W9-AOM-807

# ANNUAL REVIEW OF ECOLOGY AND SYSTEMATICS

# ANNUAL REVIEW OF ECOLOGY AND SYSTEMATICS

VOLUME 21, 1990

RICHARD F. JOHNSTON, *Editor*
University of Kansas

PETER W. FRANK, *Associate Editor*
University of Oregon

CHARLES D. MICHENER, *Associate Editor*
University of Kansas

ANNUAL REVIEWS INC. 4139 EL CAMINO WAY P.O. BOX 10139 PALO ALTO, CALIFORNIA 94303–0897

ANNUAL REVIEWS INC.
Palo Alto, California, USA



*International Standard Serial Number: 0066–4162*
*International Standard Book Number: 0–8243–1421-2*
*Library of Congress Catalog Card Number: 71-135616*



∞ The paper used in this publication meets the minimum requirements of American National Standard for Information Sciences—Permanence of Paper for Printed Library Materials, ANSI Z39.48-1984.



Typesetting by Kachina Typesetting Inc., Tempe, Arizona; John Olson, President
Typesetting Coordinator, Janis Hoffman

PRINTED AND BOUND IN THE UNITED STATES OF AMERICA

AR Annual Review of Ecology and Systematics
Volume 21, 1990

# CONTENTS



 (*continued*)

# RELATED ARTICLES FROM OTHER *ANNUAL REVIEWS*

From the *Annual Review of Earth and Planetary Sciences*, Volume 18 (1990)

*Late Precambrian and Cambrian Soft-Bodied Faunas,* S. Conway Morris
*Seafloor Hydrothermal Activity: Black Smoker Chemistry and Chimneys,* K. L. Von Damm
*The Origin and Early Evolution of Life on Earth,* J. Oró, S. L. Miller, and A. Lazcano

From the *Annual Review of Energy,* Volume 15 (1990)

*Energy Use and Acid Deposition: The View from Europe,* W. M. Stigliani and R. W. Shaw
*Energy, Greenhouse Gases, and Climate Change,* I. M. Mintzer

From the *Annual Review of Entomology,* Volume 35 (1990)

*Superparasitism as an Adaptive Strategy for Insect Parasitoids,* J. J. M. van Alphen and M. E. Visser
*Ecology and Management of the Colorado Potato Beetle,* J. D. Hare
*Properties and Potential of Natural Pesticides from the Neem Tree,* Azadirachta indica, H. Schmutterer
*Evolution of Specialization in Insect-Umbellifer Associations,* M. R. Berenbaum
*Ecological Genetics and Host Adaptation in Herbivorous Insects: The Experimental Study of Evolution in Natural and Agricultural Systems,* S. Via
*Population Biology of Planthoppers,* R. F. Denno and G. K. Roderick

From the *Annual Review of Genetics,* Volume 24 (1990)

*Genetics of Circadian Rhythms,* J. C. Hall

From the *Annual Review of Phytopathology,* Volume 28 (1990)

*The Genetics of Resistance to Plant Viruses,* R. S. S. Fraser

From the *Annual Review of Plant Physiology and Plant Molecular Biology,* Volume 41 (1990)

*Salinity Tolerance of Eukaryotic Marine Algae,* G. O. Kirst
*Cold Acclimation and Freezing Stress Tolerance: Role of Protein Metabolism,* C. L. Guy

ANNUAL REVIEWS INC. is a nonprofit scientific publisher established to promote the advancement of the sciences. Beginning in 1932 with the *Annual Review of Biochemistry*, the Company has pursued as its principal function the publication of high quality, reasonably priced *Annual Review* volumes. The volumes are organized by Editors and Editorial Committees who invite qualified authors to contribute critical articles reviewing significant developments within each major discipline. The Editor-in-Chief invites those interested in serving as future Editorial Committee members to communicate directly with him. Annual Reviews Inc. is administered by a Board of Directors, whose members serve without compensation.

John Maynard Smith

*Annu. Rev. Ecol. Syst. 1990. 21:1–12*


# THE EVOLUTION OF PROKARYOTES: DOES SEX MATTER?

*John Maynard Smith*

School of Biological Sciences, University of Sussex, Falmer, Brighton, BN1 9QG, England

KEY WORDS: transformation, recombination, sex, phylogenetic tree, penicillin

---

For many years, my lecture course on population genetics began with an account of the Hardy-Weinberg ratio. That is, I assumed that the typical evolving population could be approximated by an infinite random-mating population of sexual diploids. The effects of finite population size, demic structure, and assortative mating were treated later as departures from this ideal state, rather as a physicist might first describe an ideal gas, and then describe deviations from it due to Van der Waals forces. This clearly won't do for prokaryotes, but we are not yet clear what alternative image we should have in mind. In particular, is it sensible to see bacterial populations as composed of reproductively isolated clones? How far do processes such as conjugation, transformation, and transduction modify this picture? At the opposite extreme, is it better to take a wholly gene-centred view of bacterial evolution, and regard the bacterial cell—or, rather, the bacterial chromosome—as merely a temporary alliance of genes, analogous to a European football team, composed of players from many different countries, all liable to be transferred at any time? I think that the answers to these questions are beginning to emerge.

One of the first conclusions to emerge from a study of bacterial evolution,



0066-4162/90/1120-0001$02.00

in particular the evolution of drug resistance, was that bacterial populations can adapt to sudden changes in their environment, not, as eukaryotes do, by the selection of chromosomal mutations, either newly arisen or already present in the population, but by acquiring plasmids carrying the requisite genes. Plasmids can benefit their hosts by conferring resistance to antibiotics, drugs, or heavy metals, by producing toxins, by coding for restriction enzymes, and by utilizing novel substrates: In addition many "cryptic" plasmids exist, which probably have a function, even if unknown. Given time, plasmid-born genes can be transferred across wide taxonomic boundaries: For example, a plasmid-born $\beta$-lactamase that breaks down penicillin was first observed in bacteria related to *E. coli* and *Salmonella* but is now found in the distantly related bacterium, *Neisseria,* which it reached via an intermediate host, *Haemophilus*. This is true although, as discussed below, it is still not the common mechanism of penicillin-resistance in this genus. Of course, the gene coding for $\beta$-lactamase must have arisen in the first instance by selection and mutation, but, having once arisen, it can be acquired by a wide range of host bacteria.

When it was realized that plasmid-born genes can be transferred in this way, the idea arose that chromosomal genes might also be readily transferred between taxonomically distant species, giving rise to the "football team" model of bacterial evolution.

For example, the ability to fix nitrogen is today found in bacteria that are taxonomically diverse and phylogenetically unrelated. This patchy distribution led to the suggestion (19) that the battery of genes required to fix nitrogen has been transferred horizontally between unrelated bacteria. This suggestion, plausible enough at the time, now seems less likely. An analysis (9) of the sequences of *nif* genes, and of 16S rRNA genes from the same bacteria, shows that the two data sets are compatible with the same phylogenetic tree, suggesting that the observed distribution is to be explained, not by horizontal gene transfer, but by the loss of *nif* genes in most lineages.

This discussion of the evolution of nitrogen fixation indicates how the football team model can best be tested. If it is true, then the phylogenetic tree for a set of bacteria deduced from one class of molecules will bear no similarity to the tree deduced from a different molecule, whereas if distant transfer of genes is rare, the phylogenetic trees deduced from different molecules will be similar. The matter has recently been discussed by Woese (29), who concludes that different molecules do as a general rule give similar trees—for example, rRNA and cytochrome *c* of the purple bacteria give trees with the same topology. Of course, this does not rule out the possibility of occasional distant transfer, or that the genes serving some functions are transferred more readily than others: As Woese remarks, we do not know what fraction of functions in a bacterial cell are subject to interspecific gene transfer, or which ones they are.

How far is a tree, and the associated hierarchical classification, the appropriate image for classifying natural objects? Clearly, it is not always so—Mendeleyev's name would not be remembered if he had insisted on classifying the elements cladistically. A tree is the appropriate image only if the objects to be classified have arisen by a branching process. The oddest thing about the pattern cladists is that they wish to combine the idea, silly but not illogical, that the pattern of the living world must be fully described before questions about its evolution can be addressed, with the assumption (illogical in the absence of an evolutionary hypothesis) that the appropriate method of description is hierarchical. For the members of an asexually reproducing population, a tree is the appropriate image: If reproduction is by binary fission, the tree is dichotomously branching. For the members of a sexual population, the appropriate image (which, unhappily, cannot be drawn on a flat piece of paper) is a multidimensional net. At higher taxonomic levels, however, a tree is again the appropriate image (although dichotomous branching cannot be justified), provided that reproductively isolated species exist. This suggests a distinction between a "fractal tree," which is tree-like at all scales of magnification, down through individual reproduction and cell division to DNA replication, and a "large-scale tree," which is tree-like only at low magnification. For sexual eukaryotes, the appropriate image is a large-scale tree. Is a fractal tree appropriate for prokaryotes (at least, if we forget about their plasmids)? Surprisingly, there has been little discussion of how one might answer such a question by looking at the objects to be classified (as opposed to studying their reproduction)—but see Eigen et al (6) and Maynard Smith (11).

Before discussing the evidence, it is important to mention a third possible type of tree, which I will call a "local continuum tree." Imagine a "genus" of sexually reproducing plants (the oaks and the potentillas of North America may approximate to this image) with the following properties:

(i) Any individual can cross successfully with others that are genetically not too distant from it (if dioecious, the partners must obviously be of opposite sex), but cannot cross with genetically more distant members of the group.

(ii) There are few discontinuities, so that, even if A cannot mate with E, A can mate with B, B with C, C with D, and D with E. Hence, gene flow can occur throughout the whole taxon. However, such local continuity could be combined with occasional discontinuities, arising for accidental historical reasons, so that the living world would be divided into a number of large taxa between which gene flow could not occur but within which gene flow is possible via a series of closely related intermediates.

Why is the world not like this? It is, after all, what one would expect if the only barrier to hybridization were developmental breakdown when the two parental sets of genetic information were too disparate. Bateson (1) argued that "species" represent the different possible stable states of living matter—a view that has been resurrected recently by proponents of "laws of form." I can see little to recommend the idea. Dobzhansky sometimes wrote as if there existed a finite number of discrete ecological niches, so that "species" would represent adaptations to preexisting environmental discontinuities. I can see even less to recommend this. A third possibility lies in an intrinsic disadvantage of rarity in morphospace. This is the explanation favored in the only recent discussion of the issue known to me (2). However, my reason for raising the issue here is not so much to discuss why eukaryotes, typically, do not form a locally continuous tree, but to raise the possibility that prokaryotes in fact do so.

There is recent evidence, due largely to Selander and his colleagues (22), that the structure of bacterial populations is clonal. This comes mainly from protein electrophoreseis. For example (23), *E. coli* is highly polymorphic: 94% of loci are polymorphic, compared to 33% in humans, and mean genetic diversity per locus is 0.34–0.54, compared to 0.063 in humans. Hence, considering only this electrophoretic variability, an immense number of different genotypes could be distinguished. In practice only a small fraction of these are found, and some "electrophoretic types" (ET's) have been found repeatedly, and in different continents. This implies that recombination, resulting in crossing over between gene loci, is a relatively rare event. It does not, of course, prove that the members of a single ET are genetically identical at all loci.

This clonal pattern of variation seems to be general in prokaryotes. It is found also in bacteria that, unlike *E. coli,* habitually undergo transformation in nature—for example, *Haemophilus* (17) and *Neisseria* (13). However, as we shall see in a moment, there is equally strong evidence that recombinational events do occur in bacteria. This has led Milkman & Stoltzfus (15) to propose a modified clonal model of population structure. They suggest that a favorable chromosomal mutation will occasionally occur and will spread throughout the "species." If there were no recombination, all bacteria with the new mutation would be genetically identical (except for new mutation). But since there is some recombination, the new mutation will be associated with a "segmental clone," that is, with a chromosomal region that has not yet been broken up by recombination. The older the favorable mutation, the shorter the segmental clone, and the greater the variation that will have arisen within it by subsequent (neutral) mutation. In effect, this is an application to prokaryotes of the idea (12) that, in sexual species, each favorable mutation will create a window of genetic homozygosity by hitchhiking.

The idea of a segmental clone is attractive, but I have one reservation. This concerns the nature of recombination in prokaryotes. The model assumes that recombination has consequences similar to those of genetic crossing over in eukaryotes: that is, that it causes the replacement of one large block of genes by another. This may not be so. The evidence suggests that the important events may be much more local: A small block of DNA, of a few hundred or thousand bases, from one individual is inserted into the chromosome of another. "Local" and "global" recombination have quite different consequences for population structure. If all recombination is global, we would expect different strains of, for example, *E. coli* to have the structure of a web or net, but we would expect the genes at a single locus, if sequenced from the same set of strains, to be tree-like. In contrast, if all recombination is local, we would expect the strains to reveal a tree-like structure, but the pattern of individual genes to be net-like. Of course, the truth may lie between these extremes.

In reviewing the evidence, it is appropriate to start with *Streptococcus pneumoniae*. It was in this bug, then called *Pneumococcus,* that Griffith (7) discovered bacterial transformation and in so doing set in train the process that led to the discovery of the structure of DNA and to molecular biology. It is only recently, however, that the evolutionary significance of transformation has begun to emerge. Most bacteria evolve resistance to penicillin by acquiring a plasmid coding for $\beta$-lactamase. *Pneumococcus* has evolved resistance in a different way. Penicillin kills bacteria by binding to several high molecular weight proteins that are needed in cell wall synthesis. These proteins are misleadingly called "penicillin-binding proteins", or PBP's: Their function, of course, is not to bind to penicillin. Part of one of these proteins, PBP2B, has been sequenced from 6 sensitive and 14 resistant strains of *Pneumococcus*. The results (3) are summarized in Figure 1.

The interpretation of these results is as follows. Sensitive *Pneumococcus* are very uniform. The five "class 2" resistant strains are genetically identical to one another. They differ from the sensitive strains by the introduction, presumably by transformation, of a block of DNA, including the initial nucleotides of the sequenced region, and an unknown number of nucleotides prior to this, differing from *Pneumococcus* by 7.5% sequence divergence. The fact that the five strains, recovered from Britain and Spain over a five-year period, are identical implies that the clone, once it arose, spread without further recombination, at least in this region. The nine "class 1" strains are more complex. They are characterized by six successive amino acid substitutions: It is known from site-directed mutagenesis that these

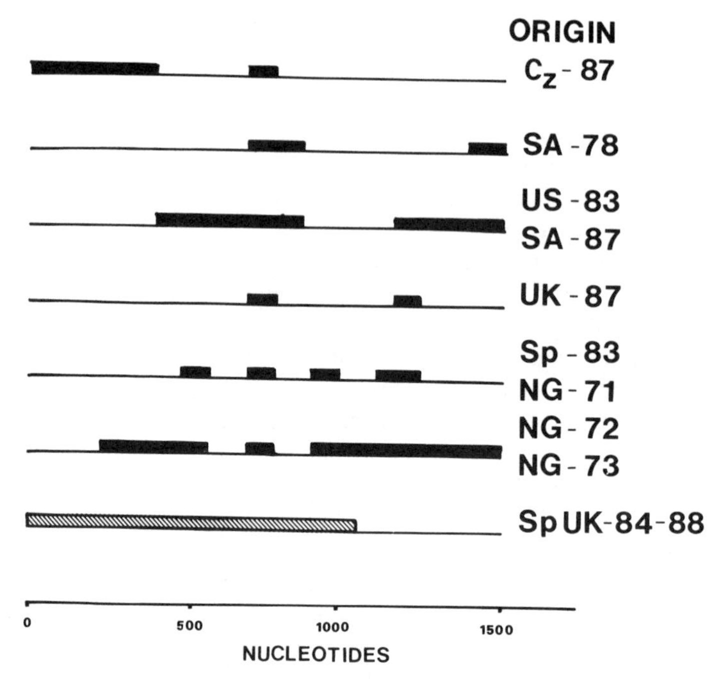


*Figure 1* Mosaic structure of PBP2B gene in penicillin-resistant strains of *Streptococcus pneumoniae*. —— , regions similar to the sensitive strains (<5% sequence divergence); ▬ , regions differing from the sensitive strains at 14% of sites (where these regions overlap in different strains, they differ from one another at less than 4% of sites); ▨ region differing from the sensitive strains at 7.5% of sites, but not resembling the other resistant strains in sequence. Places of origin: Cz, Czechoslovakia; SA, South Africa; US, United States; Sp, Spain; UK, Britain; NG, Papua New Guinea.

changes alone are sufficient to confer a high degree of resistance. However, different strains have additional blocks of introduced DNA, differing from *Pneumococcus* by 14% sequence divergence. The most likely explanation is that a block of DNA, covering the whole sequenced region, was originally introduced, by transformation, from an unknown source. This may have happened in the 1950s in Papua New Guinea, when penicillin was used extensively to treat various respiratory diseases. The varying block structures of different strains have arisen in subsequent transformation events, as resistance has spread by selection. We cannot, however, rule out the possibility that the different block structures represent different original transformation events, from the same donor species.

The species that donated the resistance gene to *S. pneumoniae* is as yet unidentified. This difficulty has been overcome in the case of penicillin resistance in *Neisseria* (25, 26). There are two pathogenic species, *N. meningitidis* and *N. gonorrhoea,* whose names are based on their extended phenotypes. The PBP2B gene has been sequenced from sensitive and resistant strains of these two species; from a naturally resistant commensal (i.e. harmless) species, *N. flavescens;* from sensitive and resistant strains of a second commensal species, *N. lactamica;* and from a resistant strain of a fifth species, *N. polysaccharae*. The results are summarized in Figure 2 and Table 1.

The following conclusions can be drawn:

(i) The sensitive strains of the two pathogenic species are very similar (1–2% divergence), although our symptoms, if humans are infected by them, are admittedly very different.

(ii) Both pathogens have evolved resistance by acquiring blocks of DNA from the naturally resistant species, *N. flavescens*. The ends of the blocks, indicating crossover points, are in some cases sufficiently similar to suggest a common origin of different resistant strains.

(iii) A second commensal species, *N. lactamica,* has acquired resistance by the introduction of DNA from a resistant strain of *N. meningitidis,* consisting in part of *N. flavescens* DNA and in part of *N. meningitidis* DNA.

(iv) Perhaps surprisingly, the sensitive *N. lactamica* strain also reveals a mosaic structure. The region between sites 728 and 1260 differs from *N. meningitidis* by 12% sequence divergence, but the next 720 nucleotide sites differ by only 2.9%.

(v) There is evidence (for example, in *N. meningitidis* strain 5) of introduced DNA from an as yet unidentified species.

(vi) One resistant strain of *N. gonorrhoea* differs from the sensitives only by the insertion of a single additional codon, which by itself is sufficient to confer resistance. It is not clear whether this was an independent mutation, or whether it represents the introduction of a short block of DNA (perhaps of the order of 20 nucleotides) from the anonymous donor species, which also has an additional codon at the same position: The amino acid is the same, but the codon is different.

The details are confusing, but it is clear that blocks of DNA have been exchanged between at least six "species" (one as yet unidentified), differing by up to 20% sequence divergence. However, the frequency of exchange has not been so high as completely to randomize the sequences, which would have destroyed all evidence of the recombinational events that have occurred. It is also likely that the frequency with which new gene sequences, arising by recombination, have been established in natural populations has been increased by the strong selection pressure imposed by penicillin. The locus is not unique, however. Similar local recombination events are known to have affected the evolution of two other genes in *Neisseria*—the *iga* gene (8), coding for an extracellular enzyme that cleaves human *IgA* protein, and the

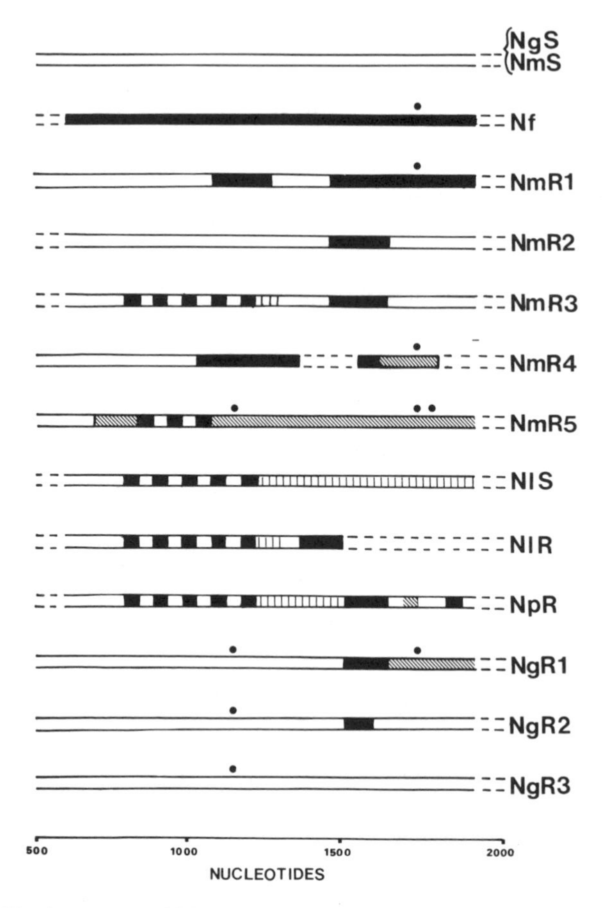


*Figure 2* Mosaic structure of PBP2B genes in *Neisseria*. NgS, NmS, penicillin-sensitive *N. gonorrhoeae* and *N. meningitidis;* Nf, *N. flavescens* (naturally resistant); NmR1–NmR5, resistant strains of *N. meningitidis;* N1S, N1R, sensitive and resistant strains of *N. lactamica;* NpR, resistant strain of *N. polysaccharae;* NgR1-NgR3, resistant strains of *N. gonorrhoeae*.

, DNA similar to sensitive *N. meningitidis*

, DNA similar to *N. flavescens*

, DNA similar to *N. lactamica,* and differing from N. meningitidis by 12% sequence divergence

, DNA similar to *N. lactamica,* and differing from N. meningitidis by 3% sequence divergence

, DNA of unknown origin

• , single additional codon

**Table 1** Mean sequence divergence (percent nucleotide sites) for the PBP2B gene in *Neisseria*

| | *N.m.* | *N.f.* | *N.l.* | Anon |
|---|---|---|---|---|
| *N. meningitidis* | — | | | |
| *N. flavescens* | 21.9 | — | | |
| *N. lactamica* | 12.0 | 20.9 | — | |
| Anonymous | 13.6 | 13.8 | 12.9 | — |

*N.gonorrhoeae* differs by 2% from *N.meningitidis*. The values for *N.lactamica* are for nucleotides 728–1260, and for "Anonymous" are for nucleotides 1259–1947 in *N. meningitidis* resistant strain 5. (unpublished data from Dr. Brian Spratt).

*pilE* gene (21), coding for the protein of the highly antigenic pili. Both these genes, too, are likely to be under strong and fluctuating selection.

It is natural to ask whether these observations on *Pneumococcus* and *Neisseria* are peculiar, either because the genes concerned have been under particularly strong selection, or because these genera have evolved a special capacity to undergo genetic transformation. Levin (10) has argued that transformation is an evolved trait, enabling a competent cell to take up single-stranded DNA from the medium, and, provided sufficient sequence homology exists, to incorporate it by homologous recombination. It is not clear whether this function has evolved because it makes possible the repair of double-stranded DNA damage (14, 30), or because it facilitates evolutionary novelty, as it certainly has done in the examples described above. As always, the mere fact that an "organ" (the machinery needed for transformation) has some effect does not by itself establish the cause of its evolution.

*E. coli* is not a bacterium competent for transformation, and there is no reason to think that the genes that have been sequenced from several strains are under specially strong selection. There are three such loci: *gnd* (5), *trp* (15), and *phoA* (4). In every case, there is evidence of a mosaic gene structure, although it is not so obvious as in the cases of *Pneumococcus* and *Neisseria*. In the latter two examples, the mosaic structure is striking, and it is easy to invent statistical tests to demonstrate its significance, although there is always an element of doubt about the exact positions of the crossover points. The *E. coli* data call for a more sophisticated analysis (20, 27). The mere fact that the nucleotide differences between two strains, A and B, are concentrated in particular regions of the gene does not demonstrate recombination: It is to be expected if mutations in particular regions of the gene are more frequent, or more likely to be selectively neutral (admittedly, it is not easy to see why silent substitutions are more likely to be neutral in one region of a gene than another). Suppose, however, that we have sequenced several strains, A,B,C.

. . . We can then list all polymorphic sites and ask of a particular pair of strains, say A and B, whether the polymorphic sites at which they differ occur in runs. Essentially, this is the approach adopted by Sawyer (20). In the case of the *phoA* gene, the mosaic structure is more obvious, and Du Bose et al (4) reconstruct the history of their eight strains, suggesting a "tree" with four crossovers. In addition to the three genes mentioned above, there is evidence of recombinational events at the *pap* gene cluster (18), and for a 3500 bp region close to the *trp* locus sequenced by Stoltzfus et al (28).

It seems, then, that local recombination is also occurring in *E. coli*, probably mediated by conjugation plasmids, or by temperate phage vectors. If most recombination events are local, they would not destroy the clonal structure revealed by electrophoresis. However, a hard question remains unanswered. It is clear that a "fractal tree" is not an appropriate image: but is the appropriate picture a "large-scale tree" or a "local continuum tree"? In other words, is there something corresponding to the "species" among eukaryotes? In sexual organisms, the species has two aspects. First, a species is a population whose members can exchange genes with one another: Second, the members of a species cannot exchange genes with members of other species. In bacteria, it seems that the evolving unit, between whose members genetic exchange is possible, is somewhat wider than the named "species," such as *S. pneumoniae* or *N. meningitidis*. Admittedly, my microbiological colleagues feel that the named entities do correspond to real groupings, but I suspect that this may mean only that they would know whether they had meningitis or the clap. To an evolutionist, the relevant entity seems to be closer to the genus *Neisseria*, or to *Escherichia* plus *Shigella*.

There remains the question whether there is anything corresponding to the isolating mechanisms found in sexual eukaryotes. The necessity for homologous recombination imposes an upper limit on the genetic distance over which at least some kinds of exchange can occur. If this were the only limit on exchange, the appropriate image would be a local continuum tree. What other isolating mechanisms are possible? Geographical isolation is not a candidate: A peculiarity of bacterial species is that their distributions are world-wide. In the case of symbiotic species, host specificity is a possibility. This is the only context in which Dobzhansky's idea of preexisting ecological niches makes some sense. However, both *Streptococcus* and *Escherichia* strains seem to cross host-species boundaries rather easily. One process that could in principle give rise to discontinuities is the action of restriction endonucleases, which can exclude foreign DNA. However, even with nucleases that recognize four-base sequences, the typical fragment length after cleavage will be 250 bp, which is long enough to explain most of the mosaic structure observed.

The evidence is at present contradictory—or I find it so. One study that suggests discontinuity is the analysis (24) of the differences between the genes of *E. coli* K12 and *Salmonella*. Some 60 genes have been sequenced in both species. Some loci show, in both species, a high bias in codon usage. These are "highly expressed genes"—that is, genes producing a lot of mRNA and protein. A plausible explanation is that, for such genes, it is selectively advantageous to use codons for which the corresponding tRNA is present in large amount. Other, less highly expressed, genes show less bias in codon usage. It turns out that, comparing the two genera, genes with a high codon bias are more similar (and, presumably, have evolved less rapidly) than genes with a low codon bias. This is precisely the result one would expect if gene substitution occurs only, or mainly, at those sites at which selection is weak. However, the result also implies that there has been little gene exchange between the two genera for a considerable time—either directly, or via intermediates as might occur on the continuum hypothesis.

Snags arise when one looks in detail at one particular locus, *gnd,* that has been sequenced from nine *E. coli* strains and from *Salmonella* (5). If one was shown these ten sequences and asked which sequence belongs to a reproductively isolated group, I do not think one could answer. There is a group of seven genes (including *E. coli* K12) that are rather similar to one another (4–6% divergence), so clearly none of these can be the odd man out. But if we compare K12, the two remaining *E. coli* strains, and *Salmonella,* all six differences are in the range 14–18%, and no strain is characterized by an unusually high number of unique nucleotides (in fact, *Salmonella* does not have the largest number of unique bases). These facts argue against discontinuity between *E. coli* and *Salmonella* (a different conclusion would be reached if the *trp* locus was examined; for these sequences *Salmonella* is an obvious outlier).

It seems that it is too early to say whether a "large-scale tree" or a "local continuum tree" is the more appropriate image of bacterial evolution, or, equivalently, whether anything corresponding to a reproductive isolating mechanism exists among bacteria. What is clear is that evolution is strongly influenced by local recombination between closely related cells: a "fractal tree" would not be an appropriate image.

### *Literature Cited*

1. Bateson, W. 1894. Materials for the study of variation. London: MacMillan
2. Bernstein, H., Byerly, H. C., Hopf, F., Michod, R. E. 1985. Sex and the emergence of species. *J. Theor. Biol.* 117: 665–90
3. Dowson, C. G., Hutchison, A., Brannigan, J. A., George, R. C., Hansman, D., Linares, J., Tomaçz, A., Maynard Smith, J., Spratt, B. G. 1989. Horizontal transfer of penicillin-binding protein genes in penicillin-resistant clinical isolates of *Streptococcus pneumoniae. Proc. Natl. Acad. Sci. USA* 86:8842–46
4. DuBose, R. F., Dykhuizen, D. E., Hartl, D. L. 1988. Genetic exchange between natural isolates of bacteria: recombination within the phoA gene of *Escherichia coli. Proc. Natl. Acad. Sci. USA* 85:7036–40
5. Dykhuizen, D. E., Green, L. 1986. DNA sequence variations, DNA phylogeny and recombination in *E. coli. Ge-*

*netics* 113: S71 and an unpublished manuscript
6. Eigen, M., Winkler-Oswatitsch, R., Dress, A. 1988. Statistical geometry in sequence space: a method of quantitative comparative sequence analysis. *Proc. Natl. Acad. Sci. USA* 85:5913–17
7. Griffith, F. 1928. Significance of Pneumococcal types. *J. Hyg. Camb.* 27: 113–59
8. Halter, R., Pohlner, J., Meyer, T. F. 1989. Mosaic-like organization of IgA protease genes in *Neisseria gonorrhoeae* generated by horizontal genetic exchange in vivo. *EMBO J.* 8:2737–44
9. Hennecke, H., Kaluza, K., Thöny, B., Fuhrmann, M., Ludwig, W., Stackebrandt, E. 1985. Concurrent evolution of nitrogenase genes and 16S rRNA in *Rhizobium* species and other nitrogen fixing bacteria. *Arch. Microbiol.* 142:342–48
10. Levin, B. R. 1988. The evolution of sex in bacteria. In *The Evolution of Sex,* ed. R. E. Michod, B. R. Levin, pp. 194–211. Sunderland, Mass: Sinauer
11. Maynard Smith, J. 1989. Trees, bundles or nets? *TREE* 4:302–4
12. Maynard Smith, J., Haigh, J. 1974. The hitch-hiking effect of a favourable gene. *Genet. Res. Camb.* 23:23–35
13. Mendelman, P. M., Caugant, D. A., Kalaitzoglou, G., Wedege, E., Chaffin, D. O., et al. 1989. Genetic diversity of Penicillin G-resistant *Neisseria meningitidis* from Spain. *Infect. Immun.* Apr. 1989:1025–29
14. Michod, R. E., Wojciechowski, M. F., Hoelzer, M. A. 1988. DNA repair and the evolution of transformation in the bacterium *Bacillus subtilis*. *Genetics* 118:31–39
15. Milkman, R., Crawford, I. P. 1983. Clustered third-base substitutions among wild strains of *E. coli*. *Science* 221:378–80
16. Milkman, R., Stoltzfus, A. 1988. Molecular evolution of the *Escherichia coli* chromosome. II. clonal segments. *Genetics* 120:359–66
17. Musser, J. M., Granoff, D. M., Pattison, P. E., Selander, R. K. 1985. A population genetic framework for the study of invasive diseases caused by serotype b strains of Haemophilus influenzae. *Proc. Natl. Acad. Sci. USA* 82:5078–82
18. Plos, K., Hull, S. I., Hull, R. A., Levin, B. R., Ørskov, I., et al. 1989. Distribution of the P-associated-pilus *(pap)* region among Escherichia coli from natural sources: evidence for horizontal gene transfer. *Infect. Immun.* 57:1604–11
19. Postgate, J. R. 1974. Evolution within nitrogen-fixing systems. In *Evolution in the Microbial World,* ed. M. Carlisle, J. Skehel, pp. 263–92. Cambridge: Cambridge Univ. Press
20. Sawyer, S. 1989. Statistical tests for detecting gene conversion. *Mol. Biol. Evol.* 6:526–38
21. Seifert, H. S., Ajioka, R. S., Marchal, C., Sparling, P. F., So, M. 1988. DNA transformation leads to pilin antigenic variation in *Neisseria gonorrhoeae*. *Nature* 336:392–95
22. Selander, R. K., Caugant, D. A., Whittam, T. S. 1987. Genetic structure and variation in natural populations of *Escherichia coli*. In Escherichia coli *and* Salmonella typhimurium: *cellular and molecular biology,* ed. F. C. Neidhart, pp. 1625–48. Washington, DC: Am. Soc. Microbiol.
23. Selander, R. K., Levin, B. R. 1980. Genetic diversity and structure in *Escherichia coli* populations. *Science* 210: 545–47
24. Sharp, P. M., Shields, D. C., Wolfe, K. H., Li, W.-H. 1989. Chromosomal location and evolutionary rate variation in enterobacterial genes. *Science* 246: 808–10
25. Spratt, B. G. 1988. Hybrid penicillin-binding proteins in penicillin-resistant strains of *Neisseria gonorrhoeae*. *Nature* 332:173–76
26. Spratt, B. G., Zhang, Q.-Y., Jones, D. M., Hutchison, A., Brannigan, J. A., Dowson, C. G. 1989. Recruitment of penicillin-binding protein genes from *Neisseria flavescens* during the emergence of penicillin resistance in *Neisseria meningitidis*. *Proc. Natl. Acad. Sci. USA* 86:8988–92
27. Stephens, J. C. 1985. Statistical methods of DNA sequence analysis: detection of intragenic recombination or gene conversion. *Mol. Biol. Evol.* 2: 539–56
28. Stoltzfus, A., Leslie, J. F., Milkman, R. 1988. Molecular evolution of the *Escherichia coli* chromosome. I. Analysis of structure and natural variation in a previously uncharacterized region between trp and tonB. *Genetics* 120:345–58
29. Woese, C. R. 1987. Bacterial evolution. *Microbial Rev.* 51:221–71
30. Wojciechowski, M. F., Hoelzer, M. A., Michod, R. E. 1989. DNA repair and the evolution of transformation in *Bacillus subtilis*. II. Role of inducible repair. *Genetics* 121:411–22

*Annu. Rev. Ecol. Syst. 1990. 21:13–55*


# SEX ALLOCATION THEORY FOR BIRDS AND MAMMALS

*Steven A. Frank*

Department of Ecology and Evolutionary Biology, University of California, Irvine, California 92717

KEY WORDS: sex ratio, bird, mammal, sex determination, evolution

---

## INTRODUCTION

Parents divide their reproductive effort into the production of sons and daughters. Darwin (43) was intrigued by the fact that parents usually split their effort so that approximately equal numbers of sons and daughters are raised. He believed that this male : female ratio had been adjusted by natural selection because he understood that the number of females set a limit on reproductive capacity. He could not, however, clearly specify how natural selection shaped the sex ratio. Fisher (51, 52) provided the explanation by noting that frequency-dependent selection stabilizes the sex ratio near equality.

Since Fisher presented his explanation, many examples of biased sex ratios have been observed in nature. For example, Hamilton (71) observed very female-biased sex ratios in parasitic wasps that mate in small groups. Hamilton explained this bias by showing that in these wasps the mating competition among brothers violates a latent assumption in Fisher's argument.

In other examples of observed biases, the explanations put forth provided new dimensions to Fisher's central theory rather than direct exceptions. The most important of these new dimensions for birds and mammals was observed by Trivers & Willard (109). They noted that, in some mammals, healthy mothers tended to produce a relatively higher proportion of sons than did unhealthy mothers. They explained this pattern of variation among families



0066-4162/90/1120-0013$02.00

by suggesting that a healthy son is reproductively more valuable than a healthy daughter and that a mother's health partly determines her offspring's vigor.

The point of these examples is to show how sex allocation theory has grown by a series of interesting observations, single-cause explanations, exceptions and new explanations. This is a natural way for a theory to grow. One problem, however, is that logical flaws slip easily into such a haphazard structure. For example, the Trivers–Willard (109) model about variation in sex ratio among families is nearly always applied simultaneously with Fisher's model for equal sex ratio at the population level. Actually, the assumptions required for the Trivers–Willard model imply that Fisher's theory cannot apply (57). The fact that these two theories are about different types of sex ratio pattern has led to the mistaken belief that the ideas can be applied independently.

The vast literature on sex allocation theory has been reviewed several times recently (12, 14, 27, 77). Several reviews of observed sex allocation patterns in birds and mammals are also available (34, 40, 75). I believe, however, that the logical structure of sex allocation theory has not been adequately reconciled with the natural history of birds and mammals. The main ideas of the theory were developed for organisms with little or no parental care, with mechanisms of sex ratio adjustment such as haplodiploidy or environmental sex determination, with no trade-offs between the sex ratio of a current brood and future reproduction, and often for simultaneous or sequential hermaphrodites.

By contrast with the types of natural history that sex allocation theory has attempted to explain, birds and mammals are characterized by extensive parental care, a sex determination mechanism that to some extent constrains sex ratio, complex mechanisms for adjusting parental investment in the sexes, variation in these mechanisms at all taxonomic levels, and a trade-off between current sex ratio and future reproduction. A complete theory for birds and mammals must show how various aspects of life history may interact in determining the patterns of selection on sex allocation. The theory viewed in total will expose some flaws in the simple single-cause explanations that typify application of sex allocation theory to birds and mammals.

## OVERVIEW

Two different perspectives on the theory must be summarized in order to provide an accurate description of its current state. The first is the series of observations and single-cause explanations that defines the history of the field. The second is the set of logical relationships among these single-cause explanations that defines the structure of the theory and the interactions that must be considered when applying the theory to real cases. For these reasons I review the theory in a loosely chronological way but use hindsight to explain ideas and provide commentary. This approach highlights the logical structure of the theory from a modern perspective and shows how some logical flaws have slipped into common usage.

In the first section of this review I recount the major single-cause forces that shape sex allocation biases. In the second section I describe how the theory has expanded and become more realistic as the ideas were first applied and then authors critically evaluated the structure of the theory in light of these applications.

Early theoretical and empirical studies of sex allocation focused mainly on adaptive significance, with less attention paid to mechanism, genetics, ontogeny, and phylogeny. To some authors this has been irksome, since in an ideal world, knowledge of phylogenetic history and of the genetic and phenotypic bases of variation must precede analyses of adaptive significance. I have delayed discussion of the bases of variation until the third section because these issues were not central to the early development of the field, which was guided by the patterns most easily observed in nature and by the simplest explanations available. I briefly summarize a few of the many fascinating recent discoveries about mechanism. Here as in other sections of the paper I cite empirical studies only to the extent that they help to understand the theory.

Up to now I have been lax about the distinction between the numbers of sons and daughters that a family produces and the relative amount of resources that is devoted to sons and daughters. From this point I use *sex ratio* for the relative number of sons and daughters and *sex allocation* for the relative amounts of energy and resources devoted to sons and daughters.

## MAJOR CONCEPTS

### *Frequency Dependence and Population-Level Patterns*

FISHER'S EQUAL ALLOCATION THEORY Darwin (43) identified the sex ratio as an interesting trait subject to natural selection, but he could offer no coherent theory as to why the sexes are generally equal in numbers. Fisher (51, 52) took up Darwin's famous challenge: "but I now see the whole problem [of sex ratio] as so intricate that it is safer to leave its solution for the future" (52, p. 158).

Fisher's argument for why the sex ratio is approximately equal is one of the most widely cited theories in evolutionary biology. The argument has been repeated in a variety of verbal and mathematical forms. I present Fisher's model in the context in which he originally described it, since it is important later to show why the idea often does not apply to birds and mammals in the manner generally accepted.

Fisher described his idea in economic metaphor—parents allocate portions of their limited reproductive energies to sons and daughters, and for each sex they get certain returns measured as genetic contribution to future generations. Because each future offspring in the population receives genes equally from its mother and father, the total genetic contribution of males and females is equal in each generation. After noting this equality in the reproductive values of males and females, Fisher (52, p. 159) concluded:

> From this it follows that the sex ratio will so adjust itself, under the influence of Natural Selection, that the total parental expenditure incurred in respect of children of each sex, shall be equal; for if this were not so and the total expenditure incurred in producing males, for instance, were less than the total expenditure incurred in producing females, then since the total reproductive value of the males is equal to that of the females, it would follow that those parents, the innate tendencies of which caused them to produce males in excess, would, for the same expenditure, produce a greater amount of reproductive value; and in consequence would be the progenitors of a larger fraction of future generations than would parents having a congenital bias towards the production of females. Selection would thus raise the sex-ratio until the expenditure upon males became equal to that upon females.

Fisher's argument is indeed compelling—reproductive profits are greater on allocation to the sex with lower total investment. The population is always pulled by frequency-dependent selection toward an equilibrium in which total allocation to the two sexes is equal. Total allocation depends on both the sex ratio and the patterns of parental investment in each son and daughter.

The implicit assumptions in Fisher's argument are often quite robust and, when met, equal allocation is a realistic prediction. Nevertheless, complications raised in the next section show that the equal allocation principle can sometimes be misleading when applied to birds and mammals. To prepare for the extensions to Fisher's theory made by later authors, let us first consider Fisher's own argument more carefully.

Suppose a parent invests some of its limited resources in a son. That son must then compete with the pool of males in his generation for a portion of the fixed genetic profits available to males, which is one half of the future population. The fraction of these fixed profits that a parent can expect by investing in sons depends on the competitive ability of the parent's sons relative to the total competitive ability of competing males in the local population. In the simplest case, suppose there are $M$ males produced by all other parents, and the parent we are considering produces $m$ sons. If all sons are equal, then our parent can expect as its fraction of the total profits in males $m/(M + m)$, or approximately $m/M$ when $M$ is much bigger than $m$ (100). This expression describes the number of sons produced by a parent relative to the population total. Note that when the number of sons produced by a parent doubles then genetic profits approximately double.

Fisher's argument is cast entirely in terms of investments and profits and is notably unconcerned with numbers of sons. In the Fisherian spirit we could view $m$ as a parent's allocation and $M$ as the total population's allocation to sons, so that expected genetic returns on investment $m$ are $m/(M + m)$, or approximately $m/M$ when $M$ is large. Likewise for females, if $f$ is a parent's investment and $F$ is everyone else's allocation, then genetic returns on $f$ are $f/(F + f)$, or $f/F$ when $F$ is large (100).

Genetic returns on a unit of investment $\epsilon$ are $\epsilon/M$ for males and $\epsilon/F$ for females. This clearly shows that when the population is currently allocating more to females, $F > M$, then investing in males gives greater returns per unit investment than investing in females. Similarly, when the population is currently allocating more to males, $M > F$, then investing in females gives greater returns per unit investment than does investing in males. Since the sex with less total investment is always more profitable, selection will constantly move the population allocation ratio toward 1 : 1 (5, 79).

When the population is at equal allocation, then genetic profits are equal for a single unit of resource invested in either males or females, $\epsilon/M = \epsilon/F$ (25, 49, 50, 87, 89). In economic language this means that the marginal returns on additional investment in males and females are equal when the population is at equilibrium. In Fisher's language, this might be said as: For if the marginal returns were not equal, and males, for instance, yielded a higher genetic return per unit investment, it would follow that those parents, the innate tendencies of which caused them to produce males in excess, would, for the same expenditure, contribute a greater amount to the genetic constitution of future populations; and in consequence would be the progenitors of a larger fraction of future generations than would parents having a congenital bias towards the production of females. Selection would thus raise the sex ratio until the marginal returns on expenditure in males became equal to the marginal returns on expenditure in females.

Fisher's theory shows the frequency dependence that occurs in all arguments about sex allocation—the current ratio of total male to female investment, $M$:$F$, affects the marginal returns per unit investment in males and females. In addition, with further assumptions implicit in Fisher's argument, frequency dependence will lead to an expected 1 : 1 allocation ratio. The theoretical extensions in the next section show that frequency dependence is a ubiquitous feature of sex allocation, but that equal allocation is not necessarily the expected result. The expected departure from equal allocation may be particularly pronounced in some birds and mammals.

CHARNOV'S NONLINEAR MODEL Charnov et al (31) used MacArthur's (87) formulation for sex ratio to study when simultaneous hermaphroditism will be favored over dioecy. In the process, they developed a model for the allocation of resources to male and female function within hermaphrodites. Nonlinear relationships between investment and returns play a key role in their analysis. Charnov (26) extended this model in a way that can be used to analyze the population sex allocation ratio in dioecious species under nonlinear returns (see also 89).

Consider how nonlinear returns may arise. Fisher's argument assumes that when a parent doubles its investment in females, $f$, that the parent will also double its genetic returns, which Fisher implicitly assumed to be $f/F$. This simple relationship between investment and profits often does not hold in birds and mammals. Imagine a red deer mother with its daughter. If the daughter has gotten little milk, then doubling the food supply may more than double the expected life-time reproductive success of that daughter. If, on the other hand, the daughter is very well fed, then doubling the food supply probably will less than double the daughter's expected life-time success. In general, when changes in investment are not directly proportional to changes in returns then the investment-return relationship is referred to as "nonlinear."

Charnov (26) showed that a nonlinear relationship between parental investment and expected genetic profits typically leads to an expected departure from equal allocation (24, 27, 57, 83, 84, 89). In particular, when the marginal return on additional investment is different for the two sexes, then equal allocation is not expected. In many polygynous birds and mammals, increasing investment in sons appears to give a rate of return different from increasing investment in daughters (33, 35, 38, 109).

Charnov (25) presented a mathematical formulation that closely follows Fisher's verbal argument and the mathematic formulation of Shaw & Mohler (100). For a particular investment in a son, a parent gets a son with a particular level of competitive ability. This competitive ability is then translated into a certain genetic profit depending on the total competitive ability of all other males in the population. The relation between investment and genetic profits may however be nonlinear in Charnov's argument, in contrast with Fisher's assumption of a linear relationship.

Charnov's (25) model is a formal statement of Fisher's insight that marginal returns on male and female investment must be equal at equilibrium (see above). The Charnov model has a very simple mathematical form that can be derived by extension of Fisher's reasoning. I first show how to derive the general result and then examine in detail an example to illustrate when a biased population sex allocation may be expected.

Assume that each of $N$ families in the population invests $m$ in sons and $f$ in daughters. For an investment in males of $m$, parents get a son with competi-

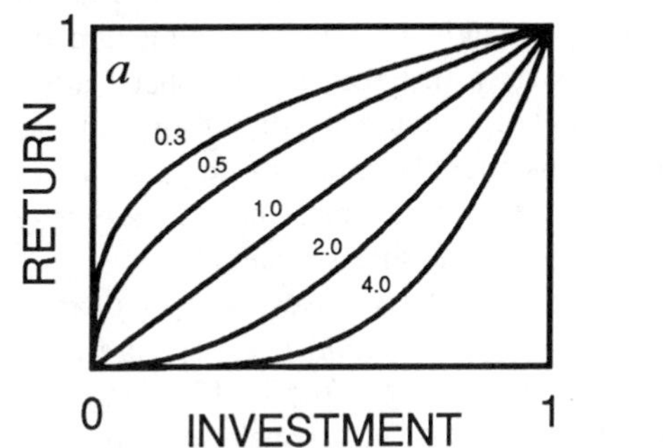


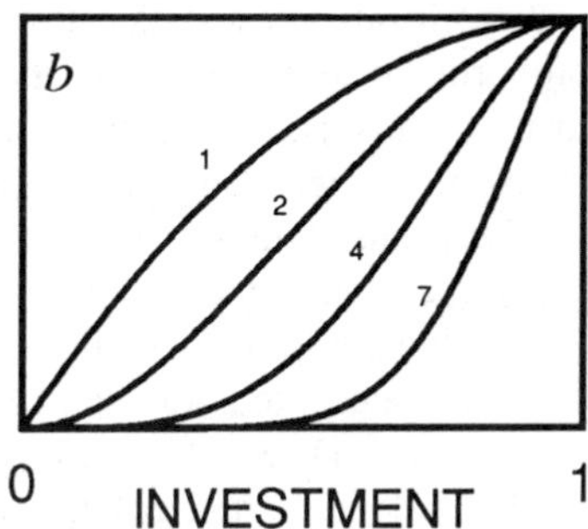


*Figure 1* The relationship between parental investment and the expected mating success of sons or fecundity of daughters. The return curves are proportional to $\int_0^k y^{s-1}(1 - y)^{t-1}dy$, where $k$ is the amount of parental investment (57). The value of $s$ is given above each curve. In $a$, $t = 1$ and the returns are proportional to $k^s$; in $b$, $t = 2$ and returns are proportional to $(s + 1)k^s - sk^{s+1}$.

tive ability $\mu(m)$ (see Figure 1, which shows a few investment-competitive ability shapes). This son will compete in a population with total male competitive ability $N\mu(m)$, so that expected genetic profits through sons will be $\mu(m)/N\mu(m)$. For an investment in females of $f$, parents get a daughter with competitive ability $\phi(f)$. This daughter will compete in a population with total female competitive ability $N\phi(f)$, so that expected genetic profits through daughters will be $\phi(f)/N\phi(f)$.

The marginal value criterion states that, at equilibrium, parents will receive the same genetic profits for investing a little bit more in either sons or daughters. As explained above, if greater profits were obtained for extra male investment, for instance, then selection would favor increased male allocation until a balance in marginal values was achieved. With a little extra investment in males, $\epsilon$, the marginal increase in profits would be $[\mu(m + \epsilon) - \mu(m)]/N\mu(m)$. The analogous expression can be written for the marginal increase in profits for females. Thus, the equality of marginal values at equilibrium guarantees that, at equilibrium,

$$\frac{\mu(m + \epsilon) - \mu(m)}{N\mu(m)} = \frac{\phi(f + \epsilon) - \phi(f)}{N\phi(f)}. \quad 1.$$

Under Fisher's explanation, returns are linear for both sexes, $\mu(m) = am$ and $\phi(f) = bf$, where $a$ and $b$ are constants. Substitution yields the condition $m = f$ at equilibrium, thus proving the equal allocation principle. Whenever $\mu$ and $\phi$ are functions with different shapes, marginal values on male and female investment are not equal when $m = f$, and the principle of equal allocation does not hold (24, 26, 57, 83).

Consider a particular example of how the relationship between investment and profits affects sex allocation. Assume that investment in a son is directly proportional to expected genetic profits from that son, as in Fisher's argument, but that genetic profits from a daughter increase at a diminishing rate as investment in that daughter increases. This might be approximately the case in a polygynous mammal such as red deer (38), where increasing male size may provide a greater rate of return than does increasing female size.

The linear returns on male investment mean that an investment by a parent of $m$ translates into a son with competitive ability of $\mu(m) = am$ (Figure 1$a$, $a = 1$ and $s = 1$), which in turn translates into a genetic profit of $m/Nm$. From the left side of Equation 1 the marginal value of further male investment is $\epsilon/Nm$. For daughters, the specific form of the diminishing returns between investment and competitive ability must be specified. Assume that the diminishing returns on female investment mean that an investment of $f$ translates into a competitive ability of $\phi(f) = b\sqrt{f}$ (see Figure 1$a$, $b = 1$ and $s = 0.5$). Genetic profit on female investment $f$ is therefore $\sqrt{f}/(N\sqrt{f})$. From the right side of Equation (1) the marginal value on further female investment is $(\sqrt{f + \epsilon} - \sqrt{f})/N\sqrt{f}$.

A simple numerical example shows that the principle of equal allocation does not apply in this case. Suppose that the population is currently allocating equal amounts of resource to males and females, $m = f = 1$. Consider how a parent can increase genetic profits with an extra investment of $\epsilon = 0.2$. For males the marginal return will be $0.2/N$, whereas for females the marginal return will be approximately $0.1/N$. When the population is allocating equally to sons and daughters a powerful selection pressure will favor an increase in allocation to sons. Now suppose that the population allocates twice as much resource to the production of sons as to the production of daughters, $m = 2$ and $f = 1$. The marginal return will be $0.1/N$ for both males and females, so a 2:1 ratio of male to female allocation must be the equilibrium. If the allocation ratio were more male biased than 2:1, we would find that extra female investment would be favored. Thus, frequency dependence is still a key feature of sex allocation, but equal allocation is not expected (26).

### *Genetic Control of Sex Ratio*

Fisher's model and its extensions discussed above depend on two implicit assumptions about the genetic control of sex allocation. First, sufficient genetic variation must exist so that any phenotypic pattern favored by natural selection can occur. For example, if all genes in the population cause parents to invest twice as much in daughters as in sons, then clearly a two to one allocation ratio will be observed no matter what selective forces occur.

The second assumption is that, from the point of view of the genes controlling sex allocation, a parent must be equally related to sons and daughters (71, 74, 82, 99; see 106 for a thorough analysis). Suppose, for example, that a matrilineally inherited cytoplasmic gene controls sex ratio (82). The success of this gene depends only on the number of daughters produced; sons do not count at all toward fitness since the gene is not transmitted to sons. Thus, a cytoplasmic gene causing its bearer to produce all daughters will have a higher fitness than will a cytoplasmic gene causing a mixture of sons and daughters. Under cytoplasmic control the population will evolve to an extremely female-biased sex ratio.

Similar complications for sex allocation arise when controlling genes are on the sex chromosomes of birds or mammals (71, 99). Mammals have *XX* females and *XY* males. If the controlling genes are on the *Y*, then selection will favor a very male-biased sex ratio because daughters do not contribute to a *Y*'s fitness, that is, with respect to the *Y*, fathers and daughters are unrelated. The situation is more complex if the controlling genes are on the *X*. If the *X* gene has its effect on sex allocation by acting in the mother, then sex allocation evolves as if controlled by haplodiploid genes (73). Under outbreeding, mother-son and mother-daughter relatedness are equal with respect to the *X*, and sex allocation evolves as if controlled by autosomal loci. If inbreeding occurs, mothers are more closely related to daughters than to sons, which favors a relatively more female-biased sex allocation than under outbreeding. If the *X* has its effects by acting in fathers, then selection favors the production of all daughters, since fathers never contribute an *X* to a son. The reverse patterns apply to birds, which have *ZZ* males and *WZ* females.

The theories reviewed below all assume unlimited genetic variety and autosomal control unless otherwise stated. The complications caused by limited genetic variety and nonautosomal control are discussed below in the section on Genetic and Phenotypic Bases of Variation.

## *Competition among Kin and Biased Sex Allocation*

LOCAL MATE COMPETITION Hamilton (71) noted that several species of small parasitic wasps have extremely female-biased sex ratios. These particular species tend to mate near where they were born. Hamilton explained the sex ratio bias by showing that when brothers compete among themselves for the limited number of mates available in a local group, then parents are favored to invest more resources in daughters than sons. In effect, since the total number of matings available in the local group is fixed by the number of females, the genetic returns to a parent for increasing allocation in males rise at a diminishing rate. If each parent is currently investing $m$ in males and there are $N$ families in the local group, then the fraction of local matings achieved by the sons in each family is $m/Nm$, and the marginal increase in the number of matings for an increase $\epsilon$ in male allocation is $(m + \epsilon)/(Nm + \epsilon) - m/Nm$

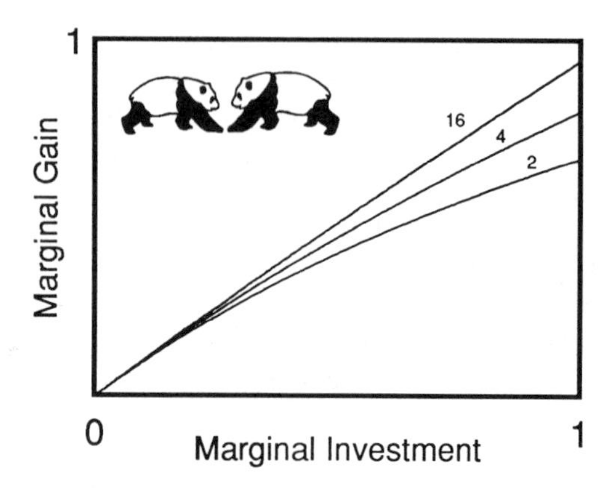


*Figure 2* These curves describe the additional mating success of sons for additional parental investment. The numbers labelling each curve are the number of families $N$ contributing sons to the local mating group. The marginal investment is described in the text as $\delta = \epsilon/m$.

or, defining $\delta = \epsilon/m$ as the marginal increase in investment, the marginal gain can be written as $\delta/(1 + \delta/N)$.

Figure 2 shows how the rate of marginal gain increases at a diminishing rate with marginal investment $\delta$ and how this effect depends on group size $N$. The decreasing rate of return on male marginal investment coupled with Charnov's general model for the role of nonlinear returns provides a simple explanation for female-biased allocation under local mate competition. In Hamilton's particular model, however, additional forces come into play (19, 105) as I describe below.

Local mate competition, as with other areas of sex allocation, has been studied with a large number of mathematical approaches (27, 53, 55, 96). I continue to formulate the theory in terms of marginal values, since that approach provides a simple way to link diverse assumptions about natural history while using the simplest mathematics.

Assume that there are $N - 1$ families in the local group, each of which allocates a fraction of their resources $m$ to sons and $f$ to daughters, where $m + f = 1$. Focus attention on the success of the $N$th family when allocating $m + \epsilon$ to sons and $f - \epsilon$ to daughters, or the success of the family through daughters when allocating $m - \epsilon$ to sons and $f + \epsilon$ to daughters. The change in expected fitness at equilibrium should be the same in either case according to the marginal value principle.

When allocating $m + \epsilon$ to sons and $f - \epsilon$ to daughters, the expected fraction of matings attained by the focal family is $(m + \epsilon)/(Nm + \epsilon)$, and the value of the females in the local group available for mating is $Nf - \epsilon$. These quantities determine the separate effects of local mating competition and local mating, respectively. When $\epsilon = 0$, the product of these quantities, which is proportional to the expected number of grandchildren through sons, is $f$, so the marginal change in grandchildren through sons when investing $m + \epsilon$ in sons and $f - \epsilon$ in daughters is proportional to:

$$\left(\frac{m + \epsilon}{Nm + \epsilon}\right)(Nf - \epsilon) - f.$$

To get the expected genetic return this quantity must be divided by the expected number of grandchildren through males in the entire population, which is proportional to $KNf$, where $K$ is the number of local mating groups each with $N$ families.

The marginal change in expected grandchildren through daughters when investing $f + \epsilon$ is proportional to $f + \epsilon - f = \epsilon$. The total number of grandchildren through daughters in the population is proportional to $KNf$. Equating the marginal values for male and female genetic return and assuming $\epsilon$ is relatively small (i.e., $\epsilon^2 \approx 0$) yields Hamilton's (71) classic formula

$$m^* = \frac{N - 1}{2N},$$

which can also be written as the ratio $m^*:f^*$ is $1 - 1/N:1 + 1/N$. This ratio form provides an easy interpretation for the forces affecting sex allocation. The genetic valuation of sons is discounted by $1/N$ for the effects of local mate competition with brothers, since $1/N$ is the probability that a male encounters a brother when competing for mates. Likewise, the genetic valuation of daughters is augmented by $1/N$ for the increase in a brother's reproduction that a sister provides through local mating between siblings, since $1/N$ is the probability of sibmating (105). The separate effects of sibmating and mate competition can be seen most clearly in a model in which females disperse before mating (19, 105). In this case there is no sibmating, but males remain at home and compete for mates with their brothers and other neighboring males. In the style of the above models one can show that the ratio $m^*:f^*$ is $1 - 1/N:1$.

LOCAL RESOURCE COMPETITION Clark (32) observed a male-biased sex ratio in bush babies. In this species young males tend to disperse whereas young females stay near their birthplace throughout life. Clark suggested, by analogy with local mate competition, that competition among sisters for limited local resources may favor parents to invest more heavily in males than females.

The theory is indeed very much like local mate competition, but there is one difference. Under the Hamilton model, there is both local mate competition among brothers and local mating between siblings (inbreeding), whereas in Clark's model there is local competition among sisters but no inbreeding since males disperse before mating.

A derivation similar to the one for the case of local mate competition can also be based on the marginal value principle. When parents allocate $m + \epsilon$ to males they obtain a fraction $(m + \epsilon)/(KNm + \epsilon)$ of the matings because competition among the dispersing males is global rather than local. When $\epsilon = 0$, this fraction of the total grandchildren is $1/KN$, so the marginal return on male investment is $(m + \epsilon)(KNm + \epsilon) - 1/KN$. When a parent allocates $f + \epsilon$ to daughters that compete among the local group of $N$ families for resources needed for reproduction, it obtains a fraction $(f + \epsilon)/(Nf + \epsilon)$ of the grandchildren through daughters within its local group. Dividing this fraction by the number of groups $K$ yields the expected proportion of grandchildren in the population by this parent. When $\epsilon = 0$ the expected proportion is $1/KN$, so the marginal gain in female genetic returns is $(f + \epsilon)/[K(Nf + \epsilon)] - 1/KN$. Equating male and female marginal values and assuming that $\epsilon$ is relatively small (i.e. $KNm + \epsilon \approx KNm$ and $\epsilon^2 \approx 0$) yields (27) the equilibrium proportion of investment in males $m^* = N/(2N - 1)$, which can also be written as $m^*:f^*$ is $1:1 - 1/N$. The latter ratio form makes it apparent that the genetic valuation of daughters is discounted by $1/N$, which is the frequency at which a female will compete with a sister for a limited resource.

Note two unrealistic assumptions in the local mate competition and local resource competition models when applied to birds and mammals. First, both the mating propensity of males and the fecundity of females are assumed to increase linearly with investment, $\mu(m) = am$ and $\phi(f) = bf$, although genetic profits increase nonlinearly because of competition among relatives. Second, the only competition between relatives and the only matings between relatives are assumed to be between siblings. Nonsiblings are assumed to be completely unrelated. These assumptions are discussed below.

## *Variation in Sex Allocation among Families*

KOLMAN'S MODEL OF SEX ALLOCATION AS A NEUTRAL TRAIT Kolman (79) and Bodmer & Edwards (5) provided the first formal model confirming Fisher's equal allocation theory. Previous analyses considered only sex ratio and implicitly assumed that expenditures per male and female offspring were equal. Kolman also stressed that any level of variation in sex allocation among families may exist at equilibrium. [This point was also briefly noted by Bodmer & Edwards (5).] For example, each family could be allocating equally to the sexes, or one half of the families could be allocating resources only to sons while the other half allocated only to daughters. Natural selection is indifferent to the variance in sex allocation among families, or, put another way, sex allocation is a neutral character at equilibrium.

Kolman's conclusion about neutrality can be derived in a straightforward way. Using the notation above, a family's fitness is proportional to $m/M + f/F$. When the population is at the Fisherian equilibrium $M = F$, fitness depends only on the family's total allocation to offspring $m + f$ and not on how the family divides this total between males and females.

The neutrality of sex allocation variance among families depends on two implicit assumptions. First, the population must be sufficiently large that an individual family's division of resources to $m$ and $f$ has a negligible effect on the population allocation ratio $M:F$. Verner (111) was the first to analyze this assumption, which is discussed in the next section. Second, lack of selection on the variance rests entirely on the assumption that returns on investment are linear for both sexes. Linearity means that $\mu(m) = am$ and $\phi(f) = bf$ so that doubling investment in a sex within a family always doubles genetic returns regardless of the initial value of investment (see above). Doubling investment in a particular offspring is unlikely to cause an exact doubling of that offspring's expected reproduction, whereas making two identical offspring does exactly double expected genetic returns.

Several later models have shown how nonlinearities can affect the expected distribution of sex allocations among families. These include the Trivers & Willard (109) model for variation in parental resources (see below) and models that consider nonlinearities induced by local mate competition (54, 58, 115) or any factor in general (57). The example given above illustrates the stabilizing effect of nonlinearities: when all families have the same amount of resource to invest, when $\mu(m) = am$ and $\phi(f) = b\sqrt{f}$, and when the population is at its equilibrium, $m^*:f^* = 2:1$, then any individual family deviating from 2:1 will suffer reduced fitness.

Fiala (49, 50) has been the only author to state explicitly that nonlinearities in "sex-specific costs" greatly reduce the expected sex allocation variation among families when there is no variation in parental resources. Because, in this context, he accepted the mistaken distinction between sex-specific costs and sex-specific returns on investment (see below), he failed to appreciate the generality of his own result as a contradiction to Kolman's model. Further discussion of Kolman's model is taken up below in the analysis of Williams's (113) paper.

VERNER'S SEX RATIO HOMEOSTASIS HYPOTHESIS Verner (111) showed that in small populations selection will tend to reduce the variance in sex ratio among families. To see how this works, consider a particular family in a small population. Assume that the equilibrium sex ratio is 1:1 and that the other families together produce a biased sex ratio. If the focal family produces a sex ratio that brings the population ratio to 1 : 1, then for all families, family fitness is independent of its sex ratio, by Kolman's argument, because $M = F$. Verner showed, however, that if the other families produce a biased sex ratio and that an allocation fraction by the focal family of $m$ in males would bring the population to 1 : 1, the focal family is in fact favored to produce a sex ratio nearer to 1 : 1 than $m$. Thus, given the sex ratios of the other families, each family is favored to produce a sex ratio nearer to 1 : 1 than the sex ratio that would produce a population ratio of 1 : 1. This process continually pushes deviants from 1 : 1 back toward equality, so that variance among families is reduced and "sex ratio homeostasis" is favored. The mechanism favoring homeostasis is subtle and can be understood by studying the calculations underlying Verner's Figure 1. Taylor & Sauer (107) have analyzed the magnitude of the homeostatic effect. Further discussion can also be found in Williams (113).

These models of homeostasis assume linearity in return functions $\mu$ and $\phi$. Nonlinearity causes selection to favor a stable and often nonzero level of sex allocation variation among families. Methods for deriving the distribution of family sex allocation ratios are described later.

THE TRIVERS-WILLARD HYPOTHESIS AND FAMILY-LEVEL BIASES Trivers & Willard (109) proposed that if one sex gains more from extra parental investment than the other, then parents with relatively more resources will bias their allocation toward the sex with the greater rate of reproductive returns. This idea has fostered much interesting research (22, 27, 33, 41, 75) and provoked some controversy over the interpretation of data (95, 113). Trivers & Willard focused primarily on the predicted positive correlation between a mother's physical condition and the proportion of sons produced. At the end of their paper they briefly outlined the more general prediction regarding the correlation between parental resources and the proportion of resources devoted to sons. I will consider the Trivers-Willard hypothesis as the more general statement about parental resources and sex allocation.

From a theoretical perspective the idea is self-evident. Incorporating the Trivers-Willard effect into a general theory of sex allocation in birds and mammals does, however, raise some important issues. The three main assumptions of the model can be expressed with the symbols used above. Consider first the assumption that one sex gains more from extra parental investment than the other. If $m$ is parental investment in males and $\mu(m)$ is the expected competitive ability and mating propensity of males with investment $m$, and if $f$ is parental investment in females and $\phi(f)$ is the expected fecundity of females with investment $f$, then the functions $\mu$ and $\phi$ must be different to satisfy the assumption that one sex gains more than the other with extra investment. Second, the assumption that parents vary in their resources available for investing in offspring means that $m + f$ varies among families.

Third, the meaning of $m$ and $f$ must be made more precise. A rough description is that the quantity $m$ means the total allocation to males within a brood and $f$ means the total allocation to females. For example, if a family has only one offspring per brood then either $m$ or $f$ must be zero. If a family has many offspring, then $m$ is the sum of the separate allocations to each male offspring and $f$ is the sum over female offspring. Selection favors an allocation strategy that depends on how parents are able to distribute resources among offspring of the same sex, how resources may be split between the sexes, and what the reproductive consequences are for each decision. Thus, the meaning of $m$ and $f$, $\mu(m)$ and $\phi(f)$, and the Trivers-Willard effect depend on the number of offspring per brood. More careful definitions of these terms and their important biological implications are considered below.

CONSEQUENCES OF THE TRIVERS-WILLARD EFFECT AT THE POPULATION LEVEL The Trivers-Willard effect depends on the assumption that a difference exists between the functions that relate investment to male and female reproductive returns. As discussed above in the section on nonlinear returns, when these functions differ then equal allocation is not expected at the population level. Thus, the Trivers-Willard effect, which is about biases within and among families, has as a corollary an expected bias in the population allocation ratio. Frank (57) presented methods for predicting the magnitude of the population-level bias.

The ways in which various authors have treated the population-level consequences of the Trivers-Willard effect form a complex historical problem. The most common approach has been to apply simultaneously both the Trivers-Willard effect and Fisher's equal allocation theory, a treatment that is logically inconsistent. Authors have often incorrectly raised Kolman's idea for the neutrality of the Fisherian equilibrium as evidence that the Trivers-Willard effect at the family level can coexist with the Fisherian effect at the population level.

## DEVELOPMENT OF THE THEORY

By the late 1970s the major ideas in sex allocation theory had been raised. Although each idea was originally formulated within the context of limiting assumptions, with hindsight the main ideas can be listed broadly as: Fisher's model for frequency dependence, which is a consequence of the joint genetic contributions of the two sexes; Charnov's marginal value model, which extends Fisher's model by allowing for nonlinearity between investment and return; Hamilton's model for competition and mating between relatives; Lewis's model for the genetic control of sex ratio; and the Trivers-Willard model for variation in sex allocation among families.

The late 1970s was also a time of rapidly growing interest in the adaptive analysis of social behavior. Sex allocation was a central platform for testing and debating this approach in evolutionary studies, beginning with Hamilton's (71, 73) papers on how interactions among relatives affect sex allocation and the evolution of fighting, Trivers & Willard's (109) paper on adaptive sex ratio in vertebrates, and Trivers & Hare's (108) effort to test theories of inclusive fitness and parent-offspring conflict through predictions about variation in sex allocation patterns in social insects.

Efforts to test sex allocation theory exposed a number of unrealistic assumptions and new problems. In response, explanations of sex allocation expanded to encompass broader assumptions and new phenomena. I here describe the development of sex allocation theory according to both its chronological development and its logical structure as seen from the present.

## *Factors Affecting Variation Among Families*

MYERS'S MODEL OF FAMILY SIZE AND SEX RATIO COMPOSITION Myers (95) extended the Trivers-Willard idea of adaptive sex ratio variation by proposing a different mechanism for generating variation. Trivers & Willard had emphasized, for stressed mothers, greater postconception mortality of sons, compared with daughters, although they mentioned that other mechanisms of variation may occur. Myers claimed that postconception mortality is an unlikely mechanism to have evolved by natural selection because it entails a loss in reproductive potential through reduction in lifetime reproductive success.

Myers proposed as an alternative mechanism that parents may be able to adjust the primary sex ratio. For example, if resource limitation causes higher offspring mortality among males than females, then stressed families that produce more daughters will have a greater number of successful offspring. Thus, Myers's idea emphasizes that sex ratio composition of families may be adjusted to maximize the number of successful offspring rather than the average reproductive potential of each offspring. Myers's critique raises the important role of the physiological, behavioral and genetic mechanisms that cause variation in sex allocation among families (see below).

WILLIAMS'S ANALYSIS OF THE DISTRIBUTION OF SEX RATIOS IN A POPULATION Williams (113) reviewed four models to explain variation in sex ratio among families. In the first part of his paper he reviewed the logic of these models and deduced contrasting expectations in terms of the variance in sex ratio among families. I summarize the main arguments here.

*The mendelian model* According to this model, the sex of each offspring in the population is determined by the random processes of meiosis and fertilization and is independent of variation in parental condition and resources. The expected variance in sex ratio among families is consistent with a binomial distribution. This prediction is difficult to test in mammals since conception sex ratios are rarely available. Birds are more promising in this regard if mortality is low before sexing.

*The adaptive model* Williams suggested that mothers in good condition in one breeding season are also likely to be in relatively good condition in the next breeding season. Likewise, those in poor condition are likely to remain in poor condition. If litter size is one, then according to the Trivers-Willard model, mothers in good condition will tend to have a series of sons and those in poor condition will tend to have a series of daughters. Thus, serial autocorrelation in maternal condition and a correlation between maternal condition and offspring sex causes a greater than binomial variance in the sex ratio among the lifetime outputs of mothers.

If litter size ranges from one to three, a purely adaptive (unconstrained) model predicts a sequence of combined litter size and sex ratio strategies according to the level of parental resources (113; Figure 3). Models concerning control of both investment per offspring and litter size trace back to Ricklefs (98) and Smith & Fretwell (101).

In the purely adaptive model shown in Figure 3, parents control number and sex of conceptions directly. The expected transitions between each litter composition depend on the return curves for sons and daughters and the distribution of parental resources in the population (57). No general qualitative trend has been demonstrated for the expected sex ratio variance among litters. Note also the weak correlation between parental resources and litter sex ratio (Figure 3).

If the conception sex ratio is not subject to modification (constrained), then some adjustment may be made by sex-biased abortion, or by sex-biased infanticide if most parental investment occurs after birth (1). In the case of constrained adjustment, the strategies favored by natural selection will depend on the particular mechanism of adjustment and cannot be predicted without further assumptions or data. Under either constrained or unconstrained adjustment, it is not obvious whether the expected variance would be greater or less than binomial. Williams (113, p. 571) suggests in his verbal model that

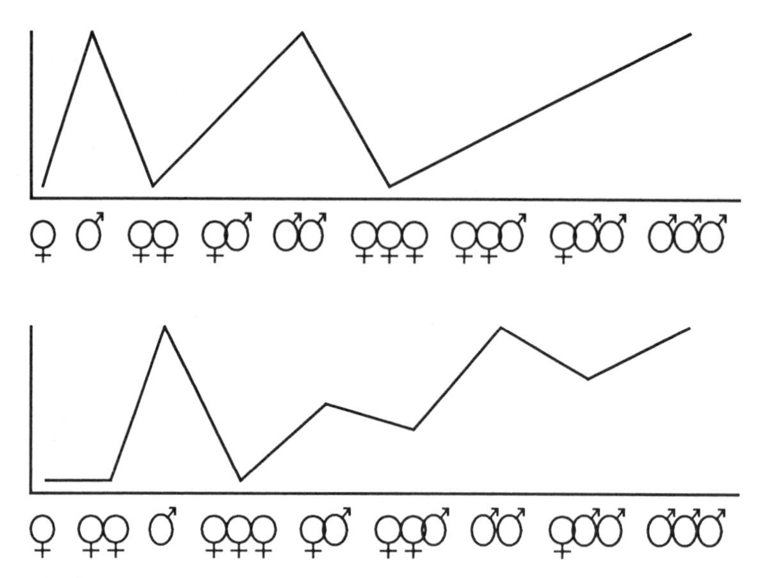

*Figure 3* The optimal number and sex of offspring in litters produced by parents with increasing amounts of resource to invest. In both panels males gain more than females from relatively high levels of investment. In the lower panel a single son can be more valuable reproductively than two daughters when given a high level of investment. The graphs in each panel show the litter sex ratio (male frequency) as a function of parental investment.

variance is expected to be greater than binomial since a preponderance of single sex broods is expected. A model to test the logic of this conjecture is feasible but would require many detailed assumptions (57). It is unclear whether a robust prediction would emerge from such an analysis.

In conclusion, the prediction from a strict interpretation of the TriversWillard model in litters with a single offspring may be a variance among lifetime broods that is greater than binomial. The predictions from a range of plausible adaptive models and larger brood sizes are, however, difficult to ascertain without numerous assumptions. In addition, the correlation between parental resources and litter sex ratio is difficult to predict in species with litter sizes greater than two (Figure 3) or in species such as humans that have a significant sharing of resources among the offspring of a series of small litters (see below).

*The neutral model* As described above, Kolman (79) had shown that the variance in sex ratio among families is a neutral character. Williams developed this notion to show that under Kolman's model the sex ratio variance among families is expected to be greater than binomial but that, in contrast with the adaptive model, the neutral model predicts no association between parental resources and sex ratio.

Kolman's model depends critically on the assumption for both sexes of linear returns between investment and reproductive returns. Since at least slight nonlinearities in the returns on male and female investment seem very likely for birds and mammals, Kolman's model and Williams's corollary are inappropriate for these organisms.

*The homeostatic model* Selection in a small population favors families that invest as closely as possible to the population optimum under certain conditions (107, 111). Williams concluded that sex ratio variation would be less than binomial if this force were operating.

Models of homeostasis in a small population depend on the assumption that Kolman's neutral equilibrium exists. When there are nonlinearities and all members of the population have the same level of resources to invest, the equilibrium is attracting rather than neutral, which induces a powerful homeostasis (small variance) of an entirely different type from Verner's. When there are nonlinearities and a variance in parental resources—the implicit assumptions whenever one analyzes the potential for adaptive sex ratio variation—the predicted sex ratio variance among families depends on many detailed assumptions (57, 67) and has yet to be worked out for any case.

In conclusion, little can be learned about adaptation and its constraints by analyzing only total sex ratio variance among families. To the extent that this approach may be useful, careful derivation of predictions under clearly specified assumptions is needed for each species considered. Discrepancies between theory and observation may point to important factors that were previously overlooked (75a).

TRADE-OFFS AMONG SIZE, SEX, AND NUMBER OF OFFSPRING Myers (95) and Williams (113) pointed out that considering both numbers and sex of offspring complicates predictions about the relationship between parental resources and sex ratio under an adaptive model (Figure 3). McGinley (92) and Gosling (67) extended these ideas and supplied some interesting data.

All of these arguments are verbal and serve to raise possible outcomes, but leave open the question of what is actually expected given particular assumptions. Frank (57) presented a series of formal models that clarifies the types of detail that must be specified and the possible outcomes. In particular, one must first specify the relationships between parental investment and reproductive potential for both male and female offspring. From these functions one can then apply a Smith & Fretwell (101) type of analysis to determine the optimal division of resources among a group of sons when given a fixed total amount for all sons, and the division among a group of daughters when given a fixed total amount for all daughters (85). This yields a pair of functions that describe the optimal relationships between investment and returns for total male and total female investment.

Given a technique for determining the optimal division of resources among males and among females, one can then calculate the optimal division of resources between the group of males and the group of females in the litter. In each family this division depends on both the family's resources and the distribution of resources among all families (57). For example, simply knowing that for a particular family 50% of the population has more resources is not sufficient to predict behavior. Predictions are of course also sensitive to the genetic variation available for selection and the physiological and behavioral mechanisms available to adjust allocation strategies (57, see below). Thus, generalizations based on verbal arguments should be treated with caution.

A CRITIQUE OF THE IDEA THAT MALES AND FEMALES EACH HAVE A PARTICULAR "COST" Many papers on sex allocation discuss the idea that males and females cost different amounts of resources to produce—for example, that male mammals are more costly to produce than female mammals. Further, in the process of developing theories, a fixed numerical value is sometimes assigned for the relative cost of producing a male or female. In reality parents across a population invest a wide range of resources in males and a different but wide range in females.

Rather than thinking that males cost more than females, it seems more appropriate to consider that each family adjusts its investments according to its resources and to the differing rates of return, given constraints such as primary sex ratio, the advantages and disadvantages of brood reduction, and the value of saving energy for future reproduction. For example, suppose parents across the population invest between 1.0 and 1.6 units in a male, and 0.7 and 1.3 in a female, and each family adjusts its allocations according to both number and sex ratio of offspring (see 57, p. 64 for a related model). A family that had 4.5 units would have several options for the number, sex, and investment in offspring and for holding some of its current energy for future reproduction. There would be little value in a model of variation among families that analyzed expected sex allocation patterns by assuming that males cost 1.3 units and females cost 1.0 units. The average relative costs of males and females may, however, be a useful quantity when making very broad phylogenetic comparisons (66a).

UNSOLVED PROBLEMS Many open questions remain in the theory itself and in the methods needed to relate theoretical predictions to observable quantities.

*The problem of currency* Theories invariably assume a unidimensional limiting resource that parents divide among offspring or save for future reproduction. In reality parents invest many types of resources. Whether theories that reduce these many dimensions to a single limiting one are robust is unknown. Further, what to measure in the field or lab in order to test predictions is often a troublesome problem. McGinley & Charnov (93) discussed the multidimensional nature of resources in allocation problems. Boomsma & Isaaks (6) discussed the problem of currency in social insects. Bull & Pease (15) developed a method that can be used to estimate the trade-off between son and daughter production and applied their method to data from a polychaete, an organism with no parental care. For species with parental care, their method requires the unlikely assumption that males and females each have a particular constant cost (see previous section). In summary, both the major theoretical questions about multidimensionality and the problems of relating theory to observation remain unsolved.

*The trade-off between current and future reproduction* The amount of parental investment in each breeding season may affect a parent's ability to invest in future offspring. For example, when a red deer mother has a son, she will skip breeding in the next season more often than after having a daughter (36). In spider monkeys, in which sons of high ranking females receive more investment than daughters, the interbirth interval is larger after a son is reared (104). In other species, such as humans, the period of parental care overlaps for sequential litters, so that investment in offspring from one litter detracts from resources available for offspring from different cohorts that are simultaneously under parental care.

*The problem of defining fecundity per investment period* Williams's (113) analysis shows that litter size plays a crucial role in determining the options available to parents and therefore in the patterns of variation expected among families (see above). Litter size may be only part of a larger problem (20, 22, 57). The parental decisions that affect genetic contribution concern trade-offs between investing in a particular offspring or using those resources to invest in other offspring from the same litter or other offspring from previous or future litters. The number and sex of offspring from past, present, and future litters among which parents can distribute limiting resources determine the options available and thus the patterns of variation that may occur. This number is the fecundity per investment period (57).

Three cases illustrate some possibilities for fecundity per investment period (57). (*a*) Current investment has no effect on past and future litters. Litter size and fecundity per investment period are identical. (*b*) Current investment affects resources available only for the next litter but, to a reasonable approximation, not more distant future pairs of litters. Fecundity per investment period is the size of sequential litters. (*c*) The period of parental care is long, and all litters must share limiting parental resources. Fecundity per investment period is the lifetime reproductive output of the parents. Humans seem the best example of this last case, which contrasts with Trivers & Willard's (109) and Williams's (113) analyses that assumed humans have a single offspring per investment period. The consequences of different fecundities per investment period are examined below.

Perhaps the most important unsolved problems concern the genetic, physiological, and behavioral mechanisms that generate sex allocation variation among families. These problems are discussed in a separate section below.

## *Variation among Families and Population Patterns*

Commentary on variation among families is usually made independently of predictions about population level patterns because most authors agree that Fisher's equal allocation theory applies. As discussed above, adaptive theories about variation among families, such as Trivers-Willard's, require a difference in the return functions between males and females. This difference usually implies that equal allocation at the population level is not a correct prediction of the theory. Whether expected departures from equality are trivial or important must be addressed explicitly. In this section I review a few key papers on population level patterns.

MAYNARD SMITH'S MODEL FOR THE CONSTRAINT OF SEX DETERMINATION Maynard Smith (89) developed three models to explain how a bias in population sex allocation may be favored by natural selection. His work was motivated in part by the observed male bias in the population allocation ratio of red deer (35, 36).

*Marginal value model* Maynard Smith began by deriving the marginal value result given in Equation 1 above (Maynard Smith's Equation 8), where for investment $m$ in males, parents receive reproductive returns $\mu(m)$ (Maynard Smith's $\psi(m)$), and for investment $f$ in females, parents receive returns $\phi(f)$. As discussed above and by Maynard Smith just following his Equation (9), the marginal value result implies a bias at the population level whenever $\mu$ and $\phi$ are different functions. Maynard Smith also points out that the direction of bias may be toward either males or females depending on the particular assumptions about $\mu$ and $\phi$. His conclusion, that at equilibrium more will be invested in males if, for a given investment, females are more likely to survive than males, depends entirely on the particular forms for $\mu$ and $\phi$ that he chose. No evidence is given that the conclusion is typical.

*Additional frequency dependence* Since Maynard Smith's goal was to explain the observed population bias in allocation toward males in red deer, he considered whether further assumptions associated with the basic marginal value model yielded a more robust conclusion. In particular, he assumed (*a*) that the return functions $\mu$ and $\phi$ are identical except (*b*) that males have an additional frequency dependent component in competitive ability or viability. One example of this frequency dependence would be that male viability depends on the total level of investment in males, for instance, male viability for a given level of investment decreases as the population-wide allocation to males increases. Using these assumptions, Maynard Smith showed that the population sex allocation ratio is expected to be biased toward males or, more generally, toward the sex with the additional frequency dependent component of viability or competitive ability. Note that this frequency dependence is distinct from the Fisherian frequency dependent competition among males that occurs in all models of sex allocation.

*Constraint of sex determination and initial investment* Maynard Smith next pointed out that these two models depend on the assumption that a parent makes the same investment in all offspring of a given sex. He suggested that parents may actually be favored to invest heavily in some offspring and little in others to the extent that they can adjust their investments by behavioral or physiological mechanisms (1, 22, 44). In the extreme case if parents know the sex of offspring before any investment has been made, then they can control their family sex ratio. Maynard Smith suggested that if parents have complete control over their family sex ratio, then Fisher's equal allocation argument applies, and we expect total investment in males and females to be equal. No model or assumptions are given under which the claimed robustness of Fisher's equal allocation actually follows. The only available models show that differences between male and female return functions as discussed in Maynard Smith's first model usually lead to biased population allocation when parents can adjust allocation to each offspring (57, see below).

Maynard Smith next examined the effects of a minimum investment in each offspring before its sex is known to the parent. This assumption was intended to match species in which control over offspring sex ratio is at least partly constrained by genetics or physiology. Under various assumptions Maynard Smith showed that parents may be favored to abandon some offspring of the sex that gains more under high levels of investment (1, 22, 44). If, for example, males gain more at high levels of investment, then under some circumstances parents will be favored to raise fewer sons than are conceived but to invest more in each son than each daughter, so that an overall bias in total investment toward males may occur. Again, a number of restrictive assumptions such as lack of variation in the amount of resources available to parents leaves the quantitative conclusions of this model open to question. This model does, however, represent a major advance because it focuses attention on the types of sex allocation variation that selection may adjust and the types of variation that are constrained by genetics and physiology.

CLUTTON-BROCK'S REVIEWS Fisher's equal allocation theory predicts that the sex ratio at the end of parental investment should be biased toward the sex with lower average investment. Clutton-Brock et al (36) and Clutton-Brock & Albon (35) found only two cases in which sufficient data were available to consider this prediction. In both red deer and northern elephant seals, more males than females were born, and males typically received considerably more milk than did the females. At the end of weaning, more males were alive in both species even though male mortality was higher. Clutton-Brock and coworkers concluded that investment was probably biased toward males. They considered and rejected Maynard Smith's (89) explanations for biased sex allocation because Maynard Smith had suggested that if certain of his unlikely assumptions did not hold, then equal allocation is expected. As discussed above and further below, the predicted population allocation ratio is unclear for these organisms under both unconstrained models and under assumptions of genetic, physiological, or behavioral constraints. It is clear, however, that equal allocation is an unlikely prediction of any realistic model for these organisms.

CHARNOV'S MODELS Charnov (26) presented the first model relating differing nonlinear returns per unit male or female investment to expected population allocation patterns. His model was developed for simultaneous hermaphrodites, but the same formulation applies to dioecious organisms. Charnov assumed in his model that all individuals had the same level of resource to invest and that at equilibrium all invested the same fraction in male and female function. The model therefore does not apply to the relationship between variation among individuals and population level patterns.

Charnov (25) presented the first formal model of sex ratio variation among individuals under differing returns per unit male or female investment. This model extends the ideas first presented verbally by Trivers & Willard (109) and applies these primarily to organisms that choose sex in a patchy environment. For example, a female parasitic wasp that lays one egg on each host may choose the sex of each offspring according to the relative host size. Charnov (25) assumed, based on available data, that males are relatively more successful than females when emerging from small hosts and that the opposite holds in large hosts. By assuming that small and large hosts exist as discrete size categories with fixed probabilities, Charnov calculated the predicted sex ratio in each size class. As expected, males are produced more frequently on small hosts and females on large hosts; the particular ratios depend on assumptions about relative male and female fitnesses and on the relative frequencies of the two host sizes. Bull (11) and Karlin & Lessard (77) have presented more rigorous population genetic analyses of this model. All of these papers concern sex ratio, and no mention is made of relative investment of resources in the two sexes. In general the predicted sex ratio is not 1:1.

In collaboration with Assem and coworkers, Charnov et al (30) presented a series of elegant experiments on parasitic wasps confirming the main predictions of Charnov's (25) model. In the 1981 paper (30) the authors made the more realistic assumption that a continuous distribution of host sizes exists. In addition, they assumed that the fitness of a female relative to a male increases steadily as host size increases. Under these assumptions a single threshold point for host size exists, below which a mother is favored to produce only sons and above which only daughters. The threshold depends on the distribution of host sizes, which is an obvious prediction of the theory and which Charnov et al (30) demonstrate by experiment. Given specific assumptions about the relative fitness of males and females as a function of host size and about the distribution of host sizes, one could in principle calculate the predicted population sex ratio. Further, if host size is taken as a measure of resources allocated to each sex, then population allocation ratios could be calculated (57, see below).

MODELS OF CONDITIONAL SEX EXPRESSION Charnov (27, pp. 140–141) showed for sequential hermaphrodites that the sex favored when relatively weak or small is expected to be more abundant in the population. Frank & Swingland (62) extended this idea to any case in which sex is conditionally expressed, including cases in which sex is environmentally determined (12, 28) or in which the sex of offspring can be manipulated by a mother in response to the amount of resources she has available for investment. Frank & Swingland (62) stressed that this theory makes a robust prediction about the greater abundance of the smaller or weaker sex, but that population sex allocation under these conditions may be biased toward either sex and that no robust prediction can be made. Thus, the usual interpretation that the cheaper sex is more abundant because of Fisher's equal allocation theory does not apply. Charnov & Bull (29) have also elaborated Charnov's (27) original model.

The most interesting prediction of this theory can be illustrated by considering red deer. The theory predicts that the sex produced when mothers are relatively weak, in this case females, should be more abundant. In fact, more males are born and more are weaned (35, 36). Two assumptions of the conditional sex expression model may be violated by red deer. First, the model assumes that a mother can freely choose offspring sex, whereas evidence suggests that although weaker red deer females do produce relatively more daughters, they are partly constrained in this regard. Second, the model assumes that investment can be made in males or females but cannot be split between offspring. If a red deer mother has energy available that she saves for future investment, then she may be dividing her resources between offspring even though she produces only one offspring per breeding season.

FRANK'S MODEL FOR INDIVIDUAL AND POPULATION SEX ALLOCATION PATTERNS Frank (57) presented quantitative predictions for sex ratio and sex allocation under Trivers-Willard type models. In the first model the assumptions are: (*a*) the amount of resources each family has is $k$, and $k$ varies according to the probability distribution $g(k)$; (*b*) returns on total male investment $m$ are $\mu(m)$ and returns on total female investment $f$ are $\phi(f)$; and (*c*) litter size is one. As described in the previous section, the predicted sex ratio is biased toward the sex produced by parents with relatively low levels of resource, but the population sex allocation ratio may be biased toward either sex. The predicted ratios depend on the male and female return functions $\mu$ and $\phi$ and are also sensitive to the shape of the resource distribution curve $g$. A variety of related assumptions were also considered.

The second model assumes that litter size may be greater than one and analyzes how parental resources are divided within a litter among offspring of the same sex and between the offspring of opposite sex. Also, as Maynard Smith (89) suggests, sex ratio at conception may be constrained, and some investment $d$ may occur before parents are able to recognize the sex of a particular offspring. The model predicts that sex-biased abortion or infanticide may occur, but the complex trade-offs discussed above between size, sex, and number of offspring prevent any general qualitative conclusions about the distribution of family strategies or the total sex ratio and sex allocation in the population.

One general conclusion did emerge from this set of models. The greater the number of offspring that must share limited parental resources, the more robust Fisher's theory for equal population sex allocation. Recall that when returns on investment are linear for both sexes, then Fisher's prediction follows. In general, the greater the difference in the shapes of the returns on male and female investment, the greater the expected departure from equal allocation. If there is only one offspring for each parental investment period, then the difference between returns on male and female investment will generally be large, and significant departures from equal allocation are expected.

If there are many offspring in each parental investment period, then returns on both male and female investment will be approximately linear because returns will scale with numbers of offspring (Figure 4). For example, if a parent has already produced 100 sons, then a 50% increase in male investment will yield 150 sons, a 50% increase in return. By contrast, if a parent has one son, a 50% increase in investment in that son or division of resources between two sons will not translate exactly into a 50% increase in return. Figure 4 illustrates the relationship between fecundity and shapes of return curves. Note that fecundity per investment period is not the same as litter size. This point was discussed in the section on *Unsolved Problems*.

SUMMARY FOR POPULATION LEVEL PATTERNS Fisherian frequency dependence is at the heart of all models of sex allocation, but equal allocation is not expected in low fecundity organisms. Departures from equal allocation increase (*a*) as fecundity declines, (*b*) as the amount of parental care increases and thus parental ability to manipulate allocation increases, and (*c*) as the intensity of sexual selection increases, and thus differences in the return curves for individual sons and daughters increase.

Not enough data are available nor has enough theory been done to provide

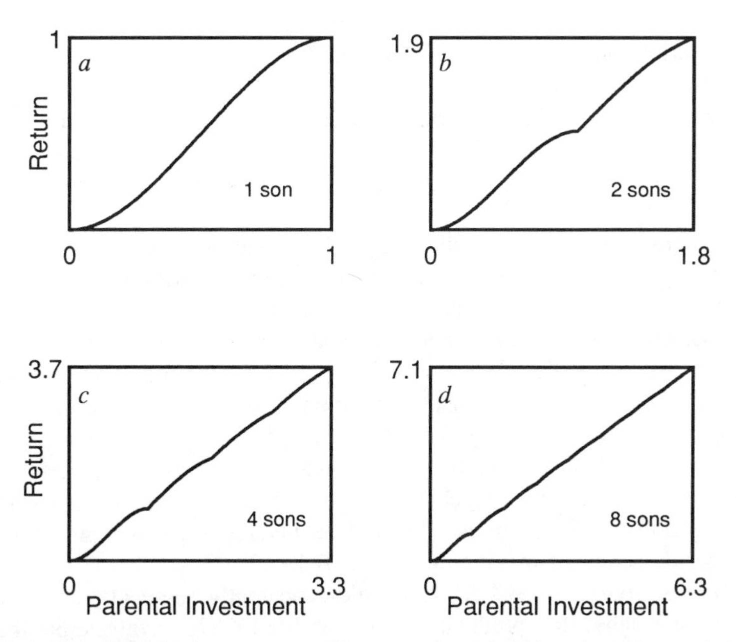


*Figure 4* The relationship between parental investment in sons within an investment period and expected total reproductive return for these sons. The relationship between investment and return for each son is nonlinear as shown in *a*. As number of sons increases the return curve rapidly approaches linearity. These curves show simultaneous optimization of size and number of sons. The return on each son is given by $h(z) = \int_0^z y(1 - y)dy$, which is also shown in Figure 1*b* with $s = 2$. In this case the return on total investment $k$ split among $n$ sons is $nh(k/n)$.

strict limits on expected departures from equal allocation under different assumptions. Based on an earlier paper (57), my own conjecture for ranges of predicted population sex allocations given as units of male investment per hundred units female investment (followed by fraction invested in males) is: between 54 (0.35) and 186 (0.65) for few offspring per period (one to two); smoothing toward a range of 67 (0.40) to 150 (0.60) fairly quickly as offspring numbers go to medium size (three to six); and then slowly approaching a robust equal allocation as the number of offspring per investment period continues to increase. These arguments assume no limits on the genetic, physiological, and behavioral variation available for natural selection to mould parental investment patterns, and the arguments assume outbreeding and no competition among relatives.

## *Extensions for Observed Genetics, Dispersions, and Demographies*

SEX CHROMOSOMES AND POPULATION CYCLES IN LEMMINGS Sex chromosome polymorphisms and segregation distortion cause strongly female-biased sex ratios in varying lemmings *(Dicrostonyx torquatus)* and wood lemmings *(Myopus schisticolor)*. Bull (12) summarized the main observations and theory; newer data and theory can be found in Gileva (65) and Bulmer (18), respectively.

These species possess *X* and *Y* chromosomes that act similarly to other mammals. In addition both species possess *X**, such that *X*X* and *X*Y* are females. In *Dicrostonyx,* the *Y* is sometimes absent so that *X*O* is female and *XO* is male.

*Myopus X*Y* females produce only *X** ova and all female broods, and *YY* zygotes are never formed (70). In *Myopus* the predicted equilibrium sex ratio is 25% males under the assumptions of equal fertility for all genotypes and normal segregation in all except *X*Y* females (4, 13). The observed sex ratio in laboratory colonies is generally below 25% (summarized in 13). Bull & Bulmer (13) pointed out that segregation distortion in favor of *Y* sperm actually may cause an increase in the expected proportion of females, but no evidence for *Y* distortion has been found in *Myopus*. Their model also demonstrated the surprising conclusion that an autosomal modifier enhancing segregation distortion of the *Y* would be favored even though it causes a further deviation of the sex ratio from 1 : 1.

*Dicrostonyx X*Y* (and *X*O*) females apparently segregate sex chromosomes normally, thus producing some inviable *YY* (*OO*) zygotes, but there is at least partial reproductive compensation for this loss in fertility (64–66). Depending on a range of likely assumptions about reproductive compensation, the predicted sex ratio varies from 36% to 42% males (13). Data from laboratory colonies tend generally to be below this predicted range. Following up on the prediction of Bull & Bulmer's (13) model for *Myopus*, Gileva (65) has shown that *Y* sperm have a segregation advantage of 0.56 ± 0.01 in *Dicrostonyx*. Bulmer (18) has developed a theory incorporating *Y* distortion and has shown that, as in *Myopus*, *Y* distortion in males actually causes a decrease in the predicted frequency of males. Bulmer (18) analyzed various assumptions based on the observed genetics that could explain the observed sex ratio and frequencies of genotypes.

Several authors have suggested that the female-biased sex ratios of lemmings are an adaptive response to their unusual population cycles (19, 91, 102); other authors have argued against the cycles as a cause of biased sex ratio (13, 18). If lemming populations are frequently broken up into small isolated demes after a population crash, and subsequently each deme expands at a rate increasing with its frequency of females, and if a deme's contribution to the population after a crash depends on its size before the crash, then female-biased sex ratios are expected. On the basis of the observed *Y* segregation distortion in *Dicrostonyx*, Bulmer (18) concluded that population cycles are unlikely to maintain the $X^*$ system because simpler mechanisms are available for sex ratio modification and because his analyses are consistent with an outbred population structure. He concluded that the $X^*$ karyotype exists because an appropriate suppressor, which would be favored by selection, has not yet arisen. Current data are not sufficient to separate these competing explanations or to suggest any compelling alternatives. All of the theories presented assume Fisherian equal allocation as a point of departure and ignore male-female differences in return curves and the low to medium fecundity per investment period.

COMPETITION AND COOPERATION AMONG KIN Competition among siblings of the same sex favors a reduction in investment in that sex (see above, 19, 32, 71, 105). Observed biases in population sex ratio have led several authors to consider kin interactions as an explanation, including the effects of cooperative rather than competitive interactions.

In red deer, population sex ratio and sex allocation both appear to be male biased (35, 36). Females tend to remain near the area in which they were born, raising the possibility that females compete for resources with sisters or their mother. Such local resource competition could explain the observed male bias. This argument is considerably weakened by the fact that elephant seals share many life history characteristics with red deer such as one offspring per litter, large sex dimorphism with bigger males, and intense male-male competition for mates. Elephant seals also have a male-biased sex ratio and allocation ratio, but they do not appear likely to experience local resource competition. Invoking local resource competition for red deer leaves unexplained the similar sex allocation pattern of elephant seals (36). Cockburn

(41, 42) summarized several other cases of male-biased sex ratios and analyzed these in the context of local resource competition. Direct evidence for female-female competition is generally lacking, but across several species male-biased dispersal tends to correlate positively with male-biased sex ratios.

Johnson (76) summarized data supporting the idea that varying levels of local resource competition could explain variation in birth sex ratios across 15 primate genera. Johnson showed that the sex ratio (frequency of males) is positively correlated with both the intensity of competition within kin groups and the level of female philopatry. The conclusion is made more convincing by Johnson's observation that in primates extreme sex ratios arise only when competition between females is intense or when sex differences in dispersal are reversed (104). Johnson also tests and rejects the idea that size dimorphism is correlated with sex ratio trends.

The model of conditional sex expression presented above provides an alternative and unexplored hypothesis for the primate data. Recall that when investment is made entirely to males or females but not both within any period, then the sex ratio is expected to be biased toward the sex that is typically produced with smaller amounts of resources. No prediction about the sex allocation ratio is made by this hypothesis. The prediction appears to be consistent with at least some of the data, since in macaques and baboons, in which daughters tend to receive more resources than do sons, the sex ratio tends to be male biased. In wild spider monkeys, sons typically receive more investment than do daughters, and the sex ratio is female biased (104). The wild spider monkey data also support the kin competition idea (104), since the males in this species are philopatric whereas females disperse. The data from red deer and elephant seals are against the conditional sex expression model, since in these species males receive greater investment and are the more numerous sex.

From a theoretical perspective a number of uncertainties remain about models of kin interactions. No models have analyzed both nonlinear returns on investment per offspring and the nonlinearities independently generated by kin competition. In primates, for example, a realistic theory must take account of: (*a*) nonlinear returns on investment per offspring and variation among parents in resources available for investment—these together form the basis for the observed Trivers-Willard variation in sex allocation among families with different resource levels; (*b*) the trade-offs among sex, size, and number of offspring in a parent's lifetime—the sex of a current offspring is known to affect future reproductive potential; and (*c*) the role of competitive and cooperative interactions among kin. From a complete theory one could analyze alternatives to the comparative explanation offered by Johnson. It would be valuable to know how alternatives might be separated, and whether one could form more precise predictions for comparisons between populations of the same species or between groups of closely related species that differ in only one or a few key ways. Finally, as discussed above, it is not valid to rely on equal allocation as an alternative and in some sense null hypothesis.

Packer & Pusey (97) observed that large cohorts of lions tend to have a more male-biased sex ratio than do small cohorts. Cohorts are groups of offspring born within one year of each other in prides of 1–18 adult females. Females in prides tend to be related, since females are philopatric and males disperse. Packer & Pusey explain the male bias in large cohorts by noting that male success depends on being a member of a successful coalition, where coalitions are typically formed from male relatives born in the same cohort. For example, in a small all male cohort (one to two) the males are less likely to form successful coalitions, whereas in large male cohorts the possibility of a successful coalition is considerably higher. Cohort size had only a slight effect on the success of an all female cohort.

Packer & Pusey (97) tested the hypothesis that males are favored in large cohorts by comparing sex ratio in prides when births are relatively more synchronous versus less synchronous. Synchrony is induced by a male takeover of the pride, so the comparison is between recently conquered and stable prides. Recently conquered prides produced a significantly higher fraction of sons (0.57) than did stable prides (0.48). Further, litter compositions of individual females after a takeover support the idea that selection favors groups of males born together. Among litters of size three or four there were a preponderance of litters with three males. Packer & Pusey discuss the trade-offs among size, sex, and number of offspring in a litter.

In all the cases discussed here, the competition and cooperation among kin includes siblings, cousins, and more distant relatives. Hamilton's (71) original model and subsequent work through the early 1980s analyzed only situations in which kin interactions were among siblings. Extensions of the mathematical theory to cover more general types of kin interactions have not uncovered any surprising qualitative conclusions—the key factors are the level of relatedness and the intensity of competition or the value of cooperation. Theoretical developments on generalized kin cooperation have been made by Frank (59) and on generalized kin competition by Frank (54, 55, 56, 58), Bulmer (16), and Taylor (106). Taylor's (106) theory is the most comprehensive, incorporating variation in genetic control of sex allocation and in complex life histories and demographies that affect the relative contribution of male and female cohorts to future generations. Frank (56, 58, 59) provided the simplest models for general kin interactions and the simplest verbal interpretations for the formal theory.

HELPERS AT THE NEST Male-biased sex ratios have been observed in adults of several cooperatively breeding bird species (8, 9, 45, 46). Female-biased mortality after fledging cannot be ruled out as an explanation, particularly since males are the philopatric sex. Brown (10, pp. 81–82) summarized evidence suggesting female-biased mortality in several cooperatively breeding species. Gowaty & Lennartz (69) have provided the only evidence showing a male-biased sex ratio among fledglings. They reported 59% males among 168 nestlings in red-cockaded woodpeckers, a species in which many nests have helpers that are mostly male. The weights of male and female nestlings were nearly equal, suggesting that sex ratio and sex allocation are approximately equivalent. Nests with older females and one or more helpers produced 54% males, whereas young females with no helpers produced 69% males (the difference is not statistically significant, but the sample size is small).

Several explanations are consistent with the data from red-cockaded woodpeckers: (*a*) Females without helpers produced more sons to increase their chance of obtaining help (69). (*b*) Older females were in some way better at producing daughters (69). There was no evidence that sex-biased mortality was affected by maternal age, nor was there evidence suggesting that females of older mothers were larger or reproductively superior to females of younger mothers. (*c*) Gowaty & Lennartz (69) suggested that, if older mothers are compared to younger mothers when each has helpers, older mothers were more likely to have sons as helpers and younger mothers were more likely to have nondescendant helpers. Sons as helpers may create intersexual competition because mother and son may not breed together but both will compete for mating at that familial nest. Gowaty & Lennartz (69) therefore favored the explanation that mothers that already have a son (older mothers) were less likely to produce more sons that would increase further the level of intersexual competition than were young mothers that were not experiencing intersexual competition. More generally, patterns of sex-biased territory inheritance may influence sex allocation in several species with helpers. (*d*) Gowaty & Lennartz (69) proposed and Emlen et al (47) quantified the idea that males were overproduced because they reduced their total costs to their parents by paying back some of the reproductive investment through helping (see also 105).

## GENETIC AND PHENOTYPIC BASES OF VARIATION

The direction of evolutionary change and the potential for adaptive modification depend on available variation. The causes of phenotypic variation in sex allocation can take several forms. Consider two extreme cases. First, alternative alleles at a single locus may cause the production of different sex ratios. Second, the population may be genetically monomorphic for a behavioral-physiological mechanism that causes parents to adjust the number, size, and sex of their offspring according to the amount of resources they have to invest. Rare genetic variants may have different parameters controlling this mechanism such that, for the same level of resources, two genetically different parents would invest differently.

Models that make testable predictions must rest on assumptions about both the selective forces acting and the genetic and phenotypic (material) bases of evolution. In this section I summarize the different types of assumption about variation. Below I consider some of the difficulties of testing theories, which necessarily rest on both selective and material assumptions.

## *Pure and Constrained ESS*

Pure evolutionarily stable strategy (ESS) models assume that individuals behave in such a way as to maximize their (single-locus) autosomal genetic contribution to future generations. This requires that all types of variants can occur and that selection has sorted among these variants. For example, in Figures 3 and 4 each parent adjusts the size, sex, and number of its offspring according to its resource level. These adjustments require that the resource level at which each transition from a smaller to a larger litter size occurs has been adjusted by selection acting on genetic variation in a behavioral-physiological mechanism.

Models of constrained ESS assume that individuals maximize their genetic contribution within the limits of certain material constraints. For example, Trivers & Willard (109) assumed that sex ratio at conception was fixed and that mothers in relatively poor condition aborted a higher frequency of male fetuses. This mechanism sets a limit on possible sex ratio variation among families. Myers (95) made the contrasting assumption that parents could manipulate conception sex ratios according to resources, which led her to a different set of predictions about sex ratio variation. Williams (113) tested a set of models with competing predictions and concluded that the sex ratio at birth in mammals is constrained by the genetic system and not subject to evolutionary modification (see above).

ESS theory and optimization theory are closely related; these approaches have been reviewed by Maynard Smith (88, 90) and, in a less genetical way, by Stephens & Krebs (103). These authors also provided discussions supporting the value of optimization models in the face of clearly unrealistic assumptions, such as unlimited genetic variation available for selection. In essence, optimization provides refutable hypotheses that can be used by field and laboratory workers to show which material aspects of variation are most constraining to adaptation and which forces of selection are most likely to have shaped the phenotypes of a particular population. ESS and optimization are theoretical methods and therefore are neither flawed nor correct. Sometimes they can be used to point out interesting lines of empirical research. When incorrectly applied or interpreted, these methods may be associated with misleading conclusions.

### *Gene Frequency Dynamics*

This approach makes explicit assumptions about the genetic and phenotypic bases of variation and then deduces the expected evolutionary outcome. The clear statement of assumptions about variation and the lack of extrinsic criteria such as optimality and adaptation make this approach an attractive alternative to the theoretically less rigorous methods of optimization and ESS. Ultimately, only a rigorous genetical model can provide a clear description for the constraining effects of particular types of material variation and a complete explanation for evolutionary pattern. It is, however, often difficult to relate the numerous assumptions and predicted gene frequency dynamics to measurable quantities.

Karlin & Lessard (77) have provided a summary of the gene frequency approach with many new extensions to the theory. From a theoretical perspective these models form a complex array based on many different assumptions about genetic control of sex allocation. A summary of these models is beyond the scope of the present paper. From the perspective of developing testable models, given the current knowledge of genetics and currently available research methods, the usable conclusions of the genetical theory are similar to the unconstrained and constrained ESS models. In particular, when variation is unconstrained—any phenotype can enter the population by mutation—then individuals allocate resources in a manner that approximately maximizes their genetic contribution. When variation is constrained, then the expected evolutionary outcomes are sensitive to the particular variety available. If the constraints concern genetic details such as dominance, epistasis, or recombination, then genetic details may sometimes be important to expected outcome (77). If the genetic details are themselves subject to extensive genetic modification, then fully genetic approaches may not yield important differences in prediction from ESS analyses (17). At present, few data bear on the types of genetic variety and on whether the dynamic models provide important and unique predictions when viewed from an empirical perspective.

### *Asymmetric Relatedness and Sex Determination*

Lewis (82) showed the interesting consequences that can occur when the genes controlling sex allocation are not equally related to sons and daughters (see above, 71, 74, 99). The most extreme case occurs when part of the genome is transmitted uniparentally, such as the strictly paternal inheritance of the Y chromosome in mammals. If the genes controlling sex allocation are on the Y then selection favors allocation only to sons and not at all to daughters. The reason is simple; a Y is never transmitted to daughters, and thus resources spent on daughters are wasted from the point of view of the controlling genes. The general theory of asymmetric relatedness and sex allocation was recently extended by Taylor (106).

Hamilton (71) noted that although a paternally inherited *Y* would favor producing only sons, selection would favor suppressors of *Y* control in the rest of the genome. Genomic competition may often be expected to lead to a stalemate between elements that are primarily matrilineal and those that are primarily patrilineal (2). When Williams (113) surveyed available data and concluded that there was no evidence for adaptive variation of sex ratio among families, he suggested that evolutionary stalemates between *X* and *Y* in mammals may have prevented variation.

The notion of evolutionary stalemate highlights a difficulty with genetic models that presume limited genetic variety: There is constant and powerful evolutionary pressure for modification of sex allocation by all parts of the genome, sometimes in conflicting directions. The qualitative features of conflicting modification pressures are fairly easy to predict (73), but a rigorous genetic theory of modification does raise a number of important complications centered around aspects of recombination and linkage disequilibrium (78), and the potential for complex evolutionary dynamics (60).

Birds and mammals provide a particularly interesting contrast with respect to sex determination and sex allocation. Most mammals have simple male heterogamety with *XY* males and *XX* females, whereas birds have simple female heterogamety with *ZZ* males and *WZ* females (12). This distinction has three interesting consequences. First, if uniparental inheritance of sex chromosomes plays an important role in sex allocation, then mammalian species would more often have male-biased sex allocation ratios than birds. There is no evidence for this at present (34, 40). Second, if the heterogametic sex were more susceptible to mortality because of the hemizygous state of its sex chromosome (70a, rejected by Trivers & Willard, 109, but see 95), then mortality under stress would be male biased in mammals and female biased in birds. Mortality under stress appears to be male biased in both birds and mammals (37), suggesting either that sex-biased parental investment (109) or sex-biased response to stress (37) are more likely explanations than heterogamety.

Third, Hamilton (72) pointed out some of the consequences of the haplodiploid type of inheritance of sex chromosomes. From the perspective of sex chromosomes, male birds value the production of brothers more highly and sisters less highly than their own offspring in terms of genetic relatedness and inclusive fitness, whereas females value their offspring at least as much as their sibs. In mammals females value sisters more highly than their own offspring. Hamilton (72) suggested that the preponderance of male helpers in birds may be associated with these asymmetries in genetic relatedness. Further, by analogy with social insects (108), if helpers control sex allocation, then to the extent that sex chromosomes control behavior, the helpers in birds will be favored to bias allocation strongly to males in their familial nests. Note that there are two distinct issues: how selection affects whether males help, and given that they help, how selection shapes the tendency of parents (61) or helpers (108) to favor males over females. Given that males help, these males will be favored to allocate more heavily to the sex to which they are more closely related independently of whether helpers are more closely related—at the genetic locus controlling the behavior—to helped offspring than to their own offspring. This is another hypothesis for the observed male-biased sex ratios in birds with predominantly male helpers (see above).

## *Observed Mechanisms of Variation*

The most often repeated observation about the genetics of sex ratio is that little variation exists (27, 88; but see 112). When considering sex allocation, however, much investment occurs after birth and sex-biased allocation and juvenile mortality can potentially be controlled by parents. Thus even if sex ratio at conception or birth were fixed, there is still the possibility of tremendous variation in sex allocation based on behavioral and physiological mechanisms which may themselves have been shaped by selection of genetic variety. I present just two mechanisms for adjusting sex allocation and then briefly discuss genetic variation for sex allocation as opposed to sex ratio. Other studies concerning parental mechanisms for controlling sex allocation have been reviewed elsewhere (22, 27, 33, 34, 40, 41, 75; see also 3, 21, 23, 63, 75a, 80, 81, 94, 114).

Bortolotti (7) studied sex-biased brood reduction in bald eagles. In this species females are about 25% larger than males, eggs hatch asynchronously, and sibling competition may result in brood reduction. Bortolotti analyzed broods with two chicks (modal brood size) and measured the hatching order, size, and sex of chicks. Labeling the four types of brood according to the hatching order and sex, and generally sampling before any mortality of eggs, of 27 broods with two chicks, the observed distribution was M-M (ten), M-F (one), F-M (nine), and F-F (seven). There were significantly fewer M-F broods than expected, and the first chick in mixed sex broods was female 90% of the time (further supporting data were reported in the paper).

Brood reduction through sibling competition depended on differences in hatching date and growth rates among the chicks and on food stress. Bortolloti analyzed growth rates and sensitivity to food stress and suggested that the paucity of M-F broods occurred because these broods were most likely to suffer brood reduction (loss of F) through sibling competition. In addition, in F-M broods under food stress, Bortolloti suggested that the female would grow well and the male could get by because of lower food requirements, yielding a robust female and a surviving male. It seems likely in this species that female size is more important for reproductive success than is male size. He also presented further discussion of brood reduction in this and other species and an analysis of some alternative hypotheses.

Gosling (68) studied sex allocation in coypus (*Myocaster coypus*), a large rodent with a polygynous mating system, female philopatry and male dispersal, and male-male competition. Males are 15% heavier than females. The adult sex ratio is 75 males per 100 females. Gestation is 19 weeks, mean litter size is 5.3, and offspring are suckled for about 8 weeks. Females breed throughout the year.

Gosling dissected 5853 adult females and collected data from 1485 that had embryos old enough to sex. He measured the mother's size; the number of embryos implanted; the number of viable embryos and their mean weight; and the number of male, female, and dead embryos. Gestation stage was inferred from previous data relating known conception date and embryo size.

The data suggest that coypu females controlled sex allocation by selectively aborting entire litters. Young females in relatively good physical condition aborted small litters of predominantly female embryos near weeks 13–14 of the 19 week gestation period. Females conceived soon after aborting a litter. The new litter size (5.82 ± 0.21 Standard Error of the mean) was significantly larger than that aborted (4.17 ± 0.32 SE). By contrast, relatively healthy females retained large litters or predominantly male litters. Neonate size was positively correlated with mother's condition and inversely related to litter size. Neonate size of an individual was positively correlated with its adult size.

In general, the likelihood that selective abortion or sex-biased infanticide would be advantageous depends on the relative costs of gestation versus lactation in mammals or egg production versus fledging chicks in birds. The coypu study shows that selective abortion may be favored under certain circumstances, but the range of conditions under which such mechanisms may or may not be favored is difficult to assess at present. Clutton-Brock et al (39) have also presented an interesting study on gestation versus lactation. They showed that the costs of gestation to the mother's subsequent survival and reproductive success are slight compared to those of lactation.

The studies of eagles and coypus show two very different mechanisms that could be used to manipulate sex allocation in a fitness-enhancing way. Discovering and characterizing the details of such mechanisms represent one of the more exciting challenges in future research. Evolutionary models would be most useful for predicting patterns of sex allocation variation, given the details of the mechanisms available to generate variation. For example, two separate populations with different distributions of resources among mothers may be expected to have different conditional responses at the phenotypic level, that is, they may be expected to have different patterns of sex allocation bias for the same resource level. Predictions could be developed by comparing the fitness (gene frequency dynamics) of competing mechanisms with differing conditional responses for litter size and sex ratio as a response to maternal condition.

The genetic influences on maternal size, litter size, sex ratio, and offspring weight are likely to be complex because of maternal effects. Falconer (48) found that selecting for increased female size in mice yielded larger litter size and smaller offspring. Genetic studies combined with analyses like Gosling's may provide interesting insights into adaptive variation in sex allocation. Marsupials and rodents are particularly promising in this regard because they are often amenable to genetic analysis and because they show a range of potentially interesting mechanisms (41, 63, 75a, 80, 81, 114).

Several difficulties may be encountered, however, when trying to make evolutionary inferences from genetic details measured on a single population at one point in time. For example, lack of observable genetic variation for the parameters controlling conditional phenotypic response does not provide strong evidence against the shaping of the phenotypic mechanism primarily by natural selection. Likewise, observed genetic variation for conditional response that is uncorrelated with other fitness traits does not provide strong evidence against genetic and physiological constraints as important factors in shaping the observed phenotypes (110).

## CONCLUSIONS

Sex allocation theory is a set of logical consequences that follow from general assumptions. As such it is, ideally, a standing pool of incontestable logic. Testable predictions can be derived for particular organisms when specific natural history assumptions are added about the relationships between investment and reproductive value, kin interactions, the distribution of resources among parents, and the genetic and phenotypic bases of variation. A test of this type of prediction determines whether the set of specific assumptions provides a good description for the forces that have shaped sex allocation. This approach to understanding sex allocation patterns must rest on a sound and complete logical structure.

The purpose of this review has been to summarize the logical structure of sex allocation theory for birds and mammals. I have emphasized some important logical flaws that have slipped into common usage and some aspects of theory that are not well understood at present. In addition, I have briefly summarized the rapidly growing body of information on the variety of proximate mechanisms that cause sex allocation variation.

The most important logical flaw in common usage is the simultaneous application of the Trivers-Willard hypothesis (109) to predict variation among families and Fisher's (51, 52) hypothesis to explain sex allocation over the total population. I reviewed the literature demonstrating that the assumptions required for application of the Trivers-Willard hypothesis imply that key assumptions of Fisher's equal allocation theory are violated. The expected quantitative departures from Fisher's equal allocation theory are not known at present, but certain qualitative expectations have been derived (57). These qualitative results highlight the importance of patterns by which parents accrue and invest resources, including the number of young per litter and the trade-off between current and future reproduction. Many important aspects of the theory have yet to be worked out for these key aspects of mammalian and avian life history.

Recent empirical work has just begun to reveal the rich set of physiological and behavioral mechanisms that parents use to adjust sex allocation in their family. Theory has, to this point, offered little in the way of predicting the types of variation that have been observed, and why certain species fail to show what would supposedly be adaptive variation. One future challenge for the theory will be to provide predictions about adaptive modification among closely related populations sharing the same genetic and phenotypic mechanisms of variation. Such predictions will help in understanding the evolutionary forces that have shaped particular mechanisms and the extent to which particular mechanisms may be subject to adaptive modification.

Sex allocation research has played a leading role in the 1980s in broadly mapping the extent and limitations of adaptive variation. As more data accumulate on the particular mechanisms of variation, sex allocation will in the 1990s provide an excellent model system for studying the evolution of behavioral plasticity under physiological and genetic constraints.

ACKNOWLEDGMENTS

I thank J. J. Bull for helpful comments on an earlier version of the manuscript. My research is supported by NSF grant BSR-9057331 and NIH grant GM42403.

## *Literature Cited*

1. Alexander, R. D. 1974. The evolution of social behavior. *Annu. Rev. Ecol. Syst.* 5:325–83
2. Alexander, R. D., Borgia, G. 1978. Group selection, altruism, and the levels of organization of life. *Annu. Rev. Ecol. Syst.* 9:449–74
3. Austad, S. N., Sunquist, M. E. 1986. Sex-ratio manipulation in the common opossum. *Nature* 324:58–60
4. Bengtsson, B. O. 1977. Evolution of the sex ratio in the wood lemming, *Myopus schisticolor*. In *Measuring Selection in Natural Populations,* ed. T. M. Fenchel, F. B. Christiansen, pp. 333–43. Berlin: Springer Verlag
5. Bodmer, W. F., Edwards, A. W. F. 1960. Natural selection and the sex ratio. *Ann. Hum. Genet.* 24:239–44
6. Boomsma, J. J., Isaaks, J. A. 1985. Energy investment and respiration in queens and males of *Lasium niger* L. *Behav. Ecol. Sociobiol.* 18:19–27
7. Bortolotti, G. R. 1986. Influence of sibling competition on nestling sex ratios of sexually dimorphic birds. *Am. Nat.* 127:495–507
8. Brown, J. L. 1978. Avian communal breeding systems. *Annu. Rev. Ecol. Syst.* 9:123–55
9. Brown, J. L. 1983. Cooperation—a biologist's dilemma. *Adv. Study Behav.* 13:1–37
10. Brown, J. L. 1987. *Helping and Communal Breeding in Birds*. Princeton, NJ: Princeton Univ. Press
11. Bull, J. J. 1981. Sex ratio evolution when fitness varies. *Heredity* 46:9–26
12. Bull, J. J. 1983. *Evolution of Sex Determining Mechanisms*. Menlo Park, Calif: Benjamin/Cummings
13. Bull, J. J., Bulmer, M. G. 1981. The evolution of *XY* females in mammals. *Heredity* 47:347–65
14. Bull, J. J., Charnov, E. L. 1988. How

fundamental are Fisherian sex ratios? *Oxf. Surv. Evol. Biol.* 5:96–135
15. Bull, J. J., Pease, C. M. 1988. Estimating relative parental investment in sons versus daughters. *J. Evol. Biol.* 1:305–15
16. Bulmer, M. G. 1986. Sex ratio theory in geographically structured populations. *Heredity* 56:69–73
17. Bulmer, M. G. 1986. Sex ratio theory. *Science* 233:1436–37
18. Bulmer, M. G. 1988. Sex ratio evolution in lemmings. *Heredity* 61:231–33
19. Bulmer, M. G., Taylor, P. D. 1980. Dispersal and the sex ratio. *Nature* 284:448–49
20. Burley, N. 1980. Clutch overlap and clutch size: Alternative and complementary reproductive tactics. *Am. Nat.* 115:223–46
21. Burley, N. 1981. Sex-ratio manipulation and selection for attractiveness. *Science* 211:721–22
22. Burley, N. 1982. Facultative sex-ratio manipulation. *Am. Nat.* 120:81–107
23. Burley, N. 1986. Sex-ratio manipulation in color-banded populations of zebra finches. *Evolution* 40:1191–1206
24. Charlesworth, D., Charlesworth, B. 1981. Allocation of resources to male and female functions in hermaphrodites. *Biol. J. Linn. Soc.* 15:57–74
25. Charnov, E. L. 1979. The genetical evolution of patterns of sexuality: Darwinian fitness. *Am. Nat.* 113:465–80
26. Charnov, E. L. 1979. Simultaneous hermaphroditism and sexual selection. *Proc. Nat. Acad. Sci. USA* 76:2480–84
27. Charnov, E. L. 1982. *The Theory of Sex Allocation*. Princeton, NJ: Princeton Univ. Press
28. Charnov, E. L., Bull, J. J. 1977. When is sex environmentally determined? *Nature* 266:828–30
29. Charnov, E. L., Bull, J. J. 1989. Non-fisherian sex ratios with sex change and environmental sex determination. *Nature* 338:148–50
30. Charnov, E. L., Los-den Hartogh, R. L., Jones, W. T., van den Assem, J. 1981. Sex ratio evolution in a variable environment. *Nature* 289:27–33
31. Charnov, E. L., Maynard Smith, J., Bull, J. J. 1976. Why be an hermaphrodite? *Nature* 263:125–126
32. Clark, A. B. 1978. Sex ratio and local resource competition in a prosiminian primate. *Science* 201:163–65
33. Clutton-Brock, T. H. 1985. Birth sex ratios and the reproductive success of sons and daughters. In *Evolution: Essays in Honour of John Maynard Smith,* ed. P. H. Harvey, M. Slatkin, P. J. Greenwood, pp. 221–35. New York: Cambridge Univ. Press
34. Clutton-Brock, T. H. 1986. Sex ratio variation in birds. *Ibis* 128:317–29
35. Clutton-Brock, T. H., Albon, S. D. 1982. Parental investment in male and female offspring in mammals. In *Current Problems in Sociobiology,* ed. King's College Sociobiology Group, pp. 223–47. New York: Cambridge Univ. Press
36. Clutton-Brock, T. H., Albon, S. D., Guinness, F. E. 1981. Parental investment in male and female offspring in polygynous mammals. *Nature* 289:487–89
37. Clutton-Brock, T. H., Albon, S. D., Guinness, F. E. 1985. Parental investment and sex differences in juvenile mortality in birds and mammals. *Nature* 313:131–33
38. Clutton-Brock, T. H., Guinness, F. E., Albon, S. D. 1982. *Red Deer: Behavior and Ecology of Two Sexes*. Chicago: Univ. Chicago Press
39. Clutton-Brock, T. H., Albon, S. D., Guinness, F. E. 1989. Fitness costs of gestation and lactation in wild mammals. *Nature* 337:260–62
40. Clutton-Brock, T. H., Iason, G. R. 1986. Sex ratio variation in mammals. *Q. Rev. Biol.* 61:339–74
41. Cockburn, A. 1989. Sex ratio variation in marsupials. *Aust. J. Zool.* In press
42. Cockburn, A., Scott, M. P., Dickman, C. R. 1985. Sex ratio and intrasexual kin competition in mammals. *Oecologia* 66:427–29
43. Darwin, C. 1871. *The Descent of Man, and Selection in Relation to Sex*. London: J. Murray
44. Dickemann, M. 1979. Female infanticide, reproductive strategies, and social stratification: a preliminary model. In *Evolutionary Biology and Human Social Behavior,* ed. N. A. Chagnon, W. Irons, pp. 321–67. North Scituate, Mass.: Duxbury
45. Emlen, S. T. 1978. The evolution of cooperative breeding in birds. In *Behavioural Ecology: An Evolutionary Approach,* ed. J. R. Krebs, N. B. Davies, pp. 245–81. Oxford: Blackwell
46. Emlen, S. T. 1984. Cooperative breeding in birds and mammals. In *Behavioural Ecology: An Evolutionary Approach,* ed. J. R. Krebs, N. B. Davies, pp. 305–39. Oxford: Blackwell. 2nd ed.
47. Emlen, S. T., Emlen, J. M., Levin, S. A. 1986. Sex-ratio selection in species

with helpers-at-the-nest. *Am. Nat.* 127: 1–8
48. Falconer, D. S. 1965. Maternal effects and selection response. In *Genetics Today, Proc. 11th International Congr. Genetics,* ed. S. J. Geerts, 3:763–74. Oxford: Pergamon
49. Fiala, K. L. 1981. Reproductive costs and the sex-ratio in red winged blackbirds. In *Natural Selection and Social Behavior,* ed. R. D. Alexander, D. W. Tinkle, pp. 198–214. New York: Chiron
50. Fiala, K. L. 1981. Sex ratio constancy in the red-winged blackbird. *Evolution* 35:898–910
51. Fisher, R. A. 1930. *The Genetical Theory of Natural Selection.* Oxford: Clarendon
52. Fisher, R. A. 1958. *The Genetical Theory of Natural Selection.* New York: Dover. 2nd ed.
53. Frank, S. A. 1983. A hierarchical view of sex-ratio patterns. *Fl. Entomol.* 66: 42–75
54. Frank, S. A. 1985. Hierarchical selection theory and sex ratios. II. On applying the theory, and a test with fig wasps. *Evolution* 39:949–64
55. Frank, S. A. 1986. Hierarchical selection theory and sex ratios I. General solutions for structured populations. *Theor. Popul. Biol.* 29:312–42
56. Frank, S. A. 1986. The genetic value of sons and daughters. *Heredity* 56:351–54
57. Frank, S. A. 1987. Individual and population sex allocation patterns. *Theor. Popul. Biol.* 31:47–74
58. Frank, S. A. 1987. Variable sex ratio among colonies of ants. *Behav. Ecol. Sociobiol.* 20:195–201
59. Frank, S. A. 1987. Demography and sex ratio in social spiders. *Evolution* 41: 1267–81
60. Frank, S. A. 1989. The evolutionary dynamics of cytoplasmic male sterility. *Am. Nat.* 133:345–76
61. Frank, S. A., Crespi, B. J. 1989. Synergism between sib-rearing and sex ratio in Hymenoptera. *Behav. Ecol. Sociobiol.* 24:155–62
62. Frank, S. A., Swingland, I. R. 1988. Sex ratio under conditional sex expression. *J. Theor. Biol.* 135:415–18
63. Fuchs, S. 1982. Optimality of parental investment: the influence of nursing on reproductive success of mother and female young house mice. *Behav. Ecol. Sociobiol.* 10:39–51
64. Gileva, E. A. 1980. Chromosomal diversity and an aberrant genetic system of sex determination in the arctic lemming, *Dicrostonyx torquatus* Pallas (1779). *Genetica* 52/53:99–103
65. Gileva, E. 1987. Meiotic drive in the sex chromosome system of the varying lemming, *Dicrostonyx torquatus* Pall. (Rodentia, Microtinae). *Heredity* 59: 383–89
66. Gileva, E. A., Benenson, I. E., Konopistevsa, L. A., Puchkov, V. F., Makaranets, I. A. 1982. XO females in the varying lemming, *Dicrostonyx torquatus:* reproductive performance and evolutionary significance. *Evolution* 36: 601–9
66a. Gomendio, M., Clutton-Brock, T. H., Albon, S. D., Guinness, F. E., Simpson, M. J. 1990. Mammalian sex ratios and variation in costs of rearing sons and daughters. *Nature* 343:261–63
67. Gosling, L. M. 1986. Biased sex ratios in stressed animals. *Am. Nat.* 127:893–96
68. Gosling, L. M. 1986. Selective abortion of entire litters in the coypu: adaptive control of offspring production in relation to quality and sex. *Am. Nat.* 127:772–95
69. Gowaty, P. A., Lennartz, M. R. 1985. Sex ratios of nestling and fledgling Red-Cockaded Woodpeckers *(Picoides borealis)* favor males. *Am. Nat.* 126: 347–53
70. Gropp, A. W. H., Frank, F., Noack, G., Fredga, K. 1976. Sex-chromosome aberrations in wood lemmings *(Myopus schisticolor). Cytogenet. Cell Genet.* 17:343–58
70a. Haldane, J. B. S. 1933. The part played by recurrent mutation in evolution. *Am. Nat.* 67:5–19
71. Hamilton, W. D. 1967. Extraordinary sex ratios. *Science* 156:477–88
72. Hamilton, W. D. 1972. Altruism and related phenomena, mainly in social insects. *Annu. Rev. Ecol. Syst.* 3:193–232
73. Hamilton, W. D. 1979. Wingless and fighting males in fig wasps and other insects. In *Reproductive Competition and Sexual Selection in Insects,* ed. M. S. Blum, N. A. Blum, pp. 167–220. New York: Academic
74. Howard, H. W. 1942. The genetics of *Armadillidium vulgare* Latr. II. Studies on the inheritance of monogeny and amphogeny. *J. Genet.* 44:143–59
75. Hrdy, S. B. 1987. Sex-biased parental investment among primates and other mammals: a critical evaluation of the Trivers-Willard hypothesis. In *Child Abuse and Neglect: Biosocial Dimensions,* ed. R. Gelles, J. Lancaster, pp. 97–147. New York: Aldine
75a. Huck, U. W., Seger, J., Lisk, R. D. 1990. Litter sex ratios in the golden hamster vary with time of mating and

litter size and are not binomially distributed. *Behav. Ecol. Sociobiol.* 26:99–109
76. Johnson, C. N. 1988. Dispersal and the sex ratio at birth in primates. *Nature* 332:726–28
77. Karlin, S., Lessard, S. 1986. *Theoretical Studies on Sex Ratio Evolution*. Princeton, NJ: Princeton Univ. Press
78. Karlin, S., McGregor, J. 1974. Towards a theory of the evolution of modifier genes. *Theor. Popul. Biol.* 5:59–105
79. Kolman, W. 1960. The mechanism of natural selection for the sex ratio. *Am. Nat.* 94:373–77
80. König, B., Markl, H. 1987. Maternal care in house mice. I. The weaning strategy as a means for parental manipulation. *Behav. Ecol. Sociobiol.* 20:1–9
81. Labov, J. B., Huck, U. W., Vaswani, P., Lisk, R. D. 1986. Sex ratio manipulation decreased growth of male offspring of undernourished golden hamsters *(Mesocricetus auratus). Behav. Ecol. Sociobiol.* 18:241–49
82. Lewis, D. 1941. Male sterility in natural populations of hermaphrodite plants. *New Phytol.* 40:56–63
83. Lloyd, D. G. 1984. Gender allocations in outcrossing cosexual plants. In *Perspectives in Plant Population Biology,* ed. R. Dirzo, J. Sarukhán, pp. 277–300. Sunderland, Mass: Sinauer
84. Lloyd, D. G. 1987. Parallels between sexual strategies and other allocations. In *The Evolution of Sex and its Consequences,* ed. S. C. Stearns, pp. 263–81. Basel: Birkhäuser Verlag
85. Lloyd, D. G. 1987. Selection of offspring size at independence and other size-versus-number strategies. *Am. Nat.* 129:800–17
86. Deleted in proof
87. MacArthur, R. H. 1965. Ecological consequences of natural selection. In *Theoretical and Mathematical Biology,* ed. T. H. Waterman, H. Morowitz, pp. 388–97. New York: Blaisdell
88. Maynard Smith, J. 1978. *The Evolution of Sex*. Cambridge: Cambridge Univ. Press
89. Maynard Smith, J. 1980. A new theory of parental investment. *Behav. Ecol. Sociobiol.* 7:247–51
90. Maynard Smith, J. 1982. *Evolution and the Theory of Games*. Cambridge: Cambridge Univ. Press
91. Maynard, Smith, J., Stenseth, N. C. 1978. On the evolutionary stability of the female-biased sex ratio in the wood lemming *(Myopus schisticolor):* the effect of inbreeding. *Heredity* 41:205–14
92. McGinley, M. A. 1984. The adaptive value of male-biased sex ratios among stressed mammals. *Am. Nat.* 124:597–99
93. McGinley, M. A., Charnov, E. L. 1988. Multiple resources and the optimal balance between size and number of offspring. *Evol. Ecol.* 2:77–84
94. Michener, G. R. 1980. Differential reproduction among female Richardson's ground squirrels and its relation to sex ratio. *Behav. Ecol. Sociobiol.* 7:173–78
95. Myers, J. H. 1978. Sex ratio adjustment under food stress: maximization of quality or numbers of offspring? *Am. Nat.* 112:381–88
96. Nunney, L. 1985. Female-biased sex ratios: individual or group selection? *Evolution* 39:349–61
97. Packer, C., Pussey, A. E. 1987. Intrasexual cooperation and the sex ratio in African lions. *Am. Nat.* 130:636–42
98. Ricklefs, R. E. 1968. On the limitation of brood size in passerine birds by the ability of adults to nourish their young. *Proc. Nat. Acad. Sci. USA* 61:847–51
99. Shaw, R. F. 1958. The theoretical genetics of the sex ratio. *Genetics* 43:149–63
100. Shaw, R. F., Mohler, J. D. 1953. The selective advantage of the sex ratio. *Am. Nat.* 87:337–42
101. Smith, C. C., Fretwell, S. D. 1974. The optimal balance between size and number of offspring. *Am. Nat.* 108:499–506
102. Stenseth, N. C. 1978. Is the female biased sex ratio in wood lemming *Myopus schisticolor* maintained by cyclic inbreeding? *Oikos* 30:83–89
103. Stephens, D. W., Krebs, J. R. 1986. *Foraging Theory*. Princeton, NJ: Princeton Univ. Press
104. Symington, M. M. 1987. Sex ratio and maternal rank in wild spider monkeys: when daughters disperse. *Behav. Ecol. Sociobiol.* 20:421–25
105. Taylor, P. D. 1981. Intra-sex and inter-sex sibling interactions as sex ratio determinants. *Nature* 291:64–66
106. Taylor, P. D. 1988. Inclusive fitness models with two sexes. *Theor. Popul. Biol.* 34:145–68
107. Taylor, P. D., Sauer, A. 1980. The selective advantage of sex-ratio homeostasis. *Am. Nat.* 116:305–10
108. Trivers, R. L., Hare, H. 1976. Haplodiploidy and the evolution of the social insects. *Science* 191:249–63
109. Trivers, R. L., Willard, D. E. 1973. Natural selection of parental ability to vary the sex ratio of offspring. *Science* 179:90–92
110. Turelli, M. 1985. Effects of pleiotropy

on predictions concerning mutation–selection balance for polygenic traits. *Genetics* 111:165–95
111. Verner, J. 1965. Selection for the sex ratio. *Am. Nat.* 99:419–22
112. Weir, J. A. 1971. Genetic control of sex ratio in mice. In *Sex Ratio at Birth–Prospects for Control,* ed. C. A. Kiddy, H. D. Hafs, pp. 43–54. Albany, NY: Am. Soc. Anim. Sci.
113. Williams, G. C. 1979. The question of adaptive variation in sex ratio in outcrossed vertebrates. *Proc. R. Soc. London, Ser. B* 205:567–80
114. Wright, S. L., Crawford, C. B., Anderson, J. L. 1988. Allocation of reproductive effort in *Mus domesticus:* responses of offspring sex ratio and quality to social density and food availability. *Behav. Ecol. Sociobiol.* 23:357–65
115. Yamaguchi, Y. 1985. Sex ratios of an aphid subject to local mate competition with variable maternal condition. *Nature* 318:460–62

*Annu. Rev. Ecol. Syst. 1990. 21:57–68*


# BIOGEOGRAPHY OF NOCTURNAL INSECTIVORES: Historical Events and Ecological Filters

*William E. Duellman*

Museum of Natural History, University of Kansas, Lawrence, Kansas 66045

*Eric R. Pianka*

Department of Zoology, University of Texas, Austin, Texas 78712–1064

KEY WORDS: biogeography, community ecology, frogs, lizards, tropics

---

## INTRODUCTION

One of the areas of overlap between systematics and ecology that has so far proven intractable to our understanding is community organization: To what extent does history constrain the structure of ecological communities? Or, how important is ecology in filtering historical elements? To clarify the distinction between the historical and the contemporary ecological perspective, consider the two extremes: (*a*) ecology is unimportant but history important, versus (*b*) ecology is important and history unimportant. (Obviously, any given situation in the real world lies somewhere between these hypothetical end points). However, a particular example can always be viewed from either perspective. For example, consider the Australian marsupial fauna. Historical: the facts that marsupials were present, but placentals failed to reach Australia are "accidental" events that profoundly shaped the entire biota. Ecological: these marsupials radiated into numerous ecological niches (arboreal, terrestrial, herbivores, megaherbivores, insectivores, and carnivores), closely paralleling adaptive radiations of placentals elsewhere. Indeed, such convergent responses demonstrate that ecological factors direct evolutionary responses.



0066-4162/90/1120-0057$02.00

Major differences in the composition of faunas in different biogeographic regions constitute a perplexing dilemma for students of community organization. We consider here certain global patterns in the diversity and geographic distribution of various nocturnal insectivores, particularly lizards and frogs. We pose three questions: (*a*) Can the disparity between species richness of nocturnal lizards in the old and new world tropics be explained on the basis of historical events? (*b*) Could the paucity of nocturnal lizard species in the New World be a result of competition with frogs (or other nocturnal insectivores)? (*c*) Could an earlier adaptive radiation of frogs have precluded gecko diversification in the neotropics? In this context we are concerned with patterns of diversity of nocturnal lizards and frogs and their occurrence in communities in the tropical parts of the Australopapuan, Ethiopian, Neotropical, and Oriental regions.

## NOCTURNAL LIZARD AND FROG DIVERSITY

### *Taxonomic Diversity*

By far, most nocturnal lizards in the world are members of the Infraorder Gekkonomorpha which contains about 860 species (25). Of the diverse groups of gekkoideans, the family Eublepharidae (5 genera, 19 species) is primarily Holarctic. One genus enters tropical Asia and one enters the neotropics as far south as Costa Rica; two genera are endemic to Africa. The family Pygopodidae is restricted to the Australopapuan region and, according to Kluge (25), contains both the nocturnal Diplodactylinae (13 genera, 84 species), some of which are arboreal and others terrestrial, and the snakelike Pygopodinae (7 genera, 31 species), most of which are crepuscular or nocturnal. The large family Gekkonidae is pantropical in distribution. The subfamily Teratoscincinae (1 genus, 4 species) occurs in the Palearctic. The subfamily Gekkoninae is pantropical. The range of the subfamily is encompassed by members of the tribe Gekkonini (62 genera, 603 species), nocturnal lizards, some of which are arboreal and others, terrestrial. Members of the tribe Sphaerodactylini are endemic to tropical America, where they have undergone a substantial adaptive radiation (5 genera and 120 species). Most species are diurnal, but some are crepuscular (46). The majority of sphaerodactylines are terrestrial, although a few species are arboreal.

Within the American tropics, there are few nocturnal lizards, all but one of which is a gekkoidean. The exception is the xantusiid *Lepidophyma flavimaculatum,* a forest floor inhabitant in Central America. Among nocturnal gekkoideans in the New World, most species are terrestrial in xeric regions (e.g. the deserts of western South America and Argentina) or semi-arid regions characterized by scrub forest or woodland-savanna (e.g. the cerrados and caatinga of Brazil), but two species are arboreal in the caatinga (47, 49). Within the tropical rainforest in South America, the only arboreal, nocturnal lizards are gekkonids. The species of *Hemidactylus* living there apparently arrived in South America via trans-Atlantic dispersal (24) Usually most species of *Hemidactylus* are found only on buildings in towns, and they are not considered to be components of natural communities. *Thecadactylus rapicauda,* a very large gecko, is the only nocturnal, arboreal lizard in neotropical rainforest communities (5, 6, 9). Five species of large, nocturnal, arboreal gekkonines occur in the West Indies, but usually only a single species is found at any given site.

A similar biogeographic situation exists among scincid lizards, virtually all of which are diurnal. Scincids are highly diverse in the Old World, but there is a distinct paucity of scincids in the Neotropics (18). Only lygosomine skinks occur in South America, where they are represented by 15 species of *Mabuya,* a genus with many species in Africa and Asia. This biogeographic parallel between scincids and gekkoideans deserves further study.

Principal families of nocturnal anurans in the American tropics are the Bufonidae, Centrolenidae, Hylidae (Hylinae and Phyllomedusinae), Leptodactylidae, and Microhylidae. The frog fauna of Australia is composed primarily of two endemic family groups, the Myobatrachidae and the Hylidae (Pelodryadinae). In Africa, major components are the Bufonidae, Hyperoliidae, Microhylidae, Ranidae, and Rhacophoridae (represented by only three species of *Chiromantis*). Tropical Asia differs from Africa by lacking hyperoliids and by having the Pelobatidae (Megophryinae) plus a diversity of rhacophorids.

## *History*

Gekkoideans represent an ancient lizard lineage thought to have arisen in Asia, probably during the Late Jurassic or Early Cretaceous; a Paleocene fossil is known from Brazil (14). The present pattern of distribution of gekkoideans in all biogeographic regions reflects Late Mesozoic and Cenozoic plate movements (25), as well as dispersal. Dispersal capabilities of some geckos are considerable, as is evidenced by their presence on many oceanic islands and frequent movements via humans.

Although the earliest separation of frogs onto different continents occurred with the original break up of Pangaea in the Jurassic, the primary tectonic factors affecting major patterns of anuran distribution were the plate movements in the Cretaceous and Cenozoic (12). Frogs of the family Pipidae are known from the Cretaceous of Africa and South America. By the early Cenozoic, the families Bufonidae, Hylidae, and Leptodactylidae also are represented in South America.

Using principles of phylogenetic systematics (19), testable hypotheses of phylogenetic relationships within monophyletic groups can be proposed. Superposition of phylogenetic cladograms on geographic areas indicates past geographic histories of areas; this approach has become known as vicariance biogeography (38, 39). Groups of organisms of approximately the same age that inhabit the same regions should show similar distributional patterns if the groups were affected by the same historical events (8). Of course, extinction of a taxon in a particular region and absence of fossils there will not support the vicariance model. Savage (40) proposed an alternative method for testing the validity of biogeographic hypotheses. He used events in earth history to predict general recurrent patterns of phylogenetic relationships; these patterns can be tested against patterns derived from phylogenetic analyses.

Unfortunately, few taxa involved have been subjected to such critical analyses. Kluge's (25) work on gekkoideans does not address relationships of Neotropical gekkonines, but Kluge (personal communication) believes that *Tarentola* in the West Indies is related to the African *Pachydactylus* and that the sister taxa of *Aristelliger* and *Thecadactylus* are probably African. With the exception of the widespread gekkonine genera *Cyrtodactylus, Gehyra, Hemidactylus,* and *Phyllodactylus* (which may not be monophyletic)—all of which are excellent dispersalists, as evidenced by their occurrence on many oceanic islands—gekkonoid evolution has paralleled earth history (Figure 1). Most notable is the endemic Pygopodidae in Australia.

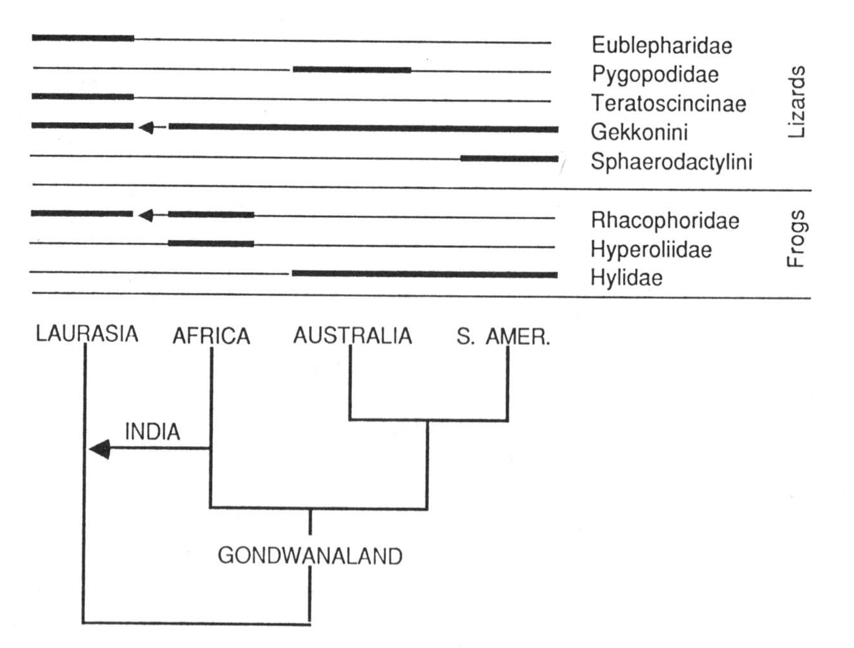


*Figure 1* Area cladogram of continents and associated groups of gekkonoid lizards and major groups of arboreal frogs. Horizontal bars indicate distributions of taxa; arrows indicate transport from Gondwanaland to Laurasia via the Indian Plate.

**Table 1** Species densities of nocturnal lizards (principally gekkonoids) and nocturnal frogs at various localities.

| Site | Lizards | | Frogs | | Reference |
|---|---|---|---|---|---|
| | Terrestrial | Arboreal | Terrestrial | Arboreal | |
| Neotropical rainforest: | | | | | |
| Chinajá, Guatemala | 1* | 1 | 3 | 7 | Duellman (10) |
| Barro Colorado Is., Panama | 1* | 1 | 6 | 17 | " |
| Belém, Brazil | 0 | 1 | 8 | 22 | " |
| Santa Cecilia, Ecuador | 0 | 1 | 17 | 52 | " |
| Cuzco Amazónico, Peru | 0 | 1 | 15 | 37 | " |
| Panguana, Peru | 0 | 1 | 17 | 34 | " |
| Neotropical dry forest: | | | | | |
| Exu, Brazil | 1 | 2 | 8 | 6 | Vitt (personal communication) |
| Asian rainforest: | | | | | |
| Labang, Borneo | 2 | 13 | 24 | 9 | Inger (21) |
| Pesu, Borneo | 0 | 6 | 34 | 17 | " |
| Nanga Tekalit, Borneo | 1 | 10 | 37 | 9 | " |
| Ulu Gombak, Malaya | 4 | 6 | 8 | 3 | " |
| Bukit Lanjan, Malaya | 3 | 5 | 19 | 11 | " |
| Sakaeret, Thailand | 3 | 7 | 10 | 5 | Inger & Colwell (23) |
| African deserts: | | | | | |
| Bloukrans, South Africa | 3 | 3 | 1 | 0 | Pianka (32) |
| Ludrille, KGNP, Botswana | 3 | 3 | 1 | 0 | " |
| Tsabong, Botswana | 3 | 2 | 1 | 0 | " |
| Australian deserts: | | | | | |
| Laverton | 10** | 2 | 1 | 9 | " |
| Red Sands | 10** | 4 | 1 | 0 | " |
| E-Area | 6** | 5 | 1 | 0 | " |

* Lepidophyma (Xantusiidae)
** Includes two non-gekkonid taxa, *Egernia* and *Eremiascincus* (Scincidae)

Frogs also have an evolutionary history associated with tectonic events. The endemic Australopapuan pygopodids are paralleled in frogs by pelodryadine hylids, whereas phyllomedusine and hyline hylids are Neotropical, hyperoliids Ethiopian, and rhacophorids principally Oriental (Figure 1).

## *Communities*

As has been noted by Cogger (3), nocturnal lizard faunas are impoverished throughout the New World, (Table 1). Even in nonrainforest areas, communities usually contain only a single species (31, 32). Comparable natural habitats in the Old World usually support much more diverse nocturnal lizard faunas; for example, a dry forest in Thailand has 10 species of nocturnal gekkonids (23). Desert habitats in Australia and southern Africa support from 3 to 10 sympatric species of nocturnal gekkoideans (33, 34). Even the cold deserts of Asia seem to support a moderate to substantial diversity of gekkoideans (1, 43, 44). The fact that this pattern applies across all habitats strongly suggests that historical biogeographic factors have played a major role in gekkonid radiations.

In contrast to gekkoideans, frogs are most diverse in the American tropics; about 45% of the 3650 species of frogs (updated from Frost 17) occur there (10). Within the neotropics, the greatest diversity of anurans is in the lowland rainforests—41 and 42 species at two sites in Costa Rica, 38–84 ($\bar{x}$ = 56) species at 10 Amazonian sites, and 35–36 ($\bar{x}$ = 50) species at three cis-Andean sites (10). These numbers are higher than for most sites in the Old World tropics—21–51 ($\bar{x}$ = 35) species at five sites in rainforest in Malaya and Borneo, and 19 and 20 species at two sites in seasonally dry forest in Thailand (21); 24 species at a site in seasonally dry forest in northern Australia (45); 49 species in a region of rainforest in Cameroon (30); and 20 species at each of two sites in seasonal rainforest in Nigeria (41).

Most frogs are nocturnal; at six sites in neotropical rainforests, 86–92% ($\bar{x}$ = 88%) of the species are nocturnal (11). In the neotropics, the majority of nocturnal arboreal frogs are members of the family Hylidae, although the small centrolenids and some *Eleutherodactylus* are arboreal. For example, at Santa Cecilia in Amazonian Ecuador, 52 of 81 (64%) species are nocturnal and arboreal (6). Generally, fewer nocturnal, arboreal frogs occur in the Old World tropics. At Foulassi, Cameroon, only 19 of 49 (38%) of frogs (all hyperoliids and rhacophorids) are in this category (30). In the Oriental region, nocturnal, arboreal frogs are rhacophorids. At sites in southeastern Asia and Borneo, all frogs are nocturnal; at five sites in rainforest, 20–50% (32%) of the species are arboreal, and at two sites in seasonally dry forest 10% and 21% of the species are arboreal (21). Numbers of species of frogs are inversely related to those of nocturnal lizards (Figure 2).

## EVALUATION OF DATA

### *Historical Events*

Evidence from phylogenetic relationships of gekkonoids and frogs and from earth history neither support nor deny the notion that the great disparity in relative numbers of frogs as compared with gekkonoids between the new and old world tropics is the result of differences in historical biogeography. The earliest known fossils of gekkonoids and hylids in the neotropics date from the Paleocene of Brazil (14, 15). No earlier fossils of either group are known. Thus, both groups apparently were present on the South American plate at the time of the break up of Gondwanaland in the Late Mesozoic.

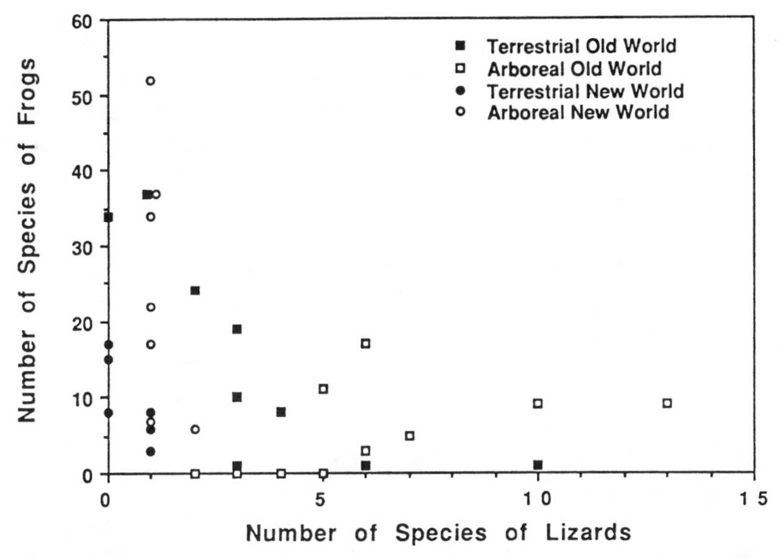


*Figure 2* Frog species density plotted against nocturnal lizard species density at seven New World (circles) and 12 Old World (squares) sites. Closed symbols represent terrestrial species; open symbols are arboreal species.

## *Frog-Lizard Competition*

If the carrying capacity of the environment limits the number of nocturnal insectivores through limited food, perches, or shelter, some sort of balance between numbers of species of nocturnal, arboreal frogs and lizards might be expected. At six sites of southeastern Asia and Borneo, Inger (21) reported 3–17 species of nocturnal, arboreal frogs and 6–13 species of arboreal gekkonoids. Ratios of species of frogs to species of lizards range from 0.40 to 2.83 ($\bar{x} = 1.25$). In contrast, at six sites in neotropical rainforests, only one species of nocturnal, arboreal gekkonoid *(Thecadactylus rapicauda)* has been found (9); at these sites the number of nocturnal, arboreal species of frogs varies from 7 to 52 ($\bar{x} = 28.2$) (10). However, as possible competitors with the large nocturnal gecko, the values for frogs in the neotropical rainforests are inflated. At these six sites the number of nocturnal, arboreal frogs having snout-vent lengths of more than 50 mm is 4–19 ($\bar{x} = 11.0$), a value that still is 8–10 times higher than those for southeastern Asia.

Abundances of frogs and lizards on the forest floor are an order of magnitude greater in the neotropics than in Indo-Malayan primary rainforest (22, 42). Presumably this disparity exists because of the comparatively lower abundance of insects (as a food supply) in the more seasonal Indo-Malayan dipterocarp forest. Nevertheless, levels of competition for food could be similar in both regions owing to the profound differential in consumer biomass.

Only one nocturnal, arboreal gecko exists on most islands in the West Indies—*Thecadactylus rapicauda* on the Lesser Antilles, *Aristelliger lar* on Hispaniola, and *Tarentola americana* on Cuba and the Bahamas. Two large, nocturnal, arboreal frogs occur on Hispaniola, but only one species is present in Cuba and the Bahamas. Thus, diversity is low, and the ratio of nocturnal, arboreal frogs to lizards in the West Indies is much closer to ratios observed in southeastern Asia than in the rainforests in Central and South America.

Daytime retreats for *Thecadactylus* include crevices under bark, cracks in tree trunks or limbs, and hollows in limbs. Although some nocturnal frogs also are found in such shelters, usually they are in more moist situations, particularly in water-filled cavities in trees or in bromeliads. Feeding sites for *Thecadactylus* are tree trunks and buttresses and large limbs, sites frequented by only a few frogs (e.g. *Osteocephalus taurinus* and *Phrynohyas venulosa*). Most nocturnal frogs in the neotropics perch on slender branches or leaves. Arthropods are principal prey of most nocturnal frogs and *Thecadactylus*. Dominant prey items identified in stomachs of *Thecadactylus* in Surinam and Ecuador were orthopterans (6, 20). Orthopterans also are common prey items of most larger treefrogs (6). However, because orthopterans are extremely abundant in neotropical forests and because feeding stations of *Thecadactylus* differ from those of most frogs, competition for food seems unlikely. Evidence for or against the existence of competition between frogs and gekkonoids for resources is hardly compelling.

Other nocturnal, insectivorous taxa that may interact with frogs and lizards include mammals, birds, and arachnids. Mouse possums *(Marmosa)* and night monkeys *(Aotus)* are abundant, nocturnal, arboreal insectivores in the American tropics; although the kinkajou *(Potos flavus)* feeds primarily on fruit at night, it also takes insects (27). Nocturnal, arboreal primates that feed on large insects also occur in the Old World (e.g. *Galago* in Africa and *Nycticebus* in southeastern Asia). Some kinds of bats glean large insects off vegetation at night. This feeding strategy is used by some phyllostomatids (e.g. *Chirotopterus, Phyllostomus,* and *Trachops*) in the neotropics and megadermatids (e.g. *Cardioderma* and *Megaderma*) in Africa. Also, small owls may well be important predators on insects. These endotherms could consume more large nocturnal insects than do the ectothermic gekkonoids and anurans, but quantitative data are not available. Also, nocturnal, arboreal arachnids prey on insects. Whip scorpions (Thelyphonidae, Amblypygidae, Uropygidae) and large predatory spiders, such as hersiliids, lycosids, and theraphosids, are abundant in the tropics throughout the world and feed on relatively large prey such as orthopterans and small vertebrates. Although we have been unable to determine if there is a differential in species diversity of arachnid insectivores between the old and new world tropics, the historical biogeography of spiders parallels that of gekkonoids and anurans (35).

### *Adaptive Radiations*

The limited fossil evidence does not support the idea that adaptive radiation on any continent of either gekkoideans or anurans preceded that of the other group. However, the diversity of living taxa certainly suggests that anuran radiation was far greater than gekkonine radiation in the neotropics. Frogs have also undergone a more extensive adaptive radiation in the neotropics than in the Old World. These differences are real unless major extinctions have occurred, but there is no basis for such speculation.

Nocturnal geckos are far more diverse and abundant in the Old World than in the New World. Lizards are generally more abundant (or at least much more conspicuous) in open habitats (e.g. deserts and savanna-woodland) than in closed-canopy forests. But this comparison of lizards as a group is unrealistic, because most lizards (gekkonoids being the principal exception) are heliophilic. The greatest diversity of nocturnal gekkonines in South America is in subarid regions, such as the cerrados and caatinga of Brazil; six species (three of which are nocturnal) occur at one site in the caatinga of northeastern Brazil (46, 47, 49, L. J. Vitt, personal communication). This number is still below that in arid regions in Africa and Australia (31, 32).

In contrast to lizards, anurans are most speciose in regions of high humidity; the richest anuran communities in the world exist in the extensive upper Amazon Basin and in the more restricted cis-Andean rainforests, where dry seasons are nonexistent or very short. Numbers of coexisting species are notably lower in monsoonal climates, such as those in tropical Africa and southeastern Asia, and decline further in deserts. Another factor contributing to the comparative paucity of arboreal anurans in the Old World tropics may be the absence there of arboreal water-holding plants (bromeliads), common diurnal retreats for neotropical anurans.

The distribution of tropical environments has changed drastically during the Late Cenozoic and Quaternary. Principally during the Pleistocene, humid tropical habitats were greatly restricted during glacial phases, whereas humid forests expanded during pluvial (or interglacial) phases. These changes have been documented in South America by Prance [(36), but see Connor (4) for a critique of interpretations]. They also are known to have occurred in Africa and Asia (16). These dramatic ecological changes have been postulated as important in determining modern distributions and patterns of speciation (7, 26, 28, 48), as well as extinctions (37).

## DISCUSSION

In the absence of evidence for large-scale extinctions of gekkonines in the neotropics or of nocturnal, arboreal anurans in the Old World tropics, we suggest that both groups had similar temporal histories, but that adaptive radiation in the two groups occurred at different rates in the Old World and in the New World. In the Old World tropics, greater climatic seasonality and the absence of bromeliads favored gecko diversification, whereas in the neotropics, less seasonal climates and the presence of bromeliads favored frog radiation.

Perhaps the difference in numbers of species of gekkonoids and frogs in the old and new world tropics reflects innate morphological and physiological design constraints of the organisms, coupled with recent climatic-ecological histories of the regions in which they live. In the extensive aseasonal rainforests in the New World, frogs have undergone a tremendous radiation; abundance of resources allows coexistence of many species. Gekkonines, most of which apparently do not tolerate environments with constantly high humidity, are much more diverse in the seasonally dry Old World tropics, where anurans are less diverse because many kinds are less capable of withstanding dry seasons. These ideas need to be tested by comparing lizard and frog communities at sites with comparable climates in Africa, Asia, and South America.

Although some evolutionary biologists (2, 13, 28, 29, 50) have included historical factors in formulating ecological hypotheses, community ecologists typically have concerned themselves with phenomena that organize communities in ecological time, such as coadjustments of patterns of resource utilization among coexisting species. Even when a coevolutionary perspective is adopted, the evolutionary time scale considered is relatively brief, such as for Pleistocene forest refugia (36). Such a "snapshot" approach provides only a brief glimpse of the factors involved in the organization of communities. Community ecology will emerge on a new level of awareness and understanding as this evolutionary time scale is broadened to include historical events and factors such as (*a*) biogeographic histories of taxa involved; (*b*) climatic histories of areas being studied; and (*c*) innate physiological tolerances and morphological design constraints of taxa involved.

## SUMMARY

The disparity in numbers of nocturnal lizards and frogs in the old and new world tropics cannot be interpreted as owing to differences in geographic histories of lineages, nor to competition between geckos and frogs. However, clearly there has been a much greater adaptive radiation of gekkonines in the Old World as compared to the New World. In contrast, frogs have diversified to a considerably greater extent in the neotropics. These differential adaptive radiations reflect the most recent climatic-ecological histories of the regions, and the historical biogeography of important biotic components (e.g. absence of bromeliads in the Old World tropics). The example of geckos and tree frogs underscores the necessity of considering historical events and biological constraints in the interpretation of patterns of community organization.

## ACKNOWLEDGMENTS

We thank Kirk Winemiller for the idea that frogs could replace nocturnal lizards in the neotropics. Albert Schwartz, Arnold G. Kluge, and Laurie J. Vitt shared their knowledge of gekkonoids. We thank Robert F. Inger, Arnold G. Kluge and Laurie J. Vitt for reading the manuscript.

### *Literature Cited*

1. Baur, A. M. 1987. [Review of] The gekkonid fauna of the U.S.S.R. and adjacent countries. *Copeia* 1987:525–27
2. Brooks, D. R. 1985. Historical ecology: a new approach to studying the evolution of ecological associations. *Ann. Missouri Bot. Gard.* 72:660–80
3. Cogger, H. G. 1987. [Review of] Ecology and natural history of desert lizards. *Q. Rev. Biol.* 62:114
4. Connor, E. F. 1986. The role of Pleistocene forest refugia in the evolution and biogeography of tropical biotas. *Trends Evol. Ecol.* 1:165–58
5. Duellman, W. E. 1963. Amphibians and reptiles of the rainforests of southern El Petén, Guatemala. *Univ. Kans. Publ. Mus. Nat. Hist.* 15:205–49
6. Duellman, W. E. 1978. The biology of an equatorial herpetofauna in Amazonian Ecuador. *Misc. Publ. Mus. Nat. Hist. Univ. Kansas* 65:1–352
7. Duellman, W. E. 1982. Quaternary climatic-ecological fluctuations in the lowland tropics: frogs and forests. In *Biological Diversification in the Tropics*. ed. G. T. Prance, pp. 389–402. New York: Columbia Univ. Press
8. Duellman, W. E. 1986. Plate tectonic, phylogenetic systematics and vicariance biogeography of anurans: methodology for unresolved problems. In *Studies in Herpetology. Proc. European Herp. Meet., Prague, 1985*, ed. Z. Rocek, pp. 59–62. Prague: Charles Univ.
9. Duellman, W. E. 1987. Lizards in an Amazonian rain forest community: resource utilization and abundance. *Natl. Geog. Res.* 3:489–500
10. Duellman, W. E. 1988. Patterns of species diversity in anuran amphibians in the American tropics. *Ann. Missouri Bot. Gard.* 75:79–104
11. Duellman, W. E. 1989. Tropical herpetofaunal communities: patterns of community organization in neotropical rainforests. In *Vertebrates in Complex Tropical Systems,* ed. M. L. Harmelin-Vivien, F. Bourliére, pp. 61–88, New York: Springer-Verlag
12. Duellman, W. E., Trueb, L. 1986. *Biology of Amphibians*. New York: McGraw-Hill
13. Endler, J. A. 1982. Problems in distinguishing historical from ecological factors in biogeography. *Am. Zool.* 22: 441–52
14. Estes, R. 1983. The fossil record and early distribution of lizards. In *Advances in Herpetology and Evolutionary Biology*. ed. A. G. J. Rhodin, K. Miyata, pp. 365–98, Cambridge, Mass: Mus. Comp. Zool. Harvard Univ.
15. Estes, R., Reig, O. A. 1973. The early fossil record of frogs: a review of the evidence. In *Evolutionary Biology of the Anurans,* ed. J. L. Vial, pp. 11–63. Columbia: Univ. Missouri Press
16. Flenley, J. R. 1979. *The Equatorial Rain Forest: A Geological History*. London: Butterworths
17. Frost, D. R. 1985. *Amphibian Species of the World*. Lawrence, Kans: Assoc. Syst. Coll.
18. Greer, A. E. 1970. A subfamilial classification of scincid lizards. *Bull. Mus. Comp. Zool. Harvard* 139:151–84
19. Hennig, W. 1966. *Phylogenetic Systematics*. Urbana Univ. Illinois Press
20. Hoogmoed, M. S. 1973. *Notes on the Herpetofauna of Surinam. IV. The Lizards and Amphisbaenians of Surinam*. The Hague: W. Junk
21. Inger, R. F. 1980. Relative abundances of frogs and lizards in forests of southeastern Asia. *Biotropica* 12:14–22
22. Inger, R. F. 1980. Densities of floor-dwelling frogs and lizards in lowland forests of southeast Asia and Central America. *Am. Nat.* 115:761–70
23. Inger, R. F., Colwell, R. K. 1977. Organization of contiguous communities of amphibians and reptiles in Thailand. *Ecol. Monogr.* 47:229–53

24. Kluge, A. G. 1969. The evolution and geographic origin of the New World *Hemidactylus mabouia-brookii* complex *(Gekkonidae, Sauria). Misc. Publ. Mus. Zool. Univ. Michigan* 138:1–78
25. Kluge, A. G. 1987. Cladistic relationships in the Gekkonoidea (Squamata, Sauria). *Misc. Publ. Mus. Zool. Univ. Michigan* 173:1–54
26. Laurent, R. B. 1973. A parallel survey of equatorial amphibians and reptiles in Africa and South America. In *Tropical Forest Ecosystems in Africa and South America: A Comparative Review*. ed. B. J. Meggers, E. S. Ayensu, D. W. Duckworth, pp. 259–66, Washington: Smithsonian Inst.
27. Nowak, R. M., Paradiso, J. L. 1983. *Walker's Mammals of the World*. Baltimore: Johns Hopkins Univ. Press.
28. Pearson, D. L. 1982. Historical factors and bird species richness. In *Biological Diversification in the Tropics*, ed. G. T. Prance, pp. 389–402. New York: Columbia Univ. Press
29. Pearson, D. L., Blum, M. S., Jones, T. H., Fales, H. M., Gonda, E., White, B. R. 1988. Historical perspective and the interpretation of ecological patterns: defensive compounds of tiger beetles (Coleoptera: Cicindelidae). *Am. Nat.* 132: 404–16
30. Perret, J.-L. 1966. Les amphibiens du Cameroun. *Zool. Jb. Syst.* 8:289–464
31. Pianka, E. R. 1985. Some intercontinental comparisons of desert lizards. *Natl. Geog. Res.* 1:490–504
32. Pianka, E. R. 1986. *Ecology and Natural History of Desert Lizards*. Princeton, NJ: Princeton Univ. Press
33. Pianka, E. R., Huey, R. B. 1978. Comparative ecology, niche segregation, and resource utilization among gekkonid lizards in the southern Kalahari. *Copeia* 1978:691–701
34. Pianka, E. R., Pianka, H. D. 1976. Comparative ecology of twelve species of nocturnal lizards (Gekkonidae) in the Western Australian desert. *Copeia* 1976:125–42
35. Platnick, N. I. 1981. Spider biogeography: past, present, and future. *Rev. Arachnol.* 3:85–96
36. Prance, G. T. 1982. *Biological Diversification in the Tropics*. New York: Columbia Univ. Press
37. Richards, P. W. 1973. Africa, the "odd man out." In *Tropical Forest Ecosystems in Africa and South America: A Comparative Review*. ed. B. J. Meggers, E. S. Ayensu, and D. W. Duckworth, pp. 21–26, Washington: Smithsonian Inst. Press
38. Rosen, D. E. 1976. A vicariance model of Caribbean biogeography. *Syst. Zool.* 24:431–64
39. Rosen, D. E. 1978. Vicariant patterns and historical explanation in biogeography. *Syst. Zool.* 27:159–88
40. Savage, J. M. 1982. The enigma of the Central American herpetofauna: dispersals or vicariance. *Ann. Missouri Bot. Gard.* 69:464–547
41. Schiøtz, A. 1963. The amphibians of Nigeria. *Vidensk. Medd. Dansk Naturh. Foren.* 125:1–92
42. Scott, N. J. 1976. The abundance and diversity of the herpetofaunas of tropical forest litter. *Biotropica* 8:41–58
43. Szczerbak, N. N. 1986. Review of the Gekkonidae in the fauna of the USSR and neighboring countries. In *Studies in Herpetology. Proc. Eropean Herp. Meetings, Prague, 1985*, ed. Z. Rocek, pp. 705–9, Prague: Charles Univ.
44. Szczerbak, N. N., M. L. Golubev. 1986. *Gekkonidae in the Fauna of the USSR and Neighboring Countries*. Kiev: Nauk. Dumka
45. Tyler, M. J., Crook, G. A., Davies, M. 1983. Reproductive biology of the frogs of the Magela Creek System, Northern Territory. *Rec. S. Australian Mus.* 18: 415–40
46. Vanzolini, P. E. 1968. Geography of South American Gekkonidae (Sauria). *Arq. Zool. São Paulo* 17:85–112
47. Vanzolini, P. E., Ramos-Costa, A. M. M., Vitt, L. J. 1980. *Repteis das Caatingas*. Rio de Janeiro: Acad. Brasil. Ciências
48. Vanzolini, P. E., Williams, E. E. 1970. South American anoles: the geographic differentiation and evolution of the *Anolis chrysolepis* species group (Sauria, Iguanidae). *Arq. Zool. São Paulo* 19:1–298
49. Vitt, L. J. 1986. Reproductive tactics of sympatric gekkonid lizards with a comment on the evolutionary and ecological consequences of invariant clutch size. *Copeia* 1986:773–786
50. Wiens, J. A. 1977. On competition and variable environments. *Am. Sci.* 65: 590–97

*Annu. Rev. Ecol. Syst. 1990. 21:69–91*


# THE END-PERMIAN MASS EXTINCTION

*D. H. Erwin*

Department of Geological Sciences, Michigan State University, East Lansing, Michigan 48824

KEY WORDS: mass extinction, end-permian extinction, global diversion, evolutionary faunas, global climate

---

## INTRODUCTION

The most severe biotic crisis of the Phanerozoic occurred at the end of the Permian 245 million years ago (Ma); the crisis eliminated 54% of marine families and perhaps as many as 96% of all marine species (65, 80, 82). This was at least twice the magnitude of the end-Ordovician extinction (the second most severe mass extinction) and eliminated far more taxa than did the end-Cretaceous mass extinction. The end-Permian extinction marks the end of marine communities dominated by the sessile, epifaunal filter-feeding articulate brachiopods, bryozoans, crinoids, and other pelmatazoan echinoderms (80, 106), but this extinction also created evolutionary opportunities which led to new, highly mobile marine organisms, particularly molluscs, and an expansion of both predators and infaunal burrowers. The new community types established during the Mesozoic radiation continue to dominate modern seas. On land the dominant floras of the Paleozoic were gradually replaced by new plant assemblages, while Permian amphibians and therapsids were replaced by new groups of therapsids and early diapsids. These changes were the most extensive reorganization of the earth's biota between the Cambrian Metazoan Radiation and the present (78, 80); the changes occurred during an interval of severe climatic, tectonic, and geophysical changes and marked geochemical shifts (31).

Determining the rate and duration of the extinction and the correlations



0066-4162/90/1120-0069$02.00

between marine and terrestrial events is complicated by a major marine regression during the upper Permian. This regression exposed most of the continental shelves, leaving few normal marine deposits outside of South China. It is unclear to what extent the regression biases our understanding of the pace of extinction, since work in South China itself gives contradictory answers about the rate of extinction. The magnitude of the regression, however, makes it difficult to perform the high-resolution biostratigraphic studies so useful in elucidating the structure of other mass extinctions. Distinguishing between catastrophic, step-wise, and mass extinctions is further complicated by preservational bias, the Signor-Lipps effect, which will tend to produce an apparently gradual extinction whatever the actual extinction pattern (86).

Despite the magnitude of this extinction, several paleontologists have argued that the changes in evolutionary faunas were underway well before the late Permian, and that the mass extinction merely accelerated an ongoing process without substantively changing the outcome (8, 42, 78, 80). The end-Permian extinction is also the earliest identified in the 26 million year (myr) mass extinctions cycle (68–70) and thus plays a significant role in discussions concerning the evolutionary role of mass extinctions. This review addresses the nature of the extinction, the physical context in which it occurred, the likely mechanisms behind it, and the importance of this mass extinction in structuring the biota.

## PERMO-TRIASSIC DIVERSITY PATTERNS

### *The Marine Extinction*

GENERAL EXTINCTION PATTERNS Global marine familial diversity declined by 54% and generic diversity by from 78 to 84% during the final two stages of the Permian (Figure 1; the latter number includes genera seen only once in the fossil record; 80, 82; see also 34, 48, 55, 71, 74). Rarefaction analysis indicates that species diversity may have dropped by 96% (65, 82), although the differential extinction patterns (53, 79) would reduce the true species extinction level. Particularly hard hit during this episode were the epifaunal, sessile filter feeding groups which dominated Paleozoic level-bottom marine communities and reef ecosystems (23). These included crinoids (98% of families became extinct), the remaining tabulate and rugose corals (96%), various articulate brachiopods (78%), bryozoans (76%), cephalopods (71%) and foraminifera (50%) (53, 79) (Figure 2). Several other groups did relatively well, including the gastropods, sponges, and bivalves. Family diversity in the classes comprising the 'Paleozoic Evolutionary Fauna' (which includes brachiopods, bryozoans and crinoids) declined by 79%, while the families that came to comprise the 'Modern Evolutionary Fauna' declined by only 27% (79).

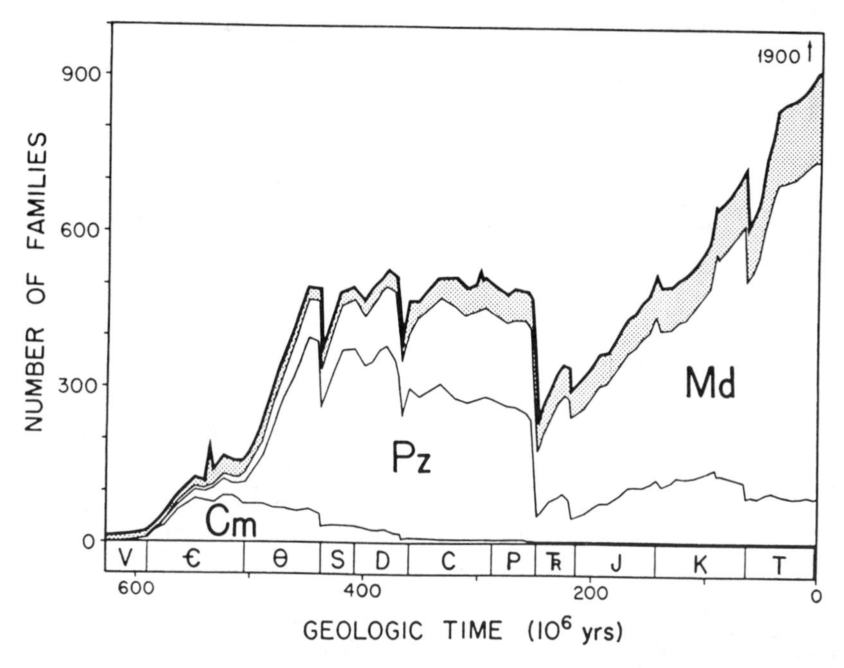


*Figure 1* Phanerozoic familial marine diversity divided into three evolutionary faunas on the basis of a Q-mode factor analysis of all extinct families. Cm, the Cambrian fauna; Pz, the Paleozoic fauna; Md, the Modern (Mesozoic-Cenozoic) fauna. The shaded area represents families not allocated by the factor analysis. Note the Paleozoic evolutionary fauna suffers a disproportionate drop during the end-Permian mass extinction. Reproduced with permission from Ref. 80.

There are several peculiar aspects of this extinction. First, the magnitude of the event was evident to paleontologists by 1841 and formed the basis for the division of the Paleozoic and Mesozoic eras (56, 57, 115). However, the erathem boundary led paleontologists to specialize on one side of the boundary or the other, and to ignore previous work on similar forms from across the boundary (2, 111). Second, many forms which disappear in the Upper Permian did not become extinct but reappeared in the Middle Triassic, after an absence of ten million years or more from the record. These so-called Lazarus-taxa (36) indicate that many refugia existed, presumably offshore islands, of which we have no record. Finally, many other lineages underwent significant morphological changes and are consequently assigned to new families even though there was no actual lineage extinction.

The extinctions within specific clades range from declines which begin by the late Guadalupian to others with a near-catastrophic extinction near the boundary. In some cases the difference is the extent of paleontologic in-

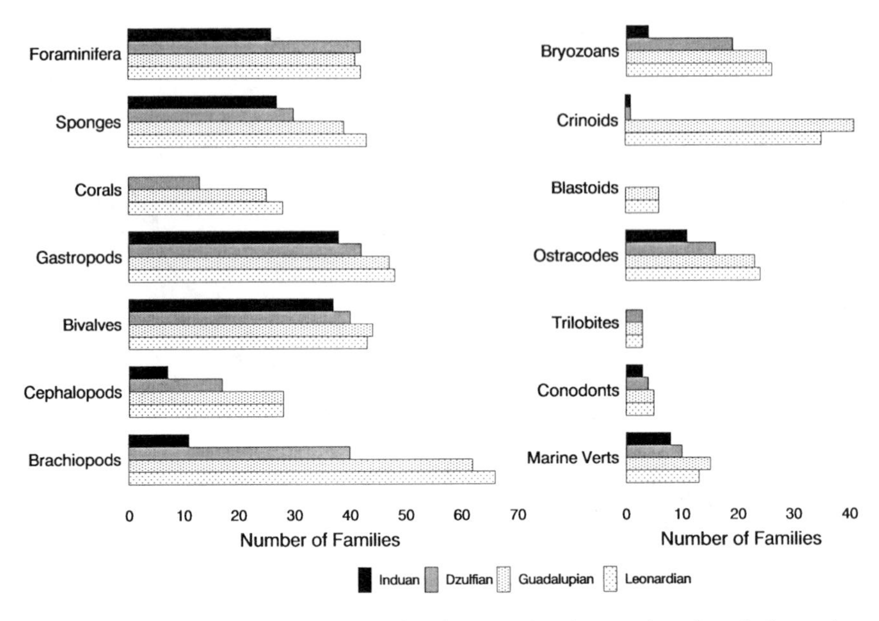


*Figure 2* Familial diversity patterns for selected groups of marine organisms from the Leonardian to Induan (earliest Triassic). Data is taken from (79) as amended. Note the higher extinction rate for the sessile, epifaunal brachiopods, bryozoans, corals, and crinoids.

vestigations in South China. Crinoids (41, 64), and tabulate and rugose corals (77) were in relative decline throughout the Permian (modern scleractinian corals did not appear until the Anisian, and their ancestors have not been identified). The bryozoans exhibit a gradual decline throughout the Guadalupian and Dzulfian, with a 76% familial extinction and a 86% generic extinction (95) during these two stages, although there is debate about the effect of this extinction at higher levels. Bryozoan workers have traditionally recognized the extinction of four orders at the end of the Permian, but more recent work suggests this may be an artifact of taxonomic practice (i.e. pseudo-extinction) (95). The foraminifera display a progressive decline during the Permian culminating in a rapid drop near the Permo-Triassic boundary. The most heavily affected taxa were architecturally complex groups, particularly those on tropical shelves (7). The fusulinids, the dominant benthic forams of the Permian were virtually wiped out.

Among the molluscan classes, the record of the Bivalvia from South China (where good latest Permian marine sections are available) displays a gradual pattern of decline beginning in the Guadalupian and culminating in the Changhsingian Stage. The extinction of bivalves, brachiopods, and some other groups may be related to differential removal of lineages with planktotrophic larval development (99, 102, 103). Modern articulate brachiopods are nonplanktotrophic, but diversity patterns of Paleozoic brachiopods suggest that both planktotrophic and nonplanktotrophic lineages existed. Nonplanktotrophic development appears to be more restrictive than planktotrophic development, which may account for the lack of success exhibited by post-Paleozoic brachiopods. Similar patterns of differential extinction may have also occurred in crinoids (93) and archaeogastropods (20, 21). In addition 75% of tropical articulate brachiopods families became extinct, but only 56% of extratropical families suffered the same fate (35), illustrating the increased extinction rates in the tropics. Conodonts, multielement phosphatic remains of a primitive, pelagic cordate, declined drastically in the Early Permian (10), but these remains exhibit no change in diversity and only a slight change in abundance across the Permo-Triassic boundary (11).

In general, although the picture is clouded by both the marine regression and the Signor-Lipps effect, the extinction begins at different times in different groups. The extinction began by the end of the Guadalupian, at least in the most affected clades, and continued into the lowest Triassic for a total duration of 5–8 myr, by far the longest duration for a mass extinction during the Phanerozoic.

BOUNDARY SECTIONS Despite the extensive marine regression, there are a number of well-preserved sections that span the interval, particularly in South China, India, and Pakistan, and which provide important information on the events at the boundary. In South China the standard position for the Permo-Triassic boundary occurs at the base of a 0.20–2.0 meter-thick sequence comprised of a very thin basal clay, a thin mudstone containing a mixture of Permian-type brachiopods and Triassic-type ammonites, a 20–30 cm thick dolomite unit with Permian brachiopods, the ammonoid *Otoceras woodwardi* and some foraminifera, and a final claystone containing the ammonoid *Ophiceras* and the bivalve *Claria,* both clear Triassic faunal elements. Permian brachiopods are very reduced in diversity and abundance in the topmost unit. The mixed faunal zones generally contain dwarfed species related to Permian species (84).

The mixture of faunas across the boundary and the lack of pronounced discontinuities suggest that deposition was fairly continuous in some sections (47, 84, but see 54, 96). However, graphic correlation among the Chinese sections, the reference stratotype at Guryul Ravine, Kashmir and sections in the Transcaucus and Iran (Sweet, personal communication 1989) indicates that the boundary is diachronous and that the mixed faunal beds in South China were deposited substantially above the boundary identified in Guryul Ravine, Kashmir. Determining the actual position of the extinction and the timing of extinction in different regions is critical to distinguishing a catastrophic from a gradual extinction, but resolving it will require additional detailed biostratigraphy.

The extinction is associated with depressed origination rates after the Guadalupian, but the drop in origination is itself insufficient to explain the magnitude of the extinction (*contra* 34, 71). Extinction rates are among the highest of the Phanerozoic (82) and peak by the end Guadalupian, well before the boundary itself. Both global diversity and individual fossil deposits remain depauperate into the Induan, and normal marine communities do not return until the latest Olenekian or Anisian, suggesting that the conditions that caused the extinction probably began by the late Guadalupian and continued into the lower Triassic. Recent work in South China supports this picture, although it does suggest that normal marine faunas may have persisted somewhat later in South China than elsewhere.

## *Terrestrial Vertebrate Extinctions*

Earlier work on extinction rates of terrestrial tetrapods suggested that the upper Permian vertebrate extinctions were not correlative with the marine extinctions and that the vertebrate extinction record was biased by the preservation of unusual faunas and by differential preservation (60, 61). However, a new compilation of global taxonomic diversity at finer stratigraphic resolution (4, 5) reveals increased extinction during the Sakmarian-Artinskian (58% of

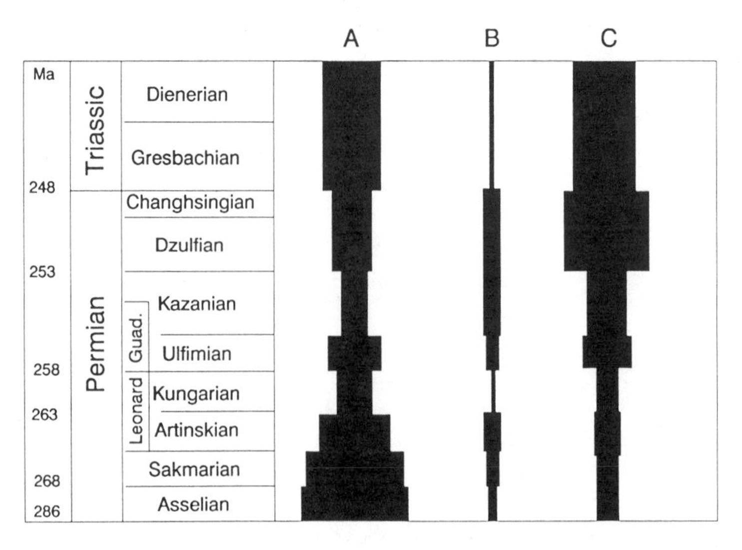


*Figure 3* Permo-Triassic terrestrial vertebrate diversity patterns. A, amphibians; B, early anapsids; C, Synapsids. Data from (4).

families), Dzulfian-Scythian (49%), and Carnian-Norian (late Triassic; 22%; Figure 3) (5, 6). These extinctions involved the progressive replacement of labyrinthodont amphibians, 'anapsids,' and early therapsids by early diapsids and new groups of therapsids (12, 61). During the end-Permian, 21 families of amniotes, largely synapsids, were eliminated. These taxa were broadly distributed geographically and spanned all size ranges. There is no evidence that the extinction was a consequence of selective preservation of unusual faunas. The peak extinction rate occurred in the earliest Dzulfian coinciding with the peak in marine extinctions.

### *Terrestrial Floral Extinctions*

Identifying a connection between the marine extinctions and the replacements in terrestrial floras has been difficult. The Paleophytic floras of equatorial regions were dominated by broad-leaved pteridosperms, cordaites, and pecopterid ferns. In high northern latitudes cordaites predominated while Glossopterid pteridosperms inhabited higher latitudes in the southern supercontinent of Gondwanaland. As climates dried out following the end of the Permo-Carboniferous glaciation, conifers began spreading into lowlands replacing pteridosperms and pteridophytes and producing mixed assemblages of Paleophytic and Mesophyic plants. In Gondwanaland, the deciduous, cold-adapted pteridosperms were unable to migrate as the climate warmed and gave way as northern pteridosperms moved south and established the Triassic Gondwanan flora (43). The Permian and Early Triassic mixed assemblages of Mesophytic and Paleophytic floral elements developed at different times in different regions (Figure 4). They generally persisted for about 5 myr within individual regions before entirely Mesophytic assemblages became established (43, 97). The new Mesophytic floras included conifers, ginkgoes, cycads, and cycadeoids and new groups of more mesic pteridophytes and pteridosperms, a transition chronicled in the palynological record by the appearance of desiccation-resistant taeniate pollen (97).

This pattern of gradual, diachronous change is clearly associated with an increase in global aridity and continental seasonality. The 50% drop in the diversity of plant families during the Permian and Triassic (43) was spread over 25 myr, and there is no evidence for a dramatic change in plant diversity at the end of the Permian. Rather, extinction was related to the ability of different groups to migrate (via dispersal) to new areas and to adapt to changing environmental conditions. Plants are generally immune to mass extinctions via catastrophic mass mortality since root systems, spores, and rhizomes remain, but they are vulnerable to both competitive displacement and climatic change (43). Thus, it comes as no surprise that sharp differences occur between terrestrial plants and marine invertebrates and vertebrates both in response to mass extinction events (44, 97).

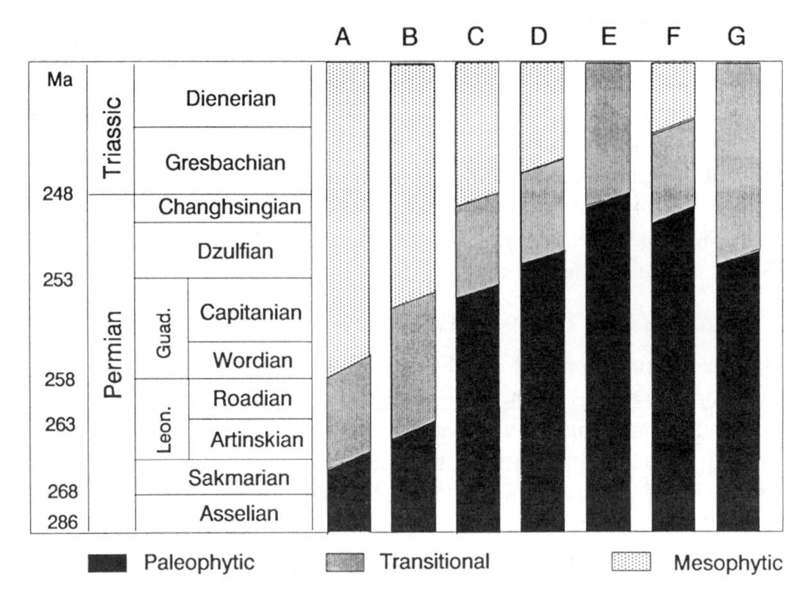


*Figure 4* Transition from Paleophytic to Mesophytic floras during the Permo-Triassic in seven different continental regions. A, North America; B, Western Europe; C, Russian platform; D, northern Eurasia; E, northern China; F, southern China; G, Australia. Redrafted from (43), used with permission.

# CHANGES IN THE PHYSICAL ENVIRONMENT

## *Marine Regression*

The Permian was a period of continuing marine regression from the Sakmarian to the Permo-Triassic boundary. The rate of regression accelerated during the Dzulfian with the maximum regression at (or very close to) the boundary itself (Figure 5). A rapid marine transgression occurred during the lowest Triassic, with the oceans quickly recovering to the pre-Dzulfian point (24, 31, 75). Estimates of the maximum regression during the Dzulfian vary from 210 meters (24) to perhaps as much as 280 meters (31). A regression of this magnitude would have reduced the percentage of continental area covered from about 40% at the end of the Guadalupian to 8–13% at the boundary, although sedimentological evidence suggests that changes in the earth's geoid may have occurred (31) that would increase the possibility of error in estimating the regression.

## *Climate*

Investigations of Cretaceous climates have confirmed that continental positions, topography, and sea level have substantial effects on climate. It comes

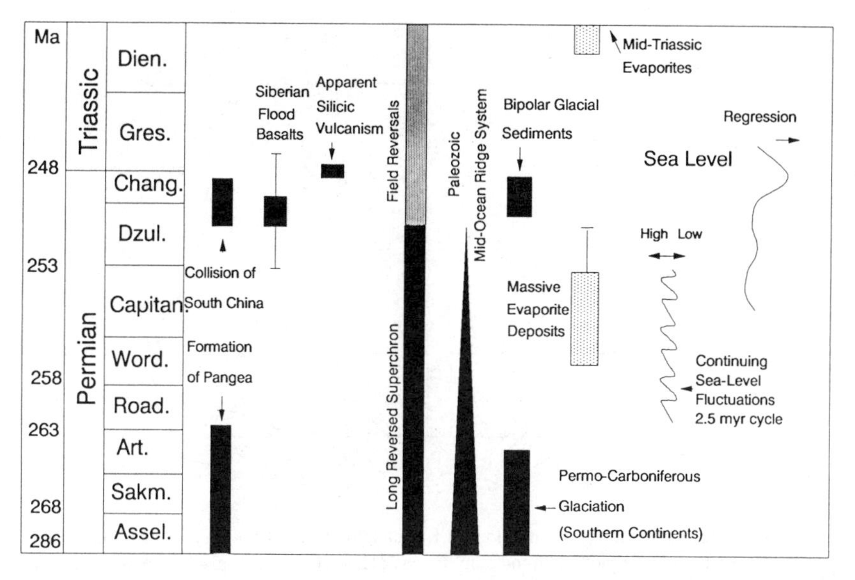


*Figure 5* Geophysical and climatic changes during the Permian and Triassic. Data from 31, 73, 89.

as no surprise that the formation of Pangea during the Permian had a significant impact on climate as well (Figure 6). Following the end of the glaciation in the Sakmarian (14, 107), Permian climates became progressively warmer and drier and by Wordian-Capitanian times massive evaporites had begun to form in low latitudes. The Permian evaporites are second only to those of the Triassic in extent during the Phanerozoic, but they seem to be less common in the latest Permian (113). Sedimentological evidence indicates that mid-Permian climates were very warm, but inequable, with high seasonality caused by the formation of Pangea (17, 62). In China, India, and Russia, coal deposits suggest a temperate-latitude region of high humidity.

At the close of the Permian the warming trend reversed, if only temporarily. Pangea rotated and drifted northward, placing Siberia in high northern latitudes, with Australia and Antarctica remaining near the South Pole (Figure 6). The distribution of land near both poles (rather than only near a single pole, as happened during the Permo-Carboniferous glaciation, but without any mass extinction) steepened the latitudinal temperature gradient and cooled global climate. Evidence of glacial activity is found in both Siberia and eastern Australia (14, 88–90). The triggering event for the glaciation may have been the cooling effects of massive vulcanism near the boundary. If polar ice has persisted throughout the Phanerozoic (25), the increased glacial activity near the Permo-Triassic boundary may represent an expansion of

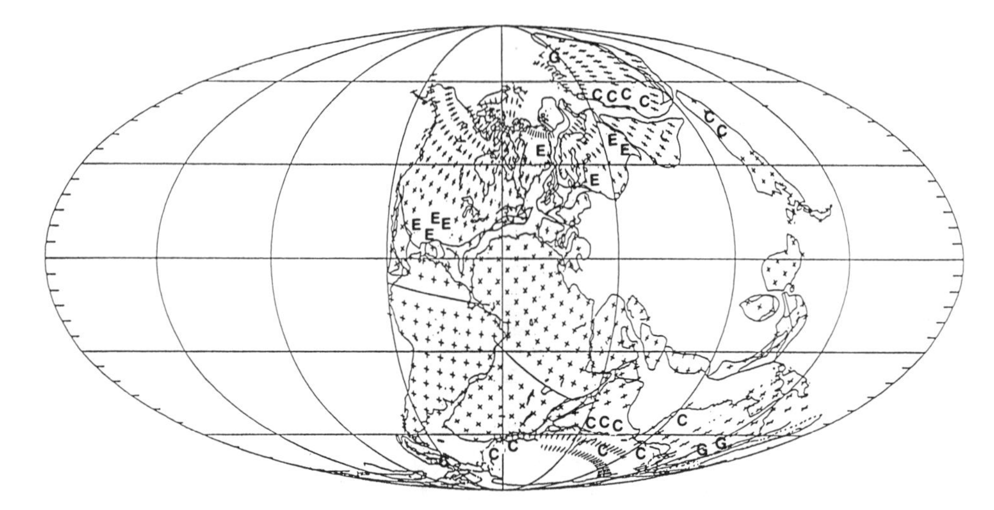

*Figure 6* Paleocontinental reconstruction for Kazanian time (255 Ma). Paleoclimatological data for the Upper Permian is plotted on the map: C, coal; E, evaporites; G, glaciers. Base map from 76a, used with permission. Paleoclimatological information from various sources.

preexisting ice sheets, rather than the onset of glaciation. A pattern of continuing sea-level fluctuations into the middle and upper Permian with an approximately 2.5 myr cycle (73) also suggests continued glacial pulses after the end of the Permo-Carboniferous glaciation, although the cyclicity differs from known Milankovich forcing periods.

The interior regions of continents lack the ameliorating effects on climate of a nearby ocean and suffer higher seasonal changes. The formation of Pangea produced extreme seasonal fluctuations, and severe monsoons in the continental interiors in both hemispheres (15, 16, 45, 62, 72). Two-dimensional energy balance models of early late Permian Pangean climates confirm that there was high seasonality with maximum mean summer temperatures of 38°C, possible daytime highs of 45°C, and a 50°C range in temperatures (15, 16). The modelling results are consistent with evidence of glacial activity in eastern Australia and Siberia and by distributions of terrestrial vertebrates. Late Permian therapsid localities are restricted to paleolatitudes of 25° to 70° north and south, with the best assemblages from South Africa and Russia. The lack of equatorial deposits is consistent with the harsh climate but may also reflect a preservational bias (62).

## *Changes in Stable Isotopes*

Some of the most exciting recent work on extinction boundaries comes from shifts in stable isotopes. The changes in carbon isotopes ($C^{12}$ and $C^{13}$), for

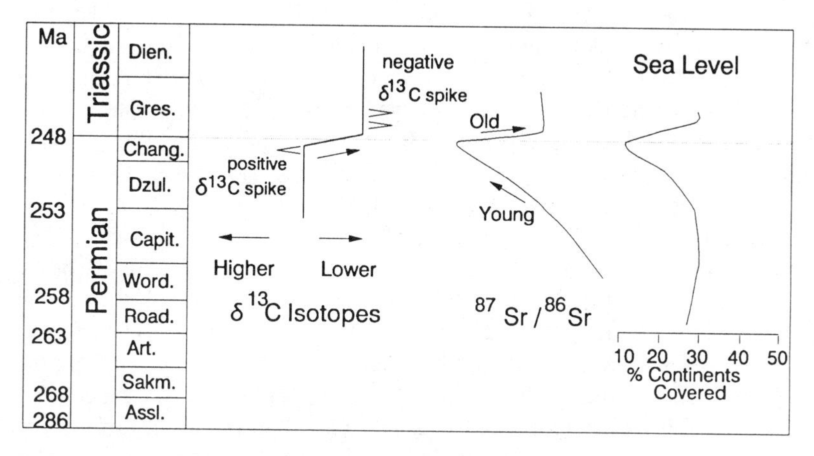


*Figure 7* Geochemical changes across the Permo-Triassic boundary. Note the overall drop in at the boundary, associated with several shorter-term spikes in $\delta C^{13}$. The $^{87}Sr/^{86}Sr$ ratio also shows a pronounced shift at the boundary, associated with the change in sea level. Redrafted from figures in 31, 59.

example, demonstrate several shifts in the rate of burial of organic carbon in sediment. These changes seem to reflect changes in global biomass. Carbon isotope measurements of brachiopod shells in Spitsbergen indicate that a period of increased organic carbon burial during the Upper Permian was followed by increased oxidation of organic carbon. A fairly complete core from a boundary section in the Austrian Alps also indicates that the extinction was preceded by several million years of oxidation of organic carbon (32). Thus, the long-term $\delta C^{13}$ signal shifts from heavy $\delta C^{13}$ values in the Permian to lighter $\delta C^{13}$ values in the Triassic (59). Immediately below the boundary there is a brief positive shift followed by several negative $\delta C^{13}$ spikes of from 2–3 ppt to 5 ppt in the earliest Triassic which appear to be associated with anoxic events (Figure 7). The negative spikes have been recorded in sections in the southern Alps (32, 51, 59), south China (9), Greenland (59), Greece, the Transcaucus, Pakistan, and Iran (32). This worldwide shift in the carbon isotopes occurred over a period of several hundred thousand to one million years (31, 32, 59).

Determining the causes of these shifts in global carbon cycling will require greater information on oceanic nutrients, actual primary productivity, and independent estimates of atmospheric oxygen. The standard explanation is that the marine regression allowed oxidation of previously buried organic matter. The subsequent decline in nutrient supply restricted marine primary productivity and caused the drop in biomass (32, 59) reflected in the $\delta C^{13}$ minima. Alternatively, the increased oxidation may have reduced both oceanic nutrients and atmospheric oxygen levels, with the extinction related to atmospheric oxygen depletion (27). Sedimentologic evidence and simulation studies suggest that the increased burial of organic material during the Carboniferous and Permian was followed by reduced burial rates as the Permian became increasingly arid. The increased oxidation of previously buried organic material may have reduced atmospheric oxygen from about 30% to less than 15% during the Upper Permian (6).

The ratio of $^{87}Sr/^{86}Sr$ drops from 0.7085 in the Lower Permian to an all-time Phanerozoic low of 0.7067 before returning in the first 10 myr of the Triassic to 0.7078 (31). Although these changes may seem small, they are an important measure of spreading rates along the mid-ocean ridges. A tectonically induced sea-level rise should be associated with a fall in $^{87}Sr/^{86}Sr$; the opposite occurs at the Permo-Triassic boundary. Holser & Margaritz offer several possible explanations for the discrepancy, but data are insufficient to choose between them (31). Changes in isotopic abundance also occur in sulfur and oxygen; together the patterns indicate dramatic changes in ocean water chemistry and perhaps in atmospheric oxygen. Unfortunately the evidence is not yet conclusive enough to aid in identifying the mechanism of extinction.

### *Geophysical Events*

Geophysical changes during the late Permian and early Triassic (Figure 5) include a massive flood basalt in Siberia of at least $1.5 \times 10^6$ km$^2$, with potassium-argon whole rock dates of $250 \pm 10$ Ma (64a). Elemental abundance patterns and the presence of bipyramidal quartz in the boundary clays in South China are consistent with an ash derived from a silica-rich (rhyolitic) magma (11, 40, 114). The thickness and geographic distribution of the clays suggests a massive eruption of $> 1000$ km$^3$ of silicic magma (114). These silica-rich clays could not have been produced by the flood basalts and require the presence of another source near South China, probably related to subduction of an oceanic plate. However, both episodes would have injected large amounts of particulates, sulfates, and other material into the atmosphere, and these would have sharply decreased global temperatures (49). It is not unreasonable to assert that these volcanic episodes provided the trigger for the late Permian glaciation once the continents had moved into a favorable position.

The upper Paleozoic was also a period of stable reversed magnetic polarity, the Kiaman Long Reversed Superchron, which ended about 250 Ma. During the early Mesozoic, magnetic field reversals occurred fairly frequently. Both the field reversals and the volcanic activity are manifestations of a changing geophysical regime which included the decline of the Paleozoic hotspot system, changes in the earth's geoid, and a decline in activity (and heat-flow) along the midocean ridge system (13, 31, 49). The correlation between magnetic field reversals and mass extinctions has been the subject of debate

(13, 49, 66), but reversals frequently occur without associated extinction episodes (66). However both the end-Permian and end-Cretaceous mass extinctions occurred at the end of long periods of stable magnetic field, and flood basalts and magnetic field reversals appear to have a periodicity of 32 +/− 1 myr (64a) and 30 myr (49), respectively. This correlation has led to suggestions that the flood basalts and field reversals are indicators of internal changes in the earth's core-mantle boundary. Although the causal link to mass extinctions remains uncertain, increased volcanic activity may produce climatic-induced mass extinctions (13, 49).

## POSSIBLE EXTINCTION MECHANISMS

A plethora of extinction mechanisms have been offered to explain the end-Permian mass extinction. These range from tectonics to extraterrestrial impacts and trace element poisoning. In the following discussion of mechanisms it is important to keep four considerations in mind: First, extinction mechanisms restricted to the marine realm may be dismissed if one accepts a causal as well as temporal connection between the terrestrial and marine extinctions. Second, the tectonic events leading to the formation of Pangea had many collateral effects, one or more of which may have been the effective cause of extinction for some groups, but the distinction between ultimate and effective causes must be made clear. Third, while the best record of the latest Permian Changhsingian Stage comes from South China, this was an isolated tectonic block, separate from Pangea, and may have a very different history from the supercontinent. Thus, data from South China must be used with caution in discussing the global extinction pattern. Finally, one must clearly distinguish between patterns associated with the extinction itself and those that were a necessary consequence of the extinction. For example, the low provinciality and depauperate marine and terrestrial vertebrate communities in the earliest Triassic were a necessary consequence of drastically reduced diversity and may not reflect an extinction caused by declining provinciality.

### *Extraterrestrial Mechanisms*

Despite several reports of Iridium concentrations of about 2.0 ppb. from the boundary clays in South China (110, 112, 115), reanalysis of the same sections revealed no anomalous iridium concentrations (11, 114). High concentrations of platinum-group metals, including iridium, have been linked to possible extraterrestrial impacts, particularly at the Cretaceous-Tertiary boundary (1). The lack of an iridium anomaly at the Permo-Triassic boundary indicates that no impact occurred there, but the lack does not, by itself, eliminate any causal connection between the end-Permian and end-Cretaceous mass extinctions. As noted earlier, the elemental abundance patterns of the boundary clays are more consistent with a volcanic than an extraterrestrial source.

Schindewolf (74) suggested that cosmic radiation caused the extinction. More recently Hatfield & Camp (29) resurrected this idea in their discussion of the movement of the solar system relative to the galactic plane, postulating that movement perpendicular to the galactic plane increased the magnetic field, and the amount of inbound cosmic radiation thus increased mutation rates and caused mass extinctions. Since the mutagenic effect of radiation is related to both intensity and duration of exposure, a lengthy period above the galactic plane would be required to explain the end-Permian extinction (52). However, the position of the solar system relative to the galactic plane is not known with certainty, and gives only a general agreement with the timing of extinction. Furthermore cosmic radiation is quickly absorbed by water, so extinction patterns should be highest for terrestrial, pelagic marine, and very shallow benthic marine organisms; these patterns might be expected to be highest in high latitudes. The available data suggest a pattern almost directly contrary to this.

## *Environmental Mechanisms*

SALINITY CHANGES The extensive Permian evaporite deposits could have created brackish oceans and the selective extinction of stenohaline taxa (22, 92). Stenohaline groups were heavily affected, but Benson (3) compared late Permian ostracode diversity patterns to the ostracode extinctions during the Miocene Messinian salinity crisis, and concluded that the maximal postulated salinity change during the Permian was insufficient to decimate the fauna. Furthermore, the greatest deposition of evaporites occurred in the Kungarian and the Middle Triassic (113), with no apparent mass extinctions during either interval.

GLOBAL COOLING Most mass extinctions occur during periods of climatic change and marine regressions. Stanley has marshalled convincing evidence that the long-term Permian trend toward increased global aridity and higher temperatures was interrupted during the Late Permian by a brief period of bipolar glaciation. This bipolar glaciation was more effective at global cooling than the unipolar Permo-Carboniferous glaciation. The resulting global cooling may have reduced the area of the tropics, causing preferential extinction of tropical taxa and other temperature-sensitive groups (88–90). I have argued above that the massive vulcanism near the boundary may have served as a trigger for the extinction. The evidence for climatic deterioration and global cooling includes progressive restriction of stenothermal taxa to low latitudes followed by a higher extinction rate, and reduced carbonate deposition into the Early Triassic. Unfortunately correlating the glacial sediments of the Kolyama block in Siberia with boundary sections in China and elsewhere is very difficult and the glaciation may have occurred well before the peak of the extinction.

CHANGES IN PRIMARY PRODUCTIVITY Terrestrially derived nutrients are vital to marine productivity, and Tappen has argued that the development of new terrestrial floras and biomass progressively sequestered more organic carbon and nutrients on land, reducing the supply to marine ecosystems (94). The reduced nutrient supply would cut marine primary productivity (63, 94) and cause a mass extinction. However, the drop in marine primary productivity recorded in the shift in stable isotopes occurred far more rapidly than Tappen's model requires. Moreover, Carboniferous terrestrial biomass changes greatly exceeded those caused by the shift to Mesophytic floras from Paleophytic, which in any case took place over a longer time span than the extinction. Thus, the drop in marine primary productivity does not appear to have been related to long-term changes in terrestrial biomass.

## *Tectonically Induced Mechanisms*

The formation of Pangea and the associated geophysical changes produced a variety of effects which could be related to the extinction. These include reduction in available shelf area for marine species, reduced biotic provinciality, and increases in climatic instability.

MacArthur & Wilson's theory of island biogeography (50) was translated to evolutionary time scales through the species-area effect. The reduction in available shelf area could induce a marine extinction if a linear (or nearly linear) relationship between species diversity and habitable area is translated during a major marine regression into increased competition and extinction (75, 85). But the species-area effect has been criticized on several grounds. It is nearly impossible to plot meaningful species-area curves for the living marine benthos, and the recent biota provides little support for a linear relationship between diversity and shelf area. The Panamic Pacific Province, for example, contains about 3000 species on a narrow shelf with no coral reefs (88) while far larger tropical regions have lower species diversities. Research on the Permian has emphasized measurements of diversity at single localities rather than across provinces, and these measurements are suspect. Investigations of diversity changes during both the Pleistocene (109) and the Eocene (28) regressions also fail to show any relationship between reduced shelf-area and extinction.

Jablonski & Flessa showed that 87% of the 276 families of molluscs, echinoderms, and coelenterates they analyzed have representatives on one or more of 22 offshore oceanic islands. Since the shelf area of an island (viewed as a cone) will increase during a regression and act as a refuge, regressions are unlikely to cause major drops in diversity (35, 36, 39). However, reduced spreading rates due to the formation of Pangea may have allowed islands to sink below the surface as the old oceanic plate cooled and thus may have reduced the number of marine islands available as refuges.

Continental collision during the formation of Pangea and the regression might have limited the number of marine faunal provinces by allowing the faunas of formerly separate continental areas to intermingle (104). This should result in a sharp drop in global marine diversity as a new equilibrium is sought. There appear to have been 14+ marine provinces in the Lower Permian, at least 8 in the Guadalupian, and 3 or less in the earliest Triassic (76, 101). However, the formation of Pangea by middle Permian times was followed by continued high provinciality and high diversity for millions of years. Recent climatic models of Pangea (described above) indicate that high latitudinal temperature gradients may have maintained high provinciality in Pangea after it formed. Valentine & Moores later modified their earlier work (100, 101) to emphasize the effects of increased environmental instability and consequent trophic resource instability (105).

The marine, terrestrial vertebrate, and terrestrial plant diversity records are all most consistent with a long period of climatic change which reached a maximum during the Upper Permian as the regression accelerated and the effects associated with the formation of Pangea were maximized. The major pulse of extinction in marine animals and terrestrial vertebrates occurred late in the Guadalupian stage, but continuing originations (at least partly a consequence of data compilation procedures) delayed the apparent maximal drop in diversity until the Dzulfian. The climatic changes would have had their greatest impact on the specialized, stenothermal, tropical taxa and the stable, well-integrated communities that suffered the greatest extinction. The initial regression must have been tectonically induced, probably by a change in the spreading rate associated with the formation of Pangea, but this recession was exacerbated by a glacio-eustatic regression in the latest Permian. Recall that South China remained a separate tectonic unit during this period and apparently maintained normal marine faunas after the extinction had begun elsewhere. The mass extinction might have been far less severe if the organisms in South China had been able to serve as a source of replenishment for Pangea. However, massive silicic vulcanism near South China intervened and caused an extinction of the faunal elements in that region. With the major regional refuge largely eliminated, the extinction continued until the waning of the glaciers in the earliest Triassic led to the rapid marine transgression and the eventual return of normal marine conditions.

## EVOLUTIONARY SIGNIFICANCE OF THE EXTINCTION

### *Alternation of Macroevolutionary Regimes*

Jablonski presented evidence from Gulf Coast (37) and Europe (38) molluscs that survival during the end-Cretaceous mass extinction was enhanced by a combination of geographic range at the species level and species-richness at the clade (generic) level. This alternation of macroevolutionary regimes, with survival during mass extinction dependent upon characteristics that cannot be selected for and that don't increase survival during background intervals, suggests that mass extinctions may impose a different selective regime upon the fauna. If this is a general feature of mass extinctions, the long-term structure of the biota may be more a function of survival of mass extinctions than the adaptive evolutionary changes which occurred between extinction events.

The analysis of distributional patterns of gastropods in the Southwestern United States and a comparison of these results to a similar-aged fauna in Malaysia provide no support for an alternation in macroevolutionary regimes during the end-Permian. Survival in marine gastropods was enhanced by broad interprovincial geographic distribution, occupation of numerous physical environments (whether achieved by few or many species) and high species richness (18, 19). There is no indication of ecological or taxonomic selectivity in extinction, which suggests that despite the variety of ecological roles adopted by late Paleozoic gastropods (from sessile filter-feeding to active predation) extinction was largely random at the species level and dependent upon the number of discrete units (populations or species) that existed. Gastropod survival both before and during the extinction was a function of broad geographic and environmental distribution and of species-richness within genera.

## *Linkage to Periodic Mass Extinctions*

The end-Permian extinction is the first in a series of events recognized by Raup & Sepkoski as forming a series with a periodicity of about 26 myr (68–70, 82). Suggestions for the forcing agent range from a companion star to the sun which periodically perturbs comets in the Oort cloud, sending them into the inner solar system (see 1, 67) to intrinsic cycles in the core-mantle boundary which produce volcanic or other events leading to climatic change and extinction (49). Since the periodicity is based on time-series data, a substantial auto-correlation problem exists that makes statistical tests of the periodic cycle difficult (30, 82).

If the cycle is real the forcing agent must have been the same for each extinction, although the proximal causes involved in single events may vary. Thus, our understanding of the end-Permian event may constrain the variety of possible forcing agents. The end-Permian mass extinction differs from the end-Cretaceous in several substantive ways: (*a*) the end-Permian was clearly not a catastrophic event and may have lasted up to 8 myr; (*b*) no iridium spike is present, nor is there any evidence of an extraterrestrial impact; (*c*) a differential extinction of planktotrophs occurred during the end-Permian event but not during the end-Cretaceous. The evidence presented above strongly implicates climatic change caused by the formation of Pangea and related geophysical events. If this scenario is correct either the end-Cretaceous and end-Permian mass extinctions have separate causes, and there is no periodicity; or the periodicity is real and the cause of both extinctions is internal tectonic changes expressed as climatic shifts; or the extraterrestrial impact during the end-Permian event coincided with an ongoing extinction event. It is difficult to choose among these alternatives, but the second appears to be the most likely.

## *The Change in Evolutionary Faunas*

One of the classic stories of the Permian extinction and the end of the Paleozoic fauna concerns the reversal of dominance relationships between the bivalves and the articulate brachiopods. In the old hagiography, competitive interaction between the two groups and the inherent competitive superiority of the bivalves led to their eventual triumph (91), with the surviving brachiopods relegated to marginal and cryptic environments. But Gould & Calloway (26) demonstrated that the late Paleozoic diversity history of both groups is *positively* correlated, and the difference in Mesozoic diversity is a direct consequence of the differential extinction of the brachiopods during the extinction. This dispute is but part of a larger controversy over the evolutionary significance of the mass extinction, and the extent to which the change in dominant taxa and in the composition of marine communities is a result of the mass extinction rather than processes already underway (e.g. 8, 21, 42, 81, 83, 106, 108). The distribution of Paleozoic marine communities along a simple, two-dimensional onshore-offshore gradient suggests that new community types become established in near-shore environments and are progressively displaced offshore by the development of new near-shore community types (81, 83). The apparent progressive displacement of the Paleozoic, brachiopod-rich communities from the near-shore by molluscan-rich communities during the Paleozoic (83) seems to confirm this pattern. However, the actual data are also consistent with a shift in community structure following the end-Devonian mass extinction and relative stasis until the close of the Paleozoic. The probability of extinction within the biota also seems to have been reset by the end-Permian extinction (106), further supporting the claim that the end-Permian mass extinction was a singular event and largely responsible for the differences between Paleozoic and post-Paleozoic marine communities.

Simulation studies of global diversity using three-phase kinetic models with a time-specific perturbation (8, 42, 80) can describe the behavior of the three evolutionary faunas. In these simulations each fauna has an intrinsic evolutionary rate and equilibrium diversity level. The inclusion of the time-specific perturbations, which simulate mass extinctions, demonstrates that the conversion from one fauna to the next would have occurred even without the extinctions. This result implies that the change in evolutionary faunas was not a consequence of the extinction, but this result can be criticized on several grounds. The most fundamental problem with both the initial onshore-offshore gradient analyses and the simulation studies is the assumption that evolutionary faunas were behaving as unitary assemblages when in fact they were the summary of diversity histories of numerous distinct lineages, and no evidence in favor of the assumption has been presented. Nonetheless the work is an interesting point of departure for further investigations of the significance of the extinction.

The pattern of morphologic innovation (21) and global marine diversity (Figure 1) is largely consistent with the pattern identified by Gould & Calloway (26): Large-scale changes in diversity dynamics within the marine biota are largely a consequence of mass extinction events. Evolutionary radiations, particularly those associated with substantive morphologic innovation, are largely restricted to low-diversity periods in earth history (the Cambrian and Ordovician), rebounds following mass extinctions, and the occupation of new environments (21). Long-term adaptive changes are a significant component of between-extinction evolutionary change (108), but these changes appear to be incapable of radically restructuring the biota.

## SUMMARY AND CONCLUSIONS

Rapid physical changes during the end-Permian included the formation of the global supercontinent of Pangea and the associated decline in a previously stable geophysical regime, climatic changes, an extensive, tectonically induced, marine regression and late Permian vulcanism. The climatic degradation and increased seasonality lead to a gradual change in terrestrial floras and, as they accelerated, to ecosystem collapse and mass extinctions in the marine realm and among terrestrial vertebrates. The complexity of the mechanisms involved in the extinction illustrates both the messiness and the contingent nature of history. There was no single cause of this extinction; rather, a series of physical events culminated in a progressive biotic collapse. But it now appears that without the apparently fortuitous massive silicic vulcanism near South China at the end of the Changhsingian Stage the fauna of South China might have survived to repopulate Pangea following the end of the glacio-eustatic marine regression and the magnitude of the extinction would have been far lower.

Whatever the cause of the extinction, the resulting shift in global marine diversity allowed the establishment of new types of marine communities and the origination of numerous new clades. Many authors have described patterns of adaptive, distributional, and compositional change in the late Paleozoic which appear to foreshadow the change in community composition associated with the end-Permian extinction. However, the available evidence suggests that the major impetus for the change was the extinction itself, and that no major shift in marine composition would have occurred without the extinction. This conclusion supports the assumption of a major role for mass extinctions in structuring biotic diversity, but it does not follow that the selective nature of a mass extinction differs radically from those prior to the extinction.

ACKNOWLEDGMENTS

I thank R. L. Anstey, M. J. Benton, C. E. Elliot, D. Jablonski, S. M. Stanley, J. W. Valentine, and T. A. Vogel for valuable discussions. This research was funded by grant BSR 87-22510 from the National Science Foundation.

## *Literature Cited*

1. Alvarez, W., Kauffman, E. G., Surlyk, F., Alvarez, L. W., Asaro, F. et al. 1984. Impact theory of mass extinctions and the invertebrate fossil record. *Science* 223:1135–41
2. Batten, R. L. 1973. The vicissitudes of the gastropods during the interval of Guadalupian-Ladinian time. See Ref. 48, pp. 596–607
3. Benson, R. H. 1984. The Phanerozoic "crisis" as viewed from the Miocene. In *Catastrophes and Earth History*, ed. W. A. Berggren, J. A. Van Couvering, pp. 437–46. Princeton, NJ: Princeton Univ. Press
4. Benton, M. 1987. Mass extinctions among families of non-marine tetrapods: the data. *Memoir. Soc. Geol. France* No. 150:21–32
5. Benton, M. J. 1988. Mass extinctions in the fossil record of reptiles: paraphyly, patchiness and periodicity (?). See Ref. 46, pp. 269–94
6. Berner, R. A. 1989. Dying, O2 and mass extinction. *Nature* 340:603–4
7. Brasier, M. D. 1988. Foraminiferid extinction and ecological collapse during global biological events. See Ref. 46, pp. 37–64
8. Carr, T. R., Kitchell, J. A. 1980. Dynamics of taxonomic diversity. *Paleobiology* 6:427–43
9. Chen, J., Shao, M., Huo, W., Yao, Y. 1984. Carbon isotope of carbonate strata at Permian-Triassic boundary in Changxing, Zhejiang. *Scientia Geologica Sinica* 1984 (1):92–93
10. Clark, D. L. 1987. Conodonts: the final fifty million years. In *Palaeobiology of Conodonts*. ed. R. J. Aldridge, pp. 165–74. Chichester: Ellis Horwood
11. Clark, D. J., Wang, C-Y., Orth, C. J., Gilmore, J. S. 1986. Conodont survival and low iridium abundances across the Permian-Triassic boundary in South China. *Science* 233:984–86
12. Colbert, E. N. 1986. Therapsids in Pangea and their contemporaries and competitors. See Ref. 33, pp. 133–45
13. Courtillot, V., Besse, J. 1987. Magnetic field reversals, polar wander and core-mantle coupling. *Science* 237:1140–47
14. Crowell, J. C. 1978. Gondwanan glaciation, cyclothems, continental positioning and climate change. *Am. J. Sci.* 278:1345–72
15. Crowley, T. J., Hyde, W. T., Short, D. A. 1989. Seasonal cycle variations on the supercontinent of Pangea. *Geology* 17:457–60
16. Crowley, T. J., North, G. R. 1988. Abrupt climate change and extinction events in earth history. *Science* 240: 996–1002
17. Dickins, J. M. 1983. Permian to Triassic changes in life. *Mem. Australasian Paleontols.* 1:297–303
18. Erwin, D. H. 1989. Regional paleoecology of Permian gastropod genera, southwestern United States and the end-Permian mass extinction. *Palaios* 4: 424–38
19. Erwin, D. H. 1990. Carboniferous-Triassic gastropod diversity patterns and the Permo-Triassic mass extinction. *Paleobiology*. In press
20. Erwin, D. H., Valentine, J. W. 1984. *Geol. Soc. Am. Abstr. with Prog.* 16(6):503 (Abstr.)
21. Erwin, D. H., Valentine, J. W., Sepkoski, J. J. Jr. 1987. A comparative study of diversification events: the early

Paleozoic vs. the Mesozoic. *Evolution* 41:1177–86
22. Fischer, A. G. 1965. Brackish oceans as the cause of the Permo-Triassic marine faunal crisis. In *Problems in Palaeoclimatology,* ed. A. E. M. Nairn, pp. 566–74. London: Interscience
23. Flugel, E., Stanley, G. D. Jr 1984. Reorganization, development and evolution of post-Permian reefs and reef organisms. *Palaeontographica Americana* 54:177–86
24. Forney, G. G. 1975. Permo-Triassic sea level change. *J. Geol.* 83:773–79
25. Frakes, L. A., Francis, J. E. 1988. A guide to Phanerozoic cold polar climates from high-latitude ice-rafting in the Cretaceous. *Nature* 333:547–49
26. Gould, S. J., Calloway, C. B. 1980. Clams and brachiopods—ships that pass in the night. *Paleobiology* 6:383–96
27. Gruszczynski, M., Halas, S., Hoffman, A., Malkowski, K. 1989. A brachiopod calcite record of the oceanic carbon and oxygen isotope shifts at the Permian/Triassic transition. *Nature* 337:64–68
28. Hansen, T. A. 1987. Extinction of late Eocene to Oligocene molluscs: relationship to shelf area, temperature changes and impact events. *Palaios* 2:69–75
29. Hatfield, C. B., Camp, M. J. 1970. Mass extinctions correlated with periodic galactic events. *Geol. Soc. Am. Bull.* 81:911–914
30. Hoffman, A. 1989. Mass extinctions: the view of a sceptic. *J. Geol. Soc. London* 146:21–35
31. Holser, W. T., Magaritz, M. 1987. Events near the Permian-Triassic boundary. *Modern Geol.* 11:155–80
32. Holser, W. T., Schonlaub, H-P., Attrep, M. Jr, Boeckelmann, K., Klein, P., et al 1989. A unique geochemical record at the Permian/Triassic boundary. *Nature* 337:39–44
33. Hotton, N. III, MacLean, P. D., Roth, J. J., Roth, E. C. eds. 1986. *Ecology and Biology of Mammal-Like Reptiles.* Washington, DC: Smithsonian Inst.
34. Hussner, M. 1983. Die Faunenwende Perm/Trias. *Geologische Rundschau* 72: 1–22
35. Jablonski, D. 1985. Marine regressions and mass extinctions: a test using the modern biota. See Ref. 98, pp. 335–54
36. Jablonski, D. 1986a. Causes and consequences of mass extinctions: a comparative approach. In *Dynamics of Extinction,* ed. D. K. Elliot, pp. 183–229. New York: Wiley
37. Jablonski, D. 1986b. Background and mass extinctions: The alternation of macroevolutionary regimes. *Science* 231:129–33
38. Jablonski, D. 1989. The biology of mass extinction: a paleontological view. *Philos. Trans. R. Soc. London B.* 325:357–68
39. Jablonski, D., Flessa, K. W. 1986. The taxonomic structure of shallow-water marine faunas: implications for Phanerozoic extinctions. *Malacologia* 27:43–66
40. Jinwen, He. 1989. Restudy of the Permian-Triassic boundary clay in Meishan, Changxing, Zhejiang, China. *Hist. Bio.* 2:73–87
41. Kier, P. M. 1965. Evolutionary trends in Paleozoic echinoderms. *J. Paleo.* 39: 436–65
42. Kitchell, J. A., Carr, T. R. 1985. Nonequilibrium model of diversification: faunal turnover dynamics. See Ref. 98, pp. 277–309
43. Knoll, A. H. 1984. Patterns of extinction in the fossil record of vascular plants. See Ref. 58, pp. 23–68
44. Knoll, A. H., Niklas, K. J. 1987. Adaptation, plant evolution, and the fossil record. *Rev. Palaeobot. Palynol.* 50:127–49
45. Kutzbach, J. E., Gallimore, R. G. 1989. Pangean climates: megamonsoons of the megacontinent. *J. Geophys. Res.* 94: 3341–57
46. Larwood, G. P., ed. *Extinction and Survival in the Fossil Record.* Oxford: Oxford Univ. Press
47. Liao, Z. T. 1980. Brachiopod assemblages from the Upper Permian and Permian-Triassic boundary Beds, South China. *Can. J. Earth Sci.* 17:289–95
48. Logan, A., Hills, L. V. 1973. The Permian and Triassic Systems and Their Mutual Boundary. *Can. Soc. Petrol. Geol.* Memoir 2. Calgary
49. Loper, D. E., McCartney, K., Buzyana, G. 1988. A model of correlated episodicity in magnetic-field reversals, climate and mass extinctions. *J. Geol.* 96:1–15
50. MacArthur, R. H., Wilson, E. O. 1967. *The Theory of Island Biogeography* Princeton, NJ: Princeton Univ. Press
51. Magaritz, M., Bar, R., Baud A., Holser, W. T. 1988. The carbon-isotope shift at the Permian/Triassic boundary in the southern Alps is gradual. *Nature* 331:337–39
52. Maxwell, W. D. 1989. The end Permian mass extinction. In: *Mass Extinctions: Processes and Evidence,* ed. S. K. Donovan, pp. 152–73. London: Belhaven
53. McKinney, M. L. 1985. Mass extinction patterns of marine invertebrate groups and some implications for a causal phenomenon. *Paleobiology* 11:227–33

54. Nakazawa, K., Bando, Y., Matsuda, T. 1980. The *Otoceras woodwardi* Zone and the time-gap at the Permian-Triassic boundary in East Asia. *Geol. Paleontol. Southeast Asia* 21:75–90
55. Newell, N. D. 1967. Revolutions in the history of life. *Geol. Soc. Am. Sp. Pap* 89:63–91
56. Newell, N. D. 1978. The search for a Paleozoic-Mesozoic boundary stratotype. *Schriftenreihe Erdwiss. Komm. Öster. Akad. Wiss.* 4:9–19
57. Newell, N. D. 1986. The Paleozoic/Mesozoic erathem boundary. *Mem. Soc. Geol. Ital.* 34:303–11
58. Nitecki, M. H., ed. 1984. *Extinctions*. Chicago: Univ. Chicago Press, pp. 354
59. Oberhansli, H., Hsu, K. J., Piasecki, S., Weissert, H. 1989. Permian-Triassic carbon-isotope anomaly in Greenland and in the southern Alps. *Hist. Biol.* 2:37–49
60. Olson, E. C. 1986. Problems of Permo-Triassic terrestrial vertebrate extinctions. *Hist. Biol.* 2:17–35
61. Padian, K., Clemens, W. A. 1985. Terrestrial vertebrate diversity: episodes and insights. See Ref. 98, pp. 41–69
62. Parrish, J. M., Parrish, J. T., Ziegler, A. M. 1986. Permian-Triassic paleogeography and paleoclimatology and implications for therapsid distribution. See Ref. 33, pp. 109–31
63. Pitrat, C. W. 1970. Phytoplankton and the late Paleozoic wave of extinction. *Palaeogeog. Palaeoclimat. Palaeoecol.* 8:49–66
64. Paul, C. R. C. 1988. Extinction and survival in the echinoderms. See Ref. 46, pp. 155–70
64a. Rampino, M. R., Strothers, R. B. 1988. Flood basalt volcanism during the past 250 million years. *Science* 241: 663–68
65. Raup, D. M. 1979. Size of the Permo-Triassic bottleneck and its evolutionary implications. *Science* 206:217–18
66. Raup, D. M. 1985. Magnetic reversals and mass extinctions. *Nature* 314:341–43
67. Raup, D. M. 1987. Mass extinction: a commentary. *Palaeontology* 30:1–13
68. Raup, D. M. and Sepkoski, J. J., Jr. 1982. Mass extinction in the marine fossil record. *Science* 215:1501–03
69. Raup, D. M., Sepkoski, J. J. Jr. 1984. Periodicity of extinctions in the geologic past. *Proc. Nat. Acad. Sci. USA* 81: 801–5
70. Raup, D. M., Sepkoski, J. J. Jr. 1986. Periodic extinction of families and genera. *Science* 231:833–36
71. Rhodes, F. H. T. 1967. Permo-Triassic Extinction. In *The Fossil Record,* ed. W. B. Harland, pp. 57–76. London: Geological Soc. London
72. Robinson, P. L. 1973. Palaeoclimatology and continental drift. In *Implications of Continental Drift to the Earth Sciences,* ed. D. H. Darlington, S. K. Runcorn, pp. 451–76. London: Academic
73. Ross, C. A., Ross, J. R. P. 1985. Late Paleozoic depositional sequences are synchronous and worldwide. *Geology* 13:194–97
74. Schindewolf, O. 1963. Neokatastrophismus. *Z. deutsch. geol. Ges. Jahrgang* 1962 114:430–57
75. Schopf, T. J. M. 1974. Permo-Triassic extinctions: relation to sea-floor spreading. *J. Geol.* 82:129–43
76. Schopf, T. J. M. 1979. The role of biogeographic provinces in regulating marine faunal diversity through geologic time. In *Historical Biogeography, Plate Tectonics and the Changing Environment,* ed. J. Gray, A. J. Boucot, pp. 449–57. Corvallis, Or: Oregon State Univ. Press
76a. Scotese, C. R. 1985. Phanerozoic reconstructions. *Paleoceonagraphic Mapping Project Project Report No. 19-1286*
77. Scrutton, C. T. 1988. Patterns of extinction and survival in Paleozoic corals. See Ref. 46, pp. 65–88
78. Sepkoski, J. J. Jr. 1981. A factor analytic description of the Phanerozoic marine fossil record. *Paleobiology* 7:36–53
79. Sepkoski, J. J. Jr. 1982. A compendium of fossil marine families. *Milwaukee Pub. Mus. Contr. Biol. Geol. No. 51.* pp. 125
80. Sepkoski, J. J. Jr. 1984. A kinetic model of Phanerozoic taxonomic diversity. III. Post-Paleozoic families and mass extinctions. *Paleobiology* 10:246–67
81. Sepkoski, J. J. Jr. 1987. Environmental trends in extinction during the Paleozoic. *Science* 235:64–65
82. Sepkoski, J. J. Jr. 1989. Periodicity in extinction and the problem of catastrophism in the history of life. *J. Geol. Soc. Lond.* 146:7–19
83. Sepkoski, J. J. Jr. and Miller, A. I. 1985. Evolutionary marine faunas and the distribution of Paleozoic benthic communities in space and time. See Ref. 98, pp. 153–89
84. Sheng, J. Z., Chen, C-Z, Wang, Y-G, Rui, L, Liao, Z-T et al. 1984. Permian-Triassic boundary in middle and Eastern Tethys. *J. Fac. Sci. Hokkaido Univ.* Ser. IV 21:133–81
85. Simberloff, D. S. 1974. Permo-Triassic

extinctions: effects of area on biotic equilibrium. *J. Geol.* 82:267–74
86. Signor, P. W. III, Lipps, J. H. 1982. Sampling bias, gradual extinction patterns, and catastrophes in the fossil record. See Ref 87. 291–96
87. Silver, L. T., Schultz, P. H., eds. 1982. *Geological Implications of Impacts of Large Asteroids and Comets on Earth.* Geol. Soc. Am. Sp. Pap. 190, pp. 528
88. Stanley, S. M. 1984. Marine mass extinctions: a dominant role for temperatures. See Ref. 58, pp. 69–117
89. Stanley, S. M. 1988a. Paleozoic mass extinctions: shared patterns suggest global cooling as a common cause. *Am. J. Sci.* 288:334–52
90. Stanley, S. M. 1988b. Climatic cooling and mass extinction of Paleozoic reef communities. *Palaios* 3:228–32
91. Steele-Petrovic, H. M. 1979. The physiological differences between articulate brachiopods and filter-feeding bivalves as a factor in the evolution of marine level-bottom communities. *Palaeontology* 22:101–34
92. Stevens, C. H. 1977. Was development of brackish oceans a factor in Permian extinctions? *Geology* 8:133–38
93. Strathman, R. R. 1978. Progressive vacating of adaptive types during the Phanerozoic. *Evolution* 32:907–14
94. Tappan, H. 1982. Extinction or survival: selectivity and causes of Phanerozoic crises. See Ref. 87, pp. 265–76
95. Taylor, P. D., Larwood, G. P. 1988. Mass extinctions and the pattern of bryozoan evolution. See Ref. 46, pp. 99–119
96. Tozer, E. T. 1979. The significance of the ammonoids *Paratirolites* and *Otoceras* in correlating the Permian-Triassic boundary beds of Iran and the People's Republic of China. *Can. J. Earth Sci.* 16:1524–32
97. Traverse, A. 1988. Plant evolution dances to a different beat. Plant and animal evolutionary mechanisms compared. *Hist. Biol.* 1:277–302
98. Valentine, J. W. ed. 1985. *Phanerozoic Diversity Patterns.* Princeton, NJ: Princeton Univ. Press. 441 pp.
99. Valentine, J. W. 1986. The Permian-Triassic extinction event and invertebrate developmental models. *Bull. Mar. Sci.* 39:607–15
100. Valentine, J. W. 1973. *Evolutionary Paleoecology of the Marine Biosphere.* Englewood Cliffs, NJ: Prentice Hall
101. Valentine, J. W., Foin, T. C., Pert, D. 1978. A provincial model of Phanerozoic marine diversity. *Paleobiology* 4:55–66
102. Valentine, J. W., Jablonski, D. 1983. Larval adaptations and patterns of brachiopod diversity in space and time. *Evolution* 37:1052–61
103. Valentine, J. W., Jablonski, D. 1986. Mass extinctions: sensitivity of marine larval types. *Proc. Nat. Acad. Sci. USA* 83:6912–14
104. Valentine, J. W., Moores, E. M. 1972. Global tectonics and the fossil record. *J. Geol.* 80:167–84
105. Valentine, J. W., Moores, E. M. 1973. Provinciality and diversity across the Permian-Triassic boundary. See Ref. 48, pp. 759–66
106. Van Valen, L. M. 1984. A resetting of Phanerozoic community evolution *Nature* 307:50–52
107. Veevers, J. J., Powell, C. M. 1987. Late Paleozoic glacial episodes in Gondwanaland reflected in transgressive-regressive depositional sequences in Euramerica. *Geol. Soc. Am. Bull.* 98:475–87
108. Vermeij, G. J. *Evolution and Escalation.* Princeton, NJ: Princeton Univ. Press. 526 pp.
109. Wise, K. P., Schopf, T. J. M. 1981. Was marine faunal diversity in the Pleistocene affected by changes in sea level?. *Paleobiology* 7:394–99
110. Xu D-Y., Ma, S-L., Chai, Z-F., Mao, X-Y., Sun, Y-Y. et al. 1985. Abundance variation of iridium and trace elements at the Permian/Triassic boundary at Shagsi in China. *Nature* 314:154–56
111. Yin, H. F. 1985. Bivalves near the Permian-Triassic boundary in South China. *J. Paleo.* 59:572–600
112. Yi Yin, S., Chai Z., Ma S., Mao Z., Xu D. et al 1984. The discovery of iridium anomaly in the Permian-Triassic boundary clay in Changxing, Zhejiang, China and its significance. In *Developments in Geosciences: Contributions to 27th Annual International Geological Congress, Moscow,* pp. 235–45. Beijing: Academica Sinica
113. Zharkov, M. A. 1981. *History of Paleozoic Salt Accumulation.* Berlin: Springer-Verlag
114. Zhou, L., Kyte, F. T. 1988. The Permian-Triassic boundary event: a geochemical study of three Chinese sections. *Earth Planet. Sci. Lett.* 90:411–21
115. Zishun, Li., Zhan Lipei, Zhu Xiufang, Zhang Jinghua, Jin Ruogu, et al. 1986. Mass extinction and geological events between Paleozoic and Mesozoic era. *Acta Geologica Sinica* 60:1–17

*Annu. Rev. Ecol. Syst. 1990. 21:93–127*


# DIGESTIVE ASSOCIATIONS BETWEEN MARINE DETRITIVORES AND BACTERIA

*Craig J. Plante, Peter A. Jumars and John A. Baross*

School of Oceanography, WB-10, University of Washington, Seattle, Washington 98195

KEY WORDS: bacteria, detritivore, digestive associations, marine

## INTRODUCTION

Although bacteria can be associated with animals in a wide variety of ways, nutritional interactions are by far the most ubiquitous. We define them simply and inclusively as interactions in which proximity allows nutrient transfer in one or both directions, with the most obvious associations involving the alimentary tract and feces, and food just prior to ingestion. Transient or indigenous, attached or free-living bacteria may be obligately or facultatively associated with their animal counterparts.

Obligate interactions of economic importance, e.g. in ruminant and termite guts, are clearly the best studied. A few other strong interactions, such as competition for food between bacteria and carnivores, are reasonably well understood intuitively. We purposely choose here a system—marine detritivory—in which intuition is a less adequate guide, and mutualism, competition, and predation all are suggested interactions. Because marine detrivores ingest bacteria-covered particles, interactions are virtually assured; we look for some methodical approach by which their existence and importance can be predicted and analyzed. The one we adopt for this review is a cost-benefit analysis that adapts and integrates disparate approaches developed for other applications. We evaluate its predictions through literature review. Then we evaluate the ecosystem consequences of our tentatively concluded interactions



0066-4162/90/1120-0093$02.00

and discuss discrepancies and gaps that require future research. Our focus is limited by space to be marine. We draw a few contrasts with terrestrial and freshwater detritivory to emphasize the differences and the reasons for them. Where we suspect that our analyses and conclusions could be extended to broader taxonomic groups more prevalent in these other environments (e.g. fungi), we broaden and loosen terminology (e.g. using "microbes" instead of "bacteria").

## *Defining Detrivitory*

Detritus is usually defined from an ecosystems perspective rather than from the perspective of a species that is a gourmet or gourmand of detritus. Most often it is defined as nonliving, particulate organic matter without regard for the fact that within the biosphere nonliving detritus and living decomposers are associated intimately. Its definition nearly always includes, either implicitly or explicitly, the observation that some kinds of organic matter degrade slowly by the totality of ecosystem processes and therefore show high standing stocks. It often includes the historically easy and gross equation of low food quality with high organic carbon content relative to components that are potentially in much shorter supply, such as biochemically available nitrogen or calories of chemical energy.

The glib definition of detritivory, then, is feeding on detritus. On closer inspection, however, the ecosystem perspective from which the definition of detritus stems is inadequate to distinguish between two radically different feeding strategies. One, a "gourmet" strategy, is ingestion of only the most digestible components of the decomposer system, especially the microbes involved in decomposition. Because of the intimate association of decomposers and the detritus they inhabit, gross inspection of gut contents is unlikely to reveal immediate distinctions between such animals and less discriminating animals that ingest detritus and its associated decomposers in bulk.

A useful distinction is Yonge's (130) idea of macrophages and microphages, the latter handling food material in bulk rather than one item at a time. The small organism searching through the detrital community and ingesting primarily relatively rich microbial biomass, then, is excluded even from the very general definition of detritivores on two counts. It feeds primarily on living matter and primarily on material that is of high bulk food value by either gross C:N indices or finer measures. We suggest that this exclusion of "gourmets" applies in general to marine meiofauna and to their terrestrial and freshwater counterparts (e.g. collembolans and free-living soil nematodes). An analogous gradation occurs among browsing herbivores. Small browsers apparently are able to be highly selective, compared to large browsers (112). Large detritivores and large browsers, then, appear to have many problems in common.

Animal size and bulk composition of ingested material clearly are easier features to identify and measure than are the constituents of ingested material that are digested and absorbed. In recognition of what was known or could be learned easily about most detritivores, a diverse group of marine scientists grappling with a useful operational definition explicitly used Sibly's (112) insightful study of browsing herbivores to redefine detrivory toward the microphagous end of the spectrum as "frequent feeding on material of low bulk food quality" (71). It should not be equated with a lack of selectivity; rather it appears that these "gourmands" of detritus that ingest up to 300 times their body dry weight per day (118) must rely on mechanical selectivity to maintain their rapid feeding rates (111). While this definition of detritivory appears to do little violence to earlier connotations based on an ecosystems perspective, there are important distinctions. Namely, there are many animals for which the bulk of ingested mass and volume is sand. It is clearly of low bulk food value by many definitions, but not by others. Near-surface, shallow-water sand, for example, may have a low C:N ratio due to the presence of phytobenthos and other colonizing microbes. In hindsight, this definition covers another area of ignorance in marine detritivory. Namely, while terrestrial and freshwater detritus originates in large measure from cellulose-rich, large, refractory packets, the origin and identity of refractory marine organic material is in serious question. Even under the open ocean, the possibility that much of the organic matter that is eventually buried in the sedimentary record comes from terrestrial sources cannot be excluded (101).

## MODEL SELECTION AND DEVELOPMENT

Our modeling approach derives from the qualitative scheme developed by Hungate. He (61) predicted the type of microbe-animal association, competition or cooperation, that should exist in the guts of vertebrates based on the "kind" or quality of food being consumed. He later (62) distinguished the combined competition-cooperation model. Cooperation is predicted if a high-carbohydrate, fibrous food is consumed whereas a competitive interaction should be found if fruit or animal food—resources rich in protein or easily digested carbohydrates—is eaten. In the former case, the animal cannot efficiently digest its food and so harbors a microbial consortium to break down ingested material. The animal lives off the metabolites and (in foregut fermenters) cellular growth of these microbes. Herbivores utilizing a rumen comprise the best examples of this strategy. This foregut fermentation chamber allows first access of ingested food to gut microbes, which are then digested by the host.

In the competition case, both animal and bacteria can efficiently digest the ingested food, and so preventive measures are required to avoid significant competition. They are especially necessary given the potential for extremely rapid growth of microorganisms. Elimination of most bacteria is accomplished by stomach acidity in many mammals. The combined competition-cooperation model is seen in hindgut fermenters. In this case the animal has first access to consumed food. Microbes in an enlarged hindgut cecum ferment undigested materials as well as sloughed cells and secretions from the animal. The apparent disadvantage of this system, especially if nitrogen is a limiting nutrient, is the inability to digest the microbial biomass. This disadvantage can be ameliorated via reingestion of feces (e.g. coprophagy in some rodents and caecotrophy in lagomorphs).

Although Hungate's ideas have been inspirational and underlie some portions of our own modeling efforts, unaltered they are insufficient for our purpose of predicting detritivore-bacteria associations. These ideas remain useful for qualitative understanding of the associations for which they are intended, but a single example shows some of their shortcomings. We have seen nonresident bacterial strains efficiently digested in the fore- and midgut of a deposit feeder, with surviving members growing at dramatic rates in the hindgut (99). It would be possible to elaborate Hungate's models to accommodate this pattern of death and growth, but we hesitate to do so. Recent experience with loop analysis (80) confirms the general problem with complex, qualitative, graph-theoretic approaches and their matrix equivalents—the outcome of perturbation and other qualitative network analyses usually are equivocal. These ambiguities cannot be removed without quantitative information (108), so we find it advantageous to bypass complex qualitative models altogether in favor of simple quantitative ones.

The obvious place to look for quantitative help is to the Lotka-Volterra equations, but the scale disparity between bacteria and detritivores makes their value dubious. These models were designed for population-population interactions and, by analogy with chemical mass action, work best where individuals of one population interact with individuals of another with a strength of interaction proportional to local abundances. Animal guts by contrast represent, in the current vernacular, a landscape (cf 47) for whole populations and communities of bacteria. More insidiously, the currency of interaction in the Lotka-Volterra equations is hidden in coefficients rather than made explicit; hence, despite their long history and abundant heuristic applications, they have not proved useful predictively with respect to kinds and intensities of interaction.

Cost-benefit analysis has the decided benefit of being explicitly predictive. We couple two of its many variants in our two-step approach. We first apply a crude level of "optimal digestion theory" (where digestion is interpreted to include absorption as well) to the separate members of the potential interaction, i.e. to detritivores and to heterotrophic microbes feeding on detritus. We note that this approach explicitly follows Wimpenny's (128) advice to make spatial and temporal variations a part of general analytic schemes involving bacterial performance. We then embark on an analysis of costs and benefits of potential associations in terms of the digestive model variables and other landscape variables identified a priori as important to bacteria.

## *Optimal Digestion Theory for Detritivores*

Optimal digestion theory (29, 97) provides a very general mass-balance and mass-flow framework in which to set the functioning of guts. The simplest gut structure and digestively the most effective one when internal fermentation is not important is tubular. Plug flow or a series of mixing cells in such a gut yields digestive products at high rates that can be sustained as long as feeding continues (cf 97). In the simplest such arrangement, the gut is not differentiated into regions, and both secretion of digestive enzymes and absorption of digestive products occur over its full length. Most generally, however, the gut can be divided into three regions on the basis of function. The first is digestive, while the second is digestive and actively absorptive. Digestion can continue in the third, but absorption is usually passive when present; the hindgut stores fecal material and thus allows feeding and digestion to be continuous while defecation can be discontinuous. We take this arrangement as primitive in the metazoan detritivore, although the very existence of hindguts may be related to microbial associations. This potential problem does not jeopardize our general conclusions, which would hold also for an animal with only fore- and midgut. Because digesta in an animal without a nonabsorptive hindgut should leave the absorptive sites when the marginal gain from them falls to the average for the whole gut (29), and because disappearance (via microbial uptake) of digestive products upstream decreases absorption rate everywhere downstream (because concentration of products drives absorption rate), the posteriormost absorptive portions are the least critical ones to protect from microbial invasion. The skeptic thus may substitute for what we call hindgut a hind section of still-absorptive midgut in the primitive detritivore.

In plug flow, there is a fixed relationship between the fractional extent of digestive conversion ($X$ in Figure 1) and gut residence time ($\tau$) (97)

$$\tau = \frac{V}{\nu} = C_{A0}\int_0^{X_{Af}} \frac{dX_A}{-r_A} \qquad 1.$$

Here the subscripts $A$ refer to a particular chemical constituent and 0 to (initial, time 0) concentration in ingested material. $V$ is gut volume, $\nu$ is volumetric throughput rate (volume time$^{-1}$) and $r$ is digestive reaction rate (moles volume$^{-1}$ time$^{-1}$).

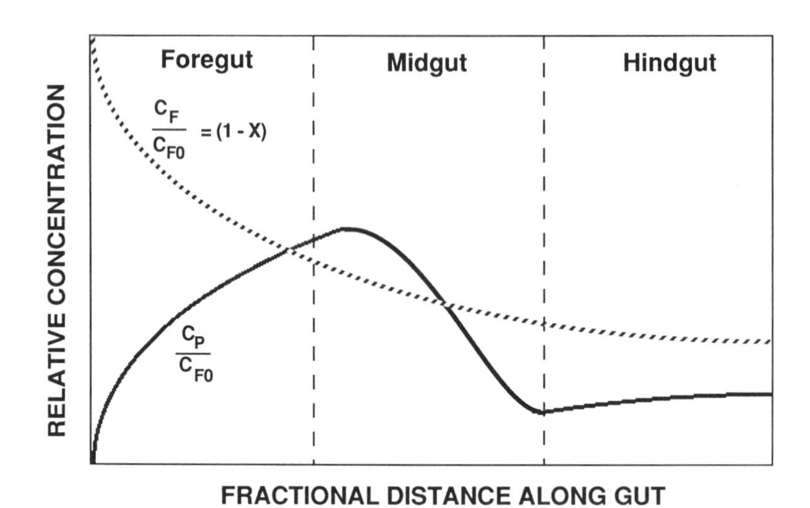


*Figure 1* Schematic pattern of particulate food concentration ($C_F$, mol vol$^{-1}$) and concentration of dissolved digestive product ($C_P$) that would be expected along the gut of a detritivore operating in plug flow to maximize its own rate of digestive absorption (after 29, 1990). Absorption is assumed to occur only in the midgut, while digestion is assumed to occur throughout the gut; both follow Michaelis-Menten kinetics. X is the extent of digestive conversion (a fraction) while $C_{F0}$ is the ingested food concentration. $C_P$ in the feces of an aquatic detritivore would diffuse out within minutes from pellets smaller than 1 mm in diameter due to the small volume and sharp gradients involved (72).

In general, digestion and the animal absorption to which it is coupled will produce a steadily decreasing concentration of the starting food material and an intermediate (in the along-gut direction) peak in digestive products. More surprisingly, an animal acting to maximize its gross rate of absorption will have an optimal retention time and will egest substantial portions of ingested food undigested and digestive products unabsorbed (29; see also Figure 1). Because it requires added gut residence time and thus necessarily a slower rate of digestive production (Equation 1 with Michaelis-Menten kinetics for digestion) more efficient absorption when food is not limiting will decrease the rates of both digestion and absorptive gain. We assume that the ingested food characterized in volumetric concentration by $C_{A0}$ for a detritivore is particulate and that the digestive product characterized by $C_P$ is dissolved.

## *Digestion by Microbes*

We know of no comparable, explicit optimal foraging models for an individual bacterium. Hence, we rely on a rough analysis for end-member cases (Figure 2) to place bacteria in the context of digestive strategies. We do not restrict ourselves to the case of bacteria attached to detritus because the possibilities for association with animal detritivores are not so restricted.

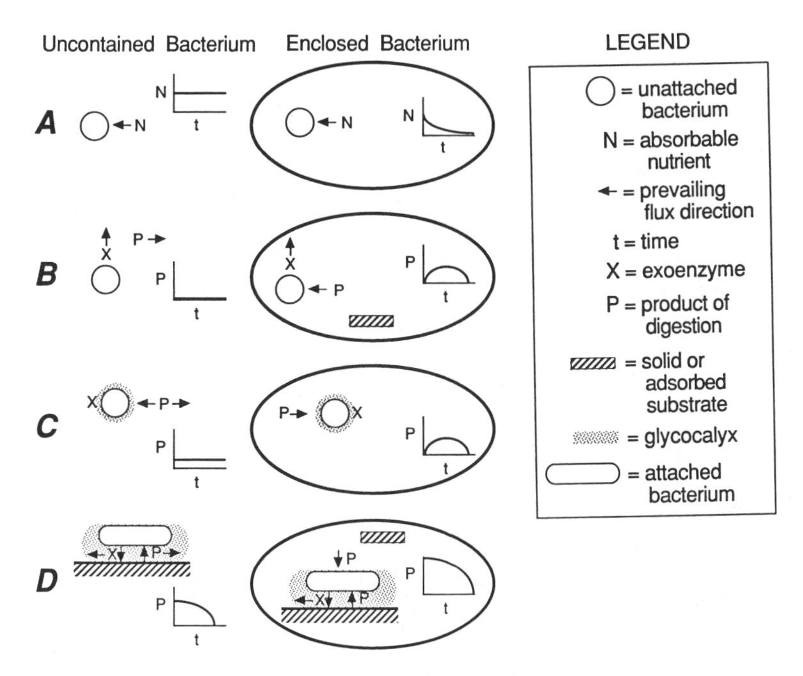


*Figure 2* Effects of enclosure on microbial digestion (via exoenzymes) and absorption across the outer membrane of the cell. Enclosure in the gut lumen occurs upon ingestion and ends upon defecation. Foraging activities of detritivores also rearrange ambient sedimentary particles and thereby both create and destroy enclosures. A: An unattached bacterium without exoenzymes depends on nutrients that can be absorbed directly; enclosure cuts off its supply. B: An unattached bacterium with freely released exoenzymes benefits from them only under enclosure. C: An unattached bacterium with exoenzymes immobilized in a glycocalyx experiences some gain in both situations. D: An attached bacterium that channels diffusion by means of its glycocalyx also gains in both cases. In cases B and D the bacterium has access to remote particulate substrates as well as dissolved substrates. Note that in these two cases, per the models of (75), enclosure of two microbial strains with different digestive products and with abilities to absorb each others' would lead readily to mutualism. It also must lead toward anoxia, thus explaining the prevalence of anaerobic consortia. Members of such consortia are likely to be ingested simultaneously by detritivores.

To place bacteria in the same digestive context as metazoans, we consider digestion via exoenzymes and absorption by active transport across the cell wall. The simplest case (Figure 2) is absorption without digestion—food acquisition without the aid of exoenzyme secretion. It is clearly the most energetically profitable, since the costs of digestion are absent. Taking an unattached bacterium as a sphere of roughly 1 $\mu$m in diameter, one can apply the diffusion equation in spherical coordinates to get some idea of constraints. Solving that equation in spherical coordinates and then integrating over the surface of the sphere, one finds that the flux of dissolved constituents to the cell surface of an individual at steady state is given as

$$4 \pi r_0 D (C_\infty - C_0), \quad 2.$$

where $r_0$ is the cell radius, $D$ is the diffusion coefficient $[L^2T^{-1}]$ for the substance of interest, $C_\infty$ is its ambient concentration, and $C_0$ is the solute concentration at the cell surface. With natural turbulence levels, this flux is virtually unaffected by the flow regime (95) for the simple reason that particles of the size and specific gravity of bacteria track flow perfectly and are too small to experience much turbulence-induced but laminar shear across their cell surfaces. The only freedom that the cell has in maximizing net uptake rate, then, is in altering cell size and uptake kinetics (the latter determining $C_0$). An expanded form of Equation 2 that incorporates uptake kinetics explicitly is given by Pasciak & Gavis (95), but for brevity we do not include it here. We simply note that for this small and unattached cell, the ensuing balance that sets $C_0$ will be of molecular diffusion of solute to the cell surface with (Michaelis-Menten) uptake kinetics. Assuming invariant uptake kinetics, the uptake-vs-time plots of Figure 2 are set, then, by the pattern of $C_\infty$ vs time.

If one considers the possibility of exoenzymatic digestion by bacteria, in an unattached state, diffusion becomes a two-edged sword. It is easy to demonstrate with three-dimensional random-walk models (39) that there is a small probability of return of a product molecule from the sending out of an exoenzyme molecule. The effective dimensionality of connected pore spaces in detritus or sediments falls below three (116), so the return probability rises. Here, time enters into the difficulties as well. If $l_f$ is the distance to a substrate molecule, the fact that diffusion times vary as distance squared gives return time a $2l_f^2$ dependence. If a released enzyme is inactivated after some time ($t_i$), particulate substrates beyond a given distance ($\alpha \sqrt{At_i}$) are unavailable. Complete enclosure in a small space (or the filling of a larger enclosure with clone members) is apparently necessary to provide strong selective pressure for such free release of exoenzymes.

Exoenzymes may be immobilized in a glycocalyx. For an unattached bacterium this situation is much improved over letting enzymes diffuse freely away, but diffusion will carry more than half of the products away from the cell rather than toward it. Bacteria attached to particles can further constrain, via the glycocalyx and the particle, diffusion geometry (Figure 2). To the degree that enzymes are prevented from diffusing away and diffusion of products can be channeled by exopolymer strands toward the microbe, exoenzymes can yield net gain. In the case of attachment to an organic particle and in the absence of enzymatic poisoning, digestion would continue until a surface inert to the enzyme complement were reached.

Attached bacteria also may benefit in enhanced solute flux—without exoenzymes—from relative motion through the surrounding fluid of the particle to which they are attached. In the extreme case of a flat object with a microbial film, boundary-layer depletion can be reduced by fluid motion. Details of cost and benefit to the bacterium depend very heavily on large and small-scale geometry (105), and thus quantitative generalizations are difficult.

Effective enclosure thus appears crucial to the success of exoenzymes. At a critical particle (volumetric) density (116), sediments become dramatically less permeable to diffusion and pore spaces become unconnected to each other. Similarly on a much more local level, natural detrital and mineral grains easily can form enclosures (e.g. Figure 25 in 2). In this case return of products resulting from diffusion of an exoenzyme can become much more likely. Significant early returns are thus more likely in small spaces until substrate concentration becomes limiting (or metabolite build-up becomes inhibiting). Further generalization again will be highly dependent upon geometry.

## *Environmental Covariables*

The stage for association is not well set by digestion of detritus alone; it requires inclusion of environmental covariables. Strong covariation exists, for example, between availability of $O_2$ and the aforementioned diffusion geometry. This difference in oxygen environments carries over to animal respiration. External surfaces of aerobic metazoans must be exposed to oxygenated fluid. Burrow- or tube-constructing animals that are large or that penetrate anoxic sediments must pump in oxygenated water to supply their respiratory needs, with numerous consequences to the surrounding sediments (2, 3). Respiratory as well as feeding currents of both benthic and planktonic animals entrain suspended microbes and expose them to solutes emanating from the animals and their feces.

Some features of the gut itself are nearly universal across environments. One is temporary freedom of microbes from a panoply of predators during gut passage or attachment to gut linings. Many chemical variables in the gut appear to covary with osmotic and water stresses across the range of terrestrial, marine, and freshwater environments and to account for variable linking of the treatment of nitrogenous wastes to the gut environment. Not only the gut environments per se, but in particular the degrees to which the gut and ambient detrital environments should differ change sharply across terrestrial, freshwater, and marine environments (Figure 3). The contrasts are most marked between marine and terrestrial systems. Decomposition of terrestrial detritus often is limited by the availability of water. Nitrogen in any form assimilable by bacteria is scarce in most terrestrial detrital systems. Guts of terrestrial detritivores guarantee water availability to resident or transient

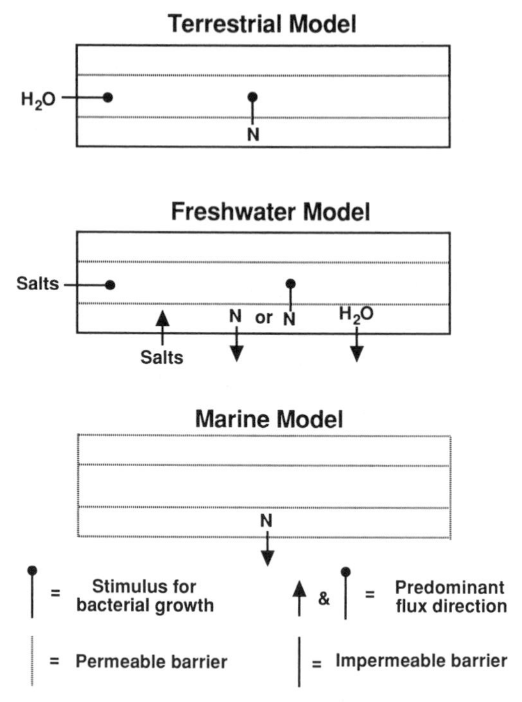


*Figure 3* Contrasts among environments of osmotic balance and nitrogenous waste excretion in detritivores. The inner "tube" (in longitudinal section) represents the gut lumen, and the oral opening is on the left. The diagram thus summarizes salient nondigestive features of the gut landscape experienced by microbes. Duration of exposure to these features is set by gut residence time ($\tau$).

microflora. Water stress in many microenvironments is so severe that detritivores control water loss not only through being relatively impermeable externally, but also through reabsorbing fluids associated with nitrogenous waste excretion; nitrogenous wastes often are excreted into the gut. While fluids are precious, recycling allows them to be used lavishly internally in terrestrial animals. Humans, for example, characteristically eat about 1 liter of solids per day, but they secrete about 7 liters of fluids into those gut contents and resorb them. The consequences for bacteria that grow in the gut are enormous. Eighty percent of the wet weight of human feces is bacterial (35).

Limited exposure of terrestrial animal guts to ambient conditions allows radical chemical departures that optimize pH for specific animal digestive enzymes and may act either as general antibacterial treatments or culture media for specialized microbes. The human "acid wash" of food in the stomach, for example, is thought to be a general antibacterial, and anaerobic conditions in the human hindgut may be an evolutionary compromise that allows containment of bacterial growth while water is resorbed, adds little respiratory demand, and can return some modest benefits as volatile fatty acids (VFAs).

The general picture in terrestrial guts, then, is of very large benefits and costs to most bacterial strains. There is little likelihood, then (since large costs and benefits rarely will sum to zero), of casual associations of bacteria with animal guts, either as residents or transients. To overgeneralize, it should be an evolutionary "love-hate" relationship.

Marine animals, in stark contrast, are highly open systems. Non-enzymatic variables, such as water availability, pH, Eh, and $O_2$ tension would be expected in the absence of microbial associations to change very little upon ingestion and during transit (Table 1), as opposed to the marked deviations from ambient values reported in some terrestrial (14) and freshwater (85) detritivore guts. Animal nitrogenous wastes rarely are excreted into the gut lumen but often are excreted through widely distributed, externally opening ducts or respiratory surfaces (with the proviso that $NH_4^+$ may still diffuse in substantial quantity into the gut). Thus, the respiratory and feeding currents of marine animals double in a nitrogen-flushing capacity. Further, in many marine systems nitrogen is available in the form of $NO_3^-$ to bacteria on detritus. Casual associations of bacteria with detritivore guts thus appear to be far more likely than in terrestrial systems.

Freshwater systems appear to be intermediate. More precisely they appear to be derived with relatively little modification from one of the other two systems, despite the obviously different osmotic problems in fresh water. The ions needed to balance the osmotic flow of water into freshwater detritivores generally are obtained both from food and from active ion transport over respiratory surfaces. As in terrestrial organisms (but for the opposite reasons), the rest of the external surfaces are often relatively impermeable. Unlike either terrestrial or marine organisms, freshwater detritivores minimize their ingestion of water. Ingestion certainly occurs, however, as does the repeatedly noted but still poorly explained phenomenon of anal intake of water (e.g. 49). Nitrogenous waste excretion is focused into the gut in some freshwater animals (notably the larval stages of insects) but not in others (e.g. crustaceans), with the difference seemingly based more on evolutionary history (terrestrial versus marine) than on present environment. Nitrogen can often be available as $NO_3^-$ in ambient water, since $PO_4^{3-}$ is more often the limiting nutrient.

## *Models of Association*

There should evolve non-chance associations (including "avoidances") between microbes and marine detritivores whenever the risks and benefits to

**Table 1** Physico-chemical conditions of marine detritivore guts[a,b]

| Animal | N | Niche | Ingested sediment | FG | MG | HG |
|---|---|---|---|---|---|---|
| ECHINODERMATA | | | | | | |
| *Molpadia intermedia* | | | | | | |
| Eh | 7 | shallow | +122 | +322 | +189 | +232 |
| pH | 7 | subtidal, | 7.2 | 7.6 | 7.5 | 7.5 |
| $[O_2]$[c] | 8 | SSDF[d] | 0[e] | 0 | 0 | 0 |
| *Brisaster latifrons* | | | | | | |
| Eh | 4 | shallow | +378 | +310 | +249 | +273 |
| pH | 3 | subtidal, | 7.5 | 7.3 | 7.4 | 7.4 |
| $[O_2]$ | 4 | SSDF? | 0 | 0 | 0 | 0 |
| *Pannychia* sp. | | | | | | |
| Eh | 3 | bathyal, | +466 | +359 | +386 | +404 |
| pH | – | SDF[d] | – | – | – | – |
| $[O_2]$ | 3 | | 49 | 26 | 26 | 30 |
| *Scotoplanes* sp. | | | | | | |
| Eh | 4 | bathyal, | +466 | +360 | +375 | +377 |
| pH | – | SDF | – | – | – | – |
| $[O_2]$ | 4 | | 49 | 8 | 3 | 4 |
| ANNELIDA | | | | | | |
| *Abarenicola pacifica* | | | | | | |
| Eh | 2 | sandy | +248 | +278 | +95 | +162 |
| pH | 2 | intertidal, | 7.1 | 7.1 | 7.1 | 6.9 |
| $[O_2]$ | 4 | SSDF | 0 | 0 | 0 | 0 |
| *Travisia foetida* | | | | | | |
| Eh | 2 | shallow | +86 | +65 | +36 | +90 |
| pH | 3 | subtidal, | 7.7 | 6.9 | 7.2 | 7.7 |
| $[O_2]$ | 2 | SSDF | 0 | 0 | 0 | 0 |
| *Eupolymnia heterobranchia* | | | | | | |
| Eh | 2 | muddy | −65 | −144 | −37 | −7 |
| pH | 1 | intertidal, | 6.9 | 6.8 | 7.6 | 7.2 |
| $[O_2]$ | 1 | SSDF? | 0 | 0 | 0 | 0 |

[a] Measured with mini- (pH) or microelectrodes ($O_2$, Eh)
[b] Mean values
[c] Concentration in mM
[d] SDF = surface deposit feeder, SSDF = subsurface deposit feeder
[e] Below detection limit (ca. 1% sat.)

either party are unequal. Note that this fitness-based analysis focuses on one, single-species population each of detritivore and microbe. What makes the detritivore-microbe system particularly interesting is that frequency of encounter of diverse microbes by a detritivore is likely to be high relative to organisms that ingest living tissues (herbivores and carnivores); decomposer microbes inhabit detritus and detritivores forage for it and them. As the ground state from which to analyze whether a mutualism or parasitism will evolve, we envisage a situation where most predation pressure and other mortality terms on the microbe population come from factors other than simple predation by the detritivore population in question. We suggest that for any marine detritus-associated microbe population, then, predation by the combination of other species of detritivores, and in particular by a diversity of smaller "gourmets" (i.e. protozoans and metazoan meiofauna), will overwhelm the negative effects of the particular detritivore population on the particular microbe population being analyzed. Thus, only two kinds of two-way association are to be contrasted explicitly, namely, mutualism and parasitism, together with their gradations into commensalisms with little effect in one direction.

To look at associations in animal guts, we follow Keeler's (75) general approach to the analysis of mycorrhizal fungi. Specifically, we define

$$w_{um} = pw_u + qw_n, \qquad 3.$$

where $w$ is fitness, the subscript $m$ denotes microbes, $u$ refers to the population fraction ($p$) that is mutualistic, and $n$ to the fraction ($q$) that may be ingested occasionally but is not mutualistic. Take $w_u = g_O + g_u$ and $w_n = g_O$ such that $g_O$ is the net population growth for the nonmutualist, background population and $g_u$ is the incremental gain from mutualism. Hence,

$$w_{um} = g_O + pg_u. \qquad 4.$$

We suggest that due to the foraging of the detritivore, encounter is likely to be less limiting in evolution of the gut-microbe association than in the development of the mycorrhizal one (allaying some of Keeler's concerns about the appropriateness of the model as applied to mycorrhizae). Hence, development of the mutualism will depend more exclusively on

$$g_u = b_u - c_u + f_u \, \mathrm{p}, \qquad 5.$$

where $b$ is the benefit and $c$ the cost of the mutualism to the microbe. The term $f$ quantifies the feedback to the microbe from the association, since microbial gain via the mutualism drives up the abundance of the detritivore and will alter $p$. Exactly the same model (with alteration of the subscript $m$ in Equation 3) applies to the detritivore. A first step in analyzing the potential for mutualism, then, is to find benefits to both parties.

Two apparently universal benefits to ingested microbes are relief from predation by other predators and exposure both to the mechanical energy (removal of diffusive limitation for particle-associated microbes) and to a reliable stream of the detritivore's food—usually higher in dissolved and particulate organic value than is the ambient medium. Costs will universally include defense from animal digestive enzymes, while terrestrial and to some extent freshwater animal guts will have added costs due to greater changes in other chemical variables (e.g. pH and Eh). The act of enclosure upon ingestion is universal, but returns to the microbe of materials digested by its exoenzymes will depend upon whether it is attached, upon whether the substrate for digestion is dissolved or particulate, on gut (enclosure) size, and upon residence time of material in the gut, as well as on the kinetics of microbial digestion. In the case of transient microbes, the time to induce exoenzyme secretion must be considered as well.

We suggest that the currency of transaction and the magnitudes of the terms can differ radically between attached and unattached (to the gut) bacteria, and thus we distinguish these two cases. We begin with the former and subdivide it further into regions of attachment, i.e. fore- mid- and hindgut. We suggest that (cf Figure 1) potential gains from dissolved products of detritivore digestion are modest in the foregut but in marine detritivores constitute the most obvious gain to microbes. Whether particulate food is available at all to gut-attached microbes via their own exoenzymes will depend critically upon the along-gut velocity of particulate food (*v*/local cross-sectional area of the gut). Since we have defined the situation as a mutualism, we must find something for the microbe to donate, and the most likely donation appears to be a digestive enzyme. Further, the costs to the detritivore may be substantial as lost potential for absorption of its own digestive products (29), escalating microbial $c_u$ via defenses against microbial attachment.

While potential gains in terms of concentrations of detritivore digestive products are greater in the midgut (Figure 1), so are the obvious costs to the detritivore. We suggest that the costs of occlusion of the detritivore's own absorptive system are so great that detritivore defenses will in general make $c_u$ insurmountable, precluding mutualism. A gift of enzymes is not as useful here, since there is less gut area remaining over which absorption can occur. Production of energy-rich products, like volatile fatty acids (VFAs), from the animal's digestive products would again (as in the foregut) appear to be at the expense of the inherent inefficiency of microbial growth (22). Hence there appears to be good reason to find the midgut relatively free of attached mutualists.

The situation in the hindgut is entirely different. There is little obvious cost to the detritivore of allowing attachment. There is substantial gain to the microbe of dissolved digestive products. The obvious microbial contribution is as VFAs that can be absorbed directly by the detritivore without an active transport system. Besides digesta, the hindgut contains the products of gut tissue ablation (that we would argue is at least in part an evolved defense against microbial fouling of the midgut), and hindgut fermentation allows partial recovery of these sloughed gut materials as VFAs.

For unattached (to the gut) microbes transiting a gut, one must take the mean of these conditions. Rather than sitting at one point in Figure 1, then, the transiting microbe follows the curves drawn in that figure. The microbe's cost of attachment to the gut wall disappears. The total time of transit may be long enough in this Lagrangian reference frame for the microbe to take local digestive advantage of the particles with which it transits. Detritivore defenses in the fore- and midgut must change in character to be effective against microbes in transit. We suggest that they will be more expensive for marine detritivores than for terrestrial and freshwater ones whose isolation of the gut from ambient conditions—by reason of both water balance and longer gut residence times than normally seen in marine detritivores without mutualists—allows radical pH and redox changes (e.g. the termite gut). It does not appear, however, that under rapid transit unattached microbes would impose any serious problem, for they would have difficulty equalling the absorptive surface area of the gut in any but the largest detritivores. Assume, for example, a cylindrical gut, spherical bacteria 1 $\mu$m in diameter, and $5 \times 10^8$ bacteria/cc of ingested detritus. We used the 10% figure of Novitsky (93a) for the proportion of bacteria in sediments that are metabolically active. For a deposit feeder of gut diameter, 1 mm, over $10^{10}$ bacteria/cc would be needed to provide equal area. At typical ambient standing stocks of bacteria, significant absorptive competition (at 10% absorptive area) would occur in guts of over 2.5 mm diam., assuming a perfectly smooth gut wall. Thus, there may be neither ready means nor reasons for general detritivore defenses. The question then becomes whether the costs of running the enzymatic gantlet are balanced by the gains in digestive products from the detritivore—in concentration of particulate organic material via selection on the part of the animal or in protection from, say, protozoan bacterivores. The cost-benefit model of Lehman (79), although developed for external encounters, is directly applicable to this analysis.

The situation grades into a one-sided mutualism or simpler commensalism (hindgut) and then into parasitism (foregut and midgut) to the extent that digestive products lost to the microbe are not recovered as VFAs. In order to save space, we forego the formal analysis of parasitism given by Keeler. The critical question clearly is whether $b_u$ for a microorganism in transit is large, and the cost in its fitness from decreased detritivore stocks (making $f_u$ negative) is smaller. It thus seems as though a one-sided benefit or a very benign form of parasitism will be common in microbes transiting detritivore guts. In terms of attached parasites, the midgut clearly is the place with the biggest benefit and cost terms.

While this form of analysis is most often applied to incipient associations, it is applicable in slightly modified form to subsequent evolution of those associations. Namely, if one takes as the ground state the present form of the association, then one can examine the effects of subsequent evolutionary perturbations (mutations) from it. Consider, for example, a simple, tubular gut of uniform diameter and an incipient hindgut mutualism, and take the animal's point of view. If the net return to the animal is limited by time for microbial growth, then evolution should cause the hindgut to lengthen or widen, with the latter being cheaper in gut-lining materials but less advantageous if diffusion of products to the gut wall is rate limiting. (Continuity demands that an increase in pipe diameter be associated with local reduction in flow velocity.) If, on the other hand, net return is limited instead or in addition by surface area for absorption of VFAs, then invagination of the gut lining may be a cheaper solution (i.e. one producing higher net gain). One can envision a chain of escalating mutualism wherein each change is repaid by added gain: The invagination eventually acquires ciliary transport that moves especially fermentable material into it. In terms of flow of material (97), such an invagination not only increases mean residence time, it also adds greatly to variance in residence time. Thus, to shift briefly to the microbe's point of view, invagination affords statistical protection from washout as would dead space in a chemostat. The marginal value theorem in turn dictates that this escalation of gut size or ramification will continue until the net gain from further expansion is not repaid.

Hindgut diverticula in turn would appear to be ideal sites for development of microbial consortia, but the animal, and hence the mutualism, might suffer from $S^{2-}$ toxicity under anaerobic conditions. Here an especially valuable donation from the microbial side would be a mechanism of detoxification (i.e. via microbial chemoautotrophy). Similar but probably weaker arguments might be made for $NH_4^+$. Any chemoautotrophy (e.g. one based on $CH_4$), in turn, might directly or indirectly provide a new source of nutrients to the detritivore. If the mutualism reached the point of cellular contact (or by diffusion even before), we suggest that $f_u$ could skyrocket from access of the bacterial consortium to oxygen provided by the detritivore to its tissues and access of the detritivore to the organic products of chemoautotrophy. If the environment, in turn, were rich in reduced solutes but poor in detrital food value, it is relatively easy to construct an apparently viable chain of events from detritivory to hindgut fermentation, to atrophy of the normal gut, to obligate mutualism. The Pogonophora appear to be likely end products of such a chain of hindgut evolutionary events, starting with a detritivore.

One can also anticipate mechanisms for reducing costs of association to the detritivore. Microbial growth requires oxidants. Where anoxia of tissues might otherwise present a problem or where provision of strong oxidants results in greater rate of gain, there appears to be no theoretical barrier to evolution of respiratory and circulatory features surrounding the region of greatest microbial activity. One can expect respiratory structures or behaviors associated with particularly active hindgut mutualisms.

This exercise thus makes it clear how radical hindgut expansions and hindgut diverticula may evolve in detritivores. Because such past chains of events are difficult to test, however, it may be more instructive to look for cases where a conceivable end product of evolution would be highly successful, but no chain of events can be found to it. For example, a termite with a foregut fermentation chamber instead of a hindgut "paunch" would appear formidable. Similarly, it is plain that hindgut fermenters are far more polyphyletic than foregut fermenters (132). Foregut fermentation, i.e. turning of particulate ingesta into dissolved products by microbial digestion before entry into the midgut, requires substantial residence time (97). Thus, an incipient mutualism of this sort requires for its development an animal with a long gut residence time—a large animal or one that specializes on especially slowly digestible material or both. An animal meeting this requirement stands, by means of the mutualism, to gain access to ingested particulate material as easily absorbed fermentation products, making its benefits transparent. The microbes, in turn, benefit from an enclosure for their exoenzymes (Figure 2). In addition, such enclosures where rapid, obligate metabolite exchanges are possible in the absence of diffusive losses, are ideal for the evolution of microbial consortia. Arguments for subsequent evolution of foregut expansions and diverticula follow those for the hindgut.

External nutritional associations can be diverse. The first dichotomy is having the microbes attached (to the detritivore body) versus unattached. Microbes associating with the external surfaces of animals, particularly with the respiratory surfaces of animals, obtain certain access to $O_2$ as one contribution to $b_u$. Defense against external attachment should result unless the return to the animal, e.g. detoxification, exceeds the tax on its respiration. It would appear that $f_u$ would be particularly large in systems where this detoxification in turn allowed the detritivore access to rich resources, e.g. chemolithotrophic products of hydrothermal vents. In this setting the feedback is intense, as the bacterial need for the attachment includes both a growing requirement for access to oxygen and a means to avoid expatriation. The likelihood of donation of chemoautotrophic production by the attached bacteria to the invertebrates is not clear, however, because the relative magnitude of its effect on $c_u$ versus $f_u$ is not obvious a priori.

Commensalism of aerobic microbes with respiratory streams of bottom-dwelling invertebrates is well established (2). It is obvious, even to the casual observer, as an oxic, reddish (from $Fe^{3+}$) halo about the tubes and burrows that penetrate anoxic sediments. The obvious benefit of such association to functionally aerobic bacteria is a stronger oxidant that provides more ATP per mole of organic matter respired than would residence in the surrounding sediment, and the scale disparity of the two associates suggests that the added water movement needed to support the oxygen demands of any one microbial population results in little cost to the invertebrate. The cost to the detritivore from $O_2$ demand distributed across all aerobic microbes may be substantial, however.

Miller et al (88) suggest that a necessary but insufficient condition for external mutualisms is that geophysical sediment movement is infrequent relative to microbial growth rate. Digestion theory coupled with Keeler's variety of cost-benefit analysis provides additional insight by taking the microbe's perspective first and then looking for positive feedbacks. For lack of space, we do not reproduce her derivations (75, pp. 113–17) for fungus-gardening ants, for they can be adopted without modification. Particularly following our suggestion of rapid microbial growth in the hindgut, microbial growth on feces is the likely beginning of such an association. One positive feedback to the detritivore from subsequent coprophagy is access to food particles with long digestive reaction times as per the arguments of Penry & Jumars (97). An added benefit underscored by Keeler is the reduced cost of foraging both in search costs and risks of predation. We have argued previously (70) that the most likely place to look for such associations among marine detritivores is in the deep sea. Keeler's analysis supports our contention that two sorts of mutualisms are likely. One is based on highly episodic inputs of labile material (phytoplanktonic detritus); Keeler's analysis shows how the caching of such material can benefit the detritivore by smoothing out the valleys between infrequent peaks of inputs at the same time that it—somewhat paradoxically—benefits the associating microbe in terms of population growth rate. The other sort is based on caching of refractory (to detritivore digestion) but energy-rich materials such as cellulose (imported plant debris) and chitin (cast exoskeletons of zooplankton). The situation here is the classic one of renewable resources (81, 83), but the gatherer of energy-rich particles in an otherwise food-poor environment fuels this microbial growth and gets its benefits. Donation of reduced nitrogenous wastes by the detritivore specifically to the microbes might be expected in a tight mutualism.

We suggest that free-living, unattached bacteria will benefit from the excretions of detritivores as sources of reduced nitrogen and from diffusion out of fecal pellets (72) as major sources of labile, dissolved organic carbon. They are entrained in feeding and respiratory streams but, because of the mechanics of particle capture, fall in a size category that runs almost no risk of being eaten by metazoans (41). Thus the quantitative risk-benefit model of Lehman (79) again can be adopted directly.

## LITERATURE-BASED EVALUATION

The intent of our modeling effort is to make predictions and guide future experimentation to test these predictions. Since very little testing of the theory presented has been completed to date, however, we are forced to use extant data, largely obtained with unrelated goals in mind, to evaluate our efforts. They constitute more a consistency check than a rigorous test.

Returns from association rest clearly on the coupled kinetics of gut passage (Equation 1) and microbial growth (Eq. 5). Of particular interest are typical detritivore gut residence times—gut residence times in detritivores not suspected to be obligate mutualists as well as in those known to harbor microbial mutualisms—and maximal bacterial growth rates. To get an idea of $\tau$ in the absence of known mutualisms, we combine characterization of median gut volume (98) and ingestion rates (23, 24) of deposit feeders to give a crude estimate of residence time of material in animal guts operating on shallow-water organic detritus (at 15°C for Cammen's results). Calculated values of order 1 hr for both *Abarenicola pacifica,* a large marine deposit feeder, and for a 20 $mm^3$ *Capitella capitata,* near the hypothetical lower size limit for a true deposit feeder (70), agree reasonably well with published values (45, 60, 99). Because there is a roughly linear relation between body size and proportion of body volume occupied by gut among deposit feeders (98) and ingestion rate scales as $weight^{0.7}$ (23, 24), there is a tendency toward longer residence time for larger animals and individuals, but it is not especially strong. For shallow-water deposit feeders a range of 0.5 hr to 2 hr encompasses most reported observations of $\tau$.

This approach is rough for a number of related reasons. The variance is large in Cammen's (23) study, with $\pm$ 1 SD spanning an order of magnitude. At least part of the reason for this spread is that residence time varies about the species mean as a function of ingested food quality (e.g. 118). The approach has the added problem that it includes whatever undefined microbial relationships already are present in the detritivores. Termites and wood-boring isopods, known with the help of microbial associates to digest extremely resistant materials, have gut residence times on the order of days (19, 58). Thus, it does not appear that the detritivores we characterize with gut residence times an order of magnitude shorter are likely to reflect comparably obligate associations. The values obtained for marine detritivores underscore the difficulty of being a bona fide detritivore at small body size; limited volume and short residence time conspire to restrict small animals to labile material. One does not know, however, whether the longer residence times characteristic of larger species provide significant access to material with slow digestive kinetics or simply allow greater efficiency of absorption of the same rapidly digested constituents that small animals must use.

The potential for association depends on the ratio of $\tau$ to microbial doubling time. Under optimal conditions with unlimited levels of dissolved carbon substrate and an assimilable source of nitrogen, bacterial growth rates can be explosive. Doubling times of under 15 min have been achieved for marine bacteria in pure cultures (122). In natural settings such as in aquatic sediments or in open-ocean and coastal waters "bulk" growth rates are considerably slower and can range from hours to days (76, 90, 91). Sedimentary bacteria can show, however, a rapid growth response to substrate manipulation or physical mixing (e.g. 91). Growth rates approaching those observed under optimal conditions in laboratory cultures have been measured in the guts of some marine detritivores (30). For example, Plante et al (99) recorded doubling times of approximately 1 hr for sedimentary bacteria both under optimal laboratory conditions (at in situ temperature) and within deposit-feeder guts.

A further restriction on mutualisms dependent upon provision of exoenzymes by microbes transiting guts would be the time needed to induce their secretion if they are not already present. In accord with our theoretical arguments (Figure 2), in water-column environments bacterial exoenzymes are scarce (26). Diffusive losses of exoenzymes in general are large in aerobic environments—they must be large to keep those environments aerobic—so exoenzymes are scarce in aerobic pore waters as well (86).

The environmental constancy offered by the gut relative to ambient detrital environments might suggest that large populations of bacteria should be found attached to the gut wall. Our cost-benefit analysis, however, points out that in the foregut—especially in view of the short residence times noted—benefits may not be sufficient to foster association, and in the midgut costs to the animal will be so great that defense mechanisms should overwhelm potentially great gains. Foreguts are, indeed, poorly utilized by microorganisms (14 and references within), and mechanisms such as harsh chemistry (14–16, 36), continuous sloughing (121), and peritrophic membranes (8, 14) are common to terrestrial, freshwater, and marine detritivore midguts and are effective in keeping tissues devoid of association (17–19, 78). Bacterial attachment to hindguts of terrestrial and freshwater detritivores is ubiquitous (8, 10, 19, 27, 28, 78, 87). Insufficiency of data precludes the same conclusion for marine detritivores, but available data do reveal hindgut associates and anterior regions free of attached bacteria (e.g. 30, 113). Net gain or loss in fitness to marine invertebrates through such interaction is currently unknown.

In terrestrial systems pH and Eh of gut compartments of both invertebrates and vertebrates often are such that microbial growth or viability is precluded within transiting material (e.g. 61). In marine detritivores direct and viable counts of ingested versus defecated material show that bacteria are indeed removed by some mechanism in the gut (8, 9, 30, 113). When counts among the gut sections have been performed, numerical reduction has been shown to occur in the foregut or anterior midgut (8, 30, 99, 113, 131). Efficiency of digestion of bacteria by detritivores, often inadvertently referred to as "assimilation efficiency," is normally quite high, often over 90% (9, 12, 56, 99). In some cases these values are underestimates since selective feeding (73, 111) or regrowth in the posterior of the gut (99) has been neglected in experimental design. The most likely explanation for bacterial removal is that cells are enzymatically digested by the animal: Appropriate enzymes are available, cell numbers and not just viabilities decrease, and bacterial carbon is assimilated. In marine deposit feeders, true assimilation efficiency has been estimated to be up to 70% via radiolabelling (57, 82). Uptake and incorporation of bacterial cell constituents indicate some benefit to the animal. The debate of direct quantitative importance of microbes to detritivore nutrition, however, will continue until limiting factors are better established.

Our cost-benefit analyses, based on predicted digestive product distributions and regional gut functions, suggest that most transient bacterial growth should be within and posterior to the midgut and that it likely continues within defecated material (albeit with the rapid loss of dissolved digestive products from pellets). Efficient digestion of protozoans and meiofauna, likely predators of bacteria, has been documented (64, 65). Experiments designed explicitly to test the predictions of digestive product distribution in the gut have yet to be performed, but the profiles of soluble proteins and carbohydrates in the guts of abyssal holothurians obtained by Sibuet et al (113) resemble those of digestive products in Figure 1. Numerous studies have shown growth, sometimes remarkably rapid, in the guts and feces of detritivores (5, 30, 74, 78, 87, 92, 94, 99, 103, 119). Once again, only a subset of these studies allows assignment of growth to sections within the gut; when done in such a manner the posterior gut regions indeed exhibit most of the growth (30, 78, 94, 99, 113).

In some instances, numbers in feces or in the hindgut are lower than in ingested materials. Attention to Keeler's models, however, shows that this observation is not sufficient to demonstrate a net loss to bacterial fitness. Numbers may initially be reduced to such an extent in the foregut (e.g. 99, 131) that growth in the hindgut, especially if transit is rapid, may not bring numbers up to original levels. Stimulated growth, however, may continue within the feces and even after disaggregation (100). Overall microbial fitness—i.e. net loss or gain to integrated population growth rates rather than short-term abundance change—needs to be estimated and may be influenced long after bacteria leave the gut. Few studies have followed the microbial response through fecal pellet production and decay. Hargrave (55), however, has demonstrated increasing metabolic activity in fecal pellets up to two days after defecation. Loss of dissolved substrates from pellets may be rapid (72), but digestion and mechanical disruption of the constituent particles clear new surfaces for microbial growth. The net benefits over time to bacteria contained in the pellets as compared to those freely suspended in fluids to which dissolved digestive products diffuse require experimental evaluation.

Bacterial growth rates in the presence of particles such as sediment grains and detritus currently cannot be measured easily or with high confidence. The reliability of the common methods (increase in abundance over time, frequency of dividing cells, and incorporation of tritiated nucleic acid precursors) has been brought into question repeatedly (e.g. 25, 89, 90, 93, 107). The few attempts to measure growth rates in detritivore guts or egesta nonetheless indicate rates that are orders of magnitude faster than in ambient detrital environments—near laboratory maxima for in situ pressures and temperatures (30, 99).

We caution that measurements of numbers, growth rates, and activities of total bacterial communities mask interactions of major biological or geochemical importance involving single strains of bacteria and detritivores. The differential effects of consumption and gut passage on various populations of microbes have been demonstrated in terrestrial (103, 117), freshwater (20, 124), and marine (32) environments. Beyond the demonstration of community changes with gut transit, little can be said about which microbes should fare best under specific circumstances. Mechanisms responsible for such community shifts remain undetermined but are likely related both to differential digestion and to varying abilities to take advantage of the benefits of gut passage. What can be said, however, is that the degree to which sediment reworking determines microbial community composition must depend heavily on the frequency of gut passage. If we equate the "rest interval" of sedimentary particles calculated by Wheatcroft et al (127) with the average period between ingestion of sedimentary bacteria by a deposit feeder, 1 to 1500 days may elapse between ingestion in various environments. It therefore appears that the effects of detritivore gut passage can be important in determining microbial community structure in at least some benthic settings.

Thus our predictions regarding incipient mutualism and commensalism within guts are supported by available data. The next question is whether the scenarios regarding further coevolution of mutualists are well met. Again, residence-time data and documentations of microbial associations are sparse, so observations of ceca or diverticula or prominent local expansions are noted as circumstantial evidence. Such information can be ambiguous, however, since structures that appear to have one function (e.g. mixing) may (also) serve another (e.g. sorting and bypass of indigestible material) (96). We also note the bias of an apparent social tabu. Structures of the foregut are drawn more frequently and in far more detail than structures near the anus.

Numerous taxa (e.g. most crustaceans, cf 14) possess midgut diverticula that sometimes have been suggested (without evidence) to harbor microbial associates. We know of no direct evidence that any such suggestion is true. On the contrary, deliberate examinations of midgut diverticula invariably have revealed them to be sites of enzyme secretion and absorption of digestive products (e.g. 17, 21) and to lack enclosed microbial associates. Thus, their enhancement of residence time and surface area for absorption and their selective entrainment of particularly digestible material appear, in accord with our theoretical suggestion of a microbially "inviolate" midgut, to be adaptations enhancing digestion by the animals' own means.

Also in accord with our predictions, the majority of ceca or other such enlargements that prolong retention are located in the hindgut, e.g. in the termite (19), cranefly larvae (78, 87), millipedes (5), mayfly larvae (28), cockroaches (27), isopods (58), and in certain holothuroids (43), irregular urchins (31) and deep-sea molluscs (1). Microbial symbiosis and fermentation have indeed been observed in the hindgut ceca of terrestrial (10, 11, 13, 119) and freshwater detritivores (77, 114), but unequivocal evidence for microbial fermentation in the guts of marine detritivores does not exist. The demonstration of fermentative mutualisms in marine herbivores (44) is, however, highly suggestive.

Prominent foregut diverticula are rare among detritivores. Models of cooperative digestive associations general to all animals would predict both foregut and hindgut diverticula for such purposes. Foregut assocations of this type are not documented in detritivores, however, in accord with our suggestions regarding residence time. While such a development would appear highly adaptive (there being no obvious reason why, for example, termite equivalents with foregut fermentation should not do better than extant termites), there appears from our theoretical considerations to be no ready way to initiate it. We suggest that long residence time would be required for initiation (to allow absorptive midgut gain from microbial digestion in the fore- and midgut), perhaps accounting for the rarity of evolution of foregut fermenters. The only suggestion that we can find of foregut diverticula among detritivores is by Penry & Jumars (98) for the deep-burrowing, large deposit feeder, *Travisia foetida*. Direct examinations of microbial associates in it and of its gut residence time and the residence times of related species (including some without diverticula) are needed to evaluate the possibility of mutualism. Decapod crustaceans would appear a likely place for foregut associations to evolve for the simple reason that gastric mills evolved for grinding large food items could serve as preadaptations adding residence time and escape from microbial washout. The crystalline style of bivalves may offer the same (110).

Experimental evidence for microbial aid in digestion among invertebrate detritivores is limited largely to the digestion of cellulose (77, 85, 114). Uptake of product from cellulose breakdown has been demonstrated in the hindgut of host detritivores (77, 85). Furthermore, these associations are reflected in hindgut expansions or diverticula. Thus, our theoretical predictions appear well met.

The strongest evidence of their contradiction, however, comes from similar tracer studies of other stream detritivore species (114) and of marine or estuarine mysid shrimp (48, 123). The stream detritivore, *Pteronarcys,* assimilates carbon from cellulose (114) but shows no obvious gut expansions or attached associates (28). Further, Sinsabaugh et al (114) have shown that absorption of the radioactive label is primarily as sugars in the midgut. Sinsabaugh et al invoke "acquired enzymes," i.e. utilization by the detritivore of enzymes acquired from ingested microbes. This interpretation contradicts our prediction that foregut associations would be rare due to kinetic constraints. It requires significant production of microbial exoenzymes and digestion of cellulose during food passage from the point of ingestion to the midgut absorptive sites (i.e. during roughly two thirds of the gut residence time). If their interpretation is correct, then we would anticipate that this captivation of enzymes resulted from the preassociation condition of detritivores specializing on refractory detritus, thereby having relatively long gut residence times compared to those we presented for marine deposit feeders. An alternative possibility cannot be ruled out, however, by the data presented. Namely, the detritivore species involved may be a "gourmet" of microbial species that are digestive specialists on the low-molecular-weight, labile sugars taken up and manufactured by cellulose-digesting bacteria. Nothing in the experimental protocol precludes the possibility that most of the cellulose digestion indicated by the presence of radioactivity in sugars absorbed by the detritivores occurs prior to ingestion. Both interpretations beg the other interesting question of what might be the source of nitrogen.

Mysids efficiently digest and assimilate cellulose with a reported gut residence time of 30 min (48). Microscopy reveals no attached associates (51), but antibiotics do inhibit the assimilation (123). One interpretation of these findings is that acquired enzymes must be responsible for cellulose breakdown. The earlier data of Foulds & Mann (48), however, appear to exclude this possiblity since initially sterile cellulose is digested and assimilated. Friesan et al (50) conclude that enzymes endogenous to the animal must be digesting cellulose and that the antimicrobials employed by Wainwright et al must have inhibited the mysid's ability to produce cellulolytic enzymes. The short residence times are still troubling, but the numerous midgut pouches (51) may allow selective retention. Endogenous cellulases in invertebrates are not extremely rare and, strangely, appear to be more common in marine than in freshwater or terrestrial detritivores (129). Perhaps the openness of marine detritivore guts has provided greater opportunity for cellulase-sharing associations to develop—which have been mistaken for endogenous production.

With respect to external associations, we do not discount the potential importance of "gardening" as originally proposed by Hylleberg (64), i.e. the stimulation of microbes by pumping in oxygen and mechanical agitation for subsequent ingestion, but we know of no published evidence that the obvious and measurable effects on the microbial community affect fitness of the supposed gardener. Hylleberg (64) studied the lugworm *Abarenicola pacifica* and speculated that its tail-toward-head pumping of water into its burrow enhances the growth of microbes in the sediments it swallows. He presented no bacterial counts or growth estimates to document the feasibility of this mechanism in enhancing food supply. A necessary and also untested corollary is that the bulk of bacterial growth that is stimulated occurs on substrates that are not available directly to the worm's digestive system. Dobbs & Whitlach (33) made a concerted effort to document gardening in another subsurface deposit feeder. While they made numerous valuable observations, they produced no convincing evidence of gardening. Bacterial growth enhancement (as in gardening) would need to occur at the feeding depth between the time that the microbial community is first influenced by the deposit feeder and the time that it is ingested. Unfortunately, the geometry and timing of site (re)visitation is not known for either of these subsurface deposit feeders. The microbial effects of pumping in of oxygen and mechanical agitation certainly can be expected to be major in what otherwise is an anoxic and quiescent environment. Thus, an effect on the microbial community—stress reactions of anaerobes and incipient growth of aerobes—would result. The time history and spatial extent of these effects relative to geometries and rates of deposit feeding need investigation as direct evidence, pro or con, of gardening. To date evidence is lacking to allow discounting of two other hypotheses. Subsurface deposit feeders may depend upon nonrenewable resources slowly accumulated with gradual, often anaerobic decomposition as material arrives by sedimentation at their feeding horizons. Alternatively, they may rely upon subduction of more recently deposited and presumably more labile organic material (106).

We have argued that gut volume is a major constraint on detritivores in general and deposit feeders in particular. Another aspect of this same volumetric constraint is on residence time. One way to alleviate both these limitations would be to carry out some of the digestion externally, with reingestion some time after egestion. Evidence for this sort of behavior in detritivores is primarily circumstantial and anecdotal, but it is at least as strong as the evidence for gardening, and there is much better reason a priori to expect a two-way interaction between microbes and detritivores. Further, gross anatomy and environmental characteristics combine to suggest that it will be found in deep-sea sipunculids and echiurans (70). Echiurans have nitrogenous waste excretion into anal sacs and are known to tend and periodically move collections of their fecal pellets (115). Nitrogenous waste excretion and egestion are closely juxtaposed in sipunculids. X-radiographs suggest smearing of burrow walls with X-ray transparent material (Figure 1 of 70), and Graf et al (53) recently have shown that bathyal sipunculids cause rapid microbial growth, presumably by subduction of newly sedimenting remains of a phytoplankton bloom, several centimeters below the sediment-water interface. Thus, there clearly is an interaction in this case between the detritivores and microbes, but its nature is still ambiguous. Jumars et al (70) have argued that if labile food arrives at the seafloor in widely separated pulses then caching should be seen, and Graf's observations appear to provide strong support of this hypothesis. The role of bacteria is unclear, however. They may be competitors for this labile material or they may be "gardened" as in fungus-ant associations (75).

Nor must caching be limited to labile material. If calories as well as available nitrogen become scarce in deep-sea sediments, then caching of structural carbohydrates and chitin would be expected. The terrestrial analogue of this sort of association would again be fungal gardening ants. While macrophytic debris apparently subducted by deposit feeders has been seen in both shallow water (106) and the deep sea (102), there is no strong evidence linking it to caching behavior or indicating particular microbial associates that might make the calories available to detritivores.

## *Geochemical Consequences*

Microbes are the primary decomposers of organic matter in nature. It has become abundantly clear, however, that the actions of detritivores accelerate decomposition in both terrestrial (6, 54, 94, 104) and aquatic (40, 42, 66) ecosystems. Mesocosm experiments reveal that benthos, chiefly deposit feeders, within shallow marine sediments in concert with microbes play a major role in enhancing primary production in the water column above by regenerating up to 100% of nutrients immobilized in deposited materials (34 and references within).

Among possible effects accelerating breakdown and mineralization are distribution of cells to sites more favorable to growth, removal of bacterial competitors and predators, enrichment by excretions or digestive products, and increase in surface area via fragmentation. In terrestrial and freshwater ecosystems detritivore fragmentation is likely the predominant factor, especially where vascular plant debris is substantial (125). Mechanical disruption is important in increasing surface area for microbial colonization and in breaking microbial barriers such as tough cell walls. The same stimulation of remineralization by mechanical disturbance has not always been observed in marine detritus (4). A high degree of autolysis is characteristic of the producers of marine detritus (chiefly algae and seaweed) relative to vascular plants

(109). In addition, phytodetritus will present high surface area without physical disruption, and seaweed, lacking the structural polymers of vascular plant cell walls, is much more prone to fragmentation by abiotic factors. Rapid decomposition of typical marine detritus is probably less dependent on invertebrate comminution. The paucity of marine shredders supports this assertion. In two situations fragmentation should be important in the marine setting, in nearshore areas where vascular plant and seaweed input is high and in places where living phytobenthos contribute significantly to the diet. The latter appear to be unavailable to bacterial colonization prior to mechanical cell rupture or autolysis (52). The lower content of available calories and nitrogen characteristic of terrestrial detritus (120) also means that a supplement food source, possibly fungi or bacteria, should be an absolute requirement.

In considering animal-microbe interactions we have focused on circumstances within the gut. In terms of geochemical importance, however, the feces may be of greater interest both because gut residence times are short relative to the lifetime of fecal materials, and chemical exchange with overlying water and sediments will occur in feces. Digestion theory predicts that substantial amounts of dissolved digestive products should be ejected in the feces, especially when food is abundant. The large concentration gradients and small size of fecal pellets should result in extremely rapid flux of solutes to surrounding waters (72). This condition, coupled with the observed enhanced mineralization in guts and feces, indicates that feces are understudied sites of major "sediment"-water exchange of dissolved organic substances and inorganic trace elements. The peritrophic membrane covering fecal pellets of some detritivores will likely affect these fluxes much as do the tube membranes of certain deposit feeders (3).

Stimulated bacterial growth and decomposition due to these dissolved products in the gut and feces will deplete oxygen and may create reducing conditions. Metabolic reactions resembling those in anaerobic sediments, e.g. sulfate reductions (67), can come to dominate in fecal pellets. Oxygen consumption by bacteria and their metabolic products, e.g. $H_2S$, will continue to remove oxygen from ambient waters. Qualitatively, the effects of a switch from aerobic to anaerobic microbial metabolism, or vice versa, are fairly well understood. In the presence of oxygen, aerobic respiration, characterized by the ability of single species to completely mineralize organic matter, will predominate. With anaerobiosis, fermentation and various types of anaerobic respiration, utilizing $(NO_3)^-$, $Mn^{4+}$, $Fe^{3+}$, $(SO_4)^{2-}$, or $CO_2$ as terminal electron acceptors, will occur together. Anaerobic processes are characterized by incomplete mineralization by any one group of microbes, so that a complex consortium is needed to achieve complete mineralization.

The quantitative aspects, i.e. rate and extent of conversion, of aerobic versus anaerobic decomposition have been debated. It is well known that aerobic growth is more efficient in terms of ATP produced per mole of substrate and that bacterial biomass production is higher in the presence of oxygen (63). Confusion between efficiency and reaction kinetics has led to the mistaken assumption that aerobic decomposition rates must exceed anaerobic ones. Additionally, numerous field studies which demonstrate that aerobic decomposition proceeds much more rapidly than does anaerobic breakdown (67, 69) have reinforced the idea that aerobic breakdown is inherently faster than anaerobic breakdown; that is, under identical conditions save for the presence or absence of oxygen, organic matter will decompose more quickly if oxygen is available. Laboratory studies, on the other hand, usually reveal that there is essentially no difference in rates (46, 126). We hold the opinion that it is not the presence of oxygen per se that seems to accelerate degradation over relevant time scales, but rather factors largely restricted to oxygenated regimes such as mechanical mixing from currents and animals. Agitation, whether due to fluid dynamic processes or bioturbation, will stimulate bacteria even without the addition of oxygen (37, 46, 48). All other things being equal, anoxic decomposition of fresh organic matter may be faster or slower than oxic decomposition depending on the nature of the substrate (59). An environment of fluctuating oxygen status, then, may show decomposition rates higher than either purely aerobic or anaerobic systems. Additionally, inhibitory end-products of anaerobic metabolism will be removed when oxygenated. This hypothesis may explain the disparate results found in field versus laboratory comparisons of aerobic and anaerobic decomposition rates since strictly aerobic detrital environments are rare in nature. Where deposit feeding is intense, fluctuating oxygen conditions will be especially prevalent due either to ingestion and defecation at different depths or oxygen consuming processes in the gut or feces.

The guts of detritivores provide a unique environment where both mechanical agitation and anaerobiosis are found. We therefore expect to find rapid decomposition comparable to rates in oxic zones yet with all the characteristics of anoxic degradation—the end-products of anaerobic respiration such as $NH_3$, $H_2$, $H_2S$, and $CH_4$, incomplete mineralization, and free fermentation products. Thus, an animal using the anaerobic degradation of its associated microbes may garner a rapid rate of gain, as do ruminants. Beyond mutualistic situations, the environmental consequences of this unique situation should motivate interest. Guts will surely be sites of high rates of anaerobic biogeochemical conversions—possibly among the highest in nature, adding credence to the speculation that the missing marine source of $CH_4$ and other end-products of anaerobic metabolism may be found in animal guts (63).

Conversely, oxygen may be supplied to the gut lumen through the ill-studied mechanisms of anal swallowing (e.g. 49), the diffusion of oxygen through the gut wall (62; C. Plante, unpublished), and possibly from the close spatial association of the respiratory trees and gut in holothuroids. Oxidation of gut contents may simply be incidental to physiological requirements unrelated to digestion, but the oxidation will affect microbial production. Deposit feeding, including events before ingestion and after defecation, may be particularly influential in geochemical cycling. The results of this sort of unsteadiness have been little explored.

## CONCLUSIONS

We conclude that the opportunity for facultative and relatively weak microbe-detritivore associations is far greater in marine than it is in terrestrial or freshwater environments. The reason is that the costs and benefits to microbial populations of association in general are both smaller. Marine animal guts on average will differ less from the detrital microbial environment in such leading chemical environmental parameters as water availability, pH, oxygen tension, Eh, and supply of inorganic but reduced nitrogen. The underlying reasons appear related to weaker or nonexistent evolutionary linkage between osmotic and digestive control in marine animals. The microbial environment of marine detritivore guts for these reasons appears well characterized spatially and temporally by digestion-absorption models based on chemical reactor theory. The major costs and benefits to microbes thus can be formulated, respectively, as risks of enzymatic digestion and local rates of supply of the products of animal digestion. Predictions can be made more precise when those products are better identified and their concentrations measured. For these same reasons of general openness to the external environment, the opportunities for invasion of detritivore guts by microbes are far more frequent than in terrestrial or even freshwater settings. Hence we anticipate that enzymatic and mechanical means of avoiding microbial attachment to the midgut will be found to be far more prevalent in the guts of marine detrivores—which do not have the general antibiotic benefits of radical pH and Eh changes. These mechanical means include encasement of gut contents in peritrophic membranes as well as frequent ablation and replacement of gut linings. On the other hand, transiting bacteria may not be important competitors, given our comparisons of absorptive area for ingested bacteria versus detritivore guts. These calculations, however, are heavily dependent on the validity of the numerous explicit assumptions. Especially important is the use of the value of 10% for proportion of active cells in ingested detritus. This number comes from one study of bacterial activity in marine sediments (93a) and may be quite different within a gut.

Once the "war of absorption" is over, the hindguts and feces of marine detrivores are excellent environments for microbial growth and may be major unstudied sites of geochemical transformation. In many detritivores they have the unusual combination of the (albeit limited) mechanical mixing of gut passage and low Eh. Digestion theory predicts that the hindguts and feces of animals operating to maximize their own rates of absorptive gain will be sites of high availability of digestive products to microbes. Micrography of crustacean hindguts and our own experiments support this idea.

Our review also highlights obvious places to look for exceptions to these generalizations. Where supplies of structural carbohydrates and proteins of relatively uniform composition are available, the strong benefits of mutualism suggest greater association. Specific substrates are cellulose from terrigenous inputs and seagrasses, various structural carbohydrates from macroalgae and chitin from crustacean molts. Conversely, we strongly doubt that the heterogeneous substrates lumped under such chemically uninformative labels as "humic substances" are amenable to rapid digestion by microbial fermentation in digestive associations. Unlike terrestrial and freshwater environments, then, there is much less apparent improvement of food quality of the dominant inputs (phytoplankton detritus) with microbial aging or animal trituration. Away from local concentrations of macrophytes, shredders constitute a much less important guild in the sea than they do in fresh water and on land.

The case for and against acquired enzymes needs particularly close scrutiny. Similarly, gain to microbes from exoenzyme release needs examination as a function of enclosure size, geometry and residence time.

Our models explicitly deal with the interactions of individual microbial populations with animals while empirical studies largely remain at the community level. With bulk measurements of bacterial standing stocks or growth rates, differential effects of gut transit on distinct populations, each playing unique biological and geochemical roles, are hidden. More specific methodologies, such as immunofluorescence techniques and molecular probes, are required.

Likely invertebrate detritivore targets for studies of associations also can be pinpointed. Obvious associations to look for are fermenters in guts of animals in or near marshes, seagrass beds, kelp beds, and river deposits. In deep water one can expect caching by detritivores of energy-rich materials. In cases where these caches are of diatom detritus one can expect associations with specific bacteria and antibiotic protection from invasion by others. In caching of chitin or cellulose, the utility of fermentative associations is clear. Sipunculids and echiurans are particularly promising for studies of close external associations because of their unique (for marine invertebrates) potential for provision of reduced inorganic nitrogen to their associates. Documented cases of associations support these ideas but are too scarce to provide much confidence. We suggest, however, that cost-benefit theory can now provide the basis for an efficient attack on the mechanisms and consequences of digestive associations of marine microbes and detritivores.

ACKNOWLEDGMENTS

We are grateful to Deborah Penry, Jody Deming, and Jim Staley for critical evaluations of the manuscript. This work was supported by NSF grant OCE 86-08157 and ONR grant N 00014-87-K0126.

## *Literature Cited*

1. Allen, J. A., Sanders, H. L. 1966. Adaptations to abyssal life as shown by the bivalve *Abra profundorum* (Smith). *Deep-Sea Res.* 13:1175–84
2. Aller, R. C. 1982. The effects of macrobenthos on chemical properties of marine sediment and overlying water. In *Animal-Sediment Relations,* ed. P. L. McCall, T. J. S. Tevesz, pp. 53–102. New York: Plenum
3. Aller, R. C., Yingst, J. Y. 1978. Biogeochemistry of tube dwellings: A study of the sedentary polychaete *Amphitrite ornata* (Leidy). *J. Mar. Res.* 36:201–54
4. Alongi, D. M. 1985. Effect of physical disturbance on population dynamics and trophic interactions among microbes and meiofauna. *J. Mar. Res.* 43:351–64
5. Anderson, J. M., Bignell, D. E. 1980. Bacteria in the food, gut contents and faeces of the litter-feeding millipede *Glomeris marginata. Soil Biol. Biochem.* 12:251–54
6. Anderson, J. M., Ineson, P., Huish, S. A. 1983. Nitrogen and cation mobilization by soil fauna feeding on leaf litter and soil organic matter from deciduous woodlands. *Soil Biol. Biochem.* 15:463–67
7. Anderson, J. M., Macfadyen, A., ed. 1976. *The Role of Terrestrial and Aquatic Organisms in Decomposition Processes* (17th Symp. of Brit. Ecol. Soc.). Oxford: Blackwell. 474 pp.
8. Austin, D. A., Baker, H. J. 1988. Fate of bacteria ingested by larvae of the freshwater Mayfly *Ephemera danica. Microb. Ecol.* 15:323–32
9. Baker, J. H., Bradnam, L. A. 1976. The role of bacteria in the nutrition of aquatic detritivores. *Oecologia* 24:94–105
10. Bayon, C. 1980. Volatile fatty acid and methane production in relation to anaerobic carbohydrate fermentation in *Oryctes nasicornis* (Coleoptera: Scarabaedae). *J. Insect Physiol.* 26:819–28
11. Bayon, C., Mathelin, J. 1980. Carbohydrate fermentation and by-product absorption studied with labelled cellulose in *Oryctes nasicornis* larvae (Coleoptera: Scarabeidae). *J. Insect Physiol.* 26:833–40
12. Berrie, A. D. 1976. See Ref. 7, pp. 323–38
13. Bignell, D. E. 1977. Some observations on the distribution of gut flora in the American cockroach *(Periplaneta americana). J. Invert. Pathol.* 29:338–43
14. Bignell, D. E. 1984. The arthropod gut as an environment for microorganisms. In *Invertebrate-Microbial Interactions,* ed. J. M., Anderson, A. D. M. Rayner, D. W. H. Walton, pp. 205–27. Cambridge: Cambridge Univ. Press
15. Bignell, D. E. 1984. Direct potentiometric determination of redox potentials of the gut content in termites *Zootermopsis nevadensis* and *Cubitermes severus* and in three other arthropods. *J. Insect Physiol.* 30:169–74
16. Bignell, D. E., Anderson, J. M. 1980. Determination of pH and oxygen status in the guts of lower and higher termites. *J. Insect Physiol.* 26:183–88
17. Bignell, D. E., Oskarsson, H., Anderson, J. M. 1980. Distribution and abundance of bacteria in the gut of a soil-feeding termite. *J. Gen. Microbiol.* 117:393–403
18. Boyle, P. J., Mitchell, R. 1978. Absence of micro-organisms in crustacean digestive tracts. *Science* 200:1157–59
19. Breznak, J. A. 1982. Intestinal microbiota of termites and other xylophagous insects. *Annu. Rev. Microbiol.* 36:323–43
20. Brinkhurst, R. O., Chua, K. E. 1969. Preliminary investigations of the exploitation of some potential nutritional resources by three sympatric tubificid oligochaetes. *J. Fish. Res. Bd. Can.* 26:2659–68
21. Buddington, R. K., Diamond, J. M. 1987. Pyloric ceca of fish: a 'new' absorptive organ. *Am. J. Physiol.* 252: 665–76
22. Calow, P. 1977. Conversion efficiencies in heterotrophic organisms. *Biol. Rev.* 52:385–409
23. Cammen, L. M. 1980. Ingestion rate: an empirical model for aquatic deposit feeders and detritivores. *Oecologia* (Berlin) 44:303–10
24. Cammen, L. M. 1987. Polychaetes. In *Animal Energetics,* ed. T. J. Pandian, F.

J. Vernberg, pp. 217–60. New York: Academic

25. Carman, K. R., Dobbs, F. C., Guckert, J. B. 1988. Consequences of thymidine catabolism for estimates of bacterial production: An example from a coastal marine sediment. *Limnol. Oceanogr.* 33:1595–1606
26. Chrost, R. J. 1989. Characterization and significance of β-glucosidase activity in lake water. *Limnol. Oceanogr.* 34:660–72
27. Cruden, D. C., Markovetz, A. J. 1987. Microbial ecology of cockroach gut. *Annu. Rev. Microbiol.* 41:617–43
28. Cummins, K. W., Klug, M. J. 1979. Feeding ecology of stream invertebrates. *Annu. Rev. Ecol. Syst.* 10:147–72
29. Dade, W. B., Jumars, P. A., Penry, D. L. 1989. Supply-side optimization: maximizing absorptive rates. In *Behavioral Mechanisms of Food Selection,* ed. R. N. Hughes, pp. 531–56. London: Springer-Verlag
30. Deming, J. W., Colwell, R. R. 1982. Barophilic growth of bacteria from intestinal tracts of deep-sea invertebrates. *Microb. Ecol.* 7:85–94
31. DeRidder, C., Jangoux, M. 1982. Digestive systems: Echinoidea. In *Echinoderm Nutrition,* ed. M. Jangoux, J. M. Lawrence, pp. 213–34. Rotterdam: Balkema
32. Dobbs, F. C., Guckert, J. B. 1988. Microbial food resources of the macrofaunal-deposit feeder *Ptychodera bahamensis* (Hemichordata: Enteropneusta). *Mar. Ecol. Prog. Ser.* 45:127–36
33. Dobbs, F. C., Whitlach, R. B. 1982. Aspects of deposit-feeding by the polychaete *Clymenella torquata. Ophelia* 21:159–66
34. Doering, P. H. 1989. On the contribution of the benthos to pelagic production. *J. Mar. Res.* 47:371–83
35. Drasar, B. S., Barrow, P. A. 1985. *Intestinal Microbiology*. Washington, DC: Am. Soc. Microbiol. 80 pp.
36. Drew, R. A. I., Courtice, A. C., Teakle, D. S. 1983. Bacteria as a natural source of food for adult fruit flies (Diptera: Tephritidae). *Oecologia* 60:279–84
37. Fallon, R. D., Brock, T. D. 1979. Decomposition of blue-green algal (cyanobacteria) blooms in Lake Mendota, Wisconsin. *Appl. Environ. Microbiol.* 37:820–30
38. Fasham, M. J., ed. 1984. *Flows of Energy in Marine Ecosystems: Theory and Practice*. New York: Plenum. 733 pp.
39. Feller, W. 1968. *An Introduction to Probability Theory and Its Applications*. New York: Wiley. 360 pp.
40. Fenchel, T. 1970. Studies on the decomposition of organic detritus from the turtle grass *Thalassia testidinum. Limnol. Oceanogr.* 15:14–20
41. Fenchel, T. 1984. See Ref. 38, pp. 301–15
42. Fenchel, T. M., Jørgensen, B. B. 1977. Detritus food chains of aquatic ecosystems: the role of bacteria. *Adv. Microb. Ecol.* 1:1–58
43. Feral, J-P., Massin, C. 1982. Digestive systems: Holothuroidea. In *Echinoderm Nutrition,* ed. M. Jangoux, J. M. Lawrence, pp. 191–212. Rotterdam: Balkema
44. Fong, J. B., Mann K. H. 1980. Role of gut flora in the transfer of amino acids through a marine food chain. *Can. J. Fish. Aquat. Sci.* 37:88–96
45. Forbes, T. L. 1989. See Ref. 84, pp. 171–200
46. Foree, E. G., McCarthy, P. D. 1970. Anaerobic decomposition of algae. *Environ. Sci. Technol.* 4:842–49
47. Forman, R. T. T., Godron, M. 1986. *Landscape Ecology*. New York: Wiley
48. Foulds, J. B., Mann, K. H. 1978. Cellulose digestion in *Mysis stenolepsis* and its ecological implications. *Limnol. Oceanogr.* 23:760–66
49. Fox, H. M. 1952. Anal and oral intake of water by crustacea. *J. Exp. Biol.* 29:583–99
50. Friesan, J. A., Mann, K. H., Novitsky, J. A. 1986. *Mysis* digests cellulose in the absence of a gut microflora. *Can. J. Zool.* 64:442–46
51. Friesan, J. A., Mann, K. H., Willison, J. H. M. 1986. Gross anatomy and fine structure of the gut of the marine mysid shrimp *Mysis stenolepsis* Smith. *Can. J. Zool.* 64:431–41
52. Golterman, H. L. 1972. The role of phytoplankton in detritus formation. *Mem. Ist. Ital. Idrobiol.* (Suppl.)29:89–103
53. Graf, G. 1989. Benthic-pelagic coupling in a deep-sea benthic community. *Nature* 341:437–39
54. Hanlon, R. D. G., Anderson, J. M. 1979. The effects of collembola grazing on microbial activity in decomposing leaf litter. *Oecologia* 38:93–99
55. Hargrave, B. T. 1976. See Ref. 7, pp. 301–21
56. Harper, R. M., Fry, J. C., Learner, M. A. 1981. A bacteriological investigation to elucidate the feeding biology of *Nais variabilis* (Oligochaete: Naididae). *Freshwater Biol.* 11:227–36
57. Harvey, R. W., Luoma, S. N. 1984. The role of bacterial exopolymer and

suspended bacteria in the nutrition of the deposit-feeding clam, *Macoma balthica*. *J. Mar. Res.* 42:957–68
58. Hassall, M., Jennings, J. B. 1975. Adaptive features of gut structure and digestive physiology in the terrestrial isopod *Philoscia muscorum* (Scopoli). *Biol. Bull.* 149:348–64
59. Henrichs, S. M., Reeburgh, W. S. 1987. Anaerobic mineralization of marine sediment organic matter: rates and the role of anaerobic processes in the oceanic carbon economy. *Geomicrobiol. J.* 5:191–237
60. Hobson, K. D. 1967. The feeding ecology of two North Pacific *Abarenicola* species (Arenicolidae, Polychaeta). *Biol. Bull.* 133:343–54
61. Hungate, R. E. 1975. The rumen microbial ecosystem. *Annu. Rev. Ecol. Syst.* 6:39–66
62. Hungate, R. E. 1976. Microbial activities related to mammalian digestion and absorption of food. In *Fiber in Human Nutrition*, ed. G. Spiller, R. Amen, pp. 131–49. New York: Plenum
63. Hungate, R. O. 1985. Anaerobic biotransformations of organic matter. In *Bacteria in Nature*, ed. E. Leadbetter, J. Poindexter, 1:39–95. New York: Plenum
64. Hylleberg, J. 1975. Selective feeding by *Abarenicola pacifica* with notes on *Abarenicola vagabunda* and a concept of gardening in lugworms. *Ophelia* 14: 113–37
65. Hylleberg, J., Gallucci, V. G. 1975. Selectivity in feeding by the deposit-feeding bivalve *Macoma nasuta*. *Mar. Biol.* 32:167–78
66. Hylleberg, J., Henriksen, K. 1980. The central role of bioturbation in sediment mineralization and element recycling. *Ophelia* suppl. 1:1–16
67. Jørgensen, B. B. 1977. Bacterial sulfate reduction within reduced microniches of oxidized marine sediments. *Mar. Biol.* 41:7–17
68. Jørgensen, B. B. 1978. A comparison of methods for the quantification of bacterial sulfate reduction in coastal marine sediments. I. Measurements with radiotracer techniques. *Geomicrobiol. J.* 1: 11–27
69. Jørgensen, B. B. 1980. Mineralization and the bacterial cycling of carbon, nitrogen and sulfur in marine sediments. In *Contemporary Microbial Ecology*, ed. D. C. Ellwood, J. N. Hedger, M. J. Latham, J. M. Lynch, J. H. Slater, pp. 239–52. London: Academic
70. Jumars, P. A., Mayer, L. M., Deming, J. D., Baross, J. A., Wheatcroft, R. A. 1990. Deep-sea deposit-feeding strategies suggested by environmental and feeding constraints. *Philos. Trans. Royal Soc. Lond.* In press
71. Jumars, P. A., Newell, R. C., Angel M. V., Fowler, S. W., Poulet, S. A., Rowe, G. T., Smetacek, V. 1984. See Ref. 38, pp. 685–93
72. Jumars, P. A., Penry, D. L., Baross, J. A., Perry, M. J., Frost, B. W. 1989. Closing the microbial loop: dissolved carbon pathway to heterotrophic bacteria from incomplete ingestion, digestion and absorption in animals. *Deep-Sea Res.* 36:483–95
73. Jumars, P. A., Self, R. F. L., Nowell, A. R. M. 1982. Mechanics of particle selection by tentaculate deposit feeders. *J. Exp. Ecol. Mar. Biol.* 64: 47–70
74. Juniper, S. K. 1981. Stimulation of bacterial activity by a deposit feeder in two New Zealand intertidal inlets. *Bull. Mar. Sci.* 31:691–701
75. Keeler, K. H. 1985. Cost:benefit models of mutualism. In *The Biology of Mutualism*, ed. Douglas H. Boucher, pp. 100–127. London: Croom Helm
76. Kemp, P. F. 1987. Potential impact on bacteria of grazing by a macrofaunal deposit-feeder, and the fate of bacterial production. *Mar. Ecol. Prog. Ser.* 36:151–61
77. Lawson, D. L., Klug, M. J. 1989. Microbial fermentation in the hindguts of two stream detritivores. *J. N. Am. Benthol. Soc.* 8:85–91
78. Klug, M. J., Kotarski, S. 1980. Ecology of the microbiota in the posterior hindgut of larval stages of the crane fly *Tipula abdominalis*. *Appl. Environ. Microbiol.* 40:408–16
79. Lehman, J. T. 1987. Selective herbivory and its role in the evolution of phytoplankton growth strategies. In *Growth and Reproductive Strategies of Freshwater Phytoplankton*, ed. C. Sandgren, pp. 369–87. London: Cambridge Univ. Press
80. Levins, R. 1975. Evolution in communities near equilibrium. In *Ecology and Evolution of Communities*, ed. M. L. Cody, J. M. Diamond, pp. 16–50. Cambridge, Mass: Belknap
81. Levinton, J. S., Lopez, G. R. 1977. A model of renewable resources and limitation of deposit-feeding benthic populations. *Oecologia* 31:177–90
82. Lopez, G. R., Cheng, I-J. 1983. Synoptic measurements of ingestion rates, ingestion selectivity, and absorption efficiency of natural foods in the deposit-feeding molluscs *Nucula annulata*

(Bivalvia) and *Hydrobia Totteni* (Gastropoda). *Mar. Ecol. Prog. Ser.* 11:55–62
83. Lopez, G. R., Levinton, J. S. 1987. Ecology of deposit-feeding animals in marine sediments. *Q. Rev. Biol.* 62: 235–60
84. Lopez, G. R., Taghon, G. L., Levinton, J. S. ed. 1989. *Ecology of Marine Deposit Feeders*. New York: Springer-Verlag. 322 pp.
85. Martin, M. M., Martin, J. S., Kukor, J. J., Merritt, R. W. 1980. The digestion of protein and carbohydrate by the stream detritivore, *Tipula abdominalis* (Diptera, Tipulidae). *Oecologia* 46:360–64
86. Mayer, L. M. 1989. Extracellular proteolytic enzyme activity in sediments of an intertidal mudflat. *Limnol. Oceanogr.* 34:973–81
87. Meitz, A. K. 1975. *Alimentary tract microbiota of aquatic invertebrates*. MS thesis. Mich. State. Univ., East Lansing, Mich. 64 pp.
88. Miller, D. C., Jumars, P. A., Nowell, A. R. M. 1984. Effects of sediment transport on deposit feeding: scaling arguments. *Limnol. Oceanogr.* 29: 1202–17
89. Moriarty, D. J. W., Pollard, P. C. 1981. DNA synthesis as a measure of bacterial productivity in seagrass sediments. *Mar. Ecol. Prog. Ser.* 5:151–56
90. Moriarity, D. J. W., Pollard, P. C. 1982. Diel variation of bacterial productivity in seagrass *(Zostera capricorni)* beds measured by rate of thymidine incorporation into DNA. *Mar. Biol.* 72:165–73
91. Moriarity, D. J. W., Pollard, P. C., Hunt, W. G., Moriarity, C. M., Wassenberg, T. J. 1985. Productivity of bacteria and microalgae and the effect of grazing by holothurians in sediments on a coral reef flat. *Mar. Biol.* 85:293–300
92. Newell, R. C. 1965. The role of detritus in the nutrition of two marine deposit feeders, the prosobranch *Hydrobia ulvae* and the bivalve *Macoma balthica*. *Zool. Soc. Lond. Proc.* 144:25–45
93. Newell, R. C., Fallon, R. D. 1982. Bacterial productivity in the water column and sediments of the Georgia (USA) coastal zone: Estimates via direct counting and parallel measurement of thymidine incorporation. *Microb. Ecol.* 8:33–46
93a. Novitsky, J. A. 1983. Heterotrophic activity throughout a vertical profile of seawater and sediment in Halifax Harbor, Canada. *Appl. Environ. Microbiol.* 45:1753–60
94. Parle, J. N. 1963. Micro-organisms in the intestines of earthworms. *J. Gen. Microbiol.* 31:1–11
95. Pasciak, W. J., Gavis, J. 1974. Transport limitation of nutrient uptake in phytoplankton. *Limnol. Oceanogr.* 19:881–88
96. Penry, D. L. P. 1989. Tests of kinematic models for deposit-feeders' guts: patterns of sediment processing by *Parastichopus californicus* (Stimpson) (Holothuroidea) and *Amphicteis scaphobranchiata* Moore (Polychaeta). *J. Exp. Mar. Biol. Ecol.* 128:127–46
97. Penry, D. L., Jumars, P. A. 1987. Modeling animal guts as chemical reactors. *Am. Nat.* 129:69–66
98. Penry, D. L., Jumars, P. A. 1990. Gut architecture, digestive constraints and feeding ecology of deposit-feeding and carnivorous polychaetes. *Oecologia.* 82:1–11
99. Plante, C. J., Jumars, P. A., Baross, J. A. 1989. Rapid bacterial growth in the hindgut of a marine deposit feeder. *Microb. Ecol.* 18:29–44
100. Porter, K. G. 1976. Enhancement of algal growth and productivity by grazing zooplankton. *Science* 192:1332–34
101. Prahl, F. G., Muehlhausen, L.A. 1989. Lipid biomarkers as geochemical tools for paleoceanographic study. In *Productivity of the Oceans: Present and Past*. Report of the Dahlem workshop on productivity of the ocean, present and past. Berlin 1988), ed. W. H. Berger, V. S. Smetacek, G. Wefer, pp. 271–89. New York: Wiley
102. Reichardt, W. 1987. Microbiological aspects of bioturbation. Proc. 22nd European Marine Biology Symposium (Barcelona, 1987), Investigacion Pesquera (Suppl.)
103. Reyes, V. G., Tiedje, J. M. 1976. Ecology of gut microbiota of *Tracheoniscus rathkei* (Crustacea, Isopoda). *Pedobiologia* 16:67–74
104. Reyes, V. G., Tiedje, J. M. 1976. Metabolism of $^{14}$C-labeled plant materials by woodlice *(Tracheoniscus rathkei* Brandt) and soil microorganisms. *Soil Biol. Biochem.* 8:103–8
105. Riber, H. H., Wetzel, R. G., 1987. Boundary-layer and internal diffusion effects on phosphorus fluxes in lake periphyton. *Limnol. Oceanogr.* 32:1181–94
106. Rice, D. L. 1986. Early diagenesis in bioadvective sediments: Relationships between the diagenesis of beryllium-7, sediment reworking rates, and the abundance of conveyor-belt deposit-feeders. *J. Mar. Res.* 44:149–84

107. Riemann, B., Nielson, P., Jeppeson, M., Fuhrman, J. A. 1984. Diel changes in bacterial biomass and growth rates in coastal environments, determined by means of thymidine incorporation into DNA, frequency of dividing cells (FDC), and microautoradiography. *Mar. Ecol. Prog. Ser.* 17:227–35
108. Roberts, F. S. 1976. *Discrete Mathematical Models.* Englewood Cliffs, NJ: Prentice-Hall 559 pp.
109. Sanders, G. W. 1976. See Ref. 7, pp. 341–73
110. Seiderer, L. J., Newell, R. C., Schultes, K., Robb, F. T., Turley, C. M. 1987. Novel bacteriolytic activity associated with the style microflora of the mussel *Mytilus edulis* (L.). *J. Exp. Mar. Biol. Ecol.* 110:213–24
111. Self, R. F. L., Jumars, P. A. 1988. Cross-phyletic patterns of particle selection by deposit feeders. *J. Mar. Res.* 46:119–43
112. Sibly, R. M. 1981. Strategies of digestion and defecation. In *Physiological Ecology: an Evolutionary Approach to Resource Use,* ed. C. R. Townsend, P. Calow, pp. 109–39. Sunderland, Mass: Sinauer
113. Sibuet, M., Khripounoff, A., Deming, J., Colwell, R., Dinet, A. 1982. Modification of the gut contents in the digestive tract of abyssal holothurians. In *Int. Echinoderms Conf.,* Tampa Bay, ed. J. M. Lawrence. Rotterdam: A. A. Balkema
114. Sinsabaugh, R. L., Linkins, A. E., Benfield, E. F. 1985. Cellulose digestion and assimilation by three leaf-shredding aquatic insects. *Ecology* 66:1464–71
115. Smith, C. R., Jumars, P. A., Demaster, D. J. 1986. In situ studies of megafaunal mounds indicate rapid sediment turnover and community response at the deep-sea floor. *Nature* 323:251–53
116. Stauffer, D. 1985. *Introduction to Percolation Theory.* London: Taylor & Francis, 124 pp.
117. Szabo, I, Marton, M., Buti, I. 1969. Intestinal microflora of the larvae of St. Mark's fly. IV. Studies of the intestinal bacterial flora of a larva-population. *Acta Microbiol. Acad. Sci, Hung.* 16:381–97
118. Taghon, G. L. 1988. The benefits and costs of deposit feeding in the polychaete *Abarenicola pacifica. Limnol. Oceanogr.* 33:1166–75
119. Taylor, E. C. 1982. Role of aerobic microbial populations in cellulose digestion by desert millipedes. *Appl. Environ. Microbiol.* 44:281–91
120. Tenore K. R., Cammen, L., Findlay, S. E. G., Phillips, N. 1982. Perspectives of research on detritus: do factors controlling the availability of detritus to macroconsumers depend on its source? *J. Mar. Res.* 40:473–90
121. Terra, W. R., Ferriera, C., de Bianchi, A. G. 1979. Distribution of digestive enzymes among the endo- and ectoperitrophic spaces and midgut cells of *Rhynchosciara* and its physiological significance. *J. Insect Physiol.* 25:487–94
122. Ulitzur, S. 1974. *Vibrio parahaemolyticus and Vibrio alginolyticus:* Short generation-time marine bacteria. *Microb. Ecol.* 1:127–35
123. Wainwright, P. F., Mann, K. H. 1982. Effect of antimicrobial substances on the ability of the mysid shrimp *Mysis stenolepis* to digest cellulose. *Mar. Ecol. Prog. Ser.* 7;309–13
124. Wavre, M., Brinkhurst, R. O. 1971. Interactions between some tubificid oligochaetes and bacteria found in the sediments of Toronto Harbour, Ontario. *J. Fish. Res. Bd. Can.* 28:335–41
125. Webster, J. R., Benfield, E. F. 1986. Vascular plant breakdown in freshwater ecosystems. *Annu. Rev. Ecol. Syst.* 17: 567–94
126. Westrich, J. T., Berner, R. A. 1984. The role of sedimentary organic matter in bacterial sulfate reduction: the G model tested. *Limnol. Oceanogr.* 29:236–49
127. Wheatcroft, R. A., Jumars, P. A., Smith, C. R., Nowell, A. R. M. 1990. A mechanistic view of the particulate biodiffusion coefficient: step lengths, rest periods and transport directions. *J. Mar. Res.* 48:177–207
128. Wimpenny, J. W. T. 1981. Spatial order in microbial ecosystems. *Biol. Rev.* 56:295–342
129. Yokoe, Y., Yasumasu, I. 1964. The distribution of cellulase in invertebrates. *Comp. Biochem. Physiol.* 13:323–38
130. Yonge, C. M. 1928. Feeding mechanisms in invertebrates. *Biol. Rev.* 3:21–76
131. Zhukova, A. I. 1963. On the quantitative significance of micro-organisms in nutrition of aquatic invertebrates. In *Symposium on Marine Microbiology,* ed. C. H. Oppenheimer, pp. 699–710. Springfield, Ill: Thomas
132. Hume, I. D., Warner, A. C. I. 1980. Evolution of microbial digestion in mammals. In *Digestive Physiology and Metabolism in Ruminanants,* ed. Y. Ruckebusch and X. Thivend, pp. 665–84. Lancaster: MTP Press

*Annu. Rev. Ecol. Syst. 1990. 21:129–66*


# MULTIVARIATE ANALYSIS IN ECOLOGY AND SYSTEMATICS: PANACEA OR PANDORA'S BOX?

*Frances C. James*

Department of Biological Science, Florida State University, Tallahassee, Florida 32306

*Charles E. McCulloch*

Biometrics Unit, Cornell University, Ithaca, New York 14853

KEY WORDS: multivariate analysis, data analysis, statistical methods

---

## INTRODUCTION

Multivariate analysis provides statistical methods for study of the joint relationships of variables in data that contain intercorrelations. Because several variables can be considered simultaneously, interpretations can be made that are not possible with univariate statistics. Applications are now common in medicine (117), agriculture (218), geology (50), the social sciences (7, 178, 193), and other disciplines. The opportunity for succinct summaries of large data sets, especially in the exploratory stages of an investigation, has contributed to an increasing interest in multivariate methods.

The first applications of multivariate analysis in ecology and systematics were in plant ecology (54, 222) and numerical taxonomy (187) more than 30 years ago. In our survey of the literature, we found 20 major summaries of recent applications. Between 1978 and 1988, books, proceedings of symposia, and reviews treated applications in ecology (73, 126, 155, 156), ordination and classification (13, 53, 67, 78, 81, 83, 90, 113, 121, 122, 159), wildlife biology (33, 213), systematics (148), and morphometrics (45, 164,



0066-4162/90/1120-0129$02.00

**Table 1** Applications of multivariate analysis in seven journals, 1983–1988. In descending order of the number of applications, the journals are *Ecology*, 128; *Oecologia*, 80; *Journal of Wildlife Management*, 76; *Evolution*, 72; *Systematic Zoology*, 55; *Oikos*, 41; *Journal of Ecology*, 35; and *Taxon*, 27.

| | |
|---|---|
| Principal components analysis | 119 |
| Linear discriminant function analysis | 100 |
| Cluster analysis | 86 |
| Multiple regression | 75 |
| Multivariate analysis of variance | 32 |
| Correspondence analysis | 32 |
| Principal coordinates analysis | 15 |
| Factor analysis | 15 |
| Canonical correlation | 13 |
| Loglinear models | 12 |
| Nonmetric multidimensional scaling | 8 |
| Multiple logistic regression | 7 |
| | 514 |

200). For the six-year period from 1983 to 1988 (Table 1), we found 514 applications in seven journals.

Clearly, it is no longer possible to gain a full understanding of ecology and systematics without some knowledge of multivariate analysis. Or, contrariwise, misunderstanding of the methods can inhibit advancement of the science (96).

Because we found misapplications and misinterpretations in our survey of recent journals, we decided to organize this review in a way that would emphasize the objectives and limitations of each of the 12 methods in common use (Table 2; Table 3 at end of chapter). Several books are available that give full explanations of the methods for biologists (53, 128, 148, 159, 164). In Table 3, we give specific references for each method. In the text we give examples of appropriate applications, and we emphasize those that led to interpretations that would not have been possible with univariate methods.

The methods can be useful at various stages of scientific inquiry (Figure 1). Rather than classifying multivariate methods as descriptive or confirmatory, we prefer to consider them all descriptive. Given appropriate sampling, 6 of the 12 methods can also be confirmatory (see inference in Table 2). Digby & Kempton (53) give numerous examples of applications that summarize the results of field experiments. Most often the methods are used in an exploratory sense, early in an investigation, when questions are still imprecise. This exploratory stage can be a very creative part of scientific work (206, pp. 23–24). It can suggest causes, which can then be formulated into research hypotheses and causal models. According to Hanson (86), by the time the

**Table 2** General objectives and limitations of multivariate analysis

| Objectives | Codes to Procedures (see Table 3) |
|---|---|
| 1. Description | All |
| 2. Prediction | MR. LDFA, MLR |
| 3. Inference | MR, MANOVA, LDFA, FA, MLR, LOGL |
| 4. Allocation | LDFA |
| 5. Classification | LDFA, MLR, CLUS |
| 6. Ordination | LDFA, PCA, PCO, FA, CANCOR, COA, NMDS |

Limitations:

1. The procedures are correlative only; they can suggest causes but derived factors (linear combinations of variables) and clusters do not necessarily reflect biological factors or clusters in nature.
2. Because patterns may have arisen by chance, their stability should be checked with multiple samples, null models, bootstrap, or jackknife.
3. Interpretation is restricted by assumptions.
4. Automatic stepwise procedures are not reliable for finding the relative importance of variables and should probably not be used at all.

theoretical hypothesis test has been defined, much of the original thinking is over. In the general scientific procedure, descriptive work, including descriptive applications of multivariate analysis, should not be relegated to a status secondary to that of experiments (28). Instead it should be refined so that research can proceed as a combination of description, modelling, and experimentation at various scales (106).

The opportunities for the misuse of multivariate methods are great. One reason we use the analogy of Pandora's box is that judgments about the results based on their interpretability can be dangerously close to circular reasoning (124, pp. 134–136; 179). The greatest danger of all is of leaping directly from the exploratory stage, or even from statistical tests based on descriptive models, to conclusions about causes, when no form of experimental design figured in the analysis. This problem is partly attributable to semantic differences between statistical and biological terminology. Statistical usage of terms like "effect" or "explanatory variable" is not meant to imply causation, so the use of terms like "effects" and "roles" in titles of papers that report descriptive research (with or without statistical inference) is misleading. Partial correlations and multiple regressions are often claimed to have sorted out alternative processes, even though such conclusions are not justified. "If . . . we choose a group of . . . phenomena with no antecedent knowledge of the causation . . . among them, then the calculation of correlation coefficients, total or partial, will not advance us a step toward evaluating the importance of the causes at work" (R. A. Fisher 1946, as quoted in reference 54, p. 432).

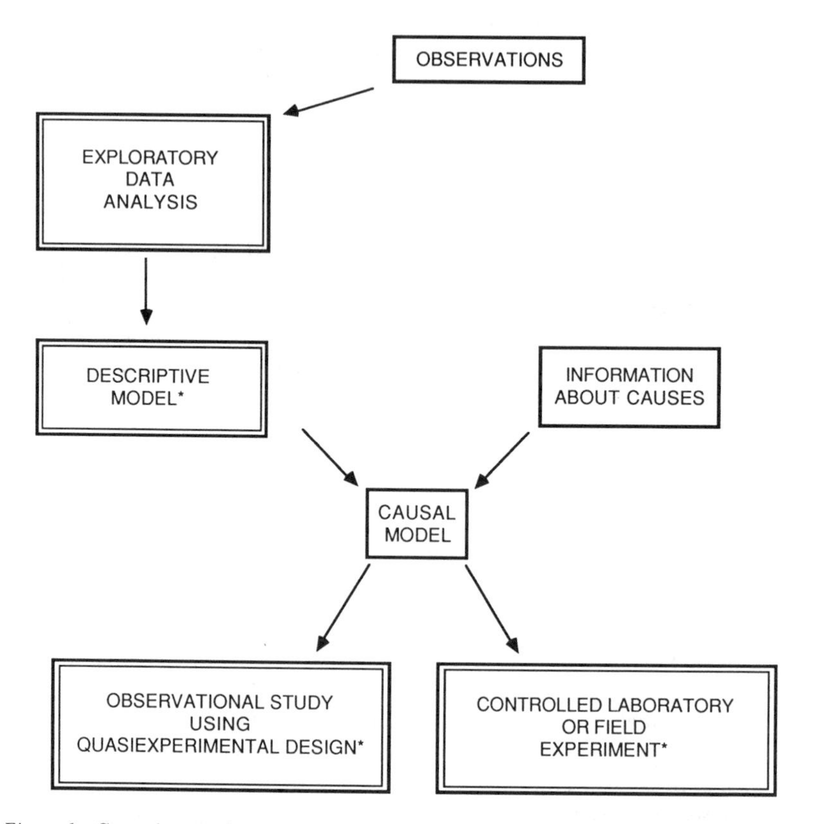


*Figure 1* General research procedure showing stages (double boxes) at which exploratory and inferential* (confirmatory) multivariate analysis may be appropriate (modified from 106).

Although this idea is familiar to biologists, it seems to get lost when they enter the realm of multivariate work.

The objective of the present review is to help the researcher navigate between the Scylla of oversimplification, such as describing complex patterns with univariate analyses (147), and the Charybdis of assuming that patterns in data necessarily reflect factors in nature, that they have a common cause, or, worse, that statistical methods alone have sorted out multiple causes.

Present understanding of the role of multivariate analysis in research affects not only the way problems are analyzed but also how they are perceived. We discuss three particularly controversial topics, and we realize that not all researchers will agree with our positions. The first is the often-cited "problem" of multicollinearity, the idea that, if correlations among variables could be removed, one could sort out their relative importance with multivariate analysis. The problem here is a confusion between the objectives of the method and the objectives of the researcher. Second, in the sections on analysis and ordination in plant ecology, we discuss the special problems that arise with indirect ordinations, such as the cases where the data are the occurrences of species in stands of vegetation. The arch pattern frequently seen in bivariate plots is not an artifact of the analysis; it is to be expected. Third, in the section on morphometrics, we explain why we argue that shape variables, which we define as ratios and proportions, should be studied directly. Of course the special properties of such variables require attention. We do not treat cladistics or the various software packages that perform multivariate analyses. In the last section, we give examples of how some basic concepts in ecology, wildlife management, and morphometrics are affected by the ways in which multivariate methods are being applied.

## SUMMARY OF METHODS: OBJECTIVES LIMITATIONS, EXAMPLES

### *Overview*

It is helpful to think of multivariate problems as studies of populations of objects about which information for more than one attribute is available (48, 169). One can describe the pattern of relationships among the objects (individuals, sampling units, quadrats, taxa) by ordination (reduction of a matrix of distances or similarities among the attributes or among the objects to one or a few dimensions) or by cluster analysis (classification of the objects into hierarchical categories on the basis of a matrix of inter-object similarities). In the former case, the objects are usually displayed in a graphic space in which the axes are gradients of combinations of the attributes. Principal components analysis is an ordination procedure of this type. It uses eigenstructure analysis of a correlation matrix or a variance-covariance matrix among the attributes. Principal coordinates analysis is a more general procedure in the sense that it starts with any type of distance matrix for distances among objects. Both principal components analysis and principal coordinates analysis are types of multidimensional scaling. Nonmetric multidimensional scaling uses the ranks of distances among objects, rather than the distances themselves. Correspondence analysis is an ordination procedure that is most appropriate for data consisting of counts (contingency tables). In this case, the distinction between objects and attributes is less relevant because they are ordinated simultaneously. Factor analysis is similar to principal components analysis in that it uses eigenstructure analysis, usually of a correlation matrix among attributes. It emphasizes the analysis of relationships among the attributes. Canonical correlation reduces the dimensions of two sets of attributes about the same set of objects so that their joint relationships can be studied.

When the objects fall into two or more groups, defined a priori, the problem is frequently to describe the differences among the groups on the basis of a set of attributes. Multivariate analysis of variance, which is often used in the analysis of experiments, can be used to test for differences among groups. Linear discriminant function analysis describes which of the attributes contribute most to the differences between the groups. When it is used as an exploratory ordination procedure, to reduce multigroup data to fewer dimensions on the basis of a set of attributes, it is called canonical variates analysis. Another objective of linear discriminant function analysis, used less frequently in ecology and systematics, is to assign new objects to previously separated groups. Multiple logistic regression permits the prediction of a binary (0, 1) attribute from a set of other attributes, which may be categorical or continuous. Its counterpart for approximately normally distributed data is multiple regression. Loglinear analysis can reveal the relationships among categorical variables. It assumes a multiplicative model, so it is linear after logarithms are taken.

Procedures 1–7 in Table 3 use linear combinations of the variables in some fashion. They are only efficient with continuous data. If the variables being analyzed are denoted by $X_1, X_2, \ldots, X_n$, then all the linear techniques find linear (additive) combinations of the variables that can be represented by:

$$L_x = b_1 X_1 + b_2 X_2 + \ldots + b_k X_k \qquad 1.$$

where $b_1, b_2, \ldots, b_n$ represent coefficients determined from the data. The way the coefficients are found is governed by the method used. For example, in principal components analysis they are chosen to make the variance $L$ as large as possible, subject to the constraint that the sum of squares of the $b$'s must be equal to one.

Linear methods are appropriate when the researcher wants to interpret optimal linear combinations of variables (e.g. principal components in principal components analysis, factors in factor analysis, and discriminant functions in linear discriminant function analysis).

The researcher applying linear methods usually assumes that the values of the variables increase or decrease regularly and that there are no interactions. If this is not the case, one should transform the variables to make them at least approximately linear (55). For example, a quadratic model can be constructed with $X_1$ as a variable $W_1$ and $X_2$ as $W_1$ squared, or interactions can be included, in which $X_3$ is $W_1$ times $W_2$ (104, 133). For some of the techniques the analysis of residuals can uncover the need for the inclusion of nonlinear terms or interactions. In multivariate analysis of variance, the nonlinearities appear in the interaction terms and may reveal biotic interactions in experimental results (see below). Presence-absence data, categorical data, and ranks are usually more efficiently handled with nonlinear models. It seldom makes sense to calculate weighted averages from these types of data, as one does with the linear methods. With nonlinear methods, the variables are combined with nonlinear functions.

The coefficient of an individual variable represents the contribution of that variable to the linear combination. Its value depends on which other variables are included in the analysis. If a different set of variables is included, the coefficients are expected to be different, the "bouncing betas" of Boyce (27).

The term "loading," often encountered in multivariate analysis, refers to the correlation of an original variable with one of the linear combinations constructed by the analysis. It tells how well a single variable could substitute for the linear combination if one had to make do with that single variable (89, p. 221). High positive or negative loadings are useful in the general interpretation of factors. However, the signs and magnitudes of the coefficients should only be interpreted jointly; it is their linear combination, not the correlations with the original variables (cf 220), that must be used to gain a proper multivariate interpretation. Rencher (162) shows how, in linear discriminant function analysis, the correlations with the original variables (loadings) lead one back to purely univariate considerations. This distinction is not important with principal components analysis because the correlations are multiples of the coefficients and their interpretations are equivalent.

Unfortunately, in observational studies, it is often difficult to provide clear descriptions of the meanings of individual coefficients. Mosteller & Tukey (146, p. 394) discuss the important idea of the construction of combinations of variables by judgment, in the context of multiple regression.

Some of the problems we found in our literature survey apply to univariate as well as to multivariate statistics. The first one is that statistical inference is being used in many cases when its use is not justified. The "alpha-level mindset" of editors leads them to expect all statements to be tested at the 0.05 level of probability (175). As a result, our journals are decorated with galaxies of misplaced stars. What the authors and editors have forgotten is that statistical inference, whether multivariate or univariate, pertains to generalization to other cases.

Confirmatory conclusions are only justified with a statistical technique if the study was conducted with appropriate sampling. It is the way the data were gathered, or how an experiment was conducted, that justifies inferences using statistical methods, not the technique itself. Inferences are justified only if the data can be regarded as a probability sample from a well-defined larger population. When this is not the case, probability values should not be reported, and the conclusions drawn should extend only to the data at hand.

The tendency to perform statistical tests when they are not justified is related to the even more general problem of when generalizations are justified. There are too many cases in which results of analyses of single study plots or single species are assumed to be representative of those for large areas or many species. More caution is warranted even in cases of widespread sampling. For example, if several vegetation variables are measured at a series of regularly spaced sites along an altitudinal gradient, the correlations among the variables will show their joint relationship to altitude, but these will differ from the correlations that would have been found had the sites been randomly selected. A principal components analysis based on the former correlations should not be interpreted as giving information about sites in general, and only limited interpretations are possible, even in an exploratory sense.

A further extension of the tendency to overinterpret data is the unjustified assignment of causation in the absence of experimentation. Papers that report the use of stepwise procedures (automatic variable selection techniques) with multiple regression, multivariate analysis of variance, linear discriminant function analysis, and multiple logistic regression to assess which variables are important are examples of the disastrous consequences of this tendency. Such judgments about the importance of variables usually carry implications about causal relationships. In the section on multiple regression, we defend our position that stepwise procedures should not be used at all.

In summary, when faced with data that contain sets of correlated variables, ecologists and systematists may prefer to interpret each variable separately. In such cases univariate methods accompanied by Bonferroni-adjusted tests (89, especially pp. 7–9, but see index; 150) may be appropriate. Often, however, the joint consideration of the variables can provide stronger conclusions than are attainable from sets of single comparisons. With proper attention to the complexities of interpretation, combinations of variables (components, factors, etc) can be meaningful. Linear methods of multivariate analysis (Table 3, 1–7) should be used when the researcher wants to interpret optimal linear combinations of variables. Otherwise, nonlinear methods (Table 3, 8–12) are more appropriate and usually more powerful. Multivariate statistics, modelling, and biological knowledge can be used in combination and may help the researcher design a crucial experiment (Figure 1).

## *Review of Methods*

Our survey of the literature revealed that the methods most commonly applied in ecology were principal components analysis, linear discriminant function analysis, and multiple regression; in systematics the order of use was cluster analysis, principal components analysis, and linear discriminant function analysis. Therefore in this section we devote most of the space to these methods.

We have included both multiple regression and multiple logistic regression even though many statisticians would not classify these methods as multivariate, a term they use only where the "response" (Y) variable rather than the "explanatory" (X) variable is multivariate. We acknowledge that, in multiple regression and multiple logistic regression, the outcome variable is univariate, but we include the topics here because many methodological issues in multiple regression carry forward to multivariate generalizations. The intercorrelations among the explanatory variables ($X$'s) in multiple regression are important to proper interpretation of the results.

MULTIPLE REGRESSION The objective of multiple regression should be either to find an equation that predicts the response variable or to interpret the coefficients as associations of one of the explanatory variables in the presence of the other explanatory variables. The coefficients ($b_1$, $b_2$, . . . , $b_k$) in Equation 1 have been determined either to maximize the correlation between $Y$ (the response variable) and $L$ (the linear combination of explanatory variables) or equivalently to minimize the sum of squared differences between $Y$ and $L$. Only in experiments where the $X$'s are controlled by the investigator can the individual coefficients of a multiple regression equation be interpreted as the effect of each variable on the $Y$ variable while the others are held constant, and only when a well-defined population of interest has been identified and randomly sampled can multiple regression provide statistically reliable predictions. Unfortunately, these conditions are rarely met. "Validation" with new, randomly collected data will be successful only when the original sample is typical of the new conditions under which validation has taken place, and this is usually a matter of guesswork.

Many workers think that, if one could eliminate multicollinearity (intercorrelations) among the X variables in a descriptive study, the predictive power and the interpretability of analyses would be improved (35). This belief has led to the practice of (*a*) screening large sets of redundant variables and removing all but one of each highly correlated set and then (*b*) entering the reduced set into a stepwise multivariate procedure, with the hope that the variables will be ranked by their importance. Statisticians have pointed out many times that this is unlikely to be the case. The procedure of screening variables may improve prediction, but it may also eliminate variables that are in fact important, and stepwise procedures are not intended to rank variables by their importance.

Many authors have documented the folly of using stepwise procedures with any multivariate method (99; 100; 139, pp. 344–357, 360–361; 215, p. 177, Fig. 8.1, pp. 195–196). One example is the reanalysis by Cochran of data from a study of the relationship between variation in sets of weather variables and the number of noctuid moths caught per night in a light trap. Stepwise forward and backward variable selection procedures did not give the same best variable as a predictor or even the same two or three variables as the best subsets of predictors (51). In another case, an investigator analyzing 13 out of 21 attributes of 155 cases of viral hepatitis used the bootstrap procedure to obtain repeated samples of the 155 cases. Of 100 stepwise regressions, only one led to the selection of the same four variables chosen by the initial stepwise regression, and it included a fifth one in addition (139, pp. 356–357). Clearly, stepwise regression is not able to select from a set of variables those that are most influential.

Wilkinson (217, p. 481) used strong language to defend his refusal to include a stepwise regression program in a recent edition of the SYSTAT manual: "For a given data set, an automatic stepwise program cannot necessarily find a) the best fitting model, b) the real model, or c) alternative plausible models. Furthermore, the order variables enter or leave a stepwise program is usually of no theoretical significance."

The best that can be hoped for, when an automatic selection method like stepwise multiple regression is used, is selection of a subset of the variables that does an adequate job of prediction (188, p. 668). However, this prediction can be achieved more reasonably without the stepwise procedure. The most reasonable solution for observational studies that have a battery of explanatory variables is to combine them into biologically meaningful groups (146), then to examine all possible subsets of regressions. The results may provide useful overall predictions, but even in this case they should not be used to rank variables by their importance. Thus, Abramsky et al (1) need not worry about field tests purported to discover interspecific competition from the values of coefficients in multiple regression equations. The method is statistically inappropriate for this purpose.

Progress toward assessing the relative importance of variables can be made by modelling, a subjective step that incorporates subject-matter knowledge into the analysis. Interactive methods (96) and methods of guided selection among candidate models (4) can incorporate reasonable biological information into the analysis (see, e.g., 37, 153, 182). This step can help develop causal hypotheses, but the testing still requires some form of experiment and outside knowledge. When controlled experiments are not feasible, quasiexperimental designs can be used to provide weak inferences about causes (32, 41, 44, 106, 111). Such designs involve either blocking, time-series models, or both.

We regret to report that, in our survey of recent journals in ecology and systematics, we could not find a single application of multiple regression to recommend as a good example. Even recent attempts to measure natural selection in the wild by means of multiple regression (119) are susceptible to the criticisms mentioned above (47, 136a). Use of a path-analytic model has been suggested as a means of adding biological information to the analysis (47, 136a), but even here, because it is not possible to break correlations among characters with experiments, it is not possible to discover whether selection is acting on individual characters. For an example of a proper application of multiple regression and subsequent discussion, see Henderson & Velleman (96) and Aitkin & Francis (2).

MULTIVARIATE ANALYSIS OF VARIANCE Multivariate analysis of variance is an inferential procedure for testing differences among groups according to the means of all the variables. It is like the usual analysis of variance except that there are multiple response variables ($Y_1, Y_2, \ldots, Y_n$). The relationship with univariate analysis of variance can be understood if MANOVA is viewed as an analysis of linear combinations of the response variables,

$$L_y = b_1 Y_1 + b_2 Y_2 + \ldots + b_k Y_k \qquad 2.$$

$L_y$ is now a single, combined, response variable. A univariate analysis of variance can be performed on $L_y$ and an F-statistic calculated to test for differences between groups. One of the suggested tests in MANOVA (Roy's maximum root test criterion) is the same as choosing the $b$'s in equation (2) to maximize the $F$-statistic and then using the maximized value of $F$ as a new test statistic. MANOVA requires that each vector of $Y$'s should be independent and that they follow a distribution that is approximately multivariate normal. A good nonmathematical introduction is available (85).

In a good example of the application of multivariate analysis of variance in ecology, a manipulative factorial experiment designed to determine processes that affect the numbers of tadpoles of several species of amphibians was conducted in artificial ponds. Predation, competition, and water level were the explanatory variables and were regulated (216). The model incorporated the explanatory variables both additively and as interactions with other variables. In one case of interaction between predation and competition, predation on newts *(Notophthalmus)* reduced the effects of competition as the pond dried up, allowing increased survival of the toad *Bufo americanus*. This result would not have been apparent from univariate analyses by species. For an application in a more evolutionary context, see Travis (204). In this paper, he used MANOVA to show that families of tadpoles grew at different rates but were not differentially susceptible to the inhibitory effects of population density.

LINEAR DISCRIMINANT FUNCTION ANALYSIS Linear discriminant function analysis can be regarded as a descriptive version of multivariate analysis of variance for two or more groups. The objective is to find linear combinations of the variables that separate the groups. In Equation 2 above they give rise to the largest F-statistics. The researcher wants to understand $L_y$ and what determines the groups to which specific data vectors belong. Linear discriminant function analysis does not formally require any assumptions, but it is the best technique for multivariate normal data when variances and covariances are the same in each group. Then the optimal combination of variables is linear. If the attributes are nonlinearly related, or the data are otherwise not multivariate normal (for example, categorical data), variances and covariances are poor summary statistics, and the technique is inefficient. An appropriate alternative, when there are only two groups, is multiple logistic regression (see below).

In a summary of applications of linear discriminant function analysis in ecology, Williams (220) warns that more attention should be paid to the assumption of equality of dispersion within groups. He also emphasizes the special problems that arise if the sample sizes are small or different (see also 34, 201, 210). Williams & Titus (221) recommend that group size be three times the number of variables, but this criterion is arbitrary. Discriminant function axes can be interpreted in either a univariate or a multivariate way (see overview). Again, the elimination of variables before the analysis and stepwise procedures should be avoided (163).

When the data are plotted on axes defined by the discriminant functions, the distances (Mahalanobis $D^2$) are measured in relation to variances and covariances. Population means may be judged far apart in cases in which the groups are similar except in one small but statistically highly significant way. This is not true of Euclidean distances in principal components space, so the two types of distances should not be interpreted in the same way (106, cf 34). Graphic presentation of the results can be clarified by the use of either concentration ellipses (43) around groups or confidence ellipses (105) around means of groups (188, pp. 594–601).

Linear discriminant function analysis can be used to summarize the results of an experiment (e.g. 91), but in both ecology and systematics it is used most often as an exploratory ordination procedure. In such cases it is called canonical variates analysis. Many descriptive uses concern resource use and the ecological niche. In the literature on wildlife management, there are applications that attempt to define the habitat of a species from quantitative samples of the vegetation taken in used and unused sites. These topics are discussed in later sections.

Some early exploratory applications of linear discriminant function analysis have made important contributions to studies of comparative morphology and functional anatomy. A good example is work comparing the shapes of the pectoral girdles (clavicles and scapulae) of mammals (8, 157). The variables were angles and indices based on the orientation of the attachments of muscles, so they were related functionally to the use of the forelimb. In Figure 2, for primates, the first discriminant function (linear combination of variables) separates the great apes, which use the forelimbs for hanging, from the quadrupedal primates. The second variate expresses an uncorrelated pattern of development that separates ground-dwellers from arboreal dwellers, some of which are quadrupedal in trees. Convergences between the suborders Anthropoidea and Prosimii and radiations within them are demonstrated simultaneously (see 164 and Figure 2), and graded patterns within groups are evident. The analysis shows, in a way that could not have been demonstrated with univariate methods or with cluster analysis, that complex adaptations of biomechanical significance can be usefully viewed as a mosaic of positions along a small number of axes of variation. Note that, although the data were unlikely to have been normally distributed, the multivariate descriptive approach was very helpful, and the 9-variable data set for 25 taxa was displayed in two dimensions.

PRINCIPAL COMPONENTS ANALYSIS Principal components analysis has been used widely in all areas of ecology and systematics. It reduces the dimensions of a single group of data by producing a smaller number of abstract variables (linear combinations of the original variables, principal components). The method is based on maximization of the variance of linear combinations of variables ($L_y$). Successive components are constructed to be uncorrelated with previous ones. Often most of the variation can be summarized with only a few components, so data with many variables can be

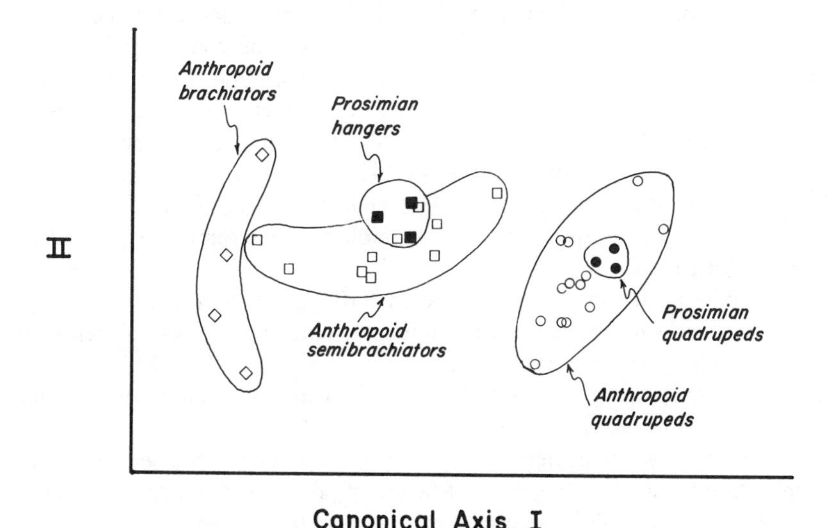


*Figure 2* Discriminant function analysis of data for the shape of the pectoral girdle (clavicle and scapula) of primates by genera (redrawn from Figure 2 of 8).

displayed effectively on a two- or three-dimensional graph that uses the components as axes.

If the original variables were not measured on the same scale, the analysis should be performed on standardized variables by the use of the correlation matrix rather than the variance-covariance matrix. Unfortunately, with the correlation matrix, the interpretation of "variance explained" or accounted for by each component is changed, because all the variables have been standardized to have a variance of one. With the variance-covariance matrix, the eigenvalues and percent of eigenvalues are equal to the variances of the components and the percent of variance explained by the components. This interpretation does not hold for analyses using the correlation matrix. When one is presenting the results of a principal components analysis, it is important to give the list of objects and attributes, the eigenvalues, and any coefficients that are interpreted and to state whether the analysis was performed on the variance-covariance or the correlation matrix.

Principal components analysis requires no formal assumptions, but in practice it is important to be aware of some of its limitations:

(*a*) Because it is based on either variances and covariances or correlations, principal components analysis is sensitive to outliers, and the coefficients of individual components are highly subject to sampling variability. One should not put too much emphasis on the exact values of the coefficients.

(*b*) When the distribution of ratios or proportions is reasonably near to normal, the analysis can be useful (see, e.g., 103, 125, 176), but without transformations principal components analysis cannot capture nonlinear relationships (135). Investigators whose data consist of counts, ratios, proportions, or percentages should check to see whether transformations might make their distribution more appropriate or whether a nonlinear approach would be preferable. Methods have been developed that incorporate the use of ratios through log transformations (140–142; see section on morphometrics).

(*c*) Mathematically orthogonal (independent) factors need not represent independent patterns in nature (14), so biological interpretations should be made with care.

(*d*) Contrary to some recommendations (101, 191), principal components analysis should not be used in a multiple-sample situation, as it then confounds within- and between-group sources of variation (60, 148, 194). In studies of geographic variation, a PCA on means by locality will give the appropriate data reduction.

A particularly interesting example of principal components analysis is its application to data for the genetic structure of present-day human populations in Europe on the basis of a correlation matrix of the frequencies of 39 alleles (5, pp. 102–108). A map on which the scores by locality for principal component 1 are contoured shows a clear gradient from the Middle East toward northwestern Europe, a pattern highly correlated with archeological evidence for the pattern of the ancient transition from hunting and gathering to agricultural societies. The analysis is compatible with the authors' demic diffusion hypothesis, which states that this major cultural change was associated with a population expansion. The genetic structure of living populations may still reflect the ancient Neolithic transition. In quantitative genetics, principal components analysis has been used to analyze genetic correlations during development (40, 205). In morphometrics, comparisons of congeneric songbirds in a space defined by principal components (123, 151) have led to useful graphic comparisons of complex forms. Little progress would have been made with any of these problems by the use of univariate statistics.

PRINCIPAL COORDINATES ANALYSIS Principal coordinates analysis begins with a matrix of distances among objects (159) and, to the extent possible, these distances are retained in a space with a reduced number of dimensions. It is the same as the technique called classical scaling by psychometricians (38, p. 190; 202). If the data are quantitative and the distances are squared distances between units in a coordinate space (Euclidean distances), a principal coordinates analysis will produce the same result as will a principal components analysis on the correlation matrix among the attributes (53).

In a good example in systematics, a matrix of Roger's genetic distances among colonizing populations of common mynahs *(Acridotheres tristis)* was expressed in a two-dimensional graphic space, and the populations in the graph were then connected with a minimum spanning tree according to their distances in the full dimensional space (16).

Another useful analysis using principal coordinates analysis was performed on a matrix of the number of interspecific contacts among 28 species of mosses (53). The procedure allowed investigators to express the associations in two dimensions, and the species were seen to occur along a shade-moisture gradient in which six habitats were clearly separated.

FACTOR ANALYSIS Basic computational similarities lead many people to regard factor analysis as a category of procedures that includes principal components analysis, but historically the two methods have had different objectives. Whereas principal components analysis is a descriptive technique for dimension reduction and summarization, factor analysis explores the resultant multivariate factors—the linear combinations of the original variables (89). The computational distinction is that, in factor analysis, the axes are rotated until they maximize correlations among the variables, and the factors need not be uncorrelated (orthogonal). The usual interpretation of the factors is that they "explain" the correlations that have been discovered among the original variables and that these factors are real factors in nature. Unfortunately, factor analysis encourages subjective overinterpretation of the data. A reading of the mythical tale about Tom Swift and his electric factor analysis machine (6) or Reyment et al (164, pp. 102–106) will persuade most people of the dangers of overinterpretation. Some newer versions of factor analysis, such as linear structural analysis (223, 224), avoid some of the problems of ordinary factor analysis.

Applications of factor analysis in systematics through 1975 have been summarized (31, pp. 135–143), and several examples have appeared in the more recent ecological literature (66a, 95, 127, 174). Q-mode factor analysis investigates the correlations among objects rather than attributes. It has been applied in an exploratory way in numerical taxonomy (185, p. 246) and morphometrics (77). The distinction between Q-mode and the more conventional R-mode analysis has been discussed by Pielou (159).

CANONICAL CORRELATION Canonical correlation is a generalization of correlation and regression that is applicable when the attributes of a single group of objects can be divided naturally into two sets (e.g. morphological variables for populations of a species at a set of sites and environmental variables associated with the same set of sites). Canonical correlation calculates overall correlations between the two sets. Linear combinations within the first set of variables, $L_1$, and within the second set, $L_2$, are considered simultaneously, and the linear combinations that maximize the correlation between $L_1$ and $L_2$ are selected. Further linear combinations are extracted that are uncorrelated with earlier ones. These are uncorrelated between sets except for paired linear combinations. Sample sizes that are small in relation to the number of variables can lead to instability, and the linear constraints imposed by the method can make interpretation difficult (198).

In spite of its limitations, canonical correlation has been useful in an exploratory sense in several ecomorphological and coevolutionary studies. One such study showed that the size of the rostrum of aphids increases and that of the tarsus decreases in proportion to the degree of pubescence of the host plant: these features could easily obscure underlying phylogenetic relationships (137). Another study explored the canonical correlation between bee and flower morphology by comparing eight species of bees according to their choice of flowers (87). Gittins (72) and Smith (183) review other examples.

MULTIPLE LOGISTIC REGRESSION Multiple logistic regression is a modification of multiple regression for the situation in which the response variable (Y) is categorical and takes one of only two values, 0 or 1. Multiple logistic regression models the log of the odds that $Y = 1$ ($\ln(\Pr(Y = 1)/\Pr(Y = 0))$) as a linear function of the independent variables, which can be continuous or categorical. The method can be used either to predict values of the response variable or to get information about particular $X$ variables and the response variable. These are some of the same goals addressed by multiple regression, and multiple logistic regression is susceptible to many of the same limitations as multiple regression. Inference of causation (e.g. 166) is not justified, and stepwise procedures should be avoided. Multiple logistic regression can be used as an alternative to two-group linear discriminant function analysis when one or more of the variables are not continuous. In this case the response variable is group membership, and the explanatory variables are those used to discriminate between the two groups. If the data are multivariate normal, linear discriminant function analysis is a more efficient procedure (56).

Multiple logistic regression is used frequently in wildlife studies, but most applications (e.g. 108, 115) use stepwise procedures. As discussed previously, this is not a reliable way to rank variables by their importance.

LOGLINEAR MODELS Loglinear analysis is an extension of the familiar chi-square analysis of two-way contingency tables (tables of counts or responses) for which there are more than two variables. If some of the variables are continuous, they must be categorized before loglinear analysis is used. The objective is simply to study the relationships among the variables. When there is a distinction between the variables, one being a response variable and the others explanatory variables, loglinear analysis is not appropriate. Fienberg (64) gives a good introduction to both loglinear models and multiple logistic regression.

There are more examples of loglinear analysis in behavior than there are in ecology (63, 94). Examples of its use in ecology include a study of population attributes in Snow Geese *(Chen caerulescens),* including interrelationships among parental morphs and the sex and cohort affiliations of the goslings (65); a study of interrelations among characteristics of fruits of the entire angiosperm tree flora of southern Africa (114); and a defense of the existence of a previously described (52) nonrandom pattern for the distribution of birds on the islands of the Bismarck Archipelago in the South Pacific Ocean (71). One excellent study combined a loglinear analysis with "causal ordering" of the variables, thereby injecting some reasonable biological information into the model for a competition hierarchy among boreal ants (211). This is a good example of how a problem can be carried forward through the research process as outlined in Figure 1. The next step would be the design of a critical experiment.

CORRESPONDENCE ANALYSIS, RECIPROCAL AVERAGING, AND DETRENDED CORRESPONDENCE ANALYSIS Correspondence analysis, which is the same as reciprocal averaging, is an ordination procedure that decomposes a two-way contingency table of counts of objects and their attributes (97, 98). The data might be the number of times various plant species occur on different quadrats, the number of times particular behaviors occur among various species, or the number of fin rays on various fish. Scores are calculated for each of the row and column categories of the table, and row and column eigenvectors show the ways in which the rows and columns deviate from what would be expected with independence. These scores are used as axes for dimension reduction, and objects and attributes are ordinated simultaneously. Because the analysis uses chi-square distances (81, p. 54) it should be based on data of counts. Continuous data such as allele frequencies, percentage of ground cover, or percentage of time spent foraging would be more efficiently handled by another method.

An excellent example of correspondence analysis is a summary of data for the distribution of 17 genera of antelope in 16 African wildlife areas (82). With supplemental information about the vegetation in these areas and about the distribution of the same species in the past, the authors were able to make inferences about the distribution of habitats in the past. In another example, an ordination of 37 lakes in the Adirondack Mountains of northern New York was found to be highly correlated with surface lakewater pH (37).

The term indirect ordination in plant ecology refers to the above class of problems, those involving a reduction of the dimensions of a table (matrix) of data for the occurrence of a set of species at a set of sites. The data may be counts, presence-absence data, or percentages. Because the species are likely to be responding in a unimodal way to underlying environmental gradients and each species is likely to have an individualistic response, their joint distribution is likely to be one of successive replacement (13). Phytosociologists have long felt that, in such cases, neither correspondence analysis nor any of the other traditional ordination procedures give reasonable results. In particular, they complain that an arch or horseshoe effect is evident in the pattern of sites in a two-dimensional ordination. Detrended correspondence analysis is an ad hoc technique intended to remove this arch (36, 67). However, it sometimes fails and can even introduce further distortion (112). A recent critique by Wartenberg et al (214) argues that detrending does not contribute to the analysis and that the arch is not an anomaly. Rather, it is an inherent property of data that represent transitions in species abundances as one passes through localities more favorable to some species and later more favorable to other species. Not even nonmetric multidimensional scaling (see below) can provide satisfactory single-dimensional ordinations in this case (214), because the relationships among the variables (species) are both nonlinear and nonmonotonic. With the indirect ordination problem, the arch in two-dimensional plots is to be expected. An unambiguous ordering along the arch would be an acceptable result.

NONMETRIC MULTIDIMENSIONAL SCALING Nonmetric multidimensional scaling is potentially a robust ordination method for reducing the dimensions of data without a priori transformations (see, e.g., 59, 112, 136, 154, and especially 214). The results are often similar to those of principal components analysis.

Like principal components analysis and principal coordinates analysis, it is a scaling technique, but with nonmetric multidimensional scaling, only the rank order of interobject distances is used. Thus the objective is to estimate nonlinear monotonic relationships. A limitation of both principal coordinates analysis and nonmetric multidimensional scaling is that interpretations must be qualitative and subjective. Because the axes are not functions of original variables, they are not very useful for formulating hypotheses about possible causal relationships. In fact with principal coordinates analysis and nonmetric multidimensional scaling, variables do not enter into the analysis; only interobject distances are used.

CLUSTER ANALYSIS With cluster analysis, objects are placed in groups according to a similarity measure and then a grouping algorithm. The reduction in the data comes from forming $g$ groups ($g$ less than $n$) out of $n$ objects. In ecology and systematics, the general term "cluster analysis" usually means agglomerative hierarchical cluster analysis. This is a set of methods that starts with a pairwise similarity matrix among objects (individuals, sites, populations, taxa; see Section on distances and similarities). The two most similar objects are joined into a group, and the similarities of this group to all other units are calculated. Repeatedly the two closest groups are combined until only a single group remains. The results are usually expressed in a dendrogram, a two-dimensional hierarchical tree diagram representing the complex multivariate relationships among the objects.

The most appropriate choice among the various algorithms for agglomerating groups depends upon the type of data and the type of representation that is desired. It has become conventional in ecology and systematics to use the UPGMA (unweighted pair-group method using averages). This method usually distributes the objects into a reasonable number of groups. It calculates differences between clusters as the average of all the point-to-point distances between a point in one cluster and a point in the other (53, 159, 185). There are also algorithms for divisive cluster analysis, in which the whole collection of objects is divided and then subdivided (67).

Cluster analysis is most appropriate for categorical rather than continuous data. It is less efficient than principal components analysis or linear discriminant function analysis when the data are vectors of correlated measurements. It has been the primary method used in phenetic taxonomy (185), in which many attributes are considered simultaneously and the objects (operational taxonomic units or OTU's) are clustered according to their overall similarity. Cluster analysis produces clusters whether or not natural groupings exist, and the results depend on both the similarity measure chosen and the algorithm used for clustering. Dendrograms codify relationships that may not really be stable in the data. They are frequently overinterpreted in both systematics and ecology. Nevertheless, as applied by Sokal et al (186) to the hypothetical caminalcules, cluster analysis can be as robust for the reconstruction of hierarchical phylogenetic relationships as are cladistic methods. Systematics relies heavily on both cluster analysis and cladistics.

## RELATED MATTERS

### *Jackknife and Bootstrap*

*Jackknifing* (146, 148, pp. 31–33) and *bootstrapping* (57, 58) are statistical techniques that resample the data in order to calculate nonparametric estimates of standard errors. They are particularly effective in two situations that arise frequently in multivariate analysis:

(*a*) in estimation of standard errors for complicated statistics for which the sampling variability is not well understood and standard formulas are not available (e.g. coefficients of principal components) and

(*b*) when the distributional assumptions necessary for the use of standard error formulas are not met (e.g. for nonnormal or skewed data).

Jackknifing and bootstrapping differ in the ways in which they resample the data and calculate standard errors. With the typical jackknifing method, each of the observations in a sample, which may be multivariate, is left out of the data set in turn, and the statistic for which one wants the standard error is recalculated. The variability in these recalculated values is used to calculate the standard error. Examples would be applications to coefficients of principal components in studies of morphometric variation (69).

With bootstrapping for a single sample, a random sample with replacement is drawn from the original sample until it is the same size as the original sample. Some of the original observations are likely to occur more than once in the bootstrap sample. The statistic is recalculated from this sample. This process is repeated, typically 200 or more times, and the standard deviation of the recalculated values is used as the standard error. Often, the bootstrap can be applied more easily to complicated situations than can the jackknife, which is mainly a single-sample technique. Applications of the jackknife and bootstrap for estimating population growth rates have been compared (134).

### *Distances and Similarities*

We use the terms *distance* and *similarity* to describe various measures of the association between pairs of objects or their attributes. Principal coordinates analysis, nonmetric multidimensional scaling, and cluster analysis require the input of a matrix of such measures. Cluster analysis operates most naturally with similarities, whereas principal coordinates analysis and nonmetric multidimensional scaling are traditionally described in terms of distances (53). With some types of data, such as immunological data (42) or DNA hybridization data (180), laboratory results are in the form of interobject distances so they can be entered directly or transformed to similarities as needed. The various distance and similarity measures have been compared (53; 149, Ch. 9; 159). The proper choice of a measure differs according to the form of the data (measurements, counts, presence-absence, frequencies), the type of standardization desired, and whether or not it is appropriate to use metric distances. The special problems that pertain to genetic distances have been discussed elsewhere (17, 61, 149, 172).

## SPECIAL PROBLEMS IN ECOLOGY AND SYSTEMATICS

We think that the present understanding of multivariate analysis among ecologists and systematists is affecting not only how they treat data but how research questions are formulated. To illustrate this point, we discuss in this section some particular issues in animal community ecology, wildlife management, ordination in plant ecology, and morphometrics.

### *Resource Use and the Niche*

Soon after it was proposed that the realized ecological niche be viewed as an area in a multidimensional resource hyperspace (102), Green (79) used linear discriminant function analysis to construct two-dimensional graphic ordinations of the relationships of bivalve molluscs in lakes in central Canada based on physical and chemical properties of the lakes. In many subsequent studies, linear discriminant function analysis has proved useful as a descriptive technique for summarizing, displaying, and comparing differences in resource use among populations (see summaries in 92 and 177).

Green (79, 80) and others have attempted a statistical test for niche size and overlap, but unfortunately, linear discriminant function analysis is not appropriate as a test of niche size. Equality of dispersion matrices is an assumption of the statistical model, but at the same time niche size is being defined by a characteristic of the dispersion matrix. Having been assumed, it cannot be tested (106, pp. 42–44). No one would expect the mean resource use of different species to be exactly the same, so the test is only of whether sample sizes in the study are sufficiently large to show these differences (see 169).

One can obtain data on resource use for each of a set of species and then express an assemblage as an ordination of their variation (43, 79, 104, and others). Or one can compare used with available resources (34). The former approach has been used to study the regeneration niche of plants (70) and to analyze interspecific associations in plant populations to get a "plant's eye view" of the biotic environment (207). In these cases the data were the species of plants that were neighbors of the species of interest. Grubb (84) used this general approach to show how species-specific "regeneration niches" vary. He suggested that this variation may contribute to the maintenance of the coexistence of both common and rare species in a plant community. This is the kind of new hypothesis, suggested partly by multivariate work, that could be tested with experiments.

## *Wildlife Management*

Wildlife biologists have maintained a good dialogue with statisticians about multivariate statistical methods (33, 213), and they are aware of the potential problems with scale, sampling, and linear methods (21). Also, they have been urged to become more experimental (173, 209).

We will give two examples of troublesome areas. First, in recent years the US Fish and Wildlife Service has supported a large program to produce predictive models of wildlife-habitat relations (212). Unfortunately, thus far, few of these models have achieved high predictive power (18, 29, 138). There are several reasons for these problems (130), not all statistical, but the issues of sampling procedures, adjustment for nonlinearities, screening variables to obtain an uncorrelated set, and the use of stepwise procedures discussed above need more attention. Even if predictive models can eventually be developed, there is no guarantee that they will be useful for management (195). That would require the additional step of causal analysis (see previous section).

An additional problem arises with studies of habitat selection, which in wildlife biology usually means the difference between occupied and available (unoccupied) habitat for a particular species. A common procedure is to measure many variables pertaining to the vegetation and its structure both at various localities where a species of interest occurs and at randomly selected locations. Then stepwise discriminant function analysis or stepwise multiple logistic regression is usually applied to examine differences between occupied and unoccupied sites and to rank the habitat variables by their "importance" (129, 165 and citations therein, 167). To see the problem with this approach, excluding the problems with stepwise procedures, recall that the linear discriminant function analysis model tests mean differences between groups. If a species were highly narrow (selective) in its habitat use, but the mean were the same as that of the average habitat, the species would be judged not to be selective by the model (see Species B in Figure 3a and 107). Also, the characteristics of the poorly defined "unused" group will always affect the result (219). Some of these problems are avoided if sites are located along principal component 1 for variation in randomly selected sites (192). An alternative is to use the first two principal components (131) for randomly selected sites and to depict concentration ellipses (188, pp. 594–601) for occupied and random sites on a graph with those components serving as axes

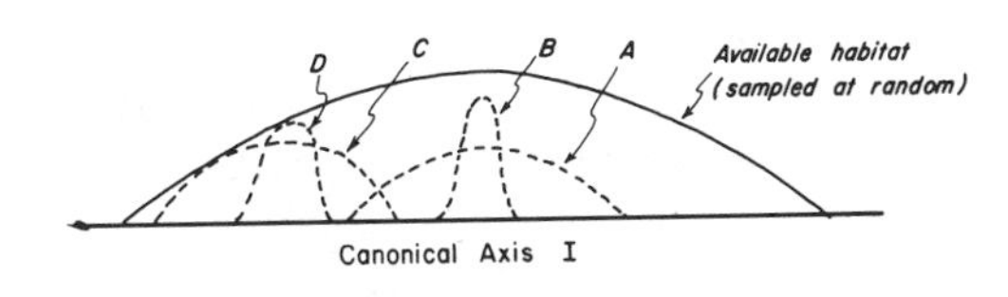




*Figure 3a* Comparisons of habitat used and habitat available for four hypothetical species (A, B, C, D). Four separate two-group linear discriminant function analysis or multiple linear regression tests between used and available habitat, one for each species, would test differences in means but not variances. A and B would not be different from habitat available; C and D would be different. However, this result is misleading because B is as selective (same variance) as D and is more selective (lower variance) than C.

*Figure 3b* Distribution of randomly selected sites in a bivariate graphic space determined by principal components I and II of their habitat characteristics. Concentration ellipses for randomly selected sites and for sites that are occupied by the species of interest indicate both the habitat used and its variance relative to the total variance.

(Figure 3b). This procedure assumes sufficient covariation in the data set for randomly selected sites that two reliable axes can be derived (152). One study that demonstrated the instability that can result otherwise attributed it to interobserver bias (76).

## *Ordination in Plant Ecology*

The most general definition of ordination is the reduction of a multivariate data set for a set of objects and their attributes so that their pattern can be seen on a continuous scale (159). Thus linear discriminant function analysis, principal components analysis, principal coordinates analysis, and nonmetric multidimensional scaling all qualify as ordination procedures (Table 2). Ordination procedures are useful for descriptions of the results of environmental perturbations and experiments (53), but they are used most often in purely observational studies. Several particularly useful reviews of the ordination literature are available (53, 112, 159).

In plant ecology, the term ordination usually refers to analyses in which the objects are stands of vegetation at study sites. When the attributes are sets of environmental variables, such as soil nutrients or quantitative measures of the structure of the vegetation, the objective is usually to find a combination of attributes that may suggest an underlying cause for a systematic pattern of the distribution of the stands, one not obvious from the geographic distribution of the stands. Austin et al (15) present some new extensions of this approach, which is called direct ordination or gradient analysis. The more common approach in plant ecology is to analyze a matrix of data for the presences and absences of species in each stand, or their actual or relative density, biomass, or cover (83), as the attributes. This is called indirect ordination. The objective is to find a systematic pattern of relationships among the stands based on the cooccurrences of their component species. The resultant ordination may subsequently be related to environmental factors (14).

If sites are being ordinated (the usual *R*-mode analysis), and they have been selected at random, inferences about patterns in a larger area are possible. If the objects and attributes are exchanged (*Q*-mode analysis), species are ordinated. The biplot (66, 196), a graphical version of principal components analysis and correspondence analysis, can provide a simultaneous view of ordinations of species and stands. The special problems that arise with indirect ordinations when the attributes do not increase or decrease regularly through the data are discussed in the section on correspondence analysis. Previous criticisms of principal components analysis as an indirect ordination technique (e.g. 67) should be reconsidered in the light of these arguments.

In recent years, principal coordinates analysis and nonmetric multidimensional scaling have been popular indirect ordination methods. Phytosociological studies that use indirect ordinations of stands by their species composition have provided succinct descriptions of stands by their species composition. We agree with Harper (88) that if the objective is to determine causes, the approach of focusing a study on the population biology of species independently and including all interspecific interactions, rather than on studying relationships among communities or among stands, should also be tried. Experiments and quasiexperiments will be required, and multivariate descriptive work at the population level, now a poorly developed field, should be important.

## *Morphometrics*

Morphometrics is the mathematical description of the form of organisms. There are many different kinds of problems in morphometric work, and even for a given problem researchers do not always agree about the best methods of analysis (46). The literature on multivariate morphometrics includes applications in growth (203) and quantitative genetics (118, 208).

For a long time the appealing graphic technique of the transformation of a grid to show its deformation when drawings of two organisms were compared (197) did not seem to be amenable to quantification. However, the study of geometric transformations of forms has been extended, and several techniques have been developed to describe geometric shape change between forms when the data are for $x,y$ coordinates for homologous landmarks (23–26, 39, 101, 158, 184). Size and shape are considered to be latent unmeasured variables, defined only after the demonstration of a global transformation between forms. Sometimes principal components analysis is used to reduce the dimensions of the result.

Mapping techniques are another set of methods designed to detect shape change among two-dimensional forms (19, 20, 181, 184). In this case the data are interpoint distances between two superimposed forms. Fourier analysis, another alternative for the description of forms that have fixed outlines, can capture shape information without using sets of homologous landmarks (161, 170). Ferson et al (62) applied linear discriminant function analysis to such shape data for two electromorph groups of the mussel *Mytilus edulis*.

A more general problem in morphometrics than the quantification of shape change among two-dimensional objects is the study of allometry, how shape changes with size during growth, or among members of a population, or among populations or taxa. Many systematists prefer conventional linear methods of multivariate analysis for this problem (148, 164). The data are standardized measurements taken on each organism. Atchley et al (12) describe the geometric and probabilistic aspects of distances among individuals (objects) in multivariate morphometric space.

If the variation in the original data is predominantly in size, the coefficients of the first principal component based on a variance-covariance matrix will be of the same sign, and that component will be highly correlated with the original variables. Size can be defined variously as this first component, as any one of the original variables, or as any combination of the original variables that is biologically reasonable (168). Principal component 1 of the correlation matrix has also been used as a size statistic (132). It is often correlated with other reasonable size measures, but we do not recommend it as a size statistic because differences in scale (size) among the variables have been removed by the construction of the correlation matrix. Similarly, a proposed method to constrain the first principal component of the correlation matrix of the logs of the measurements to be a measure of shape-free size (189) does not fully achieve its objective, because the residual variation is not interpretable as shape. A complex method proposed for the removal of within-group size in a multiple-group principal components analysis (101) removes size-related shape as well as size, and the residual variation is not necessarily uncorrelated with size (171).

With a principal components analysis on the variance-covariance matrix of log measurements, the relative magnitudes of the coefficients can often indicate whether the component contains shape information as well as size information (145). Although the first principal component often has been designated as a general size factor, it usually contains an unknown amount of allometrically related shape variation (68, 93, 140) and interpretation of the second component as shape alone is unwise (110, 190). A solution to the problem of the study of shape independently of size is to study shape directly, as either ratios or proportions, expressed as the differences between the logarithms of distances. Of course the proper mathematical treatment of shape variables requires great care, but the direct study of shape variables should play a central role in morphometric analyses.

The study of allometry, the covariation of size and shape rather than of size and size-free shape or shape orthogonal to size, has been emphasized by Mosimann (140). He shows that, if biologically reasonable size and shape variables can be defined a priori, and if the data can be assumed to be lognormally distributed, substantial mathematical theory is available for morphometric studies. The lognormal assumption can be tested (110). Log transformations do not always equalize variances (30), but equal variances among measurements are by no means required for morphometric analysis (143). Thus shape variables, which are dimensionless ratios or proportions expressed as differences between logarithms, can be analyzed directly with either univariate or multivariate methods (144, 145). In a particularly interesting example, Darroch & Mosimann (49) study shape directly in a reanalysis of Anderson's classic data set for measurements of the flowers of three species of iris, originally analyzed by R. A. Fisher. The species are well

**Table 3** Objectives and limitations of the 12 multivariate procedures used most commonly in ecology and systematics, with references.

| Procedure | Objectives and Limitations |
|---|---|
| 1. Multiple Regression (MR) | Objectives:<br>1. To predict one variable (Y, response variable) from others (X's, explanatory variables)<br>2. To investigate the association of an X variable with the Y variable in the presence of other variables<br>3. If causal models are appropriate (usually with experiments), to investigate cause and effect<br>Limitations:<br>1. Good predictability alone does not allow inference of causation.<br>2. Prediction should be carried out only in situations similar to those in which the model was derived.<br>3. Stepwise regression is usually inappropriate.<br>4. The procedure considers only linear functions of those X variables analyzed.<br>5. The procedure is intended for continuous Y variables whose values are independent; errors should be normal and sampling random for statistical inference.<br>*References:* 4, 139, 150, 215 |
| 2. Multivariate Analysis of Variance (MANOVA) | Objective:<br>1. To test for differences among two or more groups of objects according to the means of all the variables (attributes); mainly an inferential method<br>Limitation:<br>1. The procedure is intended for continuous, multivariate normal data; each vector of observations must be independent.<br>*References:* 85, 89, 109, 128, 148 |
| 3. Linear Discriminant Function Analysis (LDFA) | Objectives:<br>1. To describe multigroup situations; finds linear combinations of variables (attributes) with maximal ability to discriminate groups of objects; when used to reduce the dimensions of data, called canonical variates analysis<br>2. A linear discriminant function (equation) can be used to classify current observations or to allocate new observations to the groups<br>Limitations:<br>1. The procedure is intended mainly for continuous data; it is inefficient for data not well summarized by variances and covariances.<br>2. With linear discriminant functions, the researcher assumes equal variance-covariance matrices (identical orientation and size of concentration ellipses). |

**Table 3** (*Continued*)

| Procedure | Objectives and Limitations |
|---|---|
| | 3. Only linear combinations of the variables are considered, so th analysis will not discover nonlinear combinations.<br>4. Groups must be defined a priori.<br>*References:* 89, 109, 148, 220 |
| 4. Principal Components Analysis (PCA) | Objectives:<br>1. To describe a matrix of data consisting of objects and attribut by reducing its dimensions, usually for graphical display; to fin uncorrelated linear combinations of the original variables (attr butes) with maximal variance<br>2. To suggest new combined variables for further study<br>Limitations (see text):<br>1. The procedure is intended mainly for continuous data; it inefficient for data not well summarized by variances an covariances.<br>2. The procedure considers only linear combinations of the var ables, so it will not discover nonlinear combinations.<br>*References:* 53, 89, 109, 148, 159 |
| 5. Principal Coordinates Analysis (PCO) | Objective:<br>1. To describe the data by reducing the dimensions of a distanc matrix among objects, usually for graphical display; generalization of PCA in which non-Euclidean distances may b used<br>Limitations:<br>1. Results depend on the distance measure chosen.<br>2. The procedure produces a new coordinate system but cann indicate combinations of variables (attributes), because only th distance matrix among objects is used.<br>*References:* 53, 148, 159 |
| 6. Factor Analysis (FA) | Objectives:<br>1. To reproduce a correlation matrix among original variables b hypothesizing the existence of one or more underlying facto<br>2. To discover underlying structure in a data set by interpreting th factors<br>Limitations:<br>1. Exploratory factor analysis methods are so unstructured th interpretations are subjective.<br>2. The procedure is inefficient for data not well summarized b correlations, so it is not ideal for nonlinear relationships categorical data.<br>*References:* 54, 89, 109, 148 |

**Table 3** (*Continued*)

| Procedure | Objectives and Limitations |
|---|---|
| 7. Canonical Correlation (CANCOR) | Objective:<br>1. To analyze the correlation between two groups of variables (attributes) about the same set of objects simultaneously, rather than calculating pairwise correlations<br>Limitation:<br>1. The procedure is inefficient for data not well summarized by correlations or linear combinations, so not ideal for nonlinear relationships or categorical data.<br>*References:* 54, 89, 109, 148 |
| 8. Multiple Logistic Regression (MLR) | Objectives:<br>1. To model a dichotomous (0,1) variable (Y, response variable) as a function of other categorical or continuous variables (X's, explanatory variables), which may be categorical or continuous<br>2. To investigate the association of an X variable with the Y variable in the presence of other X variables<br>3. If causal models are appropriate (usually with experiments), to investigate cause and effect<br>4. To serve as an alternative to two group linear discriminant function analysis when the variables are categorical or otherwise not appropriate for DFA<br>Limitations:<br>1. Good predictability alone does not allow inference of causation.<br>2. Stepwise logistic regression is usually inappropriate.<br>3. The procedure considers only linear functions of those X variables analyzed.<br>4. Prediction should be carried out only in situations similar to those in which the model was estimated.<br>*References:* 64, 148 |
| 9. Loglinear Models (LOGL) | Objective:<br>1. To investigate the joint relationships among categorical variables<br>Limitations:<br>1. Variables must be categorical or made to be categorical.<br>2. When there are response and explanatory variables, techniques like logistic regression may be more appropriate.<br>*References:* 64, 148 |
| 10. Correspondence Analysis (COA) | Objectives:<br>1. To describe data consisting of counts by reducing the number of dimensions, usually for graphical display<br>2. To suggest new combined variables for further study |

**Table 3** (*Continued*)

| Procedure | Objectives and Limitations |
|---|---|
| | Limitations:<br>1. The procedure is inefficient for data that are not counts becau[se] they will not be well described by chi square distances.<br>2. The procedure is not suitable for nonlinear data; it will n[ot] discover nonlinear relationships.<br>*References:* 81, 120, 159 |
| 11. Nonmetric Multidimensional Scaling (NMDS) | Objective:<br>1. To describe data by reducing the number of dimensions, usual[ly] for graphical display; to discover nonlinear relationships<br>Limitation:<br>1. The procedure uses rank order information only.<br>*References:* 53, 54, 116, 148 |
| 12. Cluster Analysis (CLUS) | Objectives:<br>1. To classify groups of objects judged to be similar according to [a] distance or similarity measure<br>2. To reduce consideration of $n$ objects to $g$ ($g$ less than $n$) grou[ps] of objects<br>Limitations:<br>1. Results depend on the distance measure chosen.<br>2. Results depend on the algorithm chosen for forming cluster[s].<br>*References:* 53, 54, 75, 148, 159 |

discriminated by shape alone. Although these methods were developed for morphometric studies, they are applicable in other situations (e.g. 22). We think that authors who have objected to the direct use of ratios in morphometric studies (3, 9–11, 101, 160, 164, 199) have been overlooking some powerful techniques for the direct study of shape and its covariation with size.

## CONCLUSIONS

Ecologists and systematists need multivariate analysis to study the joint relationships of variables. That the methods are primarily descriptive in nature is not necessarily a disadvantage. Statistical inference may be possible, but, as with univariate analysis, without experiments even the most insightful applications can only hint at roles, processes, causes, influences, and strategies. When experiments are not feasible, quasiexperimental designs, which involve paired comparisons or time-series analysis, may be able to provide weak inferences about causes. As with univariate work, statistical inference (tests and $p$-values) should be reported only if a probability sample is taken from a well-defined larger population and if assumptions of the methods are met. Interpretations of multivariate analyses should be restricted to the joint relationships of variables, and stepwise procedures should be avoided.

We did not expect our review to have such a negative flavor, but we are forced to agree in part with the criticism that multivariate methods have opened a Pandora's box. The problem is at least partly attributable to a history of cavalier applications and interpretations. We do not think that the methods are a panacea for data analysts, but we believe that sensitive applications combined with focus on natural biological units, modelling, and an experimental approach to the analysis of causes would be a step forward. In morphometrics, few workers are taking advantage of some precise mathematical methods for the definition of size and shape and their covariation.

ACKNOWLEDGMENTS

We thank D. Burr-Doss, F. R. Gelbach, J. Rhymer, L. Marcus, L. E. Moses, R. F. Johnston, H. F. James, J. E. Mosimann, P. Frank, D. H. Johnson, J. Travis, and F. J. Rohlf for comments on various drafts of the manuscript.

## *Literature Cited*

1. Abramsky, Z., Bowers, M. A., Rosenzweig, M. L. 1986. Detecting interspecific competition in the field: testing the regression method. *Oikos* 47:199–204
2. Aitkin, M., Francis, B. 1972. Interactive regression modelling. *Biometrics* 38:511–16
3. Albrecht, G. H. 1978. Some comments on the use of ratios. *Syst. Zool.* 27:67–71
4. Allen, D. M., Cady, F. 1982. *Analyzing Experimental Data by Regression*. Belmont, Calif: Lifetime Learning
5. Ammerman, A. J., Cavalli-Sforza, L. L. 1984. *The Neolithic Transition and the Genetics of Populations in Europe*. Princeton, NJ: Princeton Univ. Press
6. Armstrong, J. S. 1967. Derivation of theory by means of factor analysis or Tom Swift and his electric factor analysis machine. *Am. Statist.* 21:17–21
7. Ashton, E. H. 1981. The Australopithecinae: their biometrical study. In *Perspectives in Primate Biology*, ed. E. H. Ashton, R. L. Holmes, pp. 67–126. New York: Academic
8. Ashton, E. H., Healy, M. J. R., Oxnard, C. E., Spence, T. F. 1965. The combination of locomotor features of the primate shoulder girdle by canonical analysis. *J. Zool., London,* 147:406–29
9. Atchley, W. R. 1983. Some genetic aspects of morphometric variation. In *Numerical Taxonomy, Series G, No. 1*, ed. J. Felsenstein, pp. 346–63. New York: Springer Verlag
10. Atchley, W. R., Anderson, D. R. 1978. Ratios and the statistical analysis of biological data. *Syst. Zool.* 27:71–78
11. Atchley, W. R., Gaskins, C. T., Anderson, D. T. 1976. Statistical properties of ratios. I. Empirical results. *Syst. Zool.* 25:137–48
12. Atchley, W. R., Nordheim, E. V., Gunsett, F. C., Crump, P. L. 1982. Geometric and probabilistic aspects of statistical distance functions. *Syst. Zool.* 31:445–60
13. Austin, M. P. 1985. Continuum concept, ordination methods, and niche theory. *Annu. Rev. Ecol. Syst.* 16:39–61
14. Austin, M. P., Noy-Meir, I. 1971. The problem of nonlinearity in ordination: experiments with two-gradient modes. *J. Ecol.* 59:763–73
15. Austin, M. P., Cunningham, R. B., Fleming, P. M. 1984. New approaches to direct gradient analysis using environmental scalars and statistical curve-fitting procedures. *Vegetatio* 55:11–27
16. Baker, A. J., Moeed, A. 1987. Rapid genetic differentiation and founder effect in colonizing populations of common mynas *(Acridotheres tristis)*. *Evolution* 41:525–38
17. Barry, D., Hartigan, J.A. 1987. Statis-

tical analysis of hominoid molecular evolution. *Statist. Sci.* 2:191–210
18. Bart, J., Petit, D. R., Linscombe, G. 1984. Field evaluation of two models developed following the Habitat Evaluation Procedures. *Trans. N. Am. Wildl. Nat. Resour. Conf.* 49:489–99
19. Benson, R. H. 1983. Biomechanical stability and sudden change in the evolution of the deep sea ostracode *Poseidonamicus*. *Paleobiology* 9:398–413
20. Benson, R. H., Chapman, R. E., Siegel, A. F. 1982. On the measurement of morphology and its change. *Paleobiology* 8:328–39
21. Best, L. B., Stauffer, D. F. 1986. Factors confounding evaluation of bird-habitat relationships. See Ref. 213, pp. 209–16
22. Boecklen, W. J. 1989. Size and shape of sawfly assemblages on arroyo willow. *Ecology* 70:1463–71
23. Bookstein, F. L. 1978. *Lecture Notes in Biomathematics. The Measurement of Biological Shape and Shape Change. Lecture Notes in Biomathematics No. 24*. Berlin: Springer-Verlag
24. Bookstein, F. L. 1982. Foundations of morphometrics. *Annu. Rev. Ecol. Syst.* 13:451–70
25. Bookstein, F. L:. 1986. Size and shape spaces for landmark data in two dimensions. *Statist. Sci.* 1:181–242
26. Bookstein, F. L., Chernoff, B., Elder, R., Humphries, J., Smith, G., Strauss, R. 1985. *Morphometrics in Evolutionary Biology, the Geometry of Size and Shape Change, with Examples from Fishes*. Special Publ. #15. Philadelphia, Pa: Acad. Nat. Sci., Philadelphia
27. Boyce, M. 1981. Robust canonical correlation of sage grouse habitat. See Ref. 33, pp. 152–59
28. Bradshaw, T., Mortimer, M. 1986. Evolution of communities. In *Community Ecology: Pattern and Process,* ed. J. Kikkawa, D. J. Anderson. Oxford: Blackwells
29. Brennan, L. A., Block, W. M., Gutierrez, R. J. 1986. The use of multivariate statistics for developing habitat suitability index models. See Ref. 213, pp. 177–82
30. Bryant, E. H. 1986. On the use of logarithms to accommodate scale. *Syst. Zool.* 35:552–59
31. Bryant, E. H., Atchley, W. R. eds. 1975. *Multivariate Statistical Methods: Within Groups Covariation*. Stroudsburg, Pa: Dowden, Hutchinson & Ross
32. Campbell, D. T., Stanley, J. C. 1966. *Experimental and Quasi-Experimental Designs for Research*. Chicago: Rand McNally
33. Capen, D. E. ed. 1981. *The Use of Multivariate Statistics in Studies of Wildlife Habitat. Rocky Mountain Forest and Range Exp. Stat. U.S. For. Serv., Gen. Tech. Rep. RM-87*. Fort Collins, Co: US Dep. Agric.
34. Carnes, B. A., Slade, N. A. 1982. Some comments on niche analysis in canonical space. *Ecology* 63:888–93
35. Carnes, B. A., Slade, N. A. 1988. The use of regression for detection of competition with multicollinear data. *Ecology* 69:1266–74
36. Chang, D. H. S., Gauch, H. G. Jr. 1986. Multivariate analysis of plant communities and environmental factors in Ngari, Tibet. *Ecology* 6:1568–75
37. Charles, D. F. 1985. Relationships between surface sediment diatom assemblages and lakewater characteristics in Adirondack lakes. *Ecology* 66:994–1011
38. Chatfield, C., Collins, A. J. 1980. *Introduction to Multivariate Analysis*. London: Chapman & Hall
39. Cheverud, J. M., Richtsmeier, J. T. 1986. Finite-element scaling applied to sexual dimorphism in Rhesus macaque *(Macaca mulatta)* facial growth. *Syst. Zool.* 35:381–99
40. Cheverud, J. M., Rutledge, J. J., Atchley, W. R. 1983. Quantitative genetics of development: genetic correlations among age-specific trait values and the evolution of ontogeny. *Evolution* 37:895–905
41. Cochran, W. G. 1983. *Planning and Analysis of Observational Studies*. New York: Wiley
42. Collier, G. E., O'Brien, S. J. 1985. A molecular phylogeny of the Felidae: immunological distance. *Evolution* 39: 473–87
43. Collins, S. L., Good, R. E. 1987. The seedling regeneration niche: habitat structure of tree seedlings in an oak-pine forest. *Oikos* 48:89–98
44. Cook, T. D., Campbell, D. T. 1979. *Quasi-experimentation: Design and Analysis Issues for Field Settings*. Boston: Houghton Mifflin
45. Corruccini, R. S. 1978. Morphometric analysis: uses and abuses. *Yearb. Phys. Anthropol.* 21:134–50
46. Corruccini, R. S. 1987. Shape in morphometrics: comparative analyses. *Am. J. Phys. Anthropol.* 73:289–303
47. Crespi, B. J., Bookstein, F. L. 1989. A path-analytic model for the measurement of selection on morphology. *Evolution* 43:18–28
48. Crovello, T. J. 1970. Analysis of char-

acter variation in ecology and systematics. *Annu. Rev. Ecol. Syst.* 1:55–98
49. Darroch, J. N., Mosimann, J. E. 1985. Canonical and principal components of shape. *Biometrika* 72:241–52
50. Davis, J. C. 1986. *Statistics and Data Analysis in Geology*. New York: Wiley. 2nd ed.
51. Dempster, A. P. 1969. *Elements of Continuous Multivariate Analysis*. Reading, Mass: Addison-Wesley
52. Diamond, J. M. 1975. Assembly of species communities. In *Ecology and Evolution of Communities*, ed. M. L. Cody, J. M. Diamond, pp. 342–444. Cambridge, Mass: Harvard Univ. Press
53. Digby, P. G. N., Kempton, R. A. 1987. *Multivariate Analysis of Ecological Communities*. New York: Chapman & Hall
54. Dillon, W. R., Goldstein, M. 1984. *Multivariate Analysis: Methods and Applications*. New York: Wiley
55. Dunn, J. 1981. Data-based transformations in multivariate analysis. See Ref. 33, pp. 93–102
56. Efron, B. 1976. The efficiency of logistic regression compared to normal discriminant analysis. *J. Am. Statist. Assoc.* 70:892–98
57. Efron, B. 1979. Bootstrap methods: another look at the jackknife. *Ann. Statist.* 7:1–26
58. Efron, B., Gong, G. 1983. A leisurely look at the bootstrap, jackknife and cross-validation. *Am. Statist.* 37:36–48
59. Faith, D. P., Minchin, P. R., Belbin, L. 1987. Compositional dissimilarity as a robust measure of ecological distance. *Vegetatio* 69:57–68
60. Felley, J. D., Hill, L. G. 1983. Multivariate assessment of environmental preferences of cyprinid fishes of the Illinois River, Oklahoma. *Am. Midl. Nat.* 109:209–21
61. Felsenstein, J. 1984. Distance methods for inferring phylogenies: a justification. *Evolution* 38:16–24
62. Ferson, S., Rohlf, F. J., Koehn, R. K. 1985. Measuring shape variation of two-dimensional outlines. *Syst. Zool.* 34:59–68
63. Fienberg, S. E. 1970. The analysis of multidimensional contingency tables. *Ecology* 51:419–33
64. Fienberg, S. E. 1980. *The Analysis of Cross-classified Categorical Data*. Cambridge, Mass.: Mass. Inst. Technol. Press. 2nd ed.
65. Findlay, C. S., Rockwell, R. F., Smith, J. A., Cooke, F. 1985. Life history studies of the Lesser Snow Goose *(Anser caerulescens caerulescens)* VI. Plumage polymorphism, assortative mating and fitness. *Evolution* 39:904–14; 39:178–79
66. Gabriel, K. R. 1971. The biplot graphic display of matrices with application to principal components analysis. *Biometrika* 58:453–67
66a. Gatz, A. J. Jr., Sale, M. J., Loar, J. M. 1987. Habitat shifts in rainbow trout: competitive influences of brown trout. *Oecologia* 74:7–19
67. Gauch, H. G. Jr. 1982. *Multivariate Analysis in Community Ecology*. Cambridge, UK: Cambridge Univ. Press
68. Gibson, A. R., Gates, M. A., Zach, R. 1976. Phenetic affinities of the Wood Thrush *Hylocichla muselina* (Aves: Turdinae). *Can. J. Zool.* 54:1679–87
69. Gibson, A. R., Baker, A. J., Moeed, A. 1984. Morphometric variation in introduced populations of the common myna *(Acridotheres tristis)*: an application of the jackknife to principal component analysis. *Syst. Zool.* 33:408–21
70. Gibson, D. J., Good, R. E. 1987. The seedling habitat of *Pinus echinata* and *Melampyrum lineare* in oak-pine forest of the New Jersey pinelands. *Oikos* 49:91–100
71. Gilpin, M. E., Diamond, J. M. 1982. Factors contributing to nonrandomness in species co-occurrences on islands. *Oecologia* 52:75–84
72. Gittins, R. 1979. Ecological applications of canonical analysis. See Ref. 156, pp. 309–535
73. Gittins, R. 1985. *Canonical Analysis: A Review with Applications in Ecology*. Berlin: Springer Verlag
74. Goodall, D. W. 1954. Objective methods for the classification of vegetation. III. An essay in the use of factor analysis. *Austr. J. Bot.* 2:304–24
75. Gordon, A. D. 1981. *Classification: Methods for the Exploratory Analysis of Multivariate Data*. London: Routledge Chapman & Hall
76. Gotfryd, A., Hansell, R. I. C. 1985. The impact of observer bias on multivariate analyses of vegetation structure. *Oikos* 45:223
77. Gould, S. J., Young, N. D. 1985. The consequences of being different: sinistral coiling in *Cerion*. *Evolution* 39:1364–79
78. Green, R. G. 1980. Multivariate approaches in ecology: the assessment of ecological similarity. *Annu. Rev. Ecol. Syst.* 11:1–14
79. Green, R. H. 1971. A multivariate statistical approach to the Hutchinsonian niche: bivalve molluscs of central Canada. *Ecology* 52:543–56
80. Green, R. H. 1974. Multivariate niche

analysis with temporally varying environmental factors. *Ecology* 55:73–83

81. Greenacre, M. J. 1984. *Theory and Application of Correspondence Analysis*. London: Academic
82. Greenacre, M. J., Vrba, E. S. 1984. Graphical display and interpretation of antelope census data in African wildlife areas, using correspondence analysis. *Ecology* 65:984–97
83. Greig-Smith, P. 1983. *Quantitative Plant Ecology*. Oxford: Blackwell. 3rd ed.
84. Grubb, P. J. 1986. Problems posed by sparse and patchily distributed species in species-rich plant communities. In *Community Ecology*, ed. J. M. Diamond, T. J. Case, pp. 207–25. New York: Harper & Row
85. Hand, D. J., Taylor, C. C. 1987. *Multivariate Analysis of Variance and Repeated Measures—a Practical Approach for Behavioral Scientists*. New York: Chapman & Hall
86. Hanson, N. R. 1958. *Patterns of Discovery*. Cambridge, UK: Cambridge Univ. Press
87. Harder, L. D. 1985. Morphology as a predictor of flower choice by bumble bees. *Ecology* 66:198–210
88. Harper, J. L. 1982. After description: the plant community as a working mechanism. In *Spec. Publ. Brit. Ecol. Soc., No. 1*, ed. E. I. Newman, pp. 11–25. Oxford: Blackwell
89. Harris, R. J. 1985. *A Primer of Multivariate Statistics*. New York: Academic. 2nd ed.
90. Hawkins, D. M. ed. 1982. *Topics in Applied Multivariate Analysis*. Cambridge, UK: Cambridge Univ. Press
91. Hawley, A. W. L. 1987. Identifying bison ratio groups by multivariate analysis of blood composition. *J. Wildl. Manag*. 51:893–900
92. Hayward, G. D., Garton, E. O. 1988. Resource partitioning among forest owls in the River of No Return Wilderness, Idaho. *Oecologia* 75:253–65
93. Healy, M. J. R., Tanner, J. M. 1981. Size and shape in relation to growth and form. *Symp. Zool. Soc. Lond. No. 46*. (1981):19–35
94. Heisey, D. M. 1985. Analyzing selection experiments with log-linear models. *Ecology* 66:1744–48
95. Henderson, C. B., Peterson, K. E., Redak, R. A. 1988. Spatial and temporal patterns in the seed bank and vegetation of a desert grassland community. *J. Ecol*. 76:717–28
96. Henderson, H. V., Velleman, 1981. Building multiple regression models interactively. *Biometrics* 37:391–411
97. Hill, M. O. 1973. Reciprocal averaging: an eigenvector method of ordination. *J. Ecol*. 61:237–49
98. Hill, M. O. 1974. Correspondence analysis: a neglected multivariate method. *J. R. Statist. Soc., Ser. C* 23:340–354
99. Hocking, R. R. 1976. The analysis and selection of variables in linear regression. *Biometrics* 32:1044
100. Hocking, R. R. 1983. Developments in linear regression methodology: 1959–1982. *Technometrics* 25:219–30
101. Humphries, J. M., Bookstein, F. L., Chernoff, B., Smith, G. R., Elder, R. L., Poss, S. G. 1981. Multivariate discrimination by shape in relation to size. *Syst. Zool*. 30:291–308
102. Hutchinson, G. E. 1968. When are species necessary? In *Population Biology and Evolution*, ed. R. C. Lewontin, pp. 177–86. Syracuse, NY: Syracuse Univ. Press
103. Jallon, J. M., David, J. R. 1987. Variations in cuticular hydrocarbons among eight species of the *D. melanogaster* subgroup. *Evolution* 41:294–302
104. James, F. C. 1971. Ordinations of habitat relationships among breeding birds. *Wilson Bull*. 83:215–36
105. James, F. C., Johnston, R. F., Wamer, N. O., Niemi, G. J., Boecklen, W. J. 1984. The Grinnelian niche of the wood thrush. *Am. Nat*. 124:17–30
106. James, F. C., McCulloch, C. E. 1985. Data analysis and the design of experiments in ornithology. In *Current Ornithology, Vol. 2*, ed. R. F. Johnston, pp. 1–63. New York: Plenum
107. Johnson, D. H. 1981. The use and misuse of statistics in wildlife habitat studies. See Ref. 33, pp. 11–19
108. Johnson, R. G., Temple, S. A. 1986. Assessing habitat quality for birds nesting in fragmented tallgrass prairies. See Ref. 213, pp. 245–49
109. Johnson, R. A., Wichern, D. W. 1988. *Applied Multivariate Statistical Analysis*. Englewood Cliffs, NJ: Prentice Hall
110. Jolicoeur, P. 1963. The multivariate generalization of the allometry equation. *Biometrics* 19:497–99
111. Kamil, A. C. 1987. Experimental design in ornithology. *Current Ornithology*. New York: Plenum
112. Kenkel, N. C., Orloci, L. 1986. Applying metric and nonmetric multidimensional scaling to ecological studies: some new results. *Ecology* 67:919–28
113. Kershaw, K. A., Looney, J. N. H.

1985. *Quantitative and Dynamic Plant Ecology*. London: Edward Arnold
114. Knight, R. S., Siegfried, W. R. 1983. Inter-relationships between type, size, colour of fruits and dispersal in southern African trees. *Oecologia* 56:405–12
115. Knopf, F. L., Sedgwick, J. A., Cannon, R. W. 1988. Guild structure of a riparian avifauna relative to seasonal cattle grazing. *J. Wildl. Manage*. 52:280–90
116. Kruskal, J. B., Wish, M. 1978. *Multidimensional Scaling*. Beverly Hills Calif: Sage
117. Lachenbruch, P. A., Clarke, W. R. 1980. Discriminant analysis and its applications in epidemiology. *Methods Inf. Med*. 19:220–26
118. Lande, R. 1979. Quantitative genetic analysis of multivariate evolution, applied to brain: body size allometry. *Evolution* 33:402–16
119. Lande, R., Arnold, S. J. 1983. The measurement of selection on correlated characters. *Evolution* 37:1210–26
120. Lebart, A., Morineau, A., Warwick, K. 1984. *Multivariate Statistical Descriptive Analysis*. New York: Wiley
121. Legendre, L., Legendre, P. 1983. *Numerical Ecology*. Amsterdam: Elsevier
122. Legendre, P., Legendre, L. eds. 1987. *Developments in Numerical Ecology*. New York: Springer Verlag
123. Leisler, B., Winkler, H. 1985. Ecomorphology. See Ref. 107, pp. 155–86
124. Levins, R., Lewontin, R. 1982. Dialectics and reductionism in ecology. In *Conceptual Issues in Ecology,* ed. E. Saarinen, pp. 107–38. Dordrecht, Holland: Reidel
125. Livezey, B. C., Humphrey, P. S. 1986. Flightlessness in steamer-ducks (Anatidae: Tachyres): its morphological bases and probable evolution. *Evolution* 40:540–58
126. Ludwig, J. A., Reynolds, J. F. 1988. *Statistical Ecology*. New York: Wiley
127. MacNally, R. C., Doolan, J. M. 1986. An empirical approach to guild structure: habitat relationships in nine species of eastern-Australian cicadas. *Oikos* 47:33–46
128. Manly, B. F. 1986. *Multivariate Statistical Methods: A Primer*. New York: Chapman & Hall
129. Mannan, R. W., Meslow, E. C. 1984. Bird populations and vegetation characteristics in managed and old-growth forests, northeastern Oregon. *J. Wildl. Manage*. 48:1219–38
130. Maurer, B. A. 1986. Predicting habitat quality for grasslands birds using density-habitat correlations. *J. Wildl. Manage*. 50:556–66
131. McCallum, D. A., Gelbach, F. R. 1988. Nest-site preferences of Flammulated Owls in western New Mexico. *The Condor* 90:653–61
132. McGillivray, W. B. 1985. Size, sexual size dimorphism, and their measurement in Great Horned Owls in Alberta. *Can. J. Zool*. 63:2364–72
133. Meents, J. K., Rice, J., Anderson, B. W., Ohmart, R. D. 1983. Nonlinear relationships between birds and vegetation. *Ecology* 64:1022–27
134. Meyer, J. S., Ingersoll, C. G., McDonald, L. L., Boyce, M. S. 1986. Estimating uncertainty in population growth rates: jackknife versus bootstrap techniques. *Ecology* 67:1156–66
135. Miles, D. B., Ricklefs, R. E. 1984. The correlation between ecology and morphology in deciduous forest passerine birds. *Ecology* 65:1629–40
136. Minchin, P. R. 1987. An evaluation of the relative robustness of techniques for ecological ordination. *Vegetatio* 69:89–107
136a. Mitchell-Olds, T., Shaw, R. G. 1987. Regression analysis of natural selection: statistical inference and biological interpretation. *Evolution* 41:1149–61
137. Moran, N. A. 1986. Morphological adaptation to host plants in *Uroleucon* (Homoptera: Aphididae). *Evolution* 40:1044–50
138. Morrison, M. L., Timossi, I. C., With, K. A. 1987. Development and testing of linear regression models predicting bird-habitat relationships. *J. Wildl. Manage*. 51:247–53
139. Moses, L. E. 1986. *Think and Explain with Statistics*. Reading, Mass: Addison-Wesley
140. Mosimann, J. E. 1970. Size allometry: size and shape variables with characterizations of the lognormal and generalized gamma distributions. *J. Am. Statist. Assoc*. 65:930–45
141. Mosimann, J. E. 1975. Statistical problems of size and shape. I. Biological applications and basic theorems. In *Statistical Distributions in Scientific Work,* ed. G. P. Patil, S. Kotz, K. Ord, pp. 187–217. Dordrecht, Holland: D. Reidel
142. Mosimann, J. E. 1975. Statistical problems of size and shape. II. Characterizations of the lognormal and gamma distributions. See Ref. 141, pp. 219–39
143. Mosimann, J. E. 1988. Size and shape analysis. In *Encyclopedia of Statistical Sciences,* Vol. 8, ed. Kotz, Johnson, pp. 497–508. New York: Wiley

144. Mosimann, J. E., James, F. C. 1979. New statistical methods for allometry with application to Florida red-winged blackbirds. *Evolution* 33:444–59
145. Mosimann, J. E., Malley, J. D. 1979. Size and shape variables. See Ref. 156, pp. 175–89
146. Mosteller, F., Tukey, J. W. 1977. *Data Analysis Including Regression, a Second Course in Statistics*. Reading, Mass: Addison-Wesley
147. Nagel, E. 1961. *The Structure of Science, Problems in the Logic of Scientific Explanation*. New York: Harcourt, Brace & World
148. Neff, W. A., Marcus, L. F. 1980. *A Survey of Multivariate Methods for Systematics*. New York: Am. Mus. Nat. Hist.
149. Nei, M. 1987. *Molecular Evolutionary Genetics*. New York: Columbia Univ. Press
150. Neter, J., Wasserman, W., Kutner, M. H. 1983. *Applied Linear Regression Models*. Homewood, Ill: Irwin
151. Niemi, G. J. 1985. Patterns of morphological evolution in bird genera of New World and Old World peatlands. *Ecology* 66:1215–28
152. Noon, B. R. 1986. Summary: biometric approaches to modeling—the researcher's viewpoint. See Ref. 213, pp. 197–201
153. Nur, N. 1984. Feeding frequencies of nestling blue tits *(Parus caeruleus)*: costs, benefits and a model of optimal feeding frequency. *Oecologia* 65:125–37
154. Oksanen, J. 1983. Ordination of boreal heath-like vegetation with principal component analysis, correspondence analysis, and multidimensional scaling. *Vegetatio* 52:181–89
155. Orloci, L. 1978. *Multivariate Analysis in Vegetation Research*. The Hague: Junk
156. Orloci, L., Rao, C. R., Stiteler, W. M. eds. 1979. *Multivariate Methods in Ecological Work*. Fairland, Md: Int. Coop. Publ.
157. Oxnard, C. E. 1968. The architecture of the shoulder in some mammals. *J. Morphol.* 126:249–90
158. Oxnard, C. E. 1980. The analysis of form: without measurement and without computers. *Am. Zool.* 20:695–705
159. Pielou, E. C. 1984. *The Interpretation of Ecological Data: A Primer on Classification and Ordination*. New York: Wiley
160. Pimentel, R. A. 1979. *Morphometrics, the Multivariate Analysis of Biological Data*. Dubuque, Ia: Kendall/Hunt
161. Read, D. W., Lestrel, P. E. 1986. Comment on uses of homologous-point measures in systematics: a reply to Bookstein et al. *Syst. Zool.* 35:241–53
162. Rencher, A. C. 1988. On the use of correlations to interpret canonical functions. *Biometrika* 75:363–65
163. Rexstad, E. A., Miller, D. D., Flather, C. H., Anderson, E. M., Hupp, J. W. Anderson, D. R. 1988. Questionable multivariate statistical inference in wildlife habitat and community studies. *J. Wildl. Manage.* 52:794–98
164. Reyment, R. A., Blacklith, R. E., Campbell, N. A. 1984. *Multivariate Morphometrics*. New York: Academic. 2nd ed.
165. Rice, J., Ohmart, R. D., Anderson, B. W. 1983. Habitat selection attributes of an avian community: a discriminant analysis investigation. *Ecol. Monogr.* 53:263–90
166. Rice, J., Anderson, B. W., Ohmart, R. D. 1984. Comparison of the importance of different habitat attributes to avian community organization. *J. Wildl. Mange.* 48:895–911
167. Rich, T. 1986. Habitat and nest-site selection by Burrowing Owls in the sagebrush steppe of Idaho. *J. Wildl. Manage.* 50:548–55
168. Rising, J. D., Somers, K. M. 1989. The measurement of overall body size in birds. *The Auk* 106:666–74
169. Rohlf, F. J. 1971. Perspectives on the application of multivariate statistics to taxonomy. *Taxon* 20:85–90
170. Rohlf, F. J., Archie, J. W. 1984. A comparison of Fourier methods for the description of wing shape in mosquitoes (Diptera: Culicidae). *Syst. Zool.* 33: 302–17
171. Rohlf, F. J., Bookstein, F. L. 1987. A comment on shearing as a method for "size correction." *Syst. Zool.* 36:356–67
172. Rohlf, F. J., Wooten, M. C. 1988. Evaluation of a restricted maximum-likelihood method for estimating phylogenetic trees using simulated allele frequency data. *Evolution* 42:581–95
173. Romesburg, H. C. 1981. Wildlife science: gaining reliable knowledge. *J. Wildl. Manage.* 45:293–313
174. Scheibe, J. S. 1987. Climate, competition, and the structure of temperate zone lizard communities. *Ecology* 68:1424–36
175. Schervish, M. J. 1987. A review of multivariate analysis. *Statist. Sci.* 2:396–433
176. Schluter, D., Grant, P. R. 1984. Ecological correlates of morphological

evolution in a Darwin's finch, *Geospiza difficilis*. *Evolution* 38:856–69
177. Schoener, T. W. 1986. Overview: Kinds of ecological communities—ecology becomes pluralistic. See Ref. 84, pp. 467–79
178. Schuerman, J. R. 1983. *Multivariate Analysis in the Human Services*. Boston: Kluwer-Nijhoff
179. Schwaegerle, K. E., Bazzaz, F. A. 1987. Differentiation among nine populations of *Phlox:* response to environmental gradients. *Ecology* 68:54–64
180. Sibley, C. G., Ahlquist, J. E. 1983. The phylogeny and classification of birds based on data of DNA-DNA hybridization. In *Current Ornithology,* Vol. 1, ed. R. F. Johnston, pp. 245–92. New York: Plenum
181. Siegel, A. F., Benson, R. H. 1982. A robust comparison of biological shapes. *Biometrics* 38:341–50
182. Simmons, R. E. 1988. Food and the deceptive acquisition of mates by polygynous male harriers. *Behav. Ecol. Sociobiol.* 23:83–92
183. Smith, K. G. 1981. Canonical correlation analysis and its use in wildlife habitat studies. See Ref. 33, pp. 80–92
184. Sneath, P. H. A. 1967. Trend-surface analysis of transformation grids. *J. Zool., Lond.,* 151:65–122
185. Sneath, P. H. A., Sokal, R. R. 1973. *Numerical Taxonomy, the Principles and Practice of Numerical Classification*. San Francisco: W. H. Freeman
186. Sokal, R. R., Fiala, K. L., Hart, G. 1984. On stability and factors determining taxonomic stability: examples from the Caminalcules and the Leptopodomorpha. *Syst. Zool.* 33:387–407
187. Sokal, R. R., Michener, C. 1958. A statistical method for evaluating systematic relationships. *Univ. Kans. Sci. Bull.* 38:1409–38
188. Sokal, R. R., Rohlf, F. J. 1981. *Biometry*. San Francisco: W. H. Freeman. 2nd ed.
189. Somers, K. M. 1986. Multivariate allometry and removal of size with principal components analysis. *Syst. Zool.* 35:359–68
190. Sprent, D. 1972. The mathematics of size and shape. *Biometrics* 28:23–37
191. Stauffer, D. F., Garten, E. O., Steinhorst, R. K. 1985. A comparison of principal components from real and random data. *Ecology* 66:1693–98
192. Stauffer, D. F., Peterson, S. R. 1985. Seasonal micro-habitat relationships of ruffed grouse in southeastern Idaho. *J. Wildl. Manage.* 49:605–10
193. Stopher, P. R., Meyburg, A. H. 1979. *Survey Sampling and Multivariate Analysis for Social Scientists and Engineers*. Lexington, Mass: Lexington Books
194. Swaine, M. D., Greig-Smith, P. 1980. An application of principal components analysis to vegetation change in permanent plots. *J. Ecol.* 68:33–41
195. Szaro, R. C. 1986. Guild management: an evaluation of avian guilds as a predictive tool. *Environ. Manage.* 10:681–88
196. Ter Braak, C. J. F. 1983. Principal components biplots and alpha and beta diversity. *Ecology* 64:454–62
197. Thompson, D'A. W. 1942. *On Growth and Form,* ed. J. T. Bonner. Cambridge, UK: Cambridge Univ. Press
198. Thorndike, R. M. 1978. *Correlational Procedures for Research*. New York: Gardner
199. Thorpe, R. S. 1983. A review of the numerical methods for recognizing and analyzing racial differentiation. See Ref. 9, pp. 404–23
200. Tissot, B. N. 1988. Multivariate analysis. In *Heterochrony in Evolution: A Multidisciplinary Approach,* ed. M. L. McKinney. New York: Plenum
201. Titus, K., Mosher, J. A., Williams, B. K. 1984. Chance-corrected classification for use in discriminant analysis: ecological applications. *Am. Midl. Nat.* 111:1–7
202. Torgerson, W. S. 1952. Multidimensional scaling: 1. Theory and method. *Psychometrika* 17:410–19
203. Travis, J. 1980. Genetic variation for larval specific growth rate in the frog *Hyla gratiosa*. *Growth* 44:167–81
204. Travis, J. 1983. Variation in development patterns of larval anurans in temporary ponds. I. Persistent variation within a *Hyla gratiosa* population. *Evolution* 37:496–512
205. Travis, J., Emerson, S. B., Blouin, M. 1987. A quantitative-genetic analysis of life-history traits in *Hyla crucifer*. *Evolution* 41:145–56
206. Tukey, J. W. 1980. We need both exploratory and confirmatory. *Am. Statist.* 34:23–25
207. Turkington, R., Harper, J. L. 1979. The growth, distribution and neighbour relationships of *Trifolium repens* in a permanent pasture. 1. Ordination, pattern and contact. *J. Ecol.* 67:201–18
208. Turrelli, M. 1988. Phenotypic evolution, constant covariances, and the maintenance of additive variance. *Evolution* 42:1342–37
209. Van Horne, B. 1986. Summary: When habitats fail as predictors, the research-

er's viewpoint. See Ref. 213, pp. 257–58

210. Van Horne, B., Ford, R. G. 1982. Niche breadth calculation based on discriminant analysis. *Ecology* 63:1172–74
211. Vepsäläinen, K., Savolainen, R. 1988. Causal reasoning in modelling multiway contingency tables. *Oikos* 53:281–85
212. Verner, J. 1986. Future trends in management of nongame wildlife: a researcher's viewpoint. In *Management of Nongame Wildlife in the Midwest: A Developing Art*, ed. J. B. Hall, L. B. Best, R. L. Clawson, pp. 149–71. Grand Rapids, Mich: Proc. Symp. Midwest Fish & Wildlife Conf.
213. Verner, J., Morrison, M. L., Ralph, C. J. eds. 1986. *Wildlife 2000, Modeling Habitat Relationships of Terrestrial Vertebrates*. Madison, Wisc: Univ. Wisc. Press
214. Wartenberg, D., Ferson, S., Rohlf, F. J. 1987. Putting things in order: a critique of detrended correspondence analysis. *Am. Nat.* 129:434–37
215. Weisberg, S. 1980. *Applied Linear Regression*. New York: Wiley
216. Wilbur, M. 1987. Regulation of structure in complex systems: experimental temporary pond communities. *Ecology* 68:1437–52
217. Wilkinson, L. 1987. *SYSTAT: The System for Statistics*. Evanston, Ill: SYSTAT
218. Williams, W. T. 1976. *Pattern Analysis in Agricultural Science*. New York: Elsevier
219. Williams, B. K. 1981. Discriminant analysis in wildlife research: theory and applications. See Ref. 33, pp. 50–71
220. Williams, B. K. 1983. Some observations on the use of discriminant analysis in ecology. *Ecology* 64:1283–91
221. Williams, B. K., Titus, K. 1988. Assessment of sampling stability in ecological applications of discriminant analysis. *Ecology* 69:1275–85
222. Williams, W. R., Lambert, J. M. 1959. Multivariate methods in plant ecology 1. Association analysis in plant communities. *J. Ecol.* 47:83–101
223. Zelditch, M. L. 1987. Evaluating models of developmental integration in the laboratory rat using confirmatory factor analysis. *Syst. Zool.* 36:368–80
224. Zelditch, M. L. 1988. Ontogenetic variation in patterns of phenotypic integration in the laboratory rat. *Evolution* 42:28–41

*Annu. Rev. Ecol. Syst. 1990. 21:167–96*


# THE RESPONSE OF NATURAL ECOSYSTEMS TO THE RISING GLOBAL $CO_2$ LEVELS

*F. A. Bazzaz*

Department of Organismic and Evolutionary Biology, Harvard University, Cambridge, Massachusetts 02138

KEY WORDS: carbon dioxide, global change, ecosystems, growth, competition

---

## INTRODUCTION

Evidence from many sources shows that the concentration of atmospheric $CO_2$ is steadily rising (61, 17). This rise is strongly correlated with the increase in global consumption of fossil fuels (104). There are also significant contributions from the clearing of forests, especially in the tropics (136, 55). Controversy continues, however, as to whether the biosphere is presently a source or a sink for carbon (see 52, 54, 56).

Despite this controversy, most scientists agree that rising $CO_2$ levels will have substantial direct and indirect effects on the biosphere (80). Because $CO_2$ is a greenhouse gas, its increase in the atmosphere may influence the earth's energy budget. Several climatologists have used general circulation models to predict changes in mean annual global temperature (58, 108). While these models differ in detail, they all predict increased global warming and substantial shifts in precipitation patterns. Recently, some scientists (60) have questioned the predictions of these models. But regardless of changes in global temperature and other climate variables, rising $CO_2$ can influence world ecosystems by direct effects on plant growth and development.

The large body of literature on the response of crops and intensively managed forests to elevated $CO_2$ is not treated in this review because there are



0066-4162/90/1120-0167$02.00

several excellent and recent reviews of it (e.g. 2, 28, 62, 127, 132 for crops, and 37, 65, 111 for trees). Instead, this review concentrates on the response of natural vegetation to elevated $CO_2$ and some of the predicted climate change. The review addresses the $CO_2$ response of individuals at the physiological level and the consequences of that response to population, community, and ecosystem levels. It must, however, be emphasized that most of the findings on the physiological and allocational response to $CO_2$ were first discovered in agricultural crops, and that much of the initial work on plants from natural ecosystems (69) tests the variation among species in these responses.

## PLANT RESPONSES AT THE PHYSIOLOGICAL LEVEL TO ELEVATED $CO_2$

Plant biologists have long known some of the effects of high $CO_2$ levels on plants, and greenhouse growers have used $CO_2$ fertilization to increase plant yield. Work on plants from natural ecosystems has lagged behind that on crops but, over the last few years, has produced a large body of information (see 120 for extensive reviews). The major emphases have been on individual physiological traits, but the consequences of these responses for the whole plant, population, and ecosystem are less understood and, in some cases, counter-intuitive. Many plant and ecosystem attributes will directly or indirectly be influenced by elevated $CO_2$ (118). Therefore, after briefly addressing physiological responses at the leaf level, I concentrate on growth and allocation, reproduction, plant-plant interactions, plant-herbivore interactions, and some ecosystem level attributes.

### *$CO_2$ and Photosynthesis*

When other environmental resources and factors are present in adequate levels, $CO_2$ can enhance photosynthesis of $C_3$ plants over a wide range of concentrations. High $CO_2$ reduces competition from $O_2$ for Rubisco, increases its activation (95), and reduces photorespiration. In contrast, in plants with the $C_4$ metabolism net photosynthetic rates rise steeply with increased $CO_2$ and level off at external $CO_2$ concentrations slightly above ambient (122).

Early studies on the response of plants to elevated $CO_2$ examined short-term responses and used tissues that were grown in near-ambient but likely quite variable $CO_2$ levels of glasshouses and growth chambers. More recent studies use plants grown under controlled $CO_2$ levels. All these studies showed an increase of photosynthetic rates with increased $CO_2$ concentrations. Measurements of photosynthetic rates of these plants grown under ambient and elevated $CO_2$ levels have shown that after a period of time some species adjust their photosynthetic rates to the $CO_2$ levels during growth (become acclimated) whereas other species show little or no adjustment (see 22, 87, 115, 121, 130, 141). The degree to which a species can adjust is probably influenced by the levels of other environmental variables and the timing of their availability (see later). Several investigators have also observed that with time plants grown at elevated $CO_2$ show a decline in photosynthetic rates. Although the reasons for this decline are not fully understood, several reasons for it have been proposed. They include: decline in carboxylation efficiency which may be caused by a decrease in the amount and activity of Rubisco (43, 105, 106); suppression of sucrose synthesis by an accumulation of starch (51, 128); inhibition of the triose-P carrier; reduction in the activity of sucrose-phosphate synthase; limitation of daytime photosynthate export from sources to sinks (36) or insufficient sinks in the plant (63). Because with acclimation there may be little overall increase in plant photosynthesis and growth, understanding acclimation to a high $CO_2$ environment is critical in assessing the long-term response of plants to the high $CO_2$ environments of the future.

From the extensive literature on the response of photosynthesis to elevated $CO_2$, the following patterns emerged: (*a*) Elevated $CO_2$ reduces or completely eliminates photorespiration; (*b*) $C_3$ plants are more responsive than $C_4$ plants to elevated $CO_2$ levels, especially those above ambient concentrations; (*c*) photosynthesis is enhanced by $CO_2$ but this enhancement may decline with time; (*d*) the response to $CO_2$ is more pronounced under high levels of other resources, especially water, nutrients, and light; (*e*) adjustment of photosynthesis during growth occurs in some species but not in others, and this adjustment may be influenced by resource availability; and (*f*) species even of the same community may differ in their response to $CO_2$ (Figure 1).

## *Dark Respiration*

Little information is available on the effects of elevated $CO_2$ concentration on dark respiration rates. Enhancement of photosynthesis may lead to increased respiration because of the increased availability of substrate for respiration. Several arctic tundra species show a substantial increase in dark respiration (88). However, there was no influence of elevated $CO_2$ on dark respiration or on light compensation in *Desmodium paniculatum* (141). There is evidence in the agronomic literature that respiration may decline at high $CO_2$ levels (4). For species from natural communities, it is not known whether the change in dark respiration is proportional to the rise in net daytime photosynthesis. Furthermore, it is unknown if both growth and maintenance respiration respond to the same degree to elevated $CO_2$. These issues are important to the understanding of the response of whole plants to the $CO_2$ rise, especially in regard to carbon gain and biomass accumulation, and they require much attention.

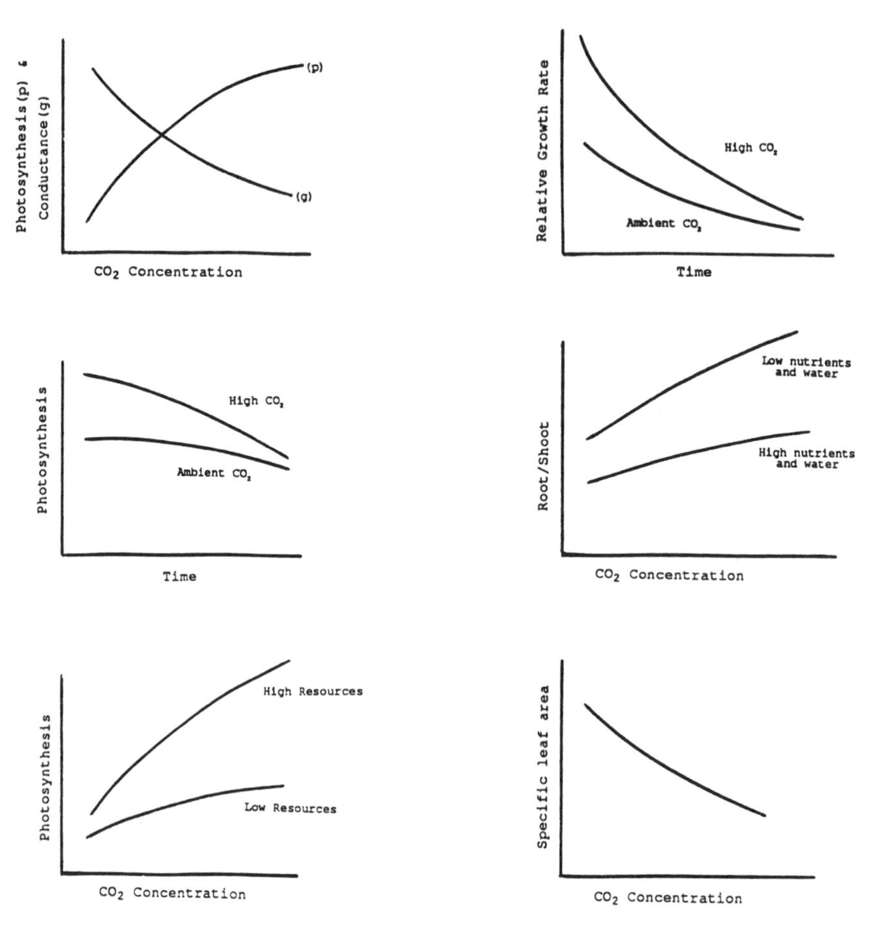


*Figure 1* General trends of response of plants to $CO_2$ concentrations.

## *Stomatal Conductance, Transpiration, and Water Use*

There is now some evidence that growth in high $CO_2$ environments causes a change in stomatal density in some species (e.g. 86). Woodward (134) has shown that stomatal density and stomatal index increased markedly as the $CO_2$ partial pressure is reduced below 340 $\mu l\ l^{-1}$. Above 340 $\mu l\ l^{-1}$ there is a slight decrease in stomatal density in several species studied (135).

Most studies have shown a decline in stomatal conductance with an increase in $CO_2$ concentrations (Figure 1). Stomatal response to $CO_2$ varies greatly among species and may be influenced by other environmental factors such as soil moisture and light levels (e.g. 125). Although strong evidence suggests that stomata respond more to internal $CO_2$ concentration than to external concentrations (e.g. 78), the mechanism by which $CO_2$ controls stomatal activities is not known (95). Therefore, explanation of the differential response of stomata to the $CO_2$ rise is not possible at this time. Transpiration rates decline as a result of decreased stomatal conductance. This decline has been shown in several studies to lead to a favorable instantaneous water use efficiency, improved plant water status, higher carbon gain and biomass accumulation, and lower season-long water consumption rate (e.g. 100). Enhancement of plant water use efficiency was observed also in plants grown in the field (102). Drought stress in plants grown at elevated $CO_2$ levels may be also ameliorated by osmo-regulation and the maintenance of higher turgor pressure (112). Lower transpiration rates should lead to higher leaf temperature under high irradiance and low windspeed conditions (57). This increase, coupled with the anticipated rise in air temperature, may have significant effects on photosynthesis and plant growth.

## *Growth and Allocation*

The critical issues that ought to be examined with regard to the effect of the rising $CO_2$ on plant growth are: (*a*) how long does the enhancement of growth continue; (*b*) how do the allocational relationships in the plant change with time under elevated $CO_2$ levels; and (*c*) how will tissue quality change over time and what are the consequences of this to herbivores, pathogens, and symbionts?

Most studies on the effects of elevated $CO_2$ show an initial enhancement in growth, and like photosynthesis, this enhancement is especially large when other resources are plentiful. In many species, however, this enhancement may decline or completely disappear in time (11, 47, 114, 123–125). Most studies have shown that there is generally an increase in allocation to roots, especially when nutrients and water are limiting (68, 75, 79, 82–84, 114, 125). There is also strong evidence that specific leaf areas (SLA) decrease with increasing $CO_2$ levels (e.g. 47). Decreased SLA in high $CO_2$-grown plants is often associated with increased starch levels in leaves and decreased N concentration. Furthermore, the concentrations of C-based secondary chemicals (e.g. phenolics) usually show no change in levels in leaves from $CO_2$-enriched plants even though the plants have greater carbon availability. Several studies (e.g. 114), especially with woody seedlings, have shown that branching increases with elevated $CO_2$. Some evidence from tree ring analysis suggests that growth in natural vegetation has been enhanced by the rising global $CO_2$ concentrations (66).

## *Phenology and Reproductive Biology*

Despite its great importance to understanding the future impact of $CO_2$ and climate change as possible agents on natural selection, there is very limited information on the effects of the rising $CO_2$ on plant reproductive biology. Most of the studies on the effects of $CO_2$ on plants of natural communities were terminated before reproduction. Because of the well-established effects of elevated $CO_2$ on plant growth, it is expected that aspects of reproduction such as flowering phenology, allocation to reproduction and to various components of reproduction, seed and flower abortion, and seed quality will also be influenced (10). Studies have shown that depending on the species, flowering time could be earlier or later under elevated $CO_2$ (24, 46). In some species these changes are only evident under unusually high levels of $CO_2$ (e.g. 117). When plants are grown in competition, significant $CO_2$ effects on flowering among the species were found only under high nutrient conditions (143), or the effects became less pronounced than when the plants were grown separately (E. G. Reekie, F. A. Bazzaz, unpublished). Differences in flower birth rate, flower longevity, and total floral display have been observed among species in the same community as well as among populations of the same species (e.g. *Phlox drummondii*) (46).

Reekie & Bazzaz (unpublished) examined the relation between $CO_2$ level and reproduction in four species from the annual community of disturbed ground in Texas. Four insect pollinated forbs with showy flowers were used. In *Gallardia pulchela,* doubling $CO_2$ reduced the time required for flowering by six days, though plant size at the time of flowering remained unchanged. In *Gaura brochycarpa,* doubling $CO_2$ also reduced time to flowering; however, these reductions do not appear to be related to increased growth at elevated $CO_2$. The response of *Lupinus texensis* was the reverse: elevated $CO_2$ increased rather than decreased the time to flowering except when the plants were given much underground space. No clear trends were found in *Oenothera laniculata*. Shifts in flowering phenology caused by $CO_2$ rise could have marked effects on community structure and regeneration, especially in communities where pollination is dependent upon animals or when the growing season could be short, as is the case in this community where drought can suddenly terminate the growing season. The combined effects of elevated $CO_2$ and other aspects of climate change, such as rising temperature, may cause large shifts in phenology such that the activities of the plants and their pollinators become decoupled.

Elevated $CO_2$ can also affect other reproductive parameters, such as seed number and size and seed nutrient content. In *Datura stramonium,* total fruit weight was higher in plants grown in elevated $CO_2$ than in plants grown in ambient $CO_2$. Plants grown in high $CO_2$ produced thicker fruit walls, which may prevent insects from laying eggs in these seeds, but, seed size was not affected (46). In *Abutilon theophrasti* grown at elevated $CO_2$ levels, total seed production did not increase, but flower number, capsule number, and seed number decreased (Figure 2). Individual seed weight was higher in plants

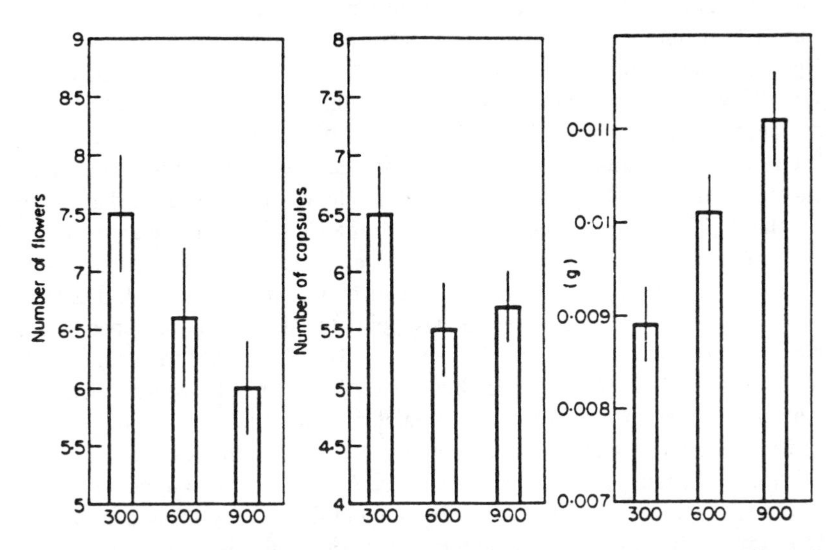


*Figure 2* The influence of $CO_2$ concentrations during growth on flower number, seed number, and mean weight of individual seed in *Abutilon theophrasti*. From (46).

grown at high $CO_2$ (46). In some species (e.g. *Ambrosia artemisiifolia*) there was also much higher N concentration in seeds from plants grown at elevated $CO_2$ concentrations (47). Because of the well-established relationship between individual seed size and nutrient content, and seedling success in nature, the effects of rising $CO_2$ and associated climate change may have great impact on the demography and evolution of natural populations.

## INTERACTION OF $CO_2$ WITH OTHER ENVIRONMENTAL FACTORS

The interaction of $CO_2$ with other plant resources has been amply demonstrated (Figure 1). The response of plants to elevated $CO_2$ is contingent upon light levels (e.g. 95, 110, 113), soil moisture (7, 138), and nutrient availability (2, 20, 50, 93, 103, 133). Several investigators have shown that the enhancing effects of $CO_2$ disappear under nitrogen and phosphorus limitation (20, 50, 133, 143). Light saturation is usually higher under elevated than under ambient $CO_2$ (126), and high $CO_2$ may compensate for low light (2). Plant response to elevated $CO_2$ is usually more strongly expressed under higher levels of these resources, in a manner consistent with predictions about the response of plants to multiple environmental resources (16, 25).

Elevated $CO_2$ may modify the effects of stress factors on plant growth. Such elevation has been shown to ameliorate effects of high salinity (19) by supplying extra energy for maintenance respiration and by the reduction in the entry of salt into the plant due to reduced transpirational pull (39, 44, 45). High $CO_2$ levels may influence the plant response to gaseous pollutants as well. Coyne & Bingham (27) have shown that reduction in stomatal conductance caused by high $CO_2$ reduces both the amount of $O_3$ entering the leaves and the resulting damage. Similarly, decreased stomatal conductance caused by high $CO_2$ reduces entry of $SO_2$ into leaves and lessens its damage in $C_3$ plants (23). $SO_2$ reduced the growth of the $C_3$ species at the ambient but not at the elevated $CO_2$ concentration. In contrast, in the $C_4$ species, $SO_2$ increased growth at the ambient $CO_2$ concentration and reduced at a high $CO_2$. The results of this experiment support the notion that $C_3$ species are more sensitive to $SO_2$ than are $C_4$ species (131). This study shows that $CO_2$ reversed the effect of $SO_2$ on $C_3$ but not $C_4$ plants, results which correlated with differences in sensitivity of stomatal conductance.

The interaction of $CO_2$ with temperature is critical to the response of plants to climate change. Acock (1) and Acock & Allen (2) present a model for the response of photosynthesis to temperature and $CO_2$. They show that at high $CO_2$ levels the optimal temperature for photosynthesis is higher than at ambient $CO_2$ and the range of optimal temperature for photosynthesis is narrower. There are, however, only a few studies that consider their joint effects. J. Coleman & F. A. Bazzaz (unpublished) examined growth and resource acquisition and allocation in response to temperature and $CO_2$ in a $C_3$ and a $C_4$ species that occur together in the field. The results show significant interactive effects on these parameters, but the strength and direction differed between the two species. In the $C_4$ species *(Amaranthus)*, final biomass was increased by $CO_2$ at 28°C but was depressed at 38°C. In the $C_3$ species *(Abutilon)*, $CO_2$ enhanced initial biomass at both temperatures, but the final biomass was not different in the two temperatures. These somewhat surprising results were explained by the amount of standing leaf areas and changed photosynthetic rates in the two species under these conditions. It was clear from this model system that the interaction between factors may be complex but could be understood by studying patterns of carbon gain and allocation.

Several of the climate models also predict that in addition to the global rise in mean annual temperature there can be an increase in temperature extremes. Furthermore, because of the generally reduced stomatal conductance under elevated $CO_2$ conditions, transpirational cooling of plant tissues will be reduced. Few studies have addressed the joint effects on plant growth of unusually high and unusually low temperature in conjunction with elevated $CO_2$. When the $C_4$ weedy grasses *Echinocloa crus-galli* and *Elusine indica* were grown in a range of temperatures and then subjected to one night of chilling at 7° C, the decline in both conductance and photosynthesis was less in plants grown under elevated $CO_2$ than in plants grown under ambient $CO_2$ levels (97). Preliminary results with *Abutilon* suggest that individuals grown at high $CO_2$ concentrations are more sensitive to heat shock than are individuals grown at ambient $CO_2$ (F. A. Bazzaz, unpublished).

## SUBSPECIFIC DIFFERENCES IN RESPONSE TO $CO_2$

Populations of the same species respond differently to $CO_2$, and these differences may be related to the $CO_2$ environment in which the plants grow (e.g. 140). However, differences among individuals of a population in response to $CO_2$ have rarely been investigated. Clearly, genetic differences among individuals in response to atmospheric $CO_2$ can affect the future of the genetic structure of the population in a changing $CO_2$ atmosphere. The studies that have examined variation among individuals have detected differences among them in response to $CO_2$. For example, Wulff & Miller (142) found that families of *Plantago lanceolata* differed in their response to $CO_2$ enrichment and to combinations of $CO_2$ and temperature treatments. They suggested the presence of genetic variability in this species in response to $CO_2$ enrichment. F. A. Bazzaz & G. Carlton (unpublished) found differences in $CO_2$ response in growth and architecture among several genotypes of *Polygonum pensylvanicum* from a single population. Garbutt & Bazzaz (46) found differences in the time of flowering, the number of flower births, and the maximum flower display among four populations of the annual *Phlox drummondii* from central Texas (Figure 3). Significant effects were also seen on plant final biomass and in the number of flowers produced per unit of plant dry weight. These responses may have significant implications for pollination success, dispersal, and establishment.

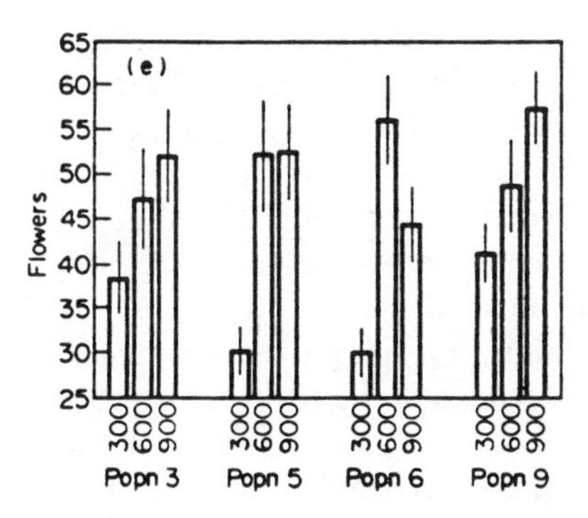


*Figure 3* Deferences in floral display among natural populations of *Phlox drummondii* in response to $CO_2$ concentration. From (46).

## PLANT RESPONSE TO $CO_2$ AT THE POPULATION LEVEL

Almost no information is available on the response to elevated $CO_2$ at the population level. But because elevated $CO_2$ affects growth, allocation, and reproduction, undoubtedly there are some effects on populations. Using our model system of the annuals *Abutilon theophrasti* and *Amaranthus retroflexus,* we investigated how the simultaneous changes in $CO_2$ and temperature affect the recruitment of seeds into the population (S. Morse, F. A. Bazzaz, unpublished). Although no differences appear in survivorship with respect to ambient $CO_2$ concentrations for either species, stand productivity was significantly affected by both $CO_2$ and temperature. In general, stand productivity increased with both $CO_2$ and temperature and was inversely proportional to the number of survivors. $CO_2$ magnified the intensity of plant-plant interactions and enhanced the growth of the remaining dominant individuals.

## INVESTIGATIONS AT THE COMMUNITY AND ECOSYSTEM LEVELS

### *Productivity*

Predictions about the changes in productivity of ecosystems are also based on the generally observed increase in plant growth under high $CO_2$ conditions. A physiologically based graphical model (Beam 82) was proposed by a group of scientists (see 119) to represent possible changes in productivity of ecosystems. To address the relationship between elevated $CO_2$ and productivity, Gates (48) suggested a modification of the B factor, described by Bacastow & Keeling (5), and proposed B' (the biotic growth factor), based on the Michaelis-Menton equation. B' is the fractional increase in net primary productivity (NPP) with a fractional increase in $CO_2$ concentration. Using data on single leaves, Gates (48) calculated B' factors for several deciduous forest tree species and showed that they could be high, ranging from 0.33–0.53. However, he also found that, depending on environmental limitations, the B' values could be small (between 0.05–0.25). Using high B' values, Gifford (49) estimated high carbon storage in the biosphere (1.65 Gt $y^{-1}$ for B' = 0.60). Several other authors (e.g. 18, 48, 54, 64, 73) have pointed out that because of the limits on plant growth already set by water and nutrient deficiency, and temperatures at the northern limits of distribution, primary productivity in natural ecosystems may not be enhanced much by the rising global $CO_2$. Furthermore, even in systems that have the potential for an increase in production, Oechel & Strain (88) show that negative feedbacks may soon lead to the elimination of any enhancement by the rising $CO_2$. For example in the chaparral, a water-limited system, increased water use efficiency may lead to enhanced productivity. However, the chaparral is a fire-prone system, and the increased accumulation of living and dead biomass may increase the frequency of fire, which in turn would reduce biomass accumulation. In contrast, Luxmoore (74) suggests a different scenario, where increased photosynthesis in a high $CO_2$ environment would increase the amount of carbon allocated to roots, resulting in increased root exudation, mycorrhizal proliferation, and increased N-fixation. Evidence also suggests increased nitrogenase activity at high $CO_2$ levels (81). These factors in turn can lead to increased water and nutrient supply to the plants and increased phytomass even in somewhat infertile habitats. The very limited evidence from field studies shows both an increase in productivity with elevated $CO_2$, especially during the first year or two (e.g. 29, 90), and no change (121). Thus, these responses to elevated $CO_2$ remain very poorly understood despite their great importance in predicting future productivity. Accurate predictions about the response of natural ecosystems to global increase in $CO_2$ levels still require much additional data on the mechanistic bases of the responses of several ecosystems (31).

## RESPONSE OF SPECIFIC ECOSYSTEMS TO ELEVATED $CO_2$

### *Graminoid-Dominated Ecosystems*

$CO_2$ AND ARCTIC TUNDRA Arctic ecosystems may be the ecosystems most sensitive to climate change (see 119). All climate models show a greater increase in mean annual temperature in these regions, compared to lower latitudes. Arctic ecosystems possess several properties that make them of particular interest to the study of $CO_2$ response (12). Because of permafrost, the active layer of the soil is shallow, and the top 10 cm of the soil contain most of the root and rhizome systems, which constitute by far most of the living biomass in this ecosystem. Up to 90% of the $CO_2$ which evolves from soil comes from root and rhizome respiration (12). Tundra soils also contain large quantities of organic matter which, being mostly in the permafrost, is normally unavailable to decomposers.

In a series of experiments with microcosms of intact cores of turf and soil of coastal arctic tundra, W. D. Billings and associates (13–15, 96) examined the effects on ecosystem carbon balance of doubling $CO_2$, increasing temperature, lowering the water table, and applying N-fertilizer. They concluded that increasing summer temperature by 4°C would reduce net ecosystem $CO_2$ uptake by half. Lowering the water table by only 5 cm and increasing temperature greatly lowered ecosystem carbon storage. In contrast, doubling $CO_2$ concentrations per se had very little effect. They suggest that warmer temperatures would extend the growing season into the short days of autumn, expose much more peat to decomposers (which become more active in the higher temperatures and the longer season), and lower the water table by high transpiration under the warm conditions. Enhanced ecosystem carbon gain caused by the release of nutrients would be more than offset by decreased insulation and the resultant lowering of the permafrost table and increased soil erosion. From these studies Billings reaches the dramatic conclusion that doubling $CO_2$ would convert the wet tundra ecosystem from a $CO_2$ sink to a $CO_2$ source.

W. Oechel and coworkers have been studying the response of arctic ecosystems to the increase in $CO_2$ and temperature using environmentally controlled greenhouses placed in situ in the tundra near Barrow, Alaska (99, 88, 121). Contrary to most results obtained on the response of single individuals, *Eriophorum* plants in situ showed little response to high levels of $CO_2$. Plants grown at the high $CO_2$ adjusted their photosynthetic rates within three weeks so that their rates were similar to those grown under ambient $CO_2$ when both were measured at $CO_2$ levels of their growth. Although there was no seasonal pattern of growth, a significant increase occurred in tillering under the high $CO_2$ conditions. When responses to elevated $CO_2$ under controlled conditions of six arctic tundra species of different growth forms were compared (in 88), most of the species had increased their photosynthetic rates on a leaf area basis, but they varied in the degree of response, and that was influenced by nutrient level. All species except *Eriophorum* had increased leaf dark respiration as well. Surprisingly, and contrary to the results from the in situ measurements, the photosynthetic rate of *Eriophorum vaginatum* was enhanced, especially under high nutrient conditions, and that enhancement was still high after 2 months of exposure to the high $CO_2$ level. Oberbauer et al (87) found that *Carex bigelowii, Betula nana,* and *Ledum palustre* responded to elevated $CO_2$ and nutrient levels. They found that nutrients enhanced growth much more than did $CO_2$ and concluded that $CO_2$ with or without nutrient limitation has little effect on the production of these species. These results point out the importance of in situ measurements to accurately assess plant response to elevated $CO_2$ concentration. Analysis of whole ecosystem response to elevated $CO_2$ and temperature from the in situ measurements shows that net $CO_2$ uptake by tussock tundra was higher at elevated $CO_2$ than at ambient $CO_2$. But, net $CO_2$ uptake was reduced by temperatures 4°C higher than ambient (53). Although the higher temperature increased conductance and consequently gross photosynthesis, higher temperatures also increased respiration to a degree that resulted in lower net $CO_2$ uptake. These authors conclude that nutrient limitation in this system lowers the ability of tundra plants to make full use of the elevated $CO_2$ concentrations.

The following conclusions emerge from work on this ecosystem: (*a*) in *Eriophorum*, the dominant species in this system, only tillering increases dramatically with rise in $CO_2$; (*b*) photosynthetic acclimation to high $CO_2$ occurs; (*c*) nutrients enhance the response to increasing $CO_2$; (*d*) species differ in the degree to which growth is enhanced by $CO_2$; (*e*) different life forms do not seem to respond differently to increase in $CO_2$; (*f*) conductance and respiration increase; and (*g*) temperature rise lowers the $CO_2$ enhancement effects. Therefore, while some general responses are similar to those observed in other ecosystems, the tundra ecosystem differs in some quite surprising ways, particularly the increase in conductance.

From the available data, the following scenario emerges: As $CO_2$ and temperature rise, thaw of permafrost increases, the growing season lengthens, decomposition of organic matter increases sharply, nutrient availability increases, net $CO_2$ uptake increases, and transpiration increases because of higher temperature and increased conductance. After a while, however, the water table recedes, photosynthesis and net ecosystem productivity decrease, and the system becomes a $CO_2$ source and a positive feedback loop would be established.

THE ESTUARINE MARSH Another in situ study of the response of graminoid ecosystem to elevated $CO_2$ has been underway in the estuarine marsh of Chesapeake Bay, Maryland, USA. Open top chambers were used by B. Drake and his associates to enclose stands of *Scirpus olneyi* ($C_3$), *Spartina patens* ($C_4$), and a combination of both species and to expose them to ambient ($350 \pm 22$ $\mu l$ $l^{-1}$) and elevated ($686 \pm 30$ $\mu l$ $l^{-1}$) $CO_2$ concentrations. Elevated $CO_2$ increased shoot density, delayed senescence, and increased biomass in *Scirpus*, the $C_3$ species, but there was no effect on *Spartina*, the $C_4$ species (30). Furthermore, *Scirpus* responded positively to elevated $CO_2$ both in pure and in mixed stands. Carbon-nitrogen relations were also examined for these species (29). While carbon percentage did not change with elevated $CO_2$ in green leaves of *Scirpus*, nitrogen was reduced by as much as 40%. Furthermore, aboveground tissue content of nitrogen on a per leaf area basis was not influenced by $CO_2$, indicating that nitrogen was allocated from storage pools. Surprisingly, litter C/N ratio was not affected by $CO_2$ level, and the authors suggested that $CO_2$ rise will not influence the rate of decomposition or N mineralization. Because of the continued input of nutrients in water from the adjacent creek into this already highly productive marsh, the authors conclude that continued exposure to high $CO_2$ levels may cause a continued increase in *Scirpus* productivity and increased dominance in this system. Thus, this situation contrasts sharply with that observed in the nutrient-limited tundra ecosystem discussed previously.

OTHER GRASSLANDS Information about the response of grasslands to elevated $CO_2$ is very limited. Smith et al (115) compared the response of four grass species from the Great Basin. High $CO_2$ resulted in increased growth, especially basal stem production, in the $C_3$ but not in the $C_4$ species. This enhancement was particularly strong for *Bromus tectorum,* an introduced weed. Since *Bromus* predisposes rangelands to burning, the authors speculated that this enhancement by high $CO_2$ levels in the future may increase the number and the severity of wildfires in this region, which could result in a change in ecosystem function. Work with Blue grama *(Bouteloua gracilis),* an important native perennial in the same region, showed that biomass and leaf area were greatly enhanced at elevated $CO_2$ levels, which is unusual for a $C_4$ plant (101).

When plants were grown individually, $CO_2$ concentration differentially influenced the growth of six species from the short annual grasslands found on serpentine soil in California (129). In competition, however, these species did not differ in their growth response to $CO_2$. The species are of small stature and presumably adapted to low nitrogen and calcium availability and to heavy metals such as Ni and Mg. Apparently, the potential for these species to respond to increased $CO_2$ concentrations may be constrained by physiological traits that enable these annuals to grow in their native, nutrient-limited environment. Furthermore, in this low-stature community with a very short growing season and nutrient limitation, competitive networks and adaptation can develop and dampen the $CO_2$ effects.

## *Regenerating Ecosystems*

The speed of the rise in $CO_2$ concentrations and the associated temperature rise will far exceed the regeneration time of many woody species in the world and their migration to new habitats (32). Thus, this rapid change would likely result in the death of many individual plants and their replacement with early successional species that, in general, are adapted to live in an environment with initially high resource levels (6). Regenerating ecosystems may be the dominant ones over much of the landscape in a high $CO_2$ world. Thus, the study of regenerating ecosystems is crucial to assessing the possible impact of global change. Our extensive knowledge of their behavior at the physiological, populational, and community levels under ambient and more recently elevated $CO_2$ may allow some predictions about their future.

## *The Early Successional Community: A Model System*

NONCOMPETITIVE RESPONSE Work in our laboratory has focused mainly on community level, using individual species responses to interpret community level responses. A major premise of the research is that the response of individuals is highly modified by the presence of other individuals in a population or a community, and that these relationships themselves would be modified by other factors in the natural environment.

Community-level investigations of $CO_2$ effects on plant growth were reported by Carlson & Bazzaz (22). The annual community of postagricultural succession and the flood plain forest community in the midwestern United States were studied. The experiments also included three crop species (corn, soybean, and sunflower) in order to compare results with the published agronomic literature. The results confirmed that species from natural communities have physiological responses similar to those of the agronomic species studied thus far. The degree of variation in response of different species even of the same community was enormous. Based on these findings, and without consideration of the associated climate change, three hypotheses about the effects of elevated $CO_2$ on plants were put forward: (*a*) Because of increased water use efficiency, plant species will be able to expand their ranges into drier habitats; (*b*) competitive interactions among species in a community may change and will result in a change in community composition and function; and (*c*) competitive interaction between crops and weeds may change. The latter hypothesis was also proposed by Patterson & Flint (92) and was later confirmed (94).

Further work with individually grown plants established the fundamental physiological and morphological basis of the response of plants to $CO_2$ and its interactions with other environmental factors. Most of these studies involved growing several species individually and studying differences among them in their photosynthetic response, growth and allocation, or some other indicator of their potential competitive success. These results were used to infer competitive outcome among species (e.g. 8). The results have also been useful in interpreting the response of communities to the rising $CO_2$ levels. We chose as a model system an early successional community of annual plants to investigate in detail aspects of the $CO_2$ response at the individual, population, and community levels. Depending on the questions asked we sometimes used all dominant species of the community and sometimes a subset of these species. This community was chosen because annual plants can be grown to maturity, so that the effects of $CO_2$ on all phases of the life cycle, including reproduction, could be studied, and also because we have accumulated much background information on this community over two decades. The community is dominated by a small number of species (five to six) and has both $C_3$ and $C_4$ plants. Comparing the response of the major species in this community to elevated $CO_2$ when the plants were grown individually (47), we found:

1. $CO_2$ concentration had little effect on the timing of seedling emergence;
2. Photosynthetic rates increased and stomatal conductance decreased with increased $CO_2$;
3. The levels of $CO_2$ during growth had no effect on photosynthetic rates;
4. Shoot water potential was less negative in plants grown at high $CO_2$;
5. Relative growth rates were enhanced by $CO_2$ early in the growth period but declined later;
6. Specific leaf area (SLA) consistently decreased with increased $CO_2$;
7. High $CO_2$ caused one species to flower earlier and one to flower later, while the rest showed no change;
8. There were significant species x $CO_2$ interactions for leaf area, leaf weight, weight of reproductive parts, and seed weight indicating species-specific response to $CO_2$; and
9. Carbon/nitrogen ratios increased with increasing $CO_2$.

The results of this experiment and others also suggest that the commonly suggested $C_3/C_4$ dichotomy does not fully explain the responses of plants to $CO_2$. For example, *Amaranthus* ($C_4$) often shows a greater increase in biomass as a result of elevated $CO_2$ than does the $C_3$ species *Abutilon theophrasti* (8, 47).

COMPETITIVE RESPONSE IN THE MODEL SYSTEM Under competitive conditions the interaction between $CO_2$ concentration and soil moisture showed that total community biomass increased with increasing $CO_2$ at both moist and dry soil moisture conditions. The contribution of each species to total community biomass was greatly influenced by $CO_2$. For example, *Polygonum pensylvanicum* contributed more at high $CO_2$ and moisture levels. In contrast, *Amaranthus retroflexus* declined under these conditions (7). These results are commensurate with the response of these species individually to $CO_2$ and moisture separately. Work on the interaction of $CO_2$ with light and nutrients (143) using all six species from this community showed that total community production reached its peak at 450 $\mu l\ l^{-1}$ $CO_2$. While total community biomass was higher under high light, relative to low light, and under high nutrients, relative to low nutrients, the response of the community to elevated $CO_2$ was affected by light level but not by nutrient availability. The relative success of some species, particularly in terms of seed biomass and reproductive allocation, was significantly altered by $CO_2$. The contribution of the $C_3$ species in this community to total production increased with $CO_2$ enrichment.

Competitive interactions and $CO_2$ have been examined in more detail using one $C_3$ and one $C_4$ plant from this community. Detailed growth analysis, patterns of leaf display, and N allocation were used to understand the mechanisms of interaction and to begin to model these interactions (11). The species were grown both individually and in competition with each other. At ambient $CO_2$ levels *Abutilon* was competitively superior to *Amaranthus* because the latter was unable to overcome the initial difference in starting capital (larger seeds and seedling). But, at elevated $CO_2$ that difference disappeared, largely because of the enhanced relative growth rate (RGR) of *Amaranthus* in high $CO_2$ (especially earlier in the growth period) which overcame the seed size advantage that *Abutilon* has over *Amaranthus*. High $CO_2$ caused an increase in root/shoot ratio in *Abutilon* and a decrease in *Amaranthus*. But *Amaranthus* had a much higher rate of N uptake per unit of root relative to *Abutilon*. Thus, the results of this experiment show that: (*a*) the response to high $CO_2$ is limited to early stages of growth; (*b*) elevated $CO_2$ greatly increased RGR in *Amaranthus;* and (*c*) although, when compared with $C_3$ plants, $C_4$ plants show a lesser enhancement of photosynthesis and net assimilation rate (NAR) with increased $CO_2$ levels, they did not "lose out" in competition with $C_3$ plants at elevated $CO_2$ concentrations.

Bazzaz & Garbutt (8) studied the influence of the identity of competing species and that of neighborhood complexity on the interaction between $CO_2$ and competition. Four species of the annual community were grown in monoculture and in all possible combinations of two, three, or four species at levels of $CO_2$. Overall, the species responded differently to $CO_2$ levels. In mixtures the species interacted strongly, and in some cases these interactions cancelled out the effects of $CO_2$. For example, there were clear differences in the responses of species in different competitive neighborhoods. All competitive arrays that had $C_3$ species in them depressed the growth of the $C_4$ species (Figure 4). The interactions between $CO_2$ and the identity of the competing species were particularly strong at the intermediate $CO_2$ level (500 $\mu l\ l^{-1}$). These findings suggested that competitive outcome will be modified by $CO_2$ and by the interaction of $CO_2$ with other environmental factors. They show that different species will behave differently in a high $CO_2$ world and that their response will depend on the identity of the competing species and perhaps on community diversity.

## *Early Perennial Stage*

The interaction between *Aster pilosus* ($C_3$) and *Andropogon virginicus* ($C_4$), important species in old-field succession, was studied by Wray & Strain (137, 138, 139). They grew the two species both separately and in competition in ambient and high $CO_2$ levels, while half of them were subjected to a drought cycle. In *Aster,* droughted plants grown at high $CO_2$ had greater leaf water potential and greater photosynthetic rates and total dry weight than did plants grown at ambient $CO_2$. In contrast, in *Andropogon* no differences appeared among $CO_2$ treatments in response to drought. In competition the differences between the species in response to elevated $CO_2$ were accentuated, and *Aster* strongly dominated *Andropogon*. These authors suggested that $CO_2$ enrich-

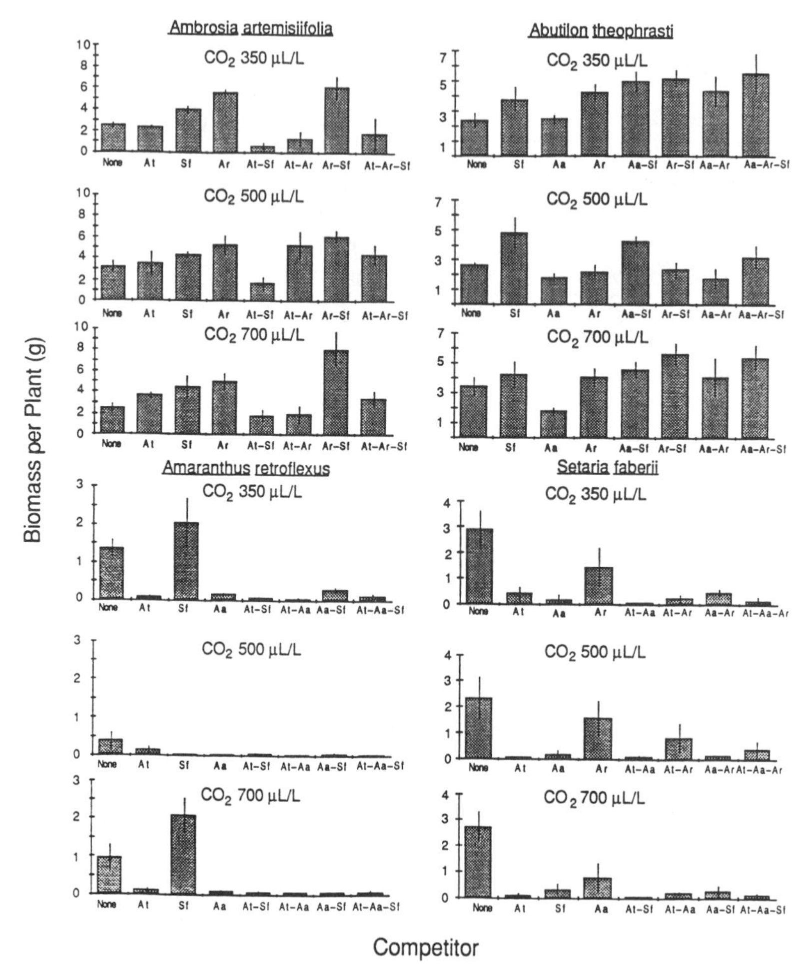


*Figure 4* The relationship between plant growth, identity and diversity of competitors, and $CO_2$ concentrations during growth in a community of annuals made up of two C3 species (*Ambrosia artemisiifolia* (Aa), and *Abutilon theophrasti* (At)) and two (C4) species (*Amaranthus retroflexus* (Ar), and *Setaria faberii* (Sf)) From (8).

ment may increase the competitive ability of *Aster* relative to *Andropogon*, allowing *Aster* to persist for longer periods during old-field succession.

## *Early Successional Trees*

Tolley & Strain (123–125), Sionit et al (114), and Fetcher et al (43) studied the response of Sweetgum *(Liquidambar styraciflua)* and loblolly pine *(Pinus taeda)*, two midsuccessional tree species, to elevated $CO_2$. They found that elevated $CO_2$ increased components of growth more in sweetgum than in loblolly pine, especially at high irradiance. Sweetgum developed more rapidly, reached maximum size earlier, and maintained height dominance relative to loblolly pine. Under drought stress high $CO_2$-grown sweetgum individuals developed internal water deficits more slowly than did those grown under ambient $CO_2$, and the seedlings maintained higher photosynthetic rates over the drying cycle. In contrast, loblolly pine seedlings had a more severe internal water deficit than did sweetgum, irrespective of $CO_2$ level. The authors concluded that sweetgum seedlings should tolerate longer exposure to low moisture, especially under high $CO_2$ conditions, and that these conditions would result in greater seedling survival on drier sites in successional fields in the piedmont. Furthermore, the height dominance and shading that sweetgum presently exerts on pine may be intensified in a high $CO_2$ environment. In the climate of the future, with high $CO_2$, the authors suggest that sweetgum could displace loblolly pine.

## *Forest Ecosystems*

TEMPERATE FORESTS Only a few studies have examined the response of tree species in a community context, and fewer still in competitive situations. Seedlings of the dominants of a floodplain forest community and of an upland deciduous forest community were grown as two groups in competition under ambient and elevated $CO_2$ concentrations (130). Photosynthetic capacity (rate of photosynthesis at saturating $CO_2$ and light) tended to decline as $CO_2$ concentration increased. Stomatal conductance also declined with an increase in $CO_2$. Nitrogen and phosphorus concentrations generally decreased as $CO_2$ increased. Overall growth of both communities was not enhanced by $CO_2$, but the relative contribution of species to the total community biomass changed in a complex way and was also influenced by light/$CO_2$ interactions.

In four cooccurring species of *Betula,* elevated $CO_2$ enhanced survivorship in yellow birch only, but nearly doubled total weight and root/shoot ratio in all species. However, differences among the species in growth response to elevated $CO_2$ were small despite the differences among the species in habitat preferance (F. Bazzaz, unpublished). The response to $CO_2$ of seven co-occurring tree species from the Northern Hardwood forests in New England was studied by F. A. Bazzaz, J. Coleman, & S. Morse, (unpublished). Seedlings of *Fagus grandifolia, Acer saccharum, Tsuga canadensis, Acer rubrum, Betula papyrifera, Prunus serotina,* and *Pinus strobus* were grown under 400 $\mu l\ l^{-1}$ and 700 $\mu l\ l^{-1}$ $CO_2$. The species differed greatly in their responses; elevated $CO_2$ significantly increased the biomass of *Fagus, Prunus, Acer saccharum,* and *Tsuga,* but only marginally that of *Betula, Acer rubrum,* and *Pinus.* Under the conditions of this experiment—relatively low light (400–700 $\mu$mole mole$^{-1}$) and high nutrients—the species that are considered more shade tolerant and late successional *(Fagus, Acer saccharum,* and *Tsuga)* showed the largest biomass increase, with high $CO_2$ levels. Furthermore, *Betula* and *Acer rubrum* grown from seed did not exhibit different responses to elevated $CO_2$ than did those individuals transplanted from the field while dormant. These results suggest that seedlings of the late successional trees in this system growing in the shade and with ample nutrients will do relatively better in a high $CO_2$ world than will early successional trees in open environments. This may be particularly important since young seedlings near the forest floor may experience a high $CO_2$ environment caused by the efflux of $CO_2$ from the soil (9). These findings, at first glance, differ from those of other studies (e.g. 86) which found that growth enhancement by elevated $CO_2$ in *Ochroma lagopus,* a fast growing pioneer species, was greater than that in *Pentaclethra macroloba,* a slower growing climax species. Furthermore, Tolley & Strain (123) found a greater enhancement of growth in the faster growing of two early successional tree species *Liquidambar styraciflua* and *Pinus taeda*. The findings of these two studies fit the general notion that early successional plants growing in open environments are able to take opportunistic advantage of available resources and that they have high growth rates (6). However, the results from the seven-species study point once again to the importance of other environmental resources in modifying the response of plants to elevated $CO_2$.

TROPICAL RAINFORESTS Reekie & Bazzaz (100) studied competition and patterns of resource use among seedlings of tropical trees under ambient and elevated $CO_2$ using five relatively fast growing early successional species from the rainforest of Mexico *(Cecropia obtusifolia, Myriocarpa longipes, Piper auritum, Senna multijuga,* and *Trichospermum mexicanum).* Elevated $CO_2$ only slightly affected photosynthesis and overall growth of the individually grown plants but greatly affected mean canopy height. Though stomatal conductance slightly declined with increased $CO_2$, leaf water potential and plant water use were relatively unaffected. However, in the competitive arrays there were marked effects of $CO_2$ on species composition, with some species decreasing and others increasing in importance. High $CO_2$ increased the mean canopy height in *Cecropia, Piper,* and *Trichospermum,* and decreased it in *Senna* (Figure 5). There were also some differences among species in allocation to roots and in the timing of that allocation. Stepwise regression analysis of several physiological and architectural measurements showed that canopy height (leaf display in the canopy) was the single most important variable determining competitive ability. Photosynthetic rates, especially in low light, and allocation to root early in the growth period were also significant. The results of this study suggest that competition for light was the major factor influencing community composition, and that $CO_2$ influenced competitive outcome largely through its effects on canopy architecture. Early in the experiment competition for nutrients was intense. This allowed *Piper,* with greater allocation to roots, to gain a competitive edge.

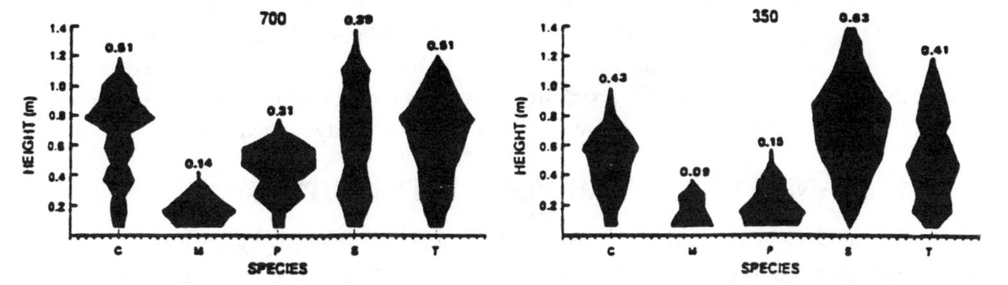


*Figure 5* Leaf area profiles, and mean canopy height of seedling of 5 fast-growing tropical rainforest trees grown at 350 (left) and 700 $\mu$l 1 (right) $CO_2$. Each unit on the horizontal axis represent 1 dm-2. The species are *Cecropia obtusifolia* (C), *Myriocarpa longipes* (M), *Piper auritum* (P), *Senna multijuga* (S), and *Trichospermum mexicanum* (T). From (100).

Very rapidly, however, the canopy closed and competition for light become more intense. Therefore, *Senna, Trichospermum,* and *Cecropia,* with their greater biomass allocation to shoots, were able to overtop the other species. *Senna* was particularly successful because of its high photosynthetic rate and tall shoot architecture. Thus, the major effect of elevated $CO_2$ on competition was through its modification of plant architecture.

## $CO_2$ AND EFFECTS ON SOIL MICROORGANISMS/PLANT ROOT INTERACTIONS

It has been hypothesized that high $CO_2$ and the resulting high availability of photosynthate will enhance root growth and root exudation in the soil. These will in turn influence plant nutrition by enlarging soil volume explored by roots and by increasing mycorrhizal colonization (119), nodulation, and nitrogen-fixing capacities (67, 74). There have been only a few tests of these ideas with plants from natural ecosystems. In *Quercus alba* seedlings grown in nutrient-poor forest soil, elevated $CO_2$ increased growth, especially of the root system (83). Much of the nitrogen was in fine roots and leaves, and the plant's efficiency of N-use was enhanced. Furthermore, elevated $CO_2$ increased uptake of P which may have also been associated with a greater proliferation of mycorrhizae and rhizosphere bacteria. The weight of new buds of seedlings grown in elevated $CO_2$ was greater than of those of seedlings grown in ambient $CO_2$, suggesting that shoot growth in the subsequent year would be enhanced (84). Seedlings of *Pinus echinata* grown in elevated $CO_2$ allocated proportionally more photosynthate to fine roots, produced larger fine root mass, and had higher mycorrhizal density than plants grown in ambient $CO_2$ (85).

Although there have been no experimental tests of the hypothesis, several authors have predicted that the rate of litter decomposition may be slower in high $CO_2$ environments (119, 130). These predictions are based on the finding in the majority of studies that the carbon-to-nitrogen ratio of tissues grown under elevated $CO_2$ levels declines and on experimental evidence that tissue with high lignin and low nitrogen content decays slowly (77).

## $CO_2$ AND PLANT-HERBIVORE INTERACTIONS

Elevated $CO_2$ concentrations within the range predicted by global change scenarios are unlikely to influence herbivores directly (e.g. 42). However, several investigators (see 119) have suggested that the tissue quality of plants grown under high $CO_2$ environments could be altered, thereby indirectly affecting insect performance. Recent experimental evidence has supported this notion. For example, most studies have demonstrated that foliar nitrogen concentrations, a limiting nutrient for insect herbivores (76), decline with increased $CO_2$ (40–42, 59, 70, 71, 130, 133). Other important nutritional factors, such as foliar carbon-based allelochemical and fiber concentrations, do not seem to be affected by elevated $CO_2$ conditions (40–42, 70, 59), and foliar water content does not change in any consistent way in higher $CO_2$ atmospheres (e.g. 71, 41). Too few systems have been examined to make any general statements about these patterns.

Insect herbivore behavior and subsequent performance are affected when they are reared on low nitrogen, high $CO_2$ grown plants. To compensate for the lower nitrogen concentrations, insect herbivores feeding on high $CO_2$ grown foliage increase their consumption rate by 20%–80% compared to those larvae feeding on low $CO_2$ grown tissue (40, 59, 70, 72). Despite this increased consumption, insect herbivore performance on high $CO_2$ grown plants is often poorer than on low $CO_2$ grown plants. Lepidopteran larval mortality increases (3, 41), and growth is often slower for larvae reared on high $CO_2$-grown plants (40–42) (Figure 6). Slower growth might reduce insect herbivore fitness in the wild due to an increased exposure to predators and parasitoids (98) and a decrease in the likelihood of their completing development in seasonal environments (e.g. 26). Reduced population numbers have also been observed for foliage-feeding herbivores on plants in enriched $CO_2$ environments in open-top chambers (21). Interactions between plants and other plant-eating organisms, such as mammals, have yet to be investigated.

## GLOBAL CHANGE AND PREDICTED CHANGES IN SPECIES RANGES

Various modelling results, based on changes in temperature caused by the increase in $CO_2$ and other greenhouse gases, have suggested a significant

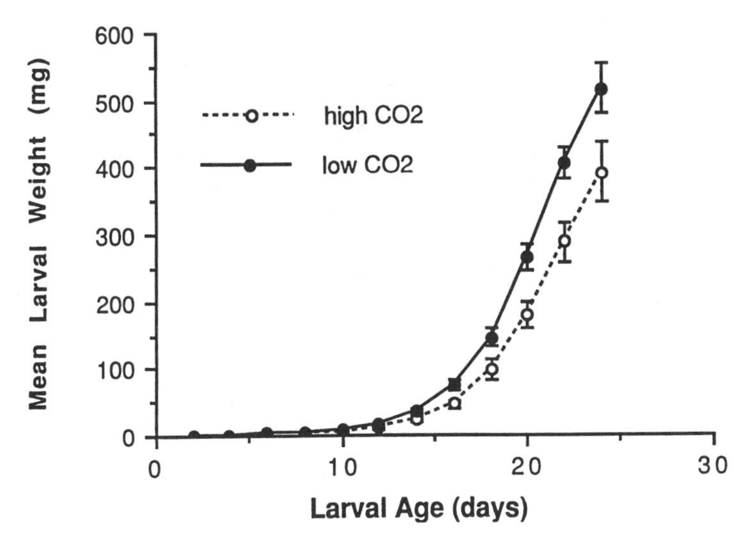


*Figure 6* Growth of larvae of the Buckeye butterfly *Junonia coenia* on *Plantago lanceolata* grown at ambient and elevated $CO_2$ concentration. From (41).

change in patterns of regional plant productivity (109, 38, 116), in the distribution ranges of some plant species (34, 33), and in species composition on a regional scale (91). For example, the range of American beech *(Fagus grandifolia)* could drastically change, and its distribution could be several hundred miles north of its current position (33). Additionally, based on the direct response to increased $CO_2$ alone and the resultant decrease in water consumption, it was also predicted that the ranges of species can expand into drier habitats (22). Of course, neither of these approaches by itself would yield definitive conclusions; the influence of both the direct and indirect effects of the rising $CO_2$ should be jointly considered. Using growth and other physiological data on the response of the weedy vines Kudzu *(Pueraria lobata),* and honeysuckle *(Lonicera japonica)* to elevated $CO_2$, and considering the indirect effects of the $CO_2$-induced climate change, Sasek & Strain (107) concluded that elevated $CO_2$ levels and increased winter minimum temperatures may allow northward and westward migration of both species, but the decreased summer precipitation may minimize the westward spread. It must be pointed out that these predictions concern only the potential for range shifts in species and do not take into account the new and potentially very effective barriers to and corridors for dispersal of propagules, nor do they consider the important factors of the changed plant/plant interactions, plant/animal interactions, and plant/microbial interactions.

## CONCLUSIONS

It is clear from this review that some general patterns of response of plants, especially at the physiological level, to the rising $CO_2$ and the associated climate change are beginning to emerge (Figure 1). Enhanced photosynthesis and growth, increased allocation to underground parts, and particularly water use efficiency have been strongly documented. However, photosynthesis and growth enhancement in some species can be of limited duration, perhaps because of shortages of sinks and the resulting simulation of photosynthates in leaves. It is also clear that $CO_2$ interacts strongly with other environmental factors, especially nutrients and temperature, to generate the response at the individual level.

Work at the community and ecosystem level has clearly shown that, in most situations, the response at the individual level may become highly modified and may not predict the response of communities. It is quite likely that the impact on productivity of ecosystems may result mainly from changes in species composition brought about by differential species response to elevated $CO_2$. The number and the identity of neighboring plants, the levels of environmental resources, the activities of herbivores, pathogens, and symbionts are crucial to the way plants respond to elevated $CO_2$. Because of the complexity of these interactions, and our limited knowledge of them, our predictions about the future impact of the rising $CO_2$ and associated climate change are very tenuous. In fact, for some ecosystems we cannot presently even predict the direction of the change that would result from increasing $CO_2$. Nevertheless, the work on a model system of annual plants, and with other assemblages, is giving us some insights into the mechanisms of the response to $CO_2$ at the community level. We are beginning to identify certain parameters that seem to explain significant amounts of the response to elevated $CO_2$. For example, initial relative plant growth rates and biomass allocation seem very important determinants of plant response to $CO_2$. Responses at the population level are essentially unknown, but that research in this area, particularly plant-animal interactions, will be of great importance in understanding the future of biological systems in a high $CO_2$ world.

### ACKNOWLEDGEMENTS

I thank J. Coleman, E. Fajer, S. Morse, K. Norweg, A. Quinn, and D. Tremmel for much help in preparing this review. I also thank the U.S. Department of Energy for their support.

## *Literature Cited*


1. Acock, B. 1980. Analysing and predicting the response of the glasshouse crop to environmental manipulation. In *Opportunities for Increasing Crop Yields*, ed. R. G. Hurd. P. V. Biscoe, and C. Dennis, pp. 131–48. London: Pitman
2. Acock, B., Allen, L. H. J. 1985. Crop responses to elevated carbon dioxide concentrations. In *Direct Effects of Increasing Carbon Dioxide on Vegetation*, ed. B. R. Strain, J. D. Cure, pp. 53–98. US Dep Energy, Washington, DC
3. Akey, D. H., Kimball, B. A. 1989. Growth and development of the beet armyworm on cotton grown in an enriched carbon dioxide atmosphere. *Southwest. Entomol.* 14:255–60
4. Amthor, J. S. 1989. *Respiration and Crop Productivity*. New York: Springer-Verlag
5. Bacastow, R., Keeling, C. D. 1973. Atmospheric carbon dioxide and radiocarbon in the natural carbon cycle: Changes from A.D. 1700 to 2070 as deduced form geochemical model. In *Carbon and the Biosphere* (CONF-720510), ed. G. M. Woodwell and E. V. Pecan. Washington, DC: Atomic Energy Com. Available from National Technical Information Service (NTIS), Springfield, Virginia
6. Bazzaz, F. A. 1979. The physiological ecology of plant succession. *Annu. Rev. Ecol. Syst.* 10:351–71
7. Bazzaz, F. A., Carlson, R. W. 1984. The response of plants to elevated $CO_2$. I. Competition among an assemblage of annuals at two levels of soil moisture. *Oecologia* 62:196–98
8. Bazzaz, F. A., Garbutt, K. 1988. The response of annuals in competitive neighborhoods: Effects of elevated $CO_2$. *Ecology* 69:937–46
9. Bazzaz, F. A., Williams, W. E., 1990. Atmospheric $CO_2$ concentrations within the canopy of a mixed forest: Implications for seedling growth. *Ecology* In press
10. Bazzaz, F. A., Garbutt, K., Williams, W. E. 1985. Effects of increased carbon dioxide concentration on plant communities. In *Direct Effects of Increasing Carbon Dioxide on Vegetation*, ed. B. R. Strain, J. D. Cure, pp. 155–70. US Dep. Energy, NTIS. Springfield, Virginia, USA
11. Bazzaz, F. A., Garbutt, K., Reekie, E. G., Williams, W. E. 1989. Using growth analysis to interpret competition between a $C_3$ and a $C_4$ annual under ambient and elevated $CO_2$. *Oecologia* 79:223–33
12. Billings, W. D., Peterson, K. M., Shaver, G. R., Trent, A. W. 1977. Root growth respiration, and carbon dioxide evolution in an arctic tundra soil. *Arctic Alpine Res.* 9:129–37
13. Billings, W. D., Luken, J. O., Mortensen, D. A., Peterson, K. M. 1982. Arctic tundra: a source or sink for atmospheric carbon dioxide in a changing environment? *Oecologia* 53:7–11
14. Billings, W. D., Luken, J. O., Mortensen, D. A., Peterson, K. M. 1983. Increasing atmospheric carbon dioxide: possible effects on arctic tundra. *Oecologia* 58:286–89
15. Billings, W. D., Peterson, K. M., Luken, J. O., Mortensen, D. A. 1984. Interaction of increasing atmospheric carbon dioxide and soil nitrogen on the carbon balance of tundra microcosms. *Oecologia* 65:26–29
16. Bloom, A. J., Chapin, F. S., Mooney, H. A. 1985. Resource limitation in plants—an economic analogy. *Annu. Rev. Ecol. Syst.* 16:363–92
17. Bolin, B. 1986. How much $CO_2$ will remain in the atmosphere? In *The Greenhouse Effect, Climatic Change and Ecosystems*, ed. B. Bolin, B. R. O. Doos, J. Jager and R. A. Warrick, pp. 93–156. Scope 29. Chichester: Wiley
18. Botkin, D. B. 1977. Forests, lakes and the anthropogenic production of carbon dioxide. *BioScience* 27:325–31
19. Bowman, W. D., Strain, B. R. 1987. Interaction between $CO_2$ enrichment and salinity stress in the $C_4$ non-halophyte *Andropogon glomeratus*. *Plant Cell. Environ.* 10:267–70
20. Brown, K., Higginbotham, K. O. 1986. Effects of carbon dioxide enrichment and nitrogen supply on growth of boreal tree seedlings. *Tree Physiology* 2:223–32
21. Butler, G. D. 1985. Populations of several insects on cotton in open-top carbon dioxide enrichment chambers. *Southwest. Entomol.* 10:264–66
22. Carlson, R. W., Bazzaz, F. A. 1980. The effects of elevated $CO_2$ concentrations on growth, photosynthesis, transpiration, and water use efficiency of plants. In *Environmental and Climatic Impact of Coal Utilization*, ed. J. J.

Singh, A. Deepak, pp. 609–23. New York: Academic
23. Carlson, R. W., Bazzaz, F. A. 1982. Photosynthetic and growth response to fumingation with $SO_2$ at elevated $CO_2$ for $C_3$ and $C_4$ plants. *Oecologia* 54:50–54
24. Carter, D. R., Peterson, K. M. 1983. Effects of a $CO_2$-enriched atmosphere on the growth and competitive interaction of a $C_3$ and a $C_4$ grass. *Oecologia* 58:188–93
25. Chapin, F. S., Bloom, A. J., Field, C. B., Waring, R. H. 1987. Plant responses to multiple environmental factors. *Bioscience* 37:49–57
26. Chew, F. S. 1975. Coevolution of pierid butterflies and their cruciferous food plants. I. The relative quality of available resources. *Oecologia* 20:117–27
27. Coyne, P. I., Bingham, G. E. 1977. Carbon dioxide correlation with oxidant air pollution in the San Bernardino mountains of California. *J. Air Pollution Control Assoc.* 27:782–84
28. Cure, J. D., Acock, B. 1986. Crop response to carbon dioxide doubling: a literature survey. *Agric. For. Meteorol.* 38:127–45
29. Curtis, P. D., Drake, B. G., Whigham, D. F. 1989. Nitrogen and carbon dynamics in $C_3$ and $C_4$ estuarine marsh plants grown under elevated $CO_2$ in situ. *Oecologia* 78:297–301
30. Curtis, P. D., Drake, B. G., Leadley, P. W., Arp, W., Whigham, D. 1989. Growth and senescence of plant communities exposed to elevated $CO_2$ concentrations on an estuarine marsh. *Oecologia* 78:20–26
31. Dahlman, R. C., Strain, B. R., Rogers, H. H. 1985. Research on the response of vegetation to elevated atmospheric carbon dioxide. *J. Environ. Qual.* 14:1–8
32. Davis, M. B. 1989. Insights from paleoecology on global change. *Bull. Ecol. Soc. Am.* 70:222–28
33. Davis, M. B. 1989. Lags in vegetation response to greenhouse warming. *Climatic Change* 15:79–82
34. Davis, M. B., Botkin, D. B. 1985. Sensitivity of cool-temperature forests and their fossil pollen record to rapid temperature change. *Quaternary Res.* 23:327–40
35. Davis, M. B., Woods, K. D., Webb, S. L., Futyma, R. P. 1986. Dispersal versus climate: expansion of *Fagus* and *Tsuga* into the Upper Great Lakes region. *Vegetatio* 67:93–103
36. DeLucia, E. H., Sasek, T. W., Strain, B. R. 1985. Photosynthetic inhibition after long-term exposure to elevated levels of atmospheric carbon dioxide. *Photosyn. Res.* 7:175–84
37. Eamus, D., Jarvis, P. G. 1989. The direct effects of increase in the global atmospheric $CO_2$ concentration on natural and commercial temperature trees and forests *Adv. Ecol. Res.* 19:1–57
38. Emanuel, W. R., Shugart, H. H., Stevenson, M. P. 1985. Climate change and the broadscale distribution of terrestrial ecosystem complexes. *Climate Change* 7:29–43
39. Enoch, H. Z., Zieslin, N., Biran, Y., Halevy, A. H., Schwartz, M., Kessler, B., Shimshi, D. 1973. Principles of $CO_2$ nutrition research. *Acta Horticult.* 32:97–118
40. Fajer, E. D. 1989. The effects of enriched $CO_2$ atmospheres on plant-insect herbivore interactions: growth responses of larvae of the specialist butterfly, Junonia coenia (Lepidoptera: Nymphalidae). *Oecologia* 81:514–20
41. Fajer, E. D., Bowers, M. D., Bazzaz, F. A. 1989. The effects of enriched carbon dioxide atmospheres on plant-insect herbivore interactions. *Science* 243: 1198–1200
42. Fajer, E. D., Bowers, M. D., Bazzaz, F. A. 1990. Enriched $CO_2$ atmospheres and the growth of the buckeye butterfly, *Junonia coenia*. Ecology. In press
43. Fetcher, N., Jaeger, C. H., Strain, B. R., Sionit, N. 1988. Long-term elevation of atmospheric $CO_2$ concentration and the carbon exchange rates of saplings of *Pinus taeda* L. and *Liquidambar styraciflua* L. *Tree Physiol.* 4:255–62
44. Gale, J. 1982. Uses of brackish and solar desalinated water in closed system agriculture. In *Biosaline Research: A Look to the Future*, ed. A. San Pietro, pp. 315–24. New York: Plenum
45. Gale, J., Zeroni, M. 1985. Cultivation of plants in brackish water in controlled environment agriculture. In *Salinity Tolerance in Plants-Strategies for crop Improvement.*, ed. R. C. Staples, G. Toenniessen, pp. 363–80. New York: Wiley Intersci.
46. Garbutt, K., Bazzaz, F. A. 1984. The effects of elevated $CO_2$ on plants. III. Flower, fruit and seed production and abortion. *New Phytol.* 98:433–46
47. Garbutt, K., William, W. E., Bazzaz, F. A. 1990. Analysis of the differential response of five annuals to elevated $CO_2$ during growth. *Ecology* 71: In press
48. Gates, D. M. 1985. Global biospheric response to increasing atmospheric carbon dioxide concentration. In *Direct Effects of Increasing Carbon Dioxide on*

*Vegetation,* ed. B. R. Strain, J. D. Cure, pp. 171–84. US Dep. Energy

49. Gifford, R. M. 1980. Carbon storage by the biosphere. In *Carbon Dioxide and Climate: Australian Research,* ed. G. I. Pearman, pp. 167–81. Austr. Acad. Sci. Canberra City, Australia
50. Goudriaan, J., de Ruiter, H. E. 1983. Plant growth in response to $CO_2$ enrichment, at two levels of nitrogen and phosphorus supply. 1. Dry matter, leaf area and development. *Neth. J. Agric. Sci.* 31:157–69
51. Guinn, G., Mauney, J. R. 1980. Analysis of $CO_2$ exchange assumptions: feedback control. In *Predicting Photosynthesis for Ecosystem Models,* ed. J. D. Hesketh, J. W. Jones, Vol. II, pp. 1–16. Boca Raton, Fla: CRC
52. Haex, A. J. C. 1984. Part one report on the $CO_2$ problem. The Hague, The Netherlands: Health Council of the Netherlands
53. Hilbert, D. W., Prudhomme, T. I., Oechel, W. C. 1987. Response of tussock tundra to elevated carbon dioxide regimes: analysis of ecosystem $CO_2$ flux through nonlinear modeling. *Oecologia* 72:466–72
54. Hobbie, J. E., Cole, J., Dungan, J., Houghton, R. A., Peterson, B. J. 1984. Role of biota in global $CO_2$ balance: the controversy. *BioScience* 34:492–98
55. Houghton, R. A. 1988. The global carbon cycle (letter to the editor). *Science* 241:1736
56. Houghton, R. A., Hobbie, J. E., Melillo, J. M., Moore, B., Peterson, B. J., Shaver, G. R., Woodwell, G. M. 1983. changes in the carbon content of terrestiral biota and soils between 1860 and 1980: A net release of $CO_2$ to the atmosphere. *Ecol. Monogr.* 53:235–62
57. Idso, S. B., Kimball, B. A., Anderson, M. G., Mauney, J. R. 1987. Effects of atmospheric $CO_2$ enrichment of plant growth: the interactive role of air temperature. *Agric. Ecosys. Environ.* 20:1–10
58. Jaeger, J. 1988. Developing policies for responding to climate change. WMO/TD-No. 225, Stockholm: World Climate Prog. Impact Stud.
59. Johnson, R. H., Lincoln, D. E. 1990. Effect of $CO_2$ on leaf chemistry, leaf nitrogen and growth responses of *Artemisia tridentata*. Oecologia. In press
60. Karr, R. A. 1989. Greenhouse skeptic out of the cold. *Science* 246:1118–19
61. Keeling, C. D. 1986. Atmospheric $CO_2$ concentrations. Mauna Loa Observatory, Hawaii 1958–1986. NDP-001/R1. Carbon dioxide information analysis center. Oak Ridge, Tenn. Oak Ridge Natl. Lab.
62. Kimball, B. A. 1986. $CO_2$ stimulation of growth and yield under environmental constraints. In *Carbon Dioxide Enrichment of Greenhouse Crops,* Vol. II, *Physiology, Yield and Economics,* ed. H. Z. Enoch, B. A. Kimball, pp. 53–67. Boca Raton, Fla: CRC
63. Koch, K. E., Jones, P. H., Avigne, W. T., Allen, L. H. 1986. Growth, dry matter partitioning, and diurnal activities of RuBP carboxylase in citrus seedlings maintained at two levels of $CO_2$. *Physiologia Plantarum* 67, 477–84
64. Kramer, P. J. 1981. Carbon dioxide concentration, photosynthesis, and dry matter production. *BioScience* 31:29–33
65. Kramer, P. J., Sionit, N. 1987. Effects of increasing carbon dioxide concentration on the physiology and growth of forest trees. In *The Greenhouse Effect, Climate Change, and U.S. Forests,* ed. W. L. Shanas, John S. Hoftman, pp. 219–46. Conservation Found.
66. LaMarche, V. J., Graybill, D. A., Fritts, H. C., Rose, M. R. 1984. Increasing atmospheric carbon dioxide: Tree ring evidence for growth enhancement in natural vegetation. *Science* 225:1019–21
67. Lamborg, M. R., Hardy, R. W. F., Paul, E. A. 1983. Microbial effects. In *$CO_2$ and Plants: The Response of Plants to Rising Levels of Atmospheric $CO_2$,* ed. E. R. Lemon, pp. 131–76. Boulder, Colo: Westview
68. Larigauderie A., Hilbert, D. W., Oechel, W. C. 1988. Effect of $CO_2$ enrichment and nitrogen availability on resource acquisition and resource allocation in a grass, *Bromus mollis. Oecologia* 77:544–49
69. Lemon, E. R. 1983. $CO_2$ and plants. AAAS Selected Symposium 84. Boulder, Colo: Westview
70. Lincoln, D. E., Couvet, D. 1989. The effect of carbon supply on allocation to allelochemicals and caterpillar consumption of peppermint. *Oecologia* 78:112–14
71. Lincoln, D. E., Sionit, N., Strain, B. R. 1984. Growth and feeding response of *Pseudoplusia includens (Lepidoptera: Noctuidae)* to host plants grown in controlled carbon dioxide atmospheres. *Environ. Entomol.* 13:1527–30
72. Lincoln, D. E., Couvet, D., Sionit, N. 1986. Response of an insect herbivore to host plants grown in carbon dioxide enriched atmospheres. *Oecologia* 69:556–60
73. Lugo, A. 1983. Influence of green

plants on the world carbon budget. In *Alternative Energy Sources* V. *Part E: Nuclear/Conservation/Environment*, ed. T. N. Veziroglu. Amsterdam, The Netherlands: Elsevier

74. Luxmoore, R. J. 1981. $CO_2$ and phytomass. *BioScience* 31:626
75. Luxmoore, R. J., O'Neill, E. G., Ells, J. M., Rogers, H. H. 1986. Nutrient-uptake and growth responses of Virginia pine to elevated atmospheric $CO_2$. *J. Environ. Qual.* 15:244–51
76. Mattson, W. T. 1980. Herbivory in relation to plant nitrogen content. *Annu. Rev. Ecol. Syst.* 11:119–61
77. Melillo, J. M., Aber, J. D., Muratore, J. F. 1982. Nitrogen and lignin control of hardwood leaf litter decomposition dynamics. *Ecology* 63:621–26
78. Mott, K. A. 1988. Do stomata respond to $CO_2$ concentrations other than intercellular? *Plant Physiol.* 86:200–203
79. Mousseau, M., Enoch, H. Z. 1989. Effect of doubling atmospheric $CO_2$ concentration on growth, dry matter distribution and $CO_2$ exchange of two-year-old sweet chestnut trees (*Castanea sativa* Mill.). *Oecologia*
80. National Academy of Science. 1988. Toward an understanding of global change. Washington, DC: NAS
81. Norby, R. J. 1987. Nodulation and nitrogenase activity in nitrogen-fixing woody plants stimulated by $CO_2$ enrichment of the atmosphere. *Physiol. Plantarum* 71:77–82
82. Norby, R. J., Luxmoore, R. J., O'Neill, E. G., Weller, D. G. 1984. *Plant responses to elevated atmospheric $CO_2$ with emphasis on belowground processes. Oak Ridge National Laboratory, Tenn. ORNL/TM-9426*
83. Norby, R. J., O'Neill, E. G., Luxmoore, R. J. 1986a. Effects of atmospheric $CO_2$ enrichment on the growth and mineral nutrition of *Quercus alba* seedlings in nutrient-poor soil. *Plant Physiol.* 82:83–89
84. Norby, R. J., Pastor, J., Melillo, J. M. 1986. Carbon-nitrogen interactions in $CO_2$–enriched white oak: physiological and long-term perspectives. *Tree Physiol.* 2:233–41
85. Norby, R. J., O'Neill, E. G., Hodd, W. G., Luxmoore, R. J. 1987. Carbon allocation, root exudation and mycorrhizal colonization. *Tree Physiol.* 3:203–10
86. Oberbauer, S. F., Strain, B. R., Fetcher, N. 1985. Effect of $CO_2$-enrichment on seedling, physiology and growth of two tropical tree species. *Physiologia Plantarum* 65:352–56
87. Oberbauer, S. F., Sionit N., Hastings S. J., Oechel W. C. 1986. Effects of $CO_2$ enrichment and nutrition on growth, photosynthesis, and nutrient concentration of Alaskan tundra plant species. *Can. J. Bot.* 64:2993–99
88. Oechel, W. C., Strain, B. R. 1985. Native species responses to increased carbon dioxide concentration. In *Direct Effects of Increasing Carbon Dioxide on Vegetation*, ed. B. R. Strain and J. D. Cure, pp. 117–54. U.S. Dep. Energy, NTIS. Springfield, Virginia
89. Osbrink, W. L. A., Trumble, J. T., Wagner, R. E. 1987. Host suitability of *Phaseolus lunata* for *Trichoplusia ni* (Lepidoptera: Noctuidae) in controlled atmospheres. *Environ. Entomol.* 16:639–44
90. Overdieck, D., Bossemeyers, D., Lieth, H. 1984. Long-term effects of an increased $CO_2$ concentration level on terrestrial plants in model-ecosystems. I. phytomass production and competition of *Trifolium repens* L. and *Lolium perenne L. Progress Biometeorol.* 3:344–52
91. Pastor, J., Post, W. M. 1988. Responses of northern forests to $CO_2$-induced climate change. *Nature* 334:55–58
92. Patterson, D. T., Flint, E. P. 1980. Potential effects of global atmospheric $CO_2$ enrichment on the growth and competitiveness of $C_3$ and $C_4$ weed and crop. *Weed Sci.* 28:71–75
93. Patterson, D. T., Flint, E. P. 1982. Interacting effects of $CO_2$ and nutrient concentration. *Weed Sci.* 30:389–94
94. Patterson, D. T., Flint, E. P., Beyers, J. L. 1984. Effects of $CO_2$ enrichment on competition between a $C_4$ weed and a $C_3$ crop. *Weed Sci.* 32:101–5
95. Pearcy, R. W., Bjorkman, O. 1983. Physiological effects. In *$CO_2$ and plants: The Response of Plants to Rising Levels of Atmospheric Carbon Dioxide*, ed. E. R. Lemon, pp. 65–106. Boulder, Colo: Westview
96. Peterson, K. M., Billings, W. D., Reynolds, D. N. 1984. Influence of water table and atmospheric $CO_2$ concentration on the carbon balance of arctic tundra. *Arctic Alpine Res.* 16:331–55
97. Potvin C. 1985. Amelioration of chilling effects by $CO_2$ enrichment. *Physiol. Veg.* 23:345–52
98. Price, P. W., Bouton, C. E., Gross, P., McPheron, B. A., Thompson, J. N., Weis, A. E. 1980. Interaction among three trophic levels: influence of plants on interactions between insect herbivores and natural enemies. *Annu. Rev. Ecol. Sys.* 11:41–65
99. Prudhomme, T. I., Oechel, W. C.,

Hastings, S. J., Lawrence, W. T. 1984. Net ecosystem gas exchange at ambient and elevated carbon dioxide concentrations in tussock tundra at Toolik Lake, Alaska: an evaluation of methods and initial results. In *The Potential Effects of Carbon Dioxide-Induced Climatic Changes in Alaska: Proceedings of a Conference School of Agricultural and Land Resources Management,* ed. J. H. McBeath, Univ. Alaska, Fairbanks, Alaska

100. Reekie, E. G., Bazzaz, F. A. 1989. Competition and patterns of resource use among seedlings of five tropical trees grown at ambient and elevated $CO_2$. *Oecologia* 79:212–22
101. Riechers, G. D., Strain, B. R. 1988. Growth of blue grama (*Bouteloua gracilis*) in response to atmospheric carbon dioxide enrichment. *Can. J. Bot.* 66: 1570–73
102. Rogers, H. H., Thomas, J. F., Bingham, G. M. 1983. Response of agronomic and forest species to elevated atmospheric carbon dioxide. *Science* 220:428–30
103. Rosenberg, N. J. 1981. The increasing $CO_2$ concentration in the atmosphere and its implication on agricultural productivity. I. Effects on photosynthesis, transportation and water use efficiency. *Climatic Change* 3:265–79
104. Rotty, R. M., Marland, G. 1986. Fossil fuel consumption: recent amounts, patterns, and trends of $CO_2$. In *The Changing Carbon Cycle: A Global Analysis,* ed. J. R. Trabalka, D. E. Reichle. New York: Springer-Verlag
105. Sage, R. F., Pearcy, R. W. 1987. The nitrogen use efficiency of $C_3$ and $C_4$ plants. I. Leaf nitrogen, growth, and biomass partitioning in *Chenopodium album* (L.) and *Amaranthus retroflexus* (L.). *Plant Physiol.* 84:954–58
106. Sage, R. F., Sharkey, T. D., Seeman, J. R. 1989. The acclimation of photosynthesis to elevated $CO_2$ in five $C_3$ species. *Plant Physiol.* In press
107. Sasek, T. W., Strain, B. R. 1990. Implications of atmospheric $CO_2$ enrichment and climatic change for the geographical distribution of two introduced vines in the U.S.A. *Climatic Change* In press
108. Schneider, S. H. 1989. The greenhouse effect: science and policy. *Science* 243:771–81
109. Shugart, H. H., Emanuel, W. R. 1985. Carbon dioxide increase: the implications at the ecosystem level. *Plant, Cell Environ.* 8:381–86
110. Sionit, N., Patterson, D. T. 1984. Responses of $C_4$ grasses to atmospheric $CO_2$ enrichment. I. Effect of irradiance. *Oecologia* 65:30–34
111. Sionit, N., Kramer, P. J. 1986. Woody plants reactions to carbon dioxide enrichment. In *Carbon Dioxide Enrichment of Greenhouse Crops,* eds. H. Z. Enoch, B. A. Kimball, Boca Raton, Fl., CRC
112. Sionit N., Hellmers H., Strain B. R. 1980. Growth and yield of wheat under $CO_2$ enrichment and water stress. *Crop Sci.* 20:687–90
113. Sionit, N., Hellmers, H., Strain, B. R. 1982. Interaction of atmospheric $CO_2$ enrichment and irradiance on plant growth. *Agron. J.* 74:721–25
114. Sionit, N., Strain, B. R., Hellmers, H., Riechers, G. H., Jaeger, C. H. 1985. Long-term atmospheric $CO_2$ enrichment effects and the growth and development of *Liquidambar styraciflua* and *Pinus taeda* seedlings. *Can. J. For. Res.* 15:468–71
115. Smith, S. P., Strain, B. R., Sharkey, T. D. 1987. Effects of $CO_2$ enrichment on four Great Basin grasses. *Functional Ecol.* 1:139–43
116. Solomon, A. M., Tharp, M. L., West, D. C., Taylor, G. E., Webb, J. M., Trimble, J. C. 1984. *Response of unmanaged forests to $CO_2$-induced climate change: available information, initial tests and data requirements. US Dep. Energy Rep. No. DOE TR009.* US Dep. Energy, Washington, DC
117. St. Omar, L., Horvath, S. M. 1983. Elevated carbon dioxide concentrations and whole plant senescence. *Ecology* 64:1311–13
118. Strain, B. R. 1985. Physiological and ecological controls on carbon sequestering in ecosystems. *Biogeochemistry* 1:219–32
119. Strain, B. R., Bazzaz, F. A. 1983. Terrestrial plant communities. In *$CO_2$ and Plants: The Response of Plants to Rising Levels of Atmospheric Carbon Dioxide,* ed. E. R. Lemon, pp. 177–222. Boulder, Colo: Westview
120. Strain, B. R., Cure, J. D. 1985. *Direct effects of increasing carbon dioxide on vegetation.* US Dep. Energy, NTIS, Springfield, Va
121. Tissue, D. T., Oechel, W. C. 1987. Response of *Eriophorum vaginatum* to elevated $CO_2$ and temperature in the Alaskan tussock tundra. *Ecology* 68:401–10
122. Tolbert, N. E., Zelitch, I. 1983. Carbon metabolism. In *$CO_2$ and Plants: The Response of Plants to Rising Levels of Atmospheric Carbon Dioxide,* ed. E. R.

Lemon, pp. 21–64. Boulder, Colo: Westview

123. Tolly, L. C., Strain, B. R. 1984a. Effects of $CO_2$ enrichment on growth of *Liquidambar styraciflua* and *Pinus taeda* seedlings under different irradiance levels. *Can. J. For. Res.* 14:343–50
124. Tolley, L. C., Strain, B. R. 1984b. Effects of atmospheric $CO_2$ enrichment and water stress on growth of *Liquidambar styraciflua* and *Pinus taeda* seedlings. *Can. J. Bot.* 62:2135–39
125. Tolley, L., Strain, B. R. 1985. Effects of $CO_2$ enrichment and water stress on gas exchange of *Liquidambar styraciflua* and *Pinus taeda* seedlings grown under different irradiance levels. *Oecologia* 65:166–72
126. Valle, R., Mishoe, J. W., Campbell, W. J., Jones, J. W., Allen, L. H., Jr. 1985. Photosynthetic response of 'Bragg' soybean leaves adapted to different $CO_2$ environments. *Crop. Sci.* 25:333–39
127. Waggoner, P. E. 1984. Agriculture and carbon dioxide. *Am. Sci.* 72:179–84
128. Walker, D. A. 1980. Regulation of starch synthesis in leaves-the role of orthophosphate. In *Physiological Aspects of Crop Productivity*. Proc. the 15th Int. Potash Inst., Bern, Switzerland
129. Williams, W. E., Garbutt, K., Bazzaz, F. A. 1988. The response of plants to elevated $CO_2$—V. Performance of an assemblage of serpentine grassland herbs. *Environ. Exp. Bot.* 28:123–30
130. Williams, W. E., Garbutt, K., Bazzaz, F. A., Vitousek, P. M. 1986. The response of plants to elevated $CO_2$ IV. Two deciduous-forest tree communities. *Oecologia* (Berlin) 69:454–59
131. Winner, W. E., Mooney, H. A. 1980. Responses of Hawaiian plants to volcanic sulfur dioxide: stomatal behavior and foliar injury. *Science* 210:789–91
132. Wittwer, S. H. 1983. Rising atmospheric $CO_2$ and crop productivity. *Hortscience* 18:667–73
133. Wong, S. C. 1979. Elevated atmospheric partial pressure of $CO_2$ and plant growth. *Oecologia* 44:68–74
134. Woodward, F. I. 1987. Stomatal numbers are sensitive to increases in $CO_2$ from pre-industrial levels. *Nature* 327:617–18
135. Woodward, F. I., Bazzaz, F. A. 1988. The responses of stomatal density of $CO_2$ partial pressure. *J. Exp. Bot.* 39:1771–81
136. Woodwell, G. M. 1988. The global carbon cycle (letter to the editor). *Science* 241:1736–37
137. Wray, S. M., Strain, B. R. 1986. Response of two old field perennials to interactions of $CO_2$ enrichment and drought stress. *Am. J. Bot.* 73:1486–91
138. Wray, S. M., Strain, B. R. 1987. Competition in old-field perennials under $CO_2$ enrichment. *Ecology* 68:1116–20
139. Wray, S. M., Strain, B. R. 1987. Interaction of age and competition under $CO_2$ enrichment. *Functional Ecol.* 1:145–49
140. Wright, R. D. 1974. Rising atmospheric $CO_2$ and photosynthesis of San Bernardino mountain plants. *Am. Midland Naturalist* 91:360–70
141. Wulff, R., Strain, B. R. 1982. Effects of carbon dioxide enrichment on growth and photosynthesis in *Desmodium paniculatum*. *Can. J. Bot.* 60:1086–91
142. Wulff, R., Miller-Alexander, H. 1985. Intraspecific variation in the response to $CO_2$ enrichment in seeds and seedlings of *Plantago lanceolata* L. *Oecologia* 66:458–60
143. Zangerl, A. R., Bazzaz, F. A. 1984. The response of plants to elevated $CO_2$. II. Competitive interactions among annual plants under varying light and nutrients. *Oecologia* 62:412–17

*Annu. Rev. Ecol. Syst. 1990. 21:197–220*


# DNA SYSTEMATICS AND EVOLUTION OF PRIMATES

*Michael M. Miyamoto*

Department of Zoology, University of Florida, Gainesville, Florida 32611

*Morris Goodman*

Department of Anatomy and Cell Biology, Wayne State University School of Medicine, Detroit, Michigan 48201

KEY WORDS: primates, DNA sequences, β-globin cluster, phylogeny, evolution

## INTRODUCTION

Modern humans *(Homo sapiens)* are one of the most extensively studied species in all areas of biology (81). A direct result of our continuing fascination with ourselves is that a tremendous amount of biological information has accumulated about our species. This trend is most obvious in the large number of DNA sequences that have been determined for humans (37, 73). As the human genome project moves ahead (8, 90, 165), the number of these DNA sequences should increase exponentially.

Comparative molecular biologists have exploited this ever-increasing data base by obtaining comparable sequences for human's closest living relatives (14, 31, 51, 57, 67, 74, 89, 105, 116, 149, 156, 161, 162). Consequently, large numbers of DNA sequences have also been collected for many other species of Primates (Class Mammalia, infraclass Eutheria), making them one of the best known groups of higher organisms (37). Analysis of the rich data base for this order has already taught us much about the evolution of DNA sequences, as well as about the phylogenetic history of humans and other primates (2, 3, 12, 14, 49, 86, 87, 95, 99, 122, 156, 160).



0066-4162/90/1120-0197$02.00

The mammalian domain of nuclear DNA encoding the $\epsilon$-, $\gamma$-, $\eta$-, $\delta$-, and $\beta$-globin genes and their flanking and intergenic sequences is usually referred to as the $\beta$-globin gene cluster (16, 21, 30, 52). This domain in mammals ranges in length from 40 to 70 kilobase pairs (kbp) of coding and noncoding genomic DNA. Presently, about 350 kbp of comparative sequence data exist on different mammalian $\beta$-globin gene clusters, including the complete determinations for human, galago, mouse, and rabbit (21, 102, 136, 153; unpublished data of D. Tagle et al). Of these 350 kbp, nearly half of the sequences have been determined in the last three years. Much of this sequence information comes from primates (51, 89).

The number of DNA sequences from the $\beta$-globin gene cluster of primates can only be expected to increase in the near future. Consequently, it is important to summarize the contributions made to the fields of DNA systematics and evolution by the many recent studies of this system (38, 39, 48, 49, 51, 52, 57, 89, 100). In our review, we first describe patterns of variation in the organization and expression of primate $\beta$-globin clusters. Next, we discuss the contributions of these investigations to a better understanding of primate phylogeny, at both the higher and the lower taxonomic levels. Finally, we review what has been learned about the evolution of DNA sequences from these comparisons of the primate $\beta$-globin system.

## BACKGROUND

### *Order Primates*

Primates constitute one of 18 or so extant orders of eutherian (placental) mammals (Table 1). The order encompasses about 170 extant species, divided into approximately 51 genera and 11 families (Table 1) (77, 157, 163). Previous studies of molecular, morphological, and other information have clearly established the monophyly of hominoids [humans, great apes (chimpanzees, gorillas, and orangutans), and gibbons]; of catarrhines (hominoids and Old World monkeys); of anthropoids [catarrhines and platyrrhines (New World monkeys and marmosets)]; of strepsirhines (lemurs and lorisoids); and of Primates as a whole (2, 10, 29, 41, 43, 51, 57, 71, 89, 107, 117, 130, 151, 152). At and above the family level, questions persist about the relationships of tarsiers to anthropoids and strepsirhines (1, 5, 11, 22, 28, 72, 117, 127, 128, 140, 152), of dwarf lemurs to lorisoids and other lemurs (41, 64, 130, 132, 141, 153), and of humans to great apes (2, 3, 47, 75). Tarsiers (infra order or semisuborder Tarsiiformes) have been traditionally united with strepsirhines, even though they may be more closely related to anthropoids (references, as above). The monophyly of Haplorhini is supported by the latter arrangement, whereas the former is consistent with a monophyletic Prosimii. Lemurs and dwarf lemurs (families Lemuridae and

Cheirogalidae, respectively) have been grouped together in the superfamily Lemuroidea by some studies. However, others have proposed that dwarf lemurs are more closely related to lorisoids than to other lemurs. The most contentious questions about primate evolution focus on the relationships of humans and great apes (genera *Pan, Gorilla,* and *Pongo*). Despite considerable effort, our closest living relative(s) has yet to be clearly identified.

## *Primate β-Globin Clusters*

Hemoglobin, whose function is transport of blood gases, is a tetrameric protein, consisting of two $\alpha$-type chains ($\zeta$ or $\alpha$) and two $\beta$-type polypeptides ($\epsilon$, $\gamma$, $\eta$, $\delta$, or $\beta$) (16, 30). The $\alpha$- and $\beta$-type chains are encoded by different gene clusters, each consisting of multiple tandemly arranged loci (21, 33, 101). Both families are composed of functional genes for embryonic and adult hemoglobin ($\zeta_2\epsilon_2$, $\alpha_2\epsilon_2$, and $\alpha_2\gamma_2$ versus $\alpha_2\delta_2$ and $\alpha_2\beta_2$, respectively), as well as one to several pseudogenes (silent loci). Comparative studies of globin protein and DNA sequences have confirmed that the $\alpha$-type and $\beta$-type hemoglobins, along with myoglobin, are all part of the same superfamily (23, 45, 49, 52, 53, 55, 56). The evolution of primate globins can be traced back to an original duplication of an ancestral gene in the stem species of gnathostomes. This duplication about 500 million years ago (MA) gave rise to myoglobin, with the duplicate counterpart leading to hemoglobin. About 450 MA in the same phyletic line, a second duplication in the ancestral hemoglobin gene resulted in its separate $\alpha$- and $\beta$-types. Soon after, the two were separated by a translocation and are now located on different chromosomes. Subsequently, additional duplications have led to the multiple, tandemly linked genes of both the $\alpha$- and $\beta$-globin families (23, 45, 49, 52, 53, 56). This picture of gnathostome myoglobin and hemoglobin evolution is well established, except for the timing of the original duplication (52, 56, 82, 83).

As deduced from comparisons of marsupials and different eutherian orders, the ancestral placental mammal condition (65–85 MA) most likely consisted of five tandemly duplicated genes linked 5' to 3': $\epsilon$-$\gamma$-$\eta$-$\delta$-$\beta$ (9, 52, 59, 61, 63, 70, 135, 158). These genes can be traced back to two separate $\beta$-related progenitors, which themselves were the result of an older tandem duplication (150–200 MA) in the stem species of marsupials and placental mammals (23, 30, 45, 52, 85). The two progenitors were most likely different in terms of their ontogenetic expression (85, 153). One of them (that leading to $\epsilon$, $\gamma$, and $\eta$) was embryonically expressed, whereas expression of the other (that ancestral to $\delta$ and $\beta$) was limited to adults.

The ancestral eutherian arrangement of 5'-$\epsilon$-$\gamma$-$\eta$-$\delta$-$\beta$-3' has been retained by tarsiers (Figure 1A) (88). In the brown lemur, a deletion between the 3'-ends of the second exons of the $\eta$- and $\delta$-loci has resulted in a unique hybrid $\eta\delta$ gene for this species (9, 52, 63, 64, 80). Catarrhines differ from the

**Table 1** (A) Classifications of living primates, as proposed from morphological and paleontological data (41, 151). These classifications are consistent with most others in the literature, with some notable exceptions (see text). (B) A molecular classification of primates (57, 89)

| A. Szalay & Delson (151) | Fleagle (41) | B. Goodman et al (57), Koop et al (89) |
|---|---|---|
| Order Primates | Order Primates | Order Primates |
| Suborder Strepsirhini | Suborder Prosimii | Suborder Strepsirhini |
| Infraorder Lemuriformes | Infraorder Lemuriformes | Superfamily Lemuroidea [lemurs |
| Superfamily Lemuroidea | Superfamily Lemuroidea | (brown lemur) and dwarf lemurs] |
| Family Lemuridae | Family Daubentoniidae | Superfamily Lorisoidea (galago |
| Superfamily Indrioidea | Family Indriidae | and lorises) |
| Family Indriidae | Family Lemuridae | Suborder Haplorhini |
| Family Daubentoniidae | Family Lepilemuridae | Semisuborder Tarsiiformes |
| Superfamily Lorisoidea | Superfamily Lorisoidea | (tarsiers) |
| Family Cheirogaleidae | Family Cheirogaleidae | Semisuborder Anthropoidea |
| Family Lorisidae | Family Galagidae | Infraorder Platyrrhini |
| Suborder Haplorhini | Family Lorisidae | [New World |
| Infraorder Tarsiiformes | Infraorder Tarsiiformes | monkeys (spider monkey and owl |
| Family Tarsiidae | Family Tarsiidae | monkey) and marmosets] |
| Infraorder Platyrrhini | Suborder Anthropoidea | Infraorder Catarrhini |
| Family Cebidae | Infraorder Platyrrhini | |
| Family Atelidae | Superfamily Ceboidea | Superfamily Cercopithecoidea [Old |
| Infraorder Catarrhini | Family Atelidae | World monkeys (baboons, maca- |
| Superfamily Cercopithecoidea | Family Callitrichidae | ques, and rhesus monkey)] |
| | Family Cebidae | Superfamily Hominoidea |
| Family Cercopithecidae | Infraorder Catarrhini | Family Hominidae |
| Superfamily Hominoidea | Superfamily | Subfamily Hylobatinae |
| Family Hominidae | Cercopithecoidea | (gibbons) |
| Subfamily Hylobatinae | Family Cercopithecidae | Subfamily Homininae |
| | Superfamily Hominoidea | Tribe Pongini (orangutans) |
| Subfamily Ponginae | Family Hylobatidae | Tribe Hominini |
| Tribe Pongini [*Pan* (= *Pan* | Family Pongidae (*Pan*, | Subtribe Gorillina |
| and *Gorilla*), *Pongo*] | *Gorilla*, and *Pongo*) | (gorillas) |
| | Family Hominidae | Subtribe Hominina (humans |
| Subfamily Homininae | *(Homo)* | and chimpanzees) |
| *(Homo)* | | |

ancestral condition by possessing two $\gamma$-loci, typically attributed to a tandem duplication 25–35 MA in their stem species (9, 38, 133, 137, 145–147). However, more recent studies of spider monkey (a platyrrhine) suggest that this species possesses the same duplication as catarrhines, but with a large deletion of exons 2 and 3 and intron 2 of $\gamma^1$ and much of the $\gamma^1$-$\gamma^2$ intergenic region (unpublished data of D. H. A. Fitch et al). If so, the duplication of primate $\gamma$-genes most likely occurred in the stem species of anthropoids and not of catarrhines.

The $\epsilon$-gene of all eutherian mammals is embryonically expressed, as is the $\gamma$-locus of galago, mouse, and rabbit (15, 32, 52, 60, 61, 70, 153, 158). The $\epsilon$- and $\gamma$-genes of these groups are therefore replaced at the beginning of the fetal stage by the adult genes ($\delta$ and/or $\beta$). This pattern of gene expression is the primitive condition for primates and other eutherian mammals. In con-

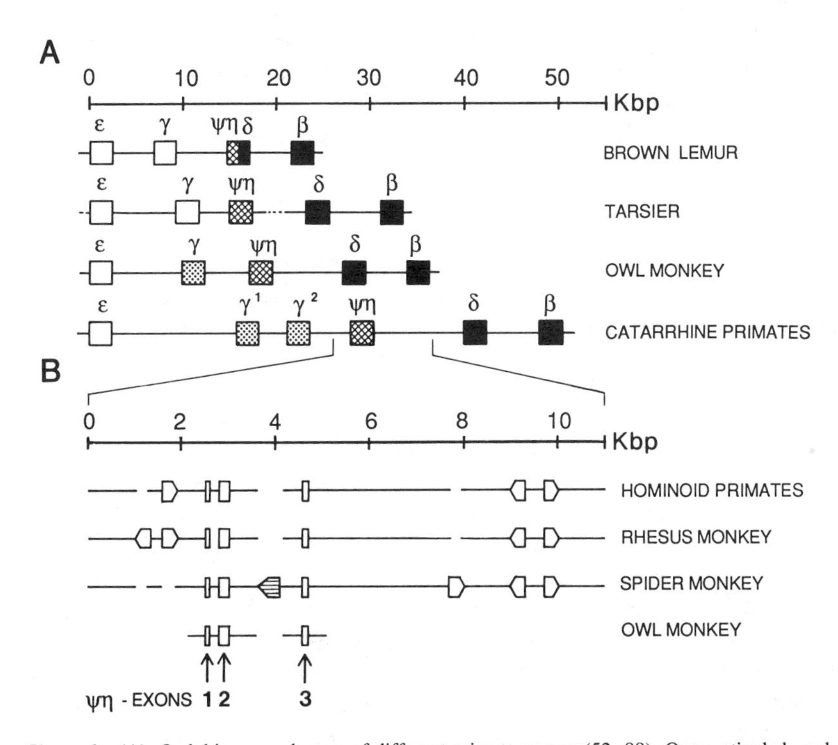


*Figure 1* (A) $\beta$-globin gene clusters of different primate groups (52, 88). Open, stippled, and closed boxes refer to genes with embryonic, fetal, and adult expression, respectively. Boxes with cross-hatching refer to the pseudogene $\psi\eta$. (B) Enlargement of the $\psi\eta$-globin gene and flanking regions of different anthropoid primates, illustrating the locations of Alu and truncated L1 elements (open and striped arrows, respectively) (51). The 5'- to 3'-direction of each element is indicated by the arrows. Only sequence from the $\psi\eta$-globin gene and immediate flanking regions is available for owl monkey (63).

trast, the $\gamma$-locus of anthropoids is delayed in its expression from the embryonic to fetal stage (16, 21, 30, 78, 101, 153). Correspondingly, expression of the $\delta$- and $\beta$-genes has been delayed until the adult stage. This ontogenetic pattern is therefore quite different from that of other placental mammals. Although thought to be embryonically expressed by early eutherians and still by extant artiodactyls, the $\eta$-locus has become a pseudogene in primates due to a series of silencing events in its stem species (i.e. a mutation in the ATG initiation codon) (19, 39, 52, 63, 86).

## DNA SYSTEMATICS

### *Phylogenetic Patterns*

A clearer picture of primate evolutionary relationships has emerged from phylogenetic analyses of $\beta$-globin sequences (39, 48, 51, 57, 89). In most cases, the maximum parsimony method of tree construction has been employed (25, 45, 50, 56). This procedure does not assume that rates of molecular evolution are constant, but only that sequence matches result more often from shared common ancestry than from parallel and back mutations. The most-parsimonious phylogeny, being closest to the original information, is retained as the best estimate of evolutionary history.

Two types of homologous comparisons are possible, when analyzing DNA sequences (40, 50, 111, 112). Paralogous comparisons involve DNA sequences that are the products of gene duplication. Such sequences are useful for reconstructing the history of genes within a family. In contrast, different orthologous sequences are those that are the result of speciation. Such sequences are important for reconstructing the sequence of cladogenesis within a group.

The evolutionary history of duplications in the globin superfamily has been based on parsimony analyses of both paralogous and orthologous sequences (7, 23, 45, 49, 50, 52, 53, 55, 56, 85). However, paralogous comparisons must be avoided when evaluating taxonomic relationships, because only orthologous sequences can provide insights into species phylogenies (40, 50, 111). Unfortunately, gene conversion has affected several $\beta$-globin genes, thereby making it difficult to isolate paralogous from orthologous sequences (38, 49, 88, 133, 145–147, 150). With detailed analysis, it has been possible to identify gene conversions in primates, using approaches to be described later. For now, we wish to emphasize that orthologues are, in most cases, easily identified in the $\beta$-globin cluster and that the following phylogenetic studies of primates are therefore based only on orthologous sequences.

Parsimony studies of primate $\beta$-globin orthologues clearly support an overall phylogenetic pattern similar to that established from morphological, paleontological, and other molecular information (Figure 2A) (24, 39, 47, 48,

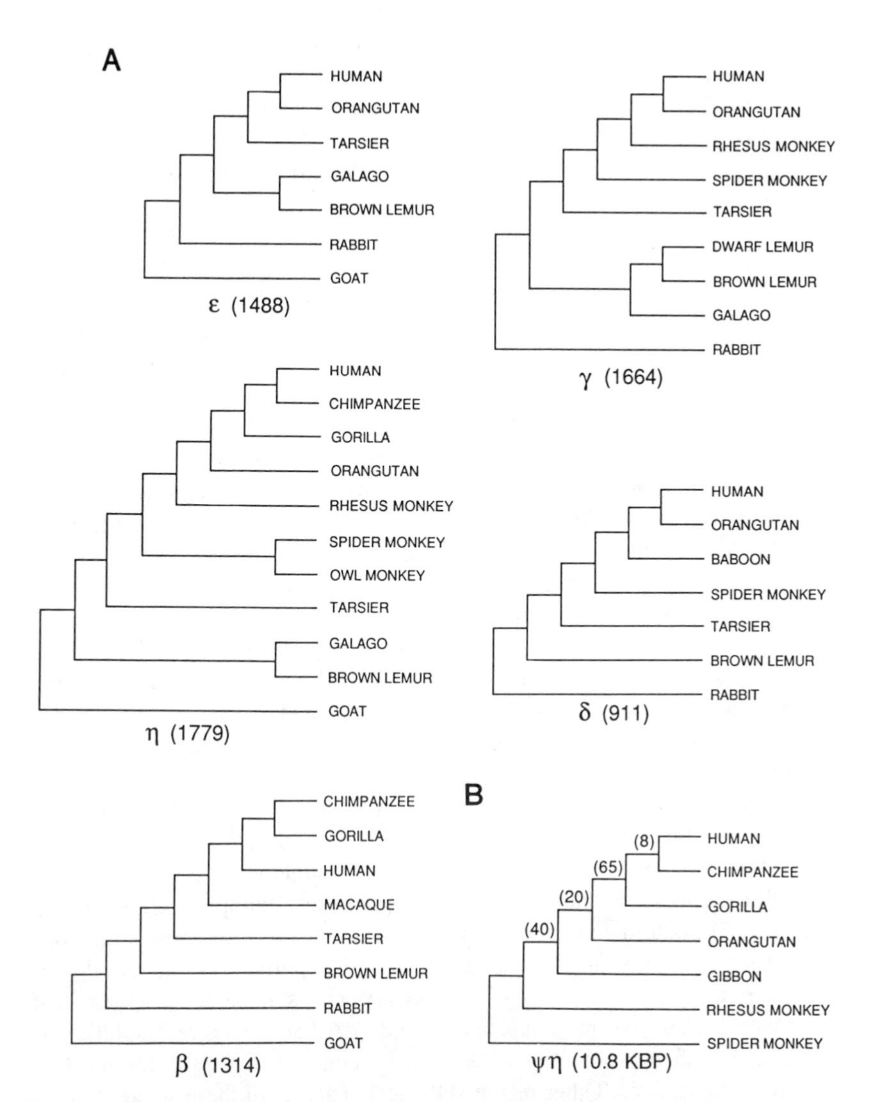


*Figure 2* (A) DNA phylogenies of primates, as determined from orthologous sequences of $\beta$-globin related genes (89). Tree scores for these most-parsimonious solutions are given in parentheses. A chimpanzee/gorilla arrangement is supported by the $\beta$-globin gene sequences, but only by a single mutation over the human/chimpanzee alternative (89). (B) Most-parsimonious phylogeny of anthropoid primates, as determined from noncoding DNA sequences of the $\psi\eta$-globin gene and flanking regions (10.8 kbp in all) (51, 57). The number of extra mutations required to undo each clade is shown in parentheses.

51, 57, 86, 89, 100, 153). These investigations provide overwhelming evidence for the monophyly of hominoids, catarrhines, anthropoids, and strepsirhines. Such conclusions are based on phylogenies: (*a*) which are supported by DNA sequences from two or more different $\beta$-globin regions; and (*b*) which are shorter than other branching arrangements by at least 69 extra mutations.

Orthologous sequences of the $\gamma^1$-globin locus (primary donor of $\gamma$-gene conversions; see below) strongly support a close relationship between dwarf lemur and brown lemur (i.e. the monophyly of Lemuroidea) (89, 153). At least 38 extra mutations must be added to the score of the most-parsimonious phylogeny to break up this superfamily. The most-parsimonious results for tarsier unite this taxon with anthropoids and not with lorisoids or lemuroids (88, 89). At least 28 extra mutations are required to undo this relationship, which supports a monophyletic Haplorhini. Furthermore, all five $\beta$-globin genes are individually congruent with this result.

Parsimony analyses of noncoding DNA sequences from the $\psi\eta$-globin gene and flanking regions (10.8 kbp in total length) have convincingly supported the monophyly of hominoids, as well as of a human/great ape clade (Figure 2B) (39, 47, 48, 51, 57, 89, 100). These results have further confirmed that orangutan is the sister group of humans and African apes (57, 86, 89, 100, 108). At least 65 extra mutations must be added to the most-parsimonious score to undo the human/African ape grouping. Interestingly, no unique mutations support the possibility of a great ape clade or human/orangutan solution, which has been proposed from morphological evidence (39, 84, 89, 100, 126, 129, 131). Clearly, these sequences provide overwhelming support for a human/African ape arrangement.

The question of the closest living human relative(s) is resolved by these sequences in favor of a human/chimpanzee grouping (39, 51, 76, 89, 100, 108, 109, 168). This solution is defined by 11 unique mutations, whereas only three diagnostic changes apiece support the chimpanzee/gorilla and human/gorilla alternatives. A recent statistical analysis of these mutations indicates that the human/chimpanzee solution is significantly supported over the other two possibilities at $p \approx 0.03$ (168). This statistic assumes: (*a*) that informative changes are independent; and (*b*) that the correct solution is associated with a probability of $\geqslant 1/3$, in contrast to $\leqslant 1/3$ for the two incorrect alternatives. Other recent statistical analyses of these sequence data have reached a similar conclusion (39).

The importance of using longer stretches of orthologous DNA to address specific systematic questions is highlighted by the $\psi\eta$-globin sequences of humans and great apes. When only the upstream 7.1-kbp region of Miyamoto et al (109) is considered, a clearly significant result in favor of human/chimpanzee is not obtained (36). The same conclusion is reached for the contiguous downstream region of noncoding DNA (3.1 kbp) obtained by

Maeda et al (100). Thus, neither set of sequences provides a sufficient number of informative positions to resolve convincingly the relationships of humans and African apes. However, when combined, the human/chimpanzee arrangement is clearly significant according to the contiguous sequences (35, 36, 168). Thus, greater numbers of informative orthologous positions are important for the resolution of specific phylogenetic questions.

Despite this statistical support, the human/chimpanzee solution may still not be the true phylogeny for higher primates (75, 108, 109). Of paramount importance in this regard is the possibility of intraspecific polymorphism confounding these phylogenetic analyses of humans and great apes (99, 108, 110, 121). For these reasons, these most-parsimonious results should be compared to those of other data sets for patterns of congruence. Unfortunately, the large and diverse body of comparative data for humans and African apes does not provide clear support for either the human/chimpanzee or chimpanzee/gorilla solution (2, 3, 58, 75, 103). Indeed, different analyses of the same data set often disagree among themselves (e.g. 2, 3, 14, 67, 76). However, these data do reemphasize that little support exists for a human/gorilla arrangement (3, 58, 75, 84). At this time, it is best to conclude that human and African ape relationships have not been decisively resolved and that more DNA sequences from different unlinked regions and individuals are now needed.

DNA hybridization studies of single-copy genomic DNA have strongly supported a human/chimpanzee arrangement (18, 138, 139). However, these results have been seriously challenged for different reasons such as the proper choice of a distance metric and the use of correction factors (92, 93, 104, 118). Nevertheless, estimates of primate sequence divergence by DNA hybridization have been shown to correspond closely with those obtained from direct sequencing of noncoding $\beta$-globin DNA (Table 2) (47, 48, 57, 86, 89, 120). This remarkable agreement between these different methods suggests (*a*) that most single-copy DNA is noncoding, and (*b*) that at least some criticisms of DNA hybridization are not of great importance.

## *Classification*

A phylogenetic classification of the order Primates is shown in Table 1B, in a synthesis of several recent studies of their $\beta$-globin clusters (47, 51, 57, 89, 108). Primates is first split into the suborders Strepsirhini and Haplorhini. Two superfamilies of Strepsirhini are recognized: Lemuroidea (lemurs and dwarf lemurs) and Lorisoidea (galago and lorises). The suborder Haplorhini is divided into the semisuborders Tarsiiformes (tarsiers) and Anthropoidea. Two infraorders of Anthropoidea are represented: Platyrrhini (New World monkeys and marmosets) and Catarrhini (Old World monkeys, apes, and humans). This classification agrees in many respects with more traditional ones

**Table 2** Estimates of primate sequence divergence based on pairwise comparisons of noncoding $\beta$-globin DNA and cross-hybridizations of total single-copy genomic DNA (11, 57, 139)

| Pairwise comparison | Percentage of sequence divergence, noncoding $\beta$-globin DNA | Delta T5OH, DNA hybridization |
|---|---|---|
| Human/Chimpanzee | 1.7 | 1.6 |
| Human/Gorilla | 1.8 | 2.3 |
| Human/Orangutan | 3.3 | 3.6 |
| Human/Gibbon | 4.3 | 4.8 |
| Human/Rhesus Monkey | 7.0 | 6.9–7.3 |
| Human/*Ateles* (i.e. spider monkey) | 10.8 | 11.2 |
| Chimpanzee/Gorilla | 1.7 | 2.2 |
| Chimpanzee/Orangutan | 3.5 | 3.6 |
| Chimpanzee/Gibbon | 4.7 | 5.1 |
| Chimpanzee/Rhesus Monkey | 7.0 | 7.3 |
| Gorilla/Orangutan | 3.5 | 3.6 |
| Gorilla/Gibbon | 4.7 | 4.5 |
| Gorilla/Rhesus Monkey | 7.2 | 7.2 |
| Orangutan/Gibbon | 4.7 | 4.9 |
| Orangutan/Rhesus Monkey | 7.3 | 7.4 |
| Gibbon/Rhesus Monkey | 7.5 | 7.1 |
| Human/Tarsier | 24.6 | 25.4 |
| Human/Galago | 28.9 | 28.0 |
| Human/Lemur | 22.6 | 22.3 |
| *Ateles*/Tarsier | 27.1 | 26.4 |
| *Ateles*/Galago | 29.7 | 30.7 |
| *Ateles*/Lemur | 25.3 | 24.5 |
| Tarsier/Galago | 30.6 | 30.2 |
| Tarsier/Lemur | 25.4 | 25.8 |
| Galago/Lemur | 21.9 | 22.1 |

based on morphological and paleontological data (Table 1A) (41, 71, 72, 130, 151, 152, 157).

However, the traditional arrangement of humans and great apes has been extensively revised (41, 130, 141, 151, 157). Instead of being assigned to separate higher taxonomic groups, humans and the great apes have been placed in the same subfamily (Homininae) on the basis of their $\beta$-globin sequences (54, 57, 58, 108). The two subfamilies of Hominidae are therefore: (*a*) Hylobatinae (gibbons); and (*b*) Homininae (humans and great apes). As African apes and humans form a natural group, they have been assigned to a tribe (Hominini) separate from that for orangutan (Pongini). The tribe Hominini has been divided into two subtribes: Gorillina for gorilla and Hominina for the putative sister groups, human and chimpanzee (57, 89).

This classification of humans and apes has clearly opted to unite these groups at the lowest taxonomic levels due to their high degrees of genetic similarity (57, 108, 120). Gibbons differ from humans and great apes in their noncoding $\beta$-globin sequences by 4.3–5.0%, whereas orangutans vary from hominines by 3.3–3.5%, and humans from African apes by 1.7–1.8% (39, 57, 86, 89, 100, 108, 109; unpublished data of W. Bailey et al). When coding regions of the $\beta$-globin cluster are considered instead, sequence divergence becomes considerably less (87, 121, 147). Clearly, higher primates are genetically very similar, which supports their union in low taxonomic levels. More importantly, this classification is now consistent with their phylogenetic relationships (69). Instead of a paraphyletic Pongidae or Ponginae for great apes, the use of separate taxa for orangutan and humans/African apes ensures that only monophyletic groups are recognized. Of course, the weakest group in the classification remains the tribe Hominina, which is based on a putative sister-taxon relationship of humans and chimpanzees (39, 108, 109).

# MOLECULAR EVOLUTION

## *Nonrandom Patterns*

Investigations of noncoding DNA commonly reveal nonrandom patterns of change, which can be related to biases in the mutational process rather than to the effects of selection (44, 94, 98). For example, studies of the 5'-flanking DNA from the $\beta$-globin gene of human and chimpanzee have shown that CpG dinucleotides and simple repeats accumulate more mutations than other types of sequence (122). The former can be related to the tendency of cytosine in a CpG pair to be methylated, thereby rendering them more vulnerable to C to T transitions (100). The latter can be attributed to replication strand slippage and mispairing, which results in deletion/insertion events.

Phylogenetic analyses of intergenic $\beta$-globin DNA have also shown that patterns of base substitution can vary between complementary strands of a DNA sequence (100, 172). For the intergenic region between the $\psi\eta$- and $\delta$-genes of anthropoid primates, it has been demonstrated that the ratio of purine to pyrimidine versus pyrimidine to purine transversions differs significantly between the two complements. Such asymmetry has been attributed to replication differences of the leading and lagging strands (172).

Noncoding regions of $\beta$-globin introns have a base composition which is rich in A and T (62%) (86). In contrast, the exons of globin genes have a base composition which is quite different (45% A + T). As calculated from base substitution frequencies (154), the $\psi\eta$-locus may eventually reach an equilibrium base composition similar to that of other noncoding DNA. At present, its base composition (51% A + T) falls between that of coding and noncoding DNA, as might be expected of a relatively young pseudogene.

## *Evolutionary Rates*

Two types of molecular clocks (global and local) have been recognized (47, 86, 169, 170). Both are based on the assumption of constant rates of change at the nucleotide level. The existence of a global clock has been challenged, as significant rate differences have been documented across major taxonomic lines (12, 96, 171). Studies of primate noncoding $\beta$-globin DNA have revealed that rates of nucleotide change are also not constant at lower taxonomic levels (39, 46, 47, 51, 86, 89, 100). There have been marked decelerations in rates from lower primates to anthropoids, from anthropoids to hominoids, and among hominoids themselves. The fastest rates are shown by lorisoids, tarsier, and the stem anthropoids, intermediate values by platyrrhines and Old World monkeys, and the slowest by hominoids. Except for stem anthropoids, values for primates are consistently less than the average neutral rate for other eutherian mammals, and sometimes by a factor of five or so (68, 86, 94). During the period of strepsirhine descent, the lorisoid rate was twice as fast as that for lemuroids (11, 89).

Even among very closely related species (e.g. hominoids), rate differences have been detected (19, 47, 67, 76, 86, 95, 109). The rate of nucleotide evolution in humans appears to be slower than that for other higher primates. This conclusion has been supported by a series of relative rate tests, using both nuclear and mitochondrial DNA sequences (95). The overall pattern of rate-deceleration leading to higher primates and ultimately to humans has been termed "the hominoid slowdown" (46). Clearly, even local molecular clocks must be used with caution (47, 86, 109). However, as cases of rate-nonconstancy are expected even under the best of circumstances due to the stochastic behavior of clocks, local molecular clocks can still be used to estimate times of divergence by compensating for species with unusual rates (95, 117, 155). In this manner, more precise estimates of divergence time can be calculated from DNA sequence data.

The $\psi\eta$-globin gene and flanking sequences of higher primates nicely illustrate these points. The 7.1-kbp upstream sequences of Miyamoto et al (109) show that human is evolving 30 to 40% slower than chimpanzee or gorilla. In contrast, the contiguous downstream region (3.1 kbp) of Maeda et al (100) suggests that the human rate is faster than those for African apes. Rates for chimpanzee and gorilla are about the same in both cases, thereby indicating that the apparent differences are attributable to human (76). These rate differences cannot be easily related to the interallelic recombination of downstream alleles reported by Maeda et al (99), because this exchange in the human lineage involved only intraspecific DNA. Furthermore, both downstream human alleles studied by them are consistent with a faster rate for human than chimpanzee or gorilla, even though only one of the two underwent recombination (100). Thus, it remains unclear why the evolutionary rate for humans varies in this case (76, 100). Nevertheless, special clock calculations can be employed to compensate for these unusual rates (95, 117). From such calculations, the human/chimpanzee split is estimated to have occurred only 500,000 to 1,000,000 years after the separation of gorilla, perhaps as recently as three to five MA (65, 108, 109, 119, 170).

Several hypotheses have been proposed to explain the hominoid slowdown (12, 46, 89, 94–96). Most often, the slowdown is related to lengthening generation times, decreased germline replication rates, and/or improved DNA repair mechanisms in higher primates. The last possibility is inconsistent with the mitochondrial DNA evidence for slowdowns in humans and other hominines (67, 95), because of the absence of a well-developed DNA repair mechanism (if any) for this genome (13). Clearly, additional research is required to determine the importance of each mechanism, as well as of ancient bottlenecks and fluctuating population size (170).

### *Gene Conversions*

Paralogous loci of multigene families are commonly more similar to each other than to their orthologous counterparts, a pattern of change which has been termed "concerted evolution" (4, 164). Many times, only parts of different paralogous genes behave in this manner, as exemplified by the duplicate $\gamma$-genes of human (137, 144, 150). The $\gamma^1$- and $\gamma^2$-polypeptides of humans differ by only one amino acid (glycine versus alanine at position 136), despite the duplication of their loci at least 35 MA (9, 38, 49, 133, 137, 145–147; unpublished data of D. H. A. Fitch et al). DNA sequences of both human $\gamma$-loci from the same haplotype revealed no differences in their upstream regions (exons 1 and 2, intron 1, and the 5'-end of intron 2), whereas many nucleotide changes were detected downstream (3'-end of intron 2 and exon 3, where the glycine/alanine substitution has occurred). Gene conversion was invoked to explain why no sequence differences existed in the upstream gene region, with a simple sequence of TG repeats in intron 2 implicated as a potential "hot spot" for the recombination.

At least 22 gene conversions among catarrhine primates have been identified since these initial studies (38, 133, 145–147). Of these, 17 have been shown to be species-specific, such as that described above for humans. Of the other five, two have occurred in the stem species of humans and apes, one in the ancestor to humans and great apes, and two in the early hominines. In eight of twelve known cases, the donor sequence is postulated to be from the $\gamma^1$-gene (i.e. the direction is biased). As several of these conversions terminate at the TG-repetitive element of intron 2, as well as in certain pyrimidine-rich and purine-rich stretches in this region, these studies have strengthened the possibility that simple sequences serve as "hot spots" for the initiation or termination of conversion events (38, 146, 147).

Gene conversions have also influenced the evolution of primate $\delta$- and $\beta$-globin genes (52, 59, 61, 88, 106). The direction of these conversions is once again biased as the donor sequence is almost always from the $\beta$-locus. However, these patterns of gene conversion differ markedly from those of the $\gamma$-loci in terms of their frequency, location, and length of converted region (38, 88, 133, 145, 146). Conversions between the $\delta$- and $\beta$-loci occur less frequently, are more restricted to conserved regions, and are shorter in length than those of the $\gamma$-genes. Such variation is consistent with the following model of gene conversion (38, 88, 146, 147, 164). Soon after duplication, conversions can be quite common and extensive, due to the perfect sequence similarity of the duplicates. However, as sequence differences accumulate, conversions become restricted to locations with fewer nucleotide differences, thereby limiting them to shorter stretches and to coding regions. The model furthermore attempts to explain the biases in directionality on the basis of differential gene expression of related loci (38, 88, 146, 147). As the $\gamma^1$- and $\beta$-loci contribute more to the composition of fetal and adult hemoglobin than their counterparts, the model proposes that stabilizing selection has preserved their sequences more than those of the $\gamma^2$- and $\delta$-genes. More research is now needed to test these hypotheses (88).

Gene conversion events have been identified by comparing the parsimony patterns of individual variable positions, either site by site or in small intervals (38, 49, 88, 145–147). Adjacent variable sites, which show a similar pattern of closer relationship between paralogous sequences than orthologous ones, are taken as evidence for the existence of a common conversion event. Converted regions are thereby isolated from other sequences, which have experienced different conversions or none at all (e.g. they are orthologous) (150). A complementary way to identify conversions has been to search for regions with unusually high or low levels of divergence, relative to their neighboring sequences (38, 59, 61, 62, 88, 106, 133, 137, 144–147). The general point is that orthology must be tested and cannot always be assumed. In recent primate studies, both approaches have been employed to confirm orthology and to test for the existence of conversion events (38, 88, 145–147, 150).

## *SINEs and LINEs of Anthropoid Primates*

SINEs and LINEs refer respectively to short and long interspersed elements, which are two categories of dispersed highly repetitive DNA (17, 27, 124, 125, 142, 166). Both have contributed greatly to the overall organization of mammalian genomes, due to their high copy numbers (greater than $10^4$ to $10^5$ per genome). As their names imply, SINEs are typically less than 500 base pairs (bp) in length, whereas LINEs can be longer than five kbp. Both SINEs and LINEs, as reflected by their high copy numbers and dispersed patterns of distribution, are types of mobile elements, whose dispersion has been related to some type of RNA intermediate and retropositioning. Both types of interspersed elements occur in primate $\beta$-globin clusters, as represented by Alu and L1 sequences, respectively (Figure 1B) (17, 21, 39, 51, 87, 99, 100, 113). Study of these primate sequences has led to a much better understanding of their patterns of expansion and potential significance.

Almost 3–6% of the total human genome is composed of Alu sequences (124, 125, 166). Phylogenetic comparisons of several Alu elements from the $\alpha$- and $\beta$-globin clusters of different anthropoid primates have revealed that many copies in these regions are the result of a single ancient (50–60 MA) expansion from one to just a few progenitors in their early stem species (87, 167). Other studies have emphasized those Alu elements that are unique to one species or anthropoid group (39, 159). Such copies provide evidence for more gradual and recent retroposon activity (27, 79, 143, 167). Clearly, more research is needed to determine what proportion of all anthropoid Alu's are due to gradual recent expansion as opposed to single ancient events. In either case, Alu repeats have been shown to be evolving at rates characteristic of noncoding DNA (6, 27, 87, 123).

A unique 5'-truncated L1 element (447 bp in length) occurs in intron 2 of the $\psi\eta$-globin locus of spider monkey (39). Comparisons of this element and the L1 consensus of human indicate that both were released from selection some time ago, even though their progenitors appear to be active and under present or recent constraints (17, 42, 66, 91, 134).

An unusual full-length L1 element of approximately 6 kbp occurs in the intergenic region between the $\epsilon$- and $\gamma^1$-genes of human (21, 113). This repeat exhibits a novel 5'-end, a 300-bp deletion at its 3'-terminus, and a 1.8-kbp insertion representing the 3'-end of a second L1 element. The original insertion of the main repeat has been estimated to be older than the catarrhine/platyrrhine split, which was then followed by the addition of its 1.8-kbp truncated element in early catarrhines. The possibility has been raised before (21) that the insertion of these elements may have been important in the switch of anthropoid $\gamma$-genes from embryonic to fetal expression (see below).

Many studies have attempted to date the origin of different interspersed elements by some type of molecular clock approach (6, 27, 87, 113, 134, 167). However, unequal rates can be even more problematic in these cases, because of the relatively short lengths of the elements themselves (39). Furthermore, gene conversion, even if uncommon, could operate to homogenize the sequences, thereby leading to inaccurate estimates of divergence time (26, 100, 113, 123, 125, 166). Consequently, it has been proposed that a much better method to determine the timing of insertions is to compare the presence or absence of individual elements among groups with different divergence times (39, 87, 123). For example, the existence of a truncated L1 element in spider monkey, but not in the related owl monkey, places the time of insertion for this sequence sometime after their cladogenesis (20–30 MA) (20, 39, 114). This estimate could be narrowed down even further by examining other New World monkeys for the presence or absence of this element.

## *Ontogenetic Expression*

The 5' to 3' order of mammalian $\beta$-globin loci follows their ontogenetic pattern of gene expression (9, 21, 30, 32, 33, 52, 61, 70, 158). Embryonic genes of the mammalian $\beta$-globin cluster are located at the 5'-end ($\epsilon$, $\gamma$, and/or $\eta$), whereas fetal and adult loci are positioned 3' ($\delta$ and $\beta$). Although no longer expressed in primates, the pseudogene $\psi\eta$ has been postulated to fill a possible regulatory role, as an important spacer between the upstream embryonic and downstream adult genes (52). The $\beta$-globin genes of strepsirhines and tarsier are expressed ontogenetically, as described above (9, 15, 52, 88, 153). However, this pattern has been changed during the evolution of anthropoid primates, as expression of their $\gamma$- and $\delta$-/$\beta$-genes has been delayed to the start of the fetal and adult stages, respectively (16, 30, 78, 101). This shift may have been facilitated by the insertion of L1 elements between the $\epsilon$- and $\gamma$-genes of the stem anthropoids (21, 113).

By comparing upstream flanking regions of different species and genes, highly conserved sequences can be identified and then studied for their possible roles as *cis*-regulating elements (38, 48, 52, 85, 88, 146, 147, 153). As noncoding flanking regions are being compared, these highly conserved sequences can be easily picked out. This approach called "phylogenetic footprinting" has led to the discovery of several known and potential *cis*-regulating factors, which may have been important in the recruitment of the $\gamma$-gene for fetal expression (38, 153). For example, the 5'-flanking region of anthropoid $\gamma$-genes contains a 26-bp invariant sequence, which differs from that of galago by seven mutations. These mutations may have been involved in the switch of the anthropoid $\gamma$-gene from embryonic to fetal expression.

The ratio of nonsynonymous to synonymous mutations in the $\gamma$-globin and other $\beta$-related genes of eutherian mammals is often less than one, indicating that most substitutions do not lead to amino acid changes (23, 49, 52, 56, 85, 153). However, a markedly elevated ratio of 3.25 has been calculated for the period when the $\gamma$-gene was being recruited for anthropoid fetal expression (49, 153). This accelerated rate of amino acid–changing substitutions has been taken as evidence of adaptive change during the remodeling of $\gamma$-globin for delayed ontogenetic use. Positive selection is therefore postulated to have played a central role during this important period of $\gamma$-globin evolution (i.e. by reducing the affinity of fetal hemoglobin for 2,3-diphosphoglycerate, an inhibitor of oxygen-binding). Once incorporated, the adaptive changes were preserved by stabilizing selection, as reflected by the return of this ratio to values of one or less. This interpretation suggests that Darwinian selection has been much more important than neutral drift in shaping the early evolution of anthropoid $\gamma$-globin genes (23, 45, 49, 55, 56, 85, 153). In contrast, the neutralist model has proposed that such rate increases are the result of reduced stabilizing selection following gene duplication (82, 83). For now, the relative roles of natural selection versus neutral drift in the evolution of $\gamma$- and other $\beta$-related genes remain unresolved.

## CONCLUSIONS

No consensus exists as to which methods are best for reconstructing phylogenies from molecular data (34, 36). Often, the validity of different approaches has been evaluated according to their abilities to recover a "known" phylogeny, one that has been generated by a simulation study (97, 115, 148). The use of simulation can be justified by arguing that the true phylogeny will always be unknown for almost all organisms. However, major patterns of primate phylogeny have been well-corroborated by many different types of data (i.e. see Figure 2), and as such, can almost be regarded as "known." Such a well-established phylogeny can be used to test the reliability of different methods, as well as to study character and organismic evolution.

Clearly, many approaches used to study primate $\beta$-globin sequences are important to fields other than evolutionary and systematic biology. For example, phylogenetic footprinting is a powerful way to identify new sequences with otherwise unknown biological importance (38, 48, 52, 85, 88, 153). Furthermore, differences among species are expected to offer novel insights about the molecular mechanisms responsible for mutation and recombination as well as human diseases (16, 21, 30, 33, 38, 39, 48, 100, 101, 137). It is in such diverse ways that the importance of comparative molecular studies of primate $\beta$-globin clusters and other systems will be fully realized.

In short, phylogenetic and evolutionary studies of primate $\beta$-related genes and intergenic regions have demonstrated the power of using DNA sequence data to address important systematic questions. These robust phylogenies have then formed the basis for investigating the tempo and mode of evolution, as well as the mutational mechanisms underlying all change. In such ways, these investigations have contributed significantly to the ever-increasing importance of comparative molecular biology to systematics and evolution.

### ACKNOWLEDGMENTS

We thank D. H. A. Fitch, B. F. Koop, D. H. Sherman, D. A. Tagle, and M. R. Tennant for their useful comments and suggestions about the manuscript.

This research was supported by grants from the National Science Foundation (BSR-8717527 and BSR-8857264) to MMM and (BSR-8607202) to MG as well as by an award from the National Institutes of Health (HL 33940) to MG.

*Literature Cited*


1. Aiello, L. C. 1986. The relationships of the Tarsiiformes: A review of the case for Haplorhini. In *Major Topics in Primate and Human Evolution,* ed. B. Wood, L. Martin, P. Andrews, pp. 47–65. Cambridge: Cambridge Univ. Press
2. Andrews, P. 1986. Molecular evidence for catarrhine evolution. In *Major Topics in Primate and Human Evolution,* ed. B. A. Wood, L. Martin, P. Andrews, pp. 107–29. Cambridge: Cambridge Univ. Press
3. Andrews, P. 1987. Aspects of hominoid phylogeny. In *Molecules and Morphology in Evolution: Conflict or Compromise?,* ed. C. Patterson, pp. 23–53. Cambridge: Cambridge Univ. Press
4. Arnheim, N. 1983. Concerted evolution of multigene families. In *Evolution of Genes and Proteins,* ed. M. Nei, R. K. Koehn, pp. 38–61. Sunderland: Sinauer
5. Baba, M., Weiss, M. L., Goodman, M., Czelusniak, J. 1982. The case of tarsier hemoglobin. *Syst. Zool.* 31:156–65
6. Bains, W. 1986. The multiple origins of human Alu sequences. *J. Mol. Evol.* 23:189–99
7. Barnabas, J., Goodman, M., Moore, G. W. 1972. Descent of mammalian alpha globin chain sequences investigated by the maximum parsimony method. *J. Mol. Biol.* 69:249–78
8. Barnhart, B. J. 1989. The Department of Energy (DOE) human genome initiative. *Genomics* 5:657–60
9. Barrie, P. A., Jeffreys, A. J., Scott, A. F. 1981. Evolution of the $\beta$-globin gene cluster in man and the primates. *J. Mol. Biol.* 149:319–36
10. Beard, K. C., Dagosto, M., Geo, D. L., Godinot, M. 1988. Interrelationships among primate higher taxa. *Nature* 331:712–14
11. Bonner, T. I., Heinemann, R., Todaro, G. J. 1980. Evolution of DNA sequences has been retarded in Malagasy primates. *Nature* 286:420–23
12. Britten, R. J. 1986. Rates of DNA sequence evolution differ between taxonomic groups. *Science* 231:1393–98
13. Brown, W. M. 1983. Evolution of animal mitochondrial DNA. In *Evolution of Genes and Proteins,* ed. M. Nei, R. K. Koehn, pp. 62–88. Sunderland: Sinauer
14. Brown, W. M., Prager, E. M., Wang, A., Wilson, A. C. 1982. Mitochondrial DNA sequences of primates: Tempo and mode of evolution. *J. Mol. Evol.* 18:225–39
15. Buettner-Janusch, J., Buettner-Janusch, V., Coppenhaver, D. 1972. Properties of the hemoglobins of newborn and adult prosimians (Prosimii: Lemuriformes and Lorisiformes). *Folia Primat.* 17:177–92
16. Bunn, H. F., Forget, B. G. 1986. *Hemoglobin: Molecular, Genetic, and Clinical Aspects.* Philadelphia: Saunders
17. Burton, F. H., Loeb, D. D., Voliva, C. F., Martin, S. L., Edgell, M. H., et al. 1986. Conservation throughout Mammalia and extensive protein-encoding capacity of the highly repeated DNA long interspersed sequence one. *J. Mol. Biol.* 187:291–304
18. Caccone, A., Powell, J. R. 1989. DNA divergence among hominoids. *Evolution* 43:925–42
19. Chang, L.-Y. E., Slightom, J. L. 1984. Isolation and nucleotide sequence analysis of the $\beta$-type globin pseudogene from human, gorilla and chimpanzee. *J. Mol. Biol.* 180:767–84
20. Ciochon, R. L., Chiarelli, A. B., eds. 1981. *Evolutionary Biology of the New World Monkeys and Continental Drift.* New York: Plenum
21. Collins, F. S., Weissman, S. 1984. The molecular genetics of human hemoglobin. *Prog. Nucleic Acid Res. Mol. Biol.* 31:315–462
22. Cronin, J. E., Sarich, V. M. 1980. Tupaiid and Archonta phylogeny: The macromolecular evidence. In *Comparative Biology and Evolutionary Relationships of Tree Shrews,* ed. W. Luckett, pp. 293–312. New York: Plenum
23. Czelusniak, J., Goodman, M., Hewett-Emmett, D., Weiss, M. L., Venta, P. J., et al. 1982. Phylogenetic origins and adaptive evolution of avian and mammalian haemoglobin genes. *Nature* 298:297–300
24. Czelusniak, J., Goodman, M., Koop, B. F., Tagle, D. A., Shoshani, J., et al. 1990. Perspectives from amino acid and nucleotide sequences on cladistic relationships among higher taxa of Eutheria. In *Current Mammalogy,* ed. H. H. Genoways, pp. 541–67. New York: Plenum

25. Czelusniak, J., Goodman, M., Moncrief, N. D., Kehoe, S. M. 1990. Maximum parsimony approach to the construction of evolutionary trees from aligned homologous sequences. *Methods Enzymol.* 183:601–15
26. Daniels, G. R., Deininger, P. L. 1983. A second major class of Alu family repeated DNA sequences in a primate genome. *Nucleic Acids Res.* 11:7595–610
27. Deininger, P. L., Daniels, G. R. 1986. The recent evolution of mammalian repetitive DNA elements. *Trends Genet.* 2:76–80
28. DeJong, W. W., Goodman, M. 1988. Anthropoid affinities of *Tarsius* supported by lens αA-crystallin sequences. *J. Hum. Evol.* 17:575–82
29. Delson, E., Rosenberger, A. L. 1980. Phyletic perspectives in platyrrhine origins and anthropoid relationships. In *Evolutionary Biology of New World Monkeys and Continental Drift,* ed. R. L. Ciochon, A. B. Chiarelli, pp. 445–58. New York: Plenum
30. Dickerson, R. E., Geis, I. 1983. *Hemoglobin: Structure, Function, Evolution, and Pathology*. Menlo Park: Benjamin/Cummings
31. Djian, P., Green, H. 1989. The involucrin gene of the orangutan: Generation of the late region as an evolutionary trend in the hominoids. *Mol. Biol. Evol.* 6:469–77
32. Edgell, M. H., Hardies, S. C., Brown, B., Voliva, C., Hill, A., et al. 1983. Evolution of the mouse β globin complex locus. In *Evolution of Genes and Proteins,* ed. M. Nei, R. K. Koehn, pp. 1–13. Sunderland: Sinauer
33. Efstratiadis, A., Posakony, J. W., Maniatis, T., Lawn, R. M., O'Connell, C., et al. 1980. The structure and evolution of the human β-globin gene family. *Cell* 21:653–68
34. Felsenstein, J. 1982. Numerical methods for inferring evolutionary trees. *Q. Rev. Biol.* 57:379–404
35. Felsenstein, J. 1985. Confidence limits on phylogenies with a molecular clock. *Syst. Zool.* 34:152–61
36. Felsenstein, J. 1988. Phylogenies from molecular sequences: Inference and reliability. *Annu. Rev. Genet.* 22:521–65
37. Fickett, J. W., Burks, C. 1989. Development of a database for nucleotide sequences. In *Mathematical Methods for DNA Sequences,* ed. M. S. Waterman, pp. 1–34. Boca Raton: CRC
38. Fitch, D. H. A., Mainone, C., Goodman, M., Slightom, J. L. 1990. Molecular history of gene conversions in the primate fetal γ-globin genes: Nucleotide sequences from the common gibbon, *Hylobates lar. J. Biol. Chem.* 265:781–93
39. Fitch, D. H. A., Mainone, C., Slightom, J. L., Goodman, M. 1988. The spider monkey ψη-globin gene and surrounding sequences: Recent or ancient insertion of LINEs and SINEs? *Genomics* 3:237–55
40. Fitch, W. M. 1970. Distinguishing homologous and analogous proteins. *Syst. Zool.* 19:99–113
41. Fleagle, J. G. 1988. *Primate Adaptation & Evolution*. San Diego: Academic
42. Fujita, A., Hattori, M., Takenaka, O., Sakaki, Y. 1987. The L1 family (*KpnI* family) sequence near the 3' end of human β-globin gene may have been derived from an active L1 sequence. *Nucleic Acids Res.* 15:4007–20
43. Gingerich, P. D. 1984. Primate evolution: Evidence from the fossil record, comparative morphology, and molecular biology. *Yearb. Phys. Anthropol.* 27:57–72
44. Gojobori, T., Li, W.-H., Graur, D. 1982. Patterns of nucleotide substitution in pseudogenes and functional genes. *J. Mol. Evol.* 18:360–69
45. Goodman, M. 1981. Decoding the pattern of protein evolution. *Progr. Biophys. Mol. Biol.* 37:105–64
46. Goodman, M. 1985. Rates of molecular evolution: The hominoid slowdown. *BioEssays* 3:9–14
47. Goodman, M. 1986. Molecular evidence on the ape subfamily Homininae. In *Evolutionary Perspectives and the New Genetics,* ed. H. Gershowitz, D. L. Rucknagel, R. E. Tashian, pp. 121–32. New York: Liss
48. Goodman, M. 1989. Emerging alliance of phylogenetic systematics and molecular biology: A new age of exploration. In *The Hierarchy of Life,* ed. B. Fernholm, K. Bremer, H. Jornvall, pp. 43–61. Amsterdam: Elsevier Sci.
49. Goodman, M., Czelusniak, J., Koop, B. F., Tagle, D. A., Slightom, J. L. 1987. Globins: A case study in molecular phylogeny. *Cold Spring Harbor Symp. Quant. Biol.* 52:875–90
50. Goodman, M., Czelusniak, J., Moore, G. W., Romero-Herrera, A. E., Matsuda, G. 1979. Fitting the gene lineage into its species lineage. A parsimony strategy illustrated by cladograms constructed from globin sequences. *Syst. Zool.* 28:132–63
51. Goodman, M., Koop, B. F., Czelusniak, J., Fitch, D. H. A., Tagle, D. A.,

et al. 1989. Molecular phylogeny of the family of apes and humans. *Genome* 31:316–35

52. Goodman, M., Koop, B. F., Czelusniak, J., Weiss, M. L., Slightom, J. L. 1984. The η-globin gene: Its long evolutionary history in the β-globin gene family of mammals. *J. Mol. Biol.* 180:803–23
53. Goodman, M., Miyamoto, M. M., Czelusniak, J. 1987. Pattern and process in vertebrate phylogeny revealed by coevolution of molecules and morphology. In *Molecules and Morphology in Evolution: Conflict or Compromise?*, ed. C. Patterson, pp. 141–76. Cambridge: Cambridge Univ. Press
54. Goodman, M., Moore, G. W. 1971. Immunodiffusion systematics of the Primates. I. The Catarrhini. *Syst. Zool.* 20:19–62
55. Goodman, M., Moore, G. W., Matsuda, G. 1975. Darwinian evolution in the genealogy of haemoglobin. *Nature* 253: 603–8
56. Goodman, M., Romero-Herrera, A. E., Dene, H., Czelusniak, J., Tashian, R. E. 1982. Amino acid sequence data on the phylogeny of primates and other eutherian mammals. In *Macromolecular Sequences in Systematic and Evolutionary Biology*, ed. M. Goodman, pp. 115–91. New York: Plenum
57. Goodman, M., Tagle, D. A., Fitch, D. H. A., Bailey, W., Czelusniak, J., et al. 1990. Primate evolution at the DNA level and a classification of hominoids. *J. Mol. Evol.* 30:260–66
58. Groves, C. P. 1986. Systematics of the great apes. In *Comparative Primate Biology*. Vol. 1: *Systematics, Evolution and Anatomy*, ed. D. R. Swindler, J. Ermin, pp. 187–217. New York: Liss
59. Hardies, S. C., Edgell, M. H., Hutchinson, C. A. III. 1984. Evolution of the mammalian β-globin cluster. *J. Biol. Chem.* 259:3748–56
60. Hardison, R. C. 1981. The nucleotide sequence of rabbit embryonic globin gene β3. *J. Biol. Chem.* 256:11780–86
61. Hardison, R. C. 1984. Comparison of the β-like globin gene families of rabbits and humans indicates that the gene cluster 5'-ϵ-γ-δ-β-3' predates the mammalian radiation. *Mol. Biol. Evol.* 1:390–410
62. Hardison, R. C., Margot, J. 1984. Rabbit globin pseudogene $\psi\beta^2$ is a hybrid of δ and β-globin gene sequences. *Mol. Biol. Evol.* 1:302–16
63. Harris, S., Barrie, P.A., Weiss, M. L., Jeffreys, A. J. 1984. The primate ψβ1 gene. An ancient β-globin pseudogene. *J. Mol. Biol.* 180:785–801
64. Harris, S., Thackeray, J. R., Jeffries, A. J., Weiss, M. L. 1986. Nucleotide sequence analysis of the lemur β-globin gene family: Evidence for major rate fluctuations in globin polypeptide evolution. *Mol. Biol. Evol.* 3:465–84
65. Hasegawa, M., Kishino, H., Yano, T. 1987. Man's place in Hominoidea as inferred from molecular clocks of DNA. *J. Mol. Evol.* 26:132–47
66. Hattori, M., Kuhara, S., Takenaka, O., Sakaki, Y. 1986. L1 family repetitive DNA sequences in primates may be derived from a sequence encoding a reverse transcriptase-related protein. *Nature* 321:625–28
67. Hayasaka, K., Gojobori, T., Horai, S. 1988. Molecular phylogeny and evolution of primate mitochondrial DNA. *Mol. Biol. Evol.* 5:626–44
68. Hayashida, H., Miyata, T. 1983. Unusual evolutionary conservation and frequent DNA segment exchange in class I genes of the major histocompatibility complex. *Proc. Natl. Acad. Sci. USA* 80:2671–75
69. Hennig, W. 1966. *Phylogenetic Systematics*. Urbana: Univ. Illinois
70. Hill, A., Hardies, S. C., Phillips, S. J., Davis, M. G., Hutchinson, C. A. III, et al. 1984. Two mouse early embryonic β-globin gene sequences. *J. Biol. Chem.* 259:3739–47
71. Hill, W. C. O. 1953. *Primates—Comparative Anatomy and Taxonomy, Volume I: Strepsirhini*. New York: Edinburgh Univ. Press
72. Hill, W. C. O. 1955. *Primates—Comparative Anatomy and Taxonomy, Volume II: Haplorhini: Tarsioidea*. New York: Edinburgh Univ. Press
73. Hillis, D. M. 1987. Molecular versus morphological approaches to systematics. *Annu. Rev. Ecol. Syst.* 18:23–42
74. Hixson, J. E., Brown, W. M. 1986. A comparison of the small ribosomal RNA genes from the mitochondrial DNA of the great apes and humans: Sequence, structure, evolution, and phylogenetic implications. *Mol. Biol. Evol.* 3:1–18
75. Holmquist, R., Miyamoto, M. M., Goodman, M. 1988. Higher-primate phylogeny—Why can't we decide? *Mol. Biol. Evol.* 5:201–16
76. Holmquist, R., Miyamoto, M. M., Goodman, M. 1988. Analysis of higher-primate phylogeny from transversion differences in nuclear and mitochondrial DNA by Lake's methods of evolutionary parsimony and operator metrics. *Mol. Biol. Evol.* 5:217–36
77. Honacki, J. H., Kinman, K. E., Koeppl, J. W. 1982. *Mammalian Species of the*

*World*. Lawrence: Assoc. Syst. Collections
78. Huisman, T. H. J., Schroeder, W. A., Keeling, M. E., Gengozian, N., Miller, A., et al. 1973. Search for non-allelic structural genes for $\gamma$-chains of fetal hemoglobin in some primates. *Biochem. Genet.* 10:309–18
79. Hwu, H. R., Roberts, J. W., Davidson, E. H., Britten, R. J. 1986. Insertion and/or deletion of many repeated DNA sequences in human and higher ape evolution. *Proc. Natl. Acad. Sci. USA* 83:3875–79
80. Jeffreys, A. J., Barrie, P. A., Harris, S., Fawcett, D. H., Nugent, Z. J., et al. 1982. Isolation and sequence analysis of a hybrid $\delta$-globin pseudogene from the brown lemur. *J. Mol. Biol.* 156:487–503
81. Jones, J. S., ed. 1990. *Cambridge Encyclopedia of the Human Species*. Cambridge: Cambridge Univ. Press. In press
82. Kimura, M. 1983. The neutral theory of molecular evolution. In *Evolution of Genes and Proteins*, ed. M. Nei, R. K. Koehn, pp. 208–33. Sunderland: Sinauer
83. Kimura, M. 1983. *The Neutral Theory of Molecular Evolution*. Cambridge: Cambridge Univ. Press
84. Kluge, A. G. 1983. Cladistics and the classification of the great apes. In *New Interpretations of Ape and Human Ancestry*, ed. R. L. Ciochon, R. S. Corruccini, pp. 151–77. New York: Plenum
85. Koop, B. F., Goodman, M. 1988. Evolutionary and developmental aspects of two hemoglobin $\beta$-chain genes ($\epsilon^M$ and $\beta^M$) of opossum. *Proc. Natl. Acad. Sci. USA* 85:3893–97
86. Koop, B. F., Goodman, M., Xu, P., Chan, K., Slightom, J. L. 1986. Primate $\eta$-globin DNA sequences and man's place among the great apes. *Nature* 319:234–38
87. Koop, B. F., Miyamoto, M. M., Embury, J. E., Goodman, M., Czelusniak, J., et al. 1986. Nucleotide sequence and evolution of the orangutan $\epsilon$ globin gene region and surrounding Alu repeats. *J. Mol. Evol.* 24:94–102
88. Koop, B. F., Siemieniak, D., Slightom, J. L., Goodman, M., Dunbar, J., et al. 1989. Tarsius $\delta$ and $\beta$ globin genes: Conversions, evolution, and systematic implications. *J. Biol. Chem.* 264:68–79
89. Koop, B. F., Tagle, D. A., Goodman, M., Slightom, J. L. 1989. A molecular view of primate phylogeny and important systematic and evolutionary questions. *Mol. Biol. Evol.* 6:580–612
90. Koshland, D. E. Jr. 1989. Sequences and consequences of the human genome. *Science* 246:189
91. Lerman, M. I., Thayer, R. E., Singer, M. F. 1983. *KpnI* family of long interspersed repeated DNA sequences in primates: Polymorphism of family members and evidence for transcription. *Proc. Natl. Acad. Sci. USA* 80:3966–70
92. Lewin, R. 1988. Conflict over DNA clock results. *Science* 241:1598–1600
93. Lewin, R. 1988. DNA clock conflict continues. *Science* 241:1756–59
94. Li, W.-H., Luo, C.-C., Wu, C.-I. 1985. Evolution of DNA sequences. In *Molecular Evolutionary Genetics*, ed. R. J. MacIntyre, pp. 1–94. New York: Plenum
95. Li, W.-H., Tanimura, M. 1987. The molecular clock runs more slowly in man than in apes and monkeys. *Nature* 326:93–96
96. Li, W.-H., Tanimura, M., Sharp, P. M. 1987. An evaluation of the molecular clock hypothesis using mammalian DNA sequences. *J. Mol. Evol.* 25:330–42
97. Li, W.-H., Wolfe, K. H., Sourdis, J., Sharp, P. M. 1987. Reconstruction of phylogenetic trees and estimation of divergence times under nonconstant rates of evolution. *Cold Spring Harb. Symp. Quant. Biol.* 52:847–56
98. Li, W.-H., Wu, C.-I., Luo, C.-C. 1984. Nonrandomness of point mutation as reflected in nucleotide substitutions in pseudogenes and its evolutionary implications. *J. Mol. Evol.* 21:58–71
99. Maeda, N., Bliska, J. B., Smithies, O. 1983. Recombination and balanced chromosome polymorphism suggested by DNA sequences 5' to the human $\delta$-globin gene. *Proc. Natl. Acad. Sci. USA* 80:5012–16
100. Maeda, N., Wu, C.-I., Bliska, J., Reneke, J. 1988. Molecular evolution of intergenic DNA in higher primates: Pattern of DNA changes, molecular clock, and evolution of repetitive sequences. *Mol. Biol. Evol.* 5:1–20
101. Maniatis, T., Fritsch, E. F., Lauer, J., Lawn, R. M. 1980. The molecular genetics of human hemoglobins. *Annu. Rev. Genet.* 14:145–78
102. Margot, J. B., Demero, G. W., Hardison, R. C. 1989. Complete nucleotide sequence of the rabbit $\beta$-like globin gene cluster. *J. Mol. Biol.* 205:15–40
103. Marks, J. 1983. Hominoid cytogenetics and evolution. *Yearb. Phys. Anthropol.* 26:131–59
104. Marks, J., Schmid, C. W., Sarich, V. M. 1988. DNA hybridization as a guide

to phylogeny: Relations of the Hominoidea. *J. Hum. Evol.* 17:769–86

105. Marks, J., Shaw, J.-P., Shen, C.-K. J. 1986. The orangutan adult $\alpha$-globin gene locus: Duplicated functional genes and a newly detected member of the primate $\alpha$-globin gene family. *Proc. Natl. Acad. Sci. USA* 83:1413–17
106. Martin, S. L., Vincent, K. A., Wilson, A. C. 1983. Rise and fall of the delta globin gene. *J. Mol. Biol.* 164:513–28
107. Miyamoto, M. M., Goodman, M. 1986. Biomolecular systematics of eutherian mammals: Phylogenetic patterns and classification. *Syst. Zool.* 35:230–40
108. Miyamoto, M. M., Koop, B. F., Slightom, J. L., Goodman, M., Tennant, M. R. 1988. Molecular systematics of higher primates: Genealogical relations and classification. *Proc. Natl. Acad. Sci. USA* 85:7627–31
109. Miyamoto, M. M., Slightom, J. L., Goodman, M. 1987. Phylogenetic relations of humans and African apes from DNA sequences in the $\psi\eta$-globin region. *Science* 238:369–73
110. Nei, M. 1986. Stochastic errors in DNA evolution and molecular phylogeny. In *Evolutionary Perspectives and the New Genetics,* ed. H. Gershowitz, D. L. Rucknagel, R. E. Tashian, pp. 133–47. New York: Liss
111. Patterson, C. 1987. Introduction. In *Molecules and Morphology in Evolution: Conflict or Compromise?,* ed. C. Patterson, pp. 1–22. Cambridge: Cambridge Univ. Press
112. Patterson, C. 1988. Homology in classical and molecular biology. *Mol. Biol. Evol.* 5:603–25
113. Rogan, P. K., Pan, J., Weissman, S. M. 1987. L1 repeat elements in the human $\epsilon$-$^{G}\gamma$-globin gene intergenic region: Sequence analysis and concerted evolution within this family. *Mol. Biol. Evol.* 4:327–42
114. Rosenberger, A. L. 1984. Fossil New World monkeys dispute the molecular clock. *J. Hum. Evol.* 13:737–42
115. Saitou, N., Imanishi, T. 1989. Relative efficiencies of the Fitch-Margoliash, maximum-parsimony, maximum-likelihood, minimum-evolution, and neighbor-joining methods of phylogenetic tree construction in obtaining the correct tree. *Mol. Biol. Evol.* 6:514–25
116. Sakoyama, Y., Hong, K.-J., Byun, S. M., Hisajima, H., Ueda, S., et al. 1987. Nucleotide sequences of immunoglobulin $\epsilon$ genes of chimpanzee and orangutan: DNA molecular clock and hominoid evolution. *Proc. Natl. Acad. Sci. USA* 84:1080–84
117. Sarich, V. M., Cronin, J. E. 1976. Molecular systematics of the primates. In *Molecular Anthropology,* ed. M. Goodman, R. E. Tashian, pp. 141–70. New York: Plenum
118. Sarich, V. M., Schmid, C. W., Marks, J. 1989. DNA hybridization as a guide to phylogenies: A critical analysis. *Cladistics* 5:3–32
119. Sarich, V. M., Wilson, A.C. 1967. Immunological time scale for hominoid evolution. *Science* 158:1200–3
120. Savatier, P., Trabuchet, G., Chebloune, Y., Faure, C., Verdier, G., et al. 1987. Nucleotide sequence of the delta-beta globin intergenic segment in the macaque: Structure and evolutionary rates in higher primates. *J. Mol. Evol.* 24:297–308
121. Savatier, P., Trabuchet, G., Chebloune, Y., Faure, C., Verdier, G., et al. 1987. Nucleotide sequence of the beta-globin genes in gorilla and macaque: The origin of nucleotide polymorphisms in human. *J. Mol. Evol.* 24:309–18
122. Savatier, P., Trabuchet, G., Faure, C., Chebloune, Y., Guoy, M., et al. 1985. Evolution of the primate $\beta$-globin gene region. High rate of variation in CpG dinucleotides and in short repeated sequences between man and chimpanzee. *J. Mol. Biol.* 182:21–29
123. Sawada, I., Willard, C., Shen, C.-K. J., Chapman, B., Wilson, A. C., et al. 1985. Evolution of Alu family repeats since the divergence of human and chimpanzee. *J. Mol. Evol.* 22:316–22
124. Schmid, C. W., Jelinek, R. 1982. The Alu family of dispersed repetitive sequences. *Science* 216:1065–70
125. Schmid, C. W., Shen, C.-K. J. 1985. The evolution of interspersed repetitive DNA sequences in mammals and other vertebrates. In *Molecular Evolutionary Genetics,* ed. R. J. MacIntyre, pp. 323–58. New York: Plenum
126. Schultz, A. H. 1963. Age changes, sex differences, and variability as factors in the classification of primates. In *Classification and Human Evolution,* ed. S. L. Washburn, pp. 85–115. Chicago: Aldine
127. Schwartz, J. H. 1978. If *Tarsius* is not a prosimian, is it a haplorhine? In *Recent Advances in Primatology,* vol. 3, ed. D. J. Chivers, K. A. Joysey, pp. 195–202. London: Academic
128. Schwartz, J. H. 1984. What is a tarsier? In *Living Fossils,* ed. N. Eldredge, S. M. Stanley, pp. 38–49. New York: Springer-Verlag
129. Schwartz, J. H. 1984. The evolutionary

relationships of man and orang-utans. *Nature* 308:501–5
130. Schwartz, J. H. 1986. Primate systematics and a classification of the order. In *Comparative Primate Biology*. Vol. 1: *Systematics, Evolution and Anatomy*, ed. D. R. Swindler, J. Ervin, pp. 1–41. New York: Liss
131. Schwartz, J. H. 1987. *The red ape: Orang-utans and human origins*. Boston: Houghton Mifflin
132. Schwartz, J. H., Tattersall, I. 1985. Evolutionary relationships of living lemurs and lorises (Mammalia, Primates) and their potential affinities with European Eocene Adapidae. *Anthropol. Pap. Am. Mus. Nat. Hist.* 60:1–100
133. Scott, A. F., Heath, P., Trusko, S., Boyer, S. H., Prass, W., et al. 1984. The sequence of gorilla fetal globin genes: Evidence for multiple gene conversions in human evolution. *Mol. Biol. Evol.* 1:371–89
134. Scott, A. F., Schmeckpeper, B. J., Abdelrazik, M., Comey, C. T., O'Hara, B., et al. 1987. Origin of the human L1 elements: Proposed progenitor genes deduced from a consensus DNA sequence. *Genomics* 1:113–25
135. Shapiro, S. G., Schon, E. A., Townes, T. M., Lingrel, J. B. 1983. Sequence and linkage of the goat $\epsilon^{I}$ and $\epsilon^{II}$ $\beta$-globin genes. *J. Mol. Biol.* 169:31–52
136. Shehee, W. R., Loeb, D. D., Adey, N. B., Burton, F. H., Casavant, N. C., et al. 1989. Nucleotide sequence of the BALB/c mouse $\beta$-globin complex. *J. Mol. Biol.* 205:41–62
137. Shen, S., Slightom, J. L., Smithies, O. 1981. A history of the human fetal globin gene duplication. *Cell* 26:191–203
138. Sibley, C. G., Ahlquist, J. E. 1984. The phylogeny of hominoid primates, as indicated by DNA-DNA hybridization. *J. Mol. Evol.* 20:2–15
139. Sibley, C. G., Ahlquist, J. E. 1987. DNA hybridization evidence of hominoid phylogeny: Results from an expanded data set. *J. Mol. Evol.* 26:99–121
140. Simons, E. L., Rasmussen, D. T. 1989. Cranial morphology of *Aegyptopithecus* and *Tarsius* and the question of the tarsier-anthropoidean clade. *Am. J. Phys. Anthr.* 79:1–23
141. Simpson, G. G. 1945. The principles of classification and a classification of mammals. *Bull. Am. Mus. Nat. Hist.* 85:1–350
142. Singer, M. F. 1982. SINEs and LINEs: Highly repeated short and long interspersed sequences in mammalian genomes. *Cell* 28:433–34
143. Slagel, V., Flemington, E., Traina-Dorge, V., Bradshaw, H., Deininger, P. 1987. Clustering and subfamily relationships of the Alu family in the human genome. *Mol. Biol. Evol.* 4:19–29
144. Slightom, J. L., Blechl, A. E., Smithies, O. 1980. Human $^{G}\gamma$- and $^{A}\gamma$-globin genes: Complete nucleotide sequences suggest that DNA can be exchanged between these duplicated genes. *Cell* 21:627–38
145. Slightom, J. L., Chang, L.-Y. E., Koop, B. F., Goodman, M. 1985. Chimpanzee fetal $^{G}\gamma$ and $^{A}\gamma$ globin gene nucleotide sequences provide further evidence of gene conversions in hominine evolution. *Mol. Biol. Evol.* 2:370–89
146. Slightom, J. L., Koop, B. F., Xu, P., Goodman, M. 1988. Rhesus fetal globin genes: Concerted gene evolution in the descent of higher primates. *J. Biol. Chem.* 263:12427–38
147. Slightom, J. L., Theisen, T. W., Koop, B. F., Goodman, M. 1987. Orangutan fetal globin genes: Nucleotide sequences reveal multiple gene conversions during hominid phylogeny. *J. Biol. Chem.* 262:7472–83
148. Sourdis, J., Nei, M. 1988. Relative efficiencies of the maximum parsimony and distance-matrix methods in obtaining the correct phylogenetic tree. *Mol. Biol. Evol.* 5:298–311
149. Spritz, R. A., Giebel, L. B. 1988. The structure and evolution of the spider monkey $\delta$-globin gene. *Mol. Biol. Evol.* 5:21–29
150. Stephens, J. C. 1985. Statistical methods of DNA sequence analysis: Detection of intragenic recombination or gene conversion. *Mol. Biol. Evol.* 2:539–56
151. Szalay, F. S., Delson, E. 1979. *Evolutionary History of the Primates*. New York: Academic
152. Szalay, F. S., Rosenberger, A. L., Dagosto, M. 1987. Diagnosis and differentiation of the order Primates. *Yearb. Phys. Anthropol.* 30:75–105
153. Tagle, D. A., Koop, B. F., Goodman, M., Slightom, J. L., Hess, D. L., et al. 1988. Embryonic $\eta$ and $\gamma$ globin genes of a prosimian primate *(Galago crassicaudatus):* Nucleotide and amino acid sequences, developmental regulation, and phylogenetic footprints. *J. Mol. Biol.* 203:439–55
154. Tajima, F., Nei, M. 1982. Biases of the estimates of DNA divergence obtained by the restriction enzyme technique. *J. Mol. Evol.* 18:115–20
155. Templeton, A. R. 1983. Phylogenetic

inference from restriction endonuclease cleavage site maps with particular reference to the evolution of humans and the apes. *Evolution* 37:221–44

156. Teumer, J., Green, H. 1989. Divergent evolution of part of the involucrin gene in the hominoids: Unique intragenic duplications in the gorilla and human. *Proc. Natl. Acad. Sci. USA* 86:1283–86
157. Thorington, R. W. Jr., Anderson, S. 1984. Primates. In *Orders and Families of Recent Mammals of the World*, ed. S. Anderson, J. K. Jones, Jr., pp. 187–217. New York: Wiley
158. Townes, T. M., Shapiro, S. G., Wernke, S. M., Lingrel, J. B. 1984. Duplication of a four-gene set during the evolution of the goat $\beta$-globin locus produced genes now expressed differentially in development. *J. Biol. Chem.* 259:1896–1900
159. Trabuchet, G., Chelbourne, Y., Savatier, P., Lachuer, J., Faure, C., et al. 1987. Recent insertion of an Alu sequence in the $\beta$-globin gene cluster of the gorilla. *J. Mol. Evol.* 25:288–91
160. Tseng, H., Green, H. 1988. Remodeling of the involucrin gene during primate evolution. *Cell* 54:491–96
161. Tseng, H., Green, H. 1989. The involucrin gene of the owl monkey: Origin of the early region. *Mol. Biol. Evol.* 6:460–68
162. Ueda, S., Watanabe, Y., Saitou, N., Omoto, K., Hayashida, H., et al. 1989. Nucleotide sequences of immunoglobulin-epsilon pseudogenes in man and apes and their phylogenetic relationships. *J. Mol. Biol.* 205:85–90
163. Walker, E. P., Warnick, F., Hamlet, S. E., Lange, K. I., Davis, M. A., et al. 1983. *Walker's Mammals of the World*, ed. R. M. Nowak, J. L. Paradiso. Vols. 1, 2. Baltimore: Johns Hopkins Univ. 4th ed.
164. Walsh, J. B. 1987. Sequence-dependent gene conversion: Can duplicated genes diverge fast enough to escape conversion? *Genetics* 117:543–57
165. Watson, J. D., Jordan, E. 1989. The human genome program at the National Institutes of Health. *Genomics* 5:654–56
166. Weiner, A. M., Deininger, P. L., Efstratiadis, A. 1986. Nonviral retroposons: Genes, pseudogenes, and transposable elements generated by the reverse flow of genetic information. *Annu. Rev. Biochem.* 55:631–61
167. Willard, C., Nguyen, H. T., Schmid, C. W. 1987. Existence of at least three distinct Alu subfamilies. *J. Mol. Evol.* 26:180–86
168. Williams, S. A., Goodman, M. 1989. A statistical test that supports a human/chimpanzee clade based on non-coding DNA sequence data. *Mol. Biol. Evol.* 6:325–30
169. Wilson, A. C., Carlson, S. S., White, T. J. 1977. Biochemical evolution. *Annu. Rev. Biochem.* 46:573–639
170. Wilson, A. C., Ochman, H., Prager, E. M. 1987. Molecular time scale for evolution. *Trends Genet.* 3:241–47
171. Wu, C.-I., Li, W.-H. 1985. Evidence for higher rates of nucleotide substitutions in rodents than in man. *Proc. Natl. Acad. Sci. USA* 82:1741–45
172. Wu, C.-I., Maeda, N. 1987. Inequality in mutation rates of the two strands of DNA. *Nature* 327:169–70

*Annu. Rev. Ecol. Syst. 1991. 21:221–41*


# COMPETITION AND PREDATION IN MARINE SOFT-SEDIMENT COMMUNITIES

*W. Herbert Wilson*[1]

Department of Zoology NJ-15, University of Washington, Seattle, Washington 98195

KEY WORDS: competition, predation, soft-sediment, infauna, succession

---

## INTRODUCTION

One of the major goals of community ecology is to understand how interactions among the resident organisms affect their distribution and abundance. For terrestrial and rocky intertidal communities competition and predation clearly can play pivotal roles in community organization (28, 47, 84). The importance of biological processes in organizing marine soft-sediment communities is not as well understood (86), despite the fact that marine soft-sediments are the most common habitat on earth. As I argue below, paradigms of community organization based on other habitats seem to offer little insight into the structure of marine soft-sediment communities. Rather than attempting to explain the failure of such paradigms, I argue that soft-sediment habitats are sufficiently different from other communities that different paradigms are needed.

Soft-sediment communities are unusual in the rate at which the nature of the physical environment can change. Most sedimentary particles are smaller than the resident organisms (the infauna). The activities of the infauna can dramatically change the nature of the environment over time periods of hours or days. For instance, burrowing infauna may increase the porosity and erodability of the sediment (102). Subsurface deposit-feeders may alter the

[1]Present address: Department of Biology, Colby College, Waterville, Maine 04901



0066-4163/90/0410-0221$02.00

vertical distribution of sediment grain sizes and change the spectrum of grain sizes by ingesting small sediment particles and egesting them as larger fecal pellets. The effects of infauna have dramatic and rapid effects on sediment biogeochemistry (2). Infaunal organisms live in, rather than on, the substratum, and their activities alter the fundamental nature of the habitat on very short temporal scales. These characteristics require a unique perspective into the nature of competitive interactions in these communities.

This review on the importance of competition and predation in structuring infaunal communities emphasizes experimental studies. Since the introduction of experimental methods to the study of soft-sediment communities (76, 103), the depth of understanding and the strength of inferences have increased greatly. The emphasis on experimental data necessitates a bias toward intertidal and shallow subtidal communities because experimental data from the deep sea have been difficult and expensive to collect.

## COMPETITION

### *Direct Interactions*

Because of the three-dimensional structure of soft-sediment habitats, vertical and horizontal partitioning of space is possible, reducing direct encounters of potential competitors with each other. Nevertheless, direct interactions are found most frequently between conspecifics or closely related species (141, 144). Agonistic interactions have been described among nereid polychaetes (104). Opheliid polychaetes apparently interact directly to maintain constant abundance per volume of sediment (112, 137). Levin (62) showed that the regular dispersion pattern of the spionid polychaete, *Pseudopolydora paucibranchiata,* is maintained by strong, direct, competitive interactions. Aggressive interactions of *P. paucibranchiata* with associated tube-building infauna result in significantly reduced foraging time for the latter species (63). Direct interactions of the polychaete *Nereis diversicolor* with the amphipod *Corophium volutator* led to declines of the amphipod (81). Croker & Hatfield (31) demonstrated strong direct effects (increased mortality and reduced reproductive effort) of one haustoriid amphipod on a second species; the inferior competitor is confined to the high intertidal zone in the field. Grant (45) demonstrated similar interactions of two haustoriids which led to vertical separation with the inferior competitor forced to occupy deeper, anoxic sediments in the presence of the superior competitor. Vertical zonation of bivalves in a California lagoon changed in apparent response to the removal of the thalassinid shrimp, *Callianassa californiensis,* implying competitive release (85). Deposit-feeding protobranch bivalves changed their position as a function of heterospecific density (64). Intertidal gastropod zonation is maintained by the displacement of *Hydrobia totteni* by *Ilyanassa obsoleta* (66). All of these interactions occur among organisms that at least attempt to establish at the same depth in the sediment.

## *Exploitative Competition*

SUSPENSION-FEEDERS Because direct competitive interactions are minimized by the three-dimensional nature of the habitat, indirect competitive mechanisms are to be expected. Peterson (87, 89, 90) has provided a wealth of experimental data on exploitative competition among suspension-feeding bivalves. Bivalves are advantageous research organisms because they can be easily marked and their growth can be accurately quantified. Identification of competitive effects can be measured through differences in growth as well as survivorship. The growth of *Sanguinolaria nuttallii* was reduced by 80% when these clams were confined with two other deep-dwelling bivalves, although no effects of a shallow-dwelling bivalve were evident on *S. nuttallii* growth. These results imply that competition for space is occurring. The interpretation of this interaction was confirmed by confining *S. nuttallii* with surrogates of the other deep-dwelling clams (valves of dead clams tied together and placed in life position in the sediment). The growth of *S. nuttallii* was depressed in the presence of the surrogate clams to an extent equal to the reduction of growth in the presence of living clams. A study of competition among two shallow-dwelling bivalves *(Protothaca staminea* and *Chione undatella*) involved the enclosure of these species at a range of densities (1/2X to 8X normal density) in the field (87). The growth rate of each species as well as the ratio of the dry weight of gonads to the dry weight of total soft parts, a measure of reproductive effort, declined as a negative function of intraspecific density. Density-dependent migration of *P. staminea* was documented. Over the two years of the study, few deaths could be attributed to competition. Interspecific interactions were usually not significant. Because the two bivalve species occupy the same depth stratum in the sediment, one infers that food is the limiting resource and that the two species differ somehow in the types of phytoplankton they consume. Experiments with Australian suspension-feeding bivalves (90) indicated that bivalves transplanted into the high intertidal grew less than conspecifics established lower in the intertidal zone. The depression of growth was not a linear function of immersion time, implying that there is density-dependent reduction of food as the tide rises. The observation that the phytoplankton can be depleted from the water column by the suspension-feeders in an intertidal mudflat (21) implies that competition for food may be occurring among suspension-feeders in a Maine mud flat.

The strength of competition may have a strong environmental component. When the bivalve *Macoma baltica* was maintained in a muddy sediment in the field, growth was found to be density-dependent, while conspecifics maintained in sand showed no density-dependence of growth (79). The difference may be related to the facultative feeding behavior of this bivalve; in muddy substrates, deposit-feeding predominates whereas suspension-feeding predominates in sandy substrates.

DEPOSIT-FEEDERS Characterization of the food resources for deposit-feeders is elusive (67). Except for differences in sizes of sediment grains ingested (124), little is known of the differential use, if any, of the various components of detritus by deposit-feeders. It is therefore impossible to enhance detritus levels in the field with any degree of confidence. As an alternative to varying food levels, soft-sediment ecologists have adopted the approach of varying densities of deposit-feeders, thereby influencing per capita resource levels. Even this approach is problematic because the feeding rate and density of deposit-feeders affect the growth of diatoms and bacteria in diverse ways; increased feeding of detritivores may enhance microfloral growth (11, 12, 30). Most of the organisms studied have been soft-bodied organisms whose growth is difficult to measure. The commonly measured responses of deposit-feeders to density are emigration rates and mortality.

For the deposit-feeding snail *Hydrobia ventrosa*, Levinton (65) showed that feeding and movement decreased as a function of increasing density and that floating, a means of dispersal, increased as density rose. Wilson (130) manipulated the densities of two species of co-occurring spionid polychaetes; the density of each species was varied independently over four experimental densities. Migration in and out of the experimental containers was strongly density-dependent; after eight weeks all containers had densities indistinguishable from ambient densities. Strong intraspecific effects were documented but interspecific effects were weak and usually not statistically significant. Ambrose (7) showed that emigration of a burrowing, deposit-feeding amphipod was density-dependent. Wilson (134) documented strong size-dependent emigration in the amphipod *Corophium volutator* with juveniles forced to emigrate from high adult densities.

Several efforts have been made to manipulate directly the detrital resource levels. Young & Young (146) added processed sewage sludge (Milorganite) to seagrass beds. Some species responded positively to Milorganite addition while others were unaffected. Dauer et al (32) added organic fertilizer to a subtidal site and documented no change in densities of members of the community, although the densities of some species in predator exclusion cages (at densities above control densities) did increase. However, it is not clear how long the enrichment persisted in the sediment. Wiltse et al (136) added urea to a salt marsh habitat in Massachusetts, which resulted in significant increases in sediment chlorophyll but no increase in infaunal abundance. The latter two studies can be cautiously interpreted to show that detritus is not limiting in the two communities. However, the uncertainty about the nature of detrital resources and the effects of the experimental treatments brings into question the realism of the enrichments.

## *Sediment-Mediated Interactions*

The rapidity with which the features of the sedimentary environment change as a function of biotic activity provides an additional means of interaction between infauna. Such sediment-mediated interactions are common in many infaunal communities. The effects of resident organisms on the sediment may render the sediment less habitable by other organisms. Such interactions were first explored by Rhoads & Young (103) who noted that sandy substrates tended to be dominated by suspension-feeders while mud substrates were dominated by deposit-feeders. Predicting that deposit feeders roil the sea bottom sufficiently to clog the filtering apparatus of suspension-feeding organisms (the trophic group amensalism hypothesis), Rhoads & Young showed that the growth of the bivalve *Mercenaria mercenaria* was, when grown near the bottom above a deposit-feeder community, significantly depressed relative to the growth of conspecifics grown at greater heights off the bottom. For the duration of the experiment (64 days), no significant mortality occurred. These data provide some support for the limitation of growth of suspension-feeders in sediments dominated by deposit-feeders. Woodin (138) shifted the focus from trophic group to mobility group; her functional groups were suspension-feeders, tube-building organisms, and bioturbators (sediment destabilizers). The effects of any group were predicted to be inimical to individuals from other functional groups. Tube-builders and bioturbators strongly alter the sedimentary environment; these functional groups are discussed below. Suspension-feeders typically have minimal effect on the nature of the sediment (141).

TUBE-BUILDERS Tube-building organisms, particularly when abundant, are predicted to retard the movement of burrowing organisms and to preempt space from large suspension-feeders (137, 138). In Washington state, Woodin (137) first demonstrated significant tube-builder effects when she noted that the larvae of tube-building polychaetes were confusing the surface of experimental cages with the sediment surface. As a result, the density of tube-builders, primarily three polychaetes, was significantly reduced inside the cages relative to unmanipulated control areas. Significantly greater numbers of a burrowing polychaete *(Armandia brevis)* were found in cages with low tube-builder density, compared to unmanipulated areas with high tube-builder density. Strong negative correlations existed between *A. brevis* abundance and combined tube-builder abundance. Laboratory experiments indicated that each *A. brevis* required a minimum volume of sediment. The interaction seen in the field is therefore interpreted as competition for space. In areas of high tube-builder abundance, burrowers have less available sediment for burrowing and feeding. Hulberg & Oliver (34, 53) challenged this interpretation, claiming from experiments in California that the larvae of the *A. brevis* preferentially settle within cages, irrespective of tube-builder abundance. The laboratory experiments of Woodin support her interpretation rather than the alternative posed by Hulberg & Oliver. Tamaki (112) showed that the larvae of *Armandia* sp. did not preferentially settle in sediments taken from within a cage although the effects of flow were not examined. He did corroborate Woodin's finding of a minimum volume requirement for each worm, but he claimed that competition for food, rather than space, was limiting. It is difficult if not impossible to separate food from space as distinct resources for a mobile infaunal species.

Other efforts to document exclusion of other functional groups by tube-builders have met with mixed results. Reise (100) found a negative relationship between burrowing capitellid polychaetes and tubicolous spionid polychaetes in some predator-exclusion experiments. Brenchley (19) showed that eelgrass *(Zostera marina)* root-mats and dense tube-mats of mixed species composition acted additively to retard the movement of burrowing deposit-feeders. In most cases, the burrowers were able to penetrate the mat. Thus, her data do not provide evidence of exclusion. Measurement of lateral movement rather than vertical burrowing would have been enlightening. Wilson (131) found that tubicolous spionid polychaetes and tanaid crustaceans coexisted with a sediment-destabilizing bivalve in laboratory experiments. Weinberg (123) found conflicting effects of tube-builders on a suspension-feeding bivalve *(Gemma gemma);* a tubicolous spionid *(Polydora ligni)* had a negative influence on bivalve abundance while a tubicolous maldanid polychaete *(Clymenella torquata)* had a positive effect. The spread of an eelgrass *(Zostera marina)* bed was associated with declines in the abundance of a burrowing crustacean *(Callianassa californiensis),* presumably because the seagrass root-mat impeded burrowing (48). Lateral movement of the bivalve *Macoma secta* was influenced more by sedimentary properties than by the dense arrays of phoronid *(Phoronopsis viridis)* tubes (34, 106). A summary of the available data indicates that the tube-builders or root-mats may be associated with diminished numbers of mobile burrowers but that burrowers are never totally excluded.

BIOTURBATORS Disruption of the sediment by animal activities (feeding, burrowing) is predicted to have negative effects on suspension-feeding and tube-building infauna. Depressed growth of suspension-feeders in the presence of bioturbating organisms was first shown for bivalves (103) and scleractinian corals (1). A suspension-feeding bivalve *(Sanguinolaria nuttallii)* increased after the experimental removal of a burrowing thalassinid shrimp,

*Callianassa californiensis* (85). Exclusion of epibenthic gastropods resulted in increases of spionid polychaetes (135) and oligochaetes and capitellid polychaetes (41). For tube-building organisms, removal of a mobile cockle *(Cerastoderma edule)* led to an increase in the tube-building amphipod crustacean, *Corophium volutator* (55). Brenchley (18) conducted a thorough laboratory analysis of the effect of sedimentation on a tube-building assemblage and showed that mortality of the tube-dwellers was a function of the size of the animal, presumably correlated positively with the ability to burrow through newly deposited sediment, as well as a function of the magnitude and frequency of the sedimentation episodes. Wilson (129) demonstrated a reduction of spionid polychaetes in areas of high sediment disturbance by the polychaete *Abarenicola pacifica*. A small spionid polychaete *(Pygospio elegans)* was reduced in the presence of *A. pacifica* relative to controls while a larger spionid *(Pseudopolydora kempi)* was unaffected. Laboratory experiments indicated that suffocation rather than emigration was the likely mechanism of reduced spionid abundances. However, emigration from intermediate levels of sediment disruption permitted local survival; despite emigration, mortality was great at high levels of bioturbation. Highsmith (49) demonstrated significant reductions in the abundance of tube-dwelling tanaids *(Leptochelia dubia)* in the presence of burrowing sand dollars *(Dendraster excentricus),* providing an explanation for the patchy distribution of these two species in the field. The deposit-feeding bivalve, *Macoma balthica,* and a cockle, *C. edule,* caused a local reduction of the small spionid, *P. elegans,* but not a larger confamilial species *(Spio filicornis)* (101). Ronan (34, 106) showed that the thalassinid shrimp, *Callianassa californiensis,* disrupted the orientation of the tubes of the phoronid *Phoronopsis viridis* in California. In Oregon, the densities of most sedentary species were reduced in the presence of *C. californiensis* (92). Laboratory experiments indicated that tube-builders could coexist with a small, mobile clam *(Transennella tantilla),* indicating that the sediment disruption of the bivalve was insufficient to affect other members of the community (131).

The available data therefore indicate that animal-mediated sediment disturbance can have significant deleterious effects on suspension-feeders and tube-builders in infaunal communities. The result of such disturbance is a complex function of the sizes of the organisms and the magnitude and frequency of the sediment disturbance (94, 129). Presently, there is insufficient information to allow the formulation of a predictive model of the effect of bioturbators on other functional groups. A major barrier to such a model is the inadequacy of functional groupings (57). Depending on animal density, sediment microflora, and the strength of ambient flow, an assemblage of tube-builders may stabilize or destabilize sediments, with correspondingly divergent effects on other members of the infauna (38, 46, 57, 68).

### *Influence of Adults on Settling Larvae and Juveniles*

ESTABLISHED COMMUNITIES In her expansion of the trophic group amensalism hypothesis (103), Woodin (138) presented the hypothesis that discrete, densely populated patches of infaunal invertebrates maintain their integrity by preventing the recruitment of larvae of other species. The mechanism of such adult-larval interactions is predicted to vary among functional groups: Bioturbators should suffocate larvae, suspension-feeders should filter larvae from the plankton, and tube-builders should exclude larvae by preemption of space and defecation on the sediment surface. Although exceptions are frequently seen to the predicted associations of functional groups (138), the concept of adult-larval interactions has had a pervasive influence in the study of marine infaunal communities. In the laboratory, Wilson (128) demonstrated that terebellid polychaetes significantly depress the survivorship of nereid polychaete larvae. Field experiments have met with mixed success in their support of the adult-larval hypothesis. Williams (126) showed reduced survivorship of clam spat *(Tapes japonica)* among high densities of conspecific adults to that in areas where adult clams were absent or at low density. Through field and laboratory experiments, Highsmith (49) showed that tanaid crustaceans *(Leptochelia dubia)* prey on the settling larvae of sand dollars *(Dendraster excentricus)*. Negative effects of organisms on the settlement and survivorship of larvae or recently settled juveniles have been demonstrated for drifting algal mats (80), nereid polychaetes (3), spionid polychaetes (69, 100, 113, 123, 125, 129), a mixed polychaete assemblage (133), phoxocephalid amphipods (82, 83), pontoporeid amphipods (39), nassariid gastropods (54), and bivalves (15, 55). Intraspecific adult-larval interactions are demonstrably strong in a nereid polychaete (58) and a corophiid amphipod (134). Despite these observations and experimental support for the importance of adult-larval interactions, other data fail to support the hypothesis. Using microcosms, Maurer (74) was unable to demonstrate any effect of varying densities of venerid bivalves on larval recruitment. In field experiments, Hines et al (51) varied the densities of a suspension-feeding bivalve *(Mya arenaria)* and a deposit-feeding bivalve *(Macoma balthica)*. Although they could demonstrate reduced larval abundances at high adult densities for both bivalves, the responses of the larvae of individual settling species were inconsistent. Positive effects of adult density were even documented for some species with planktonic larvae. Similarly, Commito (23) found that the infauna below mussel *(Mytilus edulis)* beds were not reduced; oligochaete abundance decreased in response to experimental removal of *M. edulis* (26).

Recently, Watzin (120–122) has identified meiofaunal predators, particularly turbellarians, as potent predators on settling infauna. She experimentally doubled the densities of meiofauna in experimental containers and compared settlement in those containers to settlement in cores of unmanipulated meiofauna. Although ambient meiofaunal density served as a control rather than, for instance, half-normal density or absence of meiofauna, the strengths of the meiofauna-larval effects documented suggest that these interactions may be important determinants of macrofaunal settling success.

The present data indicate that adult-larval interactions are detectable in the field and in the laboratory. However, the mechanism of the interaction is typically not determined in field experiments. Field tests are unable to distinguish between avoidance of high densities of adults by larvae and increased mortality of settling larvae by ingestion or suffocation by established adults. Woodin (143) provided field data which show reduced settlement of larvae of a spionid polychaete *(Pseudopolydora kempi)* in the presence of an arenicolid polychaete *(Abarenicola pacifica)*. Eckman (37) presented data showing that passive hydrodynamic effects from tubes or plant stalks may cause the accumulation of larvae. Similarly, Ertman & Jumars (40) showed that larvae may passively accumulate around projecting bivalve siphons. Our present level of understanding of these phenomena does not allow prediction of conditions where adult-larval interactions are important structuring forces in the community.

SUCCESSION Following a disturbance that clears a soft-sediment habitat, colonization may be quite rapid and proceed with an early pulse of opportunistic species, followed by the arrival of species whose abundance remains relatively constant (70, 71). The successional models of Connell & Slatyer (29) have provided a theoretical framework for soft-sediment succession. The models presume that early colonists may have a negative (inhibition model), a positive (facilitation model), or no effect (tolerance model) on subsequent recruits. Three major successional studies in soft-sediment habitats have been conducted in light of these theories. Zajac & Whitlatch (149, 150) followed the colonization of defaunated sediment at three tidal heights in Connecticut. Succession did not vary as a function of tidal height. Recruitment did not vary consistently as a function of the number of residents, and hence their data are most consistent with the tolerance model of succession. Gallagher et al (43) examined succession on a smaller scale on an estuarine flat in Washington as a function of several common species on the flat (polychaetes, bivalves, crustaceans) as well as simulated tubes. They monitored the dynamics of eight species over five manipulations. Twenty-four of the treatment × recruiting species combinations showed no effect of the experimental treatment, 11 showed a positive effect, and only one showed a negative influence. Although "facilitation" appears in the title of their paper, the tolerance model is most applicable to their successional dynamics. Ambrose (3) followed succession in defaunated sediments in Maine. Later recruitment varied inversely with the density of established juveniles. These studies demonstrate that all three models of succession may be operative. The data available are insufficient to allow predictions of the type of successional dynamics in a given habitat.

## PREDATION

### Methodology

Experimental analyses of the effects of predation in infaunal communities have generally involved the exclusion or enclosure of suspected predators. Comparison to unmanipulated areas allows the investigator to assess the strength of the predators. However, cages baffle currents, potentially altering sedimentation rates and the supply of phytoplankton. Some workers assess the effects of cages by comparing the sediments inside and outside of cages (16, 17, 41, 52, 97). Lack of significant differences is taken to mean that hydrodynamic artifacts are unimportant and that any effect of cages is due to exclusion of predators. An alternative method is the emplacement of partial cages (sides only, roof only, half cage) which are designed to mimic the hydrodynamic effects of cages but allow unrestricted access to predators. Any differences between unmanipulated areas and partial cages can be attributed to hydrodynamic artifact. In practice, differences between control areas and partial cages are not detected (4, 96, 111, 134–136, 140). Despite the apparent success of cages at excluding predators without inducing significant hydrodynamic artifact, there is widespread distrust of caging experiments. Working in a high energy environment, Hulberg & Oliver (53) show that the shape of cages has dramatic effects on the outcome of exclusion experiments; this work implies that hydrodynamic artifacts are large. Eckman (36) and Gallagher et al (43) show that even single tubes protruding above the sediment surface may have strong effects on local abundances of organisms. They argue that a cage is bound to have strong hydrodynamic effects. The controversial aspect of caging experiments is therefore the adequacy of partial cages as controls. Until a study of cages and partial cages is performed in a flume in a variety of flow conditions, controversy will continue over the interpretation of caging experiments. Virnstein (115, 116) offers sage advice, urging that the results of caging experiments be augmented with observational data, gut content analysis, or laboratory experiments before acceptance of predation as the process producing any experimental differences. Frid & James (42) caution that lateral migration in and out of cages may confound results. Field enclosures of predators are of limited value in my view because of altered behavior of mobile predators when their movement is restricted.

### Epibenthic Predation

A number of experiments involving exclusion of epibenthic predators resulted in increases of infauna relative to unmanipulated controls (4, 16, 17, 32, 41, 52, 58–61, 76, 78, 88, 110, 111, 114, 136). Epibenthic predation had minimal impact in several studies (14, 15, 44, 72, 73, 97). Specific epibenthic predators that have been identified as significant influences include: gastropods (88, 135), horseshoe crabs (17, 140), portunid crabs (114, 140), other decapod crustaceans (78, 117, 145–147), fish (56, 60, 61, 93, 114), and shorebirds (108, 109). Several workers have used different kinds of exclusions to separate out the effects of different predators. Using cages with differing mesh size, Reise (100) claimed that small predators (shore crabs, shrimp, gobies) have a stronger effect on infaunal abundance than do larger predators (flatfish and birds). Woodin (140) used metal stakes to exclude horseshoe crabs *(Limulus polyphemus)* from experimental areas at an intertidal flat in Virginia; this experiment permitted her to separate the strong predator effect of portunid crabs from the weaker horseshoe crab effect. Using full exclusion cages and cages whose sides were attached to floats that lifted off the bottom when the tide was in, Quammen (96) was able to separate the effects of fish (excluded only from the full cages) from the effects of shorebirds (excluded from both types of cages). Fish were found to be of minor importance, while the effect of shorebirds depended on the habitat. She also demonstrated significant effects of a shore crab. Wilson (134) used full cages and roofs to separate the effects of fish and shorebirds in the Bay of Fundy; both types of predators had significant influences on infaunal density and demography.

For studies in which significant effects of epibenthic predation are detected, two patterns consistently emerge. The first is the obvious increase in infaunal abundance and biomass in response to the exclusion of epibenthic predators. The second effect contradicts the predictions of the intermediate disturbance hypothesis which predicts that at low levels of predation, species diversity should decline as a competitive dominant monopolizes the habitat (86). The consistent pattern in epibenthic exclusions is that diversity either increases or remains unaltered. No study has documented a disproportionate rise in abundance of a competitive dominant. The effect is not unexpected because competitive exclusion rarely occurs in soft-sediment communities (see above). Overgrowth interactions which are the chief source of competitive mortality in hard-substrate marine communities do not occur among infaunal organisms (86).

### *Infaunal Predators*

Predatory infauna, primarily nemerteans and polychaetes, have received considerably less attention than epibenthic predators. However, the existing data indicate that infaunal predators may have strong impacts on community structure. The best analyzed example of infaunal predation has been conducted at intertidal sites in Maine. The predatory polychaete, *Glycera dibranchiata,* preys on the polychaete, *Nereis virens;* experimental exclusion of *G. dibranchiata* results in enhanced survivorship of *N. virens* (3, 4). In turn,

*N. virens* significantly reduces the abundance of the amphipod *Corophium volutator* (4, 22). Commito & Schrader (27) found that the addition of *N. virens* to a community where *C. volutator* is absent resulted in an increase in abundance of other infauna, rather than the predicted decrease due to prey switching by *N. virens*. They postulated that *N. virens* may be removing an intermediate predator, probably the polychaete *Nephtys incisa*, which preys on the infauna.

There are limited additional examples of the influence of infaunal predators. Roe (105) showed that a nemertean predator *(Paranemertes peregrina)* removed 14–35% of the population of a nereid polychaete each year. Reise (98) attributed a decrease in spionid polychaetes to the increased survivorship of infaunal predators within epibenthic predator exclusion cages. Schubert & Reise (110) showed that experimental additions of the predatory polychaete, *Nephtys hombergii*, resulted in significant declines of two burrowing polychaetes. Gut analyses of *N. hombergii* confirmed that these two polychaetes were the major prey items. Rönn et al (107) showed that enhancement of *Nereis diversicolor* resulted in the decrease of chironomid larvae in an oligohaline estuary. Ambrose (5) noted increased emigration by the burrowing amphipod *Rhepoxynius abronius* after the introduction of the predatory infaunal polychaete *Nephtys caeca*.

Commito & Ambrose (24, 25) surveyed the literature for evidence of infaunal predators. Ambrose (6) suggested that infaunal predators should be inserted into the current two-level model of community dynamics (epibenthic predators-infauna). Wilson (132) showed that this model requires that epibenthic predators preferentially remove infaunal predators; otherwise the advantage to the nonpredatory infauna of a decrease in epibenthic predation is exactly counterbalanced by an increase in infaunal predation. Nonetheless, both authors agree (8) that infaunal predation is likely to be of widespread importance in many infaunal communities and deserves more attention.

## *Browsing Predation*

Some predators remove only a portion of their prey. The major prey item of flatfish in the Wadden Sea are spionid palps and bivalve siphons (118). Despite the prevalence of regeneration of body parts in infaunal organisms (91, 127, 146), the importance of browsing predation in the field is poorly understood. Laboratory experiments have demonstrated that the removal of portions of polychaetes results in decreased defecation (proportional to feeding) and tube-building (142) and reduced fecundity (148), although the polychaete *Arenicola marina* tolerates weekly tail amputation with no measurable loss in growth rate (10). While most structures lost to browsing predators are exposed at the sediment surface, some infaunal predators remove portions of prey from below the sediment surface (131). Meiofaunal predators may remove portions of small macrofauna (121). Browsing predation may indirectly affect mortality. As browsing predators remove the tips of siphons of deep-dwelling bivalves, the bivalves are forced to move toward the sediment surface to make contact with the overlying water, making them more susceptible to digging predators (50).

## *Complex Trophic Interactions*

Multiple levels of predation with complex interactions are being documented in infaunal communities. The *Glycera dibranchiata–Nereis virens–Corophium volutator* interaction described above (3, 4) provides one such case. The abundance of *C. volutator* is enhanced by the presence of *G. dibranchiata* which preys on *N. virens,* a major predator of *C. volutator*. Kneib & Stiven (61) demonstrate that the size of the predators may have important ramifications on prey abundance. Using enclosures of three different sizes of the killifish *Fundulus heteroclitus* in a North Carolina salt marsh, they showed that infaunal abundances were highest in the presence of large killifish and least in the presence of small fish or no fish. Their postulate, subsequently confirmed (59), was that large killifish prey on an intermediate predator, the shrimp *Palaemonetes pugio*. Small killifish are too small to prey on the intermediate predator. Later experiments in a Georgia salt marsh (60) documented the reduction of *P. pugio* by the killifish, but only an infaunal sea anemone increased when *P. pugio* abundance declined. The anemone benefits from the presence of a higher level predator. Two levels of predators are frequently documented in seagrass beds (77, 78, 117). Exclusions of fish lead to increased survivorship of decapod crustaceans (shrimp and brachyuran crabs) which reduce infaunal abundance significantly.

The importance of size-selection by predators has received little attention by soft-sediment ecologists. Kent & Day (58) showed that flounder and sandpipers preferentially prey upon large individuals of a nereid polychaete, *Ceratonereis pseudoerythraaensis*. In the absence of epibenthic predation, adult nereids cannibalize smaller worms. Wilson (134) demonstrated that size-selective predation by shorebirds and fish on adults of the amphipod *Corophium volutator* ameliorated competition between adults and juveniles; in the absence of predators, juveniles were forced to emigrate from the vicinity of their parents' burrows.

## *Refuges from Predation*

It is apparent that some environmental features confer protection for infauna from predation. Sediment type affects the digging ability of portunid crabs and hence the susceptibility of infaunal bivalves to crab predation (9). Oystershells confer protection from digging predators (33). Seagrasses provide a deterrent to digging predators (88, 99, 100, 117). Predator exclusions in seagrass beds generally produce little increase in infaunal abundance, indicating their efficacy as refuges (111), while cages in adjacent unvegetated areas result in infaunal abundances similar to those observed within seagrass beds. Concentrations of tubes of onuphid polychaetes which project above the sediment surface confer a refuge from digging predators (69, 139, 140). Several workers have shown that shallow-dwelling species are more susceptible to epibenthic predation than are deeper-dwelling species (13, 52, 75, 114).

## CONCLUSIONS

A recurrent theme in this review is that, although we may understand the mechanisms of a particular competitive or predatory interaction, we lack the ability to generate a priori predictions of the effects of those processes. It must be appreciated that the understanding of infaunal communities has progressed rapidly in the past twenty years. Nevertheless, a unifying theory of soft-sediment community structure does not appear attainable at our present level of understanding. The great variety of sediments inhabited and modified by infaunal organisms have only begun to be studied experimentally. The distributions of infaunal organisms and the outcomes of their biological interactions are clearly affected by the nature of the physical environment (9, 20, 79) and the hydrodynamic regime (37). Experiments similar to those reviewed above performed in different habitats will provide greater insight into environment-infauna interactions. Such data alone will not, however, lead to a unified theory. I conclude by describing what I perceive to be fundamental gaps in our knowledge of how competition and predation operate in marine infaunal communities.

For competitive interactions, mortality is clearly relatively uncommon in infaunal communities. Experiments with hard-bodied invertebrates for which growth and reproduction could be quantified (87, 89, 90) are much more sensitive to competitive effects. Experiments in which the growth of populations of soft-bodied invertebrates can be monitored will be of great interest. The nature of the food of deposit-feeders is a major stumbling-block for any investigator seeking to understand exploitative competition among deposit-feeders. The microscopic analyses of sediments developed by Watling (119) seem promising. Finally, more attention needs to be given to spatial patterns. The proper spatial scale on which observations and experiments are made needs to be determined a priori by direct observation. Often the biologically relevant scale may be on the order of mm (35, 36, 40, 43, 130) whereas a typical sample (e.g., 0.01 $m^2$) is far too coarse to detect the biologically relevant patterns. Finally, further analysis of the sizes of interacting individuals seems to hold great promise for predicting the outcome of competitive encounters.

Predation is a more straightforward and better understood process in that its measurement, the disappearance of organisms, is easier to observe and to quantify. Although some infaunal communities are structured by predation, little effort has been given to understanding the behavior of predators on infaunal organisms in relation to established foraging theories (95). More data on size-selective foraging and on habitat choice would be most welcome. The mobility of many epibenthic predators and the difficulty of understanding their population dynamics are serious roadblocks to a predictive theory. Nevertheless, sufficient data on predator-prey relationships exist to begin their consideration in light of optimal foraging theory.

ACKNOWLEDGMENTS

I gratefully acknowledge the following scientists for valuable discussion and/or comments on the manuscript: W. Ambrose, B. Brown, A. H. Hines, C. H. Peterson, W. Sousa, and S. Woodin.

## *Literature Cited*

1. Aller, R. C., Dodge, R. E. 1974. Animal-sediment relations in a tropical lagoon Discovery Bay, Jamaica. *J. Mar. Res.* 32:209–30
2. Aller, R. C., Yingst, J. Y. 1985. Effects of the marine deposit-feeders *Heteromastus filiformis* (Polychaeta), *Macoma balthica* (Bivalvia), and *Tellina texana* (Bivalvia) on averaged sedimentary solute transport, reaction rates, and microbial distributions. *J. Mar. Res.* 43:615–45
3. Ambrose, Jr., W. G. 1984. Influence of residents on the development of a marine soft-bottom community. *J. Mar. Res.* 42:633–654
4. Ambrose, Jr., W. G. 1984. Influences of predatory polychaetes and epibenthic predators on the structure of a soft-bottom community in a Maine estuary. *J. Exp. Mar. Biol. Ecol.* 81:115–45
5. Ambrose, Jr. W. G. 1984. Increased emigration of the amphipod *Rhepoxynius abronius* (Barnard) and the polychaete *Nephtys caeca* (Fabricius) in the presence of invertebrate predators. *J. Exp. Mar. Biol. Ecol.* 80:67–75
6. Ambrose, Jr., W. G. 1984. Role of predatory infauna in structuring marine soft-bottom communities. *Mar. Ecol. Prog. Ser.* 17:109–15
7. Ambrose, Jr., W. G. 1986. Experimental analysis of density dependent emigration of the amphipod *Rhepoxynius abronius*. *Mar. Behav. Physiol.* 12:209–16
8. Ambrose, Jr., W. G. 1986. Importance of predatory infauna in marine soft-bottom communities: reply to Wilson. *Mar. Ecol. Prog. Ser.* 32:41–45
9. Arnold, W. S. 1984. The effects of prey size, predator size, and sediment composition on the rate of predation of the blue crab, *Callinectes sapidus* Rathbun, on the hard clam, *Mercenaria mercenaria* (Linné). *J. Exp. Mar. Biol. Ecol.* 80:207–19
10. Bergman, M. J. N., Van der Veer, H. W., Karczmarski, L. 1988. Impact of tail-nipping on mortality, growth and reproduction of *Arenicola marina*. *Nether. J. Sea Res.* 22:83–90
11. Bianchi, T. S., Levinton, J. S. 1981. Nutrition and food limitation of deposit-feeders. II. Differential effects of *Hydrobia totteni* and *Ilyanassa obsoleta* on the microbial community. *J. Mar. Res.* 39:547–56
12. Bianchi, T. S., Rice, D. L. 1988. Feeding ecology of *Leitoscoloplos fragilis*. II. Effects of worm density on benthic diatom production. *Mar. Biol.* 99:123–31
13. Blundon, J. A., Kennedy, V. S. 1982. Refuges for infaunal bivalves from blue crab, *Callinectes sapidus* (Rathbun), predation in Chesapeake Bay. *J. Exp. Mar. Biol. Ecol.* 65:67–81
14. Boates, J. S., Smith, P. C. 1979. Length-weight relationships, energy content and the effects of predation on *Corophium volutator* (Pallas) (Crustacea: Amphipoda). *Proc. Nova Scotia Inst. Sci.* 29:489–99

15. Bonsdorff, E., Mattila, J., Rönn, C., Österman, C.-S. 1986. Multidimensional interactions in shallow soft-bottom ecosystems; testing the competitive exclusion principle. *Ophelia, Suppl.* 4:37–44
16. Bottom, M. L. 1984. Effects of Laughing Gull and shorebird predation on the intertidal fauna at Cape May, New Jersey. *Est. Coast. Shelf Sci.* 18:209–20
17. Botton, M. L. 1984. The importance of predation by horseshoe crabs, *Limulus polyphemus,* to an intertidal sand flat community. *J. Mar. Res.* 42:139–61
18. Brenchley, G. A. 1981. Disturbance and community structure: an experimental study of bioturbation in marine soft-bottom environments. *J. Mar. Res.* 39:767–90
19. Brenchley, G. A. 1982. Mechanisms of spatial competition in marine soft-bottom communities. *J. Exp. Mar. Biol. Ecol.* 60:17–33
20. Brown, B. 1982. Spatial and temporal distribution of a deposit-feeding polychaete on a heterogeneous tidal flat. *J. Exp. Mar. Biol. Ecol.* 65:213–27
21. Carlson, D. J., Townsend, D. W., Hilyard, A. L., Eaton, J. F. 1984. Effect of an intertidal mudflat on plankton of the overlying water column. *Can. J. Fish. Aquat. Sci.* 41:1523–28
22. Commito, J. A. 1982. Importance of predation by infaunal polychaetes in controlling the structure of a soft-bottom community in Maine, USA. *Mar. Biol.* 68:77–81
23. Commito, J. A. 1987. Adult-larval interactions: predictions, mussels and cocoons. *Est. Coast. Shelf Sci.* 25:599–606
24. Commito, J. A., Ambrose, Jr., W. G. 1985. Multiple trophic levels in soft-bottom communities. *Mar. Ecol. Prog. Ser.* 26:289–93
25. Commito, J. A., Ambrose, Jr., W. G. 1985. Predatory infauna and trophic complexity in soft-bottom communities. *Proc. Eur. Mar. Biol. Symp., 19th, Cambridge, 1985,* 323–33
26. Commito, J. A., Boncavage, E. M. 1989. Suspension-feeders and coexisting infauna: an enhancement counterexample. *J. Exp. Mar. Biol. Ecol.* 125:33–42
27. Commito, J. A., Shrader, P. B. 1985. Benthic community response to experimental additions of the polychaete *Nereis virens. Mar. Biol.* 86:101–107
28. Connell, J. H. 1961. The influence of interspecific competition and other factors on the distribution of the barnacle *Chthamalus stellatus. Ecology* 42:710–23
29. Connell, J. H., Slatyer, R. O. 1977. Mechanisms of succession in natural communities and their role in community stability and organization. *Am. Natur.* 111:1119–44
30. Connor, M. S., Teal, J. M., Valiela, I. 1982. The effect of feeding by mud snails, *Ilyanassa obsoleta* (Say), on the structure and metabolism of a laboratory benthic algal community. *J. Exp. Mar. Biol. Ecol.* 65:29–45
30a. Coull, B. C., ed. 1977. *Ecology of Marine Benthos.* Columbia: Univ. S. Carolina Press. 467 pp.
31. Croker, R. A., Hatfield, E. B. 1980. Space partitioning and interactions in an intertidal sand-burrowing amphipod guild. *Mar. Biol.* 61:79–88
32. Dauer, D. M., Ewing, R. M., Tourtellotte, G. H., Harlan, W. T., Sourbeer, J. W., Barker, H. R. Jr. 1982. Predation, resource limitation and the structure of benthic infaunal communities of the lower Chesapeake Bay. *Int. Revue ges. Hydrobiol.* 67:477–89
33. Dauer, D. M., Tourtellotte, G. H., Ewing, R. M. 1982. Oyster shells and artificial worm tubes: the role of refuges in structuring benthic communities of the lower Chesapeake Bay. *Int. Revue ges. Hydrobiol.* 67:661–77
34. Dayton, P. K., Oliver, J. S. 1980. An evaluation of experimental analyses of population and community patterns in benthic marine environments. In *Marine Benthic Dynamics,* ed. K. R. Tenore, B. C. Coull, pp. 93–120. Columbia, Univ. S. Carolina Press. 451 pp.
35. DeWitt, T. H., Levinton, J. S. 1985. Disturbance, emigration, and refugia: how the mud snail, *Ilyanassa obsoleta* (Say), affects the habitat distribution of an epifaunal amphipod, *Microdeutopus gryllotalpa* (Costa). *J. Exp. Mar. Biol. Ecol.* 92:97–113
36. Eckman, J. E. 1979. Small-scale patterns and processes in a soft-substratum, intertidal community. *J. Mar. Res.* 37:437–57
37. Eckman, J. E. 1987. Hydrodynamic processes affecting benthic recruitment. *Limnol. Oceanogr.* 28:241–57
38. Eckman, J. E., Nowell, A. R. M., Jumars, P. A. 1981. Sediment destabilization by animal tubes. *J. Mar. Res.:* 39:361–74
39. Elmgren, R., Ankar, S., Marteleur, B., Ejdung, G. 1986. Adult interference with postlarvae in soft sediments: the *Pontoporeia-Macoma* example. *Ecology* 67:827–36
40. Ertman, S. C., Jumars, P. A. 1988. Effects of bivalve siphonal currents on

the settlement of inert particles and larvae. *J. Mar. Res.* 46:797–813
41. Frid, C. L. J., James, R. 1988. The role of epibenthic predators in structuring the marine invertebrate community of a British coastal salt marsh. *Nether. J. Sea Res.* 22:307–14
42. Frid, C. L. J. 1989. The role of recolonization processes in benthic communities, with special reference to the interpretation of predator-induced effects. *J. Exp. Mar. Biol. Ecol.* 126: 163–71
43. Gallagher, E. D., Jumars, P. A., Trueblood, D. D. 1983. Facilitation of soft-bottom benthic succession by tube builders. *Ecology* 64:1200–16
44. Gee, J. M., Warwick, R. M., Davey, J. T., George, C. L. 1985. Field experiments on the role of epibenthic predators in determining prey densities in an estuarine mudflat. *Est. Coast. Shelf Sci.* 21:429–48
45. Grant, J. 1981. Dynamics of competition among estuarine sand-burrowing amphipods. *J. Exp. Mar. Biol. Ecol.* 49:255–65
46. Grant, J., Bathmann, U. V., Mills, E. L. 1986. The interaction between benthic diatom films and sediment transport. *Est. Coast. Shelf Sci.* 23:225–38
47. Harper, J. L. 1977. *Population Biology of Plants*. New York: Academic. 892 pp.
48. Harrison, P. G. 1987. Natural expansion and experimental manipulation of seagrass (*Zostera* spp.) abundance and the response of infaunal invertebrates. *Est. Coast Shelf Sci.* 24:799–812
49. Highsmith, R. C. 1982. Induced settlement and metamorphosis of sand dollar *(Dendraster excentricus)* larvae in predator-free sites: adult sand dollar beds. *Ecology* 63:329–37
50. Hines, A. H., Posey, M. H. 1989. Complex trophic interactions in estuarine food webs: jargon, concepts and examples. *Am. Zool.* 29:27A
51. Hines, A. H., Posey, M. H., Haddon, P. J. 1989. Effects of adult suspension- and deposit-feeding bivalves on recruitment of estuarine infauna. *Veliger* 32:109–19
52. Holland, A. F., Mountford, N. K., Hiegel, M. H., Kaumeyer, K. R., Mihursky, J. A. 1980. Influence of predation on infaunal abundance in upper Chesapeake Bay, USA. *Mar. Biol.* 57:221–35
53. Hulberg, L. W., Oliver, J. S. 1980. Caging manipulations in marine soft-bottom communities: importance of animal interactions or sedimentary habitat modifications. *Can. J. Fish. Aquat. Sci.* 37:1130–39
54. Hunt, J. H., Ambrose, W. G. Jr., Peterson, C. H. 1987. Effects of the gastropod, *Ilyanassa obsoleta* (Say), and the bivalve *Mercenaria mercenaria* (L.), on larval settlement and juvenile recruitment of infauna. *J. Exp. Mar. Biol. Ecol.* 108:229–40
55. Jensen, K. T. 1985. The presence of the bivalve *Cerastoderma edule* affects migration, survival and reproduction of the amphipod *Corophium volutator*. *Mar. Ecol. Prog. Ser.* 25:269–77
56. Joyce, A. A., Weisberg, S. B. 1986. The effects of predation by the mummichog, *Fundulus heteroclitus* (L.), on the abundance and distribution of the salt marsh snail, *Melampus bidentatus* (Say). *J. Exp. Mar. Biol. Ecol.* 100: 295–306
57. Jumars, P. A., Nowell, A. R. M. 1984. Effects of benthos on sediment transport: difficulties with functional grouping. *Cont. Shelf Res.* 3:115–30
58. Kent, A. C., Day, R. W. 1983. Population dynamics of an infaunal polychaete: the effect of predators and an adult-recruit interaction. *J. Exp. Mar. Biol. Ecol.* 73:185–203
59. Kneib, R. T. 1985. Predation and disturbance by grass shrimp, *Palaemonetes pugio* Holthuis, in soft-substratum benthic invertebrate assemblages. *J. Exp. Mar. Biol. Ecol.* 93:91–102
60. Kneib, R. T. 1988. Testing for indirect effects of predation in an intertidal soft-bottom community. *Ecology* 69:1795–1805
61. Kneib, R. T., Stiven, A. E. 1982. Benthic invertebrate responses to size and density manipulations of the common mummichog, *Fundulus heteroclitus,* in an intertidal salt marsh. *Ecology* 63:1518–32
62. Levin, L. A. 1981. Dispersion, feeding behavior and competition in two spionid polychaetes. *J. Mar. Res.* 39:99–117
63. Levin, L. A. 1982. Interference interactions among tube-dwelling polychaetes in a dense infaunal assemblage. *J. Exp. Mar. Biol. Ecol.* 65:107–19
64. Levinton, J. S. 1977. See Ref. 30a, pp. 191–227
65. Levinton, J. S. 1979. The effect of density upon deposit-feeding populations: movement, feeding and floating of *Hydrobia ventrosa* Montagu (Gastropoda: Prosobranchia). *Oecologia* 43:27–39
66. Levinton, J. S., Stewart, S., DeWitt, T. H. 1985. Field and laboratory experiments on interference between *Hy-*

*drobia totteni* and *Ilyanassa obsoleta* (Gastropoda) and its possible relation to seasonal shifts in vertical mudflat zonation. *Mar. Ecol. Prog. Ser.* 22:53–58
67. Lopez, G. R., Levinton, J. S. 1987. Ecology of deposit-feeding animals in marine sediments. *Q. Rev. Biol.* 62:235–60
68. Luckenbach, M. W. 1986. Sediment stability around animal tubes: the roles of hydrodynamic processes and biotic activity. *Limnol. Oceanogr.* 31:779–87
69. Luckenbach, M. W. 1987. Effects of adult infauna on new recruits: implications for the role of biogenic refuges. *J. Exp. Mar. Biol. Ecol.* 105:197–206
70. McCall, P. L. 1977. Community patterns and adaptive strategies of the infaunal benthos of Long Island Sound. *J. Mar. Res.* 35:221–66
71. McCall, P. L. 1978. See Ref. 125a, pp. 191–219
72. Mahoney, B. M. S., Livingston, R. J. 1982. Seasonal fluctuations of benthic macrofauna in the Apalachicola estuary, Florida, USA: the role of predation. *Mar. Biol.* 69:207–13
73. Mattila, J., Bonsdorff, E. 1989. The impact of fish predation on shallow soft bottoms in brackish waters (SW Finland): an experimental study. *Nether. J. Sea Res.* 23:69–81
74. Maurer, D. 1983. The effect of an infaunal suspension feeding bivalve *Mercenaria mercenaria* (L.) on benthic recruitment. *Mar. Ecol.* 4:263–74
75. Myers, J. P., Williams, S. L., Pitelka, F. A. 1980. An experimental analysis of prey availability for sanderlings (Aves: Scolopacidae) feeding on sandy beach crustaceans. *Can. J. Zool.* 58:1564–74
76. Naqvi, S. M. Z. 1966. Effects of predation on infaunal invertebrates of Alligator Harbor, Florida. *Fla. Scient.* 29:313–21
77. Nelson, W. G. 1979. An analysis of structural pattern in an eelgrass (*Zostera marina* L.) amphipod community. *J. Exp. Mar. Biol. Ecol.* 39:231–64
78. Nelson, W. G. 1981. Experimental studies of decapod and fish predation on seagrass macrobenthos. *Mar. Ecol. Prog. Ser.* 5:141–49
79. Ólaffson, E. B. 1986. Density dependence in suspension-feeding and deposit-feeding populations of the bivalve *Macoma balthica*: a field experiment. *J. Anim. Ecol.* 55:517–26
80. Ólaffson, E. B. 1988. Inhibition of larval settlement to a soft bottom benthic community by drifting algal mats: an experimental test. *Mar. Biol.* 97:571–74
81. Ölaffson, E. B., Persson, L.-E. 1986. The interaction between *Nereis diversicolor* O. F. Müller and *Corophium volutator* Pallas as a structuring force in a shallow brackish sediment. *J. Exp. Mar. Biol. Ecol.* 103:103–17
82. Oliver, J. S., Oakden, J. M. Slattery, P. N. 1982. Phoxocephalid amphipod crustaceans as predators on larvae and juveniles in marine soft-bottom communities. *Mar. Ecol. Prog. Ser.* 7:179–84
83. Oliver, J. S., Slattery, P. N. 1985. Effects of crustacean predators on species composition and population structure of soft-bodied infuana from McMurdo Sound, Antarctica. *Ophelia* 24:155–75
84. Paine, R. T. 1966. Food web complexity and species diversity. *Am. Natur.* 100:65–75
85. Peterson, C. H. 1977. Competitive organization of the soft-bottom macrobenthic communities of southern California lagoons. *Mar. Biol.* 43:343–59
86. Peterson, C. H. 1979. Predation, competitive exclusion, and diversity in the soft-sediment benthic communities of estuaries and lagoons. In *Ecological Processes in Coastal and Marine Systems*, ed. R. J. Livingston, 244–64. New York: Plenum. 477 pp.
87. Peterson, C. H. 1982. The importance of predation and intra- and interspecific competition in the population biology of two infaunal suspension-feeding bivalves, *Protothaca staminea* and *Chione undatella*. *Ecol. Monogr.* 52:437–75
88. Peterson, C. H. 1982. Clam predation by whelks (*Busycon* spp.): experimental tests of the importance of prey size, prey density, and seagrass cover. *Mar. Biol.* 66:159–70
89. Peterson, C. H., Andre, S. V. 1980. An experimental analysis of interspecific competition among marine filter feeders in a soft-sediment environment. *Ecology* 61:129–39
90. Peterson, C. H., Black, R. 1987. Resource depletion by active suspension feeders on tidal flats: influence of local density and tidal elevation. *Limnol. Oceanogr.* 32:143–66
91. Peterson, C. H., Quammen, M. L. 1982. Siphon nipping: its importance to small fishes and its impact on growth of the bivalve *Protothaca staminea* (Conrad). *J. Exp. Mar. Biol. Ecol.* 63:249–68

92. Posey, M. H. 1986. Changes in a benthic community associated with dense beds of a burrowing deposit feeder, *Callianassa californiensis*. *Mar. Ecol. Prog. Ser.* 31:15–22
93. Posey, M. H. 1986. Predation on a burrowing shrimp: distribution and community consequences. *J. Exp. Mar. Biol. Ecol.* 103:143–61
94. Posey, M. H. 1987. Influence of relative mobilities on the composition of benthic communities. *Mar. Ecol. Prog. Ser.* 39:99–104
95. Pyke, G. H., Pulliam, H. R., Charnov, E. L. 1977. Optimal foraging: a selective review of theory and tests. *Q. Rev. Biol.* 52:137–54
96. Quammen, M. L. 1984. Predation by shorebirds, fish, and crabs on invertebrates in intertidal mudflats: an experimental test. *Ecology* 65:529–37
97. Raffaelli, D., Milne, H. 1987. An experimental investigation of the effects of shorebird and flatfish predation on estuarine invertebrates. *Est. Coast. Shelf Sci.* 24:1–13
98. Reise, K. 1977. Predator exclusion experiments in an intertidal mud flat. *Helgo. wiss. Meeresunters.* 30:263–71
99. Reise, K., 1977. Predation pressure and community structure of an intertidal soft-bottom fauna. In *Biology of Benthic Organisms,* ed. F. Keegan, P. O. Ceidigh, P. J. Boaden, pp. 513–19. New York: Pergamon. 495 pp.
100. Reise, K. 1978. Experiments on epibenthic predation in the Wadden Sea. *Helgo. wiss. Meeresunters.* 31:55–101
101. Reise, K. 1983. Biotic enrichment of intertidal sediments by experimental aggregates of the deposit-feeding bivalve *Macoma balthica*. *Mar. Ecol. Prog. Ser.* 12:229–36
102. Rhoads, D. C. 1974. Organism-sediment relations on the muddy sea floor. *Oceanogr. Mar. Biol. Ann. Rev.* 12:263–300
103. Rhoads, D. C., Young, D. K. 1970. The influences of deposit-feeding organisms on sediment stability and community trophic structure. *J. Mar. Res.* 28;150–78
104. Roe, P. 1975. Aspects of life history and of territorial behaviour in young individuals of *Platynereis bicanaliculata* and *Nereis vexillosa* (Annelida, Polychaeta). *Pac. Sci.* 29:341–48
105. Roe, P. 1976. Life history and predator-prey interactions of the nemertean *Paranemertes peregrina* Coe. *Biol. Bull.* 150:80–106
106. Ronan, T. E. Jr., 1975. *Structural and paleoecological aspects of a modern soft-sediment community: an experimental field study*. PhD thesis. Univ. Calif., Davis. 220 pp.
107. Rönn, C., Bonsdorff, E., Nelson, W. G. 1988. Predation as a mechanism of interference within infauna in shallow brackish water soft bottoms: experiments with an infauna predator, *Nereis diversicolor* O. F. Müller. *J. Exp. Mar. Biol. Ecol.* 116:143–57
108. Schneider, D. 1978. Equalisation of prey numbers by migratory shorebirds. *Nature* 271:371–72
109. Schneider, D., Harrington, B. A. 1981. Timing of shorebird migration in relation to prey depletion. *Auk* 98:801–11
110. Schubert, A., Reise, K. 1986. Predatory effects of *Nephtys hombergii* on other polychaetes in tidal sediments. *Mar. Ecol. Prog. Ser.* 34:117–24
111. Summerson, H. C., Peterson, C. H. 1984. Role of predation in organizing benthic communities of a temperate-zone seagrass bed. *Mar. Ecol. Prog. Ser.* 15:63–77
112. Tamaki, A. 1985. Detection of non-interference within a mobile polychaete species. *J. Exp. Mar. Biol. Ecol.* 90:277–87
113. Tamaki, A. 1985. Inhibition of larval recruitment of *Armandia* sp. (Polychaeta: Opheliidae) by established adults of *Pseudopolydora paucibranchiata* (Okuda) (Polychaeta: Spionidae) on an intertidal sand flat. *J. Exp. Mar. Biol. Ecol.* 87:67–82
114. Virnstein, R. W. 1977. The importance of predation by crabs and fishes on benthic infauna in Chesapeake Bay. *Ecology* 58:1199–17
115. Virnstein, R. W. 1978. See Ref. 125a, pp. 261–73
116. Virnstein, R. W. 1980. Measuring effects of predation on benthic communities in soft sediments. In *Estuarine Perspectives,* ed. M. L. Wiley, pp. 281–290. New York: Academic. 563 pp.
117. Virnstein, R. W., Mikkelsen, P. S., Cairns, K. D., Capone, M. A. 1983. Seagrass beds versus sand bottoms: the trophic importance of their associated benthic invertebrates. *Fla. Scient.* 46:363–81
118. Vlas, J. de. 1979. Annual food intake by plaice and flounder in a tidal flat area in the Dutch Wadden Sea, with special reference to consumption of regenerating parts of macrobenthic prey. *Nether. J. Sea Res.* 13:117–53
119. Watling, L. 1988. Small-scale features of marine sediments and their im-

portance to the study of deposit-feeding. *Mar. Ecol. Prog. Ser.* 47:135–44

120. Watzin, M. C. 1983. The effects of meiofauna on settling macrofauna: meiofauna may structure macrofaunal communities. *Oecologia* 59:163–66
121. Watzin, M. C. 1985. Interactions among temporary and permanent meiofauna: observations on the feeding and behavior of selected taxa. *Biol. Bull.* 169:397–416
122. Watzin, M. C. 1986. Larval settlement into marine soft-sediment systems: interactions with the meiofauna. *J. Exp. Mar. Biol. Ecol.* 98:65–113
123. Weinberg, J. R. 1984. Interactions between functional groups in soft-substrata: do species differences matter? *J. Exp. Mar. Biol. Ecol.* 80:11–28
124. Whitlatch, R. B. 1981. Animal-sediment relationships in intertidal marine benthic habitats: some determinants of deposit-feeding species diversity. *J. Exp. Mar. Biol. Ecol.* 53:31–45
125. Whitlatch, R. B., Zajac, R. N. 1985. Biotic interactions among estuarine infaunal opportunistic species. *Mar. Ecol. Prog. Ser.* 21:299–311

125a. Wiley, M. L., ed. 1978. *Estuarine Interactions.* New York: Academic. 603 pp.

126. Williams, J. G. 1980. The influence of adults on the settlement of spat of the clam, *Tapes japonica. J. Mar. Res.* 38:729–41
127. Wilson, W. H. Jr. 1979. Community structure and species diversity of the sediment reefs constructed by the maldanid polychaete *Petaloproctus socialis* (Polychaeta: Maldanidae). *J. Mar. Res.* 37:623–41
128. Wilson, W. H. Jr. 1980. A laboratory investigation of the effect of a terebellid polychaete on the survivorship of nereid polychaete larvae. *J. Exp. Mar. Biol. Ecol.* 46:73–80
129. Wilson, W. H. Jr., 1981. Sediment-mediated interactions in a densely populated infaunal assemblage: the effects of the polychaete *Abarenicola pacifica. J. Mar. Res.* 39:735–48
130. Wilson, W. H. Jr. 1983. The role of density dependence in a marine infaunal community. *Ecology* 64:295–306
131. Wilson, W. H. Jr. 1984. An experimental analysis of spatial competition in a dense infaunal community: the importance of relative effects. *Est. Coast. Shelf Sci.* 18:673–84
132. Wilson, W. H. Jr. 1986. Importance of predatory infauna in marine soft-sediment communities. *Mar. Ecol. Prog. Ser.* 32:35–40
133. Wilson, W. H. Jr. 1988. Shifting zones in a Bay of Fundy soft-sediment community: patterns and processes. *Ophelia* 29:227–45
134. Wilson, W. H. Jr. 1989. Predation and the mediation of intraspecific competition in an infaunal community in the Bay of Fundy. *J. Exp. Mar. Biol. Ecol.* 132:221–45
135. Wiltse, W. I. 1980. Effects of *Polinices duplicatus* (Gastropoda: Naticidae) on infaunal community structure at Barnstable Harbor, Massachusetts, USA. *Mar. Biol.* 56:301–10
136. Wiltse, W. I., Foreman, K. H., Teal, J. M., Valiela, I. 1984. Effects of predators and food resources on the macrobenthos of salt marsh creeks. *J. Mar. Res.* 42:923–42
137. Woodin, S. A. 1974. Polychaete abundance patterns in a marine soft-sediment environment: the importance of biological interactions. *Ecol. Monogr.* 44:171–87
138. Woodin, S. A. 1976. Adult-larval interactions in dense infaunal assemblages: patterns of abundance. *J. Mar. Res.* 34:25–41
139. Woodin, S. A. 1978. Refuges, disturbance, and community structure: a marine soft-bottom example. *Ecology* 59:274–84
140. Woodin, S. A. 1981. Disturbance and community structure in a shallow water sand flat. *Ecology* 62:1052–66
141. Woodin, S. A. 1983. Biotic interactions in Recent marine sedimentary environments. In *Biotic Interactions in Recent and Fossil Benthic Communities,* ed. M. J. S. Tevesz, P. L. McCall, pp. 3–38. New York: Plenum. 336 pp.
142. Woodin, S. A. 1984. Effects of browsing predators: activity changes in infauna following tissue loss. *Biol. Bull.* 166:558–73
143. Woodin, S. A. 1985. Effects of defecation by arenicolid polychaete adults on spionid polychaete juveniles in field experiments: selective settlement or differential mortality. *J. Exp. Mar. Biol. Ecol.* 87:119–32
144. Woodin, S. A., Jackson, J. B. C. 1979. Interphyletic competition among marine benthos. *Am. Zool.* 19:1029–43
145. Young, D. K., Young, M. W. 1977. See Ref. 30a, pp. 359–81
146. Young, D. K., Young, M. W. 1978. Regulation of species densities of seagrass-associated macrobenthos: evidence from field experiments in the Indian River estuary, Florida. *J. Mar. Res.* 36:569–93
147. Young, D. K., Buzas, M. A., Young,

M. W. 1976. Species densities of macrobenthos associated with seagrass: a field experimental study of predation. *J. Mar. Res.* 34:577–92

148. Zajac, R. N. 1985. The effects of sublethal predation on reproduction in the spionid polychaete *Polydora ligni* Webster. *J. Exp. Mar. Biol. Ecol.* 88:1–19
149. Zajac, R. N., Whitlatch, R. B. 1982a. Responses of estuarine infauna to disturbance. I. Spatial and temporal variation of initial recolonization. *Mar. Ecol. Prog. Ser.* 10:1–14
150. Zajac, R. N., Whitlatch, R. B. 1982b. Responses of estuarine infauna to disturbance. II. Spatial and temporal variation of succession. *Mar. Ecol. Prog. Ser.* 10:15–27

*Annu. Rev. Ecol. Syst. 1990. 21:243–73*


# HOST SPECIALIZATION IN PHYTOPHAGOUS INSECTS

*John Jaenike*

Department of Biology, University of Rochester, Rochester, New York 14627

KEY WORDS: behavioral ecology, diet models, herbivory, host-plant selection, niche breadth

---

## INTRODUCTION

Insects are by far the most diverse group of organisms on Earth; estimates of their current diversity range as high as 30 million species (58). A large fraction of these species feed on plants (154, 214), and Mitter et al (134) have shown that plant-feeding clades are consistently much more diverse than their nonphytophagous sister groups. By feeding at the base of the food chain, these insects have access to a potentially enormous supply of resources; these are extensively subdivided among species, as most phytophagous insects are highly host specific (55, 56, 133, 154). Investigations of the causes of host specialization in insects could therefore contribute substantially to our understanding of the origin and maintenance of diversity in this group.

The evolution of plant-insect associations has been guided to a large extent by plant chemistry in some, but not all, insect groups (26, 55, 57, 153). For instance, related species of butterflies often use plants that are chemically similar, even if taxonomically distant (23, 57, 188, 190). Cladistic analyses of papilionid butterflies and *Ophraella* leaf beetles and their host plants have revealed that host shifts are most likely to occur among chemically similar plants, even though cladograms of these insects and their host plants are not congruent (68, 131, 132). In addition, introduced plants are generally colonized by insects that feed on chemically similar plants; those with impoverished faunas are biochemically or taxonomically unusual (214, 250). Finally, within a defined set of hosts, insect herbivores have been found to



0066-4162/90/1120-0243$02.00

distribute themselves among plants according to secondary chemistry (17, 149, see also 209). Plant chemistry, therefore, plays an important role in determining the variety of plants that can be exploited by an insect species. The actual diet, however, may be significantly narrower, because it is dependent on a number of ecological variables in addition to plant chemistry.

Because adaptive evolutionary changes are guided by natural selection, yet subjected to biological constraints, it is necessary to consider both the proximate and ultimate (selective) factors affecting host use in order to understand the evolution of host specialization in insects. Here I consider some recent developments in the theory of diet breadth in insects before moving on to assess some of the proximate and selective factors affecting host use in natural populations. I conclude with a brief treatment of life history correlates of diet breadth in insects.

# MODELS OF HOST SPECIALIZATION

## *Optimality Models*

Traditional models of optimal diet breadth in insects focus on insects in a steady state condition and ask what behavior will maximize the product of an individual's realized fecundity per unit time and expected fitness of offspring. Among the predictions of most rate maximization models are: (*a*) a given host type should always be accepted or always rejected—there should be no partial preferences; (*b*) hosts should be added to the diet in order of decreasing suitability for offspring development and independently of their abundance; (*c*) monophagy is favored if the most suitable host is easy to find or if the hosts in an expanded diet would be of low suitability; and (*d*) factors restricting search time, such as short adult lifespan, favor polyphagy (32, 67, 123, 163).

Dynamic state variable models include as an integral component the state of the insect throughout the host selection process (93, 94, 124–126). Among the variables that are expected to influence behavior are age and current egg load. Older insects or those carrying many eggs should be more willing to accept low-quality hosts. While the predictions of these models are in many respects similar to those of the rate maximization models, the state variable models predict that optimally behaving insects will exhibit partial preferences for certain hosts.

## *Population Genetic Models*

Evolutionary changes in host acceptance behavior should be contingent upon the intrinsic suitabilities of various host species as well as on the extent to which host quality changes as a function of use by the insects. Suppose that oviposition behavior is governed by one genetic locus, and relative offspring fitness, assumed to be independent of the number of insects using a particular host species, is governed by another. If a newly encountered host plant (e.g. a resistant crop variety or an introduced weed) is accepted for oviposition but is less suitable for development than are ancestral hosts, then an insect population is expected to evolve behavioral avoidance or physiological tolerance of the new host, but not both (30, 76, 170). The evolutionary outcome is critically dependent on the initial genetic composition of the population.

Incorporation of a new host into an insect's diet may require changes both in loci that affect host acceptance behavior and in those that affect offspring performance (28). Because neither acceptance nor physiological tolerance is likely to evolve in the absence of the other, Rausher (170) has shown theoretically that specialization on a set of chemically similar hosts may be an evolutionary dead-end. The evolution of host specialization from generalization should be much easier, in theory, as change in either behavior or physiology is sufficient.

The genetic and optimality models discussed above predict that all members of a population will adopt similar behavioral responses to various host plants. However, if offspring fitness declines as the number of insects using a particular host increases, genetic polymorphism at a single locus governing host use can be maintained (167, 171). This leads to frequency-dependent selection in favor of individuals preferring relatively underutilized plants. Genetic polymorphism for host preference is expected to be lost under conditions of hard selection, where fitness is independent of insect density (103, 170, 217). However, even a small amount of soft selection suffices to relax greatly the conditions for maintenance of such polymorphism (85). Thus, the available models suggest that the manner of population regulation can have a major effect on the maintenance of genetic polymorphism for host use. It would be interesting to see if similar conclusions were to emerge from polygenic models of host selection, as such behavior is likely to be based on more than one locus.

### *Physiological State Models*

Models of how changing physiological state affects insect oviposition were pioneered by Singer (192, 193) and have subsequently been elaborated by others (47, 130). A general version is presented in Figure 1, where it is assumed that an insect will oviposit when the stimulus it perceives from a potential host plant exceeds some threshold. Perceived stimulus strength depends on both plant (e.g. secondary compounds) and insect (e.g. genetics, learning) qualities. As with most behaviors, the threshold is believed to be variable, declining with search time for an oviposition site, current egg load, etc. The rate at which the threshold declines can vary among insects, depending, for instance, on the rate at which eggs are produced. These models predict that immediately after oviposition, no hosts will be acceptable for oviposition. Subsequently, the top-ranked host, but no others, become acceptable during a *discrimination phase*. Other hosts become acceptable in a sequence reflecting an insect's *rank order preference* and at a rate indicative of its *specificity* for higher ranked hosts. If the top-ranked host is rarely encountered, then insects will more often accept lower ranking hosts for oviposition, resulting in a broader host range.

Courtney et al (47) predict that if insects vary in fecundity, and thus in the rate at which the threshold stimulus decreases with time, then probabilities of acceptance of different low-ranking hosts should be positively correlated. If the use of a novel host evolves by increasing the probability of acceptance of low-ranking hosts in general, then populations using such hosts should readily accept ancestral, higher-ranking hosts.

An optimally behaving insect (i.e. one behaving in a manner that maximizes fitness) should accept low-ranking hosts when its egg load is high, because the future number of host encounters is unlikely to be sufficient to ensure that all eggs will be laid on high-ranking hosts (125). Thus, optimality and physiological state models yield similar predictions regarding current egg load and acceptance of low-ranking hosts.

## PROXIMATE CAUSES OF INTRASPECIFIC VARIATION IN HOST USE

Because patterns of host use vary among conspecific insects and from one time to another within the life of an individual, the host range of a species will depend on the proximate causes of this variation. Since many insects can feed on a larger variety of plants than are actually used in nature (see next section),

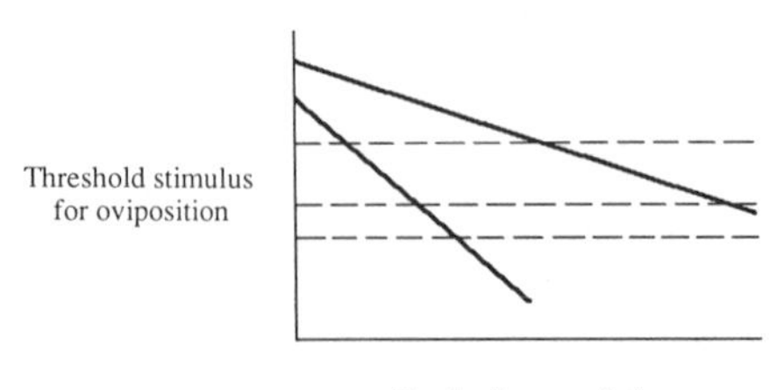


*Figure 1* Physiological state model of host acceptance behavior. Solid lines indicate the threshold stimulus necessary to elicit oviposition, which is assumed to decrease with current egg load or search time for an oviposition site. The slope of these lines depends in part on the rate at which mature eggs are produced. The two solid lines represent insects that differ in rate of egg production. Dashed lines represent the effective stimulus strength provided by potential host plants. These are functions of both plant and insect (e.g., learning and genetic variation) qualities. (Adapted from Ref. 47).

host use is often determined by adult behavior (67, 108), and that will be the primary focus of this discussion.

Interpopulation variation in host use can result simply from the occurrence of different plants in different regions within an insect's geographical range (64). This may even be the primary cause of differences in host use among allopatric species of insects (185, 186). In a number of species, however, some populations fail to use a particular plant that is used elsewhere (24, 53, 64, 137, 192, 214, 219). Because host use is the result of an interaction between plant stimuli and insect responses, such geographic variation in host use cannot, without further study, be attributed to either plant or insect characteristics.

Genetic variation among populations in response to hosts has been found in a few species. Apple populations of the apple maggot, *Rhagoletis pomonella,* are more likely to accept apple for oviposition than are hawthorn populations, yet both prefer hawthorns, the ancestral host (159). Genetically based differences in host acceptance have been found between US mainland and Virgin Island populations of the tobacco budworm, *Heliothis virescens,* (187, 232), but different mainland populations are not genetically heterogeneous. Gould (77) and Via (229) conclude that most genetic variation for host response occurs within local populations, although this finding is based on the very few species that have been studied genetically at both local and geographical scales. Variation among populations in larval adaptation has been found in several species (24, 81, 91, 92, 162, 166, 189, 228, 237). In most studies of oviposition behavior or larval performance, however, genetic variation within populations has not been assayed, thus precluding estimates of the true between-population component of variance. At this time, there is little firm evidence that geographic variation in host use is based primarily on genetic variation among insect populations.

Environmentally based variation among insects could also lead to geographical differences in host use. For instance, in areas where top-ranked hosts are abundant and oviposition frequent, thresholds for host acceptance may remain high, thus precluding use of low-ranking hosts. Where favored plants are rare or their presence is masked by associated members of the plant community (8, 206, 216, 218), thresholds for host acceptance are expected to fall, making the use of other plants more likely (138, 195, 208, 242).

The remainder of this discussion concerns genetic and nongenetic causes of variation in host use *within* populations of insects.

## *Genetic Variation*

Genetic variation for host selection has been documented in a variety of insects recently, and since this subject has been reviewed recently (70, 103, 229), the comments here serve only as supplemental notes. A genetic component to variation in host response in the field has been found in several species of *Drosophila* (12, 87, 97). For instance, isofemale strains of *D. tripunctata* differ consistently in their probabilities of capture at fruit and mushroom baits (97). Heritable variation for such behavior has also been found in *D. melanogaster:* The offspring of flies captured at a given fruit type are themselves more likely than average to be captured at such fruit (87). It should be noted that none of these studies employed natural distributions of resources—all used regular arrays of baits. If the rate of host finding is greater under such conditions, genetically based variation in host ranking and specificity is more likely to be expressed (Figure 1). With natural distributions of resources, genetic variation for host response may be effectively neutral, especially if the threshold for host acceptance is so low that individuals, regardless of their genotype, oviposit on most hosts encountered (102). No studies have yet documented genetically based differences in host response among individuals using natural arrays of resources in the field.

Patterns of host acceptance are genetically variable in many species under laboratory conditions (70, 103). Although rank order preference can vary (99), variation in specificity within a fixed rank order preference is much more common (159, 194, 215, 220, 233, 236).

Genetic variation for acceptance of low-ranking hosts may be effectively neutral if insects rarely reach a physiological state where such resources become acceptable. For example, *Drosophila buzzatii* exhibits little genetic variation for oviposition on the most preferred yeasts, but significantly more variation in acceptance of the lower-ranked yeasts (13). Similarly, in two species of *Papilio* butterflies, interstrain differences in oviposition behavior were proportionately greatest for the lowest ranking host species (220).

Components of host selection are genetically variable in most insects studied, even in monophagous species like the swallowtail *Papilio oregonius* (for exceptions, see 102, 164). The ubiquity of such variation is reminiscent of electrophoretic variation, for which adaptive explanations have been sought, largely in vain, for the past 25 years. Thus, the idea that genetic variation for host selection is neutral and nonadaptive should be considered seriously, especially in the absence of evidence that such variation affects patterns of host use in the wild.

## *Learning*

Among the nongenetic causes of variation in host use, learning has received the most attention (146). Previous experience has a substantial effect on the rate at which particular hosts are found in some butterflies (144, 147, 161, 172, 207). In the pipevine swallowtail, *Battus philenor,* this results in individuals that specialize on either narrow- or broad-leaved *Aristolochia* host plants, with the proportions of the two types governed by the rate at which they alight and oviposit on the "wrong" host and then switch their allegiance (172). In this species, infrequent encounters with the alternative host regulate the dynamics of host finding.

Learning can also affect patterns of host acceptance, occasionally effecting reversals in rank order preference. For instance, *Rhagoletis pomonella* in a field cage are more likely to oviposit on a fruit type with which they have had previous experience (145). Rank order preference for low-ranking hosts is a function of previous experience in *Drosophila melanogaster* (96). Females of *Drosophila buzzatii* exhibit an ontogeny of oviposition preference, as marked preferences for specific yeasts develop only after several days exposure to a variety of yeasts (225). Thus, the effects of learning are more varied than simply reducing the specificity of ovipositing females, as had been suggested by Courtney et al (47).

Mangel & Roitberg (127) argue that encounter rates with various resources provide information about the environment. Since optimal oviposition behavior depends on expected rates of encounter, learning allows an individual to use such information in the host selection process. For instance, as host plant density increases, *Battus philenor* females become less likely to oviposit after alighting on a potential host, perhaps as a result of increased discrimination (165).

Learning could also be advantageous if it predisposed adults to return to hosts similar to those they developed on. This could increase offspring survival if host-specific larval performance has a heritable basis. Although there is no convincing evidence to support Hopkins Host Selection Principle—the idea that larval food induces a preference for that resource as an oviposition site—an ecologically equivalent effect could be produced by early adult experience (40, 111). Early adult experience can affect subsequent capture probability in *Drosophila* (86, 100), but the extent to which this persists and influences oviposition behavior is unknown. It is intriguing that the number of fibers retained in mushroom bodies of *Drosophila* brains, which are important for olfactory learning, is dependent on early adult experience (9).

## *Physiological State*

According to both optimality and physiological state models of oviposition behavior, females should be more willing to accept low ranking hosts as current egg load or search time for an oviposition site increases (47, 125, 192, 193). Observations on a number of species support these ideas. In *Drosophila suboccidentalis* and the bruchid beetle *Callosobruchus maculatus,* more fecund individuals are more likely to accept novel or low-ranking hosts in laboratory assays (43, 234). The probability that females will oviposit on low-ranked hosts increases with the time since the last egg was laid in *Battus philenor* (151), *Rhagoletis pomonella* (181), and the checkerspot butterfly *Euphydryas editha* (193). In *E. editha,* monophagy is most likely in areas where the top-ranked host is so abundant that butterflies rarely reach a physiological state in which low-ranking hosts are accepted for oviposition (196). The tephritid fly *Dacus tryoni* matures eggs continuously in the absence of potential hosts and thus becomes increasingly likely to accept low-ranked hosts as time of deprivation increases (63). Other species of *Dacus* that do not mature eggs in the absence of preferred hosts do not become more prone to accept low-ranked hosts with time. It is noteworthy that *D. tryoni* is polyphagous, whereas the other species are host specialists (63). Deprivation does not always lead to an increase in acceptance of low-ranked hosts, as there are some insects that retain eggs until their death in the absence of preferred hosts (36, 239).

A number of environmental factors can contribute to variation in potential fecundity. 1. *Callosobruchus maculatus* beetles reared from pigeon peas are more fecund than those from azuki beans, and they are more likely to accept pigeon peas, a low-ranking host, for oviposition (233). Thus, larval host can indirectly affect adult host selection. 2. Weather is an important determinant of realized fecundity in many butterflies of temperate regions (34, 42, 49). If females cannot fly but continue to mature eggs under adverse conditions, they may become more prone to accept any hosts when the weather improves. 3. An abundance of host plants facilitates high rates of oviposition and thus could raise the threshold stimulus required for subsequent oviposition. This could account in part for *Battus philenor*'s decline in probability of oviposition with increasing host plant density (165). 4. If conspecific eggs or pheromones deter oviposition (158, 182, 191), this may result in a lower threshold stimulus for oviposition, producing an increased tendency to accept low-ranked hosts.

Oviposition per se is not the only component of host selection affected by egg load. As egg loads increase, *Battus philenor* females spend more time searching for oviposition sites (142), and *Pieris rapae* become less likely to depart from potential host plants (110).

According to dynamic state variable models of optimal oviposition behavior, older females should be more willing to accept low ranking hosts if there is not enough time to find superior hosts for all of their eggs (125). In contrast, because rates of egg production peak early in adult life in many insects (see, e.g., 249), physiological state models predict that older females, since they carry fewer eggs, should be more discriminating in such species. The effect of insect age on host acceptance has generally received little attention. In one of the few studies, *Rhagoletis pomonella* become more likely with age to accept various hosts, in accordance with expectations of optimality models (205).

A general theme of the arguments above is that individuals in good condition carry more eggs and require less stimulus to elicit oviposition, and this leads to acceptance of low-ranking hosts. Physiological stress can also have the opposite effect, if an individual's first priority is its own survival. Hoffmann & Turelli (88; see also 224) show that previously starved adults of *Drosophila melanogaster* are more likely to be captured at poor resources in the field than are healthy, well-fed individuals. Such lessened discrimination by starved flies could lead them to lay their few eggs in low-ranking hosts.

In summary, learning and the various factors influencing the threshold stimulus for oviposition appear to be the most important proximate determinants of variation in host response within natural populations of insects.

## SELECTIVE FACTORS AFFECTING HOST SPECIALIZATION

Most adaptive explanations of diet breadth in phytophagous insects focus on realized adult fecundity and offspring performance as a function of host plant. In what follows, I have grouped these explanations into density-dependent and density-independent mechanisms, where density refers to the subject species, because these can affect diet breadth in fundamentally different ways (Table 1). Other species, such as competitors and enemies, can be considered density-independent components of the host environment (105), unless their impact is directly related to the current density of the subject species.

### *Density-Independent Factors Favoring Narrow Diet Breadth*

Because plants differ in their chemical, physical, distributional, and phenological characteristics, insects are unlikely to be adapted to use most nonhost

**Table 1** Ecological factors favoring narrow versus broad diet breadth

| | Narrow | Broad |
|---|---|---|
| Density-independent | Host specific adaptations; genetically-based tradeoffs in performance<br>Interspecific competition for food or enemy-free space<br>Resistance to generalist predators<br>Similarity of some hosts to unsuitable plants<br>Host plants abundant, apparent* | Host plants rare, unpredictable, unapparent*<br>Small host plants favoring larval grazing habit<br>Risk spreading<br>Genetic correlation between preference and performance<br>Ant mutualists |
| Density-dependent | Mate finding<br>Overwhelm plant defenses | Intraspecific competition<br>Predator functional response<br>Pathogen numerical response |

*Factors influencing rate of host finding by ovipositing adults. All others, except mate finding, affect offspring fitness.

species. Even if genetic variation in ability to use nonhosts were neutral among insects using the normal host plants, genetic drift alone could prevent adaptation to nonhosts. Thus, at the grossest scale, one expects host selection behavior to be correlated with offspring performance.

In considering more restricted sets of host plants, females of some species rank hosts for oviposition in the same order, on average, as their suitability for offspring development (e.g. 1, 2, 39, 50, 54, 243, 245), but many such population-level preference-performance correlations are rather weak (e.g. 3, 33, 41, 46, 109, 112, 116, 137, 180, 185, 186, 199, 222, 242). Low correspondence between preference and performance takes two forms. On the one hand, females of many species fail to accept some plants that are suitable for larval development (109, 112, 116, 137, 180, 199). For instance, the butterfly *Papilio machaon* rejects several available plant species that are nearly as suitable for offspring survival as the normal host (242). Wiklund (242) explains this as a result of selection on larvae to be able to feed on a wide variety of plants, since their capacities for host selection are very limited, and selection on females to oviposit only on the most suitable plants for offspring development. This hypothesis applies best to species like butterflies, with mobile, long-lived adults that can be choosy in selecting oviposition sites. Some adult insects may not oviposit on introduced plants that are suitable for offspring development, because the ability to recognize them as suitable hosts has not yet evolved. The skipper *Pyrgus scriptura,* for example, fails to use a number of perfectly suitable introduced plants available in its habitat (54).

On the other hand, a number of species oviposit on plants that are poor or unsuitable for larvae (33, 39, 41, 177, 213, 242). In most cases, these plants are introduced species, and the insects presumably have not had time to evolve behavioral avoidance or physiological adaptation (33, 39, 213, 242).

Some insects are known to oviposit on unsuitable plants that are chemically similar to suitable hosts (60, 177), suggesting that females cannot distinguish the two types of plants. Levins & MacArthur (123) show that insects may be selected to avoid an entire set of plants, including both suitable and unsuitable species, if they are perceived as similar by ovipositing females. While insects may sometimes evolve physiological tolerance to previously unsuitable host plants, introduction of an unsuitable plant that is indistinguishable from current hosts could also result in behavioral avoidance of both species, although I know of no examples of this.

In each of the few cases studied, ovipositing Lepidoptera, sawflies, and aphids prefer as hosts those plants *within* a plant species that are most suitable for their offspring (122, 157, 173, 240). It would almost seem that insects are better at identifying the best individual plants within a host species than at identifying the host species that is, on average, most suitable. Most population-level studies on preference-peformance correlations across host species do not use as larval foods the individual plants selected by ovipositing females, thus diminishing the possibility of detecting adaptive oviposition behavior.

A potentially powerful explanation of host specialization in insects is that an evolutionary increase in offspring performance on one plant species entails a reduction in adaptation to other potential hosts. Evidence for such negative genetic correlations in performance across host species has now been sought in a number of species, including butterflies (115), moths (71), beetles (81, 166), flies (104, 226, 227), aphids (230, 236), and mites (65, 75). In almost every case, performance on one host species is genetically uncorrelated or positively correlated with performance on others, thus lending little support to the trade-off hypothesis.

Rausher (169) has criticized the conclusion that trade-offs are unimportant on a number of grounds. (*a*) Performance is generally tested on plants already included in the insect's natural diet. If selection had eliminated trade-offs that once existed, or if trade-offs preclude using particular sets of hosts, then none would now be found. However, even species that were tested on novel versus normal hosts showed positive correlations in performance across hosts (75, 104, 115). (*b*) Only a few components of fitness are measured, generally larval survival, development time, and pupal or adult size. Other life-history characteristics could vary as a function of host plant and be involved in trade-offs. For instance, probability of adult reproductive diapause is affected in a host-specific manner in the Colorado potato beetle (80, 91). (*c*) Subtle genetic trade-offs in performance on different hosts could be swamped by variation in general vigor in unnatural laboratory or greenhouse conditions. It is notable that in one of the few studies detecting trade-offs in performance, fitness on alternative hosts was measured in the field (230).

In the laboratory, the swamping effect of general vigor can be corrected for if performance on more than two hosts is assayed simultaneously. This has been done only in Futuyma & Philippi's (71) study of larval growth in 10 clones of the fall cankerworm, *Alsophila pometaria,* on four hosts—red maple, scarlet oak, white oak, and chestnut. Futuyma & Philippi used larval performance on the best host, scarlet oak, as a measure of general vigor, and they found some evidence of trade-offs. I have reexamined their data using a principal components analysis (Figure 2). Variation among clones in general vigor is indicated by positive factor scores on the first principal component for each of the four hosts. Clones that grow rapidly on one host grow rapidly on the others. The second principal component clearly distinguishes red maple from the other hosts, especially in their laboratory study. After correcting for variation in general vigor, clones that grow faster than average on red maple do relatively poorly on the oaks and chestnut, as one might expect on biological grounds. Thus, the only study to consider more than two hosts simultaneously does support the genetic trade-off hypothesis. Clearly, more tests using multivariate analyses of performance on several host species are warranted. It is certainly premature to dismiss the potential importance of genetically based trade-offs in performance as a general explanation of host specialization in insects.

As Janzen (105) has emphasized, "A host plant is more than its chemistry," including, from the point of view of an insect herbivore, a suite of potential competitors, mutualists, and enemies. Interspecific competition among phytophagous insects is believed to be too rare to be of much importance in restricting their feeding niches (121, 160, 214). This consensus, however, is based more on a lack of positive evidence for competition than on convincing negative evidence (114). Such competition has occasionally been demonstrated in phytophagous insects (e.g. 114, 129, 211), and it could conceivably affect host use and community structure (16, 200).

Natural enemies may affect the community structure of insects in a manner analogous to interspecific competition for resources. Competition for enemy-free space favors species that use resources or habitats where the risk of predation or parasitism is low, which may be most likely where other insect prey are rare (90, 107, 156). It is not clear how this process will influence the diet breadth of established species, as enemy pressure could favor either diet expansion to include a host where mortality is low or specialization on such a resource (221). For instance, *Drosophila* larvae that feed in amanitin-containing mushrooms are almost never attacked by parasitic nematodes (98).

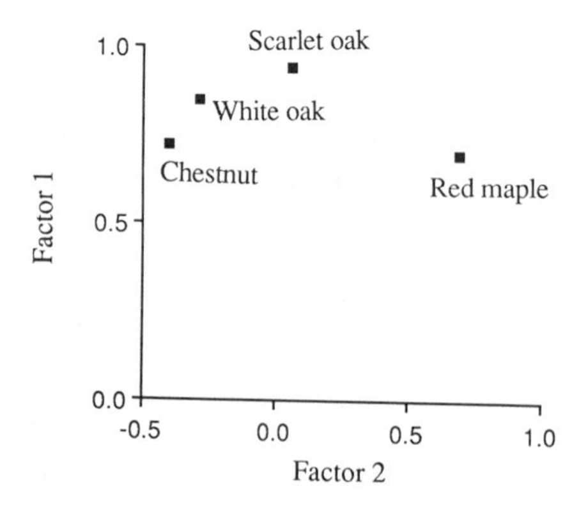


*Figure 2* Factor scores from a principal components analysis of larval growth of *Alsophila pometaria* clones on four host plants in the laboratory (data from 71). Factor 1 reflects variation among trees in overall suitability for larval growth, while factor 2 indicates that there is a trade-off in larval performance on red maple versus the other hosts.

However, because these mushrooms are generally scarce, parasite pressure probably selects for expanding the diet to use them when available rather than restricting oviposition to these resources.

Density-independent mortality from enemies can be regarded as a feature of a plant's environment and weighed as such in the equation of trade-offs in performance on different hosts. In many insects, susceptibility to predators (18, 112), parasitoids (11, 15, 29, 61, 82, 141, 155, 184, 201, 223, 231, 246), parasites (98), bacteria (10, 118), and viruses (117, 184), varies as a function of larval host plant. Such plant-conferred protection will favor host specialization if adaptation to one host reduces fitness on others. For instance, specificity of crypsis is inversely related to diet breadth in tropical moths and desert grasshoppers (105, 143).

Generalist predators, as a major cause of mortality in many insects (e.g. 49, 61, 105, 214), can have a major effect on host plant use. For example, saturniid caterpillars develop normally on nonhost plants inside predator-proof enclosures in a tropical forest understory, but outside them generalist predators rapidly consume most of these caterpillars (105). Bernays & Graham (21) hypothesize that such predators primarily consume generalist prey and thus constitute the most important selective factor favoring narrow diet breadth in phytophagous insects. In fact, a positive relation between diet breadth and susceptibility to attack by generalist predators (ants and wasps) has been found in some groups of phytophagous insects (19, 20). However, it is not clear if this is due to variation in diet breadth per se or in larval host plants, because the specialist and generalist prey were largely collected from or reared on different species of plants.

A possible example in support of Bernays & Graham's hypothesis is provided by the chrysomelid beetles *Phratora vitellinae* and *Galerucella lineola*. *P. vitellinae* exudes defensive secretions derived from salicylate-rich willows on which it specializes, whereas *G. lineola* does not produce such secretions and uses a wider variety of hosts (50). Thus, the host specialist is likely to be better protected from predators even when both beetles develop on the same host plant. In contrast to expectations of Bernays & Graham's hypothesis, no relation exists between the diet breadth and the number of parasitoid species attacking British phytophagous insects (83). In sum, there is little positive evidence to date for the notion that diet breadth per se is substantially affected by natural enemies.

Specialization on the host species, conferring highest offspring fitness, is favored only if realized female fecundity is not compromised by such diet restriction. That is, a high probability of finding the most suitable plant allows the evolution of host specialization (123). The relation between host plant apparency and insect diet breadth is discussed further in the next section.

### *Density-Independent Factors Favoring Broad Diets*

Thompson (219a) suggests that insects whose larvae feed on small host plants may have to adopt a grazing habit, requiring several plants, perhaps of different species, to complete development. Such species may be more apt to become polyphagous than those whose larvae complete development on single plants. This is reminiscent of Wiklund's (242) explanation of why larval butterflies of some species can feed successfully on more plants than are accepted for oviposition by adults.

In species that often fail to lay their full complement of eggs, such as many Lepidoptera (42, 49), selection can favor acceptance of a broader range of hosts, if the decrease in mean offspring fitness is more than offset by the increase in realized fecundity. Factors that can limit the rate of host finding relative to the number of eggs to be laid include host plant rarity or unpredictability, limited search time, low insect mobility, and limited sensory ability to detect suitable hosts. For example, gall midges, which are small and short-lived, have rather indiscriminant oviposition behavior (3), perhaps explaining why Hessian flies evolve larval adaptation to resistant wheat cultivars rather than behavioral avoidance.

Polyphagous species of insects often feed on hosts that are rare, hard to find, or unpredictable, whereas more specialized species generally use abundant, easily found plants (51, 202, 214). These behaviors would be expected under both optimality and physiological state models of oviposition. Examples of diet breadth as a function of these host characteristics have been found at the interspecific level among heliconiine butterflies that feed on passion flower vines (16, 73); various other butterflies (44, 45, 244); *Drosophila* that breed on mushrooms, flowers, and cacti (62, 95, 120, 152); beetles, Lepidoptera, and grasshoppers of the desert (31, 143); aphids (136); and flower-feeding mites (37). For instance, among mycophagous drosophilids, *Drosophila duncani* and *Mycodrosophila claytoni* specialize on a few species of long-lasting polypores, while *D. falleni* and *D. putrida* use dozens of species of more ephemeral, primarily gilled mushrooms (120). Variation in host abundance may also account for differences in diet breadth among populations of a single insect species. The butterfly *Papilio machaon* breeds on fewer plants in stable habitats, such as bogs and wet meadows, where preferred hosts are abundant, than in less stable areas like roadsides, where such hosts are rare or absent (241).

A notable counterexample to the inverse relation between diet breadth and host plant apparency is reported by Futuyma (66), who found within several groups of Lepidoptera that species using woody plants tend to have broader diets than those using herbaceous, presumably less apparent hosts. This was attributed to the greater chemical differences among herbaceous plant species. This conclusion, however, has not been sustained by subsequent studies of the relation between plant growth form and diet breadth in Lepidoptera. Analyses of 180 species of North American butterflies (84) and of all 927 species of British macrolepidoptera (72) show that the mean diet breadth of insects using woody plants is not significantly different from that of those that breed on herbaceous species. Discrepancies among these studies could be due to differences in how diet breadth was measured.

Variation in host plant apparency may explain some latitudinal trends in feeding specialization of phytophagous insects. Because plant species diversity is greatest in the tropics, individual species may be harder to find there than in temperate regions. Presumably as a result of this, tropical species of aphids (52, 55), bark and ambrosia beetles (14), and treehoppers (247) are generally less host-specific than their high-latitude counterparts. Among papilionid butterflies, however, the proportion of generalist species increases steadily from the tropics to temperate regions (190, 197), perhaps because their greater mobility and longevity allow more discrimination among potential hosts.

Just as unpredictability of host occurrence can favor polyphagy, so has it been argued that unpredictability of host suitability should favor *risk spreading,* with females laying their eggs on many host plants in order to ensure that at least some offspring survive (105, 183, 244). Such a strategy, if beneficial, could lead to the acceptance of a greater variety of host species for oviposition. Although risk spreading seems to make sense intuitively, Gillespie (74) has shown that the increase in fitness due to reduced variance in offspring production is inversely proportional to population size. Risk spreading, therefore, can lead to a significant increase in fitness only in small populations. In most populations, I would not expect this benefit to exceed the associated costs of reduced oviposition on host plants already found.

As discussed above, the impact of natural enemies may favor host specialization in phytophagous insects. However, in lycaenid butterflies, which constitute about 40% of all butterfly species, the reverse appears to be true. About one third of lycaenids are associated with ants, which protect caterpillars from predators and parasitoids (6, 7, 150). Ant-induced oviposition has been reported or suggested to occur in at least 46 species of lycaenids (150), and these mobile oviposition cues can stimulate females to accept even marginally suitable hosts (6). An important consequence is that in North America, Australia, and South Africa, lycaenids associated with ants use a greater variety of host plants than do those without such associations (150). Ants also appear to facilitate radical host shifts within lycaenid lineages (7, 150). As environmental components of plants, ants in themselves are unlikely to favor polyphagy. However, by reducing differences among plants in suitability for caterpillar development, they may tip the balance in favor of polyphagy if other factors, such as rarity of preferred hosts, do so as well.

Ants could also favor polyphagy via a density-dependent mechanism (see below) if they are a limiting resource for which caterpillars compete.

The mechanisms outlined previously in this section favor facultative polyphagy by all members of an insect population. Broad diets at the species level could also result from specialization on various plants by different subsets of an insect population. It is thus natural to ask if genetic correlations between oviposition preference and offspring performance occur within species, as this might maximize mean fitness in a polyphagous species. Such correlations could arise if preference and performance are affected by loci in linkage disequilibrium or if both are pleiotropic manifestations of the same set of genes. Positive correlations between oviposition preference and offspring performance have been reported in two species. In the agromyzid fly, *Liriomyza sativae,* the pupal weight of female, but not male, offspring was found to be relatively greater on the host plant (tomato or pea) preferred for oviposition by a female (228). However, because the females used as parents were collected as larvae in the field, an environmental component to oviposition behavior cannot be ruled out.

Singer and his colleagues (139, 194) have discovered that females of the butterfly *Euphydryas editha* vary in their propensity to accept potential host plants for oviposition. In one population, the degree of preference shown by individual females for *Collinsia parviflora* over *Plantago lanceolata* was negatively correlated with the growth rate of their offspring on *Plantago* (194). In another, females that discriminated among individual plants of *Pedicularis semibarbata* produced offspring that grew less well on plants that were rejected than on those accepted, whereas the offspring of nondiscriminating females did equally well on both types of plants (139).

Because these *Euphydryas* females were wild-caught, correlations between their oviposition behavior and offspring performance need not be genetic. Both, in fact, could be a function of a female's condition. A well-nourished female may mature eggs at a greater rate and thus be more willing to accept a greater variety of host plants. In addition, such females could produce high quality eggs, as found by Wellington (238) for western tent caterpillars, with larvae perhaps better able to survive on suboptimal hosts. In other Lepidoptera, such as the cinnabar moth, number of eggs laid per day and egg weight decrease with female age (176), which could also produce a positive correlation between probability of oviposition and offspring fitness on low quality hosts. Because genetic correlations due to linkage disequilibrium decay slowly despite random mating, the observed preference-performance correlations in *Euphydryas* could also be due to migration among genetically differentiated populations that specialize on different hosts (162, 192).

With environmental sources of variation held constant, no genetic correlations were found between female host preference and offspring performance in two large-scale experiments on *Drosophila tripunctata* and the aphid *Myzus persicae* (101, 236). Thus, there is little evidence for the existence of such correlations within sexual populations of insects. In theory, such correlations are not expected to be maintained in populations in the absence of frequency-dependent selection (103, 217).

In sum, factors like plant rarity or unpredictability that limit the rate at which host plants are found favor increased diet breadth in phytophagous insects. Risk-spreading, genetic covariance between preference and performance, and ant mutualists are probably much less important.

### *Density-Dependent Factors Favoring Narrow Diet Breadth*

If the fitness of an individual selecting a particular host plant increases with the number of other conspecific individuals selecting the same plant, then host specialization may be favored, whereas with fitness a decreasing function of conspecific density, a broader range of hosts should be used. The various density-dependent effects discussed in this and the following section act on dispersion of insects across individual host plants and only indirectly on diet breadth.

In species that mate on the host plant, it would be advantageous for individuals to choose hosts where encounters with conspecifics are most likely, especially in populations at low density. Colwell (38) argues such positive frequency-dependent selection for using a particular host species can lead to specialization on a plant that is not necessarily best for offspring development, in a manner reminiscent of Fisher's runaway sexual selection process. Host restriction has also been suggested to facilitate mate finding in monogenean parasites of fish (179). I know of no critical tests of these ideas. Data on the frequency of unmated females in natural populations would have obvious bearing on this. If this idea is correct, then in species that mate on the host plant, the fraction of unmated females should increase with the number of host plants in the diet and decrease with population density. Many insects, including those that lek on conspicuous rendezvous sites (78, 148, 203), search for new host plants after mating; problems with mate finding should have little effect on diet breadth in these species.

Aggregations of eggs or larvae may also be advantageous in some cases. Stamp (204) has summarized a variety of factors that favor egg clustering in butterflies. It may, for example, be advantageous in species with distasteful and warning-colored larvae. Aggregations are also favored in species whose larvae overwhelm the defensive responses of attacked plants (175).

### *Density-Dependent Factors Favoring Broad Diets*

Rausher (167) shows that if offspring fitness decreases with larval density on a given host plant, selection should favor females that oviposit on relatively underutilized hosts, even if these are no more suitable nutritionally than the more commonly used hosts. Such effects could be brought about by competition for resources or enemy-free space. The strength of selection to increase diet breadth should be inversely proportional to current diet breadth, as individuals are less concentrated on individual host species in more polyphagous insects. Thus, this mechanism may be less important for maintenance of diet breadth in broadly polyphagous species.

Brower (25) has argued that if enemies concentrate their foraging on plants where prey are plentiful, then polyphagy may be favored in numerically dominant species of phytophagous insects, because prey density is reduced on each host species. Decimation of high-density patches of insects by pathogens could similarly favor polyphagy. Some evidence in support of Brower's hypothesis is suggested by patterns of parasitoid attack on phytophagous insects (212). Among monophagous insect hosts, attack rates increased with host density in 12 of 44 (27%) cases, whereas such an effect was found in only 12 of 76 (16%) polyphagous species. Although this difference is not statistically significant, it does suggest a density-dependent cost to host specialization in phytophagous insects.

It has been generally believed that phytophagous insects rarely experience intraspecific competition, except during population outbreaks (e.g. 106). Strong et al (214) found evidence of juvenile competition for food in only 3 of 31 life-table analyses of survival in phytophagous insects. However, such competition has been experimentally demonstrated in natural populations (e.g. 4, 79, 211), and in these cases adult size and potential fecundity are more affected than survival to the adult stage. Such competitive effects on potential fecundity will be missed in conventional key-factor analyses.

Price et al (157) stress that analyses of insect life tables generally ignore female oviposition behavior, which can have a strong density-dependent component with major effects on population dynamics. Evidence for competition among females for suitable oviposition sites has been found in Lepidoptera, sawflies, aphids, tephritid flies, and weevils (27, 48, 158, 168, 191). Its occurrence over such a broad taxonomic range indicates that such competition may be of general importance in insects. By avoiding sites already crowded with eggs, females may be forced to oviposit on resources of lower nutritional quality for their offspring, thus bringing about a density-dependent regulation of population size. In addition, physiological state models lead one to expect that inhibition of oviposition at favored hosts should increase an individual's subsequent probability of accepting lower-ranked host species. Polyphagy as a result of competition for oviposition sites could result either from such facultative responses of all females to variable egg densities or from genetically based variation for host preference under Rausher's (167) frequency-dependence model. In the absence of evidence for genetically based variation on host use in the field, behavioral flexibility would seem to be a more likely cause of density-dependent increases in diet breadth.

Because competition for oviposition sites and the effects of larval food limitation on adult fecundity have been ignored in most studies of insect life tables, it is premature to dismiss the importance of intraspecific competition on population regulation and diet breadth in phytophagous insects.

## CORRELATES OF VARIATION IN DIET BREADTH

Associations between related groups of insects and plants, as discussed in the introduction, indicate evolutionary conservation in the types of plants used by an insect taxon. Niche breadth itself can also be evolutionarily conservative, as related groups of insects frequently differ in mean diet breadth. Among the macrolepidoptera, for instance, butterflies tend to be more host-specific than moths (26, 89). Tropical sphingid moths are largely host specific while saturniids are more polyphagous (22, 105). Holloway & Hebert (89) find that Canadian moths exhibit significant among-family heterogeneity in the frequencies of monophagy (defined as feeding on one plant family), oligophagy (feeding on 2 plant families), and polyphagy (feeding on 3 or more families). Futuyma & Moreno (69) state that some groups, such as *Calligrapha* beetles and *Yponomeuta* moths, appear to be committed to specialization per se, rather than to a particular group of plants.

In several groups of insects, it has been found that while species are relatively specialized in host use, a few have much broader diets, yielding a highly skewed frequency distribution of diet breadths. Among papilionid butterflies, *Papilio glaucus* stands out as a broad host generalist (59, 190). The vast majority of British leafhoppers and leafminers are monophagous, but a few are widely polyphagous (37). Shifts to novel hosts by pierid butterflies are restricted to species already having broad diets (42).

I have analyzed the frequency distribution of diet breadth in four insect groups within defined geographical regions: noctuid and geometrid moths in Canada (174), agromyzid flies in Britain (from 154), and treehoppers in a Costa Rican forest (247). Despite having substantial differences in mean diet breadth, these groups share a basically similar pattern. When the number of host taxa is transformed to logarithms, the frequency distribution of the number of insect species as a function of diet breadth is approximately normal, whether or not the mode of the distribution corresponds to species using only a single plant taxon (Figure 3). In other words, these groups display a *truncated lognormal distribution* of diet breadths. The fit to log-transformed data suggests that diet breadth typically changes in a proportional manner. That is, the rate of increase or decrease in diet breadth is proportional to the number of hosts a species currently uses. Already broad-niched species are more likely to expand their diets than are the more specialized species, as Courtney (42) found for pierid butterflies. One way such a normal distribution could be produced is by numerous factors contributing independently and additively to diet breadth. If truncated lognormal distributions of diet breadth

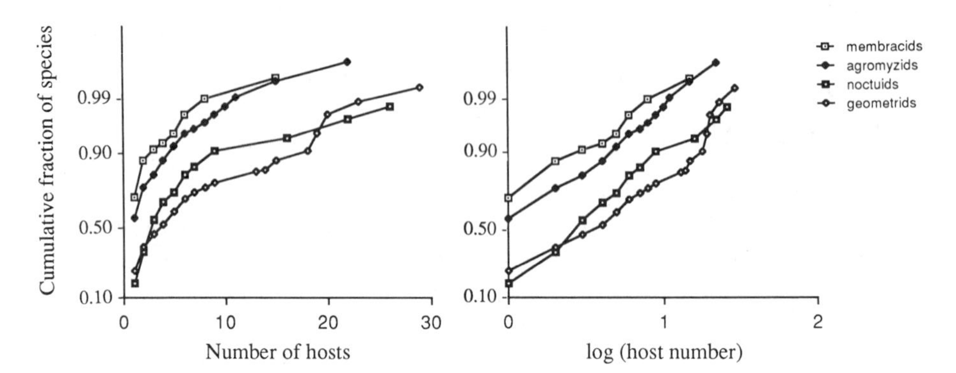


*Figure 3* Frequency distributions of host breadth within insect groups. Cumulative fraction of species are plotted on a normal probability scale as a function of number of hosts (left) and log-transformed numbers of hosts (right) in the diet of membracid treehoppers in Costa Rica (247), agromyzid flies in Great Britain (from 154), and noctuid and geometraid moths in Canada (174). A straight line, as seem for the log-transformed data, indicates a good fit to a normal distribution.

are found in other groups of insects, this could provide important insights into the evolution of host specialization in insects.

What sorts of factors may contribute to variation in diet breadth, and do they vary independently or coordinately? Diet breadth increases with body size among species of macrolepidoptera (72, 235). Wasserman & Mitter (235) suggest that larger body size may buffer individuals against a broader range of environmental stresses, thus allowing them to exploit a greater variety of plants. Broad environmental tolerances, which would be required of many high-latitude species, could allow a species to persist over wide geographical regions (210). If different host plants are used in different areas within a species' range, this could contribute to the finding that diet breadth in Lepidoptera increases with latitude (190, 198).

Mattson (128) suggests that large insect larvae should be better able than smaller individuals to use tough, nitrogen-poor leaves. Besides having a mechanical advantage in chewing such leaves, large larvae have lower metabolic rates and greater digestive efficiencies and therefore should be better able to obtain sufficient energy and nutrients from such food. If polyphagous insects tend to feed on plants with tough, nutrient-poor foliage (66), this could contribute to the correlation between insect body size and diet breadth. In contrast, Bernays & Janzen (22) find that tropical sphingid moths, which are largely host-specific, have large heads with crushing mandibles. Saturniids, a more polyphagous group, have smaller heads with mandibles that snip old, tough leaves. Within each moth family, head size and mandibular shape are correlated with diet breadth, suggesting that evolutionary changes in host specialization are not constrained by morphological inflexibility.

The relative cost of dispersal decreases with adult size in insects (178). Accordingly, large insects may have more opportunities for discrimination among potential hosts, thus facilitating host specialization in such species. However, physiological state models would lead one to expect that larger species, if they are more fecund, will accept low-ranking hosts more readily than small species. In support of the latter notion, host specialists among *Drosophila* have lower rates of egg production than do those that use a wide variety of breeding sites (5, 113, 119, 135).

Diet breadth in Lepidoptera is correlated with the stage at which insects enter diapause. Among North American species, Hayes (84) found that species diapausing late in the life cycle have broader diets than those diapausing at early stages. In Britain, species that overwinter as larvae or pupae are significantly more likely to be polyphagous than those overwintering as eggs or adults (72). In North America, but not Great Britain, diet breadth increases with the number of generations produced annually (72, 84). Finally, a variety of insects that produce oviposition-deterring pheromones tend to be host specialists, using rather ephemeral resources on long-lived host plants (182).

Clearly, a number of morphological, physiological, and behavioral variables are correlated with diet breadth in some insect groups. One may ask whether particular combinations of characters are generally found together in "adaptive syndromes" (157). Whether such character combinations are found more often than would be expected by chance remains to be seen. Hayes (84) found that among North American butterflies, host specialization, univoltinism, and diapause at early life stages tend to be weakly, but significantly intercorrelated. In contrast, Gaston & Reavey (72, p. 380) conclude that "distinct consistent life history syndromes cannot be defined for British macrolepidoptera." Perhaps intercorrelated suites of characters could be found if other traits, such as potential fecundity, were included in such analyses. The observation that diet breadth is lognormally distributed within several insect families suggests that traits affecting host specialization may be only weakly correlated at the subfamilial level.

## CONCLUSIONS

A large variety of factors have been proposed as determinants of diet breadth in phytophagous insects. In general, hypotheses that focus on maximizing mean offspring fitness predict that insects should oviposit on a narrow range of high-quality hosts if fitness is density-independent or on several hosts if density-dependent. In contrast, an adult's fecundity will generally be maximized when eggs are laid indiscriminately on many hosts. Thus, the optimal diet breadth represents a trade-off between realized adult fecundity and mean offspring fitness.

Among the hypotheses advanced to explain the prevalence of host specialization in these insects, I favor one for which there is presently little evidence: genetically based trade-offs in offspring performance on different hosts. This hypothesis has been inadequately tested to date, as most studies consider only two potential hosts and thus are often unable to distinguish genetic variation in general vigor from trade-offs in performance. The only test employing more than two hosts simultaneously (71) produced evidence that, upon reanalysis here, supports the idea of such genetically based trade-offs. This hypothesis is appealing in that it is potentially applicable to all insects; it does not invoke special ecological circumstances, such as difficulty in mate finding or interspecific competition.

Limits to the rate at which suitable hosts can be found by ovipositing females are probably the most important factors favoring polyphagy. These include host plant rarity, high insect fecundity, and limited search time. The relation between rate of host finding and diet breadth is supported by comparative data on several insect groups and is predicted by both physiological state and optimality models of insect oviposition behavior.

Within populations of oligophagous or polyphagous insects, variation among individuals in host acceptance probably is due largely to behavioral plasticity, resulting from variation in learning, physiological condition, egg load, search time for an oviposition site, and age. In the absence of evidence to the contrary, I regard local genetic variation in host use as neutral and nonadaptive, having little effect on actual patterns of host use.

ACKNOWLEDGMENTS

I wish to thank Doug Futuyma, Fred Gould, Mike Palopoli, Peter Price, Mark Rausher, Mike Singer, John Thompson, and Sara Via for very helpful comments on the paper. This work was supported in part by National Science Foundation grant BSR 89-05399.

## *Literature Cited*

1. Ahman, I. 1981. The potential of some Brassica species as host plants of the Brassica pod midge (*Dasineura brassicae* Winn.)(Dipt., Cecidomyiidae). *Ent. Tidskr.* 102:111–19
2. Ahman, I. 1984. Oviposition and larval performance of *Rhabdophaga terminalis* on *Salix* spp. with special consideration to bud size of host plants. *Entomol. Exp. Appl.* 35:129–36
3. Ahman, I. 1986. Host selection versus host suitability: how specific are herbivorous gall midges? Unpublished manuscript
4. Atkinson, W. D. 1979. A field investigation of larval competition in domestic *Drosophila*. *J. Anim. Ecol.* 48:91–102
5. Atkinson, W. D. 1979. A comparison of the reproductive strategies of domestic species of *Drosophila*. *J. Anim. Ecol.* 48:53–64
6. Atsatt, P. R. 1981. Ant-dependent food plant selection by the mistletoe butterfly

*Ogyris amaryllis* (Lycaenidae). *Oecologia* 48:60–63
7. Atsatt, P. R. 1982. Lycaenid butterflies and ants: selection for enemy-free space. *Am. Nat.* 118:638–54
8. Atsatt, P. R., O'Dowd, D. J. 1976. Plant defense guilds. *Science* 193:24–29
9. Balling, A., Technau, G. M., Heisenberg, M. 1987. Are the structural changes in adult *Drosophila* mushroom bodies memory traces? Studies on biochemical learning mutants. *J. Neurogenet.* 4:65–73
10. Barbosa, P. 1988. Natural enemies and herbivore-plant interactions: influence of plant allelochemicals and host specificity. In *Novel Aspects of Insect-Plant Interactions,* ed. P. Barbosa, D. K. Letourneau, pp. 201–29. New York: Wiley
11. Barbosa, P., Saunders, J. A. 1985. Plant allelochemicals: linkages between herbivores and their natural enemies. In *Chemically Mediated Interactions Between Plants and Other Organisms,* ed. G. A. Cooper-Driver, T. Swain, E. E. Conn, pp. 107–37. New York: Plenum

11a. Barker, J. S. F., Starmer, W. T., MacIntyre, R. 1990. *Ecological and Evolutionary Genetics of Drosophila.* New York: Plenum

12. Barker, J. S. F., Toll, G. L., East, P. D., Wilders, P. R. 1981. Attraction of *Drosophila buzzatii* and *D. aldrichi* to species of yeasts isolated from their natural environment. II. Field studies. *Aust. J. Biol. Sci.* 34:613–24
13. Barker, J. S. F., Vacek, D. C., East, P. D., Starmer, W. T. 1986. Allozyme genotypes of *Drosophila buzzatii:* feeding and oviposition preferences for microbial species, and habitat selection. *Aust. J. Biol. Sci.* 39:47–58
14. Beaver, R. A. 1979. Host specificity of temperate and tropical animals. *Nature* 281:139–41
15. Bellinger, R. G., Ravlin, W., McManus, M. L. 1988. Host plant species and parasitism of gypsy moth (Lepidoptera: Lymantriidae) egg masses by *Ooencyrtus kuvanae* (Hymenoptera: Eucurtidae). *Environ. Entomol.* 17:936–40
16. Benson, W. W. 1978. Resource partitioning in passion vines butterflies. *Evolution* 32:493–518
17. Berenbaum, M. 1981. Patterns of furanocoumarin distribution and insect herbivory in the Umbelliferae: plant chemistry and community structure. *Ecology* 62:1254–66
18. Berenbaum, M. R., Miliczky, E. 1984. Mantids and milkweed bugs: efficiency of aposematic coloration against invertebrate predators. *Am. Midl. Nat.* 111:64–68
19. Bernays, E. A. 1988. Host specificity in phytophagous insects: selection pressure from generalist predators. *Entomol. Exp. Appl.* 49:131–40
20. Bernays, E. A., Cornelius, M. L. 1989. Generalist caterpillar prey are more palatable than specialists for the generalist predator *Iridomyrmex humilis. Oecologia* 79:427–30
21. Bernays, E. A., Graham, M. 1988. On the evolution of host specificity in phytophagous arthropods. *Ecology* 69: 886–92
22. Bernays, E. A., Janzen, D. H. 1988. Saturniid and sphingid caterpillars: two ways to eat leaves. *Ecology* 69:1153–60
23. Bowers, M. D. 1983. The role of iridoid glycosides in host-plant specificity of checkerspot butterflies. *J. Chem. Ecol.* 9:475–93
24. Bowers, M. D. 1986. Population differences in larval hostplant use in the checkerspot butterfly, *Euphydryas chalcedona. Entomol. Exp. Appl.* 40:61–69
25. Brower, L. P. 1958. Bird predation and foodplant specificity in closely related procryptic insects. *Am. Nat.* 92:183–87
26. Brues, C. T. 1920. The selection of food-plants by insects, with special reference to lepidopterous larvae. *Am. Nat.* 54:313–32
27 Bultman, T. L., Faeth, S. H. 1986. Experimental evidence for intraspecific competition in a lepidopteran leafminer. *Ecology* 67:442–48
28. Bush, G. L. 1975. Sympatric speciation in phytophagous parasitic insects. In *Evolutionary Strategies of Parasitic Insects and Mites,* ed. P. W. Price, pp. 187–206. New York: Plenum
29. Campbell, B. C., Dufy, S. S. 1979. Tomatine and parasitic wasps: potential incompatibility of plant antibiosis with biological control. *Science* 205:700–2
30. Castillo-Chavez, C., Levin, S. A., Gould, F. 1988. Physiological and behavioral adaptation to varying environments: a mathematical model. *Evolution* 42:986–94
31. Cates, R. G. 1980. Feeding patterns of monophagous, oligophagous, and polyphagous insect herbivores: the effect of resource abundance and plant chemistry. *Oecologia* 46:22–31
32. Charnov, E. L., Stephens, D. W. 1988. On the evolution of host selection in solitary parasitoids. *Am. Nat.* 132:707–22
33. Chew, F. S. 1977. Coevolution of pierid butterflies and their cruciferous foodplants. II. The distribution of eggs on

potential foodplants. *Evolution* 31:568–79

34. Chew, F. S., Robbins, R. K. 1984. Egg-laying in butterflies. *Symp. R. Entomol. Soc. Lond.* 11:65–79
35. Deleted in proof
36. Claridge, M. F., Wilson, M. R. 1977. Oviposition and food plant discrimination in leafhoppers of the genus *Oncopsis*. *Ecol. Entomol.* 2:19–25
37. Colwell, R. K. 1986. Community biology and sexual selection: lessons from hummingbird flower mites. In *Community Ecology*, ed. J. Diamond, T. J. Case, pp. 406–24. New York: Harper & Row
38. Colwell, R. K. 1986. Population structure and sexual selection host fidelity in the speciation of hummingbird flower mites. In *Evolutionary Processes and Theory*, ed. S. Karlin, E. Nevo, pp. 475–95. Florida: Academic
39. Copp, N. H., Davenport, D. 1978. *Agraulis* and *Passiflora*. I. Control of specificity. *Biol. Bull.* 155:98–112
40. Corbet, S. A. 1985. Insect chemosensory responses: a chemical legacy hypothesis. *Ecol. Entomol.* 10:143–53
41. Courtney, S. P. 1981. Coevolution of pierid butterflies and their cruciferous foodplants. III. *Anthocharis cardamines* (L.) survival, development and oviposition on different hostplants. *Oecologia* 51:91–96
42. Courtney, S. P. 1986. The ecology of pierid butterflies: dynamics and interactions. *Adv. Ecol. Res.* 15:51–131
43. Courtney, S. P., Chen, G. K. 1988. Genetic and environmental variation in oviposition behaviour in the mycophagous *Drosophila suboccidentalis* Spcr. *Func. Ecol.* 2:521–28
44. Courtney, S. P., Chew, F. S. 1987. Coexistence and host use by a large community of Pierid butterflies: habitat is the template. *Oecologia* 71:210–20
45. Courtney, S. P., Forsberg, J. 1988. Host use by two pierid butterflies varies with host density. *Func. Ecol.* 2:67–75
46. Courtney, S. P., Kibota, T. T. 1990. Mother doesn't know best: selection of hosts by ovipositing insects. In *Insect-Plant Interactions*, Vol. II, ed. E. A. Bernays. Boca Raton, Florida: CRC
47. Courtney, S. P., Chen, G. K., Gardner, A. 1989. A general model for individual host selection. *Oikos* 55:55–65
48. Craig, T. P., Itami, J. K., Price, P. W. 1988. Plant wound compounds from oviposition scars used in host discrimination by a stem-galling sawfly. *J. Insect. Behav.* 1:343–56
49. Dempster, J. P. 1983. The natural control of populations of butterflies and moths. *Bio Rev.* 58:461–81
50. Denno, R. F., Larsson, S., Olmstead, K. L. 1990. Role of enemy-free space and plant quality in host-plant selection by willow beetles. *Ecology* 71:124–37
51. Dethier, V. G. 1970. Chemical interactions between plants and insects. In *Chemical Ecology*, ed. E. Sondheimer, J. B. Simeone, pp. 83–102. New York: Academic
52. Dixon, A. F. G., Kindlmann, P., Leps, J., Holman, J. 1987. Why are there so few species of aphids, especially in the tropics. *Am. Nat.* 129:580–92
53. Downey, J. C., Fuller, W. C. 1961. Variation in *Plebejus icarioides* (Lycaenidae). I. Foodplant specificity. *J. Lep. Soc.* 15:34–42
54. Dusheck, J. 1983. Larval host suitability and oviposition preference in two checkered skippers, *Pyrgus communis* and *Pyrgus scriptura* (Hesperidae). MA thesis. Univ. Calif. Davis
55. Eastop, V. F. 1973. Deductions from present day host plants of aphids and related insects. In *Insect/Plant Relationships*, ed. H. F. van Emden, pp. 157–78. Oxford: Blackwell
56. Ehrlich, P. R., Murphy, D. D. 1988. Plant chemistry and host range in insect herbivores. *Ecology* 69:908–9
57. Ehrlich, P. R., Raven, P. H. 1964. Butterflies and plants: a study in coevolution. *Evolution* 18:586–608
58. Erwin, T. L. 1982. Tropical forests: their richness in Coleoptera and other arthropod species. *Coleopt. Bull.* 36:74–75
59. Feeny, P. P. 1987. The roles of plant chemistry in associations between swallowtail butterflies and their host plants. In *Insects-Plants*, ed. V. Labeyrie et al, pp. 353–59. Dordrecht, The Netherlands: W. Junk
60. Feeny, P. 1990. Chemical constraints on the evolution of swallowtail butterflies. In *Herbivory: Tropical and Temperate Perspectives*, ed. P. W. Price et al, New York: Wiley
61. Feeny, P., W. S. Blau, and P. M. Kareiva. 1985. Larval growth and survivorship of the black swallowtail butterfly in central New York. *Ecol. Monogr.* 55:167–87
62. Fellows, D. P., Heed, W. B. 1972. Factors affecting host plant selection in desert-adapted cactiphilic *Drosophila*. *Ecology* 53:850–58
63. Fitt, G. P. 1986. The influence of a shortage of hosts on the specificity of oviposition behaviour in species of

*Dacus* (Diptera, Tephritidae). *Physiol. Entomol.* 11:133–43
64. Fox, L. R., Morrow, P. A. 1981. Specialization: species property or local phenomenon? *Science* 211:887–93
65. Fry, J. 1990. Trade-offs in fitness on different hosts: evidence from a selection experiment with the phytophagous mite *Tetranychus urticae* Kock. *Am. Nat.* In press
66. Futuyma, D. J. 1976. Food plant specialization and environmental predictability in Lepidoptera. *Am. Nat.* 110:285–92
67. Futuyma, D. J. 1983. Selective factors in the evolution of host choice by phytophagous insects. In *Herbivorous Insects,* ed. S. Ahmad, pp. 227–44. New York: Academic
68. Futuyma, D. J. 1990. The evolution of host plant associations in the leaf beetles genus *Ophraella* (Coleoptera, Chrysomelidae). II. Morphological analyses and the evolution of host associations. *Evolution.* In press
69. Futuyma, D. J., Moreno, G. 1988. The evolution of ecological specialization. *Annu. Rev. Ecol. Syst.* 19:207–33
70. Futuyma, D. J., Peterson, S. C. 1985. Genetic variation in the use of resources by insects. *Annu. Rev. Entomol.* 30:217–38
71. Futuyma, D. J., Philippi, T. E. 1987. Genetic variation and covariation in responses to host plants by *Alsophila pometaria* (Lepidoptera: Geometridae). *Evolution* 41:269–79
72. Gaston, K. J., Reavey, D. 1989. Patterns in the life histories and feeding strategies of British macrolepidoptera. *Biol. J. Linn. Soc.* 37:367–81
73. Gilbert, L. E. 1979. Development of theory in the analysis of insect-plant interactions. In *Analysis of Ecological Systems,* ed. D. J. Horn, G. R. Stairs, R. D. Mitchell, pp. 117–54. Ohio State Univ. Press: Columbus
74. Gillespie, J. H. 1974. Natural selection for within-generation variance in offspring number. *Genetics* 76:601–6
75. Gould, F. 1979. Rapid host range evolution in a population of the phytophagous mite *Tetranychus urticae* Koch. *Evolution* 33:791–802
76. Gould, F. 1984. Role of behavior in the evolution of insect adaptation to insecticides and resistant host plants. *Entomol. Soc. Am. Bull.* 30:34–41
77. Gould, F. 1990. The spatial scale of genetic variation in insect populations. Unpublished manuscript
78. Grimaldi, D. A. 1987. Phylogenetics and taxonomy of *Zygothrica* (Diptera: Drosophilidae). *Bull. Am. Mus. Nat. Hist.* 186:103–268
79. Grimaldi, D., Jaenike, J. 1984. Competition in natural populations of mycophagous *Drosophila*. *Ecology* 65:1113–20
80. Hare, J. D. 1983. Seasonal variation in plant-insect associations: utilization of *Solanum dulcamara* by *Leptinotarsa decemlineata*. *Ecology* 64:345–61
81. Hare, J. D., Kennedy, G. G. 1986. Genetic variation in plant-insect associations: survival of *Leptinotarsa decemlineata* populations on *Solanum carolinense*. *Evolution.* 40:1031–43
82. Harrington, E. A., Barbosa, P. 1978. Host habitat influences on oviposition by *Parasetigena silvestris* (R-D), a larval parasite of the gypsy moth. *Environ. Entomol.* 7:466–68
83. Hawkins, B. A., Lawton, J. H. 1987. Species richness for parasitoids of British phytophagous insects. *Nature* 326: 788–90
84. Hayes, J. L. 1982. A study of the relationships of diapause phenomena and other life history characters in temperate butterflies. *Am. Nat.* 120:160–70
85. Hedrick, P. W. 1990. Theoretical analysis of habitat selection and the maintenance of genetic variation. See Ref. 11a, pp.
86. Hoffmann, A. A. 1988. Early adult experience in *Drosophila melanogaster*. *J. Insect Physiol.* 34:197–204
87. Hoffman, A. A., O' Donnell, S. 1990. Heritable variation in resource use in *Drosophila* in the field. See Ref. 11a, pp.
88. Hoffmann, A. A., Turelli, M. 1985. Distribution of *Drosophila melanogaster* on alternative resources: effects of experience and starvation. *Am. Nat.* 126:662–79
89. Holloway, J. D., Hebert, P. D. N. 1979. Ecological and taxonomic trends in macrolepidopteran host plant selection. *Biol. J. Linn. Soc.* 11:229–51
90. Holt, R. D. 1977. Predation, apparent competition, and the structure of prey communities. *Theor. Pop. Biol.* 12:197–229
91. Horton, D. R., Capinera, J. L., Chapman, P. L. 1988. Local differences in host use by two populations of the Colorado potato beetles. *Ecology* 69:823–31
92. Hsiao, T. H. 1978. Host plant adaptations among geographic populations of the Colorado potato beetle. *Entomol. Exp. Appl.* 24:237–47
93. Iwasa, Y., Suzuki, Y., Matsuda, H. 1984. Theory of oviposition strategy of parasitoids. I. Effect of mortality and

limited egg number. *Theor. Popul. Biol.* 26:205–227

94. Jaenike, J. 1978. On optimal oviposition behavior in phytophagous insects. *Theor. Popul. Biol.* 14:350–56
95. Jaenike, J. 1978. Resource predictability and niche breadth in the *Drosophila quinaria* species group. *Evolution* 32:676–78
96. Jaenike, J. 1983. Induction of host preference in *Drosophila melanogaster*. *Oecologia* 58:320–25
97. Jaenike, J. 1985. Genetic and environmental determinants of food preference in *Drosophila tripunctata*. *Evolution* 39:362–69
98. Jaenike, J. 1985. Parasite pressure and the evolution of amanitin tolerance in *Drosophila*. *Evolution* 39:1295–1301
99. Jaenike, J. 1987. Genetics of oviposition-site preference in *Drosophila tripunctata*. *Heredity* 59:363–69
100. Jaenike, J. 1988. Effects of early experience on host selection in insects: some experimental and theoretical results. *J. Insect Behav.* 1:3–15
101. Jaenike, J. 1989. Genetic population structure of *Drosophila tripunctata:* patterns of variation and covariation of traits affecting resource use. *Evolution* 43:1467–82
102. Jaenike, J. 1990. Factors maintaining genetic variation for host preference in *Drosophila*. See Ref. 11a, pp.
103. Jaenike, J., Holt, R. D. 1990. Genetic variation for habitat preference: evidence and explanations. *Am. Nat.* In press
104. James, A. C., Jackubczak, J., Riley, M. P., Jaenike, J. 1988. On the causes of monophagy in *Drosophila quinaria*. *Evolution* 42:626–30
105. Janzen, D. H. 1985. A host plant is more than its chemistry. *Ill. Nat. Hist. Surv. Bull.* 33:141–74
106. Janzen, D. H. 1988. Ecological characteristics of a Costa Rican dry forest caterpillar fauna. *Biotropica* 20:120–35
107. Jeffries, M. J., Lawton, J. H. 1984. Enemy free space and the structure of ecological communities. *Biol. J. Linn. Soc.* 23:269–86
108. Jermy, T. 1984. Evolution of insect/host plant relationships. *Am. Nat.* 124:609–30
109. Jermy, T., Szentesi, A. 1978. The role of inhibitory stimuli in the choice of oviposition site by phytophagous insects. *Entomol. Exp. Appl.* 24:458–71
110. Jones, R. E. 1977. Movement patterns and egg distribution in cabbage butterflies. *J. Anim. Ecol.* 46:195–212
111. Jones, R. E. 1989. Host location and oviposition on plants. In *Insect Reproductive Behaviour,* ed. J. Ridell-Smith, W. Bailey,
112. Jones, R. E., Ives, P. M. 1979. The adaptiveness of searching and host selection behaviour in *Pieris rapae* (L). *Aust. J. Ecol.* 4:75–86
113. Kambysellis, M. P., Heed, W. B. 1971. Studies of oogenesis in natural populations of Drosophilidae. I. Relation of ovarian development and ecological habitats of the Hawaiian species. *Am. Nat.* 105:31–49
114. Karban, R. 1986. Interspecific competition between folivorous insects on *Erigeron glaucus*. *Ecology* 67:1063–72
115. Karowe, D. N. 1990. Predicting host range evolution: colonization of *Coronilla varia* by *Colias philodice* (Lepidoptera: Pieridae). *Evolution*. In press
116. Kearney, J. N. 1983. Selection and utilization of natural substrates as breeding sites by woodland *Drosophila* spp. *Entomol. Exp. Appl.* 33:63–70
117. Keating, S. T., Yendol, W. G., Schultz, J. C. 1988. Relationship between susceptibility of gypsy moth larvae (Lepidoptera: Lymantriidae) to a baculovirus and host plant foliage constituents. *Environ. Entomol.* 17:952–958
118. Krischik, V. A., Barbosa, P., Reichelderfer, C. F. 1988. Three trophic level interactions: allelochemicals, *Manduca sexta* (L.), and *Bacillus thuringiensis* var. *kurstaki* Berliner. *Environ. Entomol.* 17:476–82
119. Lachaise, D. 1983. Reproductive allocation in tropical Drosophilidae: further evidence on the role of breeding-site choice. *Am. Nat.* 122:132–46
120. Lacy, R. C. 1984. Predictability, toxicity, and trophic niche breadth on fungus-feeding Drosophilidae (Diptera). *Ecol. Entomol.* 9:43–54
121. Lawton, J. H., Strong, D. R. 1981. Community patterns and competition in folivorous insects. *Am. Nat.* 118:317–38
122. Leather, S. R. 1985. Oviposition preference in relation to larval growth rates and survival in the pine beauty moth, *Panolis flammea*. *Ecol. Entomol.* 10: 213–17
123. Levins, R., MacArthur, R. H. 1969. An hypothesis to explain the incidence of monophagy. *Ecology* 50:910–11
124. Mangel, M. 1987. Oviposition site selection and clutch size in insects. *J. Math. Biol.* 25:1–22
125. Mangel, M. 1989. An evolutionary interpretation of the "motivation to oviposit." *J. Evol. Biol.* 2:157–72
126. Mangel, M. 1989. Evolution of host selection in parasitoids: does the state of

the parasitoid matter? *Am. Nat.* 133: 688–05
127. Mangel, M., Roitberg, B. D. 1989. Dynamic information and host acceptance by a tephritid fruit fly. *Ecol. Entomol.* 14:181–89
128. Mattson, W. B. 1980. Herbivory in relation to plant nitrogen content. *Annu. Rev. Ecol. Syst.* 11:119–61
129. McClure, M. S., Price, P. W. 1975. Competition among sympatric *Erythroneura* leafhoppers (Homoptera: Cicadellidae) on American sycamore. *Ecology* 56:1388–97
130. Miller, J. R., Strickler, K. L. 1984. Finding and accepting host plants. In *Chemical Ecology of Insects*, ed. W. J. Bell, R. T. Carde, pp. 127–57. Sunderland, Mass: Sinauer
131. Miller, J. S. 1987. Host-plant relationships in the Papilionidae (Lepidoptera): parallel cladogenesis or colonization? *Cladistics* 3:105–20
132. Miller, J. S. 1987. Phylogenetic studies in the Papilioninae (Lepidoptera: Papilionidae). *Bull. Am. Mus. Nat. Hist.* 186:365–512
133. Mitchell, R. 1981. Insect behavior, resource exploitation, and fitness. *Annu. Rev. Entomol.* 26:73–96
134. Mitter, C., Farrell, B., Wiegmann, B. 1988. The phylogenetic study of adaptive zones: has phytophagy promoted insect diversification? *Am. Nat.* 132:107–28
135. Montague, J. R., Mangan, R. L., Starmer, W. T. 1981. Reproductive allocation in the Hawaiian Drosophilidae: egg size and number. *Am. Nat.* 118:865–71
136. Moran, N. A. 1987. Evolutionary determinants of host specificity in *Uroleucon*. In *Population Structure, Genetics and Taxonomy of Aphids and Thysanoptera*, ed. J. Holman et al, pp. 29–38. The Hague, Netherlands: SPB Academic Publ.
137. Morrow, P. A. 1977. Host specificity of insects in a community of three codominant *Eucalyptus* species. *Aust. J. Ecol.* 2:89–106
138. Murphy, D. D., Menninger, M. S., Ehrlich, P. R. 1984. Nectar source distribution as a determinant of oviposition host species in *Euphydryas chalcedona*. *Oecologia* 62:269–71
139. Ng, D. 1988. A novel level of interactions in plant-insect systems. *Nature* 334:611–13
140. Deleted in proof
141. Nordlund, D. A., Lewis, W. J., Altieri, M. A. 1988. Influence of plant-produced allelochemicals on the host/prey selection behavior of entomophagous insects. In *Novel Aspects of Insect-Plant Interactions*, ed. P. Barbosa, D. K. Letourneau, pp. 65–90. New York: Wiley
142. Odendaal, F. J. 1989. Mature egg number influences the behavior of female *Battus philenor* butterflies. *J. Insect Behav.* 2:15–25
143. Otte, D., Joern, A. 1977. On feeding patterns in desert grasshoppers and the evolution of specialized diets. *Proc. Acad. Nat. Sci. Phila.* 128:89–126
144. Papaj, D. R. 1986. Shifts in foraging behavior by a *Battus philenor* population: field evidence for switching by individual butterflies. *Behav. Ecol. Sociobiol.* 19:31–39
145. Papaj, D. R., Prokopy, R. J. 1988. The effect of prior adult experience on components of habitat preference in the apple maggot fly *(Rhagoletis pomonella)*. *Oecologia* 76:538–43
146. Papaj, D. R., Prokopy, R. J. 1989. Ecological and evolutionary aspects of learning in phytophagous insects. *Annu. Rev. Entomol.* 34:315–50
147. Papaj, D. R., Rausher, M. D. 1983. Individual variation in host location by phytophagous insects. In *Herbivorous Insects*, ed. S. Ahmad, pp. 77–124. New York: Academic
148. Parsons, P. A. 1976. Lek behavior in *Drosophila* (Hirtodrosophila) *polypori* Mallock—an Australian rainforest species. *Evolution* 31:223–25
149. Pasteels, J. M., Rowell-Rahier, M., Raupp, M. J. 1988. Plant-derived defense in chrysomelid beetles. In *Novel Aspects of Insect-Plant Interactions*, ed. P. Barbosa, D. K. Letourneaupp, pp. 235–72. New York: Wiley
150. Pierce, N. E., Elgar, M. A. 1985. The influence of ants on host plant selection by *Jalmenus evagoras*, a myrmecophilous lycaenid butterfly. *Behav. Ecol. Sociobiol.* 16:209–22
151. Pilson, D., Rausher, M. D. 1988. Clutch size adjustment by a swallowtail butterfly. *Nature* 333:361–63
152. Pipkin, S. B., Rodriguez, R. L., Leon, J. 1966. Plant host specificity among flower-feeding Neotropical Drosophila (Diptera: Drosophilidae). *Am. Nat.* 100:135–56
153. Powell, J. A. 1980. Evolution of larval food preferences in microlepidoptera. *Annu. Rev. Entomol.* 25:133–59
154. Price, P. W. 1980. *Evolutionary Biology of Parasites*. Princeton: Princeton Univ. Press
155. Price, P. W., Bouton, C. E., Gross, P., McPheron, B. A., Thompson, J. N., Weiss, A. E. 1980. Interactions among

three trophic levels: influence of plants on interactions between insect herbivores and natural enemies. *Annu. Rev. Ecol. Syst.* 11:41–65
156. Price, P. W., Westoby, M., Rice, B., Atsatt, P. R., Fritz, R. S., Thompson, J. N., K. Mobley. 1986. Parasite mediation in ecological interactions. *Annu. Rev. Ecol. Syst.* 17:487–505
157. Price, P. W., Cobb, N., Craig, T. P., Fernandes, G. W., Itami, J. K., et al. 1990. Insect herbivore population dynamics on trees and shrubs: new approaches relevant to latent and eruptive species and life table development. In *Insect-Plant Interactions,* Vol. 2., ed. E. A. Bernays. Florida: CRC Press
158. Prokopy, R. J., Roitberg, B. D., Averill, A. L. 1984. Resource partitioning. In *Chemical Ecology of Insects,* ed. W. J. Bell, R. T. Carde, pp. 301–30. Boston, Mass: Sinauer
159. Prokopy, R. J., Diehl, S. R., Coley, S. S. 1988. Behavioral evidence for host races in *Rhagoletis pomonella* flies. *Oecologia* 76:138–47
160. Rathke, B. J. 1976. Competition and coexistence within a guild of herbivorous insects. *Ecology* 57:76–87
161. Rausher, M. D. 1978. Search image for leaf shape in a butterfly. *Science* 200:1071–73
162. Rausher, M. D. 1982. Population differentiation in *Euphydryas editha* butterflies: larval adaptation to different hosts. *Evolution* 36:581–90
163. Rausher, M. D. 1983. Ecology of host-selection behavior in phytophagous insects. In *Variable Plants and Herbivores in Natural and Managed Systems,* ed. R. F. Denno, M. S. McClure, pp. 223–57. New York: Academic
164. Rausher, M. D. 1983. Conditioning and genetic variation as causes of individual variation in the oviposition behaviour of the tortoise beetle, *Deloyala guttata. Anim. Behav.* 31:743–47
165. Rausher, M. D. 1983. Alternation of oviposition behavior by *Battus philenor* butterflies in response to variation in host-plant density. *Ecology* 64:1028–34
166. Rausher, M. D. 1984. Tradeoffs in performance on different hosts: evidence from within- and between-site variation in the beetle *Deloyala guttata. Evolution* 38:582–95
167. Rausher, M. D. 1984. The evolution of habitat preference in subdivided populations. *Evolution* 38:596–608
168. Rausher, M. D. 1986. Competition, frequency-dependent selection, and diapause in *Battus philenor* butterflies. *Fla. Entomol.* 69:63–78
169. Rausher, M. D. 1988. Is coevolution dead? *Ecology* 69:898–901
170. Rausher, M. D. 1990. The evolution of habitat preference. III. The evolution of avoidance and adaptation. In *Evolution of Insect Pests: the Pattern of Variations,* ed. K. C. Kim. New York: Wiley
171. Rausher, M. D., Englander, R. 1987. The evolution of habitat preference. II. Evolutionary genetic stability under soft selection. *Theor. Popul. Biol.* 31:116–39
172. Rausher, M. D., Odendaal, F. J. 1987. Switching and the pattern of host use by *Battus philenor* butterflies. *Ecology* 68:869–77
173. Rausher, M. D., D. R. Papaj. 1983. Demographic consequences of discrimination among conspecific host plants by *Battus philenor* butterflies. *Ecology* 64:1402–10
174. Redfearn, A., Pimm, S. L. 1988. Population variability and polyphagy in herbivorous inset communities. *Ecol. Monogr.* 58:39–55
175. Rhoades, D. F. 1985. Offensive-defensive interactions between herbivores and plants: their relevance in herbivore population dynamics and ecological theory. *Am. Nat.* 125:205–38
176. Richards, L. J., Myers, J. H. 1980. Maternal influences on size and emergence time of the cinnabar moth. *Can. J. Zool.* 58:1452–57
177. Rodman, J. E., Chew, F. S. 1980. Phytochemical correlates of herbivory in a community of native and naturalized cruciferae. *Biochem. Syst. Ecol.* 8:43–50
178. Roff, D. 1977. Dispersal in dipterans: its costs and consequences. *J. Anim. Ecol.* 46:443–56
179. Rohde, K. 1979. A critical evaluation of the intrinsic and extrinsic factors responsible for niche restriction in parasites. *Am. Nat.* 114:648–71
180. Roininen, H., Tahvanainen, J. 1989. Host selection and larval performance of two willow-feeding sawflies. *Ecology* 70:129–36
181. Roitberg, B. D., Prokopy, R. J. 1983. Host deprivation influence of *Rhagoletis pomonella* to its oviposition deterring pheromone. *Physiol. Entomol.* 8:69–72
182. Roitberg, B. D., Prokopy, R. J. 1987. Insects that mark host plants. *BioScience* 37:400–6
183. Root, R. B., Kareiva, P. M. 1984. The search for resources by cabbage butterflies *(Pieris rapae):* ecological consequences and adaptive significance of

markovian movements in a patchy environment. *Ecology* 65:147–65
184. Rossiter, M. C. 1987. Use of a secondary host by non-outbreak populations of the gypsy moth. *Ecology* 68:857–68
185. Rowell, C. H. F. 1985. The feeding biology of species-rich genus of rainforest grasshoppers (*Rhachicreaga:* Orthoptera, Acrididae). I. Foodplant use and foodplant acceptance. *Oecologia* 68:87–98
186. Rowell, C. H. F. 1985. The feeding biology of a species-rich genus of rainforest grasshoppers (*Rhachicreaga:* Orthoptera, Acrididae). II. Foodplant preference and its relation to speciation. *Oecologia* 68:99–104
187. Schneider, J. C., Roush, R. T. 1986. Genetic differences in oviposition preference between two populations of *Heliothis virescens*. In *Evolutionary Genetics of Invertebrate Behavior,* ed. M. D. Huettel, pp. 163–71. New York: Plenum
188. Scott, J. A. 1986. *The Butterflies of North America*. Stanford, Calif: Stanford Univ. Press
189. Scriber, J. M. 1983. Evolution of feeding specialization, physiological efficiency, and host races in selected Papilionidae and Saturniidae. In *Variable Plants and Herbivores in Natural and Managed Systems,* ed. R. F. Denno, M. S. McClure, pp. 373–412. New York: Academic
190. Scriber, J. M. 1988. Tale of the tiger: beringeal biogeography, binomial classification, and breakfast choices in the *Papilio glaucus* complex of butterflies. In *Chemical Mediation of Coevolution,* ed. K. C. Spencer, pp. 241–301. New York: Academic
191. Shapiro, A. M. 1981. The pierid red-egg syndrome. *Am. Nat.* 117:276–94
192. Singer, M. C. 1971. Evolution of foodplant preference in the butterfly *Euphydryas editha. Evolution* 25:383–89
193. Singer, M. C. 1982. Quantification of host preference by manipulation of oviposition behavior in the butterfly *Euphydryas editha. Oecologia* 52:230–35
194. Singer, M. C., Ng, D., Thomas, C. D. 1988. Heritability of oviposition preference and its relationship to offspring performance within a single insect population. *Evolution* 42:977–85
195. Singer, M. C., Thomas, C. D., Billington, H. L., Parmesan, C. 1989. Variation among conspecific insect populations in the mechanistic basis of diet breadth. *Anim. Behav.* 37:751–59
196. Singer, M. C., Vasco, D., Parmesan, C. 1990. Distinguishing between preference and motivation in butterfly oviposition. Unpublished manuscript
197. Slansky, F. E. 1973. Latitudinal gradients in species diversity of the New World swallowtail butterflies. *J. Res. Lep.* 11:201–7
198. Slansky, F. 1976. Phagism relationships among butterflies. *J. NY Entomol Soc.* 84:91–105
199. Smiley, J. 1978. Plant chemistry and the evolution of host specificity: new evidence from *Heliconius* and *Passiflora. Science* 201:745–47
200. Smiley, J. T., Spencer, K. C. 1990. Convergent community structure in *Passiflora*-feeding insects. Unpublished manuscript
201. Smith, D. A. S. 1978. Cardiac glycosides in *Danaus chrysippus* (L.) provide some protection against an insect parasitoid. *Experientia* 34:844–45
202. Southwood, T. R. E. 1972. The insect/plant relationship—an evolutionary perspective. *Symp. R. Entomol. Soc. Lond.* 6:3–30
203. T. Spieth, H. T. 1974. Mating behavior and evolution of the Hawaiian *Drosophila*. In *Genetic Mechanisms of Speciation in Insects,* ed. M. J. D. White, pp. 94–101. Sydney: Australian and New Zealand Book Co.
204. Stamp, N. E. 1980. Egg deposition patterns in butterflies: why do some species cluster their eggs rather than lay them singly? *Am. Nat.* 115:367–80
205. Stanek, E. J, Diehl, S. R., Dgetluck, N., Stokes, M. E., Prokopy, R. J. 1987. Statistical methods for analyzing discrete responses of insects tested repeatedly. *Environ. Entomol.* 16:320–26
206. Stanton, M. L. 1983. Spatial patterns in the plant community and their effects upon insect search. In *Herbivorous Insect,* ed. S. Ahmad, pp. 125–57. New York: Academic
207. Stanton, M. L. 1983. Short-term learning and the searching accuracy of egg-laying butterflies. *Anim. Behav.* 31:33–40
208. Stanton, M. L., Cook, R. E. 1983. Sources of intraspecific variation in the hostplant seeking behavior of *Colias* butterflies. *Oecologia* 60:365–70
209. Starmer, W. T. 1981. A comparison of *Drosophila* habitats according to the physiological attributes of the associated yeast communities. *Evolution* 35:38–52
210. Stevens, G. C. 1989. The latitudinal gradient in geographical range: how so many species coexist in the tropics. *Am. Nat.* 133:240–56
211. Stiling, P. D. 1980. Competition and coexistence among *Eupteryx* leafhoppers

(Hemiptera: Cicadellidae) occurring on stinging nettles *(Urtica dioica). J. Anim. Ecol.* 49:793–805
212. Stiling, P. D. 1987. The frequency of density dependence in insect host-parasitoid systems. *Ecology* 68:844–56
213. Straatman, R. 1962. Notes on certain Lepidoptera ovipositing on plants which are toxic to their larvae, *J. Lep. Soc.* 16:99–103
214. Strong, D. L., Lawton, J. H., Southwood, R. 1984. *Insects on Plants*. Cambridge: Harvard Univ. Press
215. Tabashnik, B. E., Wheelock, H., Rainbolt, J. D., Watt, W. B. 1981. Individual variation in oviposition preference in the butterfly, *Colias eurytheme*. *Oecologia* 50:225–30
216. Tahvanainen, J. O., Root, R. B. 1972. The influence of vegetational diversity on the population ecology of a specialized herbivore, *Phyllotreta cruciferae* (Coleoptera: Chrysomelidae). *Oecologia* 10:321–46
217. Templeton, A. R., Rothman, E. D. 1981. Evolution in fine-grained environments. II. Habitat selection as a homeostatic mechanism. *Theor. Popul. Biol.* 19:326–40
218. Thiery, D., Visser, J. H. 1987. Misleading the Colorado potato beetle with an odor blend. *J. Chem. Ecol.* 13:1139–46
219. Thomas, C. D., Ng, D., Singer, M. C., Mallet, J. L. B., Parmesan, C., Billington, H. L. 1987. Incorporation of a European weed into the diet of a North American herbivore. *Evolution* 41:892–901
219a. Thompson, J. N. 1982. *Interaction and Coevolution*. New York: Wiley-Intersci.
220. Thompson, J. N. 1988. Variation in preference and specificity in monophagous and oligophagous swallowtail butterflies. *Evolution* 42:118–28
221. Thompson, J. N. 1988. Coevolution and alternative hypotheses on insect/plant interactions. *Ecology* 69:893–95
222. Thompson, J. N. 1988. Evolutionary ecology of the relationship between oviposition preference and performance of offspring in phytophagous insects. *Entomol. Exp. Appl.* 47:3–14
223. Thorpe, K. W., Barbosa, P. 1986. Effects of consumption of high and low nicotine tobacco by *Manduca sexta* (Lepidoptera: Sphingidae) on survival of gregarious endoparasitoid *Cotesia congregata* (Hymenoptera: Braconidiae). *J. Chem. Ecol.* 12:1329–37
224. Turelli, M., Hoffmann, A. A. 1988. Effects of starvation and experience on the response of *Drosophila* to alternative resources. *Oecologia* 77:497–505
225. Vacek, D. C., East, P. D., Barker, J. S. F., Soliman, M. H. 1985. Feeding and oviposition preferences of *Drosophila buzzatii* for microbial species isolated from its natural environment. *Biol. J. Linn Soc.* 24:175–87
226. Via, S. 1984. The quantitative genetics of polyphagy in an insect herbivore. I. Genotype-environment interaction in larval performance on different host plants. *Evolution* 38:881–95
227. Via, S. 1984. The quantitative genetics of polyphagy in an insect herbivore. II. Genetic correlations in larval performance within and among host plants. *Evolution* 38:896–905
228. Via, S. 1986. Genetic covariance between oviposition preference and larval performance in an insect herbivore. *Evolution* 40:778–85
229. Via, S. 1990. Ecological genetics of herbivorous insects: the experimental study of evolution in natural and agricultural systems. *Annu. Rev. Entomol.* 35:421–46
230. Via, S. 1991. The population structure of fitness in a spatial patchwork: demography of pea aphid clones from two crops in a reciprocal transplant. *Evolution*. In press
231. Vinson, S. B., Iwantsch, G. F. 1980. Host suitability for insect parasitoids. *Annu. Rev. Entomol.* 25:397–419
232. Waldvogel, M., Gould, F. 1990. Variation in oviposition preference between strains of *Heliothis virescens* (Lepidoptera: Noctuidae). *Evolution*. In press
233. Wasserman, S. S. 1986. Genetic variation in adaptation to foodplants among populations of the southern cowpea weevil, *Callosobruchus maculatus:* evolution of oviposition preference. *Entomol. Exp. Appl.* 42:201–12
234. Wasserman, S. S., Futuyma, D. J. 1981. Evolution of host plant utilization in laboratory populations of the southern cowpea weevil, *Callosobruchus maculatus* Fabricius (Coleoptera: Bruchidae). *Evolution* 35:605–17
235. Wasserman, S. S., Mitter, C. 1978. The relationship of body size to breadth of diet in some Lepidoptera. *Ecol. Entomol.* 3:155–60
236. Weber, G. 1985. Genetic variability in host plant adaptation of the green peach aphid, *Myzus persicae*. *Entomol. Exp. Appl.* 38:49–56
237. Weber, G. 1986. Ecological genetics of host plant exploitation in the green peach aphid. *Myzus persicae*. *Ent. Exp. Appl.* 40:161–68
238. Wellington, W. G. 1965. Some maternal influences on progeny quality in the

western tent caterpillar, *Malacosoma pluviale* (Dyar). *Can. Entomol.* 97:1–14

239. Weston, P. A., Keller, J. E., Miller, J. R. 1990. Influence of selective ovipositional stimulus deprivation on fecundity parameters of onion fly (Diptera: Anthomyiidae). Unpublished manuscript
240. Whitham, T. G. 1983. Host manipulation of parasites: within-plant variation as a defense against rapidly evolving pests. In *Variable Plants and Herbivores in Natural and Managed Systems,* ed. R. F. Denno, M. S. McClure, pp. 15–41. New York: Academic
241. Wiklund, C. 1974. The concept of oligophagy and the natural habitats and host plants of *Papilio machaon* L. in Fennoscandia. *Entomol. Scand.* 5:151–60
242. Wiklund, C. 1975. The evolutionary relationship between adult oviposition preferences and larval host plant range in *Papilio machaon. Oecologia* 18:185–97
243. Wiklund, C. 1981. Generalist vs. specialist oviposition behaviour in *Papilio machaon* (Lepidoptera) and functional aspects on the hierarchy of oviposition preferences. *Oikos* 36:163–70
244. Wiklund, C. 1982. Generalist versus specialist utilization of host plants among butterflies. *Proc. 5th Int. Symp. Insect-Plant Relationships,* pp. 181–91. Pudoc, Wageningen
245. Williams, K. S. 1983. The coevolution of *Euphydryas chalcedona* butterflies and their larval host plants. III. Oviposition behavior and host plant quality. *Oecologia* 56:336–40
246. Wood, T. K., Kruluts, M. K. 1990. Host plant shifts in the *Enchenopa binotata* Say complex promoted by parasitoids: an hypothesis. Unpublished manuscript
247. Wood, T. K., Olmstead, K. L. 1984. Latitudinal effects on treehopper species richness (Homoptera: Membracidae). *Ecol. Entomol.* 9:109–15
248. Deleted in proof
249. Zalucki, M. P. 1981. The effects of age and weather on egg laying in *Danaus plexippus* L. (Lepidoptera: Danaidae). *Res. Popul. Ecol.* 23:318–27
250. Zwolfer, H. 1987. Species richness, species packing, and evolution in insect-plant systems. *Ecol. Stud.* 61:301–19

*Annu. Rev. Ecol. Syst. 1990. 21:275–97*


# FUNGAL ENDOPHYTES OF GRASSES

*Keith Clay*

Department of Biology, Indiana University, Bloomington, Indiana 47405

KEY WORDS: endophyte, fungi, grasses, symbiosis, clavicipitaceae

---

## INTRODUCTION

Mutualistic interactions between species are receiving increased attention from ecologists, although research lags far behind analogous work on competition or predator-prey interactions. Most research has focused on rather showy mutualisms such as pollination or fruit dispersal and has suggested that mutualisms are more important in tropical communities than in temperate communities (67). Plant-microbial mutualisms, in contrast, have prompted little ecological research. Plant-microbial associations are more difficult to observe and manipulate than plant-animal associations. Many plants are always infected (e.g. legumes by rhizobia, forest trees by mycorrhizal fungi), so it is easy to consider the microorganisms merely as a special type of plant organ. Further, plant-microbial mutualisms historically have been outside the realm of ecology, in other areas of biology like microbiology and mycology.

Recent research has revealed a widespread mutualistic association between grasses, our most familiar and important plant family, and endophytic fungi. Asymptomatic, systemic fungi that occur intercellularly within the leaves, stems, and reproductive organs of grasses have dramatic effects on the physiology, ecology, and reproductive biology of host plants. Through the production of toxic alkaloids, endophytic fungi defend their host plants against a wide range of insect and mammalian herbivores. Poisoning of domestic livestock has spurred a great deal of research on endophytic fungi in pasture grasses. This research has shown clearly that plants benefit from



0066-4162/90/1120-0275$02.00

infection by endophytes under most circumstances. This review examines the comparative ecology of endophyte-infected and uninfected grasses and identifies areas for future research.

## CLAVICIPITACEOUS FUNGAL ENDOPHYTES OF GRASSES

Endophyte is a general term that refers to any organism that lives inside of a plant, including organisms as diverse as mycorrhizal fungi and mistletoes. Many types of microorganisms live within grasses as endophytes (106), and many other plant groups support endophytic microorganisms (23). I use endophyte in this review to refer to a particular group of closely related fungi that systemically infect grasses.

### *Taxonomy*

The endophytes considered here are members of the Ascomycete family Clavicipitaceae (tribe Balansieae) and their anamorphs (asexual derivatives). This family also includes the ergot fungi (*Claviceps* spp.) and *Cordyceps,* pathogens of other fungi or insects (47). The taxonomy of the group is based on the morphology of fruiting structures on host plants and the types of conidia and ascospores produced. Endophyte genera include *Atkinsonella* (two species), *Balansia* (approximately 15 species), *Epichloe* (less than 10 species), and *Myriogenospora* (two species). A fifth genus *Balansiopsis* was recognized by Diehl (47), but its validity has been questioned (110). Our knowledge of endophytes is based primarily on North American and European hosts; additional studies are needed on tropical and/or southern hemisphere endophytes.

The sexual state of many endophytes has never been observed, so they cannot be classified as Ascomycetes. However, these endophytes are similar to the asexual form of other species with known sexual states. The asexual endophytes thus have probably been derived from sexual ones, particularly from the species *Epichloe typhina* (34, 126). Asexual endophytes have been grouped in the form genus *Acremonium* section *Albo-lanosa* (92, 131). They do not fruit on their hosts and are transmitted through the seed of infected plants (36). Sexual and asexual endophytes have similar host ranges, growth forms in the plant, and conidial morphology in culture. Molecular data will be useful in determining phylogenetic relationships among endophyte genera and species, and between sexual and asexual forms.

### *Distribution Within Hosts*

Endophytes grow as elongate, convoluted, sparsely branched hyphae running through intercellular spaces parallel to the long axis of leaves and stems (36, 125). Within the aerial plant body, endophyte hyphae are most dense in the leaf bases surrounding the intercalary meristems of grasses. During flowering, *Acremonium* endophytes grow into ovules and become incorporated into seeds (95, 112). Endophytes are able to absorb nutrients from freely available materials found in the intercellular spaces. They do not occur in roots although they can be found in rhizomes and stolons of several hosts (31, 85).

At least four endophyte species (in the tribe Balansieae) are exceptional because they occur on the surface of young leaves and inflorescences without penetrating host tissues (78, 85). At least one of these epiphytic species, *Balansia cyperi,* occurs in the same genus as endophytic species. Technically they should not be called endophytes, but I do so here to emphasize the taxonomic affinities of the fungi rather than their particular growth form. Hyphae of epiphytic species are most dense on the surface of young, tightly appressed leaves near the meristems where the cuticle is poorly developed (78). During host flowering, the fungi proliferate and differentiate into a fruiting body surrounding the young inflorescence, aborting it in the process.

### *Endophyte Reproduction and Dispersal*

Sexual endophytes produce fruiting bodies (stromata) on the leaves or inflorescences of host plants coincident with the hosts' flowering period. Fruiting bodies generally produce both conidia and ascospores. Diehl (47) suggested that one mode of contagious spread of the fungus is infection of ovules through the germination of spores on stigmata and their growth through the style. However, the mechanism and rate of contagious spread of endophytes has not been well documented in any species (but see 73). *Atkinsonella* and *Epichloe* are heterothallic, requiring the transfer of conidia between fruiting bodies of opposite mating types before ascospores can be produced (81, 127). The conidia therefore act as spermatia or gametes in sexual reproduction. Asymptomatic, seed-borne *Acremonium* endophytes do not fruit or produce ascospores.

Endophytes are dispersed in one of three general ways. Sexual species that fruit on their hosts can infect new plants contagiously. Contagious spread is not epidemic as in many plant pathogens, but it has been observed in the field (37, 47, 73, 109). Dispersal through the seeds of its host is the only means of spread for asexual *Acremonium* endophytes, but it also occurs in plants infected by *Atkinsonella* and *Epichloe* (41, 112). Both sexual and asexual endophytes can also be disseminated through vegetative structures of host plants including rhizomes, tillers, tubers, and viviparous plantlets (30, 31, 60, 109).

## HOST RANGE OF ENDOPHYTES

A recent survey indicated that endophytes infect at least 80 grass genera and several hundred species, including several important grain crops (36). Endophytes infect grasses in most subfamilies and tribes of the grass family (36). However, endophyte species are specialized on different groups of grasses. *Epichloe* infects primarily hosts in the subfamily Pooideae, where it is a common parasite. Asexual *Acremonium* endophytes also infect grasses only from this group, supporting the hypothesis that they are derived from *Epichloe* (Table 1, classification following 55). Species in the grass genera *Agrostis, Festuca, Lolium,* and *Poa* (pooid grasses) commonly are infected by *Acremonium* endophytes, suggesting that the endophytes were present early in the history of the subfamily and have diversified concomitant with host speciation (42, 77, 128). The only tribes in the Pooideae that do not contain known hosts for *Acremonium* endophytes are very small with few species (55). Two species of *Atkinsonella* each infects only a single host genus *(Danthonia* and *Stipa),* although the two host genera are in different tribes (80, 81). Species of *Balansia* and *Myriogenospora* primarily infect panicoid grasses with the C4 photosynthetic pathway (47, 85, 109). Host ranges of individual species vary considerably, e.g. *Balansia obtecta* infects only sandbur grass *Cenchrus echinatus,* while *B. epichloe* infects many panicoid genera and species (47). Concordant with the differential distribution of C3 and C4 grasses with latitude, *Balansia*- and *Myriogenospora*-infected hosts are more common in tropical and subtropical areas, while *Acremonium-, Atkinsonella-,* and *Epichloe*-infected hosts are more frequent in temperate areas. Thus, the mutualistic seed-borne *Acremonium* endophytes predominate in temperate areas, in opposition to the oft stated generalization that mutualisms are more common in the tropics.

The host ranges and genetic similarity of asexual seed-borne endophytes may provide information about phylogenetic relationships of host grasses. Grasses infected by closely related endophytes should be more closely related themselves than grasses infected by more distantly related endophytes. The large number of species of *Festuca* and *Lolium* infected by similar seed-borne endophytes suggests close affinity of the two genera, in agreement with current taxonomic concepts (90). Endophytes might also provide evidence for the affinities of difficult genera within the grass family. A case in point is the genus *Brachyelytrum* which has alternatively been considered a pooid or bambusoid grass (90). A recent study found that this species commonly was infected by *Acremonium* or *Epichloe typhina* endophytes (42), a fact supporting its placement as a pooid grass. Fungal parasites have been used in other plant groups as indicators of host relationships (64, 114).

While most hosts of clavicipitaceous endophytes are grasses, a secondary diversification has occurred on the Cyperaceae. In some respects the sedge family represents a wetland analog to the grass family. *Balansia cyperi* infects a number of New World *Cyperus* species (30, 47, 50), including *C. rotundus,* a widespread agricultural weed (31). *Balansia cyperaceum* also infects only

**Table 1** Distribution of Balansiae and *Acremonium* endophytes in subfamilies and tribes of the Poaceae

| Grass | | Endophyte | |
|---|---|---|---|
| Subfamily | Tribe | Balansiae | Acremonium |
| Pooideae[a] | Poeae | + | + |
| | Aveneae | + | + |
| | Triticeae | + | + |
| | Meliceae | + | + |
| | Stipeae | + | + |
| | Brachyelytreae | + | + |
| | Diarrheneae | | |
| | Nardeae | | |
| | Monermeae | | |
| Panicoideae | Paniceae | + | |
| | Andropogoneae | + | |
| Chloridoideae | Eragrosteae | + | |
| | Chlorideae | + | |
| | Zoysieae | | |
| | Aeluropodeae | | |
| | Unioleae | + | |
| | Pappophoreae | | |
| | Orcuttieae | | |
| | Aristideae | + | |
| Bambusoideae | Bambuseae | + | |
| | Phareae | | |
| Oryzoideae | Oryzeae | + | |
| Arundinoideae | Arundineae | | |
| | Danthonieae | + | |
| | Centosteceae | + | |

[a] Classification follows Gould & Shaw (55)

sedges (47). There is one reported case of *Epichloe typhina* infecting a population of the rush *Juncus effusis* in New Hampshire (70). This probably represents an unusual and transitory case of host range expansion since rushes are not known to serve as hosts for other endophytes elsewhere. Thus, clavicipitaceous endophytes infect species in only three plant families, with grasses making up the greatest number of hosts, paralleling the host range of the related *Claviceps,* which also infects primarily grasses, but also sedges and rushes (20).

Many other nonclavicipitaceous fungi occur as asymptomatic endophytes that cause localized infections of grass leaves. For example, 200 fungal species have been isolated from leaves of healthy winter wheat (106) and endophytes have been isolated from other grass species (39, 75, 78). There is little evidence yet that these endophytes of grasses are biologically significant. Other plant groups also serve as hosts to a wide range of endophytic fungi of diverse taxonomic affinities (23). Nonclavicipitaceous endophytes are beyond the scope of this review; interested readers are referred to Carroll (23) and references therein.

# EFFECTS OF ENDOPHYTE INFECTION ON HOST PLANTS

## *Reproductive Biology of Infected Grasses*

Grasses infected by seed-borne *Acremonium* endophytes are similar to uninfected plants in their production of normal, healthy inflorescences. However, infected plants may produce more inflorescences and seeds than do uninfected plants, reflecting their greater vegetative vigor (33, 37). Seeds of tall fescue and perennial ryegrass germinate more rapidly and to higher levels when infected, and resulting seedlings grow faster than seedlings from uninfected seeds (26, 33). Endophyte-infected seeds also contain high concentrations of alkaloids and are less likely to be eaten by vertebrate and invertebrate seed feeders (25, 132; see also section on herbivory). However, differences in seed survival, dormancy, and germination have not been examined in field situations for any Acremonium-infected grass.

More dramatic differences between infected and uninfected plants can be observed in the reproductive biology of grasses infected by sexual Balansieae endophytes that fruit on their hosts. In many cases, infected plants do not produce seed and so are constrained to vegetative reproduction (30, 47, 72, 73). The inflorescences may be initiated normally but abort before maturity as the fungal fruiting body develops and surrounds the developing inflorescences. Examples are found in plants infected by *Epichloe, typhina, Balansia cyperi, B. obtecta,* and *B. strangulans* (47, 112). In other species infected by endophytes such as *B. epichloe* and *B. henningsiana,* inflorescence development is normally inhibited so that infected plants remain vegetative while surrounding uninfected plants are in full flower (39, 47). In these species, endophyte-infected individuals can be found easily in the field by looking for large, vigorous, nonflowering plants during the peak of the flowering season.

For grasses with well-developed means of vegetative or clonal reproduction, loss of seed production may be of little detriment. The lack of seed production may contribute to the vegetative vigor of infected plants by freeing resources to be used for growth rather than reproduction. Early work by Bradshaw, Diehl, and Harberd (21, 47, 60) suggests that grasses infected by *Epichloe* and *Balansia* endophytes tiller more profusely and spread horizontally more vigorously by means of rhizomes or stolons. Greater tillering of infected plants has been subsequently noted in several other grasses (29, 33,

39, 60). In grazed or mown situations, the prostrate vegetative growth of infected plants, relative to the more upright growth of flowering uninfected plants, may provide a selective advantage to infected plants by reducing damage. Sedges in the genus *Cyperus* produce subterranean tubers or bulbils, in addition to aerial vegetative shoots. In a greenhouse study with *C. rotundus,* plants infected with *Balansia cyperi* produced significantly more tubers and significantly fewer inflorescences than did uninfected plants (120). Tubers are the major source of reproduction in this species (seeds are rarely produced), and the fungus occurs around the buds on tubers (31). The location of the fungus on tuber buds enables it to infect new shoots or tubers that originate from that bud.

The most dramatic example of endophyte infection altering the balance between sexual and asexual reproduction occurs in those host species where infection induces vivipary, the vegetative proliferation of inflorescences (not the precocious germination of seeds as in mangroves). In the sedge *C. virens,* infection by *B. cyperi* normally causes the complete abortion of every inflorescence (30). Later in the growing season the aborted inflorescences of many infected plants become viviparous, producing up to 25 plantlets per inflorescence (30). The plantlets, which arise from dormant buds in the inflorescence and not from seed, root quickly in the marshy soils where the sedge typically occurs. Plants grown from plantlets are themselves infected with *B. cyperi* (30). Because of their large size compared to seedlings, plantlets are likely to have higher probability of establishment, in addition to the other possible advantages of endophyte infection. Viviparous plantlets have also been observed growing from aborted inflorescences of the grasses *Andropogon glomeratus* and *A. virginicus* infected by *Myriogenospora atramentosa* (30), suggesting that induced vivipary may be widespread. Sampson & Western (113) reported that infection by *E. typhina* caused *Poa bulbosa,* a viviparous grass, to produce plantlets that were themselves infected. However, uninfected plants were also viviparous. Endophyte infection is very common in grass genera where vivipary is well established (e.g. *Poa, Festuca*) (2, 36, 42, 63, 125, 128, 130, 133).

Suppression of flowering is not complete in many host species. Individual plants infected by *E. typhina* and *B. henningsiana* sometimes produce a mixture of healthy and aborted inflorescences (39, 112, 126). Aborted inflorescences are rare in *E. typhina*-infected red fescue *(F. rubra),* where the fungus is primarily transmitted by seed (112). Similar patterns of sporadic inflorescence abortion have been noted in the genera *Agrostis, Brachyelytrum, Elymus,* and *Sphenopholis,* when infected by *E. typhina* (42). These grasses apparently represent an intermediate condition between grasses such as tall fescue and perennial ryegrass where the endophyte is completely seed-borne (95, 117) and other grasses where infection causes the abortion of all inflorescences (47). In contrast, *Panicum henningsiana* infected by *B. epichloe* produces primarily aborted inflorescences and only a minority of healthy inflorescences (39). Resultant seeds are uninfected. The production of an occasional healthy inflorescence on an otherwise heavily infected plant, noted in other hosts as well (47), probably represents the "escape" of a single tiller from the systemic growth of the fungus. This is unlike the situation with *E. typhina*–infected red fescue where all inflorescences, both healthy and aborted, contain the endophyte (112).

Several host grasses have a well-developed floral dimorphism where potentially outcrossed chasmogamous flowers are produced at the apex of flowering culms while obligately self-pollinated cleistogamous flowers are produced in the lower axils of the leaf sheaths (22, 24, 28). Species of the fungus *Atkinsonella* infect several species in the genus *Danthonia* and *Stipa leucotricha,* all of which produce dimorphic flowers (29, 49, 81). Infected *Danthonia* regularly produce viable cleistogamous flowers and seeds while the apical panicles are completely aborted (29). Thus, infected plants are completely cleistogamous and are incapable of exchanging genes with uninfected members of the populations. Moreover, the cleistogamous seeds are infected by *Atkinsonella* and give rise to infected plants (41). Cleistogamous flowers and seeds in infected *S. leucotricha,* in contrast, are usually aborted although an occasional seed may be produced (K. Clay, personal observation). Other species with dimorphic reproductive systems are known to serve as hosts (22, 47), but the effect of endophyte infection on the production of chasmogamous and cleistogamous flowers and seeds has not been critically examined.

Recent theoretical and empirical work suggests that pathogens may select for sexual reproduction in host populations (58, 82, 84). The production of genetically variable progeny may make it more difficult for pathogen populations to specialize on particular genotypes, which would be easier in asexual host populations. Inflorescence-aborting Balansieae endophytes may exhibit a counteradaptation to the host strategy of producing genetically variable offspring. The fungi suppress outcrossing and enforce complete self-fertilization or asexual reproduction, promoting a more genetically uniform host population (34). For example, viviparous plantlets produced by *B. cyperi*–infected *C. virens* are genetic copies of the maternal plant and are infected by the same fungal genotype as the maternal plant. Genetic analyses of the magnitude and spatial scale of variation in natural populations containing mixtures of infected and uninfected individuals would support or refute this hypothesis.

## *Physiological Ecology of Infected Grasses*

One of the least understood aspects of the grass/fungal endophyte symbiosis is the comparative physiological ecology of infected versus uninfected plants.

Most work has focused on tall fescue, the most important pasture grass in eastern North America. Endophyte infection causes changes in host plant physiology that stimulate growth under controlled environmental conditions (33, 76). Enhanced growth of infected plants also occurred in perennial ryegrass and purple nutsedge grown in controlled environments (33, 76, 120). In tall fescue the growth advantage of endophyte-infected plants was reversed under conditions of severe nutrient stress, suggesting that the metabolic cost of supporting the endophytic fungus has a negligible detrimental effect except under extreme conditions (26).

In response to increasing irradiance, net photosynthetic rates of endophyte-infected tall fescue were found to be less than endophyte-free clones of the same genotypes, although infected clones produced more biomass and tillers (15). We have found that endophyte-infected tall fescue maintained significantly higher net photosynthetic rates at temperatures above 25°C than did uninfected plants when both were grown under well-watered conditions in the greenhouse (S. Marks and K. Clay, unpublished). Further studies have indicated that infected plants tend to have greater stomatal resistances (4, 51). These results are consistent with experimental work showing that infected tall fescue maintains higher productivity than do uninfected plants under drought conditions (4, 105, 124). Arachevaleta et al (4) showed that infected plants exhibited leaf rolling under drought stress conditions much more rapidly than did uninfected plants, and that under extreme drought conditions all uninfected plants died while all infected plants survived. Other studies have indicated that infected plants exhibited greater osmotic adjustment than uninfected plants, which allowed the infected plants to maintain higher turgor pressures under drought conditions (51, 124). Extreme environmental stresses, such as described above, may provide intermittent strong selection favoring infected plants and causing dramatic changes in population composition. The proportion of endophyte-infected tall fescue and other grasses may have increased substantially during the drought conditions experienced over much of North America in 1988.

Mechanisms by which photosynthesis and water relations might be altered by endophyte infection include changes in the plant's hormonal balance and changes in source-sink relationships. Auxin has been detected in pure cultures of the endophyte *Balansia epichloe* (99). Morphological and/or physiological changes observed in a number of host species (i.e. inhibition of flowering) are consistent with altered hormone metabolism (30, 47). Studies of photosynthate metabolism in two different host-endophyte associations have revealed that the fungi rapidly convert plant sucrose into sugar alcohols that the plants are unable to metabolize (119, 121). By continually depleting the plant's sink of available photosynthate, endophyte infection may reduce or prevent feedback inhibition of photosynthetic rates, thereby allowing higher average growth rates.

# HERBIVORY OF ENDOPHYTE-INFECTED GRASSES

## *Mammalian Herbivory*

Grasses are relatively free of toxic secondary compounds, compared to most plant families. Crop or pasture grasses infected by species of *Claviceps,* the ergot fungus, have poisoned livestock (and humans) for hundreds of years (20, 87). Many grasses toxic to mammalian herbivores in the absence of ergot infection are now known to be endophyte infected (11, 25, 93). Endophyte-infected grasses, including species in the genera *Andropogon, Festuca, Lolium, Melica, Paspalum Sporobolus,* and *Stipa* (10, 12, 25, 35, 93, 96), can be extremely toxic to a range of insect and mammalian herbivores (7, 40, 118). Animals often refuse to graze on infected grasses (19, 59, 89) and develop a number of circulatory, muscular, and neurological disorders if they do (10, 66, 93). Because most research has been conducted within an agricultural context, essentially nothing is known about how endophyte-infection affects vertebrate herbivory in natural communities. Fairly simple experiments where known infected and uninfected individuals (or ramets) of the same species are planted in random arrays in the field would provide evidence on whether differential herbivory occurs. It seems unlikely that domestic animals would react differently from wild animals in their behavioral or physiological responses to endophyte-infected grasses.

Attempts to replant heavily on endophyte-infected pastures with endophyte-free seed have been made in tall fescue pastures in the United States and perennial ryegrass pastures in New Zealand. However, eliminating the endophyte reduces the vigor of plants and their resistance to a range of biotic and abiotic stresses (4, 13, 105, 124). This is apparently the reason that endophyte-infected pasture grasses were so widely planted in the first place. Artificial selection for superior plant vigor and persistence inadvertently selected for endophyte-infected plants. Similar processes appear to occur in nature.

## *Insect Herbivory*

An important factor favoring endophyte-infected grasses in natural and agricultural communities is their enhanced resistance to insect herbivores. Field observations of fescue and ryegrass species indicate reduced herbivory of endophyte-infected plants in experimental plots and lower populations of the herbivores (54, 103, 104, 111, but see also 71, 83). Insect feeding trials and food plant preference studies conducted in the laboratory with a variety of grass species have shown that endophyte infection acts as a feeding and oviposition deterrent (61, 68) and/or reduces the survival, growth, and developmental rates of feeding insects (1, 25, 40, 74, 115). This topic has been reviewed extensively (35, 36, 118). As in the case of mammalian herbivory, evidence for insecticidal effects of endophyte infection in the field comes primarily from domesticated grasses. There is a need for careful field studies of possible differential herbivory in wild grasses similar to the work of Marquis (88).

## *Other Pests*

Endophyte infection also may protect host plants against other pests. Reduced nematode populations were associated with endophyte infection in tall fescue, both in pot culture and in field soils (97, 124). Inhibition of plant pathogens in vitro by endophyte cultures has also been documented (14, 129). Two studies have found a negative correlation between endophyte and pathogen infection in field situations. *Panicum agrostoides* infected by *Balansia henningsiana* had significantly fewer lesions of the leaf spot fungus *Alternaria triticina* than did neighboring endophyte-free plants in a natural population (39). Another study with tall fescue found reduced levels of crown rust *(Puccinia coronata)* in endophyte-infected plots (53). Possible mechanisms by which endophytes could inhibit plant pathogens include competition for resources, induction of generalized defense responses, and the production of antimicrobial compounds (32, 46).

## *Endophyte Toxins*

Alkaloids produced by endophytes within plant tissues are in part responsible for the poisonings of mammalian grazers and plant resistance to insect herbivores. Endophyte-infected grasses contain a variety of alkaloids not found in uninfected conspecifics; these alkaloids include ergot alkaloids, loline alkaloids, lolitrems, and peramine (7, 8, 38, 68, 101, 102, 107, 108). Detailed reviews on the chemistry of endophyte alkaloids are available (7, 45). Most of these alkaloids, or their precursors, have been detected in liquid cultures of the fungi (1, 5, 6, 98). Further, feeding trials using pure alkaloids have revealed insecticidal activity against several herbivores (38, 45, 68).

Alkaloid content of endophyte-infected grasses is known to vary among host species, plant parts (leaf blades, sheaths, flowering culms), within the same parts of different age, at different times of the growing season, and with fertilization and grazing history (7, 16, 17, 62, 86). In tall fescue the greatest concentration of loline alkaloids is in the seeds, suggesting a role in preventing seed predation (25). In adult plants of tall fescue, alkaloid concentrations are higher in young than in old leaves and in leaf sheaths than in leaf blades (62, 86). These patterns coincide with endophyte hyphal density, as well as with the nutritive value of the tissues (62, 65). Fungal strains also differ in the amounts and kinds of alkaloids produced in pure culture (9, 98, 100), suggesting that between-plant variation could exist in alkaloid content resulting from genetic differences between endophyte strains.

The bitter taste of many alkaloids may allow herbivores to discriminate among plants and avoid endophyte-infected individuals. Where a species is completely endophyte infected, herbivores may shift to other species. Populations differing historically in the level of herbivory would provide evidence as to whether selection favors endophyte-infected plants or plants infected with high alkaloid-producing endophyte strains where herbivory is common or chronic. Results would also bear on the question of whether grasses benefit from herbivory (18, 43, 91).

## DEMOGRAPHY OF ENDOPHYTES AND THEIR HOSTS

### *Frequency of Infection Within Populations*

As previously indicated, some grass species are never endophyte-infected, some always are infected, and others consist of mixtures of infected and uninfected individuals. However, the majority of grass species have not been examined for endophyte infection, especially grasses from tropical regions. Complete infection of all individuals in host populations occurs only in grasses infected by asexual, seed-borne *Acremonium* endophytes (42, 125), implying a strong selective advantage for infected plants. Infection can be lost by occasional production of uninfected seed or by extended seed dormancy but cannot be gained within a lineage (122). Herbarium specimens have been examined for *Acremonium* endophyte infection. High levels of infection, often 100% of specimens, were observed in many species of *Agrostis, Elymus, Festuca, Lolium, Poa, Sphenopholis,* and *Stipa,* as well as numerous other genera with fewer host species (42, 125, 128, 130). While surveys of herbarium specimens may not reflect the level of infection within populations, they provide an overview of infection levels throughout the range of the host. Further, samples are unbiased since there is no way to tell, without microscopic examination, if the plants collected were infected.

Field surveys, which have focused on pasture grasses, have also revealed very high levels of *Acremonium* endophyte infection. Neill (94, 95) found that 16 of 20 perennial ryegrass *(Lolium perenne)* varieties in New Zealand were 100% infected. High levels of endophyte infection have been found in other varieties of perennial ryegrass (54, 56, 57, 77), other species of *Lolium* (77), and in *Fesctuca rubra* and *F. longifolia* (111, 112, 123). *Festuca arundinacea,* tall fescue, has been extensively sampled in the United States. In Tennessee, 67 of 79 counties contained endophyte-infected plants (each county represented by six plants from three locations) with an overall average of 35% infection. A total of nearly 1500 samples of tall fescue from 26 states throughout the United States had an average of approximately 50% infection, with 30% of the samples containing 90% or more infected plants (116a).

Estimates of infection level have been made for several grasses infected by sexual Balansiae endophytes. In Britain up to 40% of *Agrostis tenuis* and 80% of *Dactylis glomerata* plants were infected by *Epichloe typhina* in some sites (21, 72, 73). Over one half of *Cyperus virens* plants sampled in coastal Louisiana were infected by *Balansia cyperi,* with up to 80% of plants infected in some sites (30). Similarly, the majority of *Panicum agrostoides* plants in most sites in Indiana was infected by *B. henningsiana* (39), while only 15% of *Danthonia spicata* plants in a North Carolina population were infected by *Atkinsonella hypoxylon* (29). In Sierra Leone, cultivated rice becomes infected by *Balansia pallida;* 2% of panicles were destroyed by the fungus at the most heavily infected study site (52). Anecdotal observations suggest that infection levels can be high in many other species as well (K. Clay, personal observation).

Populations of grasses completely infected by sexual endophytes that suppress host seed production might be expected to have higher extinction probabilities than ones where seed are produced, even though they could persist through vegetative reproduction alone (see 21, 60). Infection levels of *Acremonium* endophytes, in contrast, are not subject to the same constraints.

## *Dynamics of Infection*

Changes in levels of endophyte infection occur in many host populations over time. Early workers noted the tendency for *Atkinsonella* and *Epichloe* to be more prevalent in older populations (21, 47, 60, 73). *Acremonium* endophyte-infection levels in *Lolium* species were higher in older pastures than in younger pastures, and in natural populations compared to domesticated varieties (44, 77, 83, 94). In field plots of *Festuca longifolia* the percentage of endophyte infection went from less than 50% to over 90% in seven years (111).

Increasing levels of endophyte infection with increasing age of the host population may be due to contagious spread of the fungus, increased survival and/or reproduction of infected plants, or both, depending upon the ability of the endophyte to sporulate and the host plant to produce seed. For seed-borne *Acremonium* endophytes contagious spread is impossible; the fungus does not produce spores and never occurs outside the interior plant body. Increasing infection over time implies a fitness benefit to infected plants, which must live longer or produce more offspring than do endophyte-free plants. Population dynamics of endophyte infection within populations could be modeled simply by equating infection with a maternally inherited gene with two alternative alleles of different selective value.

In contrast to *Acremonium* endophytes, increasing levels of infection by *Atkinsonella* or *Epichloe* can result from contagious spread of the endophyte and/or from differential host survival. In *Dactylis glomerata* fields, levels of infection rose dramatically from the first year following establishment up through seven years. It seems likely that contagious spread is initially important since the fields were initially founded from endophyte-free seed, but increased host survival could become more important over time. In a study with marked plants in a grassland community, nearly 20% of uninfected *Sporobolus poiretti* became infected by *Balansia epichloe* in each of two years (37). Studies of other host species have found little or no contagious spread of spore-producing endophytes (3, 29, 37). Despite a low level of contagious spread, infection levels can still increase due to demographic differences between conspecific hosts and nonhosts.

## *Host Demography*

Four detailed studies of the comparative demography of endophyte-infected and uninfected plants have been conducted. Infection of *Danthonia spicata* by *Atkinsonella hypoxylon* increased from 25% to 30% of a study cohort over three years owing to greater survival of infected plants (29). Infected plants grew significantly larger but produced significantly fewer seeds than did uninfected plants. Infected plants also were superior competitors in field experiments against the cooccuring grass *Anthoxanthum odoratum* (69). In experimental populations of the sedge *C. virens* and the fungus *B. cyperi*, an initial infection level of 50% increased to 90% infection over three years owing to the greater survival of plants infected with *B. cyperi* (37). However, similar demographic experiments with *Sporobolus poiretti* revealed equivalent survival rates for *Balansia*-infected and uninfected plants (37).

Demographic studies of tall fescue, which flowers and sets seed normally when infected by endophytes, have demonstrated a nearly two-fold increase in fitness of infected plants compared to uninfected conspecifics. In field studies where ramets of known infection status were planted out into a grassland community in Louisiana, members of the infected cohort had significantly higher survival and growth rates than did uninfected plants, and a higher percentage of infected plants flowered (37). Over the long term, endophyte-infected tall fescue plants should come to dominate the population.

While increased growth of infected plants in controlled environments must result from direct effects on plant physiology, demographic and competitive differences in field situations may result from changes both in host physiology and in resistance to herbivores and other plant pests. Field competition or demographic experiments where levels of herbivory are manipulated should provide evidence on the relative importance of induced changes in host physiology and resistance to plant pests for the comparative demography of endophyte-infected plants.

## *Mutualistic and Pathogenic Endophytes*

Closely related fungi infecting various grasses can enhance host fitness by increasing host survival, growth, and reproduction, or reduce host fitness by suppressing seed production. However, even the more pathogenic endophytes like *Atkinsonella* or *Epichloe,* which abort host inflorescences, have positive effects on plant survival, growth, and resistance to herbivory. The critical difference between the two forms of endophyte infection is the production of spore-producing fruiting bodies which abort inflorescences by the more pathogenic Balansieae endophytes versus the unimpaired flowering and the concomitant seed transmission by the more mutualistic *Acremonium* endophytes (34).

A range of endophyte effects on host reproduction occurs within and among hosts of *E. typhina*. A survey of woodland grasses infected by *E. typhina* revealed that in *Glyceria striata* all flowering culms of infected plants were aborted; in *Elymus virginicus* only a minority of flowering culms were aborted (and the endophyte was seed-borne in the other culms); and in *Agrostis hiemalis* fruiting bodies were rarely produced in most populations, and the endophyte was almost entirely seed-borne (42). A similar range of variation also occurs within single host species. Sampson (112) and White (126) observed *Festuca rubra* and *E. virginicus,* respectively, and found infected plants within the same population that either (*a*) produced a high proportion of aborted culms, (*b*) a low proportion of aborted culms, or (*c*) no aborted culms, with seed transmission of the endophyte in normal culms.

These relationships illustrate a transition between mutualistic and pathogenic associations and offer the potential for determining the underlying genetic and/or environmental basis for the variation in seed transmission and inflorescence abortion (34, 126). Reciprocal inoculations of seedlings or tillers from plants in categories *a, b,* and *c* above with fungal isolates from the same categories of plants should reveal whether the observed variation is a function of plant genotype, fungal genotype, or an interaction of both. Evidence for an environmental influence could be obtained from common garden experiments where plants in categories *a–c* are cloned and ramets planted into different microenvironments. Similar responses by all clones of a genotype would suggest a genetic basis for endophyte seed transmission versus inflorescence abortion. The ecological and genetic consequences of variation in host and fungal reproductive systems for both the host and the fungus are important areas for future research that will provide insights into how symbiotic relationships move along a continuum of mutualistic and pathogenic interactions.

## CONCLUSIONS AND FUTURE DIRECTIONS

The grass-endophyte symbiosis has important implications for many areas of ecology and evolution, both applied and basic. Endophytes are significant economic problems in the beef and dairy industries because of their detrimental effects on livestock (7, 118). Simultaneously, endophytes are attractive potential biocontrol agents for reducing pest damage in nongrazed grassland communities such as lawns, athletic fields, golf courses, and roadsides (36). Turfgrass variables available on the commercial market often contain high levels of endophyte infection (56, 77, 111). Endophytes could be utilized to "vaccinate" crop grasses against pests if toxins did not accumulate in seeds, and they could be used as vectors for the genetic engineering of grasses (36).

Endophytes infecting grasses provide the best example of a defensive mutualism where the parasite protects its host against enemies (35). This association appears to be widespread, although detailed studies have been conducted only in a very few systems. Additional research needs to be conducted in several areas. We know little about the potential for herbivores to evolve tolerance or resistance to fungal alkaloids in grasses. Selection for tolerance in herbivore populations should be strongest in communities where a high percentage of individuals and species are endophyte-infected. In contrast, selection for inherent chemical defenses in grass populations would be strongest in communities where a low percentage of individuals and species are infected by endophytes. The degree of differential herbivory in mixed populations of infected and uninfected plants needs to be examined in several host species in different natural communities. It is not known whether different fungal genotypes infecting the same host population (see 80) vary in the amounts and kinds of alkaloids produced and whether hosts of the most toxic fungal genotypes are most successful when herbivory is common. There may be costs to the host, which cannot be recouped in the absence of herbivory, of supporting a high alkaloid-producing endophyte compared to one that produces smaller amounts of alkaloids. Comparisons of levels of endophyte infection and secondary defensive compounds (of either fungal or plant origin) between communities differing historically in the levels of herbivory would be valuable. Further, species that differ in the inherent nutritional content of foliage may be affected disproportionately by endophyte-infection such that the most nutritious and palatable species gain the greatest benefit from infection. The high levels of endophyte infection in pasture grasses support this hypothesis (36, 77, 118). Could our entire system of agriculture, based primarily on grasses, have originated through the accidental elimination of endophytes from grasses that, undefended, became especially nutritious or palatable (77)?

Levels of endophyte infection within and among species are known from few communities. Two community-level studies (42, 77) have revealed high levels of endophyte-infection in North American deciduous woodland communities and in central European perennial grassland communities. We do not known whether the levels of infection in these communities differ from infection levels in tropical grasslands, arctic tundra, or temperate prairie communities. Large-scale surveys in a diversity of community types throughout the world are needed to answer the basic question of how common endophytes are.

Endophyte infection of grasses affects, and can be affected by, community characteristics. A number of studies reviewed here suggest that infection levels within species increase with succession and community age (37, 105, 111). Infection-enhanced competitive abilities (69) can affect the displacement of endophyte-free species by endophyte-infected species, especially if herbivory is severe (54, 111, 116). Anecdotal report suggests community changes may be rapid. Shaw (116) reported that the endophyte-infected grass *Melica decumbens* (drunk grass) increased dramatically in South African grasslands following the introduction of European livestock, presumably because they preferentially fed on uninfected species. Similar community-level increases of endophyte-infected *Sporobolus poiretii* in heavily grazed southeastern pastures also have been reported (7). The dynamics of endophyte infection within and among host species during and following insect outbreaks or the introduction of grazers would be of particular interest.

There remain many unanswered questions regarding the population and evolutionary dynamics of endophyte infection in grass populations. The rate at which asexual seed-borne endophytes increase in host populations versus sexual Balansieae endophytes that suppress sexual reproduction of their hosts needs to be quantified. Plant survival and seed production, and the rate of contagious spread of infection, are the most important parameters to be determined. Models that reveal conditions favoring one form of transmission over the other may provide insights into how mutualistic and pathogenic associations evolve and coevolve.

The inverse relationship between endophyte sexuality and host sexuality (see 34) has important implications for gene flow and the genetic structure of both host and endophyte populations. For example, *Danthonia* grasses infected by *Atkinsonella* are reproductively isolated from uninfected plants in the population. Infected plants reproduce sexually only by self-fertilized cleistogamous seeds, which are themselves endophyte infected (29, 41). The fungus, in contrast, is cross-fertilized with other genotypes (81). Do infected plants consist of genetically similar individuals interspersed within a panmictic, uninfected subpopulation? At the other extreme are situations where the endophyte is completely asexual and seed-borne, and the host flowers normally. How many fungal genotypes exist within a single host population (80)? Do seed-borne endophytes have narrower host ranges than sexual endophytes that must contagiously infect new hosts? Are there plant genes for resistance to endophyte infection? Limited data suggest that resistance to infection, as found in many crop plants towards pathogens, does not exist in endophyte host species (79).

In conclusion, much has been learned over the last ten years about fungal endophytes of grasses, but many additional questions have been raised. Given the importance of grasses in agricultural and plant communities, the implications of endophyte symbiosis should not be underestimated. Further, endophyte-infected grasses represent ideal model systems for approaching questions about the coevolutionary origin of mutualisms from pathogenic associations, and the role of herbivory in plant population and community dynamics. We need to begin to think about many grasses not as independent organisms but as partners in a symbiosis, similar to lichens or legumes infected by N-fixing rhizobia.

ACKNOWLEDGMENTS

I wish to thank Allan Fone for his helpful comments on this manuscript. Graduate and postdoctoral students who have made major contributions to my own research on fungal endophytes include James Bier, Carol Blaney, Gregory Cheplick, Allan Fone, Adrian Leuchtmann, Margaret Maloney, Susan Marks, John Schmidt, and Mary Stovall. My research on fungal endophytes of grasses has been supported by NSF grants BSR-8400163 and BSR-8614972.

## *Literature Cited*

1. Ahmad, S., Govindarajan, S., Funk, C. R., Johnson-Cicalese, J. M. 1985. Fatality of house crickets on perennial ryegrasses infected with a fungal endophyte. *Entomol. Exp. Applic.* 39:183–90
2. Aiken, S. G., Lefkovitch, L. P., Darbyshire, S. J., Armstrong, K. C. 1988. Vegetative proliferation in inflorescences of red fescue (*Festuca rubra* s. l., Poaceae). *Can. J. Bot.* 66:1–10
3. Antonovics, J., Clay, K., Schmitt, J. 1987. The measurement of small-scale environmental heterogeneity using clonal transplants of *Anthoxanthum odoratum* and *Danthonia spicata*. *Oecologia* 71:601–7
4. Arachevaleta, M., Bacon, C. W., Hoveland, C. S., Radcliffe, D. E. 1989. Effect of tall fescue endophyte on plant response to environmental stress. *Agron. J.* 81:83–90
5. Bacon, C. W. 1985. A chemically defined medium for the growth and synthesis of ergot alkaloids by species of *Balansia*. *Mycologia* 77:418–23
6. Bacon, C. W. 1988. Procedure for isolating the endophyte from tall fescue and screening isolates for ergot alkaloids. *Appl. Environ. Microbiol.* 54: 2615–18
7. Bacon, C. W., Lyons, P. C., Porter, J. K., Robbins, J. D. 1986. Ergot toxicity from endophyte-infected grasses: a review. *Agron. J.* 78:106–16
8. Bacon, C. W., Porter, J. K., Robbins, J. D. 1975. Toxicity and occurrence of *Balansia* on grasses from toxic fescue pastures. *Appl. Microbiol.* 29:553–56
9. Bacon, C. W., Porter, J. K., Robbins, J. D. 1981. Ergot alkaloid biosynthesis by isolates of *Balansia epichloe* and *B. henningsiana*. *Can. J. Bot.* 59:2534–38
10. Bacon, C. W., Porter, J. K., Robbins, J. D. 1986. Ergot toxicity from endophyte infected weed grasses: a review. *Agron. J.* 78:106–16
11. Bacon, C. W., Porter, J. K., Robbins, J. D., Luttrell, E. S. 1977. *Epichloe typhina* from toxic tall fescue grasses. *Applied and Environ. Microbiol.* 34:576–81
12. Bacon, C. W., Siegel, M. R. 1988. Endophyte parasitism of tall fescue. *J. Prod. Agric.* 1:45–55
13. Barker, G. M., Pottinger, R. P., Addition, P. J., Prestidge, R. A. 1984. Effect of *Lolium* endophyte fungus infections on behavior of adult Argentine stem weevil. *NZ J. Agric. Res.* 27:271–77
14. Baya, B. O., Halisky, P. M., White, J. F. 1987. Inhibitory interactions between *Acremonium* spp. and the mycoflora

from seeds of *Festuca* and *Lolium*. *Phytopathology* 77:115
15. Belesky, D. P., Devine, O. J., Pallas, J. E. Jr., Stringer, W. C. 1987. Photosynthetic activity of tall fescue as influenced by a fungal endophyte. *Photosynthetica* 21:82–87
16. Belesky, D. P., Robbins, J. D., Stuedemann, J. A., Wilkinson, S. R., Devine, O. J. 1987. Fungal endophyte infection-loline derivative alkaloid concentration of grazed tall fescue. *Agron. J.* 79:217–20
17. Belesky, D. P., Stuedemann, J. A., Plattner, R. D., Wilkinson, S. R. 1988. Ergopeptine alkaloids in grazed tall fescue. *Agron. J.* 80:209–12
18. Belsky, A. J. 1986. Does herbivory benefit plants? A review of the evidence. *Am. Nat.* 127:870–92
19. Bor, N. 1960. *The Grasses of Burma, Ceylon, India, and Pakistan*. New York: Pergamon
20. Bove, F. J. 1970. *The Story of Ergot*. Basel: Karger Verlag
21. Bradshaw, A. D. 1959. Population differentiation in *Agrostis tenuis* Sibth. II. The incidence and significance of infection by *Epichloe typhina*. *New Phytol.* 58:310–15
22. Campbell, C. S., Quinn, J. A., Cheplick, G. P., Bell, T. J. 1983. Cleistogamy in grasses. *Annu. Rev. Ecol. Syst.* 14:411–41
23. Carroll, G. C. 1988. Fungal endophytes in stems and leaves: from latent pathogen to mutualistic symbiont. *Ecology* 69:2–9
24. Chase, A. 1918. Axillary cleistogenes in some American grasses. *Am. J. Bot.* 5:254–58
25. Cheplick, G. P., Clay, K. 1988. Acquired chemical defenses of grasses: the role of fungal endophytes. *Oikos* 52:309–18
26. Cheplick, G. P., Clay, K., Wray, S. 1989. Interactions between fungal endophyte infection and nutrient limitation in the grasses *Lolium perenne* and *Festuca arundinacea*. *New Phytol.* 111:89–97
27. Deleted in proof
28. Clay, K. 1982. Environmental and genetic determinants of cleistogamy in a natural population of the grass *Danthonia spicata*. *Evolution* 36:734–41
29. Clay, K. 1984. The effect of the fungus *Atkinsonella hypoxylon* (Clavicipitaceae) on the reproductive system and demography of the grass *Danthonia spicata*. *New Phytol.* 98:165–75
30. Clay, K. 1986. Induced vivipary in the sedge *Cyperus virens* and the transmission of the fungus *Balansia cyperi* (Clavicipitaceae). *Can. J. Bot.* 64:2984–88
31. Clay, K. 1986. A new disease *(Balansia cyperi)* of purple nutsedge *(Cyperus rotundus)*. *Plant Dis.* 70:597–99
32. Clay, K. 1987. The effect of fungi on the interaction between host plants and their herbivores. *Can. J. Plant Pathol.* 9:380–88
33. Clay, K. 1987. Effects of fungal endophytes on the seed and seedling biology of *Lolium perenne* and *Festuca arundinacea*. *Oecologia* 73:358–62
34. Clay, K. 1988. Clavicipitaceous fungal endophytes of grasses: Coevolution and the change from parasitism to mutualism. In *Coevolution of Fungi with Plants and Animals,* ed. D. L. Hawksworth, K. Pirozynski, pp. 79–105. London: Academic
35. Clay, K. 1988. Fungal endophytes of grasses: a defensive mutualism between plants and fungi. *Ecology* 69:10–16
36. Clay, K. 1989. Clavicipitaceous endophytes of grasses: Their potential as biocontrol agents. *Mycol. Res.* 92:1–12
37. Clay, K. 1990. Comparative demography of three graminoids infected by systemic, clavicipitaceous fungi. *Ecology*. 71:558–70
38. Clay, K., Cheplick, G. P. 1989. Effect of ergot alkaloids from fungal endophyte-infected grasses on the fall armyworm *(Spodoptera frugiperda)*. *J. Chem. Ecol.* 15:169–82
39. Clay, K., Cheplick, G. P., Wray, S. M. 1989. Impact of the fungus *Balansia henningsiana* in the grass *Panicum agrostoides:* frequency of infection, plant growth and reproduction, and resistance to pests. *Oecologia* 80:374–80
40. Clay, K., Hardy, T. N., Hammond, A. M. Jr. 1985. Fungal endophytes of grasses and their effects on an insect herbivore. *Oecologia* 66:1–6
41. Clay, K., Jones, J. P. 1984. Transmission of the fungus *Atkinsonella hypoxylon* (Clavicipitaceae) by cleistogamous seed of *Danthonia spicata* (Gramineae). *Can. J. Bot.* 62:2893–98
42. Clay, K., Leuchtmann, A. 1989. Infection of woodland grasses by fungal endophytes. *Mycologia* 81:805–11
43. Coughenour, M. B. 1985. Graminoid responses to grazing by large herbivores: adaptations, exaptations, and interacting processes. *Ann. M. Bot. Gard.* 72:852–63
44. Cunningham, J. J. 1958. Non-toxicity to animals of ryegrass endophyte and other

endophytic fungi of New Zealand grasses. *NZ J. Agric. Res.* 1:487
45. Dahlman, D. L., Eichenseer, H., Siegel, M. R. 1990. Chemical perspectives on endophyte-grass interactions and their implications to insect herbivory. In *Microorganisms, Plants and Herbivores,* ed. C. Jones, V. Krischik, P. Barbosa.
46. Dehne, H. W. 1982. Interaction between vesicular-arbuscular mycorrhizal fungi and plant pathogens. *Phytopathology* 72:1115–19
47. Diehl, W. W. 1950. *Balansia and the Balansiae in America*. Washington, DC: US Dep. Agric.
48. Deleted in proof
49. Dyksterhuis, E. J. 1945. Axillary cleistogenes in *Stipa* and their role in nature. *Ecology* 26:195–99
50. Edgerton, C. W. 1917. A new *Balansia* on *Cyperus*. *Mycologia* 2:259–61
51. Elmi, A. A., West, C. P., Turner, K. E. 1989. *Acremonium* endophyte enhances osmotic adjustment in tall fescue. *Ark. Farm Res.* 38:7
52. Fomba, S. N. 1984. Rice disease situation in mangrove and associated swamps in Sierra Leone. *Trop. Pest Manage.* 30:73–81
53. Ford, V. L., Kirkpatrick, T. L. 1989. Effects of *Acremonium coenophialum* in tall fescue on host disease and insect resistance and allelopathy to *Pinus taeda* seedlings. *Proc. Ark. Fescue Toxicosis Conf.* Special Report 140:29–34
54. Funk, C. R., Halisky, P. M., Johnson, M.C., Siegel, M. R., Stewart, A. V., et al. 1983. An endophytic fungus and resistance to sod webworms: association in *Lolium perenne*. *Bio/Technology* 1:189–91
55. Gould, F. W., Shaw, R. B. 1983. *Grass Systematics*. College Station: Texas A&M Univ. Press
56. Halisky, P. M., Funk, C. R. 1984. Fungal endophyte content of perennial ryegrass entered in the national ryegrass turf trials. *Rutgers Turfgrass Proc.* 15:178–83
57. Halisky, P. M., Funk, C. R., Vincelli, P. C. 1983. A fungal endophyte in seeds of turf-type perennial ryegrasses. *Phytopathology* 73:1343
58. Hamilton, W. D. 1980. Sex versus non-sex parasite. *Oikos* 35:282–90
59. Hance, H. F. 1876. On a mongolian grass producing intoxication in cattle. *J. Botany* 14:210–12
60. Harberd, D. J. 1961. Note on choke disease of *Festuca rubra*. *Scottish Plant Breed. Stat. Rep.* 1961:47–51
61. Hardy, T. N., Clay, K., Hammond, A. M. J. 1985. Fall armyworm (*Lepidoptera:* Noctuidae): a laboratory bioassay and larval preference study for the fungal endophyte of perennial ryegrass. *J. Econ. Entomol.* 78:571–75
62. Hardy, T. N., Clay, K., Hammond, A. M. Jr. 1986. The effect of leaf age and related factors on endophyte-mediated resistance to fall armyworm (*Lepidoptera:* Noctuidae) in tall fescue. *Environ. Entomol.* 15:1083–89
63. Harmer, R., Lee, J. A. 1978. The germination and viability of *Festuca vivipara* (L.) Sm. plantlets. *New Phytol.* 81:745–51
64. Hijwegen, T. 1979. Fungi as plant taxonomists. *Symbol. Bot. Upsalienses* 22:146–65
65. Hinton, D. M., Bacon, C. W. 1985. The distribution and ultrastructure of the endophyte of toxic tall fescue. *Can. J. Bot.* 63:36–42
66. Hoveland, C. S., Schmidt, S. P., King, C. C., Odum, J. W., Clark, E. M., et al. 1983. Steer performance and association of *Acremonium coenophialum* fungal endophyte of tall fescue. *Agron. J.* 75:821–24
67. Janzen, D. H. 1985. The natural history of mutualisms. In *The Biology of Mutualism,* ed. D. H. Boucher, pp. 40–99. New York: Oxford Univ. Press
68. Johnson, M. C., Dahlman, D. L., Siegel, M. R., Bush, L. P., Latch, G. C. M., et al. 1985. Insect feeding deterrents in endophyte-infected tall fescue. *Appl. Environ. Microbiol.* 49:568–71
69. Kelley, S. E., Clay, K. 1987. Interspecific competitive interactions and the maintenance of genotypic variation within the populations of two perennial grasses. *Evolution* 41:92–103
70. Kilpatrick, R. A., Rich, A. E., Conklin, J. G. 1961. *Juncus effusus,* a new host for *Epichloe typhina*. *Plant Dis. Rep.* 45:899
71. Kirfman, G. W., Brandenburg, R. L., Garner, G. B. 1986. Relationship between insect abundance and endophyte infestation level in tall fescue in Missouri. *J. Kans. Entomol. Soc.* 59:552–54
72. Large, E. C. 1952. Surveys for choke *(Epichloe typhina)* in cocksfoot seed crops, 1951. *Plant Pathol.* 1:23–28
73. Large, E. C. 1954. Surveys for choke (*Epichloe typhina* in cocksfoot seed crops, 1951–53. *Plant Pathol.* 3:6–11
74. Latch, G. C. M., Christensen, M. J., Gaynor, D. L. 1985. Aphid detection of

endophytic infection in tall fescue. *NZ J. Agric. Res.* 28:129–32

75. Latch, G. C. M., Christensen, M. J., Samuels, G. J. 1984. Five endophytes of *Lolium* and *Festuca* in New Zealand. *Mycotaxon* 20:535–50
76. Latch, G. C. M., Hunt, W. F., Musgrave, D. R. 1985. Endophytic fungi affect growth of perennial ryegrass. *NZ J. Agric. Res.* 28:165–68
77. Latch, G. C. M., Potter, L. R., Tyler, B. F. 1987. Incidence of endophytes in seeds from collections of *Lolium* and *Festuca* species. *Ann. Appl. Biol.* 111:59–64
78. Leuchtmann, A., Clay, K. 1988. *Atkinsonella hypoxylon* and *Balansia cyperi,* epiphytic members of the Balansiae. *Mycologia* 80:192–99
79. Leuchtmann, A., Clay, K. 1989. Experimental evidence for genetic variability for compatibility between the fungus *Atkinsonella hypoxylon* and its three host grasses. *Evolution* 43:825–34
80. Leuchtmann, A., Clay, K. 1989. Isozyme variation in the fungus *Atkinsonella hypoxylon* within and among populations of its host grasses. *Can. J. Bot.* 67:2607
81. Leuchtmann, A., Clay, K. 1989. Morphological, cultural and mating studies on *Atkinsonella,* including *A. texensis. Mycologia* 81:692–701
82. Levin, D. A. 1975. Pest pressure and recombination systems in plants. *Am. Nat.* 109:437–51
83. Lewis, G. C., Clements, R. O. 1986. A survey of ryegrass endophyte *(Adremonium loliae)* in the U.K. and its apparent ineffectuality on a seedling pest. *J. Agric. Sci.* 107:633–38
84. Lively, C. M. 1987. Evidence from a New Zealand snail for the maintenance of sex by parasitism. *Nature* 328:519–21
85. Luttrell, E. S., Bacon, C. W. 1977. Classification of *Myriogenospora* in the Clavicipitaceae. *Can. J. Bot.* 55:2090–97
86. Lyons, P. C., Plattner, R. D., Bacon, C. W. 1986. Occurrence of peptide and clavine ergot alkaloids in tall fescue grass. *Science* 232:487–89
87. Mantle, P. G. 1969. The role of alkaloids in the poisoning of mammals by sclerotia of *Claviceps* spp. *J. Stored Prod. Res.* 5:237–44
88. Marquis, R. J. 1990. Genotypic variation in leaf damage in *Piper arieianum* (Piperaceae) by a multispecies assemblage of herbivores. *Evolution* 44:104–20
89. Marsh, C. D., Clawson, A. B. 1929. Sleepy grass *(Stipa vaseyi)* as a stock-poisoning plant. *USDA Tech. Rep.* 114:1–19
90. McFarlane, T. D. 1987. Poaceae subfamily Pooideae. In *Grass Systematics and Evolution,* ed. T. R. Soderstrom, K. W. Hilu, C. Campbell, M. E. Barkworth, pp. 265–76. Washington DC: Smithsonian Inst.
91. McNaughton, S. J. 1986. Grazing lawns: on domesticated and wild grazers. *Am. Nat.* 128:937–39
92. Morgan-Jones, G., Gams, W. 1982. Notes on hyphomycetes. XLI. An endophyte of *Festuca arundinacea* and the anamorph of *Epichloe typhina,* new taxa in one of the two new sections of *Acremonium. Mycotaxon* 15:311–18
93. Mortimer, P. H., di Menna, M. E. 1985. Interactions of *Lolium* endophyte on pasture production and perennial ryegrass staggers disease. In *Trichothecenes and other Mycotoxins,* ed. J. Lacey, pp. 149–58. New York: Wiley
94. Neill, J. C. 1940. The endophyte of ryegrass (*Lolium perenne* L.). *NZ J. Science Technol.* 21:280–91
95. Neill, J. C. 1941. The endophytes of *Lolium* and *Festuca*. *NZ J. Sci. Technol.* 23:185–93
96. Nobindro, U. 1934. Grass poisoning among cattle and goats in Assam. *Indian Vet. J.* 10:235–36
97. Pederson, J. F., Rodriquez-Kabana, R., Shelby, R. A. 1988. Ryegrass cultivars and endophyte in tall fescue affect nematodes in grass and succeeding soybean. *Agron. J.* 80:811–14
98. Plowman, T. C., Leuchtmann, A., Blaney, C., Clay, K. 1990. Significance of the fungus *Balansia cyperi* infecting medicinal species of *Cyperus* (Cyperaceae) from Amazonia. *Econ. Bot.* In press
99. Porter, J. K., Bacon, C. W., Cutler, H. G., Arrendale, R. F., Robbins, J. D. 1985. In vitro auxin production by *Balansia epichloe. Phytochemistry* 24:1429–31
100. Porter, J. K., Bacon, C. W., Robbins, J. D. 1979. Lysergic acid amide derivatives from *Balansia epichloe* and *Balansia claviceps* (Clavicipitaceae). *J. Nat. Prod.* 42:309–14
101. Porter, J. K., Bacon, C. W., Robbins, J. D., Betowski, D. 1981. Ergot alkaloid identification in Clavicipitaceae systemic fungi of pasture grasses. *J. Agric. Food Chem.* 29:653–57
102. Porter, J. K., Bacon, C. W., Robbins, J. D., Himmelsbach, D. S., Higman, H. C. 1977. Indole alkaloids from *Balansia*

*epichloe* (Weese). *J. Agric. Food Chem.* 25:88–93

103. Prestidge, R. A., Gallagher, R. T. 1988. Endophyte fungus confers resistance to ryegrass: Argentine stem weevil larval studies. *Ecol. Entomol.* 13:429–35
104. Prestidge, R. A., Lauren, D. R., van der Zijpp, S. G., di Menna, M. E. 1982. An association of Lolium endophyte with ryegrass resistance to Argentine stem weevil. *Proc. NZ Weed Pest Control Conf.* 35:199–92
105. Read, J. C., Camp, B. J. 1986. The effect of fungal endophyte *Acremonium coenophialum* in tall fescue on animal performance, toxicity, and stand maintenance. *Agron. J.* 78:848–50
106. Riesen, T., Sieber, T. *Endophytische Pilze von Winterweizen* (Triticum aestivum L.). PhD Thesis. ETH, Zurich
107. Rowan, D. D., Gaynor, D. L. 1986. Isolation of feeding deterrents against Argentine stem weevil from ryegrass infected with the endophyte *Acremonium loliae*. *J. Chem. Ecol.* 12:647–58
108. Rowan, D. D., Hunt, M. B., Gaynor, D. L. 1986. Peramine, a novel insect feeding deterrent from ryegrass infected with the endophyte *Acremonium loliae*. *J. Chem. Soc. D Chem. Commun.* 142:935–36
109. Rykard, D. M., Bacon, C. W., Luttrell, E. S. 1985. Host relations of *Myriogenospora atramentosa* and *Balansia epichloe* (Clavicipitaceae). *Phytopathology* 75:950–56
110. Rykard, D. M., Luttrell, E. S., Bacon, C. W. 1984. Conidiogenesis and conidiomata in the Clavicipoideae. *Mycologia* 76:1095–103
111. Saha, D. C., Johnson-Cicalese, J. M., Halisky, P. M., Van Heemstra, M. I., Funk, C. R. 1987. Occurrence and significance of endophytic fungi in the fine fescues. *Plant Dis.* 71:1021–24
112. Sampson, K. 1933. The systematic infection of grasses by *Epichloe typhina* (Pers.) Tul. *Trans. Br. Mycol. Soc.* 18:30–47
113. Sampson, K., Western, J. H. 1954. *Diseases of British grasses and herbage legumes*. London: Cambridge Univ. Press
114. Savile, D. B. O. 1987. Use of rust fungi (Uredinales) in determining ages and relationships in Poaceae. In *Grass Systematics and Evolution*, ed. T. R. Soderstrom, K. W. Hilu, C. Cammpbell, M. E. Barkworth, pp. 168–78. Washington, DC: Smithsonian Inst.
115. Schmidt, D. 1986. La quenouille rendelle le fourrage toxique. *Rev. Suisse Agric.* 18:329–32
116. Shaw, J. 1873. On the changes going on in the vegetation of South Africa. *Bot. J. Linnean Soc.* 14:202–08

116a. Shelby, R. A., Dalrymple, L. W. 1987. Incidence and distribution of the tall fescue endophyte in the United States. *Plant Dis.* 71:783–86

117. Siegel, M. C., Latch, G. C. M., Johnson, M. C. 1985. *Acremonium* fungal endophytes of tall fescue and perennial ryegrass: significance and control. *Plant Dis.* 69:179–83
118. Siegel, M. C., Latch, G. C. M., Johnson, M. C. 1987. Fungal endophytes of grasses. *Annu. Rev. Phytopathol.* 25: 293–315
119. Smith, K. T., Bacon, C. W., Luttrell, E. S. 1985. Reciprocal translocation of carbohydrates between host and fungus in bahiagrass infected with *Myriogenospora atramentosa*. *Phytopathology* 75:407–11
120. Stovall, M. E., Clay, K. 1988. The effect of the fungus *Balansia cyperi* on the growth and reproduction of purple nutsedge, *Cyperus rotundus*. *New Phytol.* 109:351–59
121. Thrower, L. B., Lewis, D. H. 1973. Uptake of sugars by *Epichloe typhina* (Pers. ex. Fr.) Tul. in culture and from its host *Agrostis stolonifera* L. *New Phytol.* 72:501–08
122. Welty, R. E., Azevedo, M. D., Cooper, T. M. 1987. Influence of moisture content, temperature, and length of storage on seed germination and survival of endophytic fungi in seeds of tall fescue and perennial ryegrass. *Phytopathology* 77:893–900
123. Wernham, C. C. 1942. *Epichloe typhina* on imported fescue seed. *Phytopathology* 32:1093
124. West, C. P., Izekor, E., Oosterhuis, D. M., Robbins, R. T. 1988. The effect of *Acremonium coenophialum* on the growth and nematode infestation of tall fescue. *Plant Soil* 112:3–6
125. White, J. F. 1987. Widespread distribution of endophytes in the Poaceae. *Plant Dis.* 71:340–42
126. White, J. F. 1988. Endophyte-host associations in forage grasses. XI. A proposal concerning origin and evolution. *Mycologia* 80:442–46
127. White, J. F., Bultman, T. L. 1987. Endophyte-host associations in forage grasses. VIII. Heterothallism in *Epichloe typhina*. *Am. J. Bot.* 74:1716–21
128. White, J. F., Cole, G. T. 1985. Endophyte-host associations in forage grasses. I. Distribution of fungal endophytes in some species of *Lolium* and *Festuca*. *Mycologia* 77:323–27

129. White, J. F., Cole, G. T. 1985. Endophyte-host associations in forage grasses. III. In vitro inhibition of fungi by *Acremonium coenophialum*. *Mycologia* 77:487–89
130. White, J. F., Cole, G. T. 1986. Endophyte-host associations in forage grasses. V. Occurrence of fungal endophytes in certain species of *Bromus* and *Poa*. *Mycologia* 78:852–56
131. White, J. F., Morgan-Jones, G. 1987. Endophyte-host associations in forage grasses. IX. Concerning *Acremonium typhinum*, the anamorph of *Epichloe typhina*. *Mycotaxon* 29:489–500
132. Wolock-Madej, C., Clay, K. 1989. Avian seed preference and weight loss experiments: the role of fungus–infected fescue seeds. *Proc. Ind. Acad. Sci.* 105:40
133. Wycherley, P. R. 1953. The distribution of the viviparous grasses in Great Britain. *J. Ecol.* 41:275–88

*Annu. Rev. Ecol. Syst. 1990. 21:299–316*


# MORPHOMETRICS

*F. James Rohlf*

Department of Ecology and Evolution, State University of New York, Sony Brook, NY 11794-5245

KEY WORDS: shape, size, multivariate statistics, image analysis

## INTRODUCTION

Morphometrics—the quantitative description, analysis, and interpretation of shape and shape variation in biology—is a fundamental area of research. Techniques of description and comparison of shapes of structures are needed in any systematic study (whether phenetic or cladistic) that is based on the morphology of organisms. Measurements of morphological diversity are of interest in ecological and genetic studies. Ways of dealing with shape change are also important for developmental studies and for practical applications in the medical sciences.

It is impossible to cover adequately such a broad and active field. Thus, the scope of this review is limited to developments in methodology—rather than to the numerous applications of quantitative morphometric methods. Morphometrics is, I believe, in the midst of fundamental change. During such a period, disagreements are expected about issues such as the relative importance of different approaches and the interpretation of the results of different methods. While I have tried to be objective, the relative emphasis given to different approaches undoubtedly reflects my perceptions of where the field is heading.

Traditionally, of course, the variables used in morphometric analyses are distances between landmarks, and these are measured directly on the specimen. With the availability of image acquisition hardware and image analysis systems, it is useful to distinguish the problems of data acquisition (e.g. capturing information about an image in machine readable form) from those



0066-4162/90/1120-0299$02.00

of feature extraction (selecting variables and making measurements) and morphometric analysis. This review follows this structure. There are important advantages to the use of image acquisition and feature extraction techniques. For example, one does not have to decide in advance which variables should be measured. This means that one can evaluate the usefulness of alternative suites of variables without handling the original specimens again. Image enhancement techniques may also make it easier to see certain features. If measurements are made by hand, then variables other than linear distances between landmarks are difficult. As Strauss & Bookstein (108) point out, such measurements usually give very incomplete and redundant information about the shape of a structure. With more comprehensive data more powerful morphometric analyses are possible—ones that take the geometrical relationships among the variables into account. But there are alternative techniques that can be (and have been) used, and they can yield very different results. The selection among them should be based on ontogenetic, phylogenetic, or other models. There has been some progress in this area, but much remains to be done. It is, however, beyond the scope of the present review.

## DATA ACQUISITION

It is desirable to capture enough detailed information in a machine readable form that the shapes of the structures are "archived." It is then possible to consider alternative sets of measurements (perhaps automatically). To be able to do this, information about the shapes of the structures must be captured with sufficient detail that the alternative measurements one might wish to use are defined. One test of adequacy is whether or not it is possible to reconstruct the shapes of the structures of interest from the recorded information. For example, one cannot later decide to use area as a variable if the recorded information is not sufficient to allow one to reproduce the general outline. A simple method, when suitable landmarks are present, is to take distance measurements in the form of a truss (108)—a network of relatively short and nonoverlapping distance measurements that allows one to infer the geometrical arrangement of the measurements (i.e. to reconstruct the coordinates of the landmarks). This is especially useful for large specimens that cannot be placed on a digitizing tablet.

One goal in the selection of variables for archiving is efficiency—to reduce the volume of data as much as possible while retaining the ability to adequately represent the shape of a structure. Elegant examples are given by Barnsley et al (6) and Barnsley & Sloan (7). They show that the entire form of a Black Spleenwort fern frond, for example, can be expressed by very few parameters (many of which are zero), using an algorithm based on fractals. Another goal is to obtain variables, e.g. coordinates, that can easily be decoded and transformed into other kinds of variables for alternative morphometric analyses. One can also use a video framegrabber and just save entire images. With microcomputer-based image analysis systems and large hard disks, this is quite feasible (see MacLeod, 72, for a general discussion and Fink, 42, for practical issues and lists of vendors). But this is not always the best solution since image files are large and often require interpretation. One would thus have to deal with each image twice—once while digitizing and then again when locating landmarks or other features. Alternatively, the coordinates of selected landmarks and the outlines of structures of interest can be directly recorded. Outlines can also be stored compactly as chain codes and easily converted back to a list of coordinates when needed. Simple cooordinate digitizers are very effective for recording landmarks and simple outlines. Sophisticated software is not needed since the human operator makes all of the important decisions about what to measure. Video-based systems have the advantage that image enhancement algorithms can be applied to the raw image to make the features of interest easier to see. There is also the possibility of use of automatic and semi-automatic techniques to assist in the isolation of structures of interest. A disadvantage of standard video-based systems for mophometric work is their relatively low resolution (usually about $500 \times 500$ in comparison to about $2{,}000 \times 2{,}000$ for even inexpensive digitizers). Desktop scanners are also high resolution devices, but they are relatively slow and do not seem to have been used yet for image acquisition in morphometrics.

The general literature on image processing and analysis techniques is very large and cannot be reviewed adequately here. There are, however, a number of general texts (e.g. 5, 54, 84, 98) and overviews oriented toward systematics (42, 72, 92, 96).

## FEATURE EXTRACTION

This step is concerned with the selection of morphometric variables for analysis. The variables that are most convenient for archiving an image are not necessarily those that are the most appropriate for morphometric analysis. Different kinds of analyses also require different kinds of variables. The problem of selection of sets of variables is nontrivial, because different, but seemingly equally reasonable, sets of variables can yield very different results. The variables that one selects define a feature space—a multivariate space with the variables as axes and the specimens as points in this space. If one's purpose is to distinguish automatically between two populations, then what matters most is that the clouds of points representing the two populations overlap as little as possible. It matters very little whether or not the variables make sense. But in most other applications one wants distance between points in the feature space to be a reasonable measure, however defined, of morphological difference between specimens. The selection of different ways to represent a configuration of landmarks or an outline corresponds to different transformations within this space. While some alternative selections correspond to orthogonal linear transformations, which leave distances between points invariant, most alternatives correspond to oblique affine or to nonlinear transformations which can give radically different distances between the points. This is especially true in taxonomic applications. It is clear that the particular selection of variables used in a study needs to be justified.

One strategy is to select variables that describe the expected pattern of shape variation in a simple "natural" manner (for example, related in some simple way to the function or to the development of the structure). If one is interested in variation in the degree of elongation of a particular part of a structure, then it would be convenient to have a variable that directly measures such elongations. On the other hand, if one expects a structure to vary in the degree to which it is bent, then other variables would be more appropriate. In some studies one knows enough about how a structure develops (either ontogenetically or phylogenetically) that particular variables seem appropriate. A classic example is Raup's (86, 87) study of shell growth in molluscs. He developed an intuitively simple spiral model in which the shell aperture revolves around the coiling axis and expands at a rate $W$ and translates along the axis at a rate $T$. On the other hand, Ackerly (1) finds these shell parameters to have no obvious biological significance. He then develops an alternative—a growth function that specifies the magnitude of the growth vector at any given point on the margin and the rate of divergence of adjacent growth vectors. Points of growth maxima and minima are located along the margin and used as "programmatic" landmarks. Though mathematical constructs, he argues they are biologically relevant because they reflect cellular processes of mantle secretion. Thus such considerations need not lead to a unique set of variables since alternative models can usually be developed. The simplest strategy may be just to use a suite of descriptive variables that efficiently spans the universe of possible shape variation. Some examples are given below for landmark and outline data. Other types of data, such as surface texture and pattern, are more difficult to quantify and have not been used as much as morphometric studies.

### *Landmark Data*

Analyses of landmark data are usually based either on distances between selected pairs of landmarks or on the coordinates of the landmarks. Strauss & Bookstein (108) discuss the advantages of using distances in the pattern of a graph called a *truss*. They showed that discrimination between populations was improved by the use of measurements representing distances along a truss rather than in the pattern of conventional measurements used in previous studies. A number of papers have discussed the issue of whether or not the linear measurements should be used as is, log-transformed, or converted to ratios in order to correct for the effects of having specimens of different sizes (see below).

When coordinates are used as variables, the coordinates in the archive often have to be transformed because their initial origin and orientation may be with respect to the measuring device (e.g. the axes of digitizing tablet). Brower & Veinus (29) aligned their axes with respect to landmarks and then used the logarithms of the absolute values of the coordinates of the landmarks as variables. It is also possible to use constructed points such as the geometric centroid for alignment of the axes (but this is more common in studies of outline data). Other coordinate systems can also be used.

Bookstein (18–20) uses features that are simple linear transformations of the original coordinates. Assuming configurations do not differ by a reflection, he first selects a pair of landmarks ($A$ and $B$, perhaps the most distant) to serve as a baseline and then scales, translates, and rotates all specimens so that the coordinates of $A$ are (0, 0) and those of $B$ are (1, 0). Then the $x,y$-coordinates of the other landmarks are called "shape coordinates relative to the $A$-$B$ baseline" and can be used as descriptors rather than the raw coordinates. They are an improvement over linear measurements since small changes in shape can be expressed as linear combinations of shape coordinates and standard multivariate linear statistical methods can be used to detect and test hypotheses about these differences between populations. The choice of baseline is arbitrary, but the effect of different choices is mainly to translate, rotate, and rescale. When differences are small, the results of most statistical tests should be unaffected by choice of baseline.

Bookstein (23–25) suggests a very different approach—the use of "principal warps" as features with which to describe configurations of landmarks relative to a reference configuration. He showed how configurations of points can be expressed as deformations of a reference configuration (a single specimen or an average). These deformations can be decomposed into a linear part (an affine transformation) and a nonlinear part (thin-plate splines of weighted sums of the principal warps). The principal warps are geometrically independent nonlinear functions of the reference configuration of landmarks. They are eigenvectors of what Bookstein (23) calls a "bending energy matrix." The matrix is, however, based only on the reference configuration of landmarks and not on how the configuration must be "bent" to match a particular target configuration. Free of the effects of affine tranformations, the warps correspond to the geometrically independent ways in which the reference configuration could be bent. The eigenvalues corresponding to each warp are inversely related to the scales of deformation. Large eigenvalues correspond to principal warps with small-scale bending, and small eigenvalues correspond to principal warps large-scale bending. Affine transformations correspond to global features and do not involve any bending. The weights necessary to produce a thin-plate spline fitting a target configuration can be used as variables to describe the target configuration separate from any affine (uniform) size or shape change. This scheme captures information on landmarks in a very elegant manner. Since the variables are based on a continuous transformation of the landmarks, it is possible to reconstruct configurations corresponding to hypothetical points—such as sample means, points two standard deviations from the mean along PCA axis 1, etc. as routinely done with outline data (e.g. 34, 93). Further study of the properties of this feature space is needed. One obvious inconvenience is its dependence upon a reference configuration.

## *Outline Data*

Many studies are based on outlines when there are few clear landmarks in the structures of interest. Outlines are usually recorded as either a sequence of x,y-coordinates along an outline or, more compactly, as a sequence of chain codes. Chain codes are usually decoded back into coordinates, but White & Prentice (112) investigated the use of features based directly on chain codes to describe leaf outlines. Their descriptors performed poorly in discriminating between species in their study—possibly because the differences between the species did not correspond to fine details of the outline for which their chain code descriptors would be better suited.

A sequence of transformations is usually applied to the initial set of coordinates to align them or to express them in terms of special coordinate systems. The origin is placed at a landmark or, more commonly, at the centroid of the object (2). The axes are rotated to pass through a landmark or the greatest width (for example, using the principal axes of the image—41). Lohmann (70) rotated each object until it had a maximum covariance with a standard reference object. The sequence of points around the outline is also adjusted so that the first and last points correspond to homologous landmarks (the same point in the case of closed outlines). In most early studies outlines were transformed to polar coordinates relative either to a central morphological landmark or to the centroid of the object. Workers now usually use Zahn & Roskies' (116) method of expressing an outline as the slope of a tangent to the outline as a function of distance, $t$, around the outline ($t$ scaled to range from 0 to $2\pi$). The normalized form, $\phi^*(t)$, is used as a descriptor of the outline shape (e.g. 41, 93). In elliptic Fourier analysis (69), $\Delta x, \Delta y$ is expressed as a function of distance, $t$, around the outline ($t$ scaled from 0 to $2\alpha$). Kincaid & Schneider (66) described a related approach.

In eigenshape analysis (see below), the tangent slopes describing the outline are used directly as variables, but a more common approach is to fit a function to the outline and then use the parameters as variables. For example, in Fourier analysis a trigonometric function is fitted to an outline, and the Fourier coefficients are used as variables for subsequent multivariate analyses. Kaesler & Waters (63) is one of the earliest examples. Since that time, many studies have been performed using this approach. The harmonics represent an elegant, although arbitrary, decomposition of a shape into a series of orthogonal components—analogous to the decomposition of a nonlinear curve into linear, quadratic, cubic, etc components. It is doubtful that the individual harmonics themselves will have useful biological interpretations (27; but see also 38). However, since they capture the form, the coefficients may be highly correlated with variation in whatever parameters influence the shape of an organism. When there are no landmarks and the outlines cannot be aligned, information on phase angle is lost (115). When the outline is more complex, alternative methods have to be used. One choice is to use Fourier analyses of Zahn & Roskies' (116) $\phi^*$ function. An alternative is elliptic Fourier analysis (69) of the $x,y$-coordinates themselves. Other types of functions can be used in a similar way. For example, Evans et al (40) described the use of cubic splines and Engel (39), the use of Bezier curves. Rohlf (91) reviews these approaches.

Another approach is to model an outline as a probability density surface by defining the height of the surface to be a positive constant within the outline of the object and zero everywhere else. Bivariate moments can then be used as descriptors of this surface. Hu (55) and others have formulated functions called moment invariants that have the desirable property of being invariant with respect to rotation, translation, and reflection of the image. Rohlf (92) reviews the various formulations that have been proposed. But F. J. Rohlf & S. Ferson (unpublished) found their computation to be very sensitive to rounding error (denominators of some coefficients can go to zero). They also found that there were strong nonlinear relationships among some of the descriptors (different invariants were in part functions of the same moments) and that this greatly influenced the apparent relationships among the objects. White & Prentice (112) did not find this a problem in using moment invariants to discriminate between species.

Median axis transformations reduce the interiors of a structure to stick-figures or "skeletons." One definition of a skeleton is the locus of centers of circles that touch the object's edge at more than one place (12). Another is the result of thinning an object until the structure is only 1 pixel wide (84). For thin objects, skeletons often have a structure that seems biologically appropriate. For example, branch points tend to behave as landmarks (16). Applications to shape description are given by (12, 13, 16, 26, 107). An important problem is the sensitivity of the medial axis transformation to small changes in the outline. For instance, a small bump or indentation can cause drastic changes in the form of the skeleton. Different definitions of a skeleton give somewhat different results, but it is not clear which definition is most reasonable biologically.

There are many additional ways to describe outlines. How does one know which one to use, and does it make a difference? Rohlf (90) points out that the Fourier coefficients represent an orthogonal linear transformation of the data on which they are based. Thus the multivariate analyses of distances will be identical whether based on Fourier coefficients or the original variable—which may make the use of Fourier descriptors seem somewhat pointless. But usually just the lower-order harmonics are used. This has the effect both of reducing the number of descriptors one has to deal with and of smoothing the outline (which may reduce the effects of digitization error). An important advantage of fitting a function to an outline is that one can interpolate in the feature space and invert the function so that one can visualize the shapes of hypothetical objects. A problem with the fitting of functions is that the coefficients from different functions may imply different relationships among the objects, and hence subsequent multivariate analyses may give different results. For example, Rohlf (91) shows that while Fourier, cubic spline, and Bezier curve coefficients are linearly related, the transformations among them are not orthogonal. Thus, the pattern of proximity of points in the different features spaces is not the same, and multivariate analyses based on distances between points (e.g. cluster analyses and ordination analyses) need not give the same results. This is unfortunate since it means that an arbitrary choice among methods that are mathematically equivalent—in the sense that one can transform from one representation to another without error—can have a serious effect upon one's results. On the other hand, multivariate analyses relative to a within-group variance-covariance matrix (e.g. generalized distances) are invariant under linear transformations and thus will yield statistically identical results among alternative set of variables that differ only by affine transformations.

## MORPHOMETRIC ANALYSES

How should morphometric data be analyzed? There are several strategies. One is to use conventional multivariate statistical methods to analyze sets of morphometric variables. If the various assumptions can be met, this approach provides a means to perform statistical tests. Another approach is to use special methods that explicitly take into account the fact that one has landmarks in a two- or three-dimensional physical space. These are described in the subsections that follow. Present methods do not integrate information on both landmarks and outlines very well. When landmark methods are used, outlines serve only as visual references. Another problem is that morphometric methods are unable to deal with missing data or else deal with missing information in a rather arbitrary manner.

## *Multivariate Statistical Methods*

The measurement of many variables naturally leads to the use of multivariate analysis (although there has been some controversy—35, 113, 114). If one has adequate sample sizes, multivariate analyses allow one to make overall tests as well as proper a posteriori tests of sets of variables that look interesting. The reader is referred to reviews (52, 58, 80) and to useful texts such as (53, 59, 68, 85, 88) since there is not space here to review the general application of multivariate statistics in morphometrics.

Coefficients of functions fitted to outlines (e.g. Fourier, cubic spline, and Bezier coefficients) have often been used as variables in multivariate analyses. Lohmann (70) suggested the use of "empirical shape functions" rather than the use of coefficients of a priori defined functions. His eigenshape analysis is a singular-value decomposition (37, 45) of matrix, $\mathbf{A}$, of $\phi^*$-values with $p$ rows corresponding to $p$ equal-length steps around the outline and with $n$ columns corresponding to specimens. The matrix is expressed as the product $\mathbf{U}\ \Lambda\ \mathbf{V}^t$, where the columns of $\mathbf{U}$ are the shape functions, $\Lambda$ is a diagonal matrix of eigenvalues, and the columns of $\mathbf{V}$ give the relative contributions of the shape functions to the outline of each specimen. Linear combinations of columns of $\mathbf{U}$ (perhaps defined by the results of a multivariate analysis) can be interpreted as $\phi^*$-values for a hypothetical outline. Plots of these outlines are very useful as they allow one to visualize means of clusters, extremes of principal component axes, etc. Examples are given by Lohmann (70) and Lohmann & Malmgren (71), but see Full & Ehrlich (43) for critical comments. The idea of eigenshape analysis can be readily generalized to the singular-value decomposition of any data matrix rather than of a matrix of coefficients of a function fitted to the data. If the coefficients correspond to an orthogonal linear transformation of the coordinates, then the results will be statistically equivalent and the choice will depend upon computational convenience (90). If not, then a decision must be made as to which metric gives the "correct" results.

This approach can now be applied to landmark data by using principal warps. A configuration of landmarks can be described as a deformation from a reference configuration by the use of a thin-plate spline based on a weighted linear combination of the principal warps of the reference configuration. The weights can be used as variables in a multivariate analysis. The configuration corresponding to the average of these weights in a sample can then be visualized as a thin-plate spline. In a similar manner one can visualize the configurations corresponding to the results of other statistical computations made on the weights.

The results of canonical variates analysis, one of the most used methods in morphometrics, are often misinterpreted. Plots of canonical variate scores, $\mathbf{Y} = \mathbf{C}^t\ \mathbf{X}$, and canonical variate coefficients, $\mathbf{C}$, are usually superimposed so that one can see the variables on which groups differ. This is done in analogy to a biplot (44). But **Y** and **C** do not represent a decomposition of the data matrix, **X,** as required for a biplot. The matrices **C** and **X** are a decomposition of matrix **Y.** One solution is to plot the columns of $\mathbf{C}(\mathbf{C}^T\mathbf{C})^{-1}$, the least-squares inverse of **C,** rather than **C.**

Conventional multivariate analysis cannot take into account the geometrical relationships among the variables (15, 83). While a multivariate analysis does take the correlations among the variables into account, it has no knowledge of the spatial pattern of the variables on the organism—whether the variables are linear measurements, raw coordinates, or coefficients of functions. Methods that can take this additional geometrical information into account should be more powerful and allow one to see more subtle relationships in the data.

### *Geometrical Methods*

How should morphometric data be analyzed so as to take into account geometrical relationships among the variables? Several methods have been developed, and all emphasize graphical displays in terms of the original objects, synthesized hypothetical specimens, or deformations of objects—instead of simple tables of numerical results. If no true landmarks are present, then all one can do is analyze size and what might be called "pure shape." That is, one can determine only whether the outline is circular, oval, triangular, etc without any knowledge of its orientation. If a single landmark exists (e.g. a starting point on an outline), then outlines of different objects can be synchronized. This permits, for example, multivariate analyses based on the $\phi^*$ function. With two landmarks it is possible to align two outlines in a nonarbitrary manner or to study open curves. More landmarks are needed to understand in a more detailed way just how the components of an outline change.

One method is to superimpose one specimen's configuration of landmarks onto another's, rotating and scaling to fit, and then studying the differences in the positions of the landmarks. One can then see (by the lack of fit) how the relative positions of landmarks differ. The numerical technique of Hurley & Cattell (57) was first applied in morphometrics by Sneath (104). In this method one configuration of landmarks is used as a reference, and a second is scaled, translated, and rotated until the sum of the squared differences in the positions of homologous landmarks (called a Procrustes distance coefficient) is as small as possible. This is a least-squares fitting procedure. Gower (50) provided an explicit solution to this problem. This general approach has been extended in several ways. Gower (51) developed a generalized Procrustes algorithm to fit any number of configurations to a consensus configuration. Siegel & Benson (101) made the important observation that a least-squares fit usually results in a general lack of fit at most landmarks—even if the configurations are identical except for the position of a few of the landmarks. This makes the differences between two configurations seem more complex than is necessary. They proposed a nonparametric approach that they call *resistant-fit theta rho analysis,* based on robust regression techniques (99, 100). It works very well when the two organisms differ in the positions of only a few of the landmarks. Tobler (111), Goodall (46), and Bookstein & Sampson (28) present other least-squares methods with slightly different assumptions. These methods have been used in a number of studies (8–11, 32, 82, 102). Olshan et al (82) used least-squares and resistant-fit methods to compare coordinates of points distributed along a pair of outlines. Rohlf & Slice (97) extend the resistant-fit method so that any number of configurations can be superimposed. Goodall & Green (48), in their study of cell growth, suggested using affine transformations to allow for differences in uniform shape change (a stretching in orthogonal directions—28) rather than global size differences. Rohlf & Slice (97) refer to this as an oblique rather than an orthogonal rotational fit. They also extend the resistant-fit and the generalized least-squares methods to include affine transformations.

Another method for studying the relationships among more than two specimens is to use the Procrustes distance between pairs of configurations as a measure of their difference. A matrix of such distances can then be clustered or an ordination performed using principal coordinates (49) or nonmetric multidimensional scaling (67) analyses. This approach seems reasonable if the differences are like digitization error in that there are independent homogeneous small differences at all landmarks (47). However, this approach has several limitations. (*a*) The Procrustes distance coefficient defines a very complex metric (64, 65). (*b*) By analyzing only distance coefficients one considers only the magnitude of the differences and not the details of the ways in which organisms differ, which are often of most interest in a morphometric study. (*c*) The distance coefficient based on orthogonal rotations combines differences is shape due both to uniform shape changes and deformations. Bookstein (23, 25) shows that the metrics for uniform and nonuniform shape differences are incommensurate and that there is no good way to combine these into a single nonarbitrary metric.

Rather than modeling uniform shape change by simple affine transformations, and then examining the residuals to discover evidence for local and nonuniform changes, one can fit a more complex model that includes nonuniform shape change. There are two main approaches: (*a*) Fitting a continuous function that expresses one specimen as a deformation of another, and (*b*) breaking structure into a number of small regions and computing the size and shape-change parameters for each. A classic example of the former is the transformation grid of Thompson (110), and finite element analysis is an example of the latter. These approaches seem particularly appropriate when one has samples along a developmental or phylogenetic sequence.

In a finite element analysis the organism is divided into many small regions (usually triangles in two dimensions or tetrahedrons in three dimensions) based on lines or planes connecting homologous landmarks. For a given set of landmarks there will be many alternative ways in which regions can be defined. The regions should be small and compact, and the tissue they represent should be as homogeneous as possible. The differences between two organisms are analyzed by comparing each cell on a reference organism with the corresponding cell on another. For each cell one can compute descriptors such as the ratio of areas (or volumes), the principal axes of the deformation of one cell into another (the arms of a strain cross), the principal dilatations (lengths of the axes), and measures of anisotropy (ratio of the principal dilatations). Goodall & Green (48) give a good description of strain cross parameters and their estimation. These results are conveniently shown on a drawing of one of the organisms with strain crosses centered in each cell. An early example is given by Niklas (81). Discussion of this approach is given by (e.g. 17, 18, 20, 21, 33, 47, 77–79, 89, 103).

Sneath (104) explored the use of trend-surface analysis to give plots analogous to transformation grids. He computed contours of polynomial surfaces fitted to the $x$ and $y$ residuals from a Procrustes analysis. Bookstein's (14, 15) biorthogonal grid is a direct quantitative implementation of Thompson's idea. He modeled a continuous deformation of a homogeneous elastic membrane stretched across the configuration and constrained at the landmarks. But this elegant approach has been little used. The complexity of the numerical methods inhibited the development of portable software. The use of thin-plate splines as an alternative model has been proposed recently (23, 25). In this model, nonuniform shape change is a deformation like that a thin metal plate would have if the landmarks were forced into new positions corresponding to a second configuration. Since the spline is continuous, a transformation grid can be obtained by transforming a rectangular grid and plotting it. In analogy with the physical deformation of a thin metal sheet, the bending energy can be computed for a given deformation (but any energy needed to shear the sheet appropriately is not included). Bending energy is not directly useful as a measure of distance between configurations because small changes in landmarks close together can require more bending energy than do what look like much larger changes in landmarks far apart. There is also the complication that the energy needed to bend confirmation 1 into 2 is usually different from that to bend 2 into 1.

## SIZE

Size and size corrections continue to be topics of discussion. An understanding of the results of Mosimann (73), Mosimann & James (76), and Darroch & Mosimann (36) is essential for an understanding of the effects of size. Mosimann (73) considers a variable $G$ to be a size variable if the effect of multiplying linear distances in the specimen by a constant $\alpha$ is to change the value of $G$ to $\alpha G$. There are many ways such a size variable can be defined. The particular choice is important since the relationships between size and shape depend upon the size variable selected (76). No single size variable is necessarily the "correct" one to use for all organisms or for all types of analyses (22). It is often helpful to use more than one size variable in a single study to display different relationships of interest (e.g. 61, 62). But morphometric methods based on particular morphometric and statistical models should suggest which size variable or at least which class of size variables is most appropriate. Bookstein (22) gives five different definitions for size and a context within which each seems natural.

Darroch & Mosimann (36) investigate models with lognormally distributed variables. For such models, dividing by a generalized geometric mean (which is equivalent to a weighted mean of log-transformed data) yields shape variables. They show that the results of canonical variates and principal component analyses of these log-transformed shape variables do not depend on the particular generalized geometric mean used—thus the usual geometric mean can be used. Mosimann (75) and Campbell & Mosimann (31) investigated Dirichlet models in which the variables are proportions and the sum of the measurements for each specimen is the measure of size. Somers (105, 106) calls such a measure "isometric size." Mosimann & James (76) point out that since shape can be independent of at most one size variable, discussion of "isometry" must be relative to a specified size variable.

In many studies the interest in using shape variables is in the practical problem of removing the effects of size from a dataset so that one may compare samples from different populations that may have different age (and hence size) distributions. Mosimann (73, 74) shows that shape can be independent of at most one size variable. Thus, the method of constructing shape variables and the definition of size must go together in order for there to be a proper correction. Most studies, however, seem to treat these independently. The traditional use of ratios is often criticized (e.g. 3, 4). Another method is to use residuals from regressions of each variable onto a particular variable (such as total length or weight). Humphries et al (56) and Bookstein et al (26) suggest a method of "shearing" to adjust for the effects of size based on a model with size estimated as a factor. Rohlf & Bookstein (94) review this approach and suggest the use of Burnaby's (30) method if the purpose is size correction alone, instead of the estimation of a factor model. This method corresponds to the projection of the data onto a space orthogonal to a size vector. This directly ensures that the resulting variables are orthogonal to whatever variable is used to define size. The first principal component axis is often used as a size vector (60). Its use assumes that the major source of variation in the sample is size and not outliers, polymorphisms, or inadvertent mixing of different species. Somers (105, 106) suggests the use of the isometric size vector $(1, \ldots, 1)$ based on the assumption that all variables (regardless of type) contribute equally to overall size (22). Unfortunately, there were problems with the Somers algorithm (94, 105, 109).

# CONCLUSIONS

Many new tools have been developed during the past decade that greatly facilitate the capture of morphometric information and its transformation into appropriate features. New analytic techniques now also allow one to analyze shape variation and shape change, with the relationships between the features and the geometry of the organism taken into account. However, there are still many technical problems. For example, missing data are not handled very well, and comparing large numbers of diverse organisms is often awkward. Landmark and outline-based techniques are not integrated very well at present. There has also not been much work on understanding the implications of using alternative, but equally reasonable, sets of landmarks and descriptive features. While the use of alternative kinds of features and methods of analysis usually provides additional insights in morphometric studies, taxonomists would like to know which methods give the most reliable information about overall similarity and cladistic relationships. Relatively little progress has been made toward an answer to this question. Solutions to these questions cannot come from methodological studies alone. Theoretical work on the foundations of morphometrics, called for near the beginning of this last decade (17), must continue, because without theoretical work and models one has little basis for choosing among methods—no matter how ingenious they may be.

ACKNOWLEDGMENTS

The help given by Scott Ferson and Dennis Slice in providing critical comments on a draft of this paper is gratefully acknowledged. This paper is contribution number 752 from the Graduate Studies in Ecology and Evolution, State University of New York at Stony Brook. The research on which it is based was supported, in part, by a grant (BSR 8306004) from the Systematic Biology Program of the National Science Foundation.

## *Literature Cited*

1. Ackerly, S. C. 1990. Using growth functions to identify homologous landmarks on mollusc shells. See Ref. 95. In press
2. Anstey, R. L., Delmet, D. A. 1973. Fourier analyses of zooecial shapes in fossil tubular bryozoans. *Geol. Soc. Am. Bull.* 84:1753–64
3. Atchley, W. R. 1978. Ratios, regression intercepts, and the scaling of data. *Syst. Zool.* 27:78–83
4. Atchley, W. R., Gaklins, C. T., Anderson, D. 1976. Statistical properties of ratios. *Syst. Zool.* 25:137–48
5. Ballard, D. H., Brown, L. M. 1982. *Computer Vision*. Englewood, NJ: Prentice-Hall. 523 pp.
6. Barnsley, M. F., Ervin, V., Hardin, D., Lancaster, J. 1986. Solution of an inverse problem for fractals and other sets. *Proc. Natl. Acad. Sci. USA* 83:1975–77
7. Barnsley, M. F., Sloan, A. D. 1988. A

better way to compress images. *BYTE* 13(1):215–23

8. Benson, R. H. 1982. Deformation, Da Vinci's concept of form, and the analysis of events in evolutionary history. In *Palaeontology, Essential of Historical Geology,* ed. E. M. Gallitelli, pp. 241–77. STEM Mucchi, Modena, Italy
9. Benson, R. H. 1982. Comparative transformation of shape in a rapidly evolving series of structural morphotypes of the ostracode *Bradleya*. In *Fossil and Recent Ostracodes*. ed. R. H. Bate, E. Robinson, L. M. Sheppard, pp. 147–64. New York: Halstead. 350 pp.
10. Benson, R. H. 1983. Biomechanical stability and sudden change in the evolution of the deep-sea ostracode *Poseidonamicus*. *Paleobiology* 9:398–413
11. Benson, R. H., Chapman, R. E., Siegel, A. F. 1982. On the measurement of morphology and its change. *Paleobiology* 8:328–39
12. Blum, H. 1973. Biological shape and visual science (Part I). *J. Theor. Biol.* 38:205–87
13. Blum, J., Nagel, R. N. 1978. Shape description using weighted symmetric axis features. *Pattern Recog.* 10:167–80
14. Bookstein, F. L. 1977. Orthogenesis of the hominids: an exploration using biorthogonal grids. *Science* 197:901–04
15. Bookstein, F. L. 1978. *The Measurement of Biological Shape and Shape Change*. Lecture Notes in Biomathematics. Vol. 24 New York: Springer-Verlag. 191 pp.
16. Bookstein, F. L. 1979. The line skeleton. *Computer Graphics Image Process.* 11:123–37
17. Bookstein, F. L. 1982. Foundations of morphometrics. *Annu. Rev. Ecol. Syst.* 13:451–70
18. Bookstein, F. L. 1984. Tensor biometrics for changes in cranial shape. *Ann. Human. Biol.* 11:413–37
19. Bookstein, F. L. 1984. A statistical method for biological shape comparisons. *J. Theor. Biol.* 107:475–520
20. Bookstein, F. L. 1986. Size and shape spaces for landmark data in two dimensions. (With discussion and rejoinder.) *Statist. Sci.* 1:181–242
21. Bookstein, F. L. 1987. Describing a craniofacial anomaly: finite elements and the biometrics of landmark location. *Am. J. Phys. Anthropol.* 74:495–509
22. Bookstein, F. L. 1989. "Size and shape": a comment on semantics. *Syst. Zool.* 38:173–80
23. Bookstein, F. L. 1989. Principal warps: thin-plate splines and the decomposition of deformations. *IEEE Trans. Pattern Anal. Mach. Intell.* 11:567–85
24. Bookstein, F. L. 1989. Four metrics for image variation. In *Proc. XI Int. Conf. Information Processing in Medical Imaging*. ed. D. Ortendahl, J. Llacer. New York: Liss. In press
25. Bookstein, F. L. 1990. Uniform factors and relative warps: a feature space of shape variation for landmark data. *Ann. Inst. Statist. Math.* In press
26. Bookstein, F. L., Chernoff, B., Elder, R. L., Humphries, J. M. Jr., Smith, G. R., Strauss, R. E. 1985. *Morphometrics in evolutionary biology*. Spec. Publ. 15. Philadelphia: Acad. Nat. Sci. Philadelphia. 277 pp.
27. Bookstein, F. L., Strauss, R. E., Humphries, J. M., Chernoff, B., Elder, R. L., Smith, G. R. 1982. A comment upon the uses of Fourier methods in systematics. *Syst. Zool.* 31:85–92
28. Bookstein, F. L., Sampson, P. D. 1987. Statistical models for geometric components of shape change. In *Proc. Section on Statist. Graphics,* San Francisco, August 1987, pp. 18–30. Am. Statist. Assoc. Alexandria, Va. 73 pp.
29. Brower, J. C., Veinus, J. 1978. Multivariate allometry using point coordinates. *J. Paleo.* 52:1037–53
30. Burnaby, T. P. 1966. Growth-invariant discriminant functions and generalized distances. *Biometrics* 22:96–110
31. Campbell, G., Mosimann, J. E. 1987. Multivariate analysis of size and shape: modelling with the Dirichlet distribution. *Computer Science and Statistics: Proc. 19th Symp. on the Interface between Computer Science and Statistics,* Philadelphia, Penn. pp. 93–101
32. Chapman, R. E. 1990. Conventional Procrustes approaches. See Ref. 95. In press
33. Cheverud, J. M., Lewis, J. L., Lew, W. D. 1983. The measurement of form and variation in form: an application of three-dimensional quantitative morphology by finite-element methods. *Am. J. Phys. Anthropol.* 62:151–65
34. Christopher, R. A., Waters, J. A. 1974. Fourier series as a quantitative descriptor of miospore shape. *J. Paleo.* 48:697–709
35. Corruccini, R. S. 1987. Univariate versus multivariate morphometric variation: an alternate viewpoint. *Syst. Zool.* 36:396–97
36. Darroch, J. N., Mosimann, J. E. 1985. Canonical and principal components of shape. *Biometrika* 72:241–52
37. Eckart, C., Young, G. 1936. The approximation of one matrix by another

of lower rank. *Psychometrika* 1:211–318

38. Ehrlich, R. R., Pharr, Jr., R. B., Healy-Williams, N. 1983. Comments on the validity of Fourier descriptors in systematics: a reply to Bookstein et al. *Syst. Zool.* 32:202–06
39. Engel, H. 1986. A least squares method for estimation of Bezier-curves and surfaces and its applicability to multivariate analysis. *Math. Biosci.* 79:155–70
40. Evans, D. G., Schweitzer, P. N., Hanna, M. 1985. Parametric cubic splines and geologic shape descriptions. *Math. Geol.* 17:611–24
41. Ferson, S., Rohlf, F. J., Koehn, R. K. 1985. Measureing shape variation among two-dimensional outlines. *Syst. Zool.* 34:59–68
42. Fink, W. L. 1990. Data acquisition for morphometric analysis in systematic biology. See Ref. 95. In press
43. Full, W. E., Ehrlich, R. 1986. Fundamental problems associated with "eigenshape analysis" and similar "factor" analysis procedures. *J. Math. Geol.* 18:451–63
44. Gabriel, K. R. 1971. The biplot graphical display of matrices with applications to principal component analysis. *Biometrika* 58:453–67
45. Golub, G. H., Reinsch, C. 1970. Singular value decomposition and least squares solutions. *Numerical Math.* 14:403–20
46. Goodall, C. R. 1990. WLS estimators and test for shape differences in landmark data. *J. R. Statist. Soc.* In press
47. Goodall, C. R., Bose, A. 1987. Models and procrustes methods for the analysis of shape differences. Proc. Symp. on the Interface between Computer Science and Statistics, pp. 86–92. Philadelphia, Pa.
48. Goodall, C. R., Green, P. B. 1986. Quantitative analysis of surface growth. *Bot. Gaz.* 147:1–15
49. Gower, J. C. 1966. Some distance properties of latent root and vector methods used in multivariate analysis. *Biometrika* 53:325–38
50. Gower, J. C. 1971. Statistical methods of comparing different multivariate analyses of the same data. In *Mathematics in the Archaeological and Historical Sciences,* ed. F. R. Hodson, D. G. Kendall, P. Tautu, pp. 138–49. Edinburgh: Edinburgh Univ. Press. 565 pp.
51. Gower, J. C. 1975. Generalized procrustes analysis. *Psychometrika* 40:33–51
52. Gower, J. 1984. Multivariate analysis: ordination, multidimensional scaling and allied topics. In *Handbook of Applicable Mathematics,* Vol. VI: Statistics. ed. E. Lloyd, pp. 727–81. New York: Wiley. 498 pp.
53. Hand, D. J. 1981. *Discrimination and Classification*. Chichester: Wiley. 218 pp.
54. Horn, B. K. P. 1986. *Robot Vision*. Cambridge, Ma:MIT Press. 509 pp.
55. Hu, M. K. 1962. Visual pattern recognition by moment invariants. *IRE Trans. Inform. Theory* 8:179–87
56. Humphries, J. M., Bookstein, F., Chernoff, B., Smith, G. R., Elder, R. L., Poss, S. G. 1981. Multivariate discrimination by shape in relation to size. *Syst. Zool.* 30:291–308
57. Hurley, J. R., Cattell, R. B. 1962. The Procrustes program: producing direct rotation to test an hypothesized factor structure. *Behav. Sci.* 7:258–62
58. James, F. C., McCulloch, C. E. 1990. Multivariate statistical methods in ecology. *Annu. Rev. Ecol. Syst.* 21. In press
59. Johnson, R. A., Wichern, D. W. 1982. *Applied Multivariate Statistical Analysis*. Englewood Cliffs, NJ: Prentice-Hall. 607 pp.
60. Jolicoeur, P. 1963. The multivariate generalization of the allometry equation. *Biometrics* 19:497–99
61. Jungers, W. L. 1988. Relative joint size and hominoid locomotor adaptations: its implications for the evolution of hominid bipedalism. *J. Hum. Evol.* 17:247–65
62. Jungers, W. L., Cole, T. M. III, Owsley, D. W. 1988. Multivariate analysis of relative growth in the limb bones of Arikara Indians. *Growth, Dev. Aging* 52:103–7
63. Kaesler, R. L., Waters, J. A. 1972. Fourier analysis of the ostracode margin. *Geol. Soc. Am. Bull.* 83:1169–78
64. Kendall, D. G. 1981. The statistics of shape. In *Interpreting Multivariate Data,* ed. V. Barnett, pp. 75–80. New York: Wiley. 374 pp.
65. Kendall, D. G. 1984. Shape-manifolds, procrustean metrics and complex projective spaces. *Bull. London Math. Soc.* 16:81–121
66. Kincaid, D. T., Schneider, R. B. 1983. Quantification of leaf shape with a microcomputer and Fourier transform. *Can. J. Bot.* 61:2333–42
67. Kruskal, J. B. 1964. Multidimensional scaling by optimizing goodness of fit to a nonmetric hypothesis. *Psychometrika* 29:1–27
68. Krzanowski, W. J. 1988. *Principles of Multivariate Analysis: a User's Perspective*. Oxford: Oxford Press 563 pp.
69. Kuhl, F. P., Giardina, C. R. 1982. Elliptic Fourier features of a closed con-

tour. *Computer Graphics & Image Process.* 18:236–58

70. Lohmann, G. P. 1983. Eigenshape analysis of microfossils: a general morphometric: procedure for describing changes in shape. *Math. Geol.* 15:569–72
71. Lohmann, G. P., Malmgren, B. A. 1983. Equatorward migration of *Globorotalia truncatulinoides* ecophenotypes through the Late Pleistocene: gradual evolution or ocean change? *Paleobiology* 9:414–21
72. MacLeod, N. 1990. Digital images and automated image analysis systems. See Ref. 95
73. Mosimann, J. E. 1970. Size allometry: size and shape variables with characterizations of the lognormal and generalized gamma distributions. *J. Am. Statist. Assoc.* 65:930–48
74. Mosimann, J. E. 1975. Statistical problem of size and shape. I. Biological applications and basic theorems. In *Statistical Distributions in Scientific Work,* ed. G. P. Patil, S. Kotz, J. K. Ord, 2:187–217. Boston: D. Reidel. 399 pp.
75. Mosimann, J. E. 1975. Statistical problem of size and shape. II. Characterizations of the lognormal, gamma and Dirichlet distributions. In *Statistical Distributions in Scientific Work,* ed. G. P. Patil, S. Kotz, J. K. Ord, 2:219–39. Boston: D. Reidel. 399 pp.
76. Mosimann, J. E., James, F. C. 1979. New statistical methods for allometry with applications to Florida red-winged blackbirds. *Evolution* 33:444–59
77. Moss, M. L., Skalak, R., Patel, H., Moss-Salentijn, L., Vilmann, H. 1984. An allometric network model of craniofacial growth. *Am. J. Orthodontics* 85:316–32
78. Moss, M. L., Pucciarelli, H. M., Moss-Salentijn, L., Skalak, R., Bose, A., Goodall, C., Sen, K., Morgan, B., Winick, M. 1987. Effects of preweaning undernutrition on 21 day-old male rat skull form as described by the finite element method. *Gegenbaurs Morphol. Jb., Leipzig* 133:837–68
79. Moss, M. L., Skalak, R., Patel, H., Sen, K., Moss-Salentijn, L., Shinozuka, M., Vilmann, H. 1985. Finite element method modeling of craniofacial growth. *Am. J. Orthod.* 87:453–72
80. Neff, N. A., Marcus, L. F. 1980. *A Survey of Multivariate Methods for Systematics.* Privately published and Am. Mus. Nat. Hist., NY. 243 pp.
81. Niklas, K. J. 1977. Applications of finite element analyses to problems in plant morphology. *Ann. Bot.* 41:133–53
82. Olshan, A. F., Siegel, A. F., Swindler, D. R. 1982. Robust and least-squares orthogonal mapping: methods for the study of cephalofacial form and growth. *Am. J. Phys. Anthropol.* 59:131–37
83. Oxnard, C. E. 1984. *The Order of Man.* New Haven: Yale Univ. Press. 366 pp.
84. Pavlidis, T. 1982. *Algorithms for Graphics and Image Processing.* Rockville, Md: Computer Sci. Press: 416 pp.
85. Pielou, E. C. 1984. *The Interpretation of Ecological Data: a Primer on Classification and Ordination.* New York: Wiley. 263 pp.
86. Raup, D. M. 1961. The geometry of coiling in Gastropods. *Proc. Nat. Acad. Sci., USA Zoology* 47:602–09
87. Raup, D. M. 1966. Geometric analysis of shell coiling: general problems. *J. Paleontol.* 40:1178–90
88. Reyment, R. A., Blackith, R. E., Campbell, N. A. 1984. *Multivariate Morphometrics.* New York: Academic. 233 pp. 2nd ed.
89. Richtsmeier, J. T., Cheverud, J. M. 1986. Finite element scaling analysis of human craniofacial growth. *J. Craniofacial Gen. Dev. Biol.* 6:289–323
90. Rohlf, F. J. 1986. The relationships among eigenshape analysis, Fourier analysis, and the analysis of coordinates. *Math. Geol.* 18:845–54
91. Rohlf, F. J. 1990. The analysis of shape variation using ordinations of fitted functions. In *Ordinations in the Study of Morphology, Evolution and Systematics of Insects: Applications and Quantitative Genetic Rationales,* ed., J. T. Sorensen, R. G. Foottit. Amsterdam: Elsevier. In press
92. Rohlf, F. J. 1990. An overview of image processing and analysis techniques. See Ref. 95. In press
93. Rohlf, F. J., Archie, J. 1984. A comparision of Fourier methods for the description of wing shape in mosquitoes (Diptera: Culicidae). *Syst. Zool.* 33:302–17
94. Rohlf, F. J., Bookstein, F. L. 1988. A comment on shearing as a method for "size correction". *Syst. Zool.* 36:356–67
95. Rohlf, F. J., Bookstein, F. L. eds. 1990. *Proc. Michigan Morphometrics Workshop.* Ann Arbor: Dept. Zool., Univ. Michigan. In press
96. Rohlf, F. J., Ferson, S. 1983. Image analysis. In *Numerical Taxonomy,* ed. J. Felsenstein, pp. 583–99. New York: Springer-Verlag. 644 pp.
97. Rohlf, F. J., Slice, D. 1990. Extension of the Procrustes method for the optimal superimposition of landmarks. *Syst. Zool.* 39:40–59

98. Rosenfeld, A. Kak, A. C. 1982. *Digital Picture Processing,* Vol. 1. New York: Academic Press. 435 pp. 2nd ed.
99. Siegel, A. F. 1982. Robust regression using repeated medians. *Biometrika* 69:242–44
100. Siegel, A. F. 1982. Geometric data analysis: an interactive graphics program for shape comparisons. In *Modern Data Analysis,* ed. R. L. Launer, A. F. Siegel, pp. 103–22. New York: Academic. 201 pp.
101. Siegel, A. F., Benson, R. H. 1982. A robust comparison of biological shapes. *Biometrics* 38:341–50
102. Sinervo, B., McEdward, L. R. 1988. Developmental consequences of an evolutionary change in egg size: an experimental test. *Evoluiton* 42:885–99
103. Skalak, R., Dasgupta, G., Moss, M. L., Otten, E., Dullemeijer, P., Vilmann, H. 1982. A conceptual framework for the analytical description of growth. *J. Theor. Biol.* 94:555–77
104. Sneath, P. H. A. 1967. Trend-surface analysis of transformation grids. *J. Zool.* 151:65–122
105. Somers, K. M. 1986. Multivariate allometry and removal of size with principal components analysis. *Syst. Zool.* 35:359–68
106. Somers, K. M. 1989. Allometry, isometry and shape in principal components analysis. *Syst. Zool.* 38:169–73
107. Straney, D. O. 1990. Median axis methods in morphometrics. See Ref. 95. In press
108. Strauss, R. E., Bookstein, F. L. 1982. The truss: Body form reconstructions in morphometrics. *Syst. Zool.* 31:113–35
109. Sundberg, P. 1989. Shape and size-constrained principal components analysis. *Syst. Zool.* 38:166–68
110. Thompson, D. W. 1917. *On Growth and Form.* London: Cambridge. 793 pp.
111. Tobler, W. R. 1978. Comparison of plane forms. *Geograph. Anal.* 10:154–62
112. White, R. J., Prentice, H. C. 1988. Comparison of shape description methods for biological outlines. In *Classification and Related Methods of Data Analysis,* ed. H. Bock, pp. 395–402. Amsterdam: Elsevier Sci. 750 pp.
113. Willig, M. R., Owen, R. D. 1987. Univariate analyses of morphometric variation do not emulate the results of multivariate analyses. *Syst. Zool.* 36:398–400
114. Willig, M. R., Owen, R. D., Colbert, R. L. 1986. Assessment of morphometric variation in natural populations: the inadequacy of the univariate approach. *Syst. Zool.* 35:195–203
115. Younker, J. L., Ehrlich, R. 1977. Fourier biometrics: harmonic amplitudes as multivariate descriptors. *Syst. Zool.* 26:336–42
116. Zahn, C. T., Roskies, R. Z. 1972. Fourier descriptors for plane closed curves. *IEEE Trans. Comp.* C-21:260–81

*Annu. Rev. Ecol. Syst. 1990. 21:317–40*


# FUNCTIONAL MORPHOLOGY AND SYSTEMATICS: Studying Functional Patterns in an Historical Context

*George V. Lauder*

Department of Ecology and Evolutionary Biology, University of California, Irvine, California 92717

KEY WORDS: structure, function, evolution, systematics, homology

---

## INTRODUCTION

The study of function is a neglected area of systematic and historical biology. Over the last 20 years, systematic biology has expanded to include in its purview the study of many different kinds of patterns, and structural features of all kinds have been the subject of phylogenetic analyses. In addition to macroscopic structural features that have been the traditional source of characters reflecting patterns of ancestry and descent, systematists have increasingly relied on DNA base and amino acid sequences, electrophoretic banding patterns, and ontogenetic sequences of character transformation to sort out genealogical patterns (e.g. 21, 51, 63, 72, 111). Morphological features of organisms have been used by systematists as the basis for biogeographic (120, 157, 158), morphometric (17, 134), ontogenetic (1, 54, 55, 72), and quantitative genetic analyses (130) as well as for studying speciation patterns, and ecological and coevolutionary interactions (18–20, 27, 104). But data on organismal function have been both the least used and the least understood class of information about organisms in systematic biology.

There are three main reasons why the form-function relationship, long a central dichotomy in biology (124), has been so heavily weighted toward the



0066-4162/90/0410-0317$02.00

study of form in systematic and evolutionary biology. First, comparative phylogenetic analysis arose historically from a morphological tradition in the nineteenth century, while functional analysis became part of the research tradition in physiology. Since the divergence of the morphological and physiological research programs near the turn of the century (5, 6, 25), little interaction has occurred between them, and functional analysis has played a relatively small role in phylogenetic research. Second, functional analysis is often extremely time consuming, and obtaining even limited functional data for many species within a clade may take years. Thus, only rarely are functional data available for broad comparative investigations. Thirdly, the inference of function from morphology and the heuristic use of functional concepts have often been substituted for the direct experimental measurement of functional attributes in living organisms. While many general discussions have recognized the importance of data on organismal function (13–15, 41, 43–45, 62, 74, 83, 97, 137, 149, 151), a large number of papers have presented "functional analyses" based primarily on the study of structure (e.g. 11, 12, 33, 34, 59, 98, 99, 129, 136, 137). Relatively few papers have actually quantified function in living animals *and* provided direct examples in which measured functional attributes of organisms are useful for understanding problems in historical biology. Thus, despite many statements in the literature supporting the importance of functional analysis, only a small number of papers use *both* experimentally determined function and phylogenetic analysis. Many discussions of the utility of functional analysis present purely morphological data and then infer rather than measure function.

The central theme of this article is that the direct experimental measurement of organismal function in living animals provides insights, not obtainable by other means, into *(a)* the uses of structural characters, *(b)* how and why characters are distributed the way they are on a cladogram, *(c)* interactions and correlations among characters, *(d)* the nature of organismal diversity, and *(e)* historical patterns to organismal design.

## WHAT ARE FUNCTIONAL DATA?

The study of function is the study of how structures are used, and functional data are those in which the use of structural features has been directly measured. Functions are the actions of phenotypic components (34, 47, 48, 83, 84). For example, the sequence of amino acids in an enzyme constitutes structural (morphological) data on that enzyme. The maximum rate of catalysis ($V_{max}$) or the affinity of an enzyme for its substrate ($K_m$) constitutes functional data on the enzyme. Similarly, structural data on a muscle might include the morphology of the sarcoplasmic reticulum, the amino acid sequence of the myosin molecules, changes in the electrophoretic banding pattern of muscle myosins during ontogeny, the amount of pinnation in the muscle fibers, the cross-sectional area of the muscle, and the length of the muscle. Functional (or physiological) data on muscle might include the time to peak tension, maximum tetanic tension, myosin ATPase reaction rates, work done or oxygen consumption in a single contraction, patterns of electrical activity during animal movement, or how any muscular rate process changes with temperature ($Q_{10}$). I do not include measurements of muscles such as lever arms or other morphometric features as functional data, even though such information may be interpreted in a functional context and may be useful in inferring functional characteristics.

Functional characteristics may also be measured at the whole animal level. As measurements are made that integrate increasing numbers of lower level functions, functional data become difficult to distinguish from behavioral data. Examples of functional attributes at the whole organism level include resting metabolic rate, maximal running speed, footfall patterns, and preferred body temperature. Functional data may overlap measures of animal performance. Performance measures the ability of an animal to execute a behavior or the effectiveness of an animal at accomplishing a particular task (7, 39, 83, 141). Typical performance measures include maximal swimming speed, the percent of successful feeding strikes, maximal jumping distance, or the largest size of prey that can be crushed. In addition, functional data may form part of a behavioral analysis. For example, behavioral analyses of frog or bird vocalizations often include the study of specific functional attributes of the call or song.

Use of the term "function" is subject to two confusions in the literature. First, ethologists commonly use the term function as a synonym of "selective value" (10, 23, 80). As discussed elsewhere (80) the use of "function" in this way is associated with many difficulties, not the least of which is the necessity of accurately understanding selective forces on a structure before one can speak of its function.

Second, function is often confused with the role that a structure plays during the life history of an organism. Bock & von Wahlert (16) emphasized a useful distinction between the terms "function" and "biological role." *Functions* are the uses of structures, while the *biological role* of a structure reflects the task of a structure during a behavior such as mating or escape from predation. The idea of a biological role is in many ways similar to the use of the term function by ethologists.

## FUNCTION AND HISTORICAL BIOLOGY

There are five key areas in which the interplay between functional and historical analysis is likely to be particularly fruitful. These are treated seriatim below, to focus attention on the possible mutual influences of functional and historical analysis. Mayr (105, 106) has emphasized the distinction between *functional biology,* concerned primarily with understanding how organisms work and proximate causation, and *evolutionary biology,* committed particularly to studying historical pathways and ultimate causation. A major theme to emerge from a consideration of five areas of interplay discussed below is that clear areas of overlap exist between these two approaches that may provide new insights into organismal diversity and the mechanisms that have produced it. Research into the evolution of function increasingly exemplifies aspects of both "types" of biology.

### *The Evolution of Function*

Functional attributes are an important class of organismal characters. How are structures used by organisms, and how has the use of structures changed in evolution? These questions may be addressed by an analysis of both form and function in an historical context. Analysis of structural features alone will provide an incomplete picture of the nature of organismal design. Our understanding of the evolution of gastrointestinal hormones in vertebrates (140), for example, is greatly enhanced if we can, in addition to describing phylogenetic patterns to amino acid sequences, describe how changes in the structure of the active site affect function and reaction rates in each clade. Are certain changes in amino acid sequence uncorrelated with changes in function? Are particular functions retained as primitive characters despite changes in structure? Do certain structural changes at the active site permit new functions while retaining the primitive function?

Figure 1 illustrates a simple hypothetical example of a functional analysis conducted on seven species (A to G) in conjunction with a phylogenetic analysis. The branching pattern of the cladogram (Figure 1B) has been determined from a previous analysis, and structural features of a biomechanical system of muscles and bones are diagrammed in Figure 1A. Each species is studied to determine both the topology of the muscles and bones and the associated electrical activity patterns (motor patterns) of the muscles. The procedure of mapping both structural and functional attributes of the terminal species onto a cladogram allows one to make several observations about the evolution of muscle function. First, the control of bone *d* (Figure 1) evolved by a two-stage process: first ligament 6 arose, followed by ligament 7. Pathway *i* arose first, and species in clades A to E have the capability of moving bone *d* by pathway *i*. Subsequently, pathway *ii* arose by the addition of ligament 7: only clade A possesses pathway *ii*. Note that the activity patterns of muscle 5 have diverged in two clades (D - E; A - B) while clade C retains the primitive functional pattern for muscle 5. The origin of muscle 4 is congruent with its functional pattern (muscle activity pattern D), while activity patterns associated with muscle 3 show convergent evolution between clade A and clade G - F. Muscle 3 thus exhibits an incongruence between the evolution of structural and functional characteristics.

Three main benefits are to be gained from such an analysis. First, it enables one to define precisely the historical sequence of both structural and functional change that permits an understanding of how a particular biomechanical system was constructed. In any attempt to understand a mechanical system and the causal factors involved in its construction, it is useful to know the order in which a system was assembled, to help determine the interrelationships among the parts (45). In machines constructed by humans, the order of assembly may often be reasonably inferred by the way in which the parts fit together. But for biological systems the interrelationship among structural components alone rarely provides this information. Second, historical analysis of function provides the basis for an analysis of coevolutionary

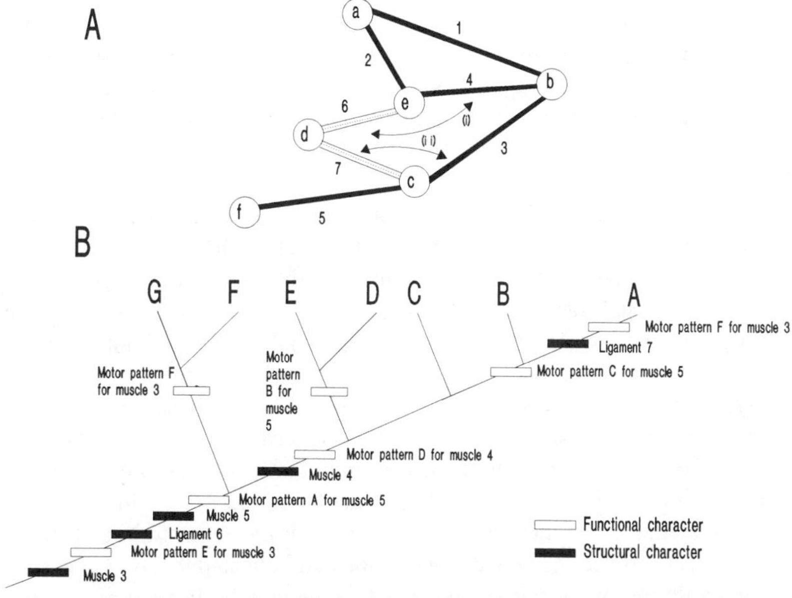


*Figure 1* Schematic diagram of a complex biomechanical system (A) and the historical sequence in which it was constructed (B). Circles indicate bones, labelled with lower-case letters, while ligaments are indicated by dashed lines and muscles by black lines. There are two mechanical pathways for moving bone d: i and ii. Pathway i involves muscle 4 and ligament 6. Pathway ii involves muscle 3 and ligament 7. Note that based on the phylogenetic analysis of (B) in which the characteristics of species A to G are mapped onto a cladogram derived from other characters, the muscle activity pattern (motor pattern) for muscle 3 is convergent between clade A and clade G. Note also that functional characters (open bars) and structural characters (solid bars) may exhibit incongruent distributions: e.g., muscle 5 arises with one motor pattern (A) and only the motor pattern is transformed (into motor pattern B and C) during subsequent cladogenesis.

patterns between structural and functional data sets: do historical changes tend to occur congruently in muscle morphology and function, for example? Third, this type of analysis allows an assessment of the conservatism (or lack thereof) in functional data as compared to structural information, and a test of the oft-stated conclusion that functional data tend to be evolutionarily plastic. These last two points are considered in more detail below in sections 2 and 4.

While considerable progress has been made in attempting analyses such as those presented in Figure 1 (22, 46, 67, 74, 80, 81, 92, 94–96, 107, 126, 148, 150), complete data sets of both structural and functional characters are not yet available for many groups (84). In part, this is due to the considerable difficulty in obtaining functional data from a variety of species. Determining mechanical properties of tissues, muscle activity patterns, rates of oxygen consumption, bone movements or deformations, or fluid pressures from three to five individuals in each of a dozen or so species is a formidable task. Many species important in phylogenetic analysis are either rare or found in inaccessible habitats, or they do not respond well to laboratory conditions, further complicating the gathering of functional data. Despite such difficulties, the development of complete case studies of structural and functional evolution would be a significant step forward in attempts to understand the evolution of organismal design.

One additional area where functional and phylogenetic analyses may interact in studies of the evolution of function lies in the framing of specific functional hypotheses. Phylogenetic analyses of morphology may suggest specific functional questions (81, 94, 96, 98, 99, 126) which may then be explored. One example was provided by an analysis of locomotion in ray-finned fishes (81). An analysis on a cladogram of the sequence of character change of muscles and bones in the tail of ray-finned fishes indicated that certain muscles arose prior to the origin of an externally symmetrical tail. This suggested that the action of these muscles might influence the way the tail was used and in fact might be able to change the function of the tail between two alternative states. A functional analysis (measuring bone strain and muscle electrical activity) confirmed this prediction and indicated that previous views on the evolution of the tail in ray-finned fishes may be in need of revision.

## *Function and Homology*

In addition to mapping functional and structural characters onto a phylogeny to conduct an historical analysis of a biomechanical system, functional characters may be used as elements of a phylogenetic analysis and may contribute to the total evidence (71) available for phylogenetic reconstruction. As such, functional characters are not different from any other attribute that might contribute to our understanding of genealogy: functions as well as structures may be synapomorphies. For example, two species of sunfishes (*Lepomis gibbosus,* the pumpkinseed, and *Lepomis microlophus,* the redear) share a synapomorphic pattern of muscle activity (76, 78) not found primitively in the sunfish clade. Although it has previously been suggested that functional characters are not suitable for systematic analysis (26, 137), recent work has shown that functional characters (such as bone movements or muscle electrical activity patterns) may be used as characters (77, 80, 128, 144) to successfully delineate monophyletic clades.

If a particular functional character (such as: "double burst pattern of electrical activity in the rectus cervicis muscle") is treated like any other character, then clearly functional characters may be homologous or convergent (just like morphological features) and may show complex patterns of phylogenetic distribution independent of their underlying structural basis. An example of muscle activity patterns treated like characters is provided by the sunfishes (Centrarchidae) discussed by Lauder (80). A further implication of treating functional characters just like structural features is that determining the homology or convergence of functional attributes is accomplished by using the same procedures applied to structural characters. Both convergent functions and structures are recognized by their incongruent distribution on a cladogram (42, 80, 109, 110), and the status of particular characters (whether functional or structural) as homologous or convergent is dependent on the topology of the cladogram. Character status may change as the cladogram changes. Figure 2 illustrates this point. In Figure 2A structure 1 and function 1 are homologous and are synapomorphies for clade A - B, based on the topology of the cladogram given. If, however, a subsequent phylogenetic analysis shows that the cladogram topology of Figure 2B is much better supported by available data (so that now clades A and B are unrelated to each other), then both structure one and function one must be interpreted as convergent between these two clades.

When functional attributes are treated like structural features, function is not useful as an a priori guide for determining homology of structures: Function does not take primacy in homology decisions. However, a number of authors have argued that functional analysis does provide an a priori guide to determining the homology of associated structural features (8, 13, 59, 60, 65, 132, 137). Under such a view, it is "impossible for nonhomologous structures to have homologous functions" (8: p. 60), and "Functions . . . should not be spoken of as homologous" (60: p. 323). Tyler (137: pp. 334, 344) has argued that "A complete determination of the probability of homology, however, is not possible on morphological grounds alone; we need a functional analysis to gauge the probability of convergence," and that "Homology applies most appropriately to the structural features, not their functions." In this view the role of functional analysis is primary, and,

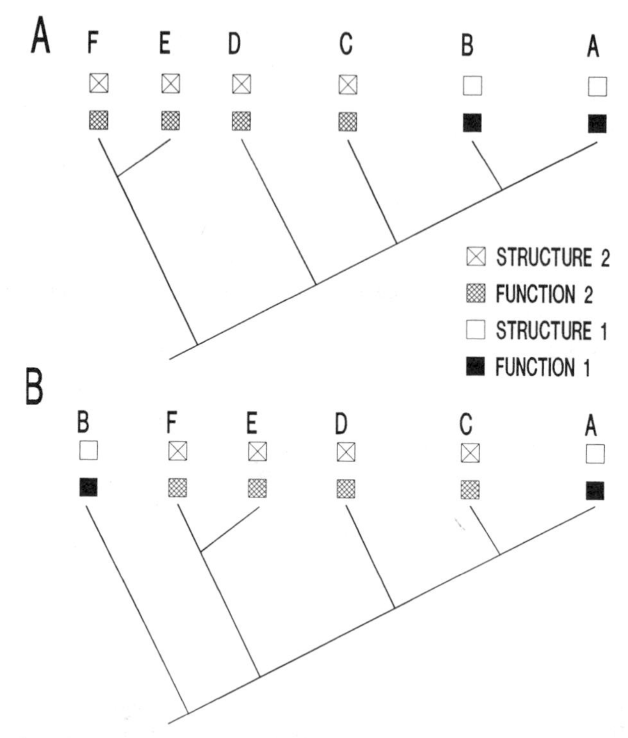


*Figure 2* Interpretation of the homology of both structures and functions depends on the topology of the cladogram of the species studied. Panel A: Both structure 1 and function 1 are homologous in clades A and B. Panel B: Given an new cladogram topology with the same pattern of character distribution, function 1 and structure 1 are considered to be convergent between clade A and clade B.

independent of the distribution of characters in other taxa, functional analysis is able to assess homology or convergence of an individual character.

The idea of a priori functional analysis as a guide to homology in systematics suffers from two primary difficulties. First, the "functional" analyses conducted to determine structural homology usually only involve morphological studies. Thus, Tyler's (137) method of "functional hierarchies" involves the heuristic use of functional ideas and hypothesized functions but not the direct measurement of function or the experimental manipulation of structures. Similarly, the methods of Bock (13, 14), Gutmann (59), and Dullemeijer (e.g., 34: p. 227) are based almost entirely on a morphological analysis. Secondly, structural characters determined in some way to be homologous by a functional analysis are not subject to refutation by the discovery of another better-corroborated cladogram. For example, if detailed a priori functional analysis is applied to two structural characters following the procedures of Tyler (137), Gutmann (59), or Dullemeijer (34), and these two structures are determined to be homologous (e.g. structure one in Figure 2A), what would be concluded if an analysis of DNA base sequences showed that cladogram B (Figure 2B) were in fact correct with a high degree of statistical probability? Would these authors conclude that structure one in clades A and B (Figure 2B) is still homologous?

## *The Analysis of Correlated Characters*

Functional analyses help to address the question of how and why characters are correlated with one another in phylogeny. Patterns of character distribution on cladograms have been the subject of increasing analysis in recent years (e.g. 31, 66, 117, 131, 150), and interest in explaining distributions of characters and determining the significance of correlated character distributions has increased also (18, 20, 24, 37, 38, 58, 66, 67, 130, 157).

One role that functional or biomechanical analysis may play in the analysis of correlated characters is to place boundary conditions on hypotheses of form-function-performance relationships (82). For example, a common goal of comparative functional analyses is to analyze morphology and function as well as the performance of the organism at particular behavioral tasks (7, 39, 101, 141). For example, the jumping performance (ability) of species in a clade of lizards, as measured by maximal jump distance, may be studied by analyzing the structure of the limb bones and muscles in an attempt to explain the evolution of interspecific differences in locomotor ability. One might hypothesize that a particular morphology (such as a long femur) or function (such as coactivation of the rectus femoris and gastrocnemius muscles) in a group of species is causally related to increased jumping performance. In Figure 3A, the clade of species A-B-C is characterized by both a long femur and increased jumping performance (solid circle and square). There is a congruent (correlated) historical origin of these characters on the cladogram. If, after an analysis of femur length and jumping distance for species A to F an incongruent origin was found (Figure 3B) such that the origin of increased bone length precedes the origin of increased jumping ability, then clearly a causal hypothesis of relationship among the two characters is not supported. Greene (58) discusses this pattern of character distribution and notes as a minimal requirement that features thought to be adaptations should have a historical origin concordant with the proposed behavioral advantage.

However, the criterion of historical concordance is not sufficient to establish a causal link between a particular morphological or functional attribute and increased performance. Typically, many derived features will characterize a clade. If, as in Figure 3C, several derived morphological attributes characterize the clade A-B-C, then how are we to choose which

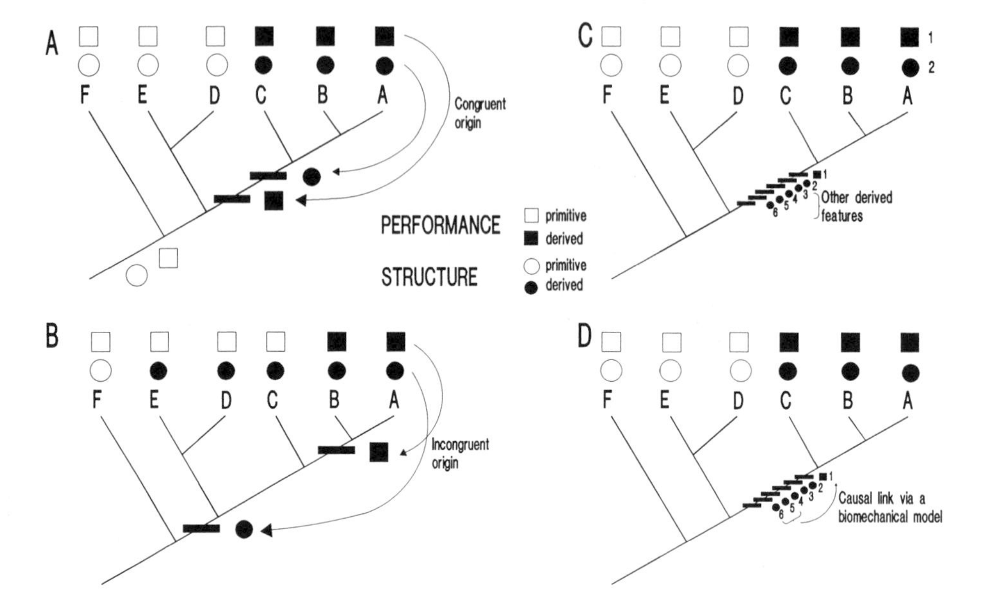


*Figure 3* Diagram to indicate how functional analysis may play a role in analyzing the pattern of characters on a cladogram. Although congruent historical origins of performance, functional, or structural characters may *suggest* a causal hypothesis of relationship, a biomechanical or functional analysis can assist in testing a causal hypothesis. Clades will often be characterized by many novelties at each level (panel C) but how are we to distinguish historically correlated characters from those sharing a causal relationship? A biomechanical or functional analysis (panel D) may assist by providing a causal model.

changes in body or limb proportions or which structural characters (features 2-6, solid circles) are *causally* related to the observed changes in performance (feature 1, solid square). Perhaps novelty 3 in Figure 3C is the morphological character "increased lever arm of the biceps muscle." Without some external model or criterion to rely on, there is no way to determine which of the five structural characters sharing a congruent (correlated) historical origin is causally related to the performance change (Figure 3C). The fact that changes in the biceps muscle are not expected to affect jumping ability in any way is only predictable from a causal biomechanical model (37, 82, 160).

A critical role of functional morphology in historical analyses of form and function is to provide an experimental, theoretical, and mechanistic basis for choosing which of several correlated characters are in fact causally related to historical changes in function or performance (Figure 3D). Functional analysis contributes to understanding at a mechanistic, proximate level why characters are historically correlated with one another (37, 82). Thus, only characters 5 and 6 (Figure 3D) might be biomechanically related to increased jumping performance. Characters 2, 3, and 4 may have nothing to do with performance character 1. As analyses of large data sets in systematics become more common, as patterns of character evolution are increasingly examined using computer programs such as MacClade and PAUP (31, 135), and as interest rises in explaining why characters have the distributions they do, it becomes increasingly important to be able to separate characters that have congruent (correlated) distribution patterns by chance from those that are causally related.

Three main types of functional analysis might be done to assist in the analysis of patterns of character distribution such as those shown in Figure 3C. First, descriptive functional studies such as quantifying patterns of movement, muscle activity, or pressure change in living animals (3, 4, 29, 30, 49, 52, 61, 68, 70, 81, 93, 114, 121, 122, 147, 159) provide a baseline of functional data that may provide a test of proposed causal links among characters. For example, if two muscles attaching to bone A had a congruent historical origin with that bone (as in characters 2, 3, and 4 in Figure 3C), measurement of muscle activity might show that only muscle 1 is electrically active during the stance phase of locomotion. An hypothesis that the correlated historical origin of muscle 2 and bone A is related to acceleration ability in species A, B, and C (Figure 3C) would be refuted as the muscle is not active during the appropriate phase of the locomotor cycle. Thus, even primarily descriptive functional analyses have the ability to provide a decisive test of historical explanations for character distribution.

A second type of functional analysis involves direct experimental modification of structures in animals to test proposed form—function—behavior relationships (e.g. 50, 69, 79, 86, 91; also see Eaton & DiDomenico—36, for a theoretical discussion of manipulation experiments). By cutting a ligament connecting two bones, for example, and monitoring behavioral or kinematic patterns before and after surgery, proposed functional hypotheses may be tested directly. If a novel muscle and increased jumping performance show a correlated historical origin in a clade (Figure 3C), then functional experiments involving cutting of the muscle tendon and the assessment of jumping performance before and after surgery provide a direct test of the causal link between the presence of the muscle and jumping ability.

A third method of assessing the significance of correlated characters is that of modelling (28, 113, 155, 160). By constructing theoretical models of morphology and using such models to generate a range of possible outputs, given known morphological inputs, understanding is gained of the significance of morphological variation in a particular mechanical system. Such models may involve mathematical descriptions of shape or direct measurements of lever arms, muscle masses, and lines of muscle action to estimate the effect of changing one anatomical link in a complex linkage system. Mechanical models of the feeding system in mammals (28, 57, 155), for example, have allowed understanding of the functional relationships among the bones and muscles and of the significance of changes in the jaw articulation at the base of the mammalian clade.

In summary, one major role of functional analysis in systematic biology is to contribute to our understanding of how characters interact within the organism. How do we *explain* patterns of character distribution? Why is it that on a particular cladogram two characters appear to have evolved together? Functional morphology by no means provides all the answers to this question as other explanations (not mutually exclusive) also exist: correlated characters may occur because of genetic linkages (40, 127, 130). But functional analysis is able to provide experimental tests, at the phenotypic level, of hypothesized functional dependencies among characters.

## *Congruence Among Classes of Characters*

Functional analyses are important for examining the extent of congruence in change among characters at different levels of biological organization. To some extent, the analysis of congruence among classes of characters is a special case of the analysis of correlated characters considered above. However, the question of the extent of congruence among different types of characters (e.g. behavioral, morphological, ecological) is an issue that is receiving increasing attention from comparative biologists (20, 24, 58, 66, 82, 101). Since the study of function provides the hierarchical link between morphological characters and performance, behavioral, and ecological characters, functional characters form an important class of information about organisms that merits detailed investigation. Table 1 summarizes one possible hierarchy of levels of design. This hierarchy illustrates seven classes of characters that could be investigated if one wanted to obtain an understanding of organismal design from the level of the nervous system to ecological interactions. A critical point is that functional information provides a link between levels 7 and 5 and between levels 5 and 3. An understanding of the morphology alone (levels 7 and 5) is not sufficient to understand the interactions among the levels of organismal design.

With a hierarchy of levels such as that depicted in Table 1 as a starting point, one can address questions about phylogenetic changes at these different levels, the extent of congruent change among levels, and questions of conservatism at any individual level. For example, within a clade are certain levels of design more evolutionarily conservative than others? Do changes in characters at two adjacent levels tend to occur in concert while characters from disparate levels show more divergent specializations? Does any one level exhibit a greater degree of homoplasy than other levels?

Most authors have viewed functional and physiological data as extremely

**Table 1** Levels of analysis that might be studied in attempts to analyze historical patterns to character transformation.

| Level of analysis | Example of patterns that might be studied |
|---|---|
| 1. Ecological | Intraspecific and interspecific resource use |
| 2. Behavioral | Sequence of behaviors used during mating |
| 3. Performance/effectiveness | Distance moved per unit time; number of prey captured per unit time |
| 4. Functional/physiological: at the level of peripheral tissues | Physiological properties of muscles and the timing of activation; kinematics of movement; enzyme kinetics; biomechanical properties of tissues |
| 5. Structural: at the level of peripheral tissues | Topology of the musculoskeletal system; tissue histology |
| 6. Functional/physiological: at the level of the nervous system | Neuronal spiking patterns; neurotransmitter modulation |
| 7. Structural: at the level of the nervous system | Neuronal morphology; patterns of neuronal circuitry |

plastic and subject to considerable homoplasy when comparisons are made across taxa. Level 4 in Table 1, function and physiology at the organismal level, is the class of characters that has been considered the most labile in evolution. Tyler (137: p. 344) asserts that functions vary extensively across clades and that therefore "functions do not constitute reliable systematic characters in their own right." Other authors (9, 64, 152) have advocated the view that physiological data are plastic and exhibit little phylogenetic coherence.

While there is no doubt that certain functional characters may vary considerably across taxa, there are no quantitative data indicating that function is any more or less variable than structural characters on average. When individual functions (such as metabolic pathways) are compared across a wide range of taxa and found to vary considerably, the fact is often ignored that many structural features of these same taxa are also extremely variable. No statement about the historical lability of function has yet been based on a phylogenetic analysis of both structural and functional data.

In fact, growing evidence exists to support the opposite view, that functional characters may often be extremely conservative: Functions may be plesiomorphic characters within a clade that are retained while associated structural features undergo considerable specialization (53, 76, 77, 80, 83, 90, 115, 125, 128, 131, 140, 144). Also, when changes do occur in functional characters, the clades diagnosed by the individual functional character states often corroborate the monophyly of clades diagnosed by structural features.

For example, Mommsen & Walsh (107) show that urea synthesis in vertebrates has a coherent phylogenetic pattern in which changes in urea synthetic pathways corroborate phylogenetic groupings based on structural features. Goslow et al (53) note that the locomotor system of vertebrates may have retained several basic functional patterns despite substantial changes in the peripheral morphology of the limbs across vertebrates. Dumont & Robertson (35) have argued that many features of neuronal circuit function are highly conserved in evolution despite significant morphological changes in peripheral structures.

Over the last ten years, research on the evolution of function in the feeding and respiratory systems of ray-finned fishes and salamanders has also shown that functional patterns may be conservative (73, 76, 77, 80, 94, 95, 100, 123, 125, 142–145, 156). Figure 4 is a schematic summary of some of the findings from this research which has focused on the evolution of muscle and bone structure and function as a case study in the evolution of levels 2, 3, 4, and 5 in Table 1. Figure 4 illustrates the fact that functional attributes, when treated phylogenetically, may be more conservative than morphological features. For example, sunfishes (Centrarchidae) differ considerably in morphology of the feeding system and are ecologically disparate in the food resources utilized (144). However, as shown in Figure 4A, the muscle activity pattern (or motor pattern) used by the jaw muscles is similar for all species that have been studied. Similarly, the pattern of muscle activity in the pharyngeal jaw muscles is conserved across sunfishes except in two species that share a derived motor pattern (Figure 4B; 76, 78). Many kinematic features associated with prey capture in ray finned fishes have been found to be conserved across phylogeny, as have basic patterns of pharyngeal jaw muscle activity (Figure 4C; 73, 90). Finally, Reilly & Lauder (115) have found that many features of the kinematics of prey transport in vertebrates are conserved across the aquatic-terrestrial transition (Figure 4D): terrestrial tongue-based prey transport uses a kinematic pattern very similar to that used by fishes for hydraulic prey transport.

Functions may also be ontogenetically conservative as shown by an analysis of muscle activity patterns during salamander ontogeny (Figure 5; 88, 89). When tiger salamanders undergo metamorphosis (during which considerable morphological change takes place in the head muscles and bones; 32, 87, 116), the pattern of muscle activity used for feeding in the water remains the same. Thus, the larval motor pattern is retained across metamorphosis, and the terrestrial motor pattern is added onto the previous functional pattern (Figure 5). Weeks & Truman (153, 154) have shown that there is considerable conservatism in the motor patterns used during metamorphosis in the tobacco hornworn *(Manduca):* Even after the prolegs and their muscles are lost at pupation, the motoneurons continue to generate the larval motor pattern.

Thus, one major conclusion to emerge from research on the ontogeny and

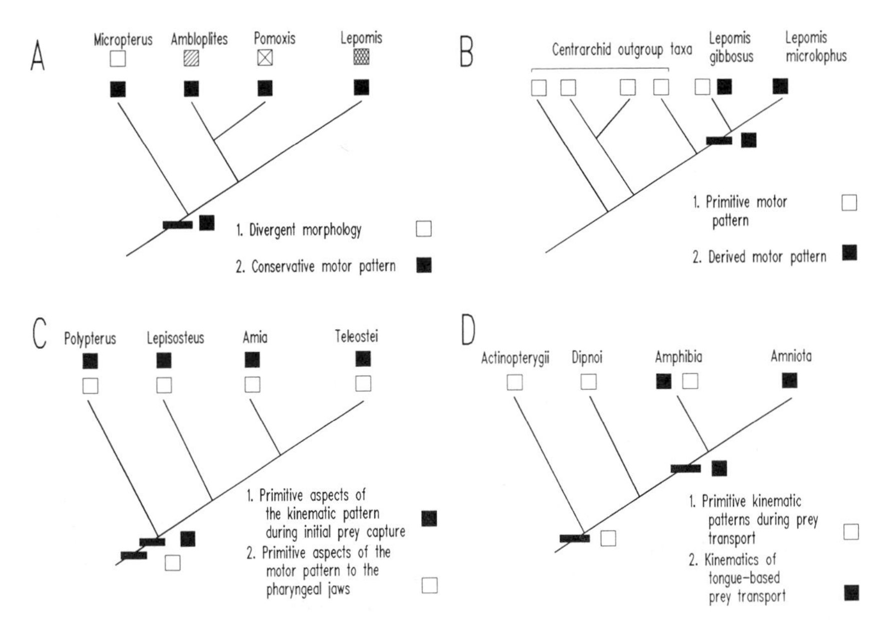


*Figure 4* Four examples of studies that provide an historical analysis of both structure and function. Panel A: Results from the research of Wainwright & Lauder (143, 144) showing that within the sunfish family (Centrarchidae) the pattern of muscle activity used during feeding is conserved while the feeding morphology has diverged considerably throughout the clade. Panel B: Results from Lauder (76, 78) showing that two species share a derived motor pattern in the pharyngeal jaw muscles. Panel C: Results from the work of Lauder (73, 77, 79, 90) to show that many functional attributes (of both initial prey capture in fishes, solid squares, and pharyngeal jaw muscle activity patterns, open squares) are conserved throughout the evolution of ray-finned fishes. Panel D: Results from Reilly & Lauder (115) to show that many aspects of the function of the jaws and tongue during prey transport are retained across the aquatic—terrestrial transition in vertebrate evolution (e.g., hyoid retraction occurs during the fast opening phase). Experimental data for amphibians show that they retain many primitive kinematic features and that they possess some derived aspects of prey transport: thus this clade is indicated by both open and filled squares.

phylogeny of muscle and bone function is that functional characters may in fact be more conservative than many structural characteristics. This will not always be the case, but it is certainly incorrect given present data to assert that functional characters are evolutionarily labile.

## *General Principles of Organismal Design*

Functional analyses contribute to defining general patterns and principles in the evolution of design. Central to any attempt to define general principles of organismal design is an understanding of how structural components of the organism are built and function (74, 82, 146). Biomechanical analysis serves

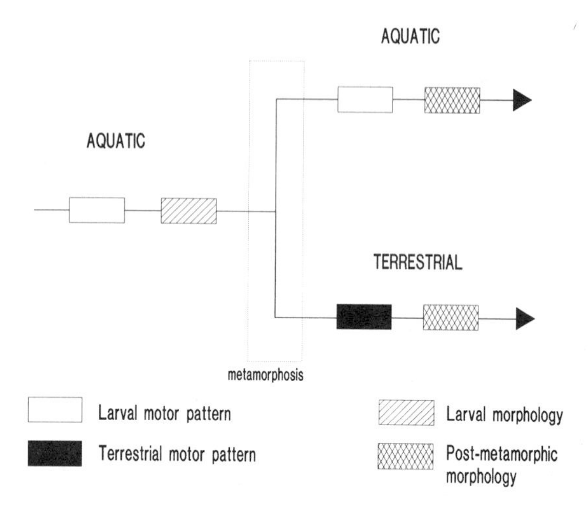


*Figure 5* Schematic diagram of ontogeny in metamorphosing tiger salamanders *(Ambystoma tigrinum)* shows that functions may be ontogenetically conserved (88, 89). At metamorphosis a muscle activity pattern used for terrestrial feeding is acquired (black box), but the larval pattern is retained and is used during aquatic feeding (open box). Retention of the larval muscle activity pattern occurs despite the many changes in morphology of the bones and muscles of the head at metamorphosis (hatched boxes).

to define the physical constraints within which organisms must work (56, 75, 147), and a phylogenetic analysis of both form and function allows the historical pathways taken during transformations of organismal design to be followed, and hypotheses about the evolution of design to be tested (74, 75, 97).

Two central contentions of my previous discussions of historical patterns to organismal design (74, 75, 82, 83, 85) are, *(a)* that there are general patterns and regularities to how structure and function change and interact in evolution and that hypotheses about these patterns may be generated and tested, and *(b)* that much of the regularity to historical patterns of form and function stems from intrinsic design features of organisms. Some authors (e.g. 133: p. 128) have argued that historical events are unique and that in the unique history of life there is no place for general patterns or "recurrent repeatable relationships." This view ignores three key facts: *(a)* that many aspects of historical change in organismal design are not unique and do occur over and over again in the history of life (e.g. segmentation of structures, helical fiber organization), *(b)* that organisms exhibit constrained ontogenies that limit the range of possible organismal design solutions to environmental problems (2, 56), and *(c)* that many characters, both structural and functional, are retained throughout speciation events (as plesiomorphies) and thus form part of the historical "burden" (118) of a clade.

One example of a repeatable historical pattern is the repetition of individual structural components of design (other examples are presented in 74, 75, 126, 138, 139). Repetition of structural elements has occurred in many groups of vertebrates, invertebrates, and plants. Vertebrae, limbs, body segments, genes, many aspects of plant structure, as well as numerous other component elements of organisms are all repeated modifications on a common structural theme. Typically, specialization of some of the repeated elements has followed phylogenetically from primitive similarity among the individual components.

What is the significance of repeated design elements for the evolution of form and function? Can we frame testable hypotheses about the evolution of structural diversity? As one example of the historical implications of repeated elements of design, consider the evolution of gene structure and function. If one asks the question: "What is the causal explanation for the diversity of proteins in the family of vertebrate hemoglobins?," it is clear that at least one proximate historical cause is gene duplication early in the evolution of hemoglobin. As has been discussed in detail elsewhere (85, 103, 108, 112, 119), gene duplication is historically permissive in that a primitive gene duplication event allows subsequent specialization of both structure and function in the second gene copy during cladogenesis. While we may have no idea of the ultimate causes of the duplication event itself (or of the selective factors, if any, that lead to gene duplication), without the additional source of structural and functional variability provided by additional copies of functionally important genes, the diversity of functions supported by the multiple variants of hemoglobin proteins would not exist: one copy of a functionally important gene is constrained from change by the critical nature of its product or regulatory function. While gene duplication is not the only mechanism responsible for protein family diversity, it is certainly a critical one for many classes of proteins. Minimally, gene duplication is sufficient for generating protein diversity; it is perhaps also necessary if other genetic mechanisms are not active for any particular family of proteins (108).

Raff et al (112) have discussed another major consequence of protein diversification through gene duplication: the ability to modulate the timing, location, and amount of protein synthesis to a greater extent than when a single-gene copy is present. Gene duplication, then, allows both structural and functional specialization in families of proteins.

Similar patterns of structural duplication are found at the organismal level in many plants and animals, and similar historical consequences may be observed. After the origin of repeated structural elements in a clade, subsequent cladogenesis is typically marked by independent specialization (both structural and functional) of at least some of the repeated elements. The evolution of the feeding system in ray-finned fishes has provided one example in which the evolution of a second biomechanical linkage system in the jaw has permitted subsequent historical divergence in the morphology and function of the feeding mechanism (73, 74, 85).

It is also possible to define a set of steps by which historical hypotheses about the evolution of morphological and functional diversity may be tested (85). Historical hypotheses are testable by quantitatively comparing ingroup and outgroup taxa with respect to a proposed novelty. As the example of gene duplication illustrates, there are common features to organismal design that appear to have general (and predictable) consequences for subsequent diversification of structure and function.

Emerson (38) has provided the best quantitative test to date of the historical effect of a morphological novelty. Emerson investigated frog pectoral girdles with the aim of assessing the extent of repeatable historical transformation in shape. Specifically, she tested the decoupling hypothesis (74)—that an increase in morphological constraint (and therefore a decrease in morphological diversity) should be associated with a reduction in the number of independent design elements. Eight phylogenetically independent cases of cartilage fusion in the frog pectoral girdle were analyzed. Emerson (38) showed that there were repeated historical changes in shape in each clade following cartilage fusion. These data provide strong support for the idea that historically regular patterns of morphological change do exist.

## SUMMARY

Functional data have been both the least used and the least understood class of data in systematics. Compared to the use of morphological features from DNA sequences to gross structural characters, patterns of distribution, and even ecological and behavioral attributes of organisms, functional characters have not been generally thought of as useful for resolving systematic and historical questions.

The current status of functional data in systematics is due to three primary factors, both historical and practical. First, current research in systematics developed primarily out of the nineteenth century morphological tradition while the analysis of organismal function was centered in physiological and experimental embryological research. At the turn of the century these two research traditions, initially complementary, diverged. While several authors have debated the severity of the divergence in research between morphologists (interested in the comparative analysis of *structure,* its development, and phylogeny) and experimental biologists (interested in *function,* physiology, or uses of structures) (5, 6, 102), there is little doubt that these two research areas diverged in the early part of this century and have remained largely separate.

Second, functional data are hard to obtain on a diversity of taxa: Gathering a range of experimental data on even one small clade is time consuming, and conducting manipulative experiments to understand causal relationships only adds to the difficulty of comparative functional analyses. Third, many so-called "functional" analyses and discussions of the import of function for systematics are in fact purely morphological. The adjective "functional" has acquired a cachet: It sounds quantitative, technical, and sophisticated. As a consequence, it is frequently used by papers in which there is no resemblance of a true functional analysis, where organismal function is directly measured and compared with measured function in other clades. The valid heuristic use of functional ideas should not be a substitute for direct measurement.

A key aim of this paper is to suggest that functional data are in fact critical to understanding five important issues in systematics. Despite difficulties in gathering functional data, the development of even a small number of well-understood case studies would greatly enhance our appreciation of *(a)* historical patterns to functional transformation, *(b)* the use of functional characters to define monophyletic clades, *(c)* the causal basis of character distributions on cladograms, *(d)* the extent to which changes in structural, functional, and behavioral characters are historically congruent, and the extent of evolutionary conservatism at any particular level of organismal design, and *(e)* general patterns and principles in the evolution of form and function.

ACKNOWLEDGMENTS

I thank Drs. Peter Wainwright, Steve Reilly, and Bruce Jayne for comments on the manuscript. Preparation of this article supported by grants NSF BSR 8520305 and DCB 8710210.

## *Literature Cited*

1. Alberch, P., Alberch, J. 1981. Heterochronic mechanisms of morphological diversification and evolutionary change in the neotropical salamander, *Bolitoglossa occidentalis* (Amphibia: Plethodontidae). *J. Morphol.* 167:249–64
2. Alberch, P., Gould, S. J., Oster, G. F., Wake, D. B. 1979. Size and shape in ontogeny and phylogeny. *Paleobiology* 5:296–317
3. Alexander, R. McN. 1975. *Biomechanics*. London: Chapman & Hall
4. Alexander, R. McN. 1983. *Animal Mechanics*. Oxford: Blackwell. 2nd ed.
5. Allen, G. 1978. *Life Science in the Twentieth Century*. Cambridge: Cambridge Univ. Press
6. Allen, G. 1981. Morphology and twentieth century biology: a response. *J. Hist. Biol.* 14:159–76
7. Arnold, S. J. 1983. Morphology, performance, and fitness. *Am. Zool.* 23:347–61
8. Atz, J. 1970. The application of the idea of homology to animal behavior. In *Development and Evolution of Behavior: Essays in Honor of T. C. Schneirla*, ed. L. Aronson, E. Tobach, D. S. Lehrman, J. S. Rosenblatt. San Francisco: W. H. Freeman
9. Barrington, E. J. W. 1975. Comparative physiology and the challenge of design. *J. Exp. Zool.* 194:271–86
10. Bertram, B. 1976. Kin selection in lions and in evolution. In *Growing Points in Ethology*, ed. P. P. G. Bateson, R. A. Hinde, pp 281–301. Cambridge: Cambridge Univ. Press
11. Bock, W. J. 1964. Kinetics of the avian skull. *J. Morphol.* 114:1–42
12. Bock, W. J. 1980. The definition and

recognition of biological adaptation. *Am. Zool.* 20:217–27
13. Bock, W. J. 1981. Functional-adaptive analysis in evolutionary classification. *Am. Zool.* 21:5–20
14. Bock W. J. 1988. The nature of explanations in morphology. *Am. Zool.* 28:205–15
15. Bock, W. J. 1989. Organisms as functional machines: a connectivity explanation. *Am. Zool.* 29:1119–32
16. Bock, W., von Wahlert, G. 1965. Adaptation and the form-function complex. *Evol.* 19:269–99
17. Bookstein, F., Chernoff, B., Elder, R., Humphries, J., Smith, G., Strauss, R. 1985. *Morphometrics in Evolutionary Biology*. Philadelphia: Acad. Nat. Sci.
18. Brooks, D. R. 1984. What's going on in evolution? A brief guide to some new ideas in evolutionary theory. *Can. J. Zool.* 61:2637–45
19. Brooks, D. R. 1985. Historical ecology: a new approach to studying the evolution of ecological associations. *Ann. Miss. Bot. Gard.* 72:660–80
20. Brooks, D. R., Wiley, E. O. 1988. *Evolution as Entropy*, Chicago: Univ. Chicago Press. 2nd ed.
21. Buth, D. 1984. The application of electrophoretic data in systematic studies. *Annu. Rev. Ecol. Syst.* 15:501–22
22. Carrier, D. R. 1987. The evolution of locomotor stamina in tetrapods: circumventing a mechanical constraint. *Paleobiology* 13:326–41
23. Clutton-Brock, T. H., Harvey, P. 1976. Evolutionary rules and primate societies. In *Growing Points in Ethology*, ed. P. P. G. Bateson, R. A. Hinde, pp 195–237. Cambridge: Cambridge Univ. Press
24. Coddington, J. 1988. Cladistic tests of adaptational hypotheses. *Cladistics* 4:3–22
25. Coleman, W. 1977. *Biology in the Nineteenth Century: Problems of Form, Function, and Transformation*. Cambridge: Cambridge Univ. Press
26. Cracraft, J. 1981. The use of functional and adaptive criteria in phylogenetic systematics. *Am. Zool.* 21:21–36
27. Cracraft, J. 1986. Origin and evolution of continental biotas: speciation and historical congruence within the Australian avifauna. *Evolution* 40:977–96
28. Crompton, A. W., Parker, P. 1978. Evolution of the mammalian masticatory apparatus. *Am. Sci.* 66:192–201
29. Crompton, A. W., Thexton, A. J., Parker, P., Hiiemae, K. 1977. The activity of the jaw and hyoid musculature in the Virginia opossum, *Didelphis virginiana*. In *The Biology of Marsupials*, ed. B. Stonehouse, G. Gilmore, pp. 287–305. New York: MacMillan
30. Cundall, D. 1983. Activity of head muscles during feeding by snakes: a comparative study. *Am. Zool.* 23:383–96
31. Donoghue, M. J. 1989. Phylogenies and the analysis of evolutionary sequences, with examples from seed plants. *Evolution* 43:1137–56
32. Duellman, W. E., Trueb, L. 1986. *Biology of Amphibians*. New York: McGraw Hill
33. Dullemeijer, P. 1972. Explanation in morphology. *Acta Biotheor.* 21:260–73
34. Dullemeijer, P. 1980. Functional morphology and evolutionary biology. *Acta Biotheor.* 29:151–250
35. Dumont, J., Robertson, R. M. 1986. Neuronal circuits: an evolutionary perspective. *Science* 233:849–53
36. Eaton, R. C., DiDomenico, R. 1985. Command and the neural causation of behavior: a theoretical analysis of the necessity and sufficiency paradigm. *Brain Behav. Evol.* 27:132–64
37. Emerson, S. 1982. Frog postcranial morphology: identification of a functional complex. *Copeia* 1982:603–13
38. Emerson, S. 1988. Testing for historical patterns of change: a case study with frog pectoral girdles. *Paleobiology* 14:174–86
39. Emerson, S., Diehl, D. 1980. Toe pad morphology and mechanisms of sticking in frogs. *Biol. J. Linn. Soc.* 13:199–216
40. Falconer, D. S. 1981. *Introduction to Quantitative Genetics*. London: Longman
41. Feder, M. E., Bennett, A. F., Burggren, W., Huey, R. B. 1987. *New Directions in Ecological Physiology*. Cambridge: Cambridge Univ. Press
42. Fink, W. L. 1988. Phylogenetic analysis and the detection of ontogenetic patterns, In *Heterochrony in Evolution*, ed. M. L. McKinney. pp. 71–91. New York: Plenum
43. Fisher, D. C. 1981. The role of functional analysis in phylogenetic inference: examples from the history of the Xiphosura. *Am. Zool.* 21:47–62
44. Fisher, D. C. 1985. Evolutionary morphology: beyond the analogous, anecdotal, and the ad hoc. *Paleobiology* 11:120–38
45. Frazzetta, T. H. 1975. *Complex Adaptations in Evolving Populations*. Sunderland, Mass: Sinauer
46. Gans, C. 1970. Strategy and sequence in the evolution of the external gas exchangers of ectothermal vertebrates. *Forma et Functio* 3:61–104

47. Gans, C. 1974. *Biomechanics, An Approach To Vertebrate Biology*. Philadelphia: J. B. Lippincott
48. Gans, C. 1988. Adaptation and the form-function relation. *Am. Zool.* 28:681–97
49. Gans, C., Gorniak, G. C. 1982. Functional morphology of lingual protrusion in marine toads *(Bufo marinus). Am J. Anat.* 163:195–222
50. Gans, C., Gorniak, G. C. 1982. How does the toad flip its tongue? Test of two hypotheses. *Science* 216:1335–37
51. Goodman, M. 1982. *Macromolecular Sequences in Systematic and Evolutionary Biology*. New York: Plenum
52. Gorniak, G. C., Rosenberg H. I., Gans, C. 1982. Mastication in the Tuatara, *Sphenodon punctatus* (Reptilia: Rhynchocephalia): structure and activity of the motor system. *J. Morphol.* 171:321–53
53. Goslow, G. E., Dial, K. P., Jenkins, F. A. 1989. The avian shoulder: an experimental approach. *Am. Zool.* 29:287–301
54. Gould, S. J. 1966. Allometry and size in ontogeny and phylogeny. *Biol. Rev.* 41:587–640
55. Gould S. J. 1977. *Ontogeny and Phylogeny*. Cambridge: Harvard Univ. Press
56. Gould, S. J. 1980. The evolutionary biology of constraint. *Daedalus* 109:39–52
57. Greaves, W. S. 1978. The jaw lever system in ungulates: a new model. *J. Zool., Lond.* 184:271–85
58. Greene, H. 1986. Diet and arboreality in the Emerald Monitor, *Varanus prasinus*, with comments on the study of adaptation. *Fieldiana* (Zool., N.S.) 31:1–12
59. Gutmann, W. F. 1981. Relationships between invertebrate phyla based on functional-mechanical analysis of the hydrostatic skeleton. *Am. Zool.* 21:63–81
60. Haas, O., Simpson, G. G. 1946. Analysis of some phylogenetic terms, with attempts at redefinition. *Proc. Am. Phil. Soc.* 90:319–49
61. Herring, S. W. 1985. The ontogeny of mammalian mastication. *Am. Zool.* 25:339–49
62. Hickman, C. S. 1988. Analysis of form and function in fossils. *Am. Zool.* 28: 775–93
63. Hillis, D. 1987. Molecular versus morphological approaches to systematics. *Annu. Rev. Ecol. Syst.* 18:23–42
64. Hoar, W. S. 1983. *General and Comparative Physiology*. New Jersey: Prentice Hall
65. Hodos, W. 1976. The concept of homology and the evolution of behavior. In *Evolution, Brain, and Behavior: Persistent Problems,* ed. R. B. Masterton, W. Hodos, H. Jerison, pp 153–67. Hillsdale, NJ: Erlbaum
66. Huey, R. 1987. Phylogeny, history, and the comparative method. In *New Directions in Ecological Physiology,* ed. M. E. Feder, A. F. Bennett, W. W. Burggren, R. B. Huey, pp. 76–101. Cambridge: Cambridge Univ. Press
67. Huey, R. B., Bennett, A. F. 1987. Phylogenetic studies of coadaptation: preferred temperatures versus optimal performance temperatures of lizards. *Evolution* 41:1098–1115
68. Jayne, B. C. 1988. Muscular mechanisms of snake locomotion: an electromyographic study of the sidewinding and concertina modes of *Crotalus cerastes, Nerodia fasciata,* and *Elaphe obsoleta*. *J. Exp. Biol.* 140:1–33
69. Jayne, B. C., Bennett, A. F. 1989. The effect of tail morphology on locomotor performance in snakes: a comparison of experimental and correlative methods. *J. Exp. Zool.* 252:126–33
70. Jenkins, F. A., Weijs, W. A. 1979. The functional anatomy of the shoulder of the Virginia opossum *Didelphis virginiana*. *J. Zool., Lond.* 188:379–410
71. Kluge, A. G. 1989. A concern for evidence and a phylogenetic hypothesis of relationships among *Epicrates* (Boidae, Serpentes). *Syst. Zool.* 38:7–25
72. Kluge, A. G., Strauss, R. E. 1985. Ontogeny and systematics. *Annu. Rev. Ecol. Syst.* 16:247–68
73. Lauder, G. V. 1980. Evolution of the feeding mechanism in primitive actinopterygian fishes: a functional anatomical analysis of *Polypterus, Lepisosteus,* and *Amia*. *J. Morphol.* 163:283–317
74. Lauder, G. V. 1981. Form and function: structural analysis in evolutionary morphology. *Paleobiology* 7:430–42
75. Lauder, G. V. 1982. Historical biology and the problem of design. *J. Theor. Biol.* 97:57–67
76. Lauder, G. V. 1983. Functional and morphological bases of trophic specialization in sunfishes (Teleostei: Centrarchidae). *J. Morphol.* 178:1–21
77. Lauder, G. V. 1983. Functional design and evolution of the pharyngeal jaw apparatus in euteleostean fishes. *Zool. J. Linn. Soc.* 77:1–38
78. Lauder, G. V. 1983. Neuromuscular patterns and the origin of trophic specialization in fishes. *Science* 219: 1235–37
79. Lauder, G. V. 1983. Prey capture hy-

drodynamics in fishes: experimental tests of two models. *J. Exp. Biol.* 104:1–13
80. Lauder, G. V. 1986. Homology, analogy, and the evolution of behavior. In *The Evolution of Behavior,* ed. M. Nitecki, J. Kitchell, pp. 9–40. Oxford: Oxford Univ. Press
81. Lauder, G. V. 1989. Caudal fin locomotion in ray-finned fishes: historical and functional analyses. *Am. Zool.* 29:85–102
82. Lauder, G. V. 1990. Biomechanics and evolution: integrating physical and historical biology in the study of complex systems, In *Biomechanics in Evolution,* ed. J. M. V. Rayner. Cambridge: Cambridge Univ. Press. In press
83. Lauder, G. V. 1990. An evolutionary perspective on the concept of efficiency: how does function evolve? In *Concepts of Efficiency in Biological Systems,* ed. R. W. Blake. Cambridge: Cambridge Univ. Press. In presss
84. Lauder, G. V., Crompton, A. W., Gans, C., Hanken, J., Liem, K. F., et al. 1989. How are feeding systems integrated and how have evolutionary innovations been introduced? Group Report #1. In *Complex Organismal Functions: Integration and Evolution in Vertebrates,* ed. D. B. Wake, G. Roth, pp. 97–115. New York: John Wiley
85. Lauder, G. V., Liem, K. F. 1989. The role of historical factors in the evolution of complex organismal functions. In *Complex Organismal Functions: Integration and Evolution in Vertebrates,* ed. D. B. Wake, G. Roth, pp. 63–78. New York: John Wiley
86. Lauder, G. V., Reilly, S. M. 1988. Functional design of the feeding mechanism in salamanders: causal bases of ontogenetic changes in function. *J. Exp. Biol.* 134:219–33
87. Lauder, G. V., Reilly, S. M. 1990. Metamorphosis of the feeding mechanism in tiger salamanders *(Ambystoma tigrinum):* the ontogeny of cranial muscle mass. *J. Zool., Lond.* In press
88. Lauder, G. V., Shaffer, H. B. 1988. The ontogeny of functional design in tiger salamanders *(Ambystoma tigrinum):* are motor patterns conserved during major morphological transformations? *J. Morphol.* 197:249–68
89. Lauder, G. V., Shaffer, H. B. 1990. Design of the aquatic vertebrate skull: major patterns and their evolutionary interpretations. In *The Vertebrate Skull,* Vol. 3, ed. J. Hanken, B. Hall. Chicago: Univ. Chicago Press. In press
90. Lauder, G. V., Wainwright, P. C. 1990. Function and history: the pharyngeal jaw apparatus in primitive ray-finned fishes. In *Systematics, Historical Ecology, and North American Freshwater Fishes,* ed. R. W. Mayden. Stanford: Stanford Univ. Press. In press
91. Liem, K. F. 1970. Comparative functional anatomy of the Nandidae (Pisces: Teleostei). *Fieldiana Zool.* 56:1–166
92. Liem, K. F. 1973. Evolutionary strategies and morphological innovations: cichlid pharyngeal jaws. *Syst. Zool.* 22:425–41
93. Liem, K. F. 1986. The pharyngeal jaw apparatus of the Embiotocidae (Teleostei): a functional and evolutionary perspective. *Copeia* 1986:311–23
94. Liem, K. F. 1988. Form and fuction of lungs: the evolution of air breathing mechanisms. *Am. Zool.* 28:739–59
95. Liem, K. F. 1989. Respiratory gas bladders in teleosts: functional conservatism and morphological diversity. *Am. Zool.* 29:333–52
96. Liem, K. F., Greenwood, P. H. 1981. A functional approach to the phylogeny of pharyngognath teleosts. *Am. Zool.* 21:83–101
97. Liem, K. F., Wake, D. B. 1985. Morphology: current approaches and concepts. In *Functional Vertebrate Morphology,* ed. M. Hildebrand, D. M. Bramble, K. F. Liem, and D. B. Wake, pp. 336–77. Cambridge: Harvard Univ. Press
98. Lombard, R. E., Wake, D. B. 1976. Tongue evolution in the lungless salamanders, family Plethodontidae. I. Introduction, theory and a general model of dynamics. *J. Morphol.* 148:265–86
99. Lombard, R. E., Wake, D. B. 1977. Tongue evolution in the lungless salamanders, family Plethodontidae. II. Function and evolutionary diversity. *J. Morphol.* 153:39–80
100. Lombard, R. E., Wake, D. B. 1986. Tongue evolution in the lungless salamanders, Family Plethodontidae. IV. Phylogeny of plethodontid salamanders and the evolution of feeding dynamics. *Syst. Zool.* 35:532–51
101. Losos, J. 1990. Concordant evolution of locomotor behavior, display rate and morphology in Anolis lizards. *Behavior.* In press
102. Maienschein, J., Rainger, R., Benson, K. R. 1981. Special section on American morphology at the turn of the century. *J. Hist. Biol.* 14:83–191
103. Markert, C. L., Shaklee, J. B., Whitt, G. S. 1975. Evolution of a gene. *Science* 189:102–14
104. Mayden, R. L. 1988. Vicariance bio-

geography, parsimony, and evolution in North American freshwater fishes. *Syst. Zool.* 37:329–55

105. Mayr, E. 1961. Cause and effect in biology. *Science* 134:1501–06
106. Mayr, E. 1982. *The Growth of Biological Thought.* Cambridge: Harvard Univ. Press
107. Mommsen, T. P., Walsh, P. J. 1989. Evolution of urea synthesis in vertebrates: the piscine connection. *Science* 243:72–5
108. Ohno, S. 1970. *Evolution by Gene Duplication.* New York: Springer-Verlag
109. Patterson, C. 1980. Cladistics. *Biologist* 27:234–40
110. Patterson, C. 1982. Morphological characters and homology, In *Problems of Phylogenetic Reconstruction,* ed. K. A. Joysey, A. E. Friday. London: Academic
111. Patterson, C. 1987. *Molecules and Morphology in Evolution: Conflict or Compromise?* London: Cambridge Univ. Press
112. Raff, E. C., Diaz, H. B., Hoyle, H. D., Hutchens, J. A., Kimble, M. et al. 1987. Origin of multiple gene families: are there both functional and regulatory constraints? In *Development as an Evolutionary Process,* ed. R. Raff, E. C. Raff, pp. 203–38. New York: Alan Liss
113. Raup, D. M. 1972. Approaches to morphologic analysis. In *Models in Paleobiology,* ed. T. J. Schopf, pp 28–44. San Francisco: W. H. Freeman
114. Reilly, S. M., Lauder, G. V. 1989. Kinetics of tongue projection in *Ambystoma tigrinum:* quantitative kinematics, muscle function and evolutionary hypotheses. *J. Morphol.* 199:223–43
115. Reilly, S. M., Lauder, G. V. 1990. The evolution of tetrapod prey transport behavior: kinematic homologies in feeding function. *Evolution.* In press
116. Reilly, S. M., Lauder, G. V. 1990. Metamorphosis of cranial design in tiger salamanders *(Ambystoma tigrinum):* a morphometric analysis of ontogenetic change. *J. Morphol.* 204:121–37
117. Ridley, M. 1983. *The Explanation of Organic Diversity.* Oxford: Clarendon
118. Riedl, R. 1978. *Order in Living Organisms.* New York: J. Wiley
119. Romero-Herrera, A. E., Lehman, H., Joysey, K. A., Friday, A. E. 1978. On the evolution of myoglobin. *Philos. Trans. R. Soc. Lond.* 283:61–163
120. Rosen, D. E. 1978. Vicariant patterns and historical explanation in biogeography. *Syst. Zool.* 27:159–88
121. Roth G. 1976. Experimental analysis of the prey catching behavior of *Hydromantes italicus* Dunn (Amphibia, Plethodontidae). *J. Comp. Physiol.* 109: 47–58
122. Roth, G. 1986. Neural mechanisms of prey recognition: an example in amphibians. In *Predator-Prey Relationships: Perspectives and Approaches From the Study of Lower Vertebrates,* ed. M. E. Feder, G. V. Lauder, pp. 42–68. Chicago: Univ. Chicago Press
123. Roth, G., Wake, D. B. 1989. Conservatism and innovation in the evolution of feeding in vertebrates. In *Complex Organismal Functions: Integration and Evolution in Vertebrates,* ed. D. B. Wake, G. Roth. pp. 7–21. New York: John Wiley
124. Russell, E. S. 1916. *Form and Function: a contribution to the history of animal morphology.* (Reprinted 1982, Univ. Chicago Press.) Chicago: Univ. Chicago Press
125. Sanderson, S. L. 1988. Variation in neuromuscular activity during prey capture by trophic specialists and generalists (Pisces: Labridae). *Brain Behav. Evol.* 32:257–68
126. Schaefer, S. A., Lauder, G. V. 1986. Historical transformation of functional design: evolutionary morphology of feeding mechanisms in loricarioid catfishes. *Syst. Zool.* 35:489–508
127. Schluter, D. 1989. Bridging population and phylogenetic approaches to the evolution of complex traits. See Ref. 151, pp. 79–95
128. Schwenk, K., Throckmorton, G. S. 1989. Functional and evolutionary morphology of lingual feeding in squamate reptiles: phylogenetics and kinematics. *J. Zool., Lond.* 219:153–75
129. Seilacher, A. 1973. Fabricational noise in adaptive morphology. *Syst. Zool.* 22:451–65
130. Shaffer, H. B. 1986. Utility of quantitative genetic parameters in character weighting. *Syst. Zool.* 35:124–34
131. Sillen-Tullberg, B. 1988. Evolution of gregariousness in aposematic butterfly larvae: a phylogenetic analysis. *Evolution* 42:293–305
132. Simpson, G. G. 1958. Behavior and evolution. In *Behavior and Evolution,* ed. A. Roe, G. G. Simpson, pp 507–35. New Haven: Yale Univ. Press.
133. Simpson, G. G. 1964. *This View of Life.* New York: Harcourt Brace & World
134. Strauss, R. E. 1985. Evolutionary allometry and variation in body form in the South American catfish genus *Corydoras* (Callichthyidae). *Syst. Zool.* 32: 381–96

135. Swofford, D. L. 1984. Phylogenetic analysis using parsimony (PAUP), version 2.3. Illinois: Ill. Nat. Hist. Survey
136. Szalay, F. S. 1981. Functional analysis and the practice of the phylogenetic method as reflected by some mammalian studies. *Am. Zool.* 21:37–45
137. Tyler, S. 1988. The role of function in determination of homology and convergence—examples from invertebrate adhesive organs. *Fortsch. Zool.* 36:331–47
138. Vermeij, G. 1973. Adaptation, versatility and evolution. *Syst. Zool.* 22:466–77
139. Vermeij, G. 1973. Biological versatility and earth history. *Proc. Natl. Acad. Sci. USA* 70:1936–38
140. Vigna, S. R. 1985. Functional evolution of gasterointestinal hormones. In *Evolutionary Biology of Primitive Fishes,* ed. R. E. Foreman, A. Gorbman, J. M. Dodd, R. Olsson. pp. 401–12. New York: Plenum
141. Wainwright, P. C. 1987. Biomechanical limits to ecological performance: mollusc-crushing by the Caribbean hogfish, *Lachnolaimus maximus* (Labridae). *J. Zool., Lond.* 213:283–97
142. Wainwright, P. C. 1989. Prey processing in haemulid fishes: patterns of variation in pharyngeal jaw muscle activity. *J. Exp. Biol.* 141:359–76
143. Wainwright, P., Lauder, G. V. 1986. Feeding biology of sunfishes: patterns of variation in prey capture. *Zool. J. Linn. Soc. Lond.* 88:217–28
144. Wainwright, P. C., Lauder, G. V. 1990. The evolution of feeding biology in sunfishes (Centrarchidae). In *Systematics, Historical Ecology, and North American Freshwater Fishes.* ed R. W. Mayden. Stanford: Stanford Univ. Press. In press
145. Wainwright, P. C., Sanford, C. P., Reilly, S. M., Lauder, G. V. 1989. Evolution of motor patterns: aquatic feeding in salamanders and ray-finned fishes. *Brain Behav. Evol.* 34:329–41
146. Wainwright, S. A. 1988. *Axis and Circumference: The Cylindrical Shape of Plants and Animals.* Cambridge: Harvard Univ. Press
147. Wainwright, S. A., Biggs, W. D., Currey, J. D., Gosline, J. M. 1976. *Mechanical Design in Organisms.* New York: John Wiley
148. Wake, D. B. 1982. Functional and developmental constraints and opportunities in the evolution of feeding systems in urodeles. In *Environmental Adaptation and Evolution,* ed. D. Mossakowski, G. Roth, pp. 51–66. New York: Gustav Fisher
149. Wake, D. B. 1982. Functional and evolutionary morphology. *Pers. Biol. Med.* 25:603–20
150. Wake, D. B., Larson, A. 1987. Multidimensional analysis on an evolving lineage. *Science* 238:42–8
151. Wake, D. B., Roth, G., eds. 1989. *Complex Organismal Functions: Integration and Evolution in Vertebrates.* New York: John Wiley
152. Waterman, T. H. 1975. Expectation and achievement in comparative physiology. *J. Exp. Zool.* 194:309–44
153. Weeks, J. C., Truman, J. W. 1984. Neural organization of peptide-activated ecdysis behaviors during the metamorphosis of *Manduca sexta* I. Conservation of the peristalsis motor pattern at the larval-pupal transformation. *J. Comp. Physiol.* A 155:407–22
154. Weeks, J. C., Truman, J. W. 1984. Neural organization of peptide-activated ecdysis behaviors during the metamorphosis of *Manduca sexta* II. Retention of the proleg motor pattern despite loss of the prolegs at pupation. *J. Comp. Physiol.* A 155:423–33
155. Weijs, W. A., Dantuma, R. 1981. Functional anatomy of the masticatory apparatus in the rabbit *(Oryctolagus cuniculus L.). Neth. J. Zool.* 31:99–147
156. Westneat, M. W., Wainwright, P. C. 1989. Feeding mechanism of *Epibulus insidiator* (Labridae: Teleostei): evolution of a novel functional system. *J. Morphol.* 202:129–50
157. Wiley, E. O. 1981. *Phylogenetics.* New York: John Wiley
158. Wiley, E. O. 1988. Vicariance biogeography. *Annu. Rev. Ecol. Syst.* 19:513–54
159. Zweers, G. 1974. Structure, movement, and myography of the feeding apparatus of the mallard *(Anas platyrhynchos L.),* a study in functional anatomy. *Neth. J. Zool.* 24:323–467
160. Zweers, G. 1979. Explanation of structure by optimization and systemization. *Neth. J. Zool.* 29:418–40

*Annu. Rev. Ecol. Syst. 1990. 21:341-72*

# FUNCTION AND PHYLOGENY OF SPIDER WEBS[1]

*William G. Eberhard*

Smithsonian Tropical Research Institute, and Escuela de Biologiá, Universidad de Costa Rica, Ciudad Universitaria, Costa Rica

KEY WORDS: spiders, webs, construction behavior, behavioral phylogeny

---

## INTRODUCTION

A general discussion of the designs and functions of the silk structures spiders use to capture prey has not been attempted since the book of Witt et al (319). Knowledge has increased so greatly in the intervening years that complete coverage is now impossible in a review article. I have cited primarily more recent and general publications. Related aspects of web biology not emphasized here are reviewed in papers on particular taxonomic groups (260), general feeding ecology (233), choice of websites (131, 232, 234), inter- and intra-specific competition (and the lack of it) (273, 275, 314), ecophysiology (192), web removal (20), sexual behavior (139, 238), vibration transmission in webs (175), neurobiology (3), attack behavior (237, 279), communication (320), energetics of web-building (227), timing of web construction (229), cues used for orientation during web construction (80, 308), structure and composition of web lines (224, 290), physical properties of silk (109, 322), and silk glands (138).

Due to the numbers of papers on orb webs, orbs are emphasized over non-orbs. The imbalance in research is strong (the ratio in the reference section is about 2:1 orb:non-orb, despite the fact that the ratio in numbers of species must be closer to 1:2 (H. W. Levi, personal communication). At least partially repairing this imbalance is a pressing need.

# IS WEB DESIGN A USEFUL TAXONOMIC CHARACTER?

Some animal constructs are useful in distinguishing closely related species (117), and it is not unusual to read that orb design is a genus- or species-specific character (17, 96, 186, 210, 235, 261, 301, 318). It is indeed true that in a local fauna species of orbweavers can often be determined from their webs (11, 96, 261).

The impression of species-specificity may usually, however, be the product of lack of information; knowing the characteristics of relatively distantly related local species, but not those of many close relatives, will give one a false impression that all are distinctive. Given the long-standing (226) and repeated documentation of substantial *intra*specific variation in at least gross web characters such as numbers of radii, spiral loops, spacing between loops, angle of web plane with vertical, web area, top-bottom asymmetry, and stabilimenta (61, 83, 85, 159, 226, 235, 267, 299), Levi's prediction (153) that species-specificity will be uncommon seems likely to be correct. Limited comparisons of 2-4 congeners in two araneid genera have shown statistical differences in some web characters (174, 235, 295, 296, 300).

Perhaps instead orb designs will prove to be useful characters at higher taxonomic levels. Although descriptions of small numbers of species have been made with this in mind (153, on *Argiope, Cyclosa*; 32, 235, 295, 299, on *Araneus),* this possibility has not yet been carefully tested. The degree of both intraspecific variation and interspecific uniformity in different web characters of numerous congeneric species will have to be determined to decide which web characters can be usefully compared. Coddington's work on theridiosomatids (29) is a step in this direction, though specific mention is made of webs of an average of only $2.5 \pm 1.3$ species/genus in the 8 web-building genera. Subsequent information has already made it necessary to substantially expand the range of webs made by one theridiosomatid genus (85). Attempts to characterize a "genus web" for *Araneus* (32, 235) do not fit the webs of some species in this large genus (97, 110). In several araneoid and uloborid genera, at least some apparently basic web characters are definitely *not* constant within the genus (*Eustala*—64, 68, 75; *Araneus*—32, 97, 110, 199, 235, 271, 291; *Alpaida*—76, 157; *Tetragnatha*—76, 155, 199, 265, 268; *Wixia*—76, 276; *Wendilgarda*—29, 85; *Uloborus*—61, 163, 166; *Philoponella*—8a, 163). Some suprageneric groupings (e.g., Cycloseae of Simon, the Synotaxidae of Forster et al) also do not reflect similarities of web design (181, 98).

It may well be that in some other groups some aspects of orb design are indeed genus-specific. For instance, all known *Cyrtophora* "orbs" are fine-meshed, nonsticky, radially organized horizontal sheets with lines converging on a hub which are hung in a mesh (161, on *cicatrosa, cylindroides,* and *monulfi;* 141, on *citricola;* 160, on *moluccensis;* 269, on *exanthematica* and *unicolor;* W. G. Eberhard, unpublished, on *nympha*). Orb characters may even link genera (e.g., *Cyrtophora* with *Mecynogea, Nephila* with *Nephilengys* and *Herennia, Argiope* with *Gea*). The modified orb of *Deinopus* may characterize the entire family (31). Web characteristics may eventually be found that will characterize even the genera mentioned above as lacking characteristic designs, or that will justify intrageneric groupings. The data and analyses necessary to demonstrate such usefulness have not yet been assembled, however.

Just as in orbs, non-orb webs also show both clear divergences within groups and dramatic convergences between groups. The related and primitive *Austrochilus,* and *Progradungula* have quite different webs (99, 112, 259), and several genera close to *Progradungula* have secondarily lost their webs (99). Web designs vary widely within the theridiid genera *Achaearanea* (e.g. 30, 52, 62, 167, 173, 269) and (to a lesser extent) *Latrodectus* (147, 282, 284) as well as the amaurobiid genus *Titanoeca* (134, 283) (*Achaearanea* may be polyphyletic, however—98). There may be species-specific differences in the designs of *Diguetia* sheet webs (200) and *Latrodectus* tangle and gumfoot webs (282, 284). On the other hand, intraspecific variation occurs in such basic aspects of web design as the presence or absence of a sheet, and presence or absence of a mesh above the sheet in the theridiid *Anelosimus jucundus* (194), and the presence or absence of sticky "gum foot" lines near the substrate in *Latrodectus geometricus* and *Theridion purcelli* (147).

Convergent evolution of horizontal aerial sheets with a mesh above and/or below is common; this design occurs in such widely separated groups as Linyphiidae (e.g. 269), Theridiidae (52, 62, 142, 162, 269, 282, 284), Cyatholipidae (115), Diguetidae (200), Pholcidae (57, 86, 162), Araneidae (269), and Uloboridae (166). Tightly woven, approximately horizontal sheets that are associated with the substrate and with a retreat, often in the form of a funnel which is positioned at one edge, also occur in a wide range of families. In some, such as Austrochilidae (99), Psechridae (240), Oxyopidae (114, 179), Amaurobiidae (111), Stiphidiidae (54, 100), the spiders run on the underside of the sheet. In others, such as Dipluridae (40), Tengellidae (252), Hahniidae (215, 269), Agelenidae (269), Lycosidae (132), and Pisauridae (188), the spider runs on the top of the sheet (some hahniids and oxyopids move about on both sides–135, 215). Shear (261) gives similar examples of convergence. Judging by the fact that the primitive sister groups of araneomorphs (hypochilids and austrochilids) move under their sheets, the tendency to run on top of the sheet is derived rather than primitive, as has been supposed by many previous authors (e.g. 134). Such evolutionary flexibility and rampant convergence suggest that attempts to use generalized webforms such as these as taxonomic characters (e.g. 178) must be treated with caution (36).

In some cases silk glands and their products rather than webs are taxonomically useful at higher levels (30). The data are still so sparse, however, that surprising discoveries are being made. For instance, drops of more or less liquid glue, thought to be confined to orb-weaving araneoids and theridiids, have been found in the webs of such diverse families as Linyphiidae (142, 178, 224), Pholcidae (9), Lycosidae (178), and Agelenidae (246). Their glandular origins and possible homologies are as yet undetermined.

Very preliminary indications suggest that chemical composition of silk is not taxonomically useful, since proportions of amino acids in ampullate gland silk show intraspecific and even intraindividual variation (323). There are substantial differences in ninhydrin reactive vs ninhydrin negative fractions of the water soluble portions of webs of one species of *Nephila* and two of *Argiope* (286, 287).

One possibly useful character that presumably results from silk chemistry is the yellow rather than the typical white color of orbs of all observed species of *Nephila* (271, on *clavata;* 26, on *senegalensis;* 2, on *edula;* 255, on *inaurata;* 162, on *clavipes;* 89 on *constricta, pilipes, turneri;* 244 on *maculata*) (it is white in the related *Nephilengys malabarensis* and *N. cruientata*–W. G. Eberhard, unpublished). In *N. clavata,* the color is present in only the webs of mature and nearly mature individuals (217). It has been claimed (46) that webs of both juvenile and adult *N. clavipes* are sometimes colorless, but this contradicts other observations (39, 41, 162, 244; W. G. Eberhard, unpublished). Yellow silk occurs in a scattered group of araneids, including *Cyclosa* sp., *Araneus expletus, Araneus* near *legonensis,* and *Cyrtophora nympha* (W. G. Eberhard, unpublished). The functional significance of the color is unknown, but the contrasting white color of wrapping and retreat silk in *A. expletus* suggests that the yellow color of orbs may be related to prey capture (see 48).

## USE OF WEBS TO DETERMINE PHYLOGENIES

While Levi (153) argued that orbs are not useful in determining phylogenies at lower taxonomic levels, there have been a number of attempts to deduce spider phylogeny from web structure. Unfortunately, some (e.g. 10, 119, 134, 144, 325) have relied on the risky combination of only a fragmentary knowledge of webs (e.g. araneid and theridiid webs mostly restricted to temperate species), and the mistaken assumption that a web design that is structurally intermediate indicates evolutionary (historical) intermediacy.

Although details of orb structure such as barrier webs, stabilimenta, free sectors, and web reduction show repeated convergences (74, 110, 153, 195), some details of the behavior patterns used in building orbs have apparently been more conservative. As noted by Ades (1), orb construction is actually very simple, in that the same motor patterns are repeated over and over; it is in the adjustments of these patterns to different conditions that the complexity and derived nature of orbs is manifest. Details such as leg positions and movements, sequences of lines laid and broken, patterns of sticky spiral attachments to radii, and patterns of orb repair have provided evidence for associating different families of araneoids (30, 74, 78, 84, 103), families of cribellates (81), for defining the family Uloboridae (30, 74, 84) and the subfamily Nephilinae (30, 74, 158, 198, 263–264—see also 240, 255), and for determining that the reduced webs of deinopids are derived from orbs (31).

Another potential source of evolutionary information is behavioral ontogeny (77, 87, 153, 319). It appears that when orb web designs change substantially as spiders mature, the designs of webs of older spiders are usually more derived (77).

## ARE ORB WEBS MONOPHYLETIC?

The old question of the monophyly of the orb design reacquired some of its controversial heat with the realization that the cribellum and calamistrum are plesiomorphic for araneomorphs (higher spiders). Thus these characters, classically used to justify the separation of the orb weaving family Uloboridae (and its possible ally Deinopidae) far from the rest of the orb weaving families (all in Araneoidae), cannot justify this separation. Coddington (30, 33–36) has argued that uloborids and/or deinopids are the sister group for araneoids, combining characters which have not been used traditionally, such as silk gland types, valves in silk gland ducts, presence and location on spinnerets of spigots through which silk fibers emerge, and details of construction behavior, with more traditional genitalic and somatic characters. Some workers are in agreement (e.g. 154), while others are not convinced (137, 178, 298).

Classifications are currently changing rapidly (37), and there are several reasons to think that it is probably too early to draw confident conclusions about monophyly. The methodological assumption of equal weight for all characters in the cladistic analyses used is likely to be misleading (e.g. 23, 184, 258). Because of the small sample available, the most complete analysis of spigot and spinneret morphology is "*at best* a first draft" (35, emphasis in original). Few species have been checked for valves in silk gland ducts (30), so it is hard to evaluate the possibility that evidence (lack of valves in a linyphiid) that is not in accord with predictions of the monophyly hypothesis can be chalked up to secondary loss (30). Homologies of different palpal sclerites are also still uncertain (34, 53, 118). Some statements about character states in current analyses are also dubious. For instance, the supposed lack of sticky spiral localization behavior in non-uloborid and deinopid cribellates (33, 36) ignores the reports of what appear to be similar exploratory tapping in *Titanoeca* (283) and *Psechrus* (79), and the very regular spacing of sticky lines (presumably resulting from localization behavior) in webs of *Fecenia angustata* (240) and *Prograndungula carraiensis* (112). Other problems yet to be solved are mentioned in 30 and 84.

Our almost incredible ignorance of the details of spinning behavior of non-orb weavers in both araneoid and cribellate families (possible outgroups) makes deductions regarding homologies and directions of behavioral transitions still uncertain (30, 84, 261). A few general descriptions exist (14, 147, 282), but to my knowledge there are only two relatively detailed behavioral studies for all of the probably >10,000 species of spiders which build non-orb webs (H. W. Levi, personal communication) (65, 102); and one concerns the obviously derived and atypical genus *Synotaxus!*

Designs of non-orb webs are also poorly documented, though there is one character that may link uloborids to dictynids and eresids. Many dictynids and allies (e.g. *Dictyna*–39, 199; *Ixeuticus* spp.–172; *Matachia*–169 and the eresid *Stegodyphus* (79, 143, 223–4) lay "blunt zig-zag" sticky lines between more or less parallel nonsticky lines, with the sticky line running along each nonsticky line a short distance before bridging back to the other nonsticky line or, occasionally (224: Fig. 67a) to a third nonsticky line. Judging from photographs, such zig-zags do not occur in the webs of amaurobiids *Titanoeca* (283), *Amaurobius* (133), the filistatid *Filistata hibernalis* (39), or the psechrids *Psechrus* and *Fecenia* (79, 240). Similar patterns, in which a sticky line hangs free, runs along a nonsticky line, hangs free, etc. (alternate between "autonomous" and "heteronomous" in the language of 224)–occur in at least the outer portions of the orbs in at least four genera of uloborids (*Uloborus, Philoponella, Tangaroa, Hyptiotes*) (61, 163, 199), and in the orb-like egg sac web of *Miagrammopes* (165). In the uloborids *Polenecia* and *Miagrammopes* sticky lines are laid for long distances along nonsticky lines (165, 311), and in some *Uloborus* they sometimes run along nonsticky frame lines (61, 163). In contrast, orb weaving araneoids almost never lay sticky silk along a nonsticky line except at the attachment itself (the "sawtooth" orb of *Eustala* is an exception–75; but comparison with other *Eustala* webs indicates this is an independent convergence). This contrast holds even for webs reduced to one or a few long sticky lines (compare the araneoid *Wendilgarda galapagensis* (85) and the uloborid *Miagrammopes* (165). Until more information is available regarding the distribution of web designs and construction behavior of nonorb builders, it will be difficult to decide whether or not this and several of the proposed synapomorphies in construction behavior linking uloborids and araneids (30) are actually plesiomorphies.

Several basic aspects of the organization of orb construction behavior may be plesiomorphies that preceded the evolution of orbs. Preliminary construction of a scaffold or skeleton which is then filled in occurs in several theridiids (65, 147, 282), agelenids (51, 293), and two pholcids (9; W. G. Eberhard, unpublished study of *Modisimus* sp.) (such a pattern is not inevitable, as the theridiid *Chrosiothes* sp. gradually extends its sheet with a crocheting-type of behavior along the edge of the sheet–W. G. Eberhard, unpublished). Placement of sticky lines following rather than preceeding nonsticky line construction is common (65, 102, 147; 282, on theridiids; 79, 240, on psechrids; 283, on an amaurobiid; 79, on eresids and a filistatid). Radial arrangement of nonsticky lines occurs, for instance, in the theridiid *Theridium* (102), hersiliids (312; W. G. Eberhard, unpublished), hypochilids (259), and an oecobiid (whose webs further resemble orbs in having the radial nonsticky lines run over an expanse of sticky silk which encircles the spider's resting place) (107). A process laying sticky lines starting at the edge of the web and working inward occurs in several cribellate families (79, 81). The apparently ancient ability to remember distances and directions moved is used to achieve regular spacing of both temporary and sticky spiral lines in the orbs of *Leucauge* (80).

Lamoral (147) found that species in three different theridiid genera start their webs by making a three-dimensional array of "radial" nonsticky lines converging on a retreat. They then interconnect these lines with shorter, nonradial nonsticky lines. Finally they lay sticky lines, in this case attached to the ground. The general resemblance to construction of the radii, temporary spiral, and sticky spiral of an orb is striking. Szlep (282) also noted radial lines laid in the horizontal platform of one theridiid, and regularly spaced but nonradial lines in that of another. She likened the vertical, gumfoot lines and the behavior employed in choosing sites where they will be laid to radii in an orb, and the regularly meshed platform where the spider waits to the hub of an orb. The order of construction (vertical lines interspersed with filling in the platform mesh) is appropriate for this comparison. Given the uncertainty as to whether theridiids had an orb-weaving ancestor, it is not clear whether these behavior patterns represent vestiges of ancestral orb construction behavior, convergences, or predecessors of orb construction in araneoids.

However the monophyly debate eventually turns out, the apparently primitive behavior of nephiline spiders indicates that even a monophyletic derivation of all orb weavers will require postulation of substantial convergences in basic aspects of exploration, radius and frame construction, and perhaps hub destruction behavior (84).

When one considers the rapid rate of discovery of new web forms and that at least three unrelated groups make nonorb webs which are nevertheless planar and have regularly spaced arrays of sticky lines (*Titanoeca*–283;

*Fecenia*–240, 255; and *Synotaxus*–65), it appears that the transition to this general webform has not been especially difficult to evolve. It is a dismaying indication of the difficulty of higher-level taxonomy to realize that even after the quantum leap forward represented by the large amounts of new data added recently and Coddington's extensive syntheses and analyses, there is so much yet to be done to obtain answers to these puzzles.

## HOW ORBS FUNCTION

### *Effects of Web Geometry*

The traditional idea is that orbs act as passive sieves in prey capture and that closer spacing of sticky lines evolved to capture smaller prey (e.g. 16, 123, 305, 317). This idea is being replaced by the realization that orbs do not act as simple filters, and that they must perform at least three different functions in prey capture: interposition of the web in the path of prey (interposition); absorption of the prey's momentum without breaking (stopping); and adhesion and/or entanglement of the prey to retain it until the spider arrives to attack (retention) (22, 42, 186, 303). The importance of the stopping and retention functions is illustrated by field observations of frequent prey escapes after hitting an orb: about 40% for *Tetragnatha elongata* (106); 61% for *T. praedonia* (326); 53% for *Metabus gravidus* (16); 17% for *Micrathena gracilis* (304); 58% for *Araneus trifolium* (205); 63% for *Argiope trifasciata* (205); about 10% for *Nephila maculata* (160) (blowflies only); 58–82% for nonadhesive sheet "orb" of *Cyrtophora moluccensis* (160) (blowflies only); and 33% in *Metepeira* colonies (301) (impacts with orbs only).

Features that improve one functional aspect of an orb can impair other functions. For instance, wider spaces between lines will result in a larger area being covered by a given volume of silk, and thus improved interposition, but will reduce stopping and retention (22, 42, 76). More closely spaced sticky lines with more or larger viscid droplets and lines with larger diameters will stop and retain prey more effectively, but make the web more visible and more easily avoided (at least during the day) (44, 46, 76). No single design is best for all prey. Those analyses of orb prey capture functions that consider only a subset of these functions (e.g. 186) can lead to unjustified conclusions ("larger webs . . . are a waste of silk"–195). Although there is a general trend for an orb to have approximately the same number of radii and sticky spiral loops (61, 76, 297, 298), some species tend to make radius-rich or radius-poor orbs. Since radii are much more effective in stopping prey (42, 76), radius-rich orbs are thought to represent adaptations for heavier, faster moving prey (42, 76).

None of the hypotheses explaining different orb geometries as adaptations to capture different prey (22, 76, 277) has yet been tested adequately. Tests will not be easy, as available prey, spider size (and thus, probably, the diameters of web lines and amounts of adhesive), and spider attack behavior must be taken into account in the field where prey are free to avoid webs. Some preliminary data are in accord with predictions. The elongate, dense vertical array of sticky lines in *Scoloderus* webs apparently increases, as predicted, captures of moths, as compared with more standard orb designs of syntopic species (276) (data for other species were not given, however). Larger prey were captured by *Micranthena gracilis,* which had more radii and more tightly spaced sticky lines, than by the sympatric and similar-sized species *Leucauge venusta* (305). Comparisons of prey captured by *M. gracilis* with prey in sticky traps also suggested that this species specializes on larger prey, especially large flies (304). Comparisons of prey from five species in a tropical forest understory showed that those with denser, more radius-rich orbs (*Micrathena, Mangora*) captured heavier prey and prey with higher predicted impact energies than those (*Leucauge, Epilineutes*) with more open, radius-poor orbs (42). The species with high energy orbs were substantially larger than the others, however, so this confirmation is inconclusive. None of these studies tested the possibility that differences in microhabitats and attack behavior were responsible for differences in prey captures.

There are other suggestive but even less complete confirmations. *Tegragnatha extensa* (199) and *T. praedonia* (271) and *Metleucauge yunohamensis* (271) make wide-meshed orbs, and *T. elongata* and *T. laboriosa* probably make similar designs typical of this genus (39). These species capture, as predicted, mainly weakly flying prey such as nematocerans and aphids (182), nematocerans (327), very small flies (106, 326), and small flies and cicadellids (152). Spider attack behavior varied with different prey, however, and some heavy prey (beetles) captured momentarily in *T. laboriosa* orbs were discarded by the spiders or allowed to escape (152). In *T. montana,* prey found in the webs (not necessarily a reliable indicator of prey attacked and consumed) were also mostly small and weakly flying dipterans (50).

The extraordinarily large variations in prey of the better studied genera *Argiope* and *Nephila* (190), which apparently have similar web designs in different species, indicate that in general apparent confirmations like those just noted should be treated with caution. For instance, in ten studies of *Argiope* (those summarized in 193 plus 1a and 177), Diptera ranged from 1.3% to 80%; Coleoptera from 0.5% to 22%; and Lepidoptera from 0.3% to 36% of the total numbers of prey. Similar variation occurred between studies of two species of *Nephila* (10–62%, 5–27%, and 0.7–24% respectively) (190, 244). In three studies of the same species (*A. aurantia*) the percent of odonates in the totals ranged from 0% to 32%. Clearly, attempts to characterize the prey taxa of at least some orb weavers are risky. The variability means that attempts to explain web design and prey capture behavior and morphology as specializations for certain prey (e.g. 205–see also below) can also be risky. Perhaps other prey characteristics (e.g. flight speed or agility, momentum on impact) are less variable, but this remains to be tested.

Some insects take evasive action to avoid orbs they are about to encounter (16, 41, 161, 193, 206, 244), and others clearly maneuver around webs [e.g. the kleptoparasitic empidid fly *Microphor* which flies along radii and sticky spiral lines as it searches for tiny prey caught in orbs (191)]. Thus, the interposition function of at least diurnal orbs is probably influenced by their visibility. Artificially increasing an orb's visibility can cause a dramatic reduction in the frequency of prey impact (44). A web's visibility is probably influenced by diameters and densities of lines, sizes and densities of balls of adhesive on the sticky spiral, and the background of the web (41, 44, 46). The importance of background has been demonstrated (44, 46); effects of higher thread density which may increase avoidance have also been studied (41), though larger species made the more dense webs used in this study, and its thicker lines (42) were probably seen more easily by the insect. The evolutionary effects of visibility on particular design characters of orbs are as yet unstudied. The lack of obvious consistent differences between designs of orbs that are used only at night (*Acacesia, Metazygia,* some *Tetragnatha, Eustala, Bertrana striolata,* some *Eriophora*) as compared with orbs built in the early morning and used during the day (e.g. many *Araneus, Mangora, Leucauge, Metabus, Argiope, Cyclosa, Micrathena, Dolichognatha*) argues, albeit inconclusively, against the selective importance of visibility on orb geometry.

Despite the fact that arachnologists often discuss "the" mesh size of an orb (e.g. 87, 177, 186, 190, 205, 305, 319), spacing between lines in a single orb varies widely. The radii are much farther apart at the edge than near the hub; sticky spirals are usually also farther apart near the edge than near the hub (61, 183, 316); and both radii and sticky spirals in the upper portions of vertical orbs are usually farther apart than those below the hub (150, 94a, 187, 309, 319). By comparing rates of prey encounter and capture in different portions of an orb it should be possible to measure the effects of these differences. This has been attempted (94a, 187), but unfortunately only for the subset of prey (mostly very small) which the spider ignored and left in the web without attacking, and which are probably of minor biological significance (205).

Movement of the web itself, as in ray spiders (Theridiosomatidae) (e.g. 29, 266), probably aids in the capture of insects with slow, tentative flight such as some nematocerous flies, though this has not been demonstrated experimentally. It is also possible that movements of typical orbs in the wind often help overcome evasive actions by prey (47). Additional measurements of orbs in nature which are not loaded in ways that increase their displacements are needed, however, to demonstrate this.

Still another important factor which must at least sometimes influence all three prey capture functions is the web's ability to resist or evade environmental stresses such as the spider's own weight, wind, rain, and falling debris (45). Production of thinner lines when spiders were weightless in space (321), and thicker lines when weights were increased by gluing weights to spiders's abdomens (25), indicates that support of the spider's weight is an important design consideration (some, however, have failed to find correlations between spider weight and diameters of draglines–7). Adjustments in web design or placement that are apparently designed to reduce wind damage have been noted in the orb weavers *Uloborus* (60), *Argiope* (94, 161), *Tetragnatha* (152), *Metepeira* (256), and *Araneus* (122) as well as the sheet weaver *Diguetia* (200); but in others, such as *Micrathena,* wind has little if any importance (5). Amounts of web damage due to wind, rain, debris, insects, and the spider's movements in the field are reported for three species (45) (criteria for distinguishing different types of damage, an apparently difficult task, were not given, however). Webs of some species accumulated damage more rapidly than those of others in the same general habitat (45).

Lower tensions improve a web's ability to stop prey, but it is not clear whether or not they also make a web better able to withstand wind stresses. A looser web will have a smaller radius of curvature under wind pressure, thus reducing the load on the supporting framework (42, 108). It will also, however, flap more in the irregular breezes typical of national websites (e.g. 104); the sudden accelerations and decelerations during snaps probably cause increased stresses (61, 148). Lower tensions will also give sticky spiral lines more freedom to swing, and thus to hit and stick to each other. Judging by photos of artificial wind damage to orbs (45), this last factor may be most important in nature (these webs were loaded, however, with powder before being exposed to wind). The marked intra-web differences in tensions on radii in *Nuctenea* and *Araneus* webs (58, 313) also suggest lesser importance of mechanical stability, as more nearly equal tensions would be needed to distribute stresses more evenly (148). Some behavioral details during orb construction seem designed to reduce tensions on both radii and sticky spirals (73, 78).

The relatively high tensions in the orbs of species such as *Araneus sexpunctatus* (145)), *Micrathena,* and *Cyclosa* (42) may represent compromises promoting greater weblife in exposed sites at the cost of some reduction in the ability to stop and hold prey. In contrast, *Meta meriana* builds in very sheltered sites, and its orb is under much less tension (145). The very slack, relatively exposed web of *Pasilobus* is, as predicted, often substantially damaged within only an hour of construction (245). Some orb weavers, such as *Micrathena* and *Wagneriana,* may try to get the best of both worlds by holding the web taut while waiting for prey, then relaxing it suddenly on impact (222; W. G. Eberhard, unpublished), as do the reduced web builders *Miagrammopes* and *Hyptiotes* (165, 207).

Defense against predators is another important characteristic of orbs and associated structures like meshes, retreats, and signal lines. These structures are discussed by Edmunds & Edmunds (89) and are not reviewed here. More data are needed to test the possibility that interposition capabilities of some orbs are improved by chemical attractants (124), as appears to be the case in a the tangle webs of social dictynid (285), and single line webs of some theridiids (71, 265a).

## *Effects of Physical Properties of Silk*

The physical properties of silk lines ultimately determine the properties of the web they form. Important properties of araneoid nonsticky (ampullate gland) silk include high tensile strength which results in increased stopping ability and high extensibility combined with low resilience, a feature that aids retention by reducing trampoline-like rebound after prey impact (109). Supercontraction of wet major ampullate gland lines (322) may increase tensions on newly laid slack lines and increase the orb's ability to survive light rain (322). The rubber-like extensibility of the sticky spiral line (flagelliform gland) probably increases prey retention by denying purchase to struggling prey (58). It is possible that the glycoproteins, amines, potassium, and inorganic phosphates in the sticky material on the sticky spiral, which sometimes constitute nearly half the weight of an orb (286, 289), may function not only to adhere to prey but in retention and/or accumulation of water in sticky balls (due to hygroscopic properties of GABamide) and prevention of bacterial degradation (254, 288).

Improved understanding of the composition and physical properties of the lines themselves will probably help to explain, in terms both of the cost to the spider (76) and the ability to capture prey, the functional significance of different orb designs. Craig's demonstration (42) of strong relations between spider size and diameters of both sticky and nonsticky lines in araneoids is an important step. Other possible examples include the stronger and more extensible frame silk of *Micrathena* and its webs, designed to capture large, fast-moving prey (42); the large and especially sticky viscid droplets of the reduced orbs of *Pasilobus* sp. (245) and, apparently, *Cyrtarachne inaequalis* (268), which probably increase retention (and whose cost may explain the reduced orbs); the low-shear radius-sticky spiral junctions in *Pasilobus* (245), and perhaps *Poecilopachys* (27) and *Wendilgarda* (85a), which may increase stopping and/or retention capacities; the sliding radius-sticky spiral connections of many araneoids (67), which probably increase a web's stopping and perhaps retention capacities; and the highly coiled lines in the sticky ball of bolas spiders (70), which increase the striking range of the spider.

## ORB DESIGNS IN RELATION TO ATTACK BEHAVIOR

Orb design and the speed with which spiders attack prey may be partially complementary. Species such as *Leucauge mariana,* which have radius-poor webs presumably adapted for only short restraint (see 42, 45 on *L. globosa*), attack very rapidly (82). Spiders such as *Micrathena* and *Gasteracantha,* which make more restraining orbs, on the other hand, attack more slowly (82, 303, 304). Some other species such as *Mangora* (W. G. Eberhard, unpublished), that attack rapidly make relatively dense orbs (42, 76).

Attack behavior probably explains the consistent tendency for the hub of a vertical orb to be nearer the top than the bottom of the web (and perhaps the associated trend for the sticky and nonsticky lines below to be closer together). Spiders reach prey below the hub more rapidly than those above it (176). Right-left displacement of the hub toward one side of the orb is also probably related in many species to running times (to run to the hub from the web's edge or a retreat in order to capture prey, and/or to run to the edge or a retreat from the hub in order to escape danger) (176). In species with retreats or preferred sides toward which spiders run, the hub is consistently closer to the side where the spider hides in both vertical orbs (e.g. 94a, 110, 143, 150, 199) and horizontal orbs (e.g. 63).

Olive (205) used data on differences in prey types and attack success with different prey in *Araneus* and *Argiope* at one site to argue that distance between sticky spiral lines may also correlate with the spider's ability to overpower prey (spiders with smaller chelicerae had more densely meshed orbs to restrain more dangerous, slowly escaping prey). Given the extreme variability in prey in different studies (above), more data are needed to test this hypothesis.

## HORIZONTAL ORBS

The nearly horizontal orientation of the orbs of some species and genera is puzzling because it probably substantially reduces both interposition (sticky trap captures were reduced by 70% when the trap was horizontal rather than vertical–22) and retention (numbers of a sepsid fly retained for more than five seconds (enough time for most attacks) were reduced by about 20% in horizontal orbs–82). Horizontal orientation probably also increases damage from rain and falling debris, though comparative data have not been gathered.

Possible advantages of horizontal orbs, most of which are as yet undocumented, include allowing spiders to run and attack more rapidly (82), reducing wind stress (60, 61), allowing web oscillations to sweep up slow moving insects flying horizontally (47), and providing access to particularly favorable microhabitats where prey are common (16). Horizontal orbs may be more difficult for prey to see, at least from below, because of background problems (44, 46) from the sky.

In some *Tetragnatha, Metabus,* and *Conoculus* (267), which build just above water surfaces, and perhaps also in the tiny anapids, symphytognathids and mysmenids that build in leaf litter (30, 78), horizontal orientation may enable spiders to build in sites particularly rich in prey (see 16 for documentation in *Metabus*). In many other groups, however (e.g. *Leucauge, Uloborus*), horizontal orbs are built at sites where vertical webs could be built.

The rapid attack hypothesis (82) is unlikely to explain the sometimes nearly horizontal orbs in short-legged, slow-attackers like *Gasteracantha cancriformis* (180). Further studies are needed with unloaded webs to determine whether web oscillations are biologically significant (above). In sum, no single explanation accounts for why horizontal orbs prevail in many genera.

## ORBS AS SELECTIVE TRAPS, AND ARTIFICIAL ORBS

Direct observations of prey striking webs consistently show that some types of insects are more reliably captured than others (e.g. 152, 160, 185, 186, 205, 304). Experiments comparing the prey captured by spiders with the prey captured in nearby sticky traps especially designed to mimic orbs (5, 21, 262, 303, 304) consistently show that spiders' biases are different from those of traps. For instance, in a species-by-species comparison, in which unusual care was taken to equalize time, site, visibility, and orientation of traps and webs (not duplicated in any other published study), only 23% of the mean squared variation in numbers of individuals of different species of spider prey was explained by trap captures (21). Data from other studies also suggest differences in both identity (164) and (in linyphiid sheet webs) size distributions of insects (130). There are a number of possible reasons why spider web and sticky trap captures differ, including differences in airflow, visibility, microhabitat, ability to stop and retain prey, and the spider's speed of response and selective attacks on different prey (21, 164, 244, 294, 303).

Unfortunately this lesson of differential selectivity has not been understood by several workers in the burgeoning field of web-spider ecology, who have attempted to measure habitat quality by counting numbers of prey "available" to orb weavers using various kinds of sticky traps. From the point of view of a given spider, however, any potential prey which it cannot capture and eat (because the prey avoids or escapes from the web, because it tastes bad, etc.) is not available, and that prey is irrelevant with respect to habitat quality for that spider; the prey is "available" only in the evolutionary sense that future changes in the spider or its web might enable it to capture this prey. Some studies interpret all trap captures as "available" prey with no corrections (12,

186, 190, 220, 229, 250, 251, 272, 274, 275, 301), while others (106, 113) include some correction factors but omit others (effect of time of day on both spiders and prey, visibility of trap, differences between prey species in ability to escape orbs, rejection of prey by spiders). Some authors (e.g. 275) then ask themselves why their trap data are not in accord with prey consumption! An obvious improvement would be to estimate the numbers of available prey by counting numbers of prey consumed by the spiders themselves in different habitats (21).

## DO DIFFERENCES IN ORB DESIGNS RESULT FROM "FINE TUNING" TO DIFFERENT PREY?

As just noted, orbs are to some degree selective traps. It is relatively clear that extreme variants on orb design, such as the asterisk web of *Wixia ectypa* or the ladder webs of *Scoloderus* and *Tylorida,* function to capture particular subsets of prey (walking insects and flying lepidopterans, respectively) (277). How far can this type of reasoning be extended? Is it probable, as is often argued (22, 42, 76, 93, 305) that different overall orb designs represent adaptations to different general sets of prey? There are several more or less direct kinds of evidence suggesting that at least the differences in details of orb designs probably do not represent fine tuning to specific subsets of prey:

1. Prey diversity within species and web types is generally very high (summary in 193; also 1a, 8a, 123, 127, 152, 177, 182, 240, 262, 302, 304 on 9 different genera). Orbweavers clearly tend to be general predators.

2. There is much intraspecific variation in web architecture related to factors other than prey type. These include: (*a*) amount or shape of available space (1, 97, 140, 149, 219, 239) (similar effects apparently occur in the theridiids *Coleosoma* and *Anelosimus*—49, 194 and the amaurobiid *Titanoeca*—283), (*b*) presence of conspecifics nearby (17, 105, 146) (may also influence space available), (*c*) lack of previous experience at a website (205), (*d*) presence or absence of water immediately below the orb (85, 267), (*e*) spider leg length (307), (*f*) amount of silk available in glands (83, 230), (*g*) time of day (perhaps also related to silk gland reserves?) (16, 228), (*h*) previous starvation (319) (also may affect silk reserves), (*i*) having ingested the previous day's orb (8) (also perhaps related to silk gland reserves), (*j*) presence of previously spun lines (105), (*k*) early experience prior to web-building (15, 18) and (*l*) weather (122, 253). Many of these factors seem unlikely to correlate with available prey or the spider's ability to capture them (no data are available, however). If one particular design was appropriate to trap the prey a spider could expect to capture, it would not seem sensible for the spider to change the designs of successive orbs.

3. Ontogenetic change in web design, with younger spiders usually mak-

ing less derived designs than older spiders (above), is not predicted in this context since fine tuning to different prey would not be likely to produce this pattern.

4. Retention capacities of an orb are extremely variable, even for a given species of prey. For instance, the coefficient of variation for retention times for 236 sepsid flies in vertical webs of adult *Metazygia* sp. was 110% (82) (A similarly large variation (c.v. = 91%) occurred in retention times of 218 sepsids in horizontal orbs of mature female *Leucauge mariana*—W. G. Eberhard, unpublished). Variation was so great that there was no statistical difference between retention in fresh orbs and that in "used" orbs, which had many segments of sticky spiral broken, or stuck to others or to radii (82). Large variations in retention time are apparently typical of both orbs and nonorbs (278).

5. Some studies suggest that websites are more important than web designs in determining which prey are captured (203, 231, 315). Wise & Barata (315) summarize other studies which also point, though less directly, to a lack of effect of web design on the sizes and kinds of prey captured by syntopic spiders. The prey of two araneids with different orb designs became more similar when the seasonal movement of one brought it into the subhabitat of the other (205). Riechert & Luczak (234) argue that (with exceptions like ladder webs) web structure plays little or no role in determining which prey taxa will be captured.

The data from all these studies share certain limitations. Different species undoubtedly built in different subhabitats, and their attack behaviors were probably not equally effective against different prey; these possible biases seem likely, however, to produce differences rather than similarities in prey capture data. Comparisons were only at high taxonomic levels (usually order), and by size categories (usually 1 mm). Similar numbers of prey in different families or orders obviously do not necessarily indicate the same species of prey were captured. As Wise & Barata (315) note, infrequent visits to some webs could have failed to document differences in smaller prey which were consumed rapidly. One study (203) used only data from prey remains left in webs rather than the prey actually fed upon by spiders.

Other studies give contradictory results, with the degree of difference in prey correlating with the degree of difference in web design (186, 201, 202; 42 for larger faster prey). In all of these studies, however, differences in prey could result from differences between species in microhabitat. In sum, a modified version of the conclusion of Riechert & Luczak (234) seems most appropriate: Differences in habitats where webs of different designs can be (are) built are probably responsible for some differences in prey captured; web design may also strongly influence prey capture, but critical data (which could come from web and spider transfer experiments combined with observations of attack and rejection behavior to take into account the possible role of microhabitat and active choice by the spider) are still lacking.

6. Attack and feeding behavior on given prey also varies intraspecifically. Most of the *Nephila clavipes* at one site ignored some types of prey in their webs which were captured at another site (123). Even the same individual may attack a given insect more or less rapidly as a result of previous experience (189, 234, 294), the availability of prey (50), or hunger (234). Digestive enzyme concentrations can change with feeding experience (189), and the amount of food extracted from a prey can vary with the size of the spider (214, 295) and hunger (295). Thus, payoffs from captures of identical prey may vary.

7. Prey captured often vary considerably in different geographic areas (123, 127), seasons (127, 177, 205), years (126), and nearby subhabitats (e.g. 16, 22, 24, 106, 177, 183, 200, 219, 244, 303, 315).

The evolutionary consequence of this multiply compounded variability is that selection on details of orb design in terms of prey captured must be weak in many species, because the effects of the details of design must only be perceptible, in terms of prey consumed, in very large samples. It seems likely that species in some groups (e.g. *Argiope, Nephila*) will prove to be extreme generalists, while others (for example, species of *Tetragnatha, Metleucauge*) will be more specialized (in this case, on light, weak, or slow moving prey). A prediction is that the webs of species limited to habitats where certain kinds of prey predominate will often be designed to deal effectively with those types of prey. Weak selection on details of web design makes it unclear whether or not minor details (e.g. differences in sticky spiral spacing near switch-back points in *Uloborus*—59; numbers of loops of hub spiral; larger spaces between loops of temporary spiral in the outer portion of the web 80, 309; and reduced distances between sticky spiral lines rather than switchbacks on shorted radii in *Micrathena* webs (222) are adaptive in terms of prey capture.

## EFFECT OF SPIDER SIZE ON WEB FUNCTION

Webs of smaller spiders, which are probably generally made with lines of smaller diameters (42) and probably with smaller amounts of adhesive (208, 214), seem to have reduced abilities to capture larger prey. Webs of smaller *Nephila* were less able to stop and retain prey for more than three seconds (13). Isolated portions of webs of immature *Metazygia* retained sepsid flies for shorter times in experimental trials than did those of adults (82). Small *Metazygia gregalis* captured much smaller prey than large individuals of the same species in webs at the same site at the same height above the ground and the same angle with the wind (21).

Several other field studies also give evidence of smaller prey for smaller individuals (12, 42, 101, 130, 177, 214, 274, 305), though the possibility that differences in microhabitats were responsible for prey differences was not eliminated. The probable importance of spider size suggests that the results of comparative studies of prey capture in webs in which spider size was not taken into account (e.g. the mix of species of very different sizes in 249) should be treated with caution.

## STABILIMENTA

A taxonomically diverse set of orb weavers add silk and/or detritus "decorations" called stabilimenta to their orbs. The following hypotheses purport to explain their function, but lack general applicability for the reasons given:

1. Web advertisement (warning off large animals such as birds which might fly through and damage the web—90, 95, 125). But many stabilimentum builders make webs in sheltered sites where birds or other large animals seldom if ever pass—e.g. *Uloborus* spp. deep in *Stegodyphus* colonies and pack rat nests (6, 60); *Lubinella* in tree buttresses and under rocky overhangs (163); *Conifaber* under prop roots of palms (166); *Argiope* in tall grass (88).

2. Prey attraction (48, 95). But stabilimenta are nearly always placed outside the trapping zone where sticky lines are present (e.g. 88, on *Gasteracantha, Argiope;* 156, 196, on *Salassina;* 271, on *Zilla;* 63, 163, on *Uloborus;* 163, 211, on *Philoponella;* 196, on *Micrathena;* 133, 199, 271, 324, on *Cyclosa;* 2, 243, on *Nephila;* 120, on *Araneus*). Stabilimenta are also often made, sometimes even more consistently and/or with more silk, on rudimentary or moulting webs which are neither designed nor used for prey capture (243, 264, on *Nephila;* 88, 95, 241, on *Argiope;* 88, 91, 92, 170, on *Gasteracantha;* W. G. Eberhard unpublished, on *Uloborus*).

3. Camouflage (hide spider or its outline) (63, 88). However the stabilimenta of some *Gasteracantha* and *Isoxya* are mostly on long frame and anchor lines which are up to 0.5 m from the hub where the gaudily colored and spiny spider rests (88, 180); the white dots in webs of *Salassina crassipina* contrast rather then blend with the spider's black or red color (162, 196).

4. Strengthening the web (241, 243). But stabilimentum silk is generally laid as unstressed, curly lines in a cottony mass, often in such a way as to preclude strengthening the web (88, 170, on *Gasteracantha;* 211, on *Philoponella;* 63, on *Uloborus*). In addition, not a single species among the many nocturnal orb weavers is known to build a stabilimentum (63, 88).

The substantial amounts of intraspecific variation in form and frequency of occurrence, typical of stabilimenta in both araneoids (88, 95, 170, 236, 247) and uloborids (63, 168, 221), are not easily explained by any but the camouflage hypothesis (95).

Some authors have argued that stabilimenta have no function and are the products of "stress" (197) or are nonfunctional vestiges (196). The common association of stabilimenta with cryptic postures in which the spider aligns its body with the stabilimentum during the day but not at night (63, 88), their strict association with daytime webs, and the repeated convergent evolution of stabilimenta argue against these interpretations. I agree with Edmunds that most stabilimenta probably function as camouflage (in some cases possibly including outline enlargement), and some as web advertisement.

## NEWLY DISCOVERED WEB FORMS

The number of known variations on both orbs and other basic web types has grown dramatically in the last 20 years. Webs show an overall pattern of exuberant diversity and frequent convergence in both orb-weaving families and others.

Many but certainly not all of the araneoid species with highly modified orb designs are of relatively small body size. Craig (43) argues that modified designs correlated evolutionarily with changes in insect sizes. Web reduction and/or loss is frequent and is often associated with increased access to prey (19, 195). In at least some cases web reduction is accompanied by substantial changes in spider morphology (212, 216).

A few modified orb designs have clear probable functions. Species in two families (Theridiosomatidae and Anapidae) have convergently evolved webs with sticky lines attached to the surfaces of streams and puddles (38, 267), presumably to capture insects in the surface film and just above it. At least three groups of species in two families (Tetragnathidae and Uloboridae) have converged on orbs with a twig running through the center (76, 233; W. G. Eberhard, unpublished observations of *Uloborus eberhardi*), presumably to hide the spider from predators. The elongate, vertical "ladder" webs (242, on a genus near *Tylorida;* 64, 276, on *Scoloderus*) probably function to capture moths.

The adaptive significance of many other variant designs is unclear. Again these include many striking convergences: the "sawtooth" orbs with radially placed sticky lines of *Polenecia* (Uloboridae) (311) and *Eustala* sp. (Araneidae) (75); the elongate webs made next to tree trunks of *Herennia ornatissima* (Tetragnathidae) (239), *Araneus atrihastula* (Araneidae) (97), and *Eustala* sp. (Araneidae) (W. G. Eberhard unpublished); loss of frame lines and drastic reduction of numbers of radii and sticky spiral loops in *Tetragnatha lauta* (Tetragnathidae) (265), *Cyrtarachne* spp. (Araneidae) (268, 269) and *Olgunius* spp. (Theridiosomatidae) (29, 30); retention of temporary spirals in finished orbs in *Nephila* (Tetragnathidae) (244) and *Phonognatha* (Araneidae) (53); reduction to a few long sticky lines diverging from a central area of nonsticky line where the spider rests in some *Miagram-*

*mopes* (Uloboridae) (165), and some webs of *Wendilgarda galapagensis* (Theridiosomatidae) (85). Such convergences suggest that similar, as yet undetermined, selective forces have operated in widely different lineages. Other equally mysterious designs are unique to certain groups, such as the nonsticky football-shaped sheet of *Paraplectanoides* (121), and the starburst, three-dimensional orbs of *Mysmena* (30, 78).

Several altered web designs are the products of radical changes in otherwise extremely conservative behavior patterns. Radius construction in "asterisk" webs of *Wixia extypa* (276) apparently differs from that of all other araneines (74); sticky "spiral" construction in the reduced orbs of *Poecilopachys* (27) and *Pasilobus* (245) (Araneidae), and *Hyptiotes* spp. (Uloboridae) (84, 171) is repeatedly interrupted and then resumed (only occasional interruptions occur in some other orbweavers—e.g. *Uloborus, Nephila, Leucauge*—W. G. Eberhard, unpublished). Movement "backward" during the construction of sawtooth orb sticky lines in *Eustala* sp. (75), toward the last site of attachment before attaching some segments of sticky spiral, is to date unique to this species. Inclusion of some temporary spiral construction and even exploratory behavior following, rather than strictly preceeding, sticky line construction in both the elongate trunk webs of *Araneus atrihastula* (Araneidae) (97) and the "high land" type of web in *Wendilgarda galapagensis* (Theridiosomatidae) (85) is again a sequence unknown, as far as I know, in undisturbed construction behavior of any other orb weaver. On the other hand, some highly modified webs, such as those of *Mysmena* spp. (Mysmenidae) (30, 78) and *Deinopus* (Deinopidae) (31), conserve typical elements of orb construction.

New designs have also been discovered in groups not closely related to orb weavers: the double sheets of the pisaurids *Pisaurina* and *Architis* (151, 188); "pseudo-orbs" with either rectangular (*Synotaxus,* 65) or radial geometries (*Titanoeca,* 283, *Fecenia,* 240); an umbrella-shaped sheet and tube or inverted cone (*Stiphidion*—54, 100; *Marplesia,* 100); and gigantic aerial planar sheets containing sticky silk that are more than 1 m in diameter in *Stegodyphus* (257). Species of *Argyrodes* (*Ariamnes*) use their simple webs in a previously undocumented way—as resting places for prey and walkways along which the attacking spider can sneak up (28, 69, 270). Perhaps the most extraordinary webs are the small sticky catching ladders of the otherwise primitive araneomorphs *Progradungula* (112) and *Macrogradungula* (99), which are sprung forward to receive prey when they are flicked backward off the substrate by the front legs. Careful study of several "well-known" webs has also revealed subtle, previously unappreciated structural details: *Pholcus phalangioides* webs have loose "screw threads" and tangles of loose silk that apparently function to retain prey (136); the somewhat similar fibrillation of fibers into subunits in webs of the diplurid *Euagrus* may entangle prey (218); and there are patterns of tension differences (of uncertain function; analogous differences aid some theridiids in finding their retreats—147) in the sheet of the linyphiid *Frontinella* (280).

Other new developments include web descriptions for previously unstudied groups (e.g. 225, on *Segestriodies;* 209, on the uloborid *Tangaroa;* 99, on austrochilids), trap construction *after* the prey has first contacted web lines in *Drymusa* (306), web construction by mature males which were previously thought incapable of web construction (66, 163, 281), lack of or reduction of webs in groups that typically build webs (56, 57, on several genera of pholcids; 4, on an agelenid; 55, on a ctenizid), and discovery of webs in genera and families previously thought to lack them (e.g. 310, on *Argyrodes;* 114, 179, on the oxyopid *Tapinillus;* 128, on the salticid *Portia* and close relatives).

Salticid webs are particularly surprising, as these spiders are relatively well-studied visial predators and have been thought to be strictly cursorial hunters. *Portia,* however, builds two types of webs, the most elaborate of which seem to function less as traps than as lookout sites for locating passing prey (especially other salticids) and perhaps also as lures for web builders of other groups (128). Salticids may be directly descended from a web-building ancestor (129). The discovery that the retreat silk of various salticids easily entangles and detains insects (116) makes it necessary to reexamine the hunting tactics of these and other cursorial spiders such as clubionids, gnaphosids, some amphinectids and anyphaenids which build similar retreats. The reverse of this, a catching web serving as a retreat (to maintain high humidity), has been documented in a linyphiid (292).

## CONCLUSIONS

Knowledge of spider webs has grown rapidly in a piecemeal fashion during the last 20 years. We are especially ignorant still of the construction behavior of nonorb builders; what little we know (79, 102, 147, 282) suggests that many webs in the process of being built may be more organized than is presently suspected, since subsequent additions can obscure the original pattern. Also needed are careful experimental tests in the field and in captivity of our relatively sophisticated ideas of the advantages and disadvantages of different orb designs. Extensive quantitative surveys of orb geometry in related genera or other taxa are needed to determine whether orb design is taxonomically useful.

### ACKNOWLEDGMENTS

I thank F. Barth, J. Coddington, C. Craig, R. Gillespie, S. Marshall, W. Nentwig, and G. Uetz for sending unpublished manuscripts and data, T. Inoue for translations from Japanese, M. Rambla and A. Arroyo for help obtaining literature, D. Mills for help with the bibliography, and J. Coddington, H. W. Levi, C. Craig, Y. D. Lubin, and M. J. West-Eberhard for criticizing a preliminary draft. I am especially grateful to H. W. Levi for his unstinting help in identifying specimens.

## *Literature Cited*

1. Ades, C. 1986. A construcao da teia geométrica como programa comportamental. *Ciencia Cult.* 38:760–75
1a. de Armas, L. F., Alayon, G. 1987. Observaciones sobre la ecologia trofica de una poblacion de *Argiope trifasciata* (Araneae: Araneidae) en el Sur de la Habana. *Poeyana* (Havana) 344:1–18
2. Austin, A. D., Anderson, D. T. 1978. Reproduction and development of the spider *Nephila edulis* (Koch) (Araneidae:Araneae). *Aust. J. Zool.* 26: 501–18
3. Barth, F. G., ed. 1985. *Neurobiology of Arachnids*. New York: Springer Verlag. 385 pp.
4. Bennett, R. G. 1985. The natural history and taxonomy of *Cicurina bryantae* (Araneae, Agelenidae). *J. Arachnol.* 13: 87–96
5. Biere, M., Uetz, G. 1981. Web orientation in the spider *Micrathena gracilis* (Araneae:Araneidae). *Ecology* 62:336–44
6. Bradoo, B. L. 1985. The primary orb web of *Uloborus ferokus* Bradoo (Araneae: Uloboridae) *Curr. Sci.* 54: 594–96
7. Brandwood, A. 1985. Mechanical properties and factors of safety of spider drag-lines. *J. Exp. Biol.* 116:141–51
8. Breed, A. L., Levine, V. D., Peakall, D. B., Witt, P. N. 1964. The fate of the intact orb web of the spider *Araneus diadematus* Cl. *Behaviour* 23:43–60
8a. Breitwitiche, R. 1989. Prey capture by a West African social spider (Uloboridae: *Philoponella* sp.). *Biotropica* 21:359–63
9. Briceño, R. D. 1985. Sticky balls in webs of the spider *Modisimus* sp. (Araneae, Pholcidae). *J. Arachnol.* 13: 267–69
10. Bristowe, W. S. 1930. Notes on the biology of spiders. I. The evolution of spiders' snares. *Ann. Mag. Nat. Hist.* ser. 10. 6:334–42
11. Bristowe, W. S. 1958. *The World of Spiders*. London: Norton. 304 pp.
12. Brown, K. 1981. Foraging ecology and niche partitioning in orb-weaving spiders. *Oecology* 50:380–85
13. Brown, S. G., Christenson, T. E. 1983. The relationships between web parameters and spiderling predatory behavior in the orb-weaver *Nephila clavipes*. *Z. Tierpsychol.* 63:241–50
14. Buche, W. 1966. Beitrage zur Okologie und Biologie Winterreifer Kleinspinnen mit besonderer Berucksichtigung der Linyphiiden *Macragus rufus rufus* (Wider), *Macrargus rufus carpenteri* (Cambridge) und *Centromerus silvaticus* (Blackwall). *Z. Morph. Okol. Tiere.* 57:329–448
15. Burch, T. L. 1979. The importance of communal experience to survival for spiderlings of *Araneus diadematus* (Araneae:Araneidae). *J. Arachnol.* 7:1–18
16. Buskirk, R. E. 1975. Coloniality, activity patterns and feeding in a tropical orb-weaving spider. *Ecology* 56:1314–28
17. Buskirk, R. E. 1986. Orb-weaving spiders in aggregations modify individual web structure. *J. Arachnol.* 14:259–65
18. Cangialosi, K. R., Uetz, G. W. 1987. Spacing in colonial spiders: effects of environment and experience. *Ethology* 76:236–46
19. Carico, J. E. 1978. Predatory behaviour in *Euryopis funebris* (Hentz) (Araneae: Theridiiadae) and the evolutionary significance of web reduction. *Symp. Zool. Soc. Lond.* 42:51–58
20. Carico, J. E. 1986. See Ref. 260, pp. 306–18
21. Castillo, J., Eberhard, W. G. 1983. The use of artificial traps to estimate prey available to web-weaving spiders. *Ecology* 64:1655–58
22. Chacón, P., Eberhard, W. G. 1980. Factors affecting numbers and kinds of prey caught in artificial spider webs, with considerations of how orb webs trap prey. *Bull. Br. Arachnol. Soc.* 5: 29–38
23. Cheetham, A. H., Hayek, L. A. C. 1988. Phylogeny reconstruction in the neogene bryozoan *Metrarabdotos:* a paleontologic evaluation of methodology. *Hist. Bio.* 165–83
24. Cherrett, J. M. 1964. The distribution of spiders on the Moor House National Nature Reserve, Westmorland. *J. Anim. Ecol.* 33:27–48

25. Christiansen, A., Baum, R., Witt, P. N. 1962. Changes in spider webs brought about by mescaline, psilocybin, and an increase in body weight. *J. Pharm. Expt. Therap.* 136:31–37
26. Clausen, I. H. S. 1987. On the biology and behaviour of *Nephila senegalensis senegalensis* (Walckenaer, 1837). *Bull. Br. Arachnol. Soc.* 7:147–50
27. Clyne, D. 1973. Notes on the web of *Poecilopachys australasia* (Griffith and Pidgeon, 1833) (Araneida:Argiopidae). *Aust. Ent. Mag.* 1:23–29
28. Clyne, D. 1979. *The Garden Jungle.* London: Collins. 1-184pp.
29. Coddington, J. A. 1986. The genera of the spider family Theridiosomatidae. *Smithson. Contrib. Zool.* 422:1–96
30. Coddington, J. A. 1986. See Ref. 260, pp. 319–63
31. Coddington, J. A. 1986. Orb webs in "non-orb weaving" ogre faced spiders (Araneae:Dinopidae): a question of genealogy, *J. Cladistics* 2:53–67
32. Coddington, J. A. 1987. Notes on spider natural history: the webs and habits of *Araneus niveus* and *A. cingulatus* (Araneae, Araneidae). *J. Arachnol.* 15: 268–70
33. Coddington, J. A. 1990. Spinneret silk spigot morphology evidence for the monophyly of orb weaving spiders, Cyrtophorinae (Araneidae), and the group Theridiidae and Nesticidae. *J. Arachnol.* In press
34. Coddington, J. A. 1990. Ontogeny and homology in the male palpus of orb weaving spiders and their potential outgroups, with comments on phylogeny (Araneoclada: Araneoidea, Deinopoidea). *Smith. Contrib. Zool.* In press.
35. Coddington, J. A. 1990. Spinneret silk spigot morphology: evidence for the monophyly of orbweaving spiders, Cyrtophorinae (Araneidae), and the group Theridiidae plus Nesticidae. In press
36. Coddington, J. A. 1990. Cladistics and spider classification: araneomorph phylogeny and the monophyly of orbweavers (Araneae:Araneomorphae:Araneoidea, Deinopoidea) *Ann. Zool. Fenn.* 26:
37. Coddington, J. A. 1990. Review of Platnick, N. I. 1989. *Advances in Spider Taxonomy. J. Arachnol.* In press
38. Coddington, J. A., Valerio, C. E. 1980. Observations on the web and behavior of *Wendilgarda* spiders (Araneae:Theridiosomatidae). *Psyche* 87:93–106
39. Comstock, J. H. 1948. *The Spider Book* (revised and edited by W. J. Gertsch). Ithaca: Cornell Univ. 729 pp.
40. Coyle, F. A. 1986. See Ref. 260, pp. 269–305
41. Craig, C. L. 1986. Orb-web visibility: the influence of insect flight behaviour and visual physiology on the evolution of web designs in Araneoidea. *Anim. Behav.* 34:54–68
42. Craig, C. L. 1987. The ecological and evolutionary interdependence between web architecture and web silk spun by orb web weaving spiders. *Biol. J. Linn. Soc.* 30:135–62
43. Craig, C. L. 1987. The significance of spider size to the diversification of spider-web architectures and spider reproductive modes. *Am. Nat.* 129:47–68
44. Craig, C. L. 1988. Insect perception of spider orb webs in three light habitats. *Funct. Ecol.* 2:277–82
45. Craig, C. L. 1989. Alternative foraging modes of orb web weaving spiders. *Biotropica* 21:257–64
46. Craig, C. L. 1990. Effects of background pattern on insect perception of webs spun by orb weaving spiders. *Anim. Behav.* In press
47. Craig, C. L., Akira, O., Andreasen, V. 1985. Effect of spider orb-web and insect oscillations on prey interception. *J. Theor. Biol.* 115:201–11
48. Craig, C. L., Bernard, G. D. 1990. Insect attraction to ultraviolet-reflecting spider webs and web decorations. Ecology. In press
49. Cutler, B. 1972. Notes on the behavior of *Coleosoma floridanum* Banks. *J. Kans. Entomol Soc.* 45:275–81
50. Dabrowska-Prot, E., Luczak, J. 1968. Studies on the incidence of mosquitoes in the food of *Tetragnatha montana* Simon and its food activity in the natural habitat. *Ekol. Polska. Ser. A* 16:843–53
51. Darchen, R. 1965. Ethologie d'une araignee sociale *Agelena cosociata* Denis. *Biol. Gabonica* 1:117–46
52. Darchen, R., Ledoux, J. C. 1978. *Achaearanea disparata,* araignée sociale du Gabon, synonyme un espece jumelle d'*A.tessellata* solitaire. *Rev. Arachnol.* 1:121–32
53. Davies, V. T. 1988. An illustrated guide to the genera of orb-weaving spiders in Australia. *Mem. Qd. Mus.* 25:273–332
54. Davies, V. T. 1988. Three new species of the spider genus *Stiphidion* (Araneae : Amaurobioidea : Stiphidiidae) from Australia. *Mem. Qd. Mus.* 25:265–71
55. Decae, A. E., Caranhac, G., Thomas, G. 1982. The supposedly unique case of *Cyrtocarenum cunicularium* (Olivier, 1811) (Araneae, Ctenizidae). *Bull. Br. Arachnol. Soc.* 5:410–19

56. Deeleman-Reinhold, C. 1986. Leaf-dwelling Pholcidae in Indo-Australian rain forests. See Ref. 86a, pp. 45–48
57. Deeleman-Reinhold, C. L. 1986. Studies on tropical Pholcidae II: Redescription of *Micromerys gracilis* Bradley and *Calapnita veriformis* Simon (Araneae, Pholcidae) and description of some related new species. *Mem. Qd. Mus.* 22: 205–24
58. Denny, M. 1976. The physical properties of spider's silk and their role in the design of orb-webs. *J. Exp. Biol.* 65: 483–506
59. Eberhard, W. G. 1969. Computer simulation of orb web construction. *Am. Zool.* 9(1):229–38
60. Eberhard, W. G. 1971. The ecology of the web of *Uloborus diversus* (Araneae:Uloboridae). *Oecologia.* 6: 328–42
61. Eberhard, W. G. 1972. The web of *Uloborus diversus* (Araneae:Uloboridae). *J. Zool., Lond.* 166:417–65
62. Eberhard, W. G. 1972. Observations on the biology of *Achaeranea tesselata* (Araneae:Theridiidae). *Psyche* 79:176–80
63. Eberhard, W. G. 1973. Stabilimenta on the webs of *Uloborus diversus* (Araneae:Uloboridae) and other spiders. *J. Zool.* 171:367–84
64. Eberhard, W. G. 1974. The "inverted ladder" orb web of *Scoloderus* sp. and the intermediate orb of *Eustala* (?) sp. (Araneidae). *J. Nat. Hist.* 9(1):93–106
65. Eberhard, W. G. 1975. "Rectangular orb" webs of *Synotaxus* (Araneae:Theridiidae). *J. Nat. Hist.* 11:501–7
66. Eberhard, W. G. 1976. The webs of newly emerged *Uloborus diversus* and of a male *Uloborus* sp. (Araneae:Uloboridae). *J. Arachnol.* 4(3):201–6
67. Eberhard, W. G. 1976. Physical properties of sticky spirals and their connections: sliding connections in orb webs. *J. Nat. Hist.* 10:481–88
68. Eberhard, W. G. 1976. Photography of orb webs in the field. *Bull. Br. Arachnol. Soc.* 3(7):200–4
69. Eberhard, W. G. 1979. *Argyrodes attenuatus:* a web that is not a snare. *Psyche* 86(4):407–413
70. Eberhard, W. G. 1980. The natural history and behavior of the bolas spider *Mastophora dizzydeani* sp. n. (Araneidae). *Psyche* 87 (3–4):143–69
71. Eberhard, W. G. 1981. The single line web of *Phoroncidia studo* Levi (Araneae:Theridiidae): a prey attractant? *J. Arachnol.* 9:229–32
72. Deleted in proof
73. Eberhard, W. G. 1981. Construction behavior and the distribution of tensions in orb webs. *Bull. Br. Arachnol. Soc.* 5(5):189–204
74. Eberhard, W. G. 1982. Behavioral characters for the higher classification of orb-weaving spiders. *Evolution* 36(5): 1067–95
75. Eberhard, W. G. 1985. The "saw-toothed" orb of *Eustala* sp., with a discussion of the ontogenetic patterns of change in web design in spiders. *Psyche* 92:105–18
76. Eberhard, W. G. 1986. See Ref. 260, pp. 70–100
77. Eberhard, W. G. 1986. Ontogenetic changes in the web of *Epeirotypus* sp. (Araneae, Theridiosomatidae). *J. Arachnol.* 143:125–28
78. Eberhard, W. G. 1987. Orb webs and construction behavior in Anapidae, Symphytognathidae, and Mysmenidae. *J. Arachnol.* 14(3):339–56
79. Eberhard, W. G. 1987. Construction behavior of non-orb weaving cribellate spiders and the evolutionary origin of orb webs. *Bull. Br. Arachnol. Soc.* 7:175–78
80. Eberhard, W. G. 1988. Memory of distances and directions moved as cues during temporary spiral construction in the spider *Leucauge mariana* (Araneae: Araneidae). *J. Ins. Behav.* 1:51–66
81. Eberhard, W. G. 1988. Combing and sticky silk attachment behavior by cribellate spiders and its taxonomic implications. *Bull. Br. Arachnol. Soc.* 7:247–51
82. Eberhard, W. G. 1989. Effects of orb web orientation and spider size on prey retention. *Bull. Br. Arachnol. Soc.* 8:45–48
83. Eberhard, W. G. 1989. Behavioral flexibility in orb web construction: effects of silk supply in different glands and spider size and weight. *J. Arachnol.* 16:295–302
84. Eberhard, W. G. 1990. Early stages of orb construction by *Philoponella vicina, Leucauge mariana,* and *Nephila clavipes* spiders (Araneae:Uloboridae and Tetragnathidae) and their phylogenetic implications. *J. Arachnol.* In press
85. Eberhard, W. G. 1990. Niche expansion in the spider *Wendilgarda galapagensis* (Araneae, Theridiosomatidae) on Cocos Island. *Rev. Biol. Tropical.* In press
85a. Eberhard, W. G. 1990. Notes on the natural history of *Wendilgarda galapagensis* (Araneae: Theridiosomatidae). *Bull. Br. Arachnol. Soc.* In press
86. Eberhard, W. G., Briceño, R. D. 1985. Behavior and ecology of four species of

*Modissimus* and *Blechroscelis* (Pholcidae). *Rev. Arachnol.* 6:29–36
86a. Eberhard, W. G., Lubin, Y. D., Robinson, B. eds. 1986. *Proceedings of the Ninth International Congress of Arachnology, Panama 1983*. Washington: Smithsonian
87. Edmunds, J. 1978. The web of *Paraneus cyrtoscapus* (Pocock, 1989) (Araneae:Araneidae) in Ghana. *Bull. Br. Arachnol. Soc.* 4:191–96
88. Edmunds, J. 1986. The stabilimenta of *Argiope flavipalpis* and *Argiope trifasciata* in West Africa, with a discussion of the function of stabilimenta. See Ref. 86a, pp. 61–72
89. Edmunds, J., Edmunds, M. 1986. The defensive mechanisms of orb weavers (Araneae:Araneidae) in Ghana, West Africa. See Ref. 86a, pp. 73–89
90. Eisner, T., Novicki, S. 1983. Spider web protection through visual advertisement: role for the stabilimentum. *Science* 219:185–87
91. Emerit, M. 1968. Contribution a l'etude de la biologie et du development du l'araignee tropicale *Gasteracantha versicolor* (Walck.) (Argiopidae). Note preliminaire. *Bull. Soc. Zool. France* 93:49–68
92. Emerit, M. 1969. *Contribution a l'etude des Gasteracanthes (Araneides, Argiopides) de Madagascar et des Iles Joisines*. PhD thesis. Univ. Monpellier, France
93. Enders, F. 1974. Vertical stratification in orb-web spiders (Araneidae, Araneae) and a consideration of other means of coexistence. *Ecology* 55:317–28
94. Enders, F. 1977. Web-site selection by orb-web spiders, particularly *Argiope aurantia* Lucas. *Anim. Behav.* 25:694–712
94a. Endo, T. 1988. Patterns of prey utilization in a web of orb-weaving spider *Araneus pinguis* (Karsch). *Res. Popul. Ecol.* 30:107–21
95. Ewer, R. R. 1972. The devices in the web of the West African spider *Argiope flavipalpis*. *J. Nat. Hist.* 6:159–67
96. Foelix, R. F. 1982. *Biology of Spiders*. Cambridge: Harvard Unv. 306 pp.
97. Forster, L. M., Forster, R. R. 1985. A derivative of the orb web and its evolutionary significance. *N. Z. J. Zool.* 12:455–65
98. Forster, R. R., Platnick, N. I., Coddington, J. A. 1990. A proposal and review of the spider family Synotaxidae (Araneae, Araneoidea), with notes on theridiid interrelationships. *Bull. Am. Mus. Nat. Hist.* 193:1–116
99. Forster, R. R., Platnick, N. I., Gray, M. R. 1987. A review of the spider superfamilies Hypochiloidea and Austrochiloidea (Araneae, Araneomorphae). *Bull. Am. Mus. Nat. Hist.* 185:1–116
100. Forster, R. R., Wilton, C. L. 1973. The spiders of New Zealand, Part IV. *Otago Mus. Bull.* 4:1–309
101. Fowler, H. G., Diehl, J. 1978. Biology of a Paraguayan colonial orb-weaver, *Eriophora bistriata* (Rengger) (Araneae, Araneidae). *Bull. Br. Arachnol. Soc.* 4:241–50
102. Freisling, J. 1961. Netz und Netzbauinstinkte bei *Theridium saxatile* Koch. *Z. Wiss. Zool.* 165:396–421
103. Fukumoto, N. 1981. Notes on the web-weaving activity (2). *Atypus* 78:17–20 (in Japanese)
104. Geiger, R. 1965. *The Climate Near the Ground*. Cambridge: Harvard Univ. 611 pp.
105. Gillespie, R. 1987. The role of prey availability in aggregative behaviour of the orb weaving spider *Tetragnatha elongata*. *Anim. Behav.* 35:675–81
106. Gillespie, R. G., Caraco, T. 1987. Risk-sensitive foraging strategies of two spider populations. *Ecology* 68:887–99
107. Glatz, L. 1967. Zur Biologie und Morphologie von *Oecobius annulipes* Lucas (Araneae, Oecobiidae). *Z. Morph. Tiere* 61:185–214
108. Gordon, J. H. 1978. *Structures: Why Things Don't Fall Down*. Sussex: Penguin. 200 pp.
109. Gosline, J. M., DeMont, M. E., Denny, M. W. 1986. The structure and properties of spider silk. *Endeavour* (NS) 10:37–43
110. Grasshoff, M., Edmunds, J. 1979. *Araneus legonensis* n. sp. (Araneidae: Araneae) from Ghana, West Africa, and its free sector web. *Bull. Br. Arachnol. Soc.* 4:303–9
111. Gray, M. 1981. A revision of the spider genus *Baiami* Lehtinen (Araneae, Amaurobioidea). *Rec. Aust. Mus.* 33: 779–802
112. Gray, M. R. 1983. The male of *Progradungula carraiensis* Forster and Gray (Araneae, Gradungulidae) with observations on the web and prey capture. *Proc. Linn. Soc. N.S.W.* 107:51–58
113. Greenstone, M. H. 1984. Determinants of web spider species diversity: vegetation structural diversity vs. prey availability. *Oecol.* 62:299–304
114. Griswold, C. E. 1986. A web-building oxyopid spider, *Tapinillus longipes* (Tac.), from Costa Rica (Abstr.) See Ref. 86a, p. 315
115. Griswold, C. E. 1987. A review of the

southern African spiders of the family Cyatholipidae Simon, 1894 (Araneae: Araneomorphae). *Ann. Natal. Mus.* 28: 499–542
116. Hallas, S. E. A., Jackson, R. R. 1986. Prey-holding abilities of the nests and webs of jumping spiders (Araneae, Salticidae). *J. Nat. Hist.* 20:881–94
117. Hansell, M. H. 1984. *Animal Architecture and Building Behaviour*. New York: Longman. 324 pp.
118. Heimer, S. 1986. From where are the Linyphiidae derived? Problems of Araneoidea phylogeny (Arachnida: Araneae). See Ref. 86a, pp. 117–20
119. Heimer, S., Nentwig, W. 1982. Thoughts on the phylogeny of the Araneoidea Latrielle, 1806. (Arachnida, Araneae). *Z. Zool. Syst. Evolutionforsch.* 20:284–95
120. Hickman, V. V. 1967. *Some Common Tasmanian Spiders.*
121. Hickman, V. V. 1975. On *Paraplectanoides crassipes* Keyserling (Araneae: Araneidae). *Bull. Br. Arachnol. Soc.* 3:166–74
122. Hieber, C. S. 1984. Orb-web orientation and modification by the spiders *Araneus diadematus* and *Araneus gemmoides* (Araneae Araneidae) in response to wind and light. *Z. Tierpsychol.* 65:250–60
123. Higgins, L. 1987. Time budget and prey of *Nephila clavipes* (Linnaeus) (Araneae, Araneidae) in southern Texas. *J. Arachnol.* 15:401–17
124. Horton, C. C. 1979. Apparent attraction of moths by webs of araneid spiders. *J. Arachnol.* 7:88
125. Horton, C. C. 1980. A defensive function for the stabilimenta of two orb weaving spiders (Araneae, Araneidae). *Psyche* 87:13–20
126. Horton, C. C., Wise, D. H. 1983. The experimental analysis of competition between syntopic species of orb-web spiders (Araneae: Araneidae). *Ecology* 64:929–44
127. Howell, F. G., Ellender, R. D. 1984. Observations on growth and diet of *Argiope aurantia* Lucas (Araneidae) in a successional habitat. *J. Arachnol.* 12: 29–36
128. Jackson, R. R. 1986. See Ref. 260, pp. 232–68
129. Jackson, R. R., Blest, A. D. 1982. The biology of *Portia fimbriata,* a web-building jumping spider (Araneae, Salticidae) from Queensland: utilization of webs and predatory versatility. *J. Zool. Lond.* 196:255–93
130. Janetos, A. C. 1983. Comparative ecology of two linyphiid spiders (Araneae, Linyphiidae). *J. Arachnol.* 11:315–22
131. Janetos, A. C. 1986. See Ref. 260, pp. 9–22
132. Job, W. 1974. Beitrage zur Biologie der fangnetz Wolfspinne *Aulonia albimana* (Walckenaer 1805). *Zool. Jb. Syst.* 101:560–608
133. Jones, D. 1983. *The Larousse Guide to Spiders*. New York: Larousse. 320 pp.
134. Kaston, B. J. 1964. The evolution of spider webs. *Am. Zool.* 4:191–207
135. Kaston, B. J. 1972. Webmaking by young *Peucetia*. *Notes Arachnol. Southw.* 3:6
136. Kirchner, W. 1986. Das Netz der Zitterspinne (*Pholcus phalangioides* Fuesslin) (Araneae:Pholcidae). *Zool. Anz.* 216:151–69
137. Kovoor, J. 1987. See Ref. 192, pp. 160–86
138. Kovoor, J., Peters, H. M. 1988. The spinning apparatus of *Polenecia producta* (Araneae, Uloboridae): structure and biochemistry. *Zoomorphology* 108: 47–59
139. Krafft, B. 1978. The recording of vibratory signals performed by spiders during courtship. *Symp. Zool. Soc. Lond.* 42: 59–67
140. Kremer, P., Leborgne, R., Pasquet, A., Krafft, B. 1987. Interactions entre femelles de *Zygiella x-notata* (Clerck) (Araneae, Araneidae): influence sur la taille des toiles. *Biol. Behav.* 12:93–99
141. Kullmann, E. 1958. Beobachtung des Netzbaues und Beitrage zur Biologie von *Cyrtophora citricola* Forskal (Araneae: Araneidae). *Zool. Jb. (Syst.)* 86:181–216
142. Kullmann, E. 1964. Neue Ergibnisse uber den Netzbau und das Sexualverhalten einiger Spinnenarten. *Z. Zool. Syst. Evolutionsforsch.* 2:41–122
143. Kullmann, E. 1971. Bemerkenswerte Konvergenzen im Verhalten cribellater und ecribellater Spinnen. *Freunde Kolner Zoo* 13:123–50
144. Kullmann, E. 1972. The convergent development of orb-webs in cribellate and ecribellate spiders. *Amer. Zool.* 12:395–405
145. Kullmann, E. 1975. Nets in Nature. In *Nets in Nature and Technics,* ed. K. Bach, pp. 319–78. Stuttgart: Fink KG. 430 pp.
146. Lahmann, E., Eberhard, W. G. 1979. La biologia de la araña colonial *Philoponella semiplumosa* (Uloboridae). *Rev. Biol. Trop.* 27:231–40
147. Lamoral, B. H. 1968. On the nest and web structure of *Latrodectus* in South Africa, and some observations on body

colouration of *L. geometricus* (Araneae: Theridiidae). *Ann. Natal. Mus.* 20:1–14
148. Langer, R. M. 1969. Elementary physics and spider webs. *Am. Zool.* 9:81–89
149. Leborgne, R., Pasquet, A. 1987. Influences of aggregative behaviour on space occupation in the spider *Zygiella x-notata* (Clerck). *Behav. Ecol. Sociobiol.* 20:203–8
150. LeGuelte, L. 1966. *Structure de la Toile de Zygiella x-notata Cl. et Facteurs que Régissent le Comportement de l'Araignée pendant la Construction de la Toile.* PhD thesis, Univ. Nancy. 77 pp.
151. Lenler-Eriksen, P. 1969. The hunting web of the young *Pisaurina mirabilis. J. Zool.*, Lond. 157:391–98
152. LeSar, C. D., Unzicker, J. D. 1978. Life history, habits, and prey preferences of *Tetragnatha laboriosa* (Araneae: Tetragnathidae). *Environ. Entomol.* 7:879–84
153. Levi, H. H. 1978. Orb-webs and phylogeny of orb-weavers. *Symp. Zool. Soc. Lond.* 42:1–15
154. Levi, H. W. 1980. Orb-webs: primitive or specialized. *Proc. Int. Arach. Congr, 8th, Vienna,* pp. 367–70
155. Levi, H. W. 1981. The American orb-weaver genera *Dolichognatha* and *Tetragnatha* North of Mexico (Araneae: Araneidae, Tetragnathinae). *Bull. Mus. Comp. Zool.* 149:271–318
156. Levi, H. W. 1986. The orb-weaver genus *Witica* (Araneae: Araneidae). *Psyche* 93:35–46
157. Levi, H. W. 1988. The neotropical orb-weaving spiders of the genus *Alpaida* (Araneae: Araneidae). *Bull. Mus. comp. Zool.* 151:365–487
158. Levi, H. W., Coddington, J. A. 1983. Progress report on the phylogeny of the orb-weaving families Araneidae and the superfamily Araneoidea (Arachnida: Araneae) (abstract). *Verh. Naturwiss. Ver. Hamburg* 26:151–54
159. Liddle, C., Putnam, J. P., Lewter, O. L., Lewis, J. Y., Bell, B., et al. 1986. Effect of 9.6-GH3 pulsed microwaves on the orb web spinning ability of the cross spider. (*Araneus diadematus*). *Bioelectro Mag.* 7:101–5
160. Lubin, Y. D. 1973. Web structure and function: the nonadhesive orb-web of *Cyrtophora moluccensis* (Doleschall) (Araneae: Araneidae). *Forma Funct.* 6:337–58
161. Lubin, Y. D. 1974. Adaptive advantages and the evolution of colony formation in *Cyrtophora* (Araneae: Araneidae). *Zool. J. Linn. Soc.* 54:321–39
162. Lubin, Y. D. 1978. Seasonal abundance and diversity of web-building spiders in relation to habitat structure on Barro Colorado Island, Panama. *J. Arachnol.* 6:31–52
163. Lubin, Y. D. 1986. See Ref. 260, pp. 132–71
164. Lubin, Y. D., Dorugl, S. 1982. Effectiveness of single-thread webs as insect traps: sticky trap models. *Bull. Br. Arachnol. Soc.* 5:399–407
165. Lubin, Y. D., Eberhard, W. G., Montgomery, G. G. 1978. Webs of *Miagrammopes* (Araneae: Uloboridae) in the Neotropics. *Psyche* 85:1–23
166. Lubin, Y. D., Opell, B. D., Eberhard, W. G., Levi, H. W. 1982. Orb plus cone webs in Uloboridae (Araneae) with a description of a new genus and four new species. *Psyche* 89:29–64
167. Main, B. 1976. *Spiders.* London: Collins. 296 pp.
168. Marples, B. J. 1962. Notes on spiders of the family Uloboridae. *Ann. Zool., Agra* 4:1–11
169. Marples, B. J. 1962. The Matachiinae, a group of cribellate spiders. *J. Linn. Soc. Zool.* 44:701–20
170. Marples, B. J. 1969. Observatations on decorated webs. *Bull. Br. Arachnol. Soc.* 1:13–18
171. Marples, M. J., Marples, B. J. 1937. Notes on the spiders *Hyptiotes paradoxus* and *Cyclosa conica. Proc. Zool. Soc. Lond.* 107:213–21
172. Marples, R. R. 1959. The dictynid spiders of New Zealand. *Trans. R. Soc. N. Z.* 87:333–61
173. Martin, D. 1974. Morphologie und Biologie der Kugelspinne *Achaearanea simulans* (Thorell, 1875) (Araneae: Theridiidae). *Mitt. Zool. Mus. Berlin* 50:251–62
174. Marusik, Y. M. 1987. Comparative studies of nets of orb-webs (Aranei, Araneidae, Tetragnathidae, Uloboridae) from the Lagodekhsky Reserve. *Vestn. Zool.* (Kiev) 1987:83–86 (in Russian).
175. Masters, W. M., Markl, H. S., Moffat, A. J. M. 1986. See Ref. 260, pp. 49–69
176. Masters, W. M., Moffat, A. J. M. 1983. A functional explanation of top-bottom asymmetry in vertical orbwebs. *Anim. Behav.* 31:1043–46
177. McReynolds, C. N., Polis, G. A. 1987. Ecomorphological factors influencing prey use by two sympatric species of orb-web spiders, *Argiope aurantia* and *Argiope trifasciata* (Araneidae). *J. Arachnol.* 15:371–83
178. Millidge, A. F. 1988. The relatives of the Linyphiidae: phylogenetic problems at the family level (Araneae). *Bull. Br. Arachnol. Soc.* 7:253–68

179. Mora, G. 1986. Use of web by *Tapinillus longipes* (Araneae: Oxyopidae). See Ref. 86a, pp. 173–75
180. Muma, M. H. 1971. Biological and behavioral notes on *Gasteracantha cancriformis* (Arachnida:Araneidae). *Fla. Entomol.* 54:345–51
181. Murphy, J., Murphy, F. 1983. The orb weaver genus *Acusilas* (Araneae, Araneidae). *Bull Br. Arachnol. Soc.* 6:115–23
182. Neet, C. R. 1986. Distribution horizontale, activite-predatrice et regime alimentaire de *Tetagnatha extensa* (L.) dans une tourbiere du Haut-Jura (Araneae, Tetragnathidae). *Bull. Soc. Ent. Suisse* 59:169–76
183. Neet, C. R. 1987. Selection de l'habitat chez l'araignée orbitele *Tetragnatha extensa* (L.) (Araneae:Tetragnathidae). *Bull. Romand Entomol.* 5:93–102
184. Neff, N. A. 1986. A rational basis for *a priori* character weighting. *Syst. Zool.* 35:110–23
185. Nentwig, W. 1982. Why do only certain insects escape from a spider's web? *Oecology* 53:412–17
186. Nentwig, W. 1983. The non-filter function of orb-webs in spiders. *Oecology* 58:418–20
187. Nentwig, W. 1985. Top-bottom asymmetry in vertical orbwebs: a functional explanation and attendant complications. *Oecology* 67:111–12
188. Nentwig, W. 1985. *Architis nitidopilosa,* a neotropical pisaurid with a permanent catching web (Araneae, Pisauridae). *Bull. Br. Arachnol. Soc.* 6:297–303
189. Nentwig, W. 1985. Spiders eat crickets artificially poisoned with KCN and change composition of their digestive fluid. *Naturwiss.* 72:545–46
190. Nentwig, W. 1985. Prey analysis of four species of tropical orb-weaving spiders (Araneae: Araneidae) and a comparison with araneids of the temperate zone. *Oecology* 66:580–94
191. Nentwig, W. 1985. Obligate kleptoparasitic behaviour of female flies at spider webs (Diptera: Empididae:Microphoridae). *Zool. Anz., Jena* 215:348–54
192. Nentwig, W. 1987. *Ecophysiology of Spiders*. New York: Springer. 448 pp.
193. Nentwig, W. 1987. See Ref. 192, pp. 249–63
194. Nentwig, W., Christenson, T. E. 1986. Natural history of the non-solitary sheetweaving spider *Anelosimus jocundus* (Araneae:Theridiidae). *Zool. J. Linn. Soc.* 87:27–35
195. Nentwig, W., Heimer, S. 1983. Orb webs and single-line webs: an economic consequence of space web reduction in spiders. *Z. Zool. Syst. Evolutionforsch.* 21:26–37
196. Nentwig, W., Heimer, S. 1987. See Ref. 192, pp. 211–25
197. Nentwig, W., Rogg, H. 1988. The cross stabilimentum of *Arigiope argentata* (Araneae:Araneidae)—nonfunctional or a nonspecific stress reaction? *Zool. Anz.* 221:248–66
198. Nentwig, W., Spiegel, H. 1986. The partial web renewal behaviour of *Nephila clavipes* (Araneae:Araneidae). *Zool. Anz.* 216:351–56
199. Nielsen, E. 1932. *The Biology of Spiders,* Vol. 2. Copenhagen: Levin & Munksgaard. 723 pp.
200. Nuessly, G. S., Goeden, R. D. 1984. Aspects of the biology and ecology of *Diguetia mojavea* Gertsch (Araneae, Diguetidae). *J. Arachnol.* 12:75–85
201. Nyffler, M., Benz, G. 1978. Die Beutespektrn der Netzspinnen *Argiope bruennichi* (Scop.), *Araneus quadratus* Cl. und *Agelena labyrinthica* (Cl.) in Odlandwiesen bei Zurich. *Rev. Susse Zool.* 85:747–57
202. Nyffler, M., Benz, G. 1979. Zur okologischen Bedeutung der Spinnen der Vegetationsschicht von Getriede- und Rapsfeldern bei Zurich (Schweiz). *Z. Ang. Entomol.* 87:348–76
203. Nyffeler, M., Dean, D. A., Sterling, W. H. 1988. Prey records of the web-building spiders *Dictyna segregata* (Dictynidae), *Theridion australe* (Theridiidae), *Tidarren haemorrhoidale* (Theridiidae), and *Frontinella pyramitela* (Linyphiidae) in a cotton agroecosystem. *Southwest. Natur.* 33: 215–18
204. Deleted in proof
205. Olive, C. 1980. Foraging specializations in orb-weaving spiders. *Ecology* 61: 1133–44
206. Olive, C. W. 1982. Behavioral response of a sit-and-wait predator to spatial variation in foraging gain. *Ecology* 63:912–20
207. Opell, B. D. 1982. Post-hatching development and web production of *Hyptiotes cavatus* (Hentz) (Araneae: Uloboridae). *J. Arachnol.* 10:185–91
208. Opell, B. D. 1982. Cribellum, calamistrum and ventral comb ontogeny in *Hyptiotes cavatus* (Hentz) (Araneae: Uloboridae). *Bull. Br. Arachnol. Soc.* 5:338–43
209. Opell, B. D. 1983. A review of the genus *Tangaroa* (Araneae, Uloboridae). *J. Arachnol.* 11:287–95

210. Opell, B. D. 1986. Webs and web-builders. *Science* 234:1593–94
211. Opell, B. D. 1987. The new species *Philoponella heredia*e and its modified orb-web (Araneae, Uloboridae). *J. Arachnol.* 15:59–63
212. Opell, B. D. 1987. The influence of web monitoring tactics on the tracheal systems of spiders in the family Uloboridae (Arachnidae, Araneida). *Zoomorph.* 107:255–59
213. Opell, B. D. 1989. Functional associations between the cribellum spinning plate and capture threads of *Miagrammopes animotus* (Araneida, Uloboridae). *Zoomorphology* 108:263–67
214. Opell, B. D. 1990. The material investment and prey capture potential of nonstereotypic spider webs. In preparation
215. Opell, B. D., Beatty, J. A. 1976. The Nearctic Hahniidae (Arachnida: Araneae). *Bull. Mus. Comp. Zool.* 147:393–433
216. Opell, B. D., Ware, A. D. 1987. Changes in visual fields associated with web reduction in the spider family Uloboridae. *J. Morph.* 192:87–100
217. Osaki, S. 1989. Seasonal change in color of spiders' silk. *Acta Arachnol.* 38:21–28
218. Palmer, J. M. 1985. The silk and silk production system of the funnel-web mygalomorph spider *Euagrus* (Araneae, Dipluridae). *J. Morph.* 186:195–207
219. Pasquet, A. 1984. Predatory-site selection and adaptation of the trap in four species of orb-weaving spiders. *Biol. Behav.* 9:3–19
220. Pasquet, A., Leborgne, R. 1984. Etude preliminaire des relations prédateur-prois chez *Zygiella x-notata* (Araneae, Argiopidae). *C. R. Soc. Biol.* 180:347–53
221. Peaslee, J. E., Peck, W. B. 1983. The biology of *Octonoba octonarius* (Muma) (Araneae: Uloboridae). *J. Arachnol.* 11: 51–67
222. Peters, H. M. 1953. Beitrage zur vergleichenden Ethologie und Okologie tropischer Webespinnen. *Z. Morph. Okol. Tiere* 42:278–306
223. Peters, H. M. 1983. Struktur und Herstellung der Fangfaden cribellater Spinnen (Arachnida: Araneae). *Verh. Naturwiss. Ver. Hamburg* 26:241–53
224. Peters, H. M. 1987. See Ref. 192, pp. 187–202
225. Platnick, N. I. 1989. A revision of the spider genus *Segestrioides* (Araneae, Diguetidae). *Am. Mus. Novit.* 2940:1–9
226. Porter, J. P. 1906. The habits, instincts, and mental powers of spiders, genera *Argiope* and *Epeira*. *Am. J. Psychol.* 17:306–57
227. Prestwich, K. N. 1977. The energetics of web-building in spiders. *Comp. Biochem. Physiol.* 57A:321–26
228. Ramousse, R., LeGuelte, L. 1984. Strategies de construction de la toile chez deux espèces d'araignées (*Araneus diadematus* et *Zygiella x-notata*). *Rev. Arachnol.* 5:255–65
229. Ramousse, R., LeGuelte, L., LeBerre, M. 1981. Organisation temporelle du comportement constructeur chez les Argiopidae. *Atti Soc. Tosc. Sci. Nat., Mem.* ser. B 88 (Suppl):159–72
230. Reed, C. F., Witt, P. N., Scarboro, M. B., Peakall, D. B. 1970. Experience and the orb web. *Dev. Psychobiol.* 3:251–65
231. Riechert, S. E., Cady, A. B. 1983. Patterns of resource use and tests for competitive release in a spider community. *Ecology* 64:899–913
232. Riechert, S. E., Gillespie, R. G. 1986. See Ref. 260, pp. 23–48
233. Riechert, S. E., Harp, J. M. 1987. Nutritional ecology of spiders. In *Nutritional Ecology of Insects, Mites, and Spiders*, ed. F. Slansky, J. G. Rodriguez, pp. 645–72. New York: Wiley
234. Riechert, S., Luczak, J. 1982. See Ref. 320, pp. 353–65
235. Risch, P. 1977. Quantitative analysis of orb web patterns in four species of spiders. *Behav. Genet.* 7:199–238
236. Robinson, B., Robinson, M. H. 1978. Developmental studies of *Argiope argentata* (Fabricius) and *Argiope acmula* (Walckenaer). *Symp. Zool. Soc. Lond.* 42:31–40
237. Robinson, M. H. 1975. The evolution of predatory behaviour in araneid spiders. In *Function and Evolution in Behaviour*, ed. G. Baerends, C. Beer, A. Manning, pp. 292–312. Oxford: Clarendon
238. Robinson, M. H. 1982. Courtship and mating behavior in spiders. *Annu. Rev. Entomol.* 27:1–20
239. Robinson, M. H., Lubin, Y. D. 1979. Specialists and generalists: the ecology and behavior of some web-building spiders from Papua New Guinea I. *Herennia ornatissima, Argiope ocyaloides* and *Arachnura Melanura*. *Pac. Ins.* 21:93–132
240. Robinson, M. H., Lubin, Y. D. 1979. Specialists and generalists: the ecology and behavior of some web building spiders from Papua New Guinea II. *Psechrus argentatus* and *Fecenia* sp. (Araneae:Psechridae). *Pac. Ins.* 21:133–64
241. Robinson, M. H., Robinson, B. 1970. The stabilimention of the orb web spi-

der. *Argiope argentata:* an improbable defence against predators. *Can. Entomol.* 102:641–55

242. Robinson, M. H., Robinson, B. 1972. The structure, possible function and origin of the remarkable ladder-web built by a New Guinea orb-web spider (Araneae:Araneidae). *J. Nat. Hist.* 6: 687–94
243. Robinson, M. H., Robinson, B. 1973. The stabilimenta of *Nephila clavipes* and the origins of stabilimentum-building in araneids. *Psyche* 80:277–88
244. Robinson, M. H., Robinson, B. 1973. Ecology and behavior of the giant wood spider *Nephila maculata* (Fabricius) in New Guinea. *Smithson. Contrib. Zool.* 149:1–76
245. Robinson, M. H., Robinson, B. 1975. Evolution beyond the orb web: the web of the araneid spider *Pasilobus* sp., its structure, operation and construction. *Zool. J. Linn. Soc.* 56:301–14
246. Rothermel, W. 1987. Spinnennetze als microskopische Praparate. *Mikrokosmos* 76:57–60
247. Rovner, J. S. 1976. Detritus stabilimenta on the webs of *Cyclosa turbinata* (Araneae, Araneidae). *J. Arachnol.* 4:215–16
248. Deleted in proof
249. Rypstra, A. L. 1981. Building a better insect trap; an experimental investigation of prey capture in a variety of spider webs. *Oecology* 52:1–6
250. Rypstra, A. L. 1985. Aggregations of *Nephila clavipes* (L.) (Araneae, Araneidae) in relation to prey availability. *J. Arachnol.* 13:71–78
251. Rypstra, A. L. 1986. Web spiders in temperate and tropical forests: relative abundance and environmental correlates. *Am. Midl. Nat.* 115:42–51
252. Santana, M., Eberhard, W. G., Bassey, G., Prestwitch, K. N., Bricẽno, R. D. 1990. Low predation rates in the field by the tropical spider *Tengella radiata* (Araneae:Tengellidae). *Biotropica.* In press
253. Schleiger, N. 1987. A clothes-peg variety of the orb-web spider *Araneus transmarinus*. *Vict. Nat.* 104:20–23
254. Schildknecht, H., Kunzelmann, P., Krauss, D., Kuhn, C. 1972. Uber die Chemie der Spinneweb, I Arthropodenabwehrstoffe, *Naturwissenschaften* 59:98–99
255. Schmidt, G. E. W. 1986. Observations on spiders from Sri Lanka and Reunion. See Ref. 86a, pp. 261–64
256. Schoener, T. W., Toft, C. A. 1983. Dispersion of a small-island population of the spider *Metepeira datona* (Araneae: Araneidae) in relation to website availability. *Behav. Ecol. Sociobiol.* 12:121–28
257. Seibt, U., Wickler, W. 1988. Bionomics and social structure of 'Family Spiders' of the genus *Stegodyphus* with special reference to the African species *S. dumicola and S. mimosarum* (Araneae, Eresidae). *Verh. Naturwiss. Ver. Hamburg* 30:255–303
258. Shaffer, H. B. 1986. Utility of quantitative genetic parameters in character weighting. *Syst. Zool.* 35:124–34
259. Shear, W. A. 1969. Observations of the predatory behavior of the spider *Hypochilus gertschi* Hoffman (Hypochilidae). *Psyche* 76:407–17
260. Shear, W. A. ed. 1986. *Spiders, Webs, Behavior and Evolution.* Palo Alto: Stanford Univ. 492 pp.
261. Shear, W. A. 1986. See Ref. 260, pp. 364–402
262. Shelley, T. E. 1984. Prey selection by the neotropical spider *Micrathena schreibersi* with notes on web-site tenacity. *Proc. Entomol. Soc. Wash.* 86:493–502
263. Shinkai, A. 1982. Web structure of *Nephila clavata* (1). *Atypus* 80:1–10 (in Japanese)
264. Shinkai, A. 1985. Comparison in the web structure between *Nephila clavata* L. Koch and *Nephila maculata* (Fabricius) (Araneae:Araneidae), and the origin of genus *Nephila*. *Acta Arachnol.* 34:11–22. (in Japanese)
265. Shinkai, A. 1988. Web structure of *Tetragnatha lauta* Yaginuma. *Kishidaia* 56:15–18 (in Japanese)

265a. Shinkai, A. 1988. Single line web of *Phoroncidia pilula* (karsch), and its prey insects. *Atypus* 92:37–39 (in Japanese)

266. Shinkai, A., Shinkai, E. 1985. The webbuilding behavior and predatory behavior of *Theridiosoma epeiroides* Bosenberg et Strand (Araneae:Theridiosomatidae) and the origin of the rayformed web. *Acta Arachnol.* 33:9–17
267. Shinkai, A., Shinkai, E. 1988. Web structure of *Conoculus lyugadinus* Komatsu (Araneae:Anapidae). *Acta Arachnol.* 37:1–12. (in Japanese)
268. Shinkai, E. 1984. *A Field Guide to the Spider of Japan.* Tokai Univ. Press: Tokai. 206 pp. (in Japanese)
269. Sinkai, E. 1989. Classification of web types in weaving spiders of Japan. *Arachnol. Pap. Pres. Yaginuma, Osaka.* 1:153–79. (in Japanese)
270. Shinkai, E., Shinkai, A. 1981. Hunters with thread. *Anima* 102:50–56. (in Japanese)
271. Shinkai, E., Takano, S. 1987. *Spiders.*

Shinrin Shobo, Ltd. 128 pp. (in Japanese)

272. Smith, D. R. 1985. Habitat use by colonies of *Philoponella republicana* (Araneae, Uloboridae). *J. Arachnol.* 13:363–73
273. Spiller, D. A. 1984. Competition between two spiders species: an experimental field study. *Ecology* 65:909–19
274. Spiller, D. A. 1986. Interspecific competition between spiders and its relevance to biological control by general predators. *Environ. Entomol.* 15:177–81
275. Spiller, D. A., Schoener, T. W. 1988. An experimental study of the effect of lizards on web-spider communities. *Ecol. Monogr.* 58:57–77
276. Stowe, M. K. 1978. Observations of two nocturnal orb weavers that build specialized webs: *Scoloderus cordatus* and *Wixia ectypa* (Araneae:Araneidae). *J. Arachnol.* 6:141–46
277. Stowe, M. K. 1986. See Ref. 260, pp. 101–31
278. Strohmenger, T., Nentwig, W. 1987. Adhesive and trapping properties of silk from different spider species. *Zool. Ans.* 218:9–16
279. Suter, R. B. 1978. *Cyclosa turbinata* (Araneae, Araneidae): prey discrimination via web-borne vibrations. *Behav. Ecol. Sociobiol.* 3:283–96
280. Suter, R. B. 1984. Web tension and gravity as cues in spider orientation. *Behav. Ecol. Sociobiol.* 16:31–36
281. Suter, R. B., Hirscheimer, A. J., Shane, C. 1987. Senescence of web construction behavior in male *Frontinella pyramitela* (Araneae, Linyphiidae). *J. Arachnol.* 15:177–83
282. Szlep, R. 1965. The web-spinning process and web-structure of *Latrodectus tredecinguttatus, L. pallidus* and *L. revivensis*. *Proc. Zool. Soc.* Lond. 145: 75–89
283. Szlep, R. 1966. Evolution of the web-spinning activities; the web spinning in *Titanoeca albomaculata* Luc. (Araneae: Amaurobiidae). *Israel J. Zool.* 15:83–88
284. Szlep, R. 1966. The web structure of *Latrodectus variolus* Walckener and *L. bishopi* Kaston. *Israel J. Zool.* 15:89–94
285. Tietjen, W. J., Ayyagari, L. R., Uetz, G. W. 1987. Symbiosis between social spiders and yeast: the role in prey attraction. *Psyche* 94:151–58
286. Tillinghast, E. K. 1984. The chemical fractionation of the orb web of *Argiope spiders*. *Ins. Biochem.* 14:115–20
287. Tillinghast, E. K., Christenson, T. E. 1984. Observations on the chemical composition of the web of *Nephila clavipes* (Araneae, Araneidae). *J. Arachnol.* 12:69–74
288. Tillinghast, E. K., Huxtable, R. S., Watson, W. H., Townley, M. A. 1987. Evidence for the presence of gabamide on the web of orb weaving spiders. *Comp. Biochem. Physiol.* 88B:457–60
289. Tillinghast, E. K., Kavanaugh, E. S., Kolbjornsen, P. H. 1981. Carbohydrates in the webs of *Argiope* spiders. *J. Morph.* 169:141–48
290. Tillinghast, E. K., Townley, M. 1987. See Ref. 192, pp. 203–10
291. Tilquin, A. 1942. *La Toile Geometrique des Araignees*. Paris: Presses Univ. 529 pp.
292. Toft, S. 1980. Humidity retaining function of the catching web of *Tapinopa longidens* (Wider) (Araneae:Linyphiidae). *Ent. Meddr.* 48:5–7
293. Tretzel, E. 1961. Biologie, Okologie und Brutpflege von *Coelotes terrestris* (Wider) (Araneae, Agelenidae) Teil 1:Biologie und Okologie. *Z. Morph. Okol. Tiere.* 49:658–745
294. Turnbull, A. L. 1960. The prey of the spider *Linyphia triangularis* (Clerck) (Araneae, Linyphidae). *Can. J. Zool.* 38:859–73
295. Turnbull, A. L. 1962. Quantitative studies of the food of *Linyphia triangularis*. *Can. Entomol.* 94:1233–49
296. Tyschenko, V. P. 1984. The catching webs of orb-weaving spiders 1. The substantiation of the method of standard webs with reference to two species of the genus *Araneus* (Aranei, Araneidae). *Zool. Zhurn.* 63:839–47 (in Russian)
297. Tyschenko, V. P. 1985. A quantitative analysis of the catching webs of orb-weaving spiders. *Proc. Zool. Inst. Leningrad* 139:17–26 (in Russian)
298. Tyshchenko, V. P. 1986. New confirmation of the convergent origin of orb webs in Cribellate and ecribellate spiders. *Doklady Akad. Nauk SSSR.* 287: 1270–73. (in Russian)
299. Tyshchenko, V. P., Marusik, Y. M. 1985. Catching webs of orb-weaving spiders. 3. Geographic variation of webs in *Araneus marmoreus* (Aranei, Araneidae). *Zool. Zhurn.* 64:1816–22 (in Russian)
300. Tyshchenko, V. P., Marusik, Y. M., Tarabaev, C. K. 1985. The catching webs of orb-weaving spiders 2. Comparative study of the webs in the genus *Nuctenea* (Aranei, Araneidae). *Zool. Zhurn.* 64:827–34 (in Russian)
301. Uetz, G. W. 1986. See Ref. 260, pp. 207–31
302. Uetz, G. W. 1990. The "ricochet effect" and prey capture. *Oecologia*. Submitted

303. Uetz, G. W., Biere, J. M. 1980. Prey of *Micrathena gracilis* (Walckenaer) (Araneae:Araneidae) in comparison with artificial webs and other trapping devices. *Bull. Br. Arachnol. Soc.* 5:101–7
304. Uetz, G. W., Hartsock, S. P. 1987. Prey selection in an orb-weaving spider: *Micrathena gracilis* (Araneae:Araneidae). *Psyche* 94:103–16
305. Uetz, G. W., Johnson, A. D., Schemske, D. W. 1978. Web placement, web structure and prey capture in orb-weavings spiders. *Bull. Br. Arachnol. Soc.* 4:141–48
306. Valerio, C. E. 1974. Prey capture by *Drymusa dinora* (Araneae, Scytodidae). *Psyche* 81:284–87
307. Vollrath, F. 1987. Altered geometry of webs in spiders with regenerated legs. *Nature* 328:247–48
308. Vollrath, F. 1988. Untangling the spider's web. *Trends Ecol. Evol.* 3:331–35
309. Vollrath, F., Mohren, W. 1985. Spiral geometry in the garden spider's orb web. *Naturwissenschaften* 72:666–67
310. Whitehouse, M. E. A. 1987. "Spider eat spider": the predatory behavior of *Romphaea* sp. from New Zealand. *J. Arachnol.* 15:355–62
311. Wiehle, H. 1931. Neue Beitrage zur Kenntnis des Fanggewebes der Spinnen aus den Familien Argiopidae, Uloboridae und Theridiidae. *Z. Morph. Okol. Tiere* 23:349–400
312. Williams, F. X. 1928. The natural history of a Nipa house with descriptions of new wasps. *Philip. J. Sci.* 35:53–118
313. Wirth, E. 1988. *Sensorische und mechanische Grundlagen des Netzbauverhaltens bei spinnen*. PhD thesis. Johann Wolfgang Goethe-Universitat, Frankfurt Am Main 85 pp.
314. Wise, D. H. 1984. The role of competition in spider communities: insights from field experiments with a model organism. In *Ecological Communities: Conceptual Issues and the Evidence*, ed. D. R. Strong, D. Simberloff, L. G. Abele, A. B.Thistle, pp. 42–53. Princeton: Princeton Univ. Press
315. Wise, D. H., Barata, J. L. 1983. Prey of two syntopic spiders with different web structures. *J. Arachnol.* 11:271–81
316. Witt, P. N. 1952. Ein einfaches Prinzip zur Deutung einiger Proportionen in Spinnenetz. *Behaviour* 4:172–89
317. Witt, P. N. 1965. Do we live in the best of all possible worlds? Spider webs suggest an answer. *Persp. Biol. Med.* 8:475–87
318. Witt, P. N., Baum, R. 1960. Changes in orb webs in spiders during growth (*Araneus diadematus* Clerck and *Neoscona vertebrata* McCook). *Behaviour* 16:309–18
319. Witt, P. N., Reed, C., Peakall, D. B. 1968. *A Spider's Web*. Springer, New York. 107 pp.
320. Witt, P. N., Rovner, J. S. 1982. *Spider Communication*. Princeton: Princeton Univ. Press. 440 pp.
321. Witt, P. N., Scarboro, M. B., Daniels, R., Peakall, D. B., Gause, R. L. 1977. Spider web building in outer space: evaluation of records from the Skylab spider experiment. *J. Arachnol.* 4:115–24
322. Work, R. W. 1985. Viscoelastic behaviour and wet supercontraction of major ampullate silk fibres of certain orb-web-building spiders (Araneae). *J. Exp. Biol.* 118:379–404
323. Work, R. W., Young, C. T. 1987. The amino acid compositions of major and minor ampullate silks of certain orb-web-building spiders (Araneae, Araneidae). *J. Arachnol.* 15:65–80
324. Yaginuma, T. 1966. Photographs of Japanese spiders. *Atypus* 41-2:1–8
325. Yaginuma, T. 1972. Evolution of spider webs. *Nat. Anim.* 2:2–6. (in Japanese)
326. Yoshida, M. 1987. Predatory behavior of *Tetragnatha praedonia* (Araneae: Tetragnathidae). *Acta Arachol.* 35:57–75
327. Yoshida, M. 1989. Predatory behavior of three Japanese species of *Metleucauge* (Araneae, Tetragnathidae). *J. Arachnol.* 17:15–25

*Annu. Rev. Ecol. Syst. 1990. 21:373–98*


# EXPERIMENTAL STUDIES OF NATURAL SELECTION IN BACTERIA

*Daniel E. Dykhuizen*

Department of Ecology and Evolution, State University of New York at Stony Brook, Stony Brook, New York 11794

KEY WORDS: natural selection, bacteria, chemostats, periodic selection, mutation rate, niche partitioning

---

## INTRODUCTION

Natural selection is the central concept in our formulation of evolutionary theory. Thus, it is important to study it and to understand its causes and effects by careful thought, observation, and experimentation. The effects—how selection changes gene frequencies—have been extensively studied by population geneticists. The causes—how genetic variation within an environment creates selective differences—have been less well studied because of many difficulties—such as understanding the development of the phenotype and defining the important components of the environment (19).

This review describes experiments with microorganisms that provide insight into the causes of natural selection and consequently the evolutionary process, insights that are difficult or impossible to obtain if evolutionary biology concentrates solely on multicellular eukaryotic organisms. A statement of the importance of studying microorganisms to increase understanding of the evolutionary process is required because of the near total exclusion of microbiology from the neo-Darwinian synthesis (61, 101). This exclusion was not intentional but occurred in part because bacterial species and their phylogenetic relationships were nearly impossible to define until recently. Consequently, microbiology has remained the least evolution-oriented of the



0066-4162/90/1120-0373$02.00

biological disciplines. However, there is an increasing interest in the evolution and ecology of microorganisms because of the possible ecological effects of various products of biotechnology (42, 90).

Microorganisms can be used to study natural selection experimentally. They are very small haploid, clonal organisms with short generation times, and they are easy to grow. Ten billion will fit into a flask containing 30 ml of defined media. Fifty to one hundred generations of bacteria can be packed into the time from Monday morning to Friday afternoon, compared to well over a year for a similar number of generations of drosophila.

The effect of the environment on the selection process will be emphasized in this review. Natural selection is distinguished from artificial selection by the role of the environment. In artificial selection, relative fitness is established by a person differentially removing phenotypes from the population. In natural selection, this function is performed by the environment. Since bacteria are small and relatively featureless, artificial selection is difficult and microbiologists manipulate the environment to select species from nature or mutants from a laboratory population, giving rise to a heuristic practice of natural selection. Since there was little conceptual interest in the causes of natural selection, few generalizations about these causes, have arisen. Generalizations similar to the postulate of Jannasch (45), however, suggest that quick growing species are selected in batch culture and slow growing species are selected in continuous culture. Reviews of mutant isolation (97) and enrichment techniques to select species (79, 83) describe the methods and practices used by microbiologists when employing natural selection.

Dykhuizen & Hartl (21) reviewed experimental studies of natural selection using microorganisms in chemostats. The present paper updates and extends that review for chromosomal genes but not for accessory elements such as plasmids, transposons, or temperate bacteriophage. The population biology of transposable elements in *Escherichia coli* has recently been reviewed (4), as have the population genetics of *E. coli* (37, 84) and of other bacteria (105). The work relating enzyme activity to fitness for the lactose pathway in *E. coli* using metabolic control theory (46) has been reviewed and discussed elsewhere (19, 35).

## *Laboratory Environments*

Three types of environments are used in the laboratory—the Petri dish, batch culture, and continuous culture. Each has advantages and disadvantages when studying natural selection.

PETRI DISH The agar surface in a petri dish provides a substrate for bacterial growth. Each cell placed on this surface grows into a colony of identical cells. Thus, a structured habitat is created, allowing spatial relationships to be maintained over many generations. By replicate plating, these relationships can be maintained indefinitely. Large spatial patterns can be visualized using indicator plates (63).

Petri dishes have been used mainly when the fitness differences are large and the advantageous types are rare. In these cases the advantageous type forms a large, visible colony or papilla. Smaller fitness differences can be measured by washing the bacteria off the agar surface and replating to estimate ratios of types. This destroys the spatial structure.

BATCH CULTURE The growth cycle in batch culture is well characterized. When a small inoculum of nongrowing cells is added to a large volume of fresh medium, initially a lag period occurs, while the cells adapt physiologically to the new conditions. Any genetic adaptation during this period will be for cells that can physiologically adapt more quickly to the new conditions. Very shortly after growth starts, the cells start to grow exponentially, at the maximum growth rate which continues until some component of the medium becomes limiting. Any genetic adaptation during this period is expected to be for cells that divide more rapidly in the presence of excess resource. Then the growth rate quickly slows and very soon stops. This very short period would be expected to select for types of genetic adaptations similar to those in chemostats (below). After growth stops, the cells remain viable for a considerable length of time. Any genetic adaptation during this period is expected to be for survival during nongrowth. This period in most experiments is too short for the cells to suffer enough death to provide much opportunity for natural selection. Even though the cell cycle is complex, most of the selection in batch culture seems to be during the growth phase. In selection experiments, usually an inoculum of 0.1 ml ($10^8$ bacteria or fewer) is added to 9.9 ml of fresh medium in a culture tube. This results in 100-fold increase in cell numbers or 6.64 generations. Populations can be transferred serially day after day, for any desired number of generations.

Batch cultures or serial transfer cultures are different from dish cultures in that all the cells in the culture are physically isolated from each other, and they are all in the same stage of the growth cycle. Serial transfer cultures are easy to handle, and many different cultures can be maintained concurrently. Using serial transfer cultures, high repeatability is obtained, and selection coefficients as low as 0.5% per generation can be resolved as significantly different from zero (71). This precision is equal to that achieved in continuous cultures.

CONTINUOUS CULTURE Continuous cultures are distinguished from batch cultures because they are open systems in which nutrients are continually added and spent media plus cells removed. Various types of continuous cultures can be built (21, 79, 100). The most common one is the chemostat, a culture vessel to which fresh media is added at a constant rate and from which exhausted medium and living cells are removed at the same rate, such that a constant culture volume ($V$) is maintained. The rate of addition of media ($f$) is such that the dilution rate ($D = f/V$) is less than the maximum growth rate of the bacteria. The medium is constructed so that one of the added components, the limiting nutrient, is added in amounts that limit the number of bacteria. As the limiting nutrient is exhausted by growth, growth slows (41). Since fresh medium is continuously added, growth never stops. Instead the population is maintained near carrying capacity (K-type population) with competition for the limiting nutrient. The relationship between growth rate ($u$) and the concentration of the limiting nutrient ($S$) is adequately described by the conventional Michaelis saturation equation:

$$u(S) = u_{max}\, S/(K_S + S),$$

where $u_{max}$ is the theoretical maximum growth rate, and $K_S$ is the saturation constant. Nutrient is usually converted into cell density ($X$) by a constant yield coefficient ($y$), so $X = yS$. Thus, the chemostat equations become:

$$dS(t)/dt = S_o \cdot D - S(t) \cdot D - u(S)X(t)/y$$

$$dX(t)/dt = u(S) \cdot X(t) - D \cdot X(t),$$

where $S_o$ is the concentration of limiting nutrient in the fresh medium. At equilibrium, $u = D$ and $X = y(S_o - S)$. Thus, the population density can be set by the concentration of the limiting nutrient in the medium and the growth rate by the rate of addition of the limiting nutrient. The chemostat is very useful to hold the growth rate at some particular point between the maximal rate and zero. The growth rate can be held at the maximum in a turbidostat (10). The lag phase can be mimicked in continuous culture by pulsing the limiting nutrient (52). In distinction from batch culture, continuous culture holds the cells in the same physiological state all the time.

For both batch culture and continuous culture, selection coefficients can be calculated by

$$ln[x_1(t)/x_2(t)] = ln[x_1(0)/x_2(0)] + st,$$

where $x_1(t)$ and $x_2(t)$ represent the relevant density or number of the two competing strains at time $t$, where $t$ can be measured either in generations or hours, and $s$, the selection coefficient, which is then measured either per generation or per hour. Various neutral markers, such as the ability to ferment lactose or resistance to phage T5, are used to distinguish the two strains. These markers are neutral only in certain environments. In others, such as lactose limitation for the *lac* mutants or nitrogen limitation instead of sugar limitation for T5 resistance, the markers are selected. With the use of neutral markers that are as easy to score as those above, accurate selective coefficients can be estimated for subtle genetic differences like thermal stability variants.

SELECTION AND THE LABORATORY ENVIRONMENT The characteristics of selection are different in these various environments. Some strains of *E. coli* produce a toxin, a colicin, to which they themselves are resistant but to which other strains are sensitive. In a nonstructured environment, either batch culture (12) or chemostats (39), colicinogenic and sensitive *E. coli* show frequency dependent selection such that neither at low frequencies can invade a population of the other. In a structured environment like a Petri plate, a single coliciogenic cell can invade a population of sensitive cells (12). On the other hand, one cannot select for resistance to bacteriophage T2 in *E. coli* on plates, but one can select resistance quite easily in a chemostat (54).

Batch cultures and chemostats represent different types of environment. While batch culture provides a cyclic boom-bust environment, continuous culture provides a constant, nutrient-limited environment. When a specific medium is used to enrich particular organisms—for example, the ability to utilize thiosulfate as a source of energy—batch culture selects for opportunistic, quickly growing (*r* type) or zymogenous organisms which usually can utilize many other resources for growth, while the chemostat selects for slower growing, more specialized (*k* type) or autochthonous organisms which often can utilize only thiosulfate for growth (45).

## *Periodic Selection*

Since experimental populations of microorganisms are large, often containing $10^{10}$ individuals, advantageous mutations are expected each generation. The consequences of this were first investigated by Atwood et al (6) and were named periodic selection because, periodically, advantageous mutations eliminated the original type. These population changeovers are observed by tracking the replacement of the resident population. A mutant phenotype that is selectively neutral, like resistance to phage T5 (in a phage-free environment) or reversion to lactose fermentation (when a sugar other than lactose is used for growth), should increase linearly in a population of constant size because such mutations occur at a constant rate. The rate of increase is equal to the mutation rate (Figure 1), or $dq/dt = ut$ where $q$ is the frequency of the mutant, $u$ is the mutation rate, and $t$ is time. An advantageous mutation will most likely happen in a cell that does not have the neutral mutation because, although they accumulate with time, these cells occur at a very low frequency. The strain with the advantageous mutation will replace the original strain at the rate of selection, which will be log linear (Figure 1). The frequency of the neutral mutation drops as the strain with the advantageous mutation replaces the less fit strain. This is because the strain with the advantagous mutation, $P_1$, arose free of the neutral mutation after neutral mutations had already accumulated for a time in the previous population, $P_0$. $P_1$ simply has not had as much time to accumulate neutral mutations. The frequency of the neutral mutation will increase in the $P_1$ strain at the same rate as before until another strain, $P_{12}$, with another advantageous mutation, replaces it and the frequency of the neutral mutation drops again and the cycle is repeated.

THE CLASSICAL MODEL A model of the above process was suggested by Atwood et al (6) and developed mathematically by Moser (65). It assumes that adaptive mutations will accumulate sequentially, and between the adaptive shifts the population is virtually monomorphic (Figure 1). In the original population $P_0$, an advantageous mutation arises that replaces $P_0$ to form population $P_1$. One individual of this population has an advantageous mutation (mutation 2), and this strain replaces $P_1$ to form population $P_{12}$, which is

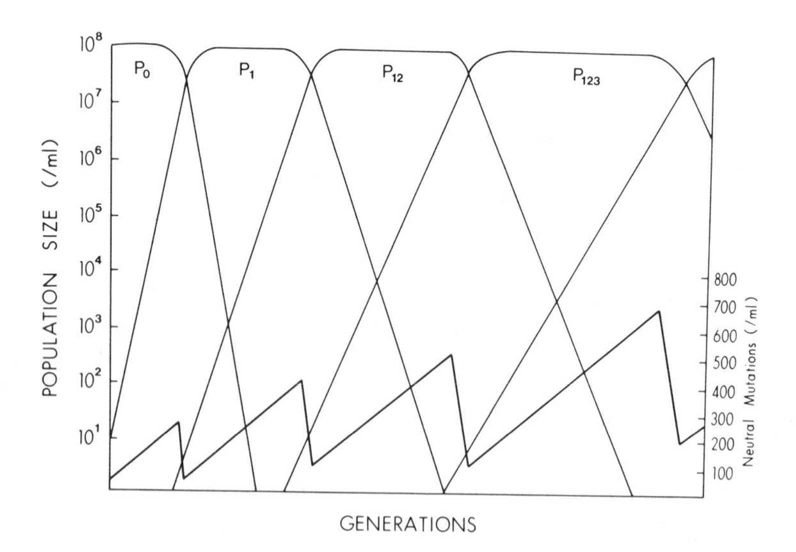


*Figure 1* The population dynamics ot periodic selection as traditionally viewed. The dark line at bottom represents the numbers of neutral mutations. The other lines are the numbers of the various populations that replace one another. The subscripts on the P are the advantageous mutations carried in that population.

two mutational steps different from the original strain. Likewise, in this population an advantageous mutation will arise (mutation 3), which will take over to form population $P_{123}$ and so on.

The time between population turnovers consists of two parts—a waiting time for the advantagous mutation to appear, and one in which this mutation increases in frequency by selection, from one cell, to replace the original population. Since the populations are large, the next advantageous mutation happens as the previous turnover is happening. So the waiting time for advantageous mutations is short, compared to the time it takes for the advantageous mutation to replace the original population. If the original frequency is $10^{-6}$ and the selection is 0.1 per generation, it will take 138 generations for the mutation to rise to 50%. It will take longer if the starting frequency is lower or the selection coefficient less.

This model has formed the basis for our thinking about periodic selection (49, 57) and has led to such interesting questions as: (*a*) Will the time between population turnovers get longer? (*b*) If one set up parallel populations, would the same mutations be selected at about the same time? Or, how regular is the evolutionary process?

This model resembles Fisher's model of the adaptive process (23; see pp. 135–37 of Ref. 48). Fisher assumes that there is a single optimal genotype in a constant environment. Advantageous mutations bring the population closer to this state, implying that the fitness increase of subsequent advantagous mutations will be less. After enough adaptive mutations are incorporated, the population will be so close to the fitness peak that further improvements are virtually impossible and population turnovers will cease. Thus Fisher's model predicts that the time between population turnovers will get longer and that parallel populations would arrive at the same fitness. This model also predicts that fitness will be transitive: If population B displaces A and C displaces B, then C will displace A. In addition, we expect that the differential in fitness of C over A will be larger than that of B over A. Fisher's model of the adaptive process is usually contrasted with Wright's model. Wright (104 and references therein) suggested there is much nontransitivity in fitness. Gene pools are coadapted, and frequency dependent selection will be important. Thus, there are many different optimal genotypes whose optimality depends upon the starting conditions. These various optimal genotypes need not be equally fit if set in competition with each other.

BATCH CULTURE In this and the next section, most of the studies of periodic selection are described. Much experimental detail is given because it is not clear how important initial conditions and the particularities of the environment are for the evolutionary outcome. The original experiments of Atwood et al (6) used serial dilutions (0.5 ml or $2.5 \times 10^8$ cells into 50 ml of fresh minimal medium every 12 hours). The neutral marker was reversion to histidine prototrophy ($h^+$) from histidine auxotrophy ($h^-$). The first population turnover took place between 200 and 300 generations. The $h^+$ cells isolated as they were being eliminated during the turnover were still selectively equivalent to the original $h^-_o$ cells, while the advantageous $h^-_1$ were competitively superior. The $h^+$ cells isolated a short time later were of equal fitness with the $h^-_1$ and were assumed to be derived from them. Two more rounds of advantageous mutations were isolated, $h^+_{12}$ and $h^+_{123}$, each giving a yet higher maximum growth rate. The fitness of the advantageous mutations isolated from different cultures at the same round were similar on the medium used for isolation, but very different on other media, suggesting they represent mutations in different genes.

Luckinbill (59) used the same strains and media, repeating the experiments of Atwood et al except for the culture conditions. He diluted one group of cultures 100-fold every 5–6 hours (*r* populations), so that the culture was constantly growing at its maximal rate; the other group was diluted 10-fold twice daily (*K* populations). The population thus cycled often between growth and nongrowth. Periodic selection was seen after 200–300 generations, with the time of onset more variable in the *K* populations, even though the effective population size was larger in these because more cells were carried over in each transfer. Advantageous mutations were isolated from 3 *r* populations and 3 *K* populations. These mutations had different fitnesses when compared to each other, but the order of strains ranked by the fitness in one culture condition was the same as the other, showing no trade-off in the adaptation to rapid maximal growth, compared to frequent cyclic growth. On complex medium, the advantageous mutations were generally selected against compared to the original strain, confirming the results of Atwood et al (6). L. S. Luckinbill (personal communication) showed that the adapted strains produce much more acetate during growth, and that the genetic change responsible for this excess acetate production and the adaptation in four strains, including two obtained from Atwood, mapped at 3 minutes, presumably in pyruvate dehydrogenase (*aceE* and *F*, 7). Since all the adapted strains show the elevated levels of acetate excretion, Luckinbill proposes that all adaptive mutations in the first round of selection are variations in the same complex enzyme, and the differences in growth on complex media are pleiotrophic effects of different mutations in this gene.

Are mutations of the same gene selected even in different genetic backgrounds? This question has not been answered. However, Lenski (55) repeated the experiment of Atwood et al using a different isolate of *E. coli* with different markers and saw periodic selection within 400 generations. He showed that alleles of the same gene are selected whether the background contains a maladaptive mutation or not. It is interesting that the advantageous mutation provides greater increase in fitness for the strain with the maladaptive mutation.

Another long-term serial dilution experiment (0.1 ml or $3 \times 10^6$ cells into 9.9 ml each day) was started with 12 replicate populations derived from a common ancestor and run for 2000 generations to test the classical model of periodic selection (56). Half the populations could ferment arabinose ($ara^+$) and half could not ($ara^-$). Rather than follow periodic selection using a neutral marker, and isolate clones after each turnover, workers isolated a single clone from each population each 100 generations. Successive clones could thus have the same fitness. Each clone was competed against the original strain to estimate the fitness increase of the population over the time before isolation. The average fitness of the 12 populations increased over time at a decreasing rate.

At each 500 generation interval, multiple clones were taken to analyze the intrapopulational variance. There was no evidence of significant intrapopulational variance in fitness, suggesting that the populations are homogeneous most of the time. The among-population variance was significant but did not increase with time. The data from this experiment could be satisfactorily modeled given these assumptions: random mutations, sequential adaptive substitutions with a waiting time for replacement, decreasing rate of fitness improvement, and eventual convergence of all populations to the same fitness. The model worked well when three substitutions were assumed over the 2000 generations, which is reasonable if the first one occurred between 200 and 300 generations and the rate of incorporation slowed over time. Overall, experiments using batch culture—in which most of the selection seems to be for maximal growth rate—support the classical model of periodic selection.

CONTINUOUS CULTURE Generally, the results of periodic selection experiments using chemostats do not support the classical model. The population changeovers happen in rapid succession, as if either the second advantageous mutation happened in the strain of the first while it was still rare or mutations are not being incorporated sequentially (49).

Dykhuizen & Hartl (20) presented additional evidence against sequential replacement of mutations. They removed populations from glucose-limited chemostats every 100 hr for 500 hr and charted the course of increase in competitive ability by competing each population with the population isolated 100 hours previously (Table 1). Fitness increased every hundred hours in every culture, which was not expected.

This experiment supported the classical model in that fitness seems to be transitive and there is a single adaptive peak. When the 500-hr cultures from chemostats 1 and 2 were competed, the culture from chemostat 1 was favored

**Table 1** The estimated selective increase (per hour) for 100 hour intervals in four parallel chemostat cultures.

| Time (hour) | Fitness increase over period | | | |
|---|---|---|---|---|
| Chemostat number | 1 | 2 | 3 | 4 |
| generation time (hour) | 2.5 | 2.5 | 5.0 | 5.0 |
| 0–100 | .1290[a] | .1389 | .0976 | .0531 |
| 100–200 | .0150 | .0073 | .0548 | .0902 |
| 200–300 | .0065 | .0057 | .0293 | .0172 |
| 300–400 | .0457 | .0280 | .0099 | .0159 |
| 400–500 | .0053 | .0136 | .0048 | .0017 |
| 0–500 (sum) | .2016 | .1935 | .1964 | .1781 |

[a] Values estimated from Table 1 of Ref. 36

($s$ = 0.0071 $hr^{-1}$) and likewise when 500-hr cultures from chemostats 3 and 4 were competed, the culture from chemostat 3 was favored ($s$ = .0142 $hr^{-1}$). These outcomes are predicted from the fitnesses summed over the 500 hours (Table 1). Little of the adaptation was specific to the generation time at which the cultures evolved (20). Thus, it was not surprising, given summed fitnesses on Table 1, that competition of 500-hour cultures from chemostats 2 and 3 run at 5.0 hours only slightly favored the one from chemostat 3 ($s$ = 0.00811 $hr^{-1}$) and competition of cultures from chemostats 1 and 4 run at 2.5 hr. strongly favored the one from chemostat 1 (s = 0.0512 $hr^{-1}$).

There is another way of visualizing periodic selection. If T5-resistant and T5-sensitive bacteria are mixed in equal proportions in a phage-free, glucose-limited chemostat, then the frequency should shift during periodic selection. If only one mutation were selected, then half the time it would appear in a T5-resistant cell and half the time in a T5-sensitive one. Thus, after the advantageous mutation took over, one genotype would dominate the population and the other would reappear only by mutation. This happens in batch culture (55, 59) but not in chemostats (36). Usually there is only a small change in frequency, often less than a factor of two (36; D. E. Dykhuizen, unpublished observations). This suggests that advantageous mutations are frequent, appearing multiple times in both strains, but it does not distinguish between many different mutations of a single gene or many mutations in many genes.

Eleven populations of the yeast, *Saccharomyces cerevisiae,* in glucose-limited chemostats (3.5 hr/generation) were run for up to 300 generations to test the rate of evolution in six haploid and five diploid populations (76–78). Clones were isolated after each periodic selection event as determined by the decrease in frequency of a neutral mutation. On the average, periodic selection events occurred every 37 generations. This rate remained constant over the 300 generations and was the same in both haploid and diploid populations

(76). Paquin & Adams (77) calculated the rate per cell division, which suggests that the frequency of adaptive mutations is higher in diploids than haploids since there are fewer diploids in a chemostat ($3 \times 10^9$) than haploids ($5 \times 10^9$) (76). With populations of this size, favorable mutations are expected every generation, and most of the time between periodic selection events would be due to the time for selective replacement. Thus the correct parameterization is in terms of events per generation and not in terms of per cell generation.

The result suggests that the fitness advantage of selected mutations is about equal in haploid and diploid populations. This result is surprising only if it is assumed that most advantageous mutations are recessive. However, mutations which give small differences in activity will be codominant even when the amorphic mutations are recessive (47).

Population turnovers every 37 generations are too fast. The average fitness increase was 0.09/generation (76). If we again take the population turnover when the advantageous strain has risen to 50% of the population, then with a selection coefficient of 9%, this strain was 3% of the population 37 generations before, at the previous turnover. This implies that the culture is heterogeneous and that advantageous mutations will not be selected sequentially, i.e. the population from the third turnover will not contain the mutation selected in the second.

Paquin & Adams (78) showed that fitness increases are not transitive. From one haploid and one diploid chemostat, they isolated five clones, representing the population after each population turnover, and mixed these clones to create a population representing that turnover. Each of these populations is competitively superior to the population from the previous turnover (Table 2). However, when each population was competed with the original strain, there were large reversals in fitness. In three cases, the evolved strain was even less fit than the original strain (Table 2). These data suggest a number of conclusions. First, not all adaptation increases the ability to recover and metabolize glucose, the limiting nutrient in these experiments. Adams et al (3) confirmed this for another series of adaptive clones. Second, this provides additional evidence that an adaptive strain is not always derived from the previous one. For example, adaptive strain 3 in the haploid cells of Table 2 could have an advantage over strain 2 because of a released toxin and strain 4 could be a derivative of 2 which is resistant to this toxin.

Adams & Oellar (2) have proposed a clever way to visualize the heterogeneity of these cultures. Yeast contains multiple copies of the mobile element, Ty. When the element moves into new sites, it will give new bands when the yeast DNA is probed with Ty during Southern analysis. Clones can thus be followed in the population. They saw 12 new bands from 100 clones analyzed. Most of the new bands were seen in only a single clone. This was

**Table 2** Estimated fitness increase (per gen) after each periodic selection event for two chemostat cultures[a].

| Population number turnover | Diploid | | Haploid | |
|---|---|---|---|---|
| | previous | original | previous | original |
| 1 | 0.18[b] | 0.18 | 0.10 | 0.10 |
| 2 | 0.05 | 0.08 | 0.05 | 0.11 |
| 3 | 0.10 | 0.13 | 0.14 | −0.16 |
| 4 | 0.12 | −0.03 | 0.08 | 0.13 |
| 5 | 0.05 | 0.07 | | |
| 6 | 0.03 | −0.23 | | |

[a] The fitness differences were for competition experiments between five clones isolated after a periodic selection event with the five clones isolated after the previous event and with the original strain before inoculation of the chemostat.
[b] data from Ref. 76

interpreted as evidence of culture heterogeneity. However, since the transposition rate per genome per generation is about $10^{-2}$ to $10^{-3}$ (9), the expected frequency of cells containing transpositions in a chemostat will be about 25 times this rate or 2.5% to 25%. This factor of 25 relates the per generation mutation rate to the frequency of neutral mutations in the chemostat. Thus, the frequency of transpositions is the expected frequency of a neutral mutation and does not imply culture heterogeneity. However, with intensive sampling, the method could illuminate the clonal structure of these populations.

THE BAROQUE MODEL Adams & Oellar (2) have suggested that the classical model of periodic selection is too simple and should be replaced by a more complex model.

With a base pair mutation rate of $2 \times 10^{-10}$ and $10^{10}$ cells, nearly every possible single step mutation may be produced every generation. Since an *E. coli* has about $4.5 \times 10^6$ base pairs, this represents $13.5 \times 10^6$ possible different single step mutations of which roughly $10^7$ would be present every generation. Many of these either will be selected against or will be selectively neutral (say half and half). Even if only 0.01% are advantageous, this is roughly 1000 different advantageous mutations each generation. Even if most are lost by drift, they would be produced again the next generation. The selection coefficients for these advantageous mutations would range from small to fairly large, and all would be increasing in frequency. Chao & Cox (11) estimated that the rate for advantageous mutations is $5 \times 10^{-9}$ or 5 advantageous mutations equivalents per generation. Most of the approximately 1000 different advantageous mutations generated per generation will not contribute much to this number because they will still be rare by the time of the population turnover. Only those with the largest selection coefficients—

those increasing fastest—dominate and change the culture conditions, changing the selection coefficients of all the other advantageous mutations. Presumably, most would now be selected against and decrease in frequency, but others could still be advantageous in the changed environment. These later ones would soon dominate, replacing the others. However, this takeover would not be easily visualized by a decrease in the neutral mutation because lineages for both turnovers have been extant for approximately equal times. This model can explain many of the anomalous results presented above. However, reality is even more complex than we have postulated as is shown in the next section.

BIOTIC INTERACTIONS Helling et al (40) ran glucose-limited chemostats (3.5 hr/generation) with *E. coli* for up to 1867 generations. There was no apparent decrease in the frequency of population turnovers during this period. However, after 800 generations, population turnovers became difficult to visualize using a neutral mutation; this fact suggested cultural heterogeneity. After 100 to 350 generations variants appeared that formed small colonies when plated. These small colony or S variants fluctuated in frequency from 5–80%. One chemostat, where the frequency of these S variants was particularly high (99% at generation 765), was investigated in detail. Four clones were isolated: two S variants and two normal. The clones giving large colonies grew like the initial strain in flasks with excess glucose. The two S variants differed from each other as well as from the others. One S variant, CV103, grew significantly slower and exhibited a final yield of about 55% of the other strains. The other S variant, CV116, exhibited a complex growth pattern, showing variation in growth rate, but it finally achieved the same yield as the non-S variants. When the original chemostat culture at generation 765 was analyzed, 60% of the clones were like CV103 and 40% like CV116. When these two strains were competed in a chemostat, they reached an equilibrium, with CV103, the strain that grows more slowly and has a lower yield, dominant. Thus there is coexistence of two "species" in the simplest environment possible—a homogenous, constant environment with one limiting resource. The coexistance can be explained. CV103 outcompetes CV116 for glucose, but CV103 excretes compounds which are used by CV116 for growth. Thus, through evolutionary change, an environment which seemingly could support only one "species" was divided to support more. If phenomena like this are common, a balanced model, with multiple coexisting adaptive states, will have to be proposed.

### *The Effect of Environmental Complexity*

Bacteria grown on mixed sugars in batch culture will often use the sugars sequentially (diauxic growth), with the substrate producing the most rapid growth used first (34). In chemostats with low dilution rates, both substrates are consumed at the same time. As the dilution rates increase to the maximum, the sugars will be preferentially underutilized (87, 88).

For the purposes of this review, I am conflating population genetics and population ecology because noncrossing clones of *E. coli* can be formally treated as two different species, and conversely, two different species can be thought of as different clones of the same species. If two resources are added to the chemostat along with two species, the species can coexist if they are growth limited by different nonsubstitutable resources such as phosphate and silicate (91) or by different substitutable resources such as glucose and *p*-hydroxybenzoate (81). Mixed substrate utilization has been reviewed by Gottschal (24). The most studied case is when one species or strain (the specialist) can use only one resource and is competitively dominant on it, and the other species or strain (the generalist) can use both (18, 25, 51, 53). Of the studies referenced above, one used isogenic strains of *E. coli* (18), and the others used different species. The result of these experiments is that a relatively small amount of the resource used by only the generalist allows it to dominate. For example, a generalist that is selectively inferior on the shared resource, maltose ($s = 5.3\%\ gen^{-1}$), maintains an equilibrium frequency of 60% if only 10% of the total sugar is sugar metabolized only by the generalist (18). The generalist consumes half of the maltose.

Experiments by Gottchall and his coworkers (25–29) used three species, two specialists and a generalist. The specialists are *Thiobacillus neopolitanus,* an autotroph, which can grow only on thiosulfate and a *Spirillum,* a heterotroph, which can grow only on acetate. The generalist was *Thiobacillus* A2, a mixotroph, which could grow on either resource. In two-species mixtures, the generalist was eliminated by the heterotroph on acetate. With small amounts of thiosulfate included, an equilibrium was established, until the molar ratio of thiosulfate to acetate became greater than .66. Then the heterotroph was eliminated. On thiosulfate, *Thiobacillus neopolitanus* dominated but did not eliminate the generalist (Figure 2). The generalist remained in the population because the specialist excreted glycollate which the generalist could use as an alternative energy source. As the molar ratio of thiosulfate to acetate became less than 3.7, the autotroph was eliminated. With a relatively small amount of unique resource, a species that is competitively inferior on the major resource can thus dominate and even exclude the other species. This is important for an understanding of the evolution of cometabolism (see below).

What happens when three species compete on two resources? Armstrong & McGehee (5) have demonstrated theoretically that three consumer species can not coexist at fixed densities on two resources. At least one species will be eliminated. Two-species competitions have shown that an equilibrium with one specialist and the generalist is impossible in the region where the ratio of resources is between 3.7 and .66 (25). The two remaining possibilities are that

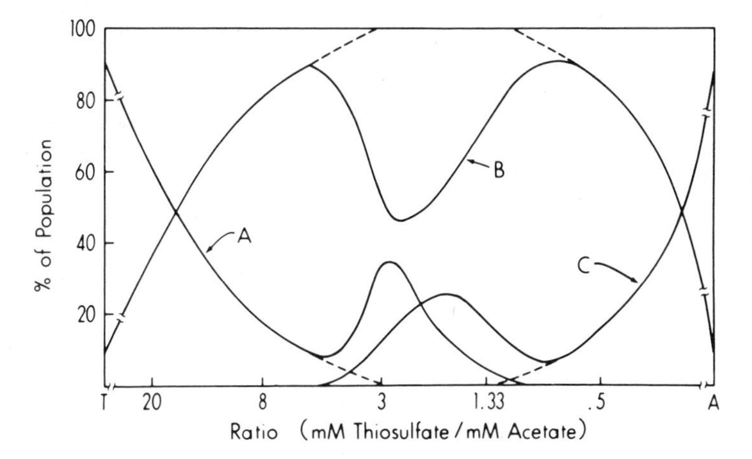


*Figure 2* The lines represent the equilibrium percentage of the three species. The A line represents *Thiobacillus neopolitanus;* the B line, *Thiobacillus* A2; and the C line a *Spirillum* species. The dotted lines represent the expected frequencies from the two species competitions where they deviate from the observed results of the three species competitions. On the abscissa, the T represents the experiment with only thiosulfate in the medium and the A, only acetate in the medium.

the generalist eliminates both specialists or that the specialists eliminate the generalist. The former is predicted theoretically using measured parameter values (29). However, neither happens (Figure 2). At ratios of about 2.5 to 3.0, there seems to be an equilibrium of about 40% generalist, 35% autotrophic specialist, and 25% heterotrophic specialist. This surprising result should be investigated further. Possible explanations are: (*a*) The species are not coexisting. Rather, the rates of elimination are so slow that it only appears that there is coexistence. (*b*) There are really three distinct resources—thiosulfate, acetate, and the excreted glycollate. (*c*) The resources are being treated in different ways by different species and thus are becoming effectively three resources (1). These would be thiosulfate alone, acetate alone, and the combination of them. The yield of the generalist on both resources was 30% higher than predicted from the sum of the yields on the separate substrates (26), implying the last explanation may be most important.

Additional experiments using varied environments (27) changed the dilution rate to 0.05 (g = 14 hr) from 0.075 for the experiments in the above paragraph. In all these experiments (27), two media were used, alternating between one and the other every four hours. The cycle time of the environmental variation was thus less than the generation time. When the substrate alteration was between pure acetate and pure thiosulfate, which would give a ratio of 4.0 if the media were combined (Figure 2), the generalist was eliminated and the two specialists were maintained in equal numbers. This result supports the hypothesis that the generalist uses the two resources together differently than the way it uses each separately. In the other two experiments, both media contained both resources but at different ratios. In the first experiment the alteration was between a medium with a ratio of 2.2 and one with a ratio of 6.4, to give a combined ratio of 5.16. The two specialists rose to about 45%, and the generalist dropped to about 10%. This compared to a frequency of about 90% for the generalist in a constant environment. In the second experiment, the alteration was between a medium with a ratio of 5.44 and one with a ratio of 0.54, giving a combined ratio of 2.13. This seemed to establish a long-term cycling. It is theoretically possible to obtain coexistence of three species on two resources if the species numbers are oscillating (62) or if the environment varies over time (58). Grover (30) tried to maintain two species of diatoms on limiting phosphorus by regularly pulsing the chemostat with phosphorus. This only slowed the rate of elimination of the inferior species.

## *Neutrality and Selection*

Chemostats were used to study the selective effects of naturally occurring allozyme polymorphisms in *E. coli*. Significant selection could not be detected except in novel environments (21, 37, 38). Of 29 tests of different allozymes at five loci, not one showed significant selection in glucose-limited chemostats, while of 24 tests of allozymes at 4 loci, 9 showed significant selection in chemostats limited for a sugar other than glucose. Since so many natural substrates contain glucose or are converted to glucose metabolites within the cell, while many of the other sugars tested seem to be rare in the environment, we postulated that all the allozymes were effectively neutral in the natural environment (38). The distribution of amino acid changes in 6-phosphogluconate dehydrogenase suggests that many of the allozymes are under weak selection (of the order of $10^{-7}$—84). Estimates of the effective population size indicate a value around $10^8$ to $10^9$ in nature (60, 74, 85). Mutants for which $s$ lies between about $10/N_e$ and $100/N_e$ are not nearly neutral in the usual sense (75)—it is very unlikely they will be fixed in the population. However, it is very likely they will be fixed in the population. However, it is very likely that they can drift to a significant fraction (>1%) of the population. It is interesting in this context that using the computer simulation method of Watterson (98), David Haymer (unpublished results) tested the electrophoretic data of Selander & Levin (86) for neutrality and found that 18 out of 20 loci showed an excess of the most frequent allele, even though none showed a significant deviation from neutrality by the Ewens criterion (22). This suggests that many of the rare alleles are detrimental, but with a very small selection coefficient ($100/N_e > s > 10/N_e$).

When the DNA sequences from a number of genes from *E. coli* and *Salmonella typhimuium* are compared, the rate of amino acid substitution is much less than the rate of change for synonymous substitutions. The ratio of nonsynonymous to synonymous nucleotide changes is 0.05 for bacteria, compared to the ratio of 0.20 for mammals (72, 73). This suggests that most amino acid changes are detrimental. If there is no selection on amino acid changes, the ratio would be about 2, so only about 2.5% of the changes in bacteria and 10% of the changes in mammals are sufficiently neutral or advantageous to be incorporated. The retardation in the rate of nonsynonymous changes in bacteria is probably larger than four-fold, compared to mammals, because there is more constraint on synonymous change (72).

The effective population size for many mammals is estimated to be $10^4$ to $10^5$ (67, 99). This smaller effective population size implies that a higher proportion of the amino acid changes in mammals are effectively neutral, not that mammalian proteins are less constrained. This reasoning implies that only about 1% of the amino acid changes have a fitness effect of less than $10^{-9}$, 10% have a fitness effect of less than $10^{-5}$, and the others have a larger effect. Of course, this varies considerably from protein to protein.

Estimation of the distribution of selection coefficients among newly arising mutants can also be done experimentally (17). We started with a series of nonsense mutations in the beta-galactosidase *(lacZ)* gene of *E. coli*. These mutations prematurely terminate synthesis of the protein, creating a truncated and usually inactive protein. A number of these nonsense mutations were also polar, meaning the distal genes on the polycistronic messenger RNA are not synthesized. The lactose permease is distal to beta-galactosidase. Any revertant that substitutes an amino acid in place of the polar nonsense mutation will allow synthesis of the permease even if the beta-galactosidase is completely inactive. Since the lactose permease can also transport melibiose, selection for melibiose transport allows selection for permease activity without selecting for function of the beta-galactosidase. This technique allows the estimation of the number of amino acid replacements in *lacZ* which inactivate the enzyme. Six polar nonsense mutations were studied, and the proportion of amino acid replacements that were inactive was 0% for three and 12%, 21%, and 96% for the other three, to give an unweighted average of about 20% (17). A beta-galactosidase mutation was judged to be active if it could be detected on indicator plates, a test which requires only about 0.5% of wild-type activity (33). Twenty-five mutations of beta-galactosidase which give a functional enzyme that is phenotypically different from wild-type were completed against wild-type in lactose-limited chemostats. Only three of these variants could be shown to be selectively different from wild-type (17). One of the three was even selectively superior to wild-type ($s = 6 \times 10^{-3}$/gen).

The limit of resolution in the chemostat is $s = 4 \times 10^{-3}$/gen. We think that the 22 mutations for which we cannot detect a significant selection coefficient would be, at most, weakly selected in nature. The intensity of selection on the lactose operon in the mosaic of natural environments is likely to be considerably less than in a lactose-limited chemostat. Using the observation that about 5% of naturally occurring *E. coli* are Lac−, Dean et al (17) estimated that the limit of resolution in a chemostat of $4 \times 10^{-3}$ is comparable to a selection coefficient of $3.2 \times 10^{-7}$ in nature. This suggests that most amino acid changes in most genes will have a small effect on fitness; whether or not they replace the present amino acid will depend on the effective population size.

## *Acquisitive Evolution*

Acquisitive or directed evolution refers to the evolution of new traits or metabolic abilities by alteration of existing genetic material. This field has been well reviewed in a recent book (64). Generally, enzymes used for new metabolic functions are recruited from other pathways that metabolize similar compounds. These enzymes usually handle the new compounds poorly compared to the related compounds normally metabolized. Mutations in the regulatory system are required so that the enzymes can be synthesized constitutively, since the new compound is not an inducer of the operon. The inefficiency of these recruited enzymes clearly limits growth. Growth rate is usually increased by producing more enzyme rather than improved enzyme. Mutations that improve the enzyme by changing kinetic parameters seem to be quite rare.

One system in which there was considerable evolutionary change of kinetic parameters is *ebg*, studied by Hall (31). The *ebg* (evolved beta-galactosidase) operon is located on the opposite side of the *E. coli* chromosome from the *lac* operon and requires at least two mutations before it can function as a lactamase in a strain deleted for *lacZ*, beta-galactosidase. One mutation is in the repressor, allowing the enzyme to be synthesized constitutively. The other changes the kinetic parameters of the enzyme. These later mutations fall into three classes, called I, II, and IV. Sequencing of three class-I and three class-II mutations has shown that all class-I mutations change a G to an A at position 1566 and all class-II mutations change a G to an A or T at position 4223 (32). Class-IV mutations are simply the combination of a class-I and a class-II mutation. This suggests that only a very limited number of possible mutations improve the kinetic functioning of the enzyme. In this case, changes in only 2 bases out of 3090 improve the function of the enzyme. If this result is generally true, it has profound consequences for our view of molecular evolution. This finding suggests that enzyme activity as determined by kinetic parameters rather than amount of enzyme is limited to a few discrete values. Enzyme activity in this sense would then be a highly constrained trait. However, these mutations were selected as papilla on Petri dishes. Other environments may select mutations with smaller effects and show that there are a larger number of possible advantageous changes.

## *Selection for Mutation Rate*

Mutations of certain genes cause an increased mutation rate and have been referred to as mutator genes. In chemostats, strains carrying mutator genes are selected. This has been shown for the mutation, *mutT*1, (14) which gives an increased rate of A:T to G:C transversions; mutations of *mutS* (93), *mutL* (92), and *mutH* (69), all three of which promote transitions and frame shifts and are part of the mismatch repair system (50); and mutations of DNA polymerase I, *polA,* (95), which increase the frequency of frameshift and deletion mutations. Even though mutations at the *dam* locus elevate mutation rate, these are selected against due to various pleiotropic effects (96).

Mutator genes increase the frequency of all mutations—advantageous, disadvantageous, and neutral. The most studied mutator, *mutT*1 or Treffer's mutator, causes a 100- to 1000-fold increase in the mutation rate of transversions converting AT pairs into CG pairs. This strong bias in the mutation rate increased the percent CG in *E. coli* by 0.2–0.5% after only 1000 generations (15). Other mutators preferentially convert CG pairs into AT pairs (70, 80).

Strains carrying a mutator gene increase in frequency in chemostats because the mutator increases the chance of undergoing a favorable mutation. If there are $10^8$ mutator cells and $10^{10}$ nonmutator, and the mutation rate to favorable mutations is $10^{-7}$ and $10^{-9}$, respectively, then there will be 10 advantageous mutations in the cells carrying the mutator and 10 in other cells. After periodic selection, the progeny of these 20 cells dominate, and the frequency of mutator strains increases from 1% to 50%. The evidence for this, the mutation hypothesis, rather than the metabolic hypothesis which postulates that the advantage of the mutator is due to some unknown metabolic effect of the mutator itself, is twofold. First, the mutation hypothesis predicts that there will be little fitness difference (a lag) until the frequency of the mutator increases rapidly during a period of periodic selection, while the metabolic hypothesis predicts immediate and continuous selection. Chao & Cox (11) have shown that there is a lag of about 70 generations during which there seems to be a slight amount of selection against the mutator, *mutT*1, presumably because of the elevated frequency of disadvantageous mutations, before the mutator increased rapidly. Another lag between periodic selection events is not seen (13). Second, the mutation hypothesis predicts frequency dependent selection and the metabolic hypothesis does not. If the number of mutator cells is so low that no advantageous mutation arises in them before advantageous mutations arising in the wild-type replace the population, the cells carrying the mutators will be eliminated. This is analogous to the frequency dependence of mutations used by Atwood et al to monitor periodic selection. Chao & Cox (11) showed that when the frequency of *mutT*1 is less than $7 \times 10^{-5}$, it is lost from the population and that when it is greater than $10^{-4}$, it dominates. Given that there were about $2 \times 10^{10}$ cells in each chemostat population, about $2 \times 10^{6}$ mutator cells are required for a sufficiently advantageous mutation to arise in one cell to compete successfully with the advantageous mutations arising in the wild-type.

Troebner & Piechocki (94) maintained a single chemostat containing a mutator (*mutT*) for 2200 generations. A strain isolated from this chemostat had a lowered mutation rate, roughly intermediate between wild-type and the mutator. This lower mutation rate was caused by a second site mutation since the original *mutT*1 allele could be isolated from the strain, suggesting that the optimal mutation rate is lower than the mutation rate produced by *mutT*1.

The frequency of mutator strains in nature varies from about 2% to about .5%, depending on the study (94). This suggests that even for a cosmopolitan species like *E. coli,* there are some environments where fitness increases are common enough that mutators have an advantage.

## *Evolutionary Explanations*

In the following three sections, I wish to show that results from the laboratory experiments can be used as elements in evolutionary stories. Even though "just so" stories are out of fashion now, we still have to use the available knowledge to make sense of our world.

COMETABOLISM Many xenobiotic compounds, man-made compounds foreign to microbial systems, are possible additional resources for microorganisms. Xenobiotic compounds could be utilized by enzymes recruited for these novel substrates from other pathways by acquisitive evolution. However, it has proven difficult to isolate organisms that grow on many of these compounds, even when these compounds are obviously being modified or degraded by organisms. The ability of microorganisms to modify or degrade compounds that they can not utilize as the sole nutrient for growth is called cometabolism (16). Cometabolism can be either detrimental or beneficial to the organism. It is often detrimental, costing the organism more to modify the compound than is obtained. It is postulated that the xenobiotic compounds are metabolized simply because of secondary reactions of important enzymes, (16). However, it may be postulated that there will be rapid evolution to metabolize xenobiotic compounds, since they represent an underutilized resource allowing a competitive advantage similar to the advantage the unique resource offers the generalist.

G+C CONTENT Among eubacteria, the mean guanine plus cytosine (G+C) content of the genomic DNA varies from approximately 25% to 75%, with *E. coli* at about 50%. The genomic G+C content is a phylogenetic character, with closely related organisms showing similar base composition (43, 44). Within an organism, the various pieces of DNA show similar G+C content since they are unimodal with a relatively narrow range in CsCl gradients. Sueoka (89) has proposed that directional mutational pressure has created this divergence in G+C content. This idea is supported by sequence data (66). Since many mutator genes, e.g. *mutT,* show strong directional mutational pressure, it may be postulated that many of the major eubacterial groups had a period of rapid evolution where high mutation rate was advantageous and that afterwards, the rate was reduced by secondary mutations, leaving the directional bias as a pleiotropic character.

SIGNATURE SEQUENCES In his fine review on bacterial evolution, Woese (101) stresses that ribosomal RNA is a most useful molecule for phylogenic analysis. If any molecule changes in a clocklike manner, it should. It has an important and constant function (protein synthesis) which implies that constraints will not change, and there is little opportunity for selection of advantageous mutations. Changes will thus be introduced by chance where the probability of any particular change is a function of the mutation rate, the selection coefficient against the change, and the effective population size (48). This would lead to the predictions that branch lengths would be roughly equal and that the distribution of rare, improbable changes or signatures would be randomly scattered throughout the phylogenetic tree. These predictions are not realized. Branch lengths of certain groups are too long—for example, the Eubacteria relative to Archaebacteria or the mycoplasms relative to the other groups of Eubacteria. Rare or unique changes seem to be found more often in rapidly evolving groups (102) and at the base of major groups (103). The elevated rates and increased frequency of signature sequences may reflect periods of rapid evolution prior to adaptive radiations (101). The function of the ribosomal RNA must remain both important and constant even during periods of rapid evolution. Thus, the signature sequences in rRNA cannot be developed by inactivation and reactivation by second site mutations. These periods may represent rapid evolution into new niches where survival is much more important than competitive ability. Any mutation which improved survivability would be selected. This situation would select-for high mutation rate even more than chemostat cultures. There would be small population size and intense selection on the part of the genome adapting to the new environment with relaxed selection on the rest. Under these conditions, rapid and unusual changes in the ribosomal sequence are expected. If species of sexually reproducing diploids commonly evolve in small isolated demes, the successful groups could have elevated mutation rates since those demes with elevated mutation rates are more likely to produce the required series of advantageous mutations. Then signature sequences are expected in the rRNA and in the genome DNA, unrelated to the adaptive changes, in rapidly evolving lineages of Eukaryotes. Because of the elevated mutation rate, the whole genome will show a faster rate of change, not just the genes which are incorporating the adaptive changes.

## *Outlook and Summary*

In discussing the causes of natural selection I have emphasized the environment. The structures of the environment in batch and continuous culture are sufficiently different that in batch culture a Fisherian view of evolution seems adequate while in continuous culture a view of evolution more akin to that of Wright seems to be required. I am postulating that environments where most of the selection is for density independent fitness promote a Fisherian type of evolution, and environments where most of the selection is for density dependent fitness promote a Wrightian type of evolution.

In chemostats, where density dependent selection is important, more than $n$ species coexist on $n$ resources. The experiment by Helling et al (40) showed the evolution of a single clone into two distinct clones which used the single limiting resource differently, allowing coexistence. With bacterial systems the experimental investigation of what is meant by the term "resource" is possible.

Bacterial systems allow one to study the importance of various types of genetic variation in creating differences in fitness. The data from experiments with bacteria suggest that most base pair or amino acid mutations which do not eliminate function completely are either nearly neutral or under very weak purifying selection. Advantageous mutations are rare. Thus, during periods of adaptation to new environments, higher rates of mutation are advantageous. Higher rates of mutation will produce a faster-paced molecular clock in all genes. If this increased mutation rate is combined with small effective population size, signature changes will be produced randomly in the genome. Looking for and using signature changes should be incorporated into the methodology for constructing phylogenies from molecular data.

ACKNOWLEDGMENTS

I thank R. Lenski, L. Luckinbill, and H. Ochman for sharing unpublished work and D. Futuyma and J. Kim for reading a draft of this paper. This work was supported by NIH grant GM30201 and NSF grant BSR-8614997 and is contribution number 753 from the Graduate Studies in Ecology and Evolution, State University of New York at Stony Brook.

## Literature Cited


1. Abrams, P. A. 1988. How should resources be counted? *Theor. Pop. Biol.* 33:226–42
2. Adams, J., Oeller, P. W. 1986. Structure of evolving populations of *Saccharomyces cerevisiae:* Adaptive changes are frequently associated with sequence alterations involving mobile elements belonging to the Ty family. *Proc. Natl. Acad. Sci. USA* 83:7124–27
3. Adams, J., Paquin, C., Oeller, P. W., Lee, L. W. 1985. Physiological characterization of adaptive clones in evolving populations of yeast, *Saccharomyces cerevisiae. Genetics* 110:175–85
4. Ajioka, J. W., Hartl, D. L. 1989. Population dynamics of transposable elements. See Ref. 8, pp. 939–58
5. Armstrong, R. A., McGehee, R. 1980. Competitive exclusion. *Am. Nat.* 115: 151–70
6. Atwood, K. C., Schneider, L. K., Ryan, F. J. 1951. Selective mechanisms in bacteria. *Cold Spring Harbor Symp. Quant. Biol.* 16:345–55
7. Bachmann, B. J. 1987. Linkage map of *Escherichia coli* K-12, edition 7. See Ref. 68, pp. 807–76
8. Berg, D. E., Howe, M. M. ed. 1989. *Mobile DNA.* Washington DC: Am. Soc. Microbiol.
9. Boeke, J. D. 1989. Transposable elements in *Saccharomyces cerevisiae.* See Ref. 8, pp. 335–74
10. Bryson, V., Szybalski, W. 1952. Microbial selection. *Science* 116:45–51
11. Chao, L., Cox, E. C. 1983. Competition between high and low mutating strains of *Escherichia coli. Evolution* 37:125–34
12. Chao, L., Levin, B. R. 1981. Structured habitats and the evolution of anticompetitor toxins in bacteria. *Proc. Natl. Acad. Sci. USA* 78:6324–28
13. Chao, L., Ramsdell, G. 1985. The effects of wall populations on coexistence of bacteria in the liquid phase of chemostat cultures. *J. Gen. Microbiol.* 131:1229–36
14. Cox, E. C., Gibson, T. C. 1974. Selection for high mutation rates in chemostats. *Genetics* 77:169–84
15. Cox, E. C., Yanofsky, C. 1967. Altered base ratios in the DNA of an *Escherichia coli* mutator strain. *Proc. Natl. Acad. Sci. USA* 58:1895–1902
16. Dalton, H., Stirling, D. I. 1982. Cometabolism. *Philos. Trans. R. Soc. Lond. B* 297:481–96
17. Dean, A. M., Dykhuizen, D. E., Hartl, D. L. 1988. Fitness effects of amino acid replacements in the beta-galactosidase of *Escherichia coli. Mol. Biol. Evol.* 5:469–85
18. Dykhuizen, D., Davies, M. 1980. An experimental model: Bacterial specialists and generalists competing in chemostats. *Ecology* 61:1213–27
19. Dykhuizen, D. E., Dean, A. M. 1990. Enzyme activity and fitness: evolution in solution. *Trends Ecol. Evol.* 5:257–62
20. Dykhuizen, D., Hartl, D. 1981. Evolution of competitive ability in *Escherichia coli. Evolution* 35:581–94
21. Dykhuizen, D. E., Hartl, D. L. 1983. Selection in chemostats. *Microbiol. Rev.* 47:150–68
22. Ewens, W. J. 1980. *Mathematical Population Genetics.* New York: Springer Verlag
23. Fisher, R. A. 1930. *The Genetical Theory of Natural Selection.* Republished 1958. New York: Dover
24. Gottschal, J. C. 1986. Mixed substrate utilization by mixed cultures. See Ref. 82, pp. 261–92
25. Gottschal, J. C., De Vries, S., Kuenen, J. G. 1979. Competition between the facultatively chemolithotrophic *Thiobacillus* A2, an obligately chemolithotrophic *Thiobacillus* and a heterotrophic *Spirillum* for inorganic and organic substrates. *Arch. Microbiol.* 121:241–49
26. Gottschal, J. C., Kuenen, J. G. 1980. Mixotrophic growth of *Thiobacillus* A2 on acetate and thiosulfate as growth limiting substrates in the chemostat. *Arch. Microbiol.* 126:33–42
27. Gottschal, J. C., Nanninga, H. J., Kuenen, J. G. 1981. Growth of *Thiobacillus* A2 under alternating growth conditions in the chemostat. *J. Gen. Microbiol.* 126:85–96
28. Gottschal, J. C., Pol, A., Kuenen, J. G. 1981. Metabolic flexibility of *Thiobacillus* A2 during substrate transitions in the chemostat. *Arch. Microbiol.* 129:23–28
29. Gottschal, J. C. Thingstad, T. F. 1982. Mathematical description of competition between two and three bacterial species under dual substrate limitation in the chemostat: A comparison with experimental data. *Biotechnol. Bioeng.* 24:1403–18
30. Grover, J. P., 1988. Dynamics of competition in a variable environment: Experiments with two diatom species. *Ecology* 69:408–17
31. Hall, B. G. 1983. Evolution of new

metabolic functions in laboratory organisms. In *Evolution of Genes and Proteins,* ed. M. Nei and R. Koehn, pp. 234–57. Sunderland, Mass: Sinauer
32. Hall, B. G., Betts, P. W., Wootton, J. C. 1989. DNA sequence analysis of artificially evolved *ebq* enzyme and *ebq* repressor genes. *Genetics* 123:635–48
33. Hall, B. G., Clark, N. D. 1977. Regulation of newly evolved enzymes. III. Evolution of the *ebq* repressor during selection for enhanced lactase activity. *Genetics* 85:193–201
34. Harder, W., Dijkhuizen, L. 1976. Mixed substrate utilization. In *Continuous Culture 6: Applications and New Fields,* ed. A. C. R. Dean, D. C. Ellwood, C. G. T. Evans, and J. Melling, pp. 297–314. Chichester: Ellis Howard
35. Hartl, D. L. 1989. The physiology of weak selection. *Genome* 31:183–89
36. Hartl, D., Dykhuizen, D. 1979. A selectively driven molecular clock. *Nature* 281:230–31
37. Hartl, D. L., Dykhuizen, D. E. 1984. The population genetics of *Escherichia coli. Annu. Rev. Genet.* 18:31–68
38. Hartl, D. L., Dykhuizen, D. E. 1985. The neutral theory and the molecular basis of preadaptation. In *Population Genetics and Molecular Evolution,* ed. T. Ohta and K. Aoki, pp. 107–24. Tokyo: Japan Sci. Societies
39. Helling, R. B., Kinney, T., Adams, J. 1981. The maintenance of plasmid-containing organisms in populations of *Escherichia coli J. Gen. Microbiol.* 123:129–41
40. Helling, R. B., Vargas, C. N., Adams, J. 1987. Evolution of *Escherichia coli* during growth in a constant environment. *Genetics* 116:349–58
41. Herbert, D., Elsworth, R., Telling, R. C. 1956. The continuous culture of bacteria: a theoretical and experimental study. *J. Gen. Microbiol.* 14:601–22
42. Hodgson, J., Sugden, A. M. 1988. *Planned Release of Genetically Engineered Organisms* (Trends in Biotechnology/Trends in Ecology and Evolution Special Publication). Cambridge: Elsevier
43. Holt, J. G., ed. 1977. *The Shorter Bergey's Manual of Determinative Bacteriology.* Baltimore: Williams & Wilkins. *8th ed.*
44. Hori, H., Osawa, S. 1986. Evolutionary change in 5S rRNA secondary structure and a phylogenetic tree of 352 5S rRNA species. *BioSystems* 19:163–72
45. Jannasch, H. W. 1967. Enrichments of aquatic bacteria in continuous culture. *Archiv. Mikrobiol.* 56:355–69
46. Kacser, H., Burns, J. A. 1973. The control of flux. *Symp. Soc. Exp. Biol.* 27:65–104
47. Kacser, H., Burns, J. A. 1981. The molecular basis of dominance. *Genetics* 97:639–66
48. Kimura, M. 1983. *The Neutral Theory of Molecular Evolution.* Cambridge: Cambridge Univ. Press
49. Kubitschek, H. E. 1974. Operation of selection pressure on microbial populations. *Symp. Soc. Gen. Microbiol.* 24:105–30
50. Kushner, S. R. 1987. DNA Repair. See Ref 68, pp. 1044–53
51. Laanbroek, H. J., Smit, A. J., Klein Nulend, G., Veldkamp, H. 1979. Competition for L-glutamate between specialized and versatile *Clostridium* species. *Arch. Microbiol.* 120:61–66
52. Leegwater, M. P. M. 1983. *Microbial Reactivity: Its Relevance to Growth in Natural and Artificial Environments.* Utrecht: Drukkerij Elinkwijk BV
53. Legan, J. D., Owens, J. D., Chilvers, G. A. 1987. Competition between specialist and generalist methylotrophic bacteria for an intermittent supply of methylamine. *J. Gen. Microbiol.* 133: 1061–73
54. Lenski, R. E. 1984. Two-step resistance by *Escherichia coli* B to bacteriophage T2. *Genetics* 107:1–7
55. Lenski, R. E. 1988. Experimental studies of pleiotropy and epistasis in *Escherichia coli* II. Compensation for maladaptive effects associated with resistance to virus T4. *Evolution* 42:433–40
56. Lenski, R. E., Rose, M. R., Simpson, S. C., Tadler, S. C. 1990. Long-term experimental evolution in *Escherichia coli.* I Adaptation and divergence during 2000 generations. *Am. Nat.* Submitted
57. Levin, B. R. 1981. Periodic selection, infectious gene exchange and the genetic structure of *E. coli* populations. *Genetics* 99:1–23
58. Levins, R. 1979. Coexistance in a variable environment. *Am. Nat.* 114:765–83
59. Luckinbill, L. S. 1984. An experimental analysis of a life history theory. *Ecology* 65:1170–84
60. Maruyama, T., Kimura, M. 1980. Genetic variability and effective population size when local extinction and recolonization of subpopulations are frequent. *Proc. Natl. Acad. Sci. USA* 77:6710–14
61. Mayr, E. 1982. *The Growth of Biologi-*

*cal Thought. Diversity, Evolution and Inheritance*. Cambridge, Mass: Belknap
62. McGehee, R., Armstrong, R. A. 1977. Some mathematical problems concerning the ecological principle of competitive exclusion. *J. Differ. Equations* 23:30–52
63. Miller, J. H. 1972. *Experiments in Molecular Genetics*. Cold Spring Harbor: Cold Spring Harbor Laboratory
64. Mortlock, R. P. ed. 1984. *Microorganisms as Model Systems for Studying Evolution*. New York: Plenum
65. Moser, H. 1958. *The Dynamics of Bacterial Populations Maintained in the Chemostat*. Washington DC: Carnegie Inst. Washington Publ. 614
66. Muto, A., Osawa, S. 1987. The guanine and cytosine content of genomic DNA and bacterial evolution. *Proc. Natl. Acad. Sci. USA* 84:166–69
67. Nei, M., Graur, D. 1984. Extent of protein polymorphism and the neutral mutation theory. *Evol. Biol.* 17:73–118
68. Neidhardt, F. C., Ingraham, J. L., Low, K. B., Magasanik, B., Schaechter, M., Umbarger, H. E. ed. 1987. Escherichia coli *and* Salmonella typhimurium: *Cellular and Molecular Biology*. Washington, DC: Am. Soc. Microbiol.
69. Nestmann, E. R., Hill, R. F. 1973. Population changes in continuously growing mutator cultures of *Escherichia coli*. *Genetics* 73:41–44
70. Nghiem, Y., Cabrera, M., Cupples, C. G., Miller, J. H. 1988. The *mutY* gene: A mutator locus in *Escherichia coli* that generates G.C→T.A transversions. *Proc. Natl. Acad. Sci. USA* 85:2709–13
71. Nguyen, T. N. M., Phan, Q. G., Duong, L. P., Bertrand, K. P., Lenski, R. E. 1989. Effects of carriage and expression of the Tn*10* tetracycline-resistance operon on the fitness of *Escherichia coli* K12. *Mol. Biol. Evol.* 6:213–25
72. Ochman, H. 1990. Patterns of nucleotide evolution in bacteria and eukaryotes: Glyceraldehyde-3-phosphate dehydrogenase. *Mol. Biol. Evol.* Submitted
73. Ochman, H., Wilson, A. C. 1987. Evolution in bacteria: Evidence for a universal substitution rate in cellular genomes. *J. Mol. Evol.* 26:74–86
74. Ochman, H., Wilson, A. C. 1987. Evolutionary history of enteric bacteria. See Ref. 68, pp. 1649–54
75. Ohta, T. 1974. Mutation pressure as the main cause of molecular evolution and polymorphism. *Nature* 252:351–54
76. Paquin, C. M. E. 1982. *Characterization and rate of occurence of adaptive mutations in haploid and diploid populations of the yeast* Saccharomyces cerevisiae. PhD thesis. Univ. Mich., Ann Arbor. 81 pp.
77. Paquin, C., Adams, J. 1983. Frequency of fixation of adaptive mutations is higher in evolving diploid than haploid yeast populations. *Nature* 302:495–500
78. Paquin, C. E., Adams, J. 1983. Relative fitness can decrease in evolving asexual populations of *S. cerevisiae*. *Nature* 306:368–71
79. Parkes, R. J. 1982. Methods for enriching, isolating, and analysing microbial communities in laboratory systems. In *Microbial Interactions and Communities,* Vol 1., ed. A. T. Bull, J. H. Slater, pp. 45–102. London:Academic
80. Piechocki, R., Kupper, D., Quinõnes, A., Langhammer, R. 1986. Mutational specificity of a proof-reading defective *Escherichia coli dnaQ49* mutator. *Mol. Gen. Genet.* 202:162–68
81. Pirt, S. J. 1975. *Principles of Microbe and Cell Cultivation*. Oxford: Blackwell
82. Poindexter, J. S., Leadbetter, E. R. ed. 1986. *Bacteria in Nature: Methods and Special Applications in Bacterial Ecology*. New York: Plenum
83. Poindexter, J. S., Leadbetter, E. R. 1986. Enrichment cultures in bacterial ecology. See Ref. 82, pp. 229–60
84. Sawyer, S. A., Dykhuizen, D. E., Hartl, D. L. 1987. Confidence interval for the number of selectively neutral amino acid polymorphisms. *Proc. Natl. Acad. Sci. USA* 84:6225–28
85. Selander, R. K., Caugant, D. A., Whittam, T. S. 1987. Genetic Structure and variation in natural populations of *Escherichia coli*. See Ref. 68, pp. 1625–48
86. Selander, R. K., Levin, B. R. 1980. Genetic diversity and structure in populations of *Escherichia coli*. *Science* 210:545–47
87. Silver, R. S., Mateles, R. L. 1969. Control of mixed substrate utilization in continuous cultures of *Escherichia coli*. *J. Bacteriol.* 97:535–43
88. Smith, M. E., Bull, A. T. 1976. Studies of the utilization of coconut water waste for the production of food yeast *Saccharomyces fragilis*. *J. Appl. Bacteriol.* 41:81–95
89. Sueoka, N. 1962. On the genetic basis of the variation and heterogenity of DNA base composition. *Proc. Natl. Acad. Sci. USA* 48:582–92
90. Tiedje, J. M., Colwell, R. K., Grossman, Y. L., Hodson, R. E., Lenski, R.

E., et al. 1989. The planned introduction of genetically engineered organisms: Ecological considerations and recommendations. *Ecology* 70:298–315

91. Tilman, D. 1977. Resource competition between planktonic algae: An experimental and theoretical approach. *Ecology* 58:338–48
92. Troebner, W., Piechocki, R. 1981. Competition growth between *Escherichia coli mutL* and *mut*$^+$ in continuously growing cultures. *Z. Allg. Mikrobiol.* 21:347–49
93. Troebner, W., Piechocki, R. 1984. Competition between isogenic *mutS* and *mut*$^+$ population of *Escherichia coli:* K12 in continuously growing cultures. *Mol. Gen. Genet.* 198:175–76
94. Troebner, W., Piechocki, R. 1984. Selection against hypermutability in *Escherichia coli* during long term evolution. *Mol. Gen. Genet.* 198:177–78
95. Troebner, W., Piechocki, R. 1985. Selective advantage of *polA1* mutator over *polA*$^+$ strains of *Escherichia coli* in a chemostat. *Naturwissenschaften* 72: 377–78
96. Troebner, W., Piechocki, R. 1985. Competition between the *dam* mutator and the isogenic wild-type of *Escherichia coli. Mutation Res.* 144:145–49
97. Vinopal, R. T. 1987. Selectable phenotypes. See Ref. 68, pp. 990–1015
98. Waterson, G. A. 1978. The homozygosity test of neutrality. *Genetics* 88:405–17
99. Wilson, A. C., Cann, R. L., Carr, S. M., George, M., Gyllensten, U. B., et al. 1985. Mitochondrial DNA and two perspectives on evolutionary genetics. *Biol. J. Linn. Soc.* 26:373–400
100. Wimpenny, J. W. T. 1982. Responses of microorganisms to physical and chemical gradients. *Philos. Trans. R. Soc. Lond. B* 297:497–515
101. Woese, C. R. 1987. Bacterial evolution. *Microbiol. Rev.* 51:221–71
102. Woese, C. R., Stackebrandt, E., Ludwig, W. 1985. What are mycoplasmas: The relationship of tempo and mode in bacterial evolution. *J. Mol. Evol.* 21:305–16
103. Woese, C. R., Stackebrandt, E., Macke, T. J., Fox, G. E. 1985. A phylogenetic definition of the major eubacterial taxa. *System. Appl. Microbiol.* 6:143–51
104. Wright, S. 1980. Genic and organismic selection. *Evolution* 34:825–43
105. Young, J. P. W. 1989. The population genetics of bacteria. In *Genetics of Bacterial Diversity*, ed. D. A. Hopwood, K. F. Chater, pp. 417–38. London: Academic

*Annu. Rev. Ecol. Syst. 1990. 21:399–422*


# PLANT-POLLINATOR INTERACTIONS IN TROPICAL RAIN FORESTS

*K. S. Bawa*

Department of Biology, University of Massachusetts, Boston, Massachusetts 02125

KEY WORDS: pollination, tropical rain forests, speciation, coevolution

## INTRODUCTION

Plant-pollinator interactions in tropical lowland rain forests (TLRF) offer unique opportunities to address several problems of current evolutionary and ecological interest. First, conspecifics of many tree species are spatially isolated and self-incompatible or dioecious (11, 16, 24, 36, 61). Thus selection for long-distance pollen flow may be more intense in TLRF than in any other community (78), making it possible to study the patterns of pollen flow that perhaps are not observed anywhere else. Second, in tree species longevity combined with intense pressure from competitors, predators, and pathogens as well as abiotic agents places a high premium on genetic recombination (78, 91), which may also select for larger pollen (and seed) shadows not generally encountered in other communities. Third, the high species richness of TLRF correlated in part with the richness of pollination mechanisms (3, 5, 11, 25) offers an unusual opportunity to examine the role of plant-pollinator interactions in plant speciation (33, 133). Fourth, the wide range of specialization in plant-pollinator interactions at various taxonomic levels provides rich material for an assessment of factors promoting coevolution (47). Fifth, the ubiquitousness of biotic pollination in almost all plant species in TLRF (80) makes it a unique community to study the effects of plant-pollinator interactions in the structure and organization of communities. Finally, the multitude of plant-pollinator interactions permits an analysis of the role of mutualistic interactions in maintaining stability in complex communities (57).



0066-4162/90/1120-0399$02.00

The question of community stability, apart from its theoretical importance, is a central issue in conservation biology.

It should be noted at the outset that the unusual importance of studying pollination systems in TLRF is matched by unusual difficulties encountered in gathering basic information. Canopy trees that define the structure and properties of TLRF present, because of their height, logistical difficulties for empirical and observational work not generally found in other ecosystems (100, 108, 110). Furthermore, in a given TLRF, hundreds of plant species together with thousands of pollinator species form a complex web of relationships difficult to unravel without a concerted effort lasting many years. Although a wealth of information exists about certain systems, e.g. figs and fig wasps (81, 152) and orchids and orchid bees (1, 41, 42), the data for a particular species assemblage are not from one site. One of the few exceptions is Stile's data set for *Heliconia* species and their hummingbird pollinators (135, 137). Much of the available information about pollination systems in TLRF at the level of particular "guilds" or communities is from scattered studies undertaken at sites throughout the tropics.

Plant-pollinator interactions in TLRF have been used as paradigms to study coevolution (47), gene flow (11, 24, 79), evolution of plant sexual systems (16, 20), and community stability (74) and can be thus reviewed in several different contexts. Here I first recapitulate the diversity of pollination systems in TLRF, based on recent work in the lowlands of Central America. The focus, unless specified otherwise, is on *lowland* rain forests. Montane forests are considered in a separate section; coastal mangrove forests are excluded due to the paucity of data (149). I then discuss plant-pollinator interactions in the context of gene flow and speciation, two topics central to the issue of species richness of tropical communities. I conclude with a brief commentary on the effects of disruption in plant-pollinator interactions on community stability and the maintenance of biodiversity, two topics, again, of much current interest.

The paper complements two other recent reviews of the subject (22, 90). Related topics that have been lately reviewed particularly in the context of tropical wet forests are: flowering phenology (18, 29), plant reproductive systems (24, 90), harvest of floral resources (82), pollinator specialization and coevolution (47, 74, 75), and the role of pollinators in the evolution of sexual systems of plants (16, 20).

## MODES OF POLLINATION

### *Diversity*

The diverse range of pollination systems found in angiosperms can be encountered in its entirety in most TLRF. It is well known that wind pollination is rare, but not absent in TLRF (21). On the basis of studies on trees (25), I estimate that approximately 98% to 99% of all flowering plant species in TLRF are pollinated by animals. Biotic pollen vectors range all the way from one- to two-millimeter-long fig wasps (152) to flying foxes with a wingspan of two meters (38). Although flowers may receive a wide range of visitors, members of only one or two main classes, usually of the same order, act as effective vectors (25, 128). The subject of specialization and constancy is treated elsewhere in the paper. In the discussion below, reference to a particular pollinator implies that it is the primary pollen vector for a given plant species (or a group of plant species). The commentary is concerned with the ecological aspects of various pollination systems, rather than with the evolution of particular morphological, anatomical, or behavioral traits associated with various plant-pollinator interactions. Prance (115) describes detailed case studies for many of the major pollination modes considered below.

POLLINATION BY VERTEBRATES Bats, some nonflying mammals, and birds are the only pollen vectors known among vertebrates.

*Bats* Examples of bat pollination can be found in many families, but this mode of pollination is particularly common or well studied in the Bombacaceae (13, 14, 43, 98), and the genera *Passiflora* in the Passifloraceae (125, 126), *Parkia* in the Mimosaceae (73), and *Bauhinia* in the Caesalpiniaceae (66). Of the two orders of Chiroptera to which bats belong, only Microchiroptera, in which nectarivory is of relatively recent origin, are found in the neotropics; Megachiroptera, some of which are exclusively vegetarian, are restricted to the old world (13).

The number of plant species pollinated by bats or the number of bat species involved as pollen vectors is not known for any tropical wet forest. In a tropical lowland dry deciduous forest with approximately 150 tree species, 7 species of bats were found to carry pollen of 13 species of trees over a one-year period (65). Several species of bats apparently serviced a given plant species, and a particular bat species utilized the nectar and pollen of many plant species.

The floral syndromes of pollination by bats are well documented (13). In general, flowers open at dusk or soon after, are large, white or pale yellow in color, have a musky odor, and produce large quantities of nectar. However, in some species, flowers are small (3–5 mm across), but borne in dense clusters (135). In a community-wide study, Opler found the highest amount of nectar in a bat-pollinated species (107). On a per flower basis, bat pollination is perhaps energetically most expensive, but its benefits may be in the form of a long pollen shadow because bats forage over long distances (65, 132).

Bat-pollinated species may flower massively for a few days or bear a few flowers every day for several months (65).

*Nonflying mammals* Sussman & Raven (143) have presented circumstantial evidence for pollination of several tree species in Madagascar by lemurs, especially in areas where bat-pollination is absent or rare. They are uncertain about any specific adaptations involved in pollination by nonflying mammals and consider the system to be a relict that has survived from ancient times in certain areas. Janson et al (77) have also implicated nonflying mammals, e.g. opossums, kinkajous, and monkeys, in the pollination of several tree species such as *Cieba pentandra, Ochroma pyramidale,* and *Quararibea cordata* (all in the Bombacaceae) in an Amazonian forest. However, two of these three species *(C. pentandra* and *O. pyramidale)* are known to be bat-pollinated, and the nonflying mammals, the presumed pollinators, do destroy a number of flowers. The evidence for the effective transfer of pollen by nonflying mammals is indirect and weak.

Substantial evidence for pollination by nonflying mammals in tropical wet forests exists only for *Mabea occidentalis* (Euphorbiaceae), a small tree in the Central American lowland forests. The red woolly opossum, *Caluromys derbianus,* is a common visitor to the inflorescences of *M. occidentalis,* the flowers of which are also visited by noctuid and pyralid moths, Cerambycid beetles, Trigona bees, and bats (135). However, the inflorescences are "clearly adapted" to pollination by bats (135).

Pollination by rodents has been implicated for an epiphytic species of *Blakea* (Melastomataceae) in a Costa Rican montane forest (96), but the example remains to be explored in detail.

Clearly, the nectar-rich flowers or inflorescences with nocturnal anthesis, pollinated by moths or bats, are exploited by nonflying mammals. Inevitably, these mammals will be found visiting the night-blooming flowers and moving from one plant to another. However, such observations are not enough to suggest that the nonflying mammals transfer significant amounts of pollen from one plant to another. The contributions of these flower visitors to fruit and seed set must be measured against the frequent damage to the flowers they presumably pollinate (77). However, the availability of such flowers may be an important factor in the life cycle of nonflying mammals, if the flowers provide critical resources during periods of drought or low fruit abundance. The nature of interactions between nonflying mammals and flowers, and the consequences of these interactions for both mammals and plants, remain unexplored.

*Birds* In Central America, hummingbirds constitute the major group of bird pollinators. Pollination by hummingbirds is common in Acanthaceae,

Bromeliaceae, Gesneriaceae, Marantaceae, Musaceae, Rubiaceae, and Zingiberaceae. In a Costa Rican lowland rain forest with approximately 1800 flowering plant species, hummingbirds have been recorded to collect nectar regularly from 42 species and occasionally from another 27 species (137). None of the species in this community has been observed to be pollinated by nonhovering birds; however, several species of *Erythrina* in Central America are known to be pollinated by passerine birds (48, 103, 134, 148).

Stiles (136, 137) has reviewed the ecology of hummingbird pollination in a TLRF, and Feinsinger (46) in a montane rain forest. In the lowland forest, nine species of *Heliconia* are pollinated by nine species of hummingbirds, but there is no species specificity; each species of *Heliconia* is visited by more than one species of hummingbird, and each hummingbird species visits more than one species of *Heliconia* (136). Flowering patterns of *Heliconia* species are staggered in time (137). The staggered blooming has been attributed to competition for pollinators (137, 139), but the idea remains debatable (37, 111a, 140).

Although a very large number of birds pollinate many plant species in the old-world tropics, much of the available information is anecdotal and descriptive (2, 112–115). Coevolution between flower-visiting birds and flowers on a global basis has been reviewed by Stiles (138).

POLLINATION BY INVERTEBRATES The vast majority of plant species in tropical rain forests are pollinated by insects.

*Bees* Among insects, bees constitute perhaps the most important group in number and diversity of plant species pollinated. In the neotropical lowland rain forests, the vast majority of species in many common families such as Burseraceae, Euphorbiaceae, Clusiaceae, Fabaceae, Flacourtiaceae, Lecythidaceae, Melastomataceae, Orchidaceae, and Sapotaceae are pollinated by bees. The bee pollination system is particularly predominant in canopy trees (25).

The number and diversity of bee species that act as pollen vectors is equally great (122). Approximately 70 species have been recorded to visit the flowers of a single tree of *Andira inermis* (52) in a seasonal forest and 26 species as visiting the flowers of *Dipteryx panamensis* (104) in an aseasonal forest in Costa Rica. Both tree species are in the Fabaceae.

In general, based on size, two types of bees may be distinguished: medium to large-sized bees of the families Andrenidae, Apidae, Anthophoridae, Halictidae, and Megachilidae; and the small-sized bees in Apidae (tribe Apini), Halictidae, and Megachilidae (25). The former appear more prevalent on the canopy flowers and the latter on the understory flowers. The medium- to large-sized bees constitute a very heterogeneous group. The brightly colored orchid bees of the tribe Euglossini in Apidae tend to forage singly or in small groups, primarily in the understory and subcanopy. In contrast, many anthophorids forage in large aggregations, mainly, but not entirely, in the canopy (G. W. Frankie, personal communication). Species pollinated by these bees often flower massively.

The diversity of bee-pollinated trees is so large that no generalizations can be made with respect to flower morphology or flowering pattern. Bee flowers, especially those that are pollinated by small bees can be relatively small and inconspicuous, and white, pale, or green in color. Flowers pollinated by medium-sized to large bees, may however be relatively large, brightly colored, and morphologically specialized as in many species of the Bignoniaceae, Fabaceae, Melastomataceae and Orchidaceae. Flowering may extend from a few days in some species to several months in others (54).

*Moths* Moth pollination is particularly prevalent in the Rubiaceae (25, 63). The heavily scented, white or pale flowers with narrow floral tubes and nocturnal anthesis in many species of such families as Apocynaceae, Meliaceae, Mimosaceae, and Solanaceae suggest that moths are also important pollen vectors in these groups. Moth-pollinated trees are mostly found in the understory and subcanopy.

Pollinating moths may be distinguished into two broad categories: (a) the large sphinx moths and (b) small moths in the Noctuidiae, and possibly in other families. Virtually nothing is known about the biology of the interactions that the latter group have with tropical trees. Even for sphinx moths, the available information is largely derived from Gottsberger's (58) work in Brazil, Nilsson and associates' investigations in Madagascar (104), and Haber & Frankie's (63) comprehensive study of sphinx moth–pollinated plants in a dry deciduous forest of Costa Rica.

The flowers pollinated by sphinx moths are generally white or pale yellow in color with deep corolla tubes; the flowers open in the late afternoon or after dark, are sweet scented, and offer nectar as the main reward to the moths (44, 63). Two types of flowers may be distinguished: tubular flowers with narrow corolla tubes terminated by four to six corolla lobes, and brush type of flowers with reduced corolla and many exerted stamens (63).

Although sphinx moth–pollinated plants may be found to flower at all times of the year, Frankie and coworkers (50, 63) have observed peak flowering during the wet season. They explain this seasonality by saying the plants serve as sources of food not only for adults but also for the larvae. The larvae depend upon leaves which in the case of most plant species in the dry deciduous forest are borne only during the wet season. The sphinx moth–pollinated species may flower in highly synchronous episodes lasting only four to five days, or they may bloom for as long as ten months (63; see also 32).

Moth pollination system is one of the most common but the least studied systems in tropical lowland rain forests.

*Beetles* Beetles constitute an important group of pollen vectors, next in importance perhaps only to bees and moths. Beetle pollination is particularly common in Annonaceae, Araceae (127, 155, 156), Cyclanthaceae (26), Lauraceae, Myristicaceae (10, 76), and Palmae (27, 67, 68, 102). The beetles involved are diverse, from weevils two millimeters in length to scarabs that are two centimeters long (26, 27, 127).

Very little is known about the ecology of interactions between beetles and flowers. Recent studies on aroids (155, 156), Cyclanthaceae (26), palms (27), and *Myristica* (9) are among the most well-documented cases of beetle pollination. Beach (26, 27) has shown the complex nature of interactions among beetles and other flower visitors such as fruit flies and weevils in the pollination of *Pejibaye* palm. Flowers pollinated by beetles range from small as in the Myristicaceae (9) to several centimeters across as in the Annonaceae (127). Flowers when small may be borne on large inflorescences that in the Araceae are enclosed by bracts (156). Nocturnal anthesis is characteristic of the system, which is driven by strong odors (127).

Recent studies in Australian rain forests indicate that in some communities up to one quarter of all plant species may be pollinated by beetles (76). Beetles pollinate plants of all life forms and in all the strata of the forests in such communities. The Australian studies suggest that overall beetles may be third in importance, after bees and moths, in the number of plant species that they pollinate in rain forests.

*Butterflies* Many species of butterflies are common visitors to the flowers of a diverse array of species with brightly colored corollas (or other appendages), especially in the Boragiaceae, Rubiaceae, and Vochysiaceae. However, this pollination system is among the least studied in tropical rain forests.

*Wasps* The mutualistic relationship between the cosmopolitan genus *Ficus* and wasps is well known and has been a subject of some recent reviews (81, 152). Apart from the agaonid wasps, a diverse array of wasps are found among the insects visiting generalized flowers of such taxa as Anacardiaceae, Burseraceae, Simaroubaceae, and others. However, the extent to which such species transfer pollen is not known. Curiously, the type of specialized relationship found between fig wasps and figs has not been reported for other wasps and plants or other insects and plants in tropical lowland rain forests.

*Large Flies* Fly pollination appears to be widespread in Sterculiaceae (115, 154). Many species of *Aristolochia* (115) and *Rafflesia* (28) are also pollinated by flies. Little is known about the ecology of fly pollination. Beaman et al's (28) recent study of *Rafflesia* represents one of the few well-documented examples.

*Other insects* Pollination by thrips has been reported in some species of Myristicaceae (25) and Dipterocarpaceae (6). Curiously, although ants are abundant in TLRF and are known to pollinate plants in other regions (71, 86), pollination by ants is unknown in TLRF.

In summary, there are groups of species pollinated by the same class of pollinators, but little is known about the structure of these groups and the factors that influence the number and the diversity of the interacting species. Another general feature is the presence of a large number of species with a relatively generalist mode of pollination among trees. Such species with small, white, pale yellow or green, shallow flowers may account for up to *31%* of all species, and most seem to be collectively pollinated by a diverse array of small insects (25). What selects for specialized and generalized modes of pollination in the same community? The two modes might differ with respect to energetic costs, including the cost of defending flowers from predators and nectar robbers. They might also differ with respect to reliability of pollination, distances over which pollen is dispersed, and the manner in which they influence male and female components of fitness.

## *Spatial Distribution*

The distribution of various pollination systems in TLRF appears to be nonrandom (25). In particular, systems based on medium-sized to large bees and small diverse bees primarily occur in the canopy, and those based on hummingbirds sphingid moths, and beetles in the understory (Table 1, see also 25). Enough data to evaluate the distribution of other systems do not exist. The vegetation in tropical forests is often differentiated into more than two vertical strata. Indeed, Kress & Beach (90) have organized data on plant-pollinator interactions under three strata. As our knowledge of plant-pollinator interactions at the community level increases, it may be feasible to find evidence for nonrandom distribution at a finer vertical scale.

The nonrandom distribution of plant-pollinator interactions may be expected on the basis of vertical stratification of animal communities in general (130). Plant taxa, particularly at the generic and the familial levels, are also often distributed in particular strata. Examples include dipterocarps, which occur primarily in the canopy, and the Rubiaceae, which are generally confined to the subcanopy or understory. Although the associations between a particular plant taxon and a particular pollinator vector could also contribute to the observed patterns, the origin of spatial correlations of such associations remains unexplained. It should be interesting to determine the extent to which

**Table 1** Frequencies of different pollination systems

| | Forest stratum | |
|---|---|---|
| | Canopy[1] | Subcanopy & Understorey[2] |
| Pollination type | Percent of species (N) | Percent of species (N) |
| Bat | 3.8 (2) | 3.6 (8) |
| Hummingbird | 1.9 (1) | 17.7 (39) |
| Medium-sized to large bee | 44.2 (23) | 21.8 (48) |
| Small bee | 7.7 (4) | 16.8 (37) |
| Beetle | — | 15.5 (34) |
| Butterfly | 1.9 (1) | 4.5 (10) |
| Moth | 13.5 (7) | 7.3 (16) |
| Wasp | 3.8 (2) | 1.8 (4) |
| Small diverse insect | 23.1 (12) | 7.7 (17) |
| Wind | — | 3.2 (7) |
| TOTAL | 100% (52) | 100% (220) |

[1] Data from Bawa et al (25).
[2] Data from Kress and Beach (90).

plants and pollinators reciprocally influence their abundance in various vertical strata.

The diversity of pollination systems seems to be the highest in the understory (25). This might simply be a reflection of the diversity of plant species in that stratum. The significance of this observation is explored in another section.

## *Tropical-Temperate Zone Comparisons*

There are four major differences between pollination systems of TLRF and the north temperate zone forests. First, in aseasonal TLRF, pollination at the community level occurs throughout the year, though there may be well-defined peaks in flowering during certain times of year (51). By contrast, flowering in the north temperate zone is mostly confined to late spring and summer (118). Second, flowers of plants in TLRF generally last a day or two, whereas the mean longevity of flowers in the north temperate zone communities may extend to 7 days (116). Unpredictable conditions for pollination due to uncertain weather have been cited as one of the factors influencing longer flower longevity of the temperate zone plants (116). Third, pollination by vertebrates is almost nonexistent in forest communities in the north temperate zone, though birds and rodents constitute an important group of pollinators in the temperate zone Australia (49, 124) and in south Africa (15, 124). Finally, the proportion of wind-pollinated plants steadily increases as one moves from the equatorial region, reaching 80–100% among trees in some of the northernmost latitudes (119).

## *Tropical Montane Rain Forests*

Pollination systems in tropical montane rain forests appear to differ from those in the lowland rain forests in at least two respects. First, pollination by hummingbirds in the neotropics is more common in the montane than in the lowland rain forests (39, 47). For example, in a cloud forest of Costa Rica with 600 flowering plant species, hummingbirds have been observed to visit flowers of 100 species (47); a comparable figure for a lowland rain forest is less than 70 out of 1800 species of flowering plants (137). Second, the pollination system involving small generalist insects also appears to be more widespread in cloud forests (131, 145). Low temperatures in montane forests may limit the activity of bees and may explain their displacement by the hummingbird (39) and the generalist insect pollination system.

The number of bat species in all feeding guilds is known to decrease with an increase in altitude (59a). However, it is not known if the tropical montane forests have disproportionately fewer species of bat-pollinated plants than the lowland forests.

Montane forests also differ from the lowland forests with respect to sexual and breeding systems (Table 2). Proportionately, many more tree species are self-compatible in high altitude forests (69, 131, 145). Unpredictable weather conditions for pollination in the generally cold and wet environments in montane forests have been invoked to explain the high incidence of self-compatibility. Lack of strong selection for outcrossing or direct selection for homozygosity could also explain the preponderance of self-compatibility. However, the proportion of dioecious species in montane forests is similar to or exceeds that found in TLRF (131, 145). The prevalence of two somewhat opposite modes of reproduction in the same community defies an easy

**Table 2** Distribution of self-compatible, self-incompatible, and dioecious tree species in tropical lowland and montane rain forests.

| Forest Type | Percentage self-compatible[1] species | Percentage self-incompatible[1] species | Percentage dioecious species[2] | References |
|---|---|---|---|---|
| Tropical lowland rain forest, Costa Rica | 20 | 80 | 23 | 24 |
| Montane forest, Venezuela | 62 | 28 | 31 | 131 |
| Montane forest, Jamaica | 85 | 15 | 21 | 145 |

[1] Expressed as percentage of hermaphroditic species tested for self-incompatibility.
[2] Expressed as percentage of all tree species.

explanation. The preponderance of dioecious species in montane forests may not necessarily be due to selection for outcrossing in these ecosystems, but may be due to other advantages associated with dioecy (16, 17).

Finally, the individual flowers in montane forests on an average last 2–8 days more than in lowland rain forests (141); the longer life span is consistent with the notion of unpredictability in pollination, a possibility suggested above.

## *Southeast Asian Lowland Rain Forests*

The aseasonal southeast Asian rain forests are well known for the irregular, supraannual flowering at the community level (4). These forests also differ from most neotropical rain forests in having members of a single family, Dipterocarpaceae, dominating the canopy (12). Another unusual feature in some of these forests is that many species of the dominant genus *Shorea* may be pollinated by thrips (6). The thrips complete their life cycle during the flowering of congeneric sympatric species. Species of *Shorea* in south Asia are pollinated by bees, *Apis dorsata* and *A. indica* (40). The impact of irregular, supraannual flowering on the stability of the pollinator fauna, thrips and nonthrips remains unexplored. The temporal fluctuations in the abundance of pollinators may be responsible for the evolution of apomixis, which has been reported in some trees in southeast Asian rain forests (62, 85).

## *Global Patterns*

Do rain forests in different parts of the world differ with respect to the proportion of various pollination systems? Such differences might be expected on the basis of differences in geographical distribution of plants and animals. For example, hummingbirds which dominate the bird pollination systems in the neotropics are confined to the new world. Although other birds serve as pollen vectors in the old world tropics, their role in terms of species pollinated is not as well documented as in Central American forests. Bats have coevolved with plants over a much longer geological time scale in the paleotropics than in the neotropics, and completely vegetarian bats, as mentioned earlier, are confined to the old world (13). But we do not know whether the proportion of plants pollinated by bats in the paleotropics is greater than in the neotropics. Irvine & Armstrong (76) suggest that the frequency of beetle pollination is much greater in Australian than in Central American forests, but the latter have not been completely surveyed and the reported differences may be a sampling artifact. Finally, for bees, the most dominant pollen vectors in all TLRF, Roubik (122) reports similar patterns for the old and the new world tropics with respect to the proportion of species numbers in major families.

## SPECIALIZATION

The issue of specialization is critical to the discussion of the role of plant-pollinator interactions in speciation and community stability, as we note later.

On the basis of existing evidence, species in tropical lowland rain forests may be distinguished into three categories. To the first category belong species like figs that are extremely specialized in their pollinator requirements and the pollinating fig wasps which are very host specific (117, 152). However, the type of specific interaction exemplified by figs and fig wasps appears rare in tropical lowland rain forests.

The second category is exemplified by orchids and orchid bees and other plants that, as a taxonomic group, are pollinated by a particular assemblage of animals. Many species of orchids in the neotropics are pollinated by male euglossine bees (41, 42, 123). Each species of orchid may be visited by one or two species (1, 47). Similarly bees of any one species of euglossines may visit as many as nine different species of orchids, but most confine their visits to only one or two species. However, the euglossine bees also gather resources from many other plant species. (47). Similarly a majority of 20 species of *Dalechampia* (Euphorbiaceae) are pollinated by female euglossine bees only, and each species is visited by one or two species of the bees (7, 8).

Preliminary observations suggest that the type of specialization exemplified by orchids and orchid bees also exists in other plant-pollinator groups. For example, each of the five species of angraecoid orchids in Madagascar is pollinated by one species of hawkmoth, *Panogena lingens* (104). Several species of the Araceae and Annonaceae are pollinated by one or two species of scarab beetles (155; G. Schatz, personal communication).

At the next level of decreasing specialization are examples like the *Heliconia* and the hummingbirds. The nine sympatric species of the genus *Heliconia* (Musaceae) in a tropical lowland rain forest are visited by nine species of hummingbirds (136). Although as many as eight species have been recorded visiting one species of *Heliconia,* most species are predominantly visited by one or two species of hummingbirds. A group of species in Lecythidaceae are pollinated by euglossine bees, and the geographic ranges of both groups are known to coincide (101). In bat-pollinated species, evidence from tropical dry deciduous forests suggests that a given plant species is visited by several species of bats (65) though examples exist of a plant species being pollinated by a particular species of bat (59).

A large number of species in tropical lowland rain forests are pollinated by medium-sized to large bees, as mentioned earlier. Current data from several tree species in the Fabaceae indicate that flowers of most species, though displaying great morphological complexity for pollination by a specific group of bees, are visited by a large number of bee species. For example, as stated earlier, Perry & Starrett (109) reported 19 species of bees visiting the flowers of a single large canopy tree of *Dipteryx panamensis*. In a dry deciduous forest, Frankie et al (52) captured 70 species of bees on the flowers of *Andira inermis,* also a large tree. Even though the diversity of bee species is very high, it is possible that only one or a few species constitute the effective group of pollinators.

To the third category belong species that apparently have a generalist mode of pollination. These species bear small flowers in which pollen and nectar are accessible to a wide range of small insects, such as bees, butterflies, beetles, flies, and wasps which collectively visit the flowers (25). It is not known if the visitors differ in their effectiveness as pollinators. In one such species, *Calathea ovandensis,* ten species of Hymenoptera and Lepidoptera were found to visit flowers, but Schemske & Horvitz (128) showed that one species of Hymenoptera was responsible for 66% and another species for 14% of all fruit set; collectively five species of Hymenoptera accounted for 99% of the fruit set. Even species that appear to exploit a wide range of pollen vectors may thus in practice be pollinated effectively by only one or two species.

In general, the type of specialized, almost one-to-one relationship that exists between figs and fig wasps is an exception rather than the norm in the tropics (47). Nonetheless in a majority of species, pollination systems are specialized to the extent that a given plant species is pollinated by one or a few species belonging to the same taxonomic group (e.g. euglossine or other bees, hummingbirds, scarab beetles, bats, etc). A further level of specialization may exist, but studies to evaluate the relative effectiveness of various flower visitors in achieving pollen dispersal and pollen deposition are lacking.

## POLLEN FLOW

Tropical forest plants with their diverse patterns of dispersion and modes of pollination provide an ideal material to compare the effectiveness of various pollinators in long-distance pollen flow. Nevertheless, little is known about dispersal of pollen in TLRF. Several lines of evidence, however, suggest that pollen flow in tree species may be extensive. First, most species are either self-incompatible or dioecious (11, 16, 24, 36, 61). Apomixis is known in some species of the south-east Asian Dipterocarp forests (62, 85), but the true extent of apomixis within individuals and species or among populations has not been determined. Second, studies based on mark-recapture techniques indicate that bees (52, 79) and hawkmoths (93) forage over long distances and have the potential for pollen flow among widely spaced conspecifics. Third, direct observations of flight patterns also reveal that some pollinators—bats, for example—forage over distances of many kilometers (65, 132). On the other hand, territorial hummingbirds move pollen over restricted distances, though nonterritorial hummingbirds forage over long distances (92). Data derived from the studies of sexual systems, breeding systems, and flight patterns of pollen vectors, however, provide an estimate only of potential pollen flow. Realized pollen flow may not correspond with potential flow; there is evidence that in some species crosses involving conspecifics that are many hundred meters apart yield more fruits than among individuals that are relatively close to each other (89).

Genetic markers offer considerable promise in understanding the patterns of realized pollen flow in tropical rain forests (23, 31, 105, 106). Recent studies of mating systems based on progeny arrays of individual trees, utilizing genetic markers, have revealed a high degree of outcrossing and indicated potential for extensive pollen flow in several large canopy trees (21, 95, 96). Similarly, analysis of the population genetic structure of several species indicates low values for gene flow among populations (64). Both electrophoretic markers (105, 106) and the DNA "fingerprinting" (121) have the potential in the future considerably to enhance our understanding of pollen flow within and among populations.

The linkage of conspecifics by means of pollen flow over a large area in canopy and subcanopy tree species does not negate the possibility of restricted pollen flow and the potential for local genetic differentiation due to inbreeding in other taxa. In contrast to trees, many herbs and shrubs in TRLF are self-compatible; also the frequency of dioecy in the understory plants is only half of that encountered in tree species (90). It has been suggested that pollen flow in understory plants may be generally restricted (90), and that such taxa, especially in montane forests, may be largely inbred (131, 145). Many herbs, especially epiphytes, are also patchily distributed. Localized gene dispersal due to limited pollen flow in such species could result in subdivision of the populations. However, the neighborhood size that determines the potential for subdivision within a population is a function of both the distance over which pollen is dispersed and the density of individuals (153). Although pollen flow in understory plants may be localized, such plants, because of their small size, have much higher densities than canopy trees. Thus, the reduction in neighborhood size relative to those in trees may not be as great as expected from gene flow alone.

Fedorov (45) argued that a main contributor to the origin and maintenance of many closely related species in the humid tropics may be inbreeding combined with drift. In the absence of population genetic data, it is difficult to evaluate the validity of this argument. However, results of two population genetic studies are consistent with the notion of inbreeding and genetic drift in understory plants. A preliminary survey of genetic variation in some species of *Piper*, one of the most species-rich understory genera in Costa Rica, shows little genetic diversity within populations (70) and indicates inbreeding. Similarly, low levels of genetic variation within populations, but high levels among populations, in a species complex of shrubs in the Gesneriaceae have been revealed (144). Autogamy and self-compatibility seem prevalent in several other taxa of herbs and shrubs, e.g. Marcgraviaceae (56), Ericaceae (97), and Melastomotaceae (120). Gentry (55) provides indirect evidence for the importance of drift and inbreeding as factors in speciation of many tropical plants.

In summary, although there is considerable evidence for outcrossing and long-distance pollen flow in several species, some data suggest that many taxa, primarily herbs and shrubs, may be highly inbred.

## POLLINATION AND SPECIATION

To what extent do plant-pollinator interactions contribute to species richness of TLRF? Although the role of pollinators as isolating mechanisms is well known (60, 142), the part that plant-pollinator interactions might have played in speciation in tropical communities has not been adequately evaluated (72). As I have argued elsewhere (19), ecological interactions between plants and animals by themselves or in combination with other factors may promote speciation in several ways. The following arguments are from Bawa (19).

First, a founder population can be reproductively isolated from the parental species if it interacts with a pollinator that has no or little interaction with the ancestral species. Geographical ranges of plants and their pollinators are often dissimilar (G. Stiles, personal communication). Thus, plants with a slightly variant floral morphology may be exposed to a different assemblage of pollinators. The floral variants in small, isolated, founder populations may be "fixed" not necessarily by genetic drift but by a new set of pollinators.

Second, differentiation of plant populations with a variant floral morphology may also lead to differentiation of host-specific pollinators; plant-pollinator interactions do lead to cospeciation more often than other kinds of mutualisms (133, 147). There is no direct evidence for cospeciation of plants and their pollinators from tropical rain forests, but it is suspected to have occurred in figs and pollinating fig wasps (117, 151).

Third, West-Eberhard (150) has argued that under selection for success in intraspecific competition (including competition for mates), characters important in the outcome of competition can undergo quick change, leading to rapid population divergence and speciation. Thus, the combined effects of plant-pollinator coevolution and sexual selection can accelerate speciation. West-Eberhard suggests that competition for mates via pollinators in groups with extremely specific pollinators could be a significant diversifying force in plant evolution. Apparently, in plants, the floral variants may arise first as a result of sexual selection and then rapidly spread and be isolated by specific pollinators.

Fourth, plant-pollinator coevolution in combination with sexual selection in pollinators rather than in plants, as in the previous example, may enhance the rate of population divergence and hence speciation (87). Kiester et al (87) suggest that in orchids, sexual selection in euglossine bees, based on variation in their mating behavior due to variation in chemical odors collected from flowers, may lead to genetic instability in bee populations. This instability in conjunction with the selection by the orchid bees on floral characters may result in explosive cospeciation.

Finally, genetic drift, alone, or in conjunction with inbreeding combined with coevolution and sexual selection in plants or pollinators, or both, should accelerate speciation. Both West-Eberhard (150) and Kiester et al (87) assign a major role to drift in their models which consider the combined effects of coevolution and sexual selection in speciation. It is generally accepted that inbreeding has the potential to cause rapid population divergence (153).

The validity of the arguments above depends upon the prevalence of high specificity between plants and pollinators, sexual selection, genetic drift in plants, and inbreeding. I have already reviewed the evidence of plant-pollinator specificity as well as genetic drift and inbreeding in tropical rain forests. A review of sexual selection is beyond the scope of this paper, but arguments for the operation of sexual selection in tropical plants have been made before (16, 32).

Overall, specificity in plant-pollinator coevolution is critical in promoting continued cycles of speciation (87). Accordingly, high specificity should be positively correlated with species richness at various taxonomic levels. *Ficus* is the largest genus in the Moraceae. It also has the most specialized mode of pollination in the family. Locally, *Ficus* usually has a greater number of species than any other genus of the Moraceae. In the neotropics, many genera in such families as the Annonaceae, Lauraceae, and Rubiaceae display high specificity as well as considerable species richness. However, specificity in plant-pollinator interactions may not be a characteristic feature of all the species-rich genera because factors other than plant-pollination coevolution also play a role in speciation (133).

## COMMUNITY STABILITY

Stability, as defined here, refers to the ability of all populations to return to equilibrium following perturbation (111). It has been generally asserted that the abundance of obligate mutualisms in tropical rain forests makes such communities prone to instability (53, 99). Another viewpoint is that the evolution of obligate mutualisms requires stringent conditions, and such mutualisms are relatively rare in natural communities (74). According to this viewpoint, mutualistic interactions often involve a diverse array of species. For example, a given species of plant may be pollinated by a wide variety of animals, and conversely a particular pollinator species may use floral resources of a wide variety of species. Selective pressures exercised by interacting species on each other are thus highly diffuse and often asymmetrical. As a result, removal of one of the interacting species is not likely to have a significant effect on the stability of the system. Obviously specificity to a large extent determines the effect of the disruption of a mutualistic interaction on community stability.

The disruption of a mutualistic interaction can influence stability in two ways. First, the effect may be direct, as, for example, the loss of one of the interacting partners in species-specific interactions may lead to the extinction of the other. Second, the effect may be indirect. For instance, the loss of a fig species and its pollinating wasp species may also lead to a loss of the nonpollinating wasp species that parasitize the pollinating species. Such an effect, referred to as the ripple effect (147), can extend through a large part of the community, depending upon the number of interacting species and the strength of the interactions. Apart from specificity, the importance of a species as a critical resource may be a primary determinant of the consequences of a disruption in plant-pollinator mutualism. Fluctuations in populations of keystone mutualists (57, 146) that provide resources when other resources are scarce or not available are expected to have a drastic effect on the community. Fig trees are presumed to be keystone resources because they provide fruits to a large number of primates and birds when the overall abundance of fruits in the community is low (146). A disruption of the pollination system in figs thus has consequences not only for figs and pollinating and nonpollinating fig wasps, but also for a large segment of the frugivore community. The disappearance of species like figs then could have a ripple effect throughout the community. Gilbert (57) and Howe (74) provide other examples.

There are two major problems in assessing how community stability may be influenced by a breakdown in plant-pollinator interactions. First, our understanding of the way in which pollinators interact with plants is very elementary. Flowers not only provide food to the pollinators but also act as sites of mating and predator avoidance (129). Furthermore, vegetative parts of the plants whose flowers are used as sources of food or mating sites by adults may provide food for the larvae. Adults may thus use a wide variety of species as sources of pollen and nectar, but only one species as a larval host. The extinction of this larval host will result in the extinction of the pollinator species and may also affect the host species serviced by the pollinator; but these effects may not be anticipated if the focus of attention is the interaction between the adults and the flowers. Second, although a number of qualitative models have explored the impact of perturbations in plant-pollinator interactions on community stability (30, 57) there is no formal treatment of the subject.

The effect of habitat fragmentation on plant-pollinator interactions and consequently on community stability is another area of increasing concern. Insect abundance as well as diversity is known to decrease with a decrease in the size of the habitat (88). Furthermore, by altering light regimes and other microclimatic conditions (94) edge effects may also influence the composition and foraging of pollinators. Changes in composition and abundance of specific pollinators in small forested areas have been shown to result in lowered seed set in plants (84).

Small, isolated habitats may also lack habitat heterogeneity to support pollinator populations all year round. Nectarivorous bats on a diurnal basis (132) and moths on a seasonal basis (83) have been shown to utilize resources from distinct habitats, often involving different vegetation types or zones, separated by several kilometers. Clearly, plant-pollinator interactions can be severely disrupted in small, isolated, fragments of vegetation. The consequences for plants may be not only lowered reproductive output, but also altered patterns of pollen flow. Gene flow among small fragmented habitats via pollen (and seed) may be curtailed. The resulting inbreeding may further decrease fruit set. Changes in pollination and mating systems may thus act synergistically to lower reproductive output. However, the effects of such changes as manifested themselves in decreased regeneration may remain obscure for a long time.

## CONCLUDING REMARKS

Plant-pollinator interactions provide model systems to address a wide variety of ecological and evolutionary questions. Here I have briefly explored their role in microevolution and speciation of tropical forest plants, community structure, and community stability. Basic information about the natural history of plant-pollinator interactions in tropical lowland rain forests is obviously limited. Nevertheless, enough is known that precise, testable, hypotheses can be formulated. Among the subjects reviewed here, the following require special attention.

First, it is apparent that there are well-defined "guilds" of pollinators with an associated set of "host" plants in TLRF. A detailed study of these "guilds" and of the plants with which they interact, along the lines of Stiles' (136, 137) work on hummingbird—plant interactions, is essential for understanding the structure and organization of a particular class of plant-pollinator interactions and their relative role in maintaining the overall organization of the whole community.

Second, empirical studies are needed to determine the specificity of plant-pollinator interactions. Flowers of tropical forest plants receive a diverse array of visitors, but only a few act as effective pollinators. Schemske & Horvitz's (128) study provides a model for similar work on other species. The determination of specificity is critical to our notions of community stability and to evaluate the degree of coevolution between plants and their pollinators.

Third, pollination systems in plants are known to be the primary determinants of population genetic structure (95). The diversity of pollination systems combined with diverse patterns of dispersion and distribution of plants provides a novel material for comparative evaluation of the roles of pollinators and plant density in the genetic structure and microevolution of plant populations. Information on pollen flow and population genetic structure are also critical to the adequate conservation and management of forest genetic resources.

Fourth, in order to understand the role of plant-pollinator interactions in speciation, one needs to study such interactions in genera with a large number of sympatric species (19). It is particularly important to investigate the origin of variation in floral characters involved in sexual selection, especially in peripheral populations where incipient speciation is likely to occur (34, 35).

Fifth, the effect of forest fragmentation on plant-pollinator interactions is likely to assume special significance as efforts to conserve biodiversity in nature reserves continue to gain momentum. Extreme fragmentation and isolation of habitats can drastically affect major mutualistic interactions (83). The maintenance of biodiversity in fragmented reserves would require a knowledge of the dynamics of the key mutualistic interactions. In the vast majority of tropical rain forest reserves, virtually nothing is known about the basic plant-pollinator (and other plant-animal) interactions.

ACKNOWLEDGMENTS

Richard Primack provided useful commentary on an earlier draft of this manuscript. This research was supported by the US National Science Foundation, the Guggenheim Foundation and the University of Massachusetts Faculty Development Fund.

## *Literature Cited*

1. Ackerman, J. D. 1983. Specificity and the mutual dependency of the orchid-euglossine bee interaction. *Biol. J. Linn. Soc.* 20:301–14
2. Ali, S. A. 1932. Flower birds and bird flowers in India. *J. Bombay Nat. Hist. Soc.* 35:573–605
3. Appanah, S. 1981. Pollination in Malaysian primary forests. *Malaysian For.* 44:37–42
4. Appanah, S. 1985. General flowering in the climax rain forests of South-East Asia. *J. Trop. Ecol.* 1:225–40
5. Appanah, S. 1990. See Ref. 22
6. Appanah, S., Chan, H. T. 1981. Thrips: the pollinators of some dipterocarps. *Malaysian For.* 44:234–52
7. Armbruster, W. S. 1986. Reproductive interactions between sympatric *Dalechampia* species: are natural assemblages "random" or organized? *Ecology* 67:522–33

8. Armbruster, W. S., Webster, G. L. 1979. Pollination of two species of *Dalechampia* (Euphorbiaceae) in Mexico by euglossine bees. *Biotropica* 11: 278–83
9. Armstrong, J. E., Drummond, B. A. 1986. Floral biology of *Myristica fragrans* Houtt. (Myristicaceae), the nutmeg of commerce. *Biotropica* 18:32–38
10. Armstrong, J. E., Irvine, A. K. 1989. Floral biology of *Myristica insipida* R. Br. (Myristicaceae), a distinctive beetle pollination syndrome. *Am. J. Bot.* 76: 86–94
11. Ashton, P. S. 1969. Speciation among tropical forest trees: some deductions in the light of recent evidence. *Biol. J. Linn. Soc.* 1:155–96
12. Ashton, P. S. 1988. Dipterocarp biology as a window to the understanding of tropical forest structure. *Annu. Rev. Ecol. Syst.* 19:347–70
13. Baker, H. G. 1973. Evolutionary relationships between flowering plants and animals in American and African tropical forests. In *Tropical Forest Ecosystems in Africa and South America: A Comparative Review,* ed B. J. Meggers, E. S. Ayensu, D. Duckworth, pp. 145–59. Washington, DC: Smithsonian Inst. Press
14. Baker, H. G., Cruden, R. W., Baker, I. 1971. Minor parasitism in pollination biology and its community function: the case of *Cieba Acuminata. Bioscience* 21:1127–29
15. Baker, H. A., Oliver, E. G. H. 1967. *Ericas in Southern Africa.* Cape Town: Purnell
16. Bawa, K. S. 1980. Evolution of dioecy in flowering plants. *Annu. Rev. Ecol. Syst.* 11:15–39
17. Bawa, K. S. 1982. Outcrossing and the incidence of dioecism in island floras. *Am. Nat.* 119:866–71
18. Bawa, K. S. 1983. Patterns of flowering in tropical plants. In *Handbook of Experimental Pollination Biology,* ed. C. E. Jones, R. J. Little, pp. 394–410. New York: Van Nostrand, Reinhold
19. Bawa, K. S. 1989. Mating systems, genetic differentiation and speciation in tropical rain forest plants. *Biotropica.* (In review)
20. Bawa, K. S., Beach, J. H. 1981. Evolution of sexual systems in flowering plants. *Ann. Mo. Bot. Gard.* 62:254–74
21. Bawa, K. S., Crisp. J. E. 1980. Wind pollination in the understorey of a rain forest in Costa Rica. *J. Ecol.* 68:871–76
22. Bawa, K. S., Hadley, M., eds., 1990. *Reproductive Ecology of Tropical Forest Plants.* Carnforth, England: Parthenon
23. Bawa, K. S., O'Malley D. M. 1987. Estudios geneticos y de systemas de cruzamiento en algunas especies arboreas de bosques tropicales. *Rev. Biol. Trop.* 35(Suppl. 1):177–88
24. Bawa, K. S., Perry, D. R., Beach, J. H. 1985. Reproductive biology of tropical lowland rain forest trees. I. Sexual systems and self-incompatibility mechanisms. *Am. J. Bot.* 72:331–45
25. Bawa, K. S., Perry, D. R., Bullock, S. H., Coville, R. E., Grayum, M. H. 1985. Reproductive biology of tropical lowland rain forest trees. II. Pollination mechanisms. *Am. J. Bot.* 72:346–56
26. Beach, J. H. 1982. Beetle pollination of *Cyclanthus bipartitus* (Cyclanthaceae). *Am. J. Bot.* 69:1074–81
27. Beach, J. H. 1984. The reproductive biology of the peach or "Pejibaye" palm *(Bactris gasipaes)* and a wild cogener *(B. perschiana)* in the Atlantic lowlands of Costa Rica. *Principes* 28:107–19
28. Beaman, R. S., Decker, P. J., Beaman, J. H. 1988. Pollination of *Rafflesia* (Rafflesiaceae). *Am. J. Bot.* 75:1148–62
29. Borchert, R. 1983. Phenology and control of flowering in tropical trees. *Biotropica* 15:81–89
30. Boucher, D. H., ed. 1985. *The Biology of Mutalisms: Ecology and Evolution.* London: Croom & Helm
31. Buckley, D. P., O'Malley, D. P., Apsit, V., Prance, G. T., Bawa, K. S. 1988. Genetics of Brazil nut (*Bertholletia excelsa* Humb. & Bonpl.: Lecythidaceae). 1. Genetic variation in natural populations. *Theor. Appl. Genet.* 76:923–28
32. Bullock, S. H., Bawa, K. S. 1981. Sexual dimorphism and the annual flowering pattern in *Jacaratia dolichaula* (D. Smith) Woodson (Caricaceae) in a Costa Rican rain forest. *Ecology* 62:1494–1504
33. Burger, W. C. 1981. Why are there so many kinds of flowering plants? *Bioscience* 31:572–81
34. Carson, H. L. 1985. Unification of speciation theory in plants and animals. *Syst. Bot.* 10:380–90
35. Carson, H. L. 1987. The genetic system, the deme, and the origin of species. *Annu. Rev. Genet.* 21:405–23
36. Chan, H. T. 1981. Reproductive biology of some Malaysian Dipterocarps. III. Breeding systems. *Malaysian For.* 44:28–34
37. Cole, B. J. 1981. Overlap, regularity and flowering phenologies. *Am. Nat.* 117:993–97
38. Cox, P. A. 1984. Chiropterophily and ornithophily in *Frecycinetia* (Panda-

naceae) in Samoa. *Plant Syst. Evol.* 144:277–90
39. Cruden, R. W. 1972. Pollinators in high-elevation ecosystems: relative effectiveness of birds and bees. *Science* 176:1439–40
40. Dayanandan, S., Attygalla, D. N. C., Abegunasekera, A. W. W. L., Gunatilleke, I. A. U. N., Gunatilleke, C. V. S. 1990. See Ref. 22
41. Dodson, C. H. 1975. Coevolution of orchids and bees. In *Coevolution of Animals and Plants,* ed. L. Gilbert, P. Raven, pp. 91–99. Austin: Univ. Texas Press
42. Dressler, R. L. 1968. Pollination by euglossine bees. *Evolution* 22:202–10
43. Equiarte, L., Rio, C. M., Arita, H. 1987. El nectar y el polen como recursos: el papel ecologica de los visitantes a las flores de *Pseudobombax ellipticum* (H.B.K.) Dugand. *Biotropica* 19:74–82
44. Faegri, K., van Der Pijl, L. 1971. *Principles of Pollination Ecology*. Oxford: Pergamon. 248 pp.
45. Fedorov, A. A. 1966. The structure of the tropical rain forest and speciation in the humid tropics. *J. Ecol.* 54:1–11
46. Feinsinger, P. 1978. Ecological interactions between plants and hummingbirds in a successional tropical community *Ecol. Monogr.* 48:269–87
47. Feinsinger, P. 1983. Coevolution and pollination. In *Coevolution,* ed. D. J. Futuyma, M. Slatkin, pp. 282–310. Sunderland, Mass: Sinauer
48. Feinsinger, P., Linhart, Y. B., Swarm, L. A., Wolfe, J. A. 1979. Aspects of the pollination biology of three *Erythrina* species on Trinidad and Tobago. *Ann. Mo. Mot. Gard.* 66:451–71
49. Ford, H. A., Paton, D. C., Forde, N. 1979. Birds as pollinators of Australian plants. *N.Z. J. Bot.* 17:509–19
50. Frankie, G. W. 1975. Tropical forest phenology and pollinator plant coevolution. In *Coevolution of Animals and Plants,* ed. L. E. Gilbert, P. H. Raven, pp. 282–310. Austin: Univ. Texas Press
51. Frankie, G. W., Baker, H. G., Opler, P. A. 1974. Comparative phenological studies of trees in tropical wet and dry forests in the lowlands of Costa Rica. *J. Ecol.* 62:881–919
52. Frankie, G. W., Opler, P. A., Bawa, K. S. 1976. Foraging behavior of solitary bees: implications for outcrossing of a neotropical forest tree species. *J. Ecol.* 64:1049–57
53. Futuyma, D. J. 1973. Community structure and stability in constant environments. *Am. Nat.* 107:443–46
54. Gentry, A. 1974. Flowering phenology and diversity in Bignoniaceae. *Biotropica* 6:64–68
55. Gentry, A. 1982. Neotropical floristic diversity: phytogeographical connections between Central and South America. Pleistocene climatic fluctuations or an accident of Andean orogeny? *Ann. Mo. Bot. Gard.* 69:557–93
56. Gentry, A. 1989. See Ref. 72, pp. 113–34
57. Gilbert, L. E. 1980. Food web organization and conservation of neotropical diversity. In *Conservation Biology,* ed. M. E. Soule, B. A. Wilcox, pp. 11–34. Sunderland, Mass.: Sinauer
58. Gottsberger, I. S., Gottsberger, G. 1975. Über Sphingophile Angiospermen Brasiliens. *Plant Syst. Evol.* 123:157–84
59. Gould, E. 1978. Foraging behavior of Malaysian nectar-feeding bats. *Biotropica* 10:184–93
59a. Graham, G. L. 1983. Changes in bat species diversity along an elevational gradient up the Peruvian Andes. *J. Mamm.* 64:559–571
60. Grant, V. 1949. Pollination systems as isolating mechanisms in angiosperms. *Evolution* 3:82–97
61. Ha, C. O., Sands, V. E., Soepadmo, E., Jong, K. 1988. Reproductive patterns of selected understorey trees in the Malaysian rain forest: the sexual species. *Bot. J. Linn. Soc.* 97:295–316
62. Ha, C. O., Sands, V. E., Soepadmo, E., Jong, K. 1988. Reproductive patterns of selected understorey trees in the Malaysian rain forest: the apomictic species. *Bot. J. Linn. Soc.* 97:317–31
63. Haber, W. A., Frankie, G. W. 1989. A tropical hawkmoth community: Costa Rican dry forest Sphingidae. *Biotropica* 21:155–72
64. Hamrick, J. L., Loveless, M. D. 1989. Genetic structure of tropical tree populations: associations with reproductive biology. In *The Evolutionary Ecology of Plants,* ed. J. H. Bock, Y. B. Linhart, pp. 129–49. San Francisco: Westview
65. Heithaus, E. R., Fleming, T. H., Opler, P. A. 1975. Foraging patterns and resource utilization in seven species of bats in a seasonal tropical forest. *Ecology* 56:841–54
66. Heithaus, E. R., Opler, P. A., Fleming, T. H. 1974. Bat activity and pollination of *Bauhinia pauletia:* plant-pollinator coevolution. *Ecology* 55:412–19
67. Henderson, A. H. 1985. Pollination of *Socratea exorrhiza* and *Iriartia ventricosa. Principes* 29:64–71
68. Henderson, A. H. 1986. A review of pollination studies in the *Palmae. Bot. Rev.* 52:221–59

69. Hernandez, H. M., Abud, Y. C. 1987. Notas sobre la ecologia reproductiva de arboles en un bosque mefofilo de montana en Michoacan, Mexico. *Bol. Soc. Bot. Mex.* 47:5–35
70. Heywood, J. S., Fleming, T. H. 1986. Patterns of allozyme variation in three Costa Rican species of *Piper*. *Biotropica* 18:208–13
71. Hickman, J. C. 1974. Pollination by ants: a low energy system. *Science* 184:1290–92
72. Holm-Nielsen, L. B., Nielsen, I. C., Basslev, H., eds. 1989. *Tropical Forest: Botanical Dynamics, Speciation and Diversity*. New York: Academic. 380 pp.
73. Hopkins, H. C. 1984. Floral biology and pollination ecology of the neotropical species of *Parkia*. *J. Ecol.* 72:1–23
74. Howe, H. F. 1983. Constraints on the evolution of mutualisms. *Am. Nat.* 123:764–77
75. Howe, H. F., Westley, L. C. 1988. *Ecological Relationships of Plants and Animals*. New York: Oxford Univ. Press
76. Irvine, T. K., Armstrong, J. E. 1990. See Ref. 22
77. Janson, C. H., Terborgh, J., Emmons, L. H. 1981. Non-flying mammals as pollinating agents in the Amazonian forest. *Biotropica* 13 (Suppl.): 1–6
78. Janzen, D. H. 1970. Herbivores and the number of tree species in tropical forests. *Am. Nat.* 104:501–28
79. Janzen, D. H. 1971. Euglossine bees as long distance pollinators of tropical plants. *Science* 171:203–5
80. Janzen, D. H. 1975. *Ecology of Plants in the Tropics*. London: Edward Arnold
81. Janzen, D. H. 1979. How to be a fig. *Annu. Rev. Ecol. Syst.* 10:13–51
82. Janzen, D. H. 1985. The natural history of mutualisms. See Ref. 30, pp. 39–99
83. Janzen, D. H. 1987. Insect diversity of a Costa Rican dry forest: why keep it, and how? *Biol. J. Linn. Soc.* 30:343–56
84. Jennersten, O. 1988. Pollination in Dianthus deltoides (Caryophyllaceae): Effects of habitat fragmentation on visitation and seed set. *Conserv. Biol.* 2:359–66
85. Kaur, A., Ha, C. D., Jong, K., Sands, V. E., Chan, H. T., et al. 1978. Apomixis may be widespread among trees of the climax rain forest. *Nature* 270:440–41
86. Kevan, P. G., Baker, H. G. 1983. Insects as flower visitors and pollinators. *Annu. Rev. Entomol.* 28:407–53
87. Kiester, A. R., Lande, R., Schemske, D. W. 1984. Models of coevolution and speciation in plants and their pollinators. *Am. Nat.* 124:220–43
88. Klein, B. C. 1989. Effects of forest fragmentation on dung and carrion beetle communities in central Amazonia. *Ecology* 70:1715–25
89. Koptur, S. T. 1984. Outcrossing and pollinator limitation of fruit set; breeding systems of neotropical Inga trees (Fabaceae: Mimosoideae). *Evolution* 38:1130–43
90. Kress, W. J., Beach, J. H. 1990. Flowering plant reproductive systems at La Selva Biological Station. Ms.
91. Levin, D. A. 1975. Pest pressure and recombination systems in plants. *Am. Nat.* 109:437–51
92. Linhart, Y. B. 1973. Ecological and behavioral determinants of pollen dispersal in hummingbird pollinated *Heliconia*. *Am. Nat.* 107:115–23
93. Linhart, Y. B., Mendenhall, J. A. 1977. Pollen dispersal by hawkmoths in a *Lindenia rivalis* Benth. population in Belize. *Biotropica* 9:143
94. Lovejoy, T. E., Bieergaard, R. O. Jr., Rylands, A. B., Malcolm, J. R., Quintela, C. E., et al. 1986. Edge and other effects of isolation on Amazon forest fragments. In *Conservation Biology*, ed. M. E. Soule, pp. 257–85. Sunderland, Mass: Sinauer
95. Loveless, M. D., Hamrick, J. L. 1984. Ecological determinants of genetic structure in plant populations. *Annu. Rev. Ecol. Syst.* 15:65–90
96. Lumer, C. 1980. Rodent pollination of Blakea (Melastomataceae) in a Costa Rican cloud forest. *Brittonia* 32:512–17
97. Luteyn, J. L. 1989. See Ref. 72, pp. 297–307
98. Marshall, A. G. 1983. Bats, flowers and fruit: evolutionary relationships in the old world. *Biol. J. Linn. Soc.* 20:115–35
99. May, R. 1973. *Stability and Complexity in Model Ecosystems*. Princeton, NJ: Princeton Univ. Press
100. Mitchell, A. 1982. *Reaching the Rain Forest Roof: A Handbook on Techniques of Access and Study in the Canopy*. Leeds, UK: Leeds Philos. Lit. Soc.
101. Mori, S. A. 1989. See Ref. 72, pp. 319–32
102. Mori Urpi, J. 1982. Pollination en *Bactris gasipaes* H.B.K. (Palmae). *Rev. Biol. Trop.* 28:153–74
103. Neill, D. A. 1988. Experimental studies on species relationships in *Erythrina* (Leguminosae: Papilionoideae). *Ann. Mo. Bot. Gard.* 75:886–969
104. Nilsson, L. A., Jonsson, L., Ralison, L., Randrianjohany, E. 1987. Angraceoid orchids and hawkmoths in Central Madagascar: specialized pollination

systems and generalist foragers. *Biotropica* 19:310–18
105. O'Malley, D. M., Bawa, K. S. 1987. Mating system of a tropical rain forest tree species. *Am. J. Bot.* 74:1143–49
106. O'Malley, D. M., Buckley, D. P., Prance, G. T., Bawa, K. S. 1988. Genetics of Brazil nut (*Bertholletia excelsa* Humb. Bonpl.: Lecythidaceae).2. Mating system. *Theor. Appl. Genet.* 76: 929–32
107. Opler, P. A. 1983. Nectar production in a tropical ecosystem. In *The Biology of Nectaries,* ed. B. Bentley, T. Elias, pp. 30–79. New York: Columbia Univ. Press
108. Perry, D. R. 1978. A method of access into the crown of emergent and canopy trees. *Biotropica* 10:155–57
109. Perry, D. R., Starrett, A. 1980. The pollination ecology and blooming strategy of a neotropical emergent tree, *Dipteryx panamensis*. *Biotropica* 12:307–13
110. Perry, D. R., Williams, J. 1981. The tropical rain forest canopy: a method for providing total access. *Biotropica* 13:283–85
111. Pimm, S. L. 1986. Community stability and structure. In *Conservation Biology,* ed. M. E. Soule, pp. 309–30. Sunderland, Mass: Sinauer
111a. Poole, R. W., Rathcke, B. J. 1979. Regularity, randomness, and seggregation in flowering phenologies. *Science* 203:470–71
112. Porsch, O. 1934. Säugetiere als Blumenausbeuter und die Frage der Saugetierblume. I. *Biol. Gen.* 10:657–85
113. Porsch, O. 1935. Säugetiere als Blumenausbeuter und die Frage der Saugetiereblume. II. *Biol. Gen.* 11:171–88
114. Porsch, O. 1936. Säugetiere als Blumenausbeuter und die Frage der Saugetierblume. III. *Biol. Gen.* 12:1–21
115. Prance, G. T. 1985. The pollination of Amazonian plants. In *Key Environments: Amazonia,* ed. G. T. Prance, T. E. Lovejoy, pp. 166–91. Pergamon
116. Primack, R. B. 1985. Longevity of individual flowers. *Annu. Rev. Ecol. Syst.* 16:15–37
117. Ramirez, W. 1970. Host specificity of fig wasps (Agaonidae). *Evolution* 24:680–91
118. Rathcke, B., Lacey, E. P. 1985. Phenological patterns of terrestrial plants. *Annu. Rev. Ecol. Syst.* 16:179–214
119. Regal, R. J. 1982. Pollination by wind and animals: ecology of geographic patterns. *Annu. Rev. Ecol. Syst.* 13:497–24
120. Renner, S. 1986. The neotropical epiphytic Melastomataceae: phytogeographic patterns, fruit types and floral biology. *Selbyana* 9:104–11
121. Rogstad, S. H., Paton, J. C. II, Schaal, B. A. 1988. M 13 repeat probe detects DNA minisatellite-like sequences in gymnosperms and angiosperms. *Proc. Natl. Acad. Sci. USA* 85:9176–78
122. Roubik, D. W. 1989. *Ecology and Natural History of Tropical Bees.* Cambridge: Cambridge Univ. Press. 514 pp.
123. Roubik, D. W., Ackerman, J. D. 1987. Long term ecology of euglossine orchid bees (Apidae: Euglossini) in Panama. *Oecologia* 73:321–33
124. Rourke, J., Wiens, D. 1977. Convergent floral evolution in South African and Australian Proteaceae and its possible bearing on pollination by non flying mammals. *Ann. Mo. Bot. Gard.* 64:1–17
125. Sazima, M., Sazima, I. 1978. Bat pollination of the passion flowers, *Passiflora mucronata,* in southeastern Brazil. *Biotropica* 10:100–9
126. Sazima, M., Sazima, I. 1987. Additional observations on *Passiflora mucronata,* the bat-pollinated passion flower. *Cienc. Cult.* 39:310–12
127. Schatz, G. 1990. See Ref. 22
128. Schemske, D. W., Horvitz, C. C. 1984. Variation among floral visitors in pollination ability: a precondition for mutualism specialization. *Science* 225: 519–21
129. Simpson, B. B., Neff, J. L. 1981. Floral reward, alternatives to pollen and nectar. *Ann. Mo. Bot. Gard.* 68:301–22
130. Smith, A. P. 1973. Stratification of temperate and tropical forests. *Am. Nat.* 107:671–83
131. Sobrevila, C., Arroyo, M. T. K. 1982. Breeding systems in a montane tropical cloud forest in Venezuela. *Plant Syst. Evol.* 140:19–38
132. Start, A. N., Marshall, A. G. 1976. Nectarivorous bats as pollinators of trees in west Malaysia. In *Tropical Trees: Variation, Breeding and Conservation,* ed. J. Burley, B. T. Styles, pp. 141–50. New York: Academic
133. Stebbins, G. L. 1981. Why are there so many species of flowering plants? *Bioscience* 31:573–77
134. Steiner, K. E. 1979. Passerine pollination of *Erythrina megistophylla* Diels. (Fabaceae). *Ann. Mo. Bot. Gard.* 66: 490–502
135. Steiner, K. 1981. Nectarivory and potential pollination by a neotropical marsupial. *Ann. Mo. Bot. Gard.* 68: 505–13

136. Stiles, F. G. 1975. Ecology, flowering phenology and hummingbird pollination of some Costa Rican Heliconia species. *Ecology* 56:285–10
137. Stiles, F. G. 1978. Temporal organization of flowering among the hummingbird food plants of a tropical wet forest. *Biotropica* 10:194–10
138. Stiles, F. G. 1981. Geographical aspects of bird-flower coevolution, with particular reference to Central America. *Ann. Mo. Bot. Gard.* 68:323–51
139. Stiles, F. G. 1977. Coadapted competitors: the flowering seasons of hummingbird pollinated plants in a tropical forest. *Science* 196:1177–78
140. Stiles, F. G. 1979. Reply to Poole and Rathcke. *Science* 203:471
141. Stratton, D. A. 1989. Longevity of individual flowers in a Costa Rican cloud forest: ecological correlates and phylogenetic constraints. *Biotropica* 21:308–18
142. Straw, R. M. 1956. Floral isolation in *Penstimon*. *Am. Nat.* 90:47–63
143. Sussman, R. W., Raven, P. H. 1978. Pollination by lemurs and marsupials: an archaic coevolutionary system. *Science* 200:731–36
144. Sytsma, K. J., Schaal, B. A. 1985. Genetic variation, differentiation and evolution in a species complex of tropical shrubs based on isozymic data. *Evolution* 39:582–93
145. Tanner, E. V. J. 1982. Species diversity and reproductive mechanisms in Jamaican trees. *Biol. J. Linn. Soc.* 18:263–78
146. Terborgh, J. 1986. Keystone plant resources in the tropical forest. In *Conservation Biology*, ed. M. E. Soule, pp. 330–44. Sunderland, Mass: Sinauer
147. Thompson, J. N. 1982. *Interaction and Coevolution*. New York: Wiley
148. Toledo, V. M. 1977. Pollination of some rain forest plants by nonhovering birds in Veracruz, Mexico. *Biotropica* 9:262–67
149. Tomlinson, P. B., Primack, R. B., Bunt, J. 1979. Preliminary observations of floral morphology in mangrove Rhizophoraceae. *Biotropica* 11:256–77
150. West-Eberhard, M. J. 1983. Sexual selection, social competition and speciation. *Q. Rev. Biol.* 58:155–83
151. White, M. J. D. 1978. *Modes of Speciation*. San Francisco: Freeman
152. Wiebes, J. T. 1979. Coevolution of figs and their insect pollinators. *Annu. Rev. Ecol. Syst.* 10:1–12
153. Wright, S. 1969. *Evolution and the Genetics of Populations*, Vol. II. Chicago: Univ. Chicago Press
154. Young, A. M. 1982. Effects of shade cover and availability of midge breeding sites on pollinating midge populations and fruit set in two cocoa farms. *J. Appl. Ecol.* 19:47–63
155. Young, H. J. 1986. Beetle pollination of *Dieffenbachia longispatha* (Araceae). *Am. J. Bot.* 73:931–44
156. Young, H. J. 1990. See Ref. 22

*Annu. Rev. Ecol Syst. 1990. 21:423–47*


# THE ECOLOGY AND ECONOMICS OF STORAGE IN PLANTS

*F. Stuart Chapin III*

Department of Integrative Biology, University of California, Berkeley, California 94720

*Ernst-Detlef Schulze*

Lehrstuhl für Pflanzenökologie der Üniversitat Bayreuth, Universitatsstrasse 30, D-8580 Bayreuth, Federal Republic of Germany

*Harold A. Mooney*

Department of Biological Sciences, Stanford University, Stanford, California 94305

KEY WORDS: accumulation, growth, recycling, reserve, storage

---

## INTRODUCTION

Storage is a characteristic feature of most plants, particularly perennials, and the subject has been thoroughly reviewed according to its chemistry and physiology (9, 40, 59, 80, 133, 139). However, in ecology much of the information on storage is based on observation rather than experimentation, and experiments often fail to confirm common perceptions of the nature and dynamics of stored reserves. For example, clipping studies show that not all carbohydrates are available to the plant, even though they are considered to be stored reserves. In this review we suggest criteria for defining storage in ecological and eonomic contexts in order to examine the costs and benefits of storage. We then evaluate the evidence for, and ecological importance of, different types of storage. We discuss storage in relation to vegetative growth



0066-4162/90/1120-0423$02.00

and reproduction, but we ignore storage in seeds and fruits in this review (41,85) because the purposes and constraints on storage differ somewhat between vegetative and reproductive tissues.

## WHAT IS STORAGE?

Storage is a major plant function, along with acquisition, transport, growth, defense, and reproduction. The term storage is confusing, however, because it is seldom defined explicitly and has been used differently in various disciplines. We define storage broadly as *resources that build up in the plant and can be mobilized in the future to support biosynthesis* for growth or other plant functions. We recognize three general classes of storage: accumulation, reserve formation, and recycling (Figure 1).

1. *Accumulation* is the increase in compounds that do not directly promote growth. It occurs because resource supply exceeds demands for growth and maintenance (96).
2. *Reserve formation* involves the metabolically regulated compartmentation or synthesis of storage compounds from resources that might otherwise directly promote growth. Reserve formation directly competes for resources with growth and defense.
3. *Recycling* is the reutilization of compounds whose immediate physiological function contributes to growth or defense but which can subsequently be broken down to support future growth. In the absence of recycling, these compounds would be lost as litter.

Biochemists generally define storage more restrictively as specific compounds that do not directly promote growth but which may be mobilized in the future to support structural biosynthesis (e.g. daily starch storage in leaves). This includes accumulation and reserve formation (Figure 1). By contrast, whole-plant physiologists and ecologists may include reserves and recycling

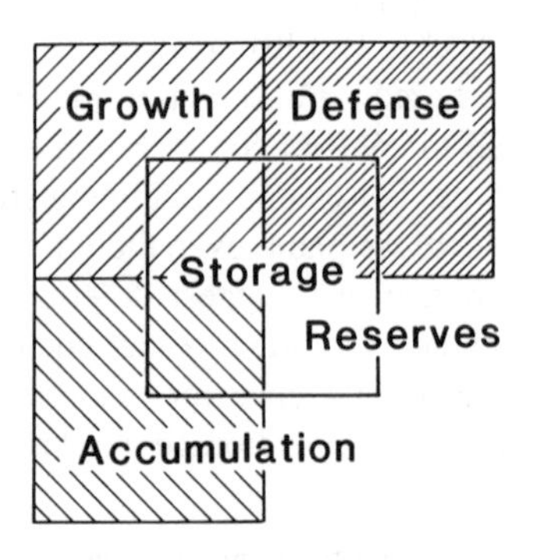


*Figure 1* Storage includes reserves and components of defense, growth, and accumulation.

as storage but exclude accumulation (96). Some defensive compounds turn over and have the potential to support future structural biosynthesis (36, 62). Our definition of storage encompasses both the biochemical and the ecological definitions of storage but emphasizes the potential of stores to contribute to future growth.

Accumulation occurs when acquisition exceeds inputs to growth (and associated defense and reserve storage; flux 1 > flux 2 in Figure 2). Accumulated compounds can be lost from the plant or can contribute to future growth (flux 6). Reserves are formed when acquisition is partitioned among growth, defense, and stored reserves (flux 2 = fluxes 3, 4, and 5 in Figure 2). The partitioning to reserves (flux 5) therefore competes with growth (flux 3) and defense (flux 4). Accumulation, reserves, and defense can subsequently support growth (fluxes 6, 7, and 8, respectively). Recycling involves breakdown of components of growth to form a pool of recycled materials that supports additional growth (flux 9). Storage is mobilized through the sum of fluxes 6, 7, 8, and 9 (Figure 2). Those resources that are not mobilized from growth, reserves, accumulation, and defense are lost as litter.

The role of storage must be evaluated in the whole-plant context and in light of alternative patterns of allocation (34, 100). In particular, we must clearly define growth and its controls. We define *growth* in a restricted sense as the buildup of those components of biomass that directly promote further acquisition and transport of resources. Growth includes structure (e.g. cellu-

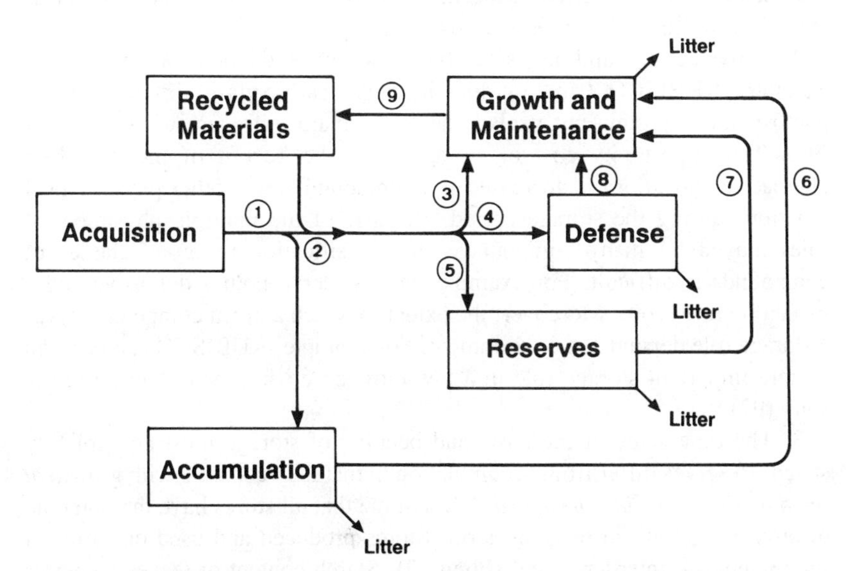


*Figure 2* Interrelationships among pools (boxes) and fluxes (numbered arrows) associated with storage. See text for explanation.

lose), biochemical machinery (e.g. functional enzymes), and small pools of metabolic intermediates (e.g. cytosolic sucrose). It excludes compounds whose major function is storage (e.g. starch, vacuolar sucrose) or defense (e.g. tannins and phenolics), because allocation to these pools competes with growth at the time of allocation. Growth by our definition also excludes compounds that have accumulated due to an excess of supply over demand and which, therefore, do not promote growth at the time they are produced or accumulated (e.g. nitrate), or compounds which have no growth-promoting function (e.g. heavy metals). We prefer to talk about growth in a restricted sense rather than total growth (e.g. biomass accumulation), because the latter includes storage and defense and would lead to circular reasoning when we define reserves and defense as being formed in competition with growth.

Growth and allocation can be considered at several levels: whole-plant allocation to organs, functional allocation to sources and sinks, and biochemical allocation to specific compounds (34, 100). In the context of storage we prefer the biochemical rather than the anatomical allocation scheme, because storage is generally distributed throughout the plant. There are four potential sources of confusion in our concept of storage that must be addressed:

1. A given compound may be formed by one or more types of storage processes. For instance, starch may be synthesized when carbon gain exceeds the carbon demands for growth (accumulation) or when the plant partitions carbon between growth and reserves. Amino acids, especially proline, may accumulate as osmoticants during drought, but amino acids can also act as overwinter nitrogen reserves (31, 64, 111).

2. A given compound may serve both a storage and a nonstorage role. For example, RUBISCO (ribulose bis-phosphate carboxylase) is an essential photosynthetic protein but is also one of the major nitrogen stores in leaves (96). The opportunity cost of storage (e.g. the benefit of the next best alternative allocation) is decreased if a compound serves other physiological functions during the storage period. Because of the multiple physiological roles played by many compounds, the classification of broad classes of compounds is difficult. For example, tannins serve both a defensive and a metabolic role (160). Moreover, the extent to which a given compound serves a storage role depends on environment. For example, RUBISCO can perform a more important storage role in a low-nitrogen environment than on fertile soils (97).

3. The time scale of the costs and benefits of storage must be explicitly stated. Reserves differ from accumulation in that they compete with growth *at the time the stores are produced*. We assume that all stores have the potential to promote growth in the long term. Stores produced and used on different time scales are interdependent (Figure 3). Starch content of leaves increases during the day and decreases at night to support growth. Any daily cycle of

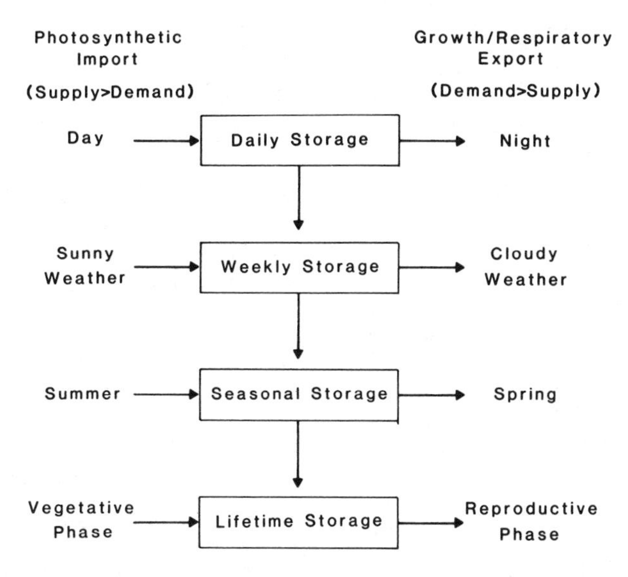


*Figure 3* Causes of changes in storage occurring over different time scales.

storage components contributes to storage pools that may be drawn down over various time scales (e.g. daily, weekly, seasonally) to support future growth or reproduction. The ultimate benefit of stores in supporting reproduction depends on net balances of reserve production and use occurring at several shorter time scales.

4. Two distinct measures are useful in describing the production and use of stores. Allocation in a storage organ can be characterized by concentration, providing allocation ratios remain constant during the period of interest. For example, a constant starch concentration in a growing potato could reflect a constant proportional allocation to starch vs other components of the potato. However, allocation at the whole-plant level is best measured by the pool size of stored reserves as a fraction of total growth. Moreover, the pool size of stored reserves is probably the best measure of the potential of stores to contribute to future growth. Thus, both concentration and pool size are useful in describing the processes that control formation and use of stores.

## AN ECONOMIC ANALOGY OF STORAGE

Microeconomic theory predicts how a business firm or plant should allocate resources to maximize profit (the excess of revenues over cost; 15, 16). Here we discuss those aspects of the economic analogy that relate to storage by plants. Although the analogy between plants and business firms has limitations, it provides a framework to evaluate the costs and benefits of storage.

A business firm earns profits when its revenues exceed costs. The firm can grow by investing profits and savings in additional inputs. If the economy is perfectly stable and predictable, and the output of the firm is small relative to market demand, a profitable firm can grow exponentially by reinvesting all profits in additional inputs, thereby saving nothing. Any savings detract from the exponential growth rate that the firm can realize. Similarly, investment by a plant in growth results in a compounding of the investment in terms of new resources gained, and any stored reserves detract from the potential exponential growth rate of the plant.

## *Reserve Storage*

Although reserve storage (savings) detracts from growth, all successful firms and plants invest in some savings (i.e. have some resources that are not used in a productive function). Three main reasons exist for saving:

1. *Asynchrony of supply and demand.* Some firms, such as those that process tomatoes, experience large asynchronies in anticipated supply and demand. Such firms must save to purchase and process large quantities of tomatoes when they are cheap. Asynchrony of supply and demand is the rule rather than the exception for plants, and the growth demand may be supported largely by stored reserves. The greater the asynchrony of supply and demand, the greater should be the expected storage reserve.

2. *Risk aversion.* Firms save primarily to minimize risk of a large catastrophic loss such as from a fire. In the economic realm this risk is averted by purchasing insurance from another company, in which case the business firm invests less in insurance than it expects to regain in the event of catastrophic loss. By contrast, plants can only use internally stored compounds to recover from catastrophic events like fire or grazing. In this case the plant must store internally more resources than are to be used in recovery from catastrophe because of respiratory losses. There is a continuum from risk aversion to asynchrony of supply and demand. This continuum depends on the predictability of the timing and magnitude of demand. The greater the risk (high probability of a large or frequent loss), the more a firm or plant should save. The alternative to storage is to shorten the life cycle to minimize the risk of catastrophe.

3. *Change in type of product.* Firms occasionally undergo large changes in patterns of production (e.g. change from producing guns to producing butter). These expensive one-time investments can be accomplished only if the firm has substantial savings or can borrow money. If plants change patterns of production from a vegetative to a reproductive state, such large bursts of biosynthesis may be supported largely by internally stored reserves (savings).

The best way to evaluate the costs and benefits of savings is through estimation of opportunity costs of storage—i.e. the benefit achieved from the most favorable alternative pattern of allocation. Plants should store if the opportunity cost is less than the benefit achieved by storing now and using resources to support growth at a future date. For example, a plant may store carbohydrates during mild water stress and initiate growth again after the rain rather than allocate carbohydrates to additional root growth with the danger of exhausting the water resource. Obviously, the opportunity cost of storage depends critically on the time scale over which the calculation is made. In the context of fitness, it should be evaluated over the lifetime of the individual in the environment of interest. Short-term analysis might produce quite different conclusions, because catastrophic losses are less likely to occur over a short than over a long time interval. Opportunity costs depend on environment because the relative benefit of different allocations is environmentally determined.

### Accumulation

The economics of accumulation are quite different from those of storing reserves, because accumulation does not directly compete with growth and therefore has a lower opportunity cost. There are two economic analogs of accumulation: by-products and inventory. By-products are materials that accumulate as an inevitable consequence of the productive process; they may be either useless, in which case they are expensive wastes, or they may be useful, in which case they can be sold and contribute to profit. Useful by-products are components of inventory, i.e. the materials and products held in stock by a firm for future production and sale. Inventories that result from a decline in demand (sales or growth) are disadvantageous, whereas those resulting from abundant (cheap) supply are beneficial to plants and firms. Plants differ from business firms in that they can control demand more tightly than supply, whereas firms have tighter control over supply. Consequently, within certain limits, inventories are generally good for plants but bad for firms.

Wastes are by-products of production and are costly to store and dispose of. For instance, plants accumulate heavy metals and salt as a result of cumulative transpiration in metal-contaminated or saline environments. These leaves become inefficient due to waste accumulation and must therefore be replaced. Such plants typically have low transpiration rates (and therefore low photosynthetic rates) which minimize waste accumulation (4, 18).

### Recycling

When firms undergo major changes in patterns of production, they convert as much of the equipment and buildings as is profitable to the new function. Plants undergo a dramatic change from vegetative to reproductive growth, requiring large resource inputs into new structures and biosynthetic equipment. This is supported in large part by recycling materials from ageing tissues. The opportunity cost of recycling a structure is the additional growth that could be achieved by retaining and continuing to use that structure. We expect the plant to recycle as much resource as possible, because any remaining resources are lost in litter.

## STORAGE COMPOUNDS

Accumulation and reserve storage occurs primarily in vacuoles and plastids, because this prevents degradation of stores by isolating them from other metabolic pathways. Vacuoles also protect cellular machinery from potentially toxic metabolites (1, 17, 89). Accumulation is the least expensive and probably most common mode of storage in plants.

Carbon accumulates primarily as starch, fructosans, and sucrose, depending on species and plant part (Table 1: 9, 66, 115, 133, 155, 159). Carbon accumulation occurs under conditions of high light, low nutrients, or mild water or salt stress (25, 63). These carbon stores subsequently support growth after the stress is alleviated (93). Starch storage occurs in plastids, so its synthesis and breakdown are tightly coupled to photosynthesis (9, 139).

Other carbon compounds are generally less important storage products: In response to high carbon supply, organic acids increase in some species and decline in others (39). Soluble phenolics and hydrolyzable tannins can increase in response to carbon surplus and exhibit turnover (20, 21, 83), but it is unclear whether their breakdown products can be mobilized to support growth. Lipids, which were once thought to play a major storage role in vegetative tissues of shrubs, are primarily cutins, waxes, and antiherbivore resins that cannot be broken down to support biosynthesis (30, 31, 67, 145). However, many trees build up lipids (fat trees; 80, 158, 159), which can be storage (10) or nonstorage lipids (105). Lipids are usually quantitatively less important than carbohydrates as an energy store (105). The possible use of hemicellulose as stores is equivocal (31, 155, 159). Of the other classes of compounds that could conceivably build up under conditions of carbon surplus, some are insenstitive to carbon supply (cellulose), and others increase but are not broken down (lignin, condensed tannins, terpene resins, calcium oxalate; 19, 20, 21, 81, 151). These compounds therefore cannot serve a storage function.

Nitrogen builds up in concentration and pool size under conditions of high nitrogen or low light, primarily as specialized storage proteins, amino acids (especially arginine, glutamine, and asparagine), and nitrate, depending on the species (Table 1; 28, 30, 40, 89, 111, 141, 143, 146, 149). Some of the RUBISCO which builds up under high-nitrogen conditions is inactive and

**Table 1** The role that major chemical fractions in plants play in storage and in nonstorage aspects of growth and accumulation.

| | Components of Storage | | | | |
|---|---|---|---|---|---|
| | Accumulation | | | Growth | |
| | Nonstorage | Storage | Stored Reserve | Recycled | Nonrecycled |
| Carbon fractions | | | | | |
| Polysaccharide | − | + | + | − | − |
| Sugar | − | + | + | − | − |
| Organic acid | − | + | ? | + | − |
| Phenolics | + | ? | ? | ? | − |
| Tannins | + | ? | ? | ? | + |
| Hemicellulose | − | ? | − | ? | + |
| Lipids | + | ? | − | − | − |
| Cellulose | − | − | − | − | + |
| Lignin | + | − | − | − | + |
| Nitrogen fractions | | | | | |
| Protein | − | + | + | + | − |
| Amino acid | − | + | + | + | − |
| Alkaloid | − | + | − | + | − |
| Nitrate | − | + | − | + | − |
| Nucleic acid | − | + | ? | + | − |
| Phosphorus fractions | | | | | |
| Phosphate | − | + | − | + | − |
| Polyphosphate | − | + | + | − | − |
| Phospholipid | − | + | + | + | − |
| Nucleic acid | − | + | ? | + | − |
| Sugar phosphate | − | + | + | + | − |
| Ions | | | | | |
| Potassium | − | + | − | + | − |
| Calcium | + | − | − | − | + |
| Magnesium | + | + | − | + | + |
| Salt | + | − | − | − | + |
| Heavy metals | + | − | − | − | + |

therefore should be considered accumulation (96). Some alkaloid-producing plants increase alkaloid content under conditions of nitrogen surplus, but others do not (147). Under conditions of nitrogen stress, these alkaloids are broken down to support growth. Proline often increases in response to drought or salinity, because it serves as an osmotically active, nontoxic nitrogen store (96).

Phosphorus is stored as inorganic phosphate or polyphosphate, and to a lesser extent as ribonucleic acids and phospholipids, depending on the species (13, 14, 26, 28, 30, 74, 104). Some compounds (e.g. ribonucleic acids) contain carbon, nitrogen, and phosphorus and consequently can serve a storage role for all three of these elements. However, ribonucleic acid constitutes a larger proportion of the plant's total phosphorus than of its nitrogen or carbon store, and amino acids constitute a larger proportion of the plant's total nitrogen than of its carbon store. Thus, the potential importance of different classes of compounds in storage is largely as described in Table 1.

Mineral ions other than phosphorus can increase in plants either when supply exceeds demand (luxury uptake) or because the plant fails to exclude them (46, 57), as in the case of potassium in xylem-tapping mistletoes (43, 125, 130). Some are essential nutrients required by plants (e.g. potassium, calcium, and magnesium). Others serve as osmoticants in halophytes (sodium chloride). Some ions are stored and support future growth (e.g. potassium, and magnesium). Others are shed in litter when the tissue senesces (e.g. toxic heavy metals, calcium, and sodium chloride). Inorganic ions never form a reserve in the sense that they are not sequestered in competition with growth (14, 46, 57).

## STORAGE AT DIFFERENT TIME SCALES

The costs and benefits of storage must ultimately be evaluated in terms of contributions to fitness over the lifetime of the organism. However, individuals seldom differ only in storage, making it difficult to estimate the contribution of storage to fitness. It is, therefore, useful to estimate costs and benefits of storage over shorter time scales, particularly for long-lived plants.

### *Daily Storage*

Leaves of most plants store starch and/or vacuolar sucrose during the day and break starch down for export at night. The demand by the plant for carbohydrate determines export rate from a well-lighted leaf. Partitioning between starch and sugar is enzymatically regulated and feeds back to control photosynthesis (137). Manipulations of photoperiod, nutrition, and source/sink activities demonstrate that regulation of this storage reserve results in a nearly constant export rate throughout the 24-hr cycle (6, 33, 52, 53, 72, 116, 140). However, under natural conditions, hourly and daily fluctuations in carbon supply or growth demand cause variations in leaf starch content. In this short-term sense, starch acts as an overflow (131, 138, 159). Plants export recent photosynthate before mobilizing starch stores, suggesting an opportunity cost to starch storage (131).

Mutants that cannot produce starch accumulate sugars in the leaves during the day. The higher sugar concentration in turn leads to a higher leaf respiration rate, causing overall plant growth to decline (140). This demonstrates that, over a 24 hr cycle, starch storage has a low opportunity cost, e.g. the plant gains more by storing starch than by leaving soluble sugars in the cytoplasm. Consequently, plants should store starch during the day for use at night rather than using some alternative allocation. Mutants that cannot store starch provide a system in which the fitness consequences of starch storage could be directly measured.

Nitrate concentration in leaves also shows a diurnal pattern in many species, increasing at night and decreasing during the day (58, 94, 127), because nitrate reduction is closely coupled to photosynthesis. During the day leaves often have excess reducing power and can reduce nitrate with little or no decline in carbon reduction (16). Thus, the daily cycle of nitrate storage appears to have a low opportunity cost. The alternative of nitrate reduction in roots would be energetically more expensive (16). Avoiding uptake of nitrate under conditions where it cannot be immediately assimilated is another alternative to nitrate storage. However, this might well have a negative effect on total plant nitrogen gain and therefore on fitness in a competitive environment.

Daily water storage is generally of little ecological importance, because the daily turnover of water by a leaf may be ten times its water content. There is no tissue in which water can be sequestered from the main transport path in amounts large enough to cope with this turnover, except in large trees. In *Picea abies,* water storage causes water flow to begin in the crown two hours before it begins in the base of the tree (122). Even this water storage is small compared to total daily water turnover. It would be expensive for a plant to produce a structure large enough to store enough water for more than a few hours' use. In other words, the opportunity cost of water storage is high, and most plants invest little or nothing in a water-storing structure.

### *Short-Term Fluctuations*

Manipulation of light, nutrients, and water clearly indicates that storage levels are sensitive to short-term changes in environment. Under natural conditions the level of carbohydrates declines during cloudy weather when respiratory and growth demands exceed net carbon gain (51). These reserves build up again under conditions favoring photosynthesis. The opportunity cost of this short-term storage might be either a larger leaf biomass to supply carbon dependably even during periods of cloudy weather, or pronounced reductions in growth rate during cloudy weather. In both cases the expected benefit is probably less than observed patterns of fluctuation with weather.

Nutrient availability is highly pulsed. Plants accumulate nitrate in response to pulses of nutrient availability (78, 127) and use these accumulated stores to support continued growth when nutrient availability declines (25, 32, 78, 94). The opportunity cost of this short-term nitrate storage is probably small because it reflects accumulation in excess of immediate demand rather than competition with growth (96). This opportunity cost is further reduced because nitrate may be essential as a counter-ion during uptake of cations.

Short-term water storage can occur in trunks of trees (128, 150). However, in succulents where short-term changes in water content are most pronounced, this storage is either a consequence of salt accumulation, or it is related to acid metabolism. It thus serves other functions than just storage and is not readily available for transpiration.

## *Seasonal Storage*

ACCUMULATION AND RESERVES In perennial plants growing in a seasonal environment, stores of carbon, nitrogen, and phosphorus decline when growth is most rapid and recover when growth stops and/or when senescence recycles leaf nutrients back to storage organs (27, 29, 31, 40, 70, 76, 102, 122, 124, 133, 141, 146, 155, 159). Current acquisition is used before stores when plants have access to both (60, 141, 146), which suggests an opportunity cost to storage. Although seasonal storage is generally viewed as a carefully regulated reserve storage, the pattern could also be interpreted as accumulation in response to supply and demand (105, 146, 159). Only a few studies distinguish between these possibilities.

Sugar beet is one species that clearly builds storage reserves in competition with growth. This species maintains a nearly constant partitioning of carbon between sucrose storage and root growth over a spectrum of light and nutrient conditions ranging from optimal to moderately growth-limiting (95, 152). However, outside the "normal" range of light and nutrient supply, sugar beet stores little sugar at low light or high nitrogen (86, 148). Grafting experiments also demonstrate that reserve storage competes with growth. Sugar beet, which allocates strongly to storage, decreases shoot growth when grafted to shoots of a leafy variety of the same species (chard), whereas chard roots, which have a small capacity for storage, cause grafted sugar beet shoots to grow larger than normal (117). Thus, in crops that have been bred for storage, allocation to storage is maintained under growth-limiting conditions, and this competition causes a decline in growth. However, under conditions of extreme carbon limitation, carbon storage declines even more than does growth. Much less is known about controls over carbon storage in wild species.

Nutrient reserves are also stored in competition with growth. In the biennial *Arctium tomentosum,* the proportional allocation of nitrogen to storage and growth and the final concentration of nitrogen in storage organs were the same for shaded and fertilized plants, plants with reduced leaf area, and control plants (64, 136). This indicates that the increase in asparagine, arginine, and proline in the hypocotyl competed with growth, according to a preset genetic program. In this species, growth rate and carbon balance determine the size of the store (i.e. hypocotyl size), and nitrogen status determines the extent to which this storage organ is filled with nitrogen. Because both carbon and nitrogen are allocated to the hypocotyl over a range of conditions in which either carbon or nitrogen become limiting to growth, this is a clear example of carbon and nitrogen reserve storage in competition with growth.

In perennial rhizomatous plants it is more difficult to distinguish between reserve storage and accumulation of stores, because growth of new stores and use of existing stores may occur simultaneously. In the nitrophilous *Urtica dioica* growing in full sun, or in various degrees of shade that were limiting to growth, old rhizomes accumulated starch during the growing season in all treatments, indicating reserve storage in competition with growth (142). These stores were broken down in autumn, when leaves were shed and new rhizomes started to grow. At the same time rhizomes accumulated amino acids that were recycled from senescing leaves or were acquired by nitrate uptake and assimilation. The assimilation process further depleted carbohydrate stores. In this species patterns of nitrogen storage could reflect either accumulation or reserve storage but clearly depended on recycling.

Seasonal patterns of storage pools in relation to growth suggest that an increase of stores in late season may compete with growth. Nutrient-limited individuals of the tundra sedge *Eriophorum vaginatum* greatly diminish their growth rate when reserve accumulation begins in late summer. If provided with added nutrients, both growth and reserve accumulation (in the form of sugars, arginine, and sugar phosphates) continue simultaneously (31). In a tundra environment where spring growth begins before soils thaw, nutrient storage is essential to support spring growth and therefore has a low opportunity cost.

The examples described above indicate that seasonal storage often reflects reserves that are stored in competition with growth (i.e. money in the bank). However, pool sizes and dynamics of stores also reflect accumulation. For instance, nutrient concentrations in storage organs generally increase more in response to fertilization than do concentrations in vegetative tissues (132, 146, 149). Unfavorable conditions for growth (i.e. low demand) at high altitude or latitude also cause accumulation of nitrogen in tundra plants (79). The extent to which seasonal fluctuations in stores reflect accumulation vs reserves formed in competition with growth remains uncertain, particularly in wild plants.

RECYCLING Other than some organic acids, few non-nitrogenous organic compounds are recycled from senescing leaves (143). During senescence, only about 10% of the leaf weight is lost in respired or resorbed material (28, 54), and most of this can be accounted for by recycled nitrogenous compounds (54). This fact supports the idea that non-nutrient-containing compounds are not recycled to any large extent. By contrast, about half the nitrogen, phosphorus, and potassium are recycled from a senescing leaf to support new growth (28). Thus, recycling constitutes a large store of nutrients but a small store of carbon.

Recycling allows nutrients that have been used to support previous growth to be reused. In *Arctium* the same molecule of reduced nitrogen can be reutilized as many as six times in a single growing season: from hypocotyl to rosette leaves to leaves on the flowering stalk to the flower stalk to the flowers and finally to the seeds (64). Even within the canopy of the same individual, recycling takes place from old leaves to young leaves and from shaded parts of the canopy to more sun-lit leaves, maximizing carbon gain (50, 69, 76). Similarly, carbohydrates from the heartwood can be recycled to support growth of outer tree rings (70).

The widespread forest decline in Europe shows the importance of recycling (123). Here acid rain has caused a dramatic increase in plant nitrogen and a decline in cation availability. The flush of spring growth in Norway spruce depends primarily on stored nitrogen and magnesium (49), because soils are too cold to allow much uptake at this time. In trees stressed by acid rain, new growth depletes the magnesium stored in older leaves, causing them to become chlorotic and die (108). Wood growth occurs after foliage development has depleted magnesium stores and is directly limited by magnesium deficiency. When recycling is prevented by clipping off leaf buds to prevent spring growth, trees remain green and have magnesium concentrations in old foliage that are comparable to those of healthy trees (129, 153).

In other cases recycling of nutrients from old leaves has no major effect on nutrient supply to young leaves. In several evergreen species, ranging from arctic to mediterranean, removal of old leaves prior to senescence had no effect on the nutrient pool size of young leaves (75). However, growth of new leaves was reduced in defoliated plants, suggesting that leaf growth in these species was limited more by carbon supply than by recycling of nutrients. Clearly, more experimental studies are needed to demonstrate the conditions under which recycling constitutes an important storage process.

## STORAGE AND RECOVERY FROM CATASTROPHE

In perennial plants there is no clear relationship among species between the amount of stored carbohydrate and capacity for regrowth after grazing (144). Moreover, most plants fail to use a large proportion of their stored carbohydrate in response to clipping, raising questions of whether these reserves are completely accessible to the plant (22, 38, 47, 90, 98, 121, 144, 155). Similarly, even though regrowth after fire depletes carbohydrate reserves, substantial carbohydrate concentrations remain (113). Such studies generally emphasize the magnitude of decline in carbohydrate reserves in response to clipping, but they fail to point out that substantial carbohydrates remain at levels of clipping that greatly repress regrowth. In fact, if plants are repeatedly clipped and kept in the dark to prevent photosynthesis, they generally cease regrowth well before carbohydrate reserves are exhausted (121).

We suggest four potential causes for the failure of plants to use all carbohydrate reserves following catastrophe. (*a*) Intense defoliation may deplete nitrogen and phosphorus reserves more strongly than carbon reserves (3, 24), so that nutrients rather than carbon may limit regrowth. However, grazing usually causes tissue nutrient concentrations to increase (73), suggesting that grazing usually depletes carbon more than nutrient stores or that high root-shoot ratio of grazed plants enables the plant to meet nutrient demands more readily than carbon demands. (*b*) Some carbohydrate stores may become inaccessible to the plant with time because they are in dead cells and cannot be retrieved (158). (*c*) These clipping experiments may not have provided the appropriate cue to trigger mobilization of the stores. (*d*) Our chemical measures of stores may include some nonstorage forms such as breakdown products of hemicellulose. Because of the uncertainty in estimating the pool size of stores available to the plant after catastrophe, it is presently impossible to estimate their opportunity cost. A comparison of the clipping response of individuals that differed only in the magnitude of stored reserves (due to differences in carbohydrate or nutrient status) or in genetic potential to store could provide insight into the opportunity costs of storage for recovery from grazing. The observation that plants use concurrent photosynthate rather than stores to support regrowth, when both are available, suggests an opportunity cost to storage in support of recovery from catastrophe (155).

In semi-arid environments, fire has led to quite different strategies for recovery from catastrophe. In California, woody species recover from fire as sprouters, i.e. species which activate dormant buds for growth following fire; sprouters contain specialized tissues in their root (lignotubers) which store large quantities of starch (77). By contrast, seeders, which recolonize from seed following fire, lack such storage tissues. The observation that both strategies are well represented in this ecosystem suggests that the opportunity cost of lignotuber formation (in terms of reduced reproductive output) must be substantial.

## STORAGE FOR REPRODUCTION

If stores are important to reproduction, we expect strong depletion of stores when the plant switches to the reproductive mode, particularly in monocarpic plants. Optimality models of allocation predict that reserve storage in annuals will be best developed when the switch to reproduction occurs late in the season, particularly in monocarpic species, because such species have a greater probability of loss of productive potential and recycling stores to herbivores (35). Field data with *Hemizonia* conform to this prediction.

When annuals are grown under optimal conditions, less than 25% (generally less than 5%) of seed carbon comes from stores (11, 12, 48, 112, 118, 157). The remaining carbon comes from concurrent photosynthesis. The reproductive structures themselves contribute as much as 30–65% of their carbon requirement through photosynthesis (7, 8, 48, 119). This may explain the close relation between seed yield and evapotranspiration (126). Thus, under optimal conditions these crops rely primarily on current photosynthate to support seed production. By contrast, 50–90% of the nitrogen and phosphorus in seeds of annual plants is recycled from vegetative tissues rather than taken from concurrent uptake (11). Thus, as with autumn leaf senescence, recycling is a much more important source of nutrients than of carbon to support reproduction. However, under conditions of drought severe enough to restrict photosynthesis, carbon for grain growth comes mainly from stores (12, 56, 110). Similarly, plants under nutrient stress draw proportionally more on stores of nitrogen and phosphorus than do plants growing under optimal nutrition (5, 11, 156). Thus, the opportunity cost of storage may be less under conditions of low-resource supply. There is remarkably little evidence on the extent to which wild plants draw on nutrient stores to support reproduction.

Biennials are an excellent example of the importance of reserves for supporting reproduction at the end of the life cycle. In *Arctium,* reserves accumulated during the first year support rapid vegetative growth at the beginning of the second year. At seed filling, more than 70% of the total N in the plant is recycled into seed production (64). Stored reserves also have an indirect effect on nutrient supply to reproduction. Large stores in *Arctium* support production of large rosettes which exclude competing individuals and provide access to a larger soil pool of mineralized nitrogen than in the first year. For this reason, *Arctium* absorbs two thirds of its nitrogen in the second year, when it has a larger rosette despite a constant or decreasing root biomass (64).

In biennials, rosette size is a good predictor of the quantity of stored reserves, which in turn is a good predictor of seed output (154). Biennials that delay reproduction and remain vegetative to attain a larger size depend more strongly on stores but achieve greater reproductive output (68). This dependence on stores is particularly important in infertile and dry environments (55, 68, but see 120), again supporting our economic prediction that the opportunity cost of storage is reduced in low-resource environments.

In perennials there is clear evidence of the importance of reserves in supporting reproduction. In an extreme case, *Aesculus californicus,* a drought-deciduous tree, produces large fruits which develop after leaves have been shed from the trees. Its fruits must, therefore, draw their carbon entirely from stored reserves (101, 103). Reproductive branches show delayed bud break and less growth the following year than do nonreproductive branches, indicating that this reserve allocation to reproduction competes with future growth (106). Similarly, mast-cropping conifers (37, 44, 87, 91, 92), biennially bearing fruit trees (23, 65, 122), and some herbaceous species (2, 84, 99, 134, 135) show marked declines in carbon reserves and growth following heavy reproduction. In most of these cases, carbohydrate reserves are drawn down more strongly by reproduction than is nitrogen. The nitrogen and phosphorus that support reproduction come largely from recycling of nutrients from senescing leaves (11, 156). If grazing or browsing depletes storage reserves, this often causes a decline in reproduction (42, 82, 107). This effect is particularly pronounced in females of dioecious species which are more dependent on stored reserves for reproduction than are males (45). This impact of grazing on reproduction is most pronounced in infertile soils (107), again suggesting that the dependence of reproduction on reserve storage is most pronounced (low opportunity cost) in low-resource environments.

In other cases, reproduction shows no clear relationship to reserves. For example, in two arctic sedge species, reserves were drawn down no more strongly to support reproductive development than to support normal spring shoot growth (88). Similarly, in many perennial grasses and herbs, extent of reproduction had little or no influence on vegetative growth (71, 114, 120). Clearly the importance of reserves in supporting reproduction varies among species and deserves further study.

## ADAPTIVE PATTERNS IN STORAGE

Drought-deciduous plants show large seasonal variations in carbohydrate stores whereas co-occurring evergreen species do not (103). Similarly, in tundra, deciduous species show more pronounced seasonal fluctuations in carbohydrate and nutrient stores than do co-occurring evergreens (27, 30). The greater dependence on storage probably reflects the lower opportunity cost (benefit of an alternative allocation) in species which experience a large asynchrony in resource supply and demand, as predicted by our economic assumptions.

Species adapted to low resource supply have an inherently low growth rate even under conditions of high resource supply (25, 61, 109). These plants accumulate larger nutrient stores in response to a pulse of nutrients than do plants with a high growth rate, because rapid growth dilutes the nutrient pool over a larger biomass (25). This nutrient storage (luxury consumption) by species adapted to low-resource environments suggests that the opportunity cost of such storage is lower here than in environments which support more rapid growth.

## CONCLUSIONS

1. Economics provides a qualitative framework to evaluate the adaptive significance of storage. Concurrent uptake is used before stores during daily and seasonal cycles and to support reproduction or recovery from grazing, presumably because of the greater cost of storage. Carbohydrate and nitrate contents of plants show greater daily fluctuation than does water because they have a lower opportunity cost of daily storage. Seasonal and lifetime storage of carbon and nutrients has a lower opportunity cost (and is therefore better developed) in plants of low- (compared to high-) resource environments. The greater allocation to storage in monocarpic plants (e.g. many biennials) compared to polycarpic plants is consistent with the greater risk associated with the monocarpic strategy. Greater storage by deciduous than by evergreen species reflects the greater asynchrony of supply and demand experienced by deciduous species. Storage for reproduction supports a large change in pattern of production at the end of a plant's life, particularly for monocarpic plants; this storage for reproduction appears to be much more pronounced for nutrients than for carbon. Future experiments will be necessary to test these hypotheses more rigorously and to provide a more quantitative assessment of the adaptive value of storage.

2. Distinguishing among the components of storage (accumulation, reserve, and recycling) allows estimation of the cost of storage in different environments. Accumulation is most pronounced in species with inherently slow growth rates. Reserve storage has been critically demonstrated in only a few studies, making it difficult to detect any broadscale ecological pattern in reserve storage. Recycling stores are unimportant for carbon but critical for nitrogen, phosphorus, and potassium. The dependence of growth on recycling may be greater in infertile environments, suggesting that the opportunity cost of recycling is higher in fertile sites.

3. In order to distinguish among different components of storage and to relate these to controls over growth, defense, and reproduction, field experiments which perturb these processes (e.g. shading and fertilization) will be necessary. Even then, however, it will be difficult to dissect the causes of the coordinated whole-plant response. An alternative is to manipulate the storage properties of plants through comparisons of storage mutants or closely related ecotypes that differ only in storage characteristics. A combination of these approaches may be particularly fruitful.

4. An appreciation of time scale is critical to an understanding of the causes and benefits of storage. The cause of storage (reserves vs accumulation) must be evaluated at the time of allocation. The benefit of storage should be measured in terms of fitness and thus should incorporate probabilities of survival and reproduction over the life of the individual. In long-lived species

where fitness is difficult to measure, the benefits of storage can be approximated over the length of repeatable cycles such as a day, season, or cycle of mast cropping.

ACKNOWLEDGMENTS

We thank J. Armstrong, D. Bishop, A. Bloom, M. Chapin, N. Chiariello, R. Dickson, K. Fichtner, B. Gartner, H. Heilmeier, P. Millard, P. Rich, T. Steinlein, E. Steudle, and M. Stitt for constructive criticism of the ideas presented in this review.

*Literature Cited*

1. Alibert, G., Boudet, A. M., Canut, H., Rataboul, P. 1985. Protoplasts in studies of vacuolar storage compounds. In *The Physiological Properties of Plant Protoplasts*, ed. P. E. Pilet, pp. 105–21. Berlin: Springer-Verlag
2. Antonovics, J. 1980. Concepts of resource allocation and partitioning in plants. In *Limits to Action: The Allocation of Individual Behavior*, ed. J. E. R. Staddon, pp. 1–25. New York: Academic
3. Archer, S. R., Tieszen, L. L. 1986. Plant response defoliation: hierarchical considerations. In *Grazing Research at Northern Latitudes*, ed. O. Gudmundsson, pp. 45–59. New York: Plenum
4. Ball, M. C. 1988. Ecophysiology of mangroves. *Trends Ecol. Evol.* 2:129–42
5. Batten, G., Wardlaw, I. F. 1987. Senescence and grain development in wheat plants grown with contrasting phosphorus regimes. *Aust. J. Plant Physiol.* 14:253–65
6. Baysdorfer, C., Robinson, J. M. 1985. Sucrose and starch synthesis in spinach plants grown under long and short photosynthetic periods. *Plant Physiol.* 79:838–42
7. Bazzaz, F. A., Carlson, R. W., Harper, J. L. 1979. Contribution to reproductive effort by photosynthesis of flowers and fruits. *Nature* 279:554–55
8. Bazzaz, F. A., Reekie, E. G. 1985. The meaning and measurement of reproductive effort in plants. In *Studies on Plant Demography: A Festschrift for John L. Harper*, ed. J. White, pp. 373–87. London: Academic
9. Beck, E., Ziegler, P. 1989. Biosynthesis and degradation of starch in higher plants. *Annu. Rev. Plant Physiol. Plant Mol. Biol.* 40: 95–117
10. Beeson, R. C., Proebsting, W. M. 1988. Carbon metabolism in scions of Colorado blue spruce. II. Carbon storage compounds. *J. Am. Soc. Hortic. Sci.* 113:800–5
11. Below, F. E., Christensen, L. E., Reed, A. J., Hageman, R. H. 1981. Availability of reduced N and carbohydrates for ear development of maize. *Plant Physiol.* 68:1186–90
12. Bidinger, F., Musgrave, R. B., Fisher, R. A. 1977. Contribution of stored preanthesis assimilate to grain yield in wheat and barley. *Nature* 270:431–33
13. Bieleski, R. L. 1968. Effect of phosphorus deficiency on levels of phosphorus compounds in *Spirodela*. *Plant Physiol.* 43:1309–16
14. Bieleski, R. L. 1973. Phosphate pools, phosphate transport, and phosphate availability. *Annu. Rev. Plant Physiol.* 24:225–52
15. Bloom, A. 1986. Plant economics. *Trends Ecol. Evol.* 1:98–100
16. Bloom, A. J., Chapin, F. S. III, Mooney, H. A. 1985. Resource limitation in plants—an economic analogy. *Annu. Rev. Ecol. Syst.* 16:363–92
17. Boller, T., Wiemken, A. 1986. Dynamics of vacuolar compartmentation. *Annu. Rev. Plant Physiol.* 37:137–64
18. Bradshaw, A. D. 1969. An ecologist's viewpoint. In *Ecological Aspects of the Mineral Nutrition of Plants*, ed. I. H. Rorison, pp. 415–27. Oxford: Blackwell
19. Bryant, J. P. 1987. Feltleaf willow-snowshoe hare interactions. Plant carbon/nutrient balance and its implications for floodplain succession. *Ecology* 68:1319–27
20. Bryant, J. P., Chapin, F. S. III, Reichardt, P. B., Clausen, T. P. 1987. Response of winter chemical defense in Alaska paper birch and green alder to manipulation of plant carbon/nutrient balance. *Oecologia* 72:510–14

21. Bryant, J. P., Clausen, T. P., Reichardt, P. B., McCarthy, M. C., Werner, R. A. 1987. Effect of nitrogen fertilization upon the secondary chemistry and nutritional value of quaking aspen (*Populus tremuloides* Mich). Leaves for the large aspen tortrix (*Choristoneura conflictana* (Walker)). *Oecologia* 73:513–17
22. Buwai, M., Trlica, M. J. 1977. Multiple defoliation effects on herbage yield, vigor and total nonstructural carbohydrates of five range species. *J. Range Manage*. 30:164–71
23. Chandler, W. J. 1957. *Deciduous Orchards*. Philadelphia: Lea & Febiger
24. Chapin, F. S. III 1977. Nutrient/carbon costs associated with tundra adaptations to a cold nutrient-poor environment. In *Proceedings of the Circumpolar Conference on Northern Ecology*, pp. I183–94. Ottawa: Res. Council Canada
25. Chapin, F. S. III. 1980. The mineral nutrition of wild plants. *Annu. Rev. Ecol. Syst*. 11:233–60
26. Chapin, F. S. III, Follet, J. M., O'Connor, K. F. 1982. Growth, phosphate absorption and phosphorus chemical fractions in two *Chionochloa* species. *J. Ecol*. 70:305–21
27. Chapin, F. S. III, Johnson, D. A., McKendrick, J. D. 1980. Seasonal movement of nutrients in plants of differing growth form in an Alaskan tundra ecosystem: Implications for herbivory. *J. Ecol*. 68:189–209
28. Chapin, F. S. III, Kedrowski, R. A. 1983. Seasonal changes in nitrogen and phosphorus fractions and autumn retranslocation in evergreen and deciduous taiga trees. *Ecology* 64:373–91
29. Chapin, F. S. III, McKendrick, J. D., Johnson, D. A. 1986. Seasonal changes in carbon fractions in Alaskan tundra plants of differing growth form: Implications for herbivory. *J. Ecol*. 76:707–32
30. Chapin, F. S. III, Shaver, G. R. 1989. Differences in carbon and nutrient fractions among arctic growth forms. *Oecologia* 77:506–14
31. Chapin, F. S. III, Shaver, G. R., Kedrowski, R. A. 1986. Environmental controls over carbon, nitrogen, and phosphorus fractions in *Eriophorum vaginatum* in Alaskan tussock tundra. *J. Ecol*. 74:167–95
32. Chapin, F. S. III, Walter, C. H. S., Clarkson, D. T. 1988. Growth response of barley and tomato to nitrogen stress and its control by abscisic acid, water relations and photosynthesis. *Planta* 173:352–66
33. Chatterton, N. J., Silvius, J. E. 1981. Photosynthate partioning into starch in soybean leaves. II. Irradiance level and daily photosynthetic period duration effects. *Plant Physiol*.67:257–60
34. Chiariello, N. R., Mooney, H. A., Williams, K. 1989. Growth, carbon allocation and cost of plant tissue. In *Plant Physiological Ecology: Field Methods and Instrumentation*, ed. R. W. Pearcy, J. R. Ehleringer, H. A. Mooney, P. W. Rundel, pp. 327–65. London: Chapman & Hall
35. Chiariello, N., Roughgarden, J. 1984. Storage allocation in seasonal races of a grassland annual: optional versus actual allocation. *Ecology* 65:1290–1301
36. Coley, P. D., Bryant, J. P., Chapin, F. S. III. 1985. Resource availability and plant anti-herbivore defense. *Science* 230:895–99
37. Daubenmire, R. 1960. A seven-year study of cone production as related to xylem layers and temperature in *Pinus ponderosa*. *Am. Midl. Nat*. 64:187–93
38. Davidson, J. L., Milthorpe, F. L. 1966. The effect of defoliation on the carbon balance in *Dactylis glomerata*. *Ann. Bot*. 30:185–98
39. Dickson, R. E. 1987. Diurnal changes in leaf chemical constituents and $^{14}C$ partitioning in cottonwood. *Tree Physiol*. 3:157–71
40. Dickson, R. E. 1990. Carbon and nitrogen allocation in trees. In *Integrated Response of Plants to Stress*, ed. H. A. Mooney, W. E. Winner, E. J. Pell. New York: Academic
41. Donald, C. M., Hamblin, J. 1976. The biological yield and harvest index of cereals as agronomic and plant breeding criteria. *Adv. Agron*. 28:361–405
42. Edwards, J. 1985. Effects of herbivory by moose on flower and fruit production by *Aralia nudicaulis*. *J. Ecol*. 73:861–68
43. Ehleringer, J. R., Schulze, E.-D. 1985. Mineral concentrations in an autoparasitic *Phoradendron californicum* growing on a parasitic *P. californicum* and its host, *Cercideum floridum*. *Am. J. Bot*. 72:568–71
44. Eis, S., Garman, E. H., Ebell, L. F. 1965. Relation between cone production and diameter increment of Douglas-fir (*Pseudotsuga menziesii* (Mirb.) Franco), grand fir (*Abies grandis* (Dougl.) Lindl.), and western white pine (*Pinus monticola* Dougl.). *Can. J. Bot*. 43: 1553–59
45. Elmqvist, T., Ericson, L., Danell, K., Salomonson, A. 1987. Flowering, shoot production, and vole bark herbivory in a boreal willow. *Ecology* 68:1623–29

46. Epstein, E., 1972. *Mineral Nutrition of Plants: Principles and Perspectives.* New York: Wiley. 412 pp.
47. Ericsson, A., Larsson, S., Tenow, O. 1980. Effects of early and late season defoliation on growth and carbohydrate dynamics in Scots pine. *J. Appl. Ecol.* 17:747–69
48. Evans, L. T., Rawson, H. M., 1970. Photosynthesis and respiration by the flag leaf and components of the ear during grain development in wheat. *Aust. J. Biol. Sci.* 23:245–59
49. Fagerstöm, T. 1977. Growth in Scots pine (*Pinus silvestris* L.). Mechanism of response to nitrogen. *Oecologia* 26:305–15
50. Field, C. 1983. Allocating leaf nitrogen for the maximization of carbon gain: Leaf age as a control on the allocation program. *Oecologia* 56:341–47
51. Fonda, R. W., Bliss, L. C. 1966. Annual carbohydrate cycle of alpine plants on Mt. Washington, New Hampshire. *Bull. Torrey Bot. Club* 93:268–77
52. Fondy, B. R., Geiger, D. R. 1985. Diurnal changes in allocation of newly fixed carbon in exporting sugar beet leaves. *Plant Physiol.* 78:753–57
53. Fox, T. C., Geiger, D. R. 1984. Effects of decreased net carbon exchange on carbohydrate metabolism in sugar beet source leaves. *Plant Physiol.* 76:763–68
54. Fries, N. 1952. Variations in the content of phosphorus, nucleic acids and adenine in the leaves of some deciduous trees during the autumn. *Plant Soil* 4:29–42
55. Gadgil, M., Bossert, W. H. 1970. Life history consequences of natural selection. *Am. Nat.* 104:1–24
56. Gallagher, J. N., Biscoe, P. V., Hunter, B. 1976. Effects of drought on grain growth. *Nature* 264:541–42
57. Gauch, H. G. 1972. *Inorganic Plant Nutrition.* Stroudsburg: Dowden, Hutchinson, and Ross. 488 pp.
58. Gebauer, G., Melzer, A., Rehder, H. 1984. Nitrogen content and nitrate reductase activity in *Rumex obtusifolius* L. I. Differences in organs and diurnal changes. *Oecologia* 63:136–42
59. Geiger, D. F. 1990. Allocation of recently fixed and of reserve carbon in relation to stress. In *Integrated Response of Plants to Stress,* ed. H. A. Mooney, W. E. Winner, E. J. Pell. New York: Academic
60. Greenway, H., Gunn, A. 1966. Phosphorus retranslocation in *Hordum vulgare* during early tillering. *Planta* 71:43–67
61. Grime, J. P. 1977. Evidence for the existence of three primary strategies in plants and its relevance to ecological and evolutionary theory. *Am. Nat.* 111: 1169–94
62. Gulmon, S. L., Mooney, H. A. 1986. Costs of defense on plant productivity. In *On the Economy of Plant Form and Function,* ed. T. J. Givnish, R. Robichaux, pp. 681–98. Cambridge: Cambridge Univ. Press
63. Hajibagheri, M. A., Flowers, T. 1985. Salt tolerance in the halophyte *Sueda maritima* (L.) Dum. The influence of the salinity of the culture solution on leaf starch and phosphate content. *Plant Cell Environ.* 8:261–67
64. Heilmeier, H., Schulze, E.-D., Whale, D. M. 1986. Carbon and nitrogen partitioning in the biennial monocarp *Arctium tomentosum* Mill. *Oecologia* 70:466–67
65. Heim, G., Landsberg, J. J., Wilson, R. L., Brian, P. 1979. Ecophysiology of apple trees: Dry matter production and partitioning by young golden delicious trees in France and England. *J. Appl. Ecol.* 16:179–94
66. Hendry, G. 1987. The ecological significance of fructan in a contemporary flora. *New Phytol.* 106(Suppl.):201–10
67. Hetherington, A. M., Hunter, M. J. S., Crawford, R. M. M. 1984. Evidence contrary to the existence of storage lipids in leaves of plants inhabiting cold climates. *Plant Cell Environ.* 7:223–27
68. Hirose, T., Kachi, N. 1982. Critical plant size for flowering in biennials with special reference to their distribution in a sand dune system. *Oecologia* 55:281–84
69. Hirose, T., Werger, M. J. A. 1987. Maximizing daily canopy photosynthesis with respect to the leaf nitrogen allocation pattern in the canopy. *Oecologia* 72:520–26
70. Höll, W. 1985. Seasonal fluctuation of reserve materials in the trunkwood of spruce (*Picea abies* (L.) Karst.) *J. Plant Physiol.* 117:355–62
71. Horvitz, C. C., Schemske, D. W. 1988. Demographic cost of reproduction in a neotropical herb: an experimental field study. *Ecology* 69:1741–45
72. Huber, S. C., Israel, D. W. 1982. Biochemical basis for partitioning of photosynthetically fixed carbon between starch and sucrose in soybean (*Glycine max* Merr.) leaves. *Plant Physiol.* 69: 691–96
73. Jameson, D. A. 1963. Responses of individual plants to harvesting. *Bot. Rev.* 29:532–94
74. Jeffrey, D. W. 1968. Phosphate nutrition of Australian heath plants. II. The

formation of polyphosphate by five heath species. *Aust. J. Bot.* 16:603–13
75. Jonasson, S. 1989. Implications of leaf longevity, leaf nutrient re-absorption and translocation for the resource economy of five evergreen plant species. *Oikos* 56:121–31
76. Jonasson, S., Chapin, F. S. III. 1985. Significance of sequential leaf development for nutrient balance of the cotton sedge, *Eriophorum vaginatum* L. *Oecologia* 67:511–18
77. Keeley, J. E., Zedler, P. H. 1978. Reproduction in chaparral shrubs after fire: A comparison of sprouting and seeding strategies. *Am. Midl. Nat.* 99: 142–61
78. Koch, G. W., Schulze, E. D., Percival, F., Mooney, H. A., Chu, C. 1989. The nitrogen balance of *Raphanus sativus* x *raphanistrum* plants. II. Growth, nitrogen redistribution and photosynthesis under $NO_3$ deprivation.
79. Körner, C. 1989. The nutritional status of plants from high altitudes. A worldwide comparison. *Oecologia* 81:379–91
80. Kozlowski, T. T., Keller, T. 1966. Food relations of woody plants. *Bot. Rev.* 32:293–382
81. Kristic, B., Gebauer, G., Saric, M. 1986. Specific response of sugar beet cultivars to different nitrogen forms. *Z. Pflanz. Bodenk.* 149:561–71
82. Laine, K., Henttonen, H. 1983. The role of plant production in microtine cycles in northern Fennoscandia. *Oikos* 40: 407–18
83. Larsson, S., Wiren, A., Lundgren, L., Ericsson, T. 1986. Effects of light and nutrient stress on leaf phenolic chemistry in *Salix dasyclados* susceptibility to *Galerucella lineola* (Coleoptera). *Oikos* 47:205–10
84. Law, R. 1978. The costs of reproduction in an annual meadow grass. *Am. Nat.* 113:3–16
85. Levin, D. A. 1974. The oil content of seeds: an ecological perspective. *Am. Nat.* 108:193–206
86. Loomis, R. S., Worker, G. F. Jr. 1963. Responses of the sugar beet to low moisture at two levels of nitrogen nutrition. *Agron. J.* 55:509–15
87. Lowry, W. P. 1966. Apparent meteorological requirements for abundant cone crop in Douglas-fir. *For. Sci.* 12:185–92
88. Mark, A. F., Chapin, F. S. III. 1988. Seasonal control over allocation to reproduction in a tussock-forming and a rhizomatous species of *Eriophorum* in central Alaska. *Oecologia* 78:27–34
89. Matile, P. 1978. Biochemistry and function of vacuoles. *Annu. Rev. Plant Physiol.* 29:193–213
90. Mattheis, P. J., Tieszen, L. L., Lewis, M. C. 1976. Responses of *Dupontia fischeri* to simulated lemming grazing in an Alaskan arctic tundra. *Ann. Bot.* 40: 179–97
91. Matthews, J. D. 1955. The influence of weather on the frequency of beech mast years in England. *Forestry* 28:107–16
92. Matthews, J. D. 1963. Factors affecting the production of seed by forest trees. *For. Abs.* 24(1):i–xiii
93. McCree, K. J., Kallsen, C. E., Richardson, S. G. 1984. Carbon balance of sorghum plants during osmotic adjustment to water stress. *Plant Physiol.* 76:898–902
94. Melzer, A., Gebauer, G., Rehder, H. 1984. Nitrogen content and nitrate reductase activity in *Rumex obtusifolius* L. II. Responses to nitrate starvation and nitrogen fertilization. *Oecologia* 63: 380–85
95. Milford, G. F. J., Thorne, G. N. 1973. The effect of light and temperature late in the season on the growth of sugar beet. *Ann. Appl. Biol.* 75:419–25
96. Millard, P. 1988. The accumulation and storage of nitrogen by herbaceous plants. *Plant Cell Environ.* 11:1–8
97. Millard, P., Thomson, C. 1989. The effect of autumn senescence of leaves on the internal cycling of nitrogen for the spring growth of apple trees. *J. Exp. Bot.* 40:1285–89
98. Milthorpe, F. L., Davidson, J. L. 1966. Physiological aspects of regrowth in grasses. In *The Growth of Cereals and Grasses*, ed. F. L. Milthorpe, J. D. Ivin, pp. 241–54. London: Butterworths
99. Montalvo, A. M., Ackerman, J. D. 1987. Limitations to fruit production in *Ionopsis utricularioides* (Orchidaceae). *Biotropica* 19:24–31
100. Mooney, H. A. 1972. The carbon balance of plants. *Annu. Rev. Ecol. Syst.* 3:315–46
101. Mooney, H. A., Bartholomew, B. 1974. Comparative carbon balance and reproductive modes of two Californean *Aesculus* species. *Bot. Gaz.* 135:306–13
102. Mooney, H. A., Billings, W. D. 1960. The annual carbohydrate cycle of alpine plants as related to growth. *Am. J. Bot.* 47:594–98
103. Mooney, H. A., Hayes, R. I. 1973. Carbohydrate storage cycles in two California Mediterranean-climate trees. *Flora* 162:295–304
104. Nassery, H. 1969. Polyphosphate formation in the roots of *Deschampsia*

*flexuosa* and *Urtica dioica*. *New Phytol.* 68:21–23

105. Nelson, E. A., Dickson, R. E. 1981. Accumulation of food reserves in cottonwood stems during dormancy induction. *Can. J. For. Res.* 11:145–54
106. Newell, E. A. 1987. *The costs of reproduction in* Aesculus californica, *the California buckeye tree*. PhD Thesis. Stanford Univ., Stanford, Calif.
107. Oksanen, L., Ericson, L. 1987. Dynamics of tundra and taiga populations of herbaceous plants in relation to the Tihomirov-Fretwell and Kalela-Tast hypotheses. *Oikos* 50:381–88
108. Oren, R., Schulze, E.-D. 1989. Nutritional disharmony and forest decline: A conceptual model. *Ecol. Studies* 77: 425–43
109. Parsons, R. F. 1968. The significance of growth-rate comparisons for plant ecology. *Am. Nat.* 102:595–97
110. Passioura, J. B. 1976. Physiology of grain yield in wheat growing on stored water. *Aust. J. Plant Physiol.* 3:559–65
111. Pate, J. S. 1983. Patterns of nitrogen metabolism in higher plants. In *Nitrogen as an Ecological Factor*, ed. J. A. Lee, S. McNeill, I. H. Rorison, pp. 225–55. Oxford: Blackwell
112. Pate, J. S., Atkins, C. A., Perry, M. W. 1980. Significance of photosynthate produced at different stages of growth as carbon source for fruit filling and seed reserve accumulation in *Lupinus angustifolius* L. *Aust. J. Plant Physiol.* 7:283–97
113. Payton, I. J., Brasch, D. J. 1978. Growth and nonstructural carbohydrate reserves in *Chionochloa rigida* and *C. macra*, and their short-term response to fire. *N. Z. J. Bot.* 16:435–60
114. Pitelka, L. F., Hansen, S. B., Ashmun, J. W. 1985. Population biology of *Clintonia borealis*. *I.* Ramet and patch dynamics. *J. Ecol.* 73:169–83
115. Pollock, C. J. 1986. Fructans and the metabolism of sucrose in vascular plants. *New Phytol.* 104:1–24
116. Rao, M., Fredeen, A. L., Terry, N. 1990. Leaf phosphate status, photosynthesis, and carbon partitioning in sugar beet. III. Diurnal changes in carbon partitioning and carbon export. *Plant Physiol.* 92:29–36
117. Rappoport, H. F., Loomis, R. S. 1985. Interaction of storage root and shoot in grafted sugarbeet and chard. *Crop Sci.* 25:1079–84
118. Rawson, H. M., Evans, L. T. 1971. The contribution of stem reserves to grain development in a range of wheat cultivars of different height. *Aust. J. Agric. Res.* 22:851–63
119. Reekie, E. G., Bazzaz, F. A. 1987. Reproductive effort in plants. 1. Carbon allocation to reproduction. *Am. Nat.* 129:876–96
120. Reekie, E. G., Bazzaz, F. A. 1987. Reproductive effort in plants. 3. Effect of reproduction on vegetative activity. *Am. Nat.* 129:907–19
121. Richards, J. H., Caldwell, M. M. 1985. Soluble carbohydrates, concurrent photosynthesis and efficiency in regrowth following defoliation: a field study with *Agropyron* species. *J. Appl. Ecol.* 22:907–20
122. Schulze, E.-D. 1982. Plant life forms and their carbon, water and nutrient relations. *Encycl. Plant Physiol.* N. S. 12B:616–76
123. Schulze, E.-D. 1989. Air pollution and forest decline in a spruce *(Picea abies)* forest. *Science* 244:776–83
124. Schulze, E.-D. Chapin, F. S. III. 1987. Plant specialization to environments of different resource availability. In *Potentials and Limitations in Ecosystem Analysis*, ed. E.-D. Schulze, H. Zwolfer, pp. 120–48. Berlin: Springer-Verlag
125. Schulze, E.-D., Ehleringer, J. R. 1984. The effect of nitrogen supply on growth and water-use efficiency of xylem tapping mistletoes. *Planta* 162:268–75
126. Schulze, E.-D., Hall, A. E. 1972. Short-term and long-term effects of drought on steady-state and time-integrated plant processes: transpiration, carbon dioxide assimilation, biomass production and seed yield. In *Physiological Processes Limiting Plant Productivity*, pp. 217–35. Borough Green: Butterworths
127. Schulze, E.-D., Koch, G. W., Percival, F., Mooney, H. A., Chu, C. 1985. The nitrogen balance of *Raphanus sativus* x *raphanistrum* plants. I. Daily nitrogen use under high nitrate supply. *Plant Cell Environ.* 8:713–20
128. Schulze, E.-D., Mooney, H. A., Bullock, S. H., Mendoza, A. 1988. Water contents of wood of tropical deciduous forest species during the dry season. *Bol. Soc. Bot. Mexico* 48:113–18
129. Schulze, E.-D., Oren, R., Lange, O. L. 1989. Nutrient relations of trees in healthy and declining Norway spruce stands. *Ecol. Studies* 77:392–417
130. Schulze, E.-D., Turner, N. C., Glatzel, G. 1984. Carbon, water and nutrient relations of two mistletoes and their hosts. A hypothesis. *Plant Cell Environ.* 7: 293–99

131. Servaites, J. C., Tucci, M. A., Geiger, D. R. 1987. Glyphosate effects on carbon assimilation, ribulose bisphosphate carboxylase activity, and metabolite levels in sugar beet leaves. *Plant Physiol.* 85:370–74
132. Shaver, G. R., Chapin, F. S. III, Gartner, B. L. 1986. Factors limiting growth and biomass accumulation of *Eriophorum vaginatum* L. in Alaskan tussock tundra. *J. Ecol.* 74:257–78
133. Smith, D. 1973. The nonstructural carbohydrates. In *Chemistry and Biochemistry of Herbage,* vol. 1., ed. G. W. Butler, R. W. Bailey, pp. 105–55. London: Academic
134. Snow, A. A., Wigham, D. F. 1989. Costs of flower and fruit production in *Tipaularia discolor* (Orchidaceae). *Ecology* 70:1286–93
135. Sohn, J. J., Policansky, D. 1977. The costs of reproduction in the mayapple *Podophyllum peltatum* (Berberidaceae). *Ecology* 58:1366–74
136. Steinlein, T. 1987. *Die Bedeutung des Stickstoffs für Stoffproduktion and Stoffverteilung bei der Biennen Pflanze* Arctium tomentosum. MA thesis. Univ. Bayreuth
137. Stitt, M. 1988. Control analysis of photosynthetic sucrose synthesis: assignment of elasticity coefficients and flux control coefficients to the cytosolic fructose-1, 6-bisphosphatase and sucrose phosphate synthase. *Philos. Trans. R. Soc. London* 323:327–38
138. Stitt, M., Huber, S., Kerr, P. 1987. Control of photosynthetic sucrose formation. *Biochem. Plants* 10:327–408
139. Stitt, M., Quick, W. P. 1989. Photosynthetic carbon partitioning: its regulation and possibilities for manipulation. *Physiol. Plant.*
140. Schulze, W., Stitt, M., Schulze, E.-D. Neuhaus, H., Fichtner, K. 1990. A quantification of the significance of assimilatory starch for growth of Arabidopsis. *Plant Phys.* In press
141. Taylor, B. K. 1967. The nitrogen nutrition of the peach tree. I. Seasonal changes in nitrogenous constituents in mature trees. *Aust. J. Biol. Sci.* 20:379–87
142. Teckelmann, M., Schulze, E.-D. 1990. Carbon, water and nitrogen relations of the nitrophilous perennial *Urtica dioica* in contrasting habitats. *Oecologia.* In press
143. Titus, J. S., Kang, S.-M. 1982. Nitrogen metabolism, translocation, and recycling in apple trees. *Hort. Rev.* 4:204–46
144. Trlica, M. J. Jr., Cook, C. W. 1971. Defoliation effects on carbohydrate reserves of desert species. *J. Range Manage.* 24:418–25
145. Tschager, A., Hilscher, H., Franz, S., Kull, U., Larcher, W. 1982. Jahreszeitliche dynamik der fettspeicherung von *Loiseleuria procumbens* und anderen Ericaceen der alpinen zwergstrauchheide. *Acta Oecol.* 3:119–34
146. Tromp, J. 1970. Storage and mobilization of nitrogenous compounds in apple trees with special reference to arginine. In *Physiology of Tree Crops,* ed. L. C. Luckwill, C. V. Cutting, pp. 143–59, New York: Academic
147. Tso, T. C., Jeffrey, R. N. 1961. Biochemical studies on tobacco alkaloids. IV. The dynamic state of nicotine supplied to *N. rustica. Arch. Biochem. Biophys.* 92:253–56
148. Ulrich, A. 1955. Influence of night temperature and nitrogen nutrition on the growth, sucrose accumulation and leaf minerals of sugar beet plants. *Plant Physiol.* 30:250–57
149. van den Driessche, R. 1974. Prediction of mineral nutrient status of trees by foliar analysis. *Bot. Rev.* 40:347–94
150. Waring, R. H., Running, S. W. 1978. Sapwood water storage: its contribution to transpiration and effect upon water conductance through stems of old-growth Douglas-fir. *Plant Cell Environ.* 1:131–40
151. Waring, R. H., McDonald, A. J. S., Larsson, S., Ericsson, T., Wiren, A., et al. 1985. Differences in chemical composition of plants grown at constant relative growth rates with stable mineral nutrition. *Oecologia* 66:157–60
152. Watson, D. J., Motomatsu, T., Loach, K., Milford, G. F. J. 1972. Effects of shading and of seasonal differences in weather on the growth, sugar content and sugar yield of sugar-beet crops. *Ann. Appl. Biol.* 71:159–85
153. Weikert, R. M., Wedler, M., Lippert, M., Schramel, P., Lange, O. L. 1989. Photosynthetic performance, chloroplast pigments, and mineral content of various needle age classes of spruce *(Picea abies)* with and without the new flush: an experimental approach for analyzing forest decline phenomena. *Trees* 3: 161–72
154. Werner, P. A. 1975. Predictions of fate from rosette size in teasel (*Dipsacus fullonum* L.) *Oecologia* 20:197–201
155. White, L. M. 1973. Carbohydrate reserves of grasses: a review. *J. Range Manage.* 26:13–18

156. Williams, R. F. 1948. The effects of phosphorus supply on the rates of intake of phosphorus and nitrogen and upon certain aspects of phosphorus metabolism in gramineous plants. *Aust. J. Sci. Res. Ser. B* 1:333–61
157. Yamagata, M., Kouchi, H., Yoneyama, T. 1987. Partitioning and utilization of photosynthate produced at different growth states after anthesis in soybean (*Glycine max* L. Merr.): Analysis by long-term $^{13}C$-labelling experiments. *J. Exp. Bot.* 38:1247–59
158. Ziegler, H. 1964. Storage, mobilization and distribution of reserve material in trees. In *The Formation of Wood in Forest Trees,* ed. M. H. Zimmermann, pp. 307–20. New York: Academic
159. Zimmermann, M. H. 1971. Storage, mobilization and circulation of assimilates. In *Trees: Structure and Function,* ed. M. H. Zimmermann, C. L. Brown, pp. 307–22. New York: Springer-Verlag
160. Zucker, W. V. 1983. Tannins: Does structure determine function? An ecological perspective. *Am. Nat.* 121:335–65

*Annu. Rev. Ecol. Syst. 1990. 21:449–80*


# EVOLUTION OF DISPERSAL: Theoretical Models and Empirical Tests Using Birds and Mammals

*Michael L. Johnson*[1,2], *Michael S. Gaines*[2]

[1]Kansas Biological Survey, Foley Hall, University of Kansas, Lawrence, Kansas 66047

[2]Department of Systematics and Ecology, Haworth Hall, University of Kansas, Lawrence, Kansas 66045

KEY WORDS: birds, dispersal, evolution, mammals, philopatry

## INTRODUCTION

Dispersal is a life history trait that, viewed from both an ecological and a genetic perspective, has profound effects on populations. Dispersal is a demographic process that needs to be considered by population ecologists who want to understand the distribution and abundance of organisms. Population geneticists are interested in the genetic structure of populations and must assess how dispersal, with its potential for gene flow, changes allele frequencies within and among populations.

A fundamental question related to ecological and genetic aspects of dispersal is: What are the mechanisms for the evolution of dispersal? In recent years a prodigious number of mathematical models have described the evolution of dispersal. We believe that it is a propitious time to examine these models with respect to their underlying assumptions (Table 1), critical parameters, and predictions. Our objective is to elucidate common themes as well as differences in these models, and to present them in a manner comprehensible to individuals who lack extensive mathematical training, so that others might be encouraged to perform critical experiments in the field. We conclude the review with a summary of empirical work that uses data on birds and mammals to test some of these models.



0066-4162/90/1120–0449$02.00

**Table 1** Assumptions and critical parameters from some selected mathematical models. Due to spac limitations, not all assumptions are included for every model. Models grouped together share the assumptio listed, but do not necessarily share all of their assumptions.

| Model | Assumptions | Critical parameters | Reference |
|---|---|---|---|
| Hamilton-May | Parthenogenesis<br>One adult per site<br>Non-overlapping generations<br>Dispersal determined by parent<br>Site occupant chosen at random from potential colonists | Probability of surviving a dispersal episode | 76 |
| Comins-Hamilton-May | Reproduction, competition, migration are stochastic<br>Empty sites possible<br>Many individuals per site<br>Sites extirpated with fixed probability | Probability of surviving a dispersal episode<br>Site extinction rate<br>Number of individuals per site | 40 |
| Taylor | Infinite population size<br>Sexual reproduction<br>Discrete, non-overlapping generations<br>Many individuals per site | Cost of dispersal (survival)<br>Number of individuals per site | 188 |
| Frank | Infinite number of patches<br>Discrete non-overlapping generations<br>Females produce equal number of offspring<br>Many individuals per site | Cost of dispersal (survival)<br>Coefficient of relatedness among members of the population | 62 |
| Motro | Infinite population size<br>Discrete, non-overlapping generations<br>Single occupant per site<br>Site occupant chosen at random | Probability of surviving a dispersal episode | 136 |
| Asmussen | Discrete generation<br>Modifier gene influences dispersal, not fitness directly<br>Density dependent dispersal | Fitness in patches | 5 |
| Bull et al. | Dispersal between two sites<br>Discrete generations<br>Fitness determined by migratory history through survival<br>Fitness parameters constant over generations | Fitness in patches | 26 |
| Hastings-Holt | Population growth rate represented by differential equations<br>Dispersal by passive diffusion<br>Spatial variation only | Fitness in patches | 81, 88 |
| Waser et al | Population initially outbred<br>Inbreeding results in a reduction in number of offspring produced<br>Dispersal involves cost | Cost of inbreeding<br>Cost of dispersal | 13, 33, 1 |
| Van Valen | Discrete generations<br>Sites extirpated with fixed probability<br>Existing populations have equal probability of producing successful colonists | Probability of site extinction<br>Probability of dispersal | 71 |

## DEFINITION OF TERMS

Of the many definitions of dispersal proposed, we use Howard's (91) definition as "the permanent movement an individual makes from its birthsite to the place where it reproduces or would have reproduced had it survived and found a mate." This definition of natal dispersal is essentially the same as that of Endler (54). An additional type of movement is breeding dispersal, or the movement from one home range to another between attempts at reproduction, disregarding whether reproduction is successful. The distinction between natal and breeding dispersal was also made by Greenwood & Harvey (73). We include no minimum distance requirement for dispersal, either in actual distance or in home range diameters, because such a priori limitation is arbitrary. For example, Shields (176) used a distance of ten home ranges to distinguish philopatry from dispersal, while Waser & Jones (200) implicitly considered movement the distance of one home range to be dispersal. Harvey et al (80) found that juvenile great tits dispersed between four and seven home ranges before attempting to breed; these great tits would be considered philopatric by Shields and dispersers by Waser & Jones. We also distinguish between ecological and genetic dispersal. *Ecological dispersal* is simply the movement of an individual from one place to another; *genetic dispersal* refers specifically to individuals moving from one breeding population to another and successfully breeding there, resulting in the potential for changes in allele frequency (26). This distinction is not always made in the literature. However, if dispersers do not leave offspring in the new population, it is difficult to construct an argument for the evolution of dispersal. Genetic dispersal includes ecological dispersal, but ecological dispersal need not imply genetic dispersal.

*Migration* is often defined as directional movement made by individuals or (more commonly) groups during specific times within a generation (64). These movements may or may not result in a return to the point of origin. However, *migration* and *dispersal* are often used interchangeably in the theoretical models. In fact, Baker (7) defined migration as "the act of moving from one spatial unit to another." We use this more general definition of migration, and we use migration and *dispersal* interchangeably.

In our review we codify evolutionary models into two basic types, models invoking group selection and models using individual selection. The mathematical models, in general, focus on genetic dispersal; the remaining models focus on ecological dispersal. Even though this distinction can be critical when designing experiments to test these models, we present the models without categorizing them as genetic or ecological. Also, some models were developed implicitly or explicitly for natal dispersal; others may model either natal or breeding dispersal. The distinction is usually clear, and we make no attempt to assign the models to either category.

# GROUP SELECTION MODELS

Early speculations about the evolution of dispersal were often couched in terms of group selection. Although never formally developed in detail, many authors hypothesized that dispersal evolved from selection taking place at the level of the species or population (e.g. 23, 91, 97, 183). The benefits to the population or species were assumed to be population regulation and range expansion.

## *Wynne-Edwards's Model*

Wynne-Edwards (210) viewed emigration as the most universally available means of reducing population density, and he suggested that emigration was a direct result of populations increasing above the optimal density. Wynne-Edwards felt that emigration fulfilled the need for extending or restocking the range of the species, or for exchanging genes between neighboring groups that may have become genetically impoverished. He suggested that the young of the population should be the pioneering stock. Interestingly, while Wynne-Edwards viewed emigration as occurring for the good of the species, the mechanism he proposed to determine which individuals dispersed could be based on individual selection. His hypothesis was that dominant individuals forced subordinant individuals from the population, and that the chances for reproduction of the subordinant individuals were better in a new location.

## *Van Valen's Model*

The only explicit quantitative treatment of dispersal and group selection is Van Valen's (192) simulation model, in which he assumed that dispersal outside of the population resulted in death and was selected against, but that this emigration could result in the reestablishment of extinct populations. Van Valen (192) clearly recognized the conditions for the maintenance of dispersal by individual selection when he stated, "in order for within population selection to do anything other than minimize dispersal, the probability of survival of an individual disperser, wherever it goes, must be as large as that of a nondisperser." He dismissed this possibility as unlikely except for populations with locally high densities and ephemeral habitats. Each population had a constant probability of extinction, and so the persistence of the population was not dependent on the frequency of dispersers in the population. Thus, his model operated only when differential establishment of populations resulted from dispersal, not from differential extinction. Van Valen varied the extinction rate of the populations and the proportion of the offspring that remained in the population. Using a three-genotype, two-allele system, he found that he was able to maintain the dispersal allele with a probability of population extinction as low as .05 per generation. In general,

Van Valen found that the optimal dispersal rate was about equal to the probability of local extinction. Empirical validation of his theory would require a knowledge of two parameters, the probability of local extinction, and the probability of an individual dispersing.

### *Altruistic Trait Models*

A large number of general theoretical models exist that describe the evolution of altruistic traits by the process of group selection. One of the findings of many of these models is that for group selection to be effective, little or no migration or gene flow should occur between demes (3, 42, 57, 101, 112, 124, 177, but see 194, 195). Dispersal could be an altruistic act given the assumption that successful dispersal occurs only within a short time during the formation of the migrant or propagule pool (see above models for details about the formation of migrant pools) and that it results in differential colonization of groups. As an altruistic act, dispersal at any other time would serve to reduce the size of the local population, providing more resources for those that do not disperse. This was the basic idea of Wynne-Edwards. Dispersal would result in both differential extinction and formation of populations.

An interesting question arises as to whether the processes of group and individual selection could result in the same dispersal rate. If dispersal occurred only during the formation of the migrant pool, successful dispersal of individuals would result in differential formation of populations. The problem that immediately presents itself is whether the rate of dispersal that maximizes individual fitness also maximizes group fitness (see below).

## INDIVIDUAL SELECTION MODELS

By far the largest group of models, both mathematical and verbal, hypothesize that selection occurring at the level of the individual is responsible for the evolution of dispersal. Different types of models exist within this category. A large number of models deal with Evolutionarily Stable Strategies (ESS) of dispersal, where the analysis attempts to solve for the evolutionarily stable dispersal rate of the population. Another set of models, also mathematical, derive the optimal dispersal rate without resorting to the mathematics of ESSs. Several genetic models investigating the maintenance of dispersal polymorphisms fall into this category. A third set of models, usually verbal, attempt to intuit the conditions that lead to the evolution of dispersal. Finally, a large group of models assume an ultimate mechanism for the evolution of dispersal (e.g. inbreeding avoidance) and then attempt to explain why a particular subgroup, e.g. one sex or age class, is the predominant disperser or disperses the greater distance.

## *Evolutionarily Stable Strategy Models*

A large number of models use the concept of Evolutionarily (sometimes Evolutionary) Stable Strategies (ESS). Maynard Smith and his coworkers originated the concept of ESS in a series of papers that used game theory to analyze outcomes of differing behavioral strategies that could be used by members of a population (e.g. 123, 125–128). ESS models basically use a set of simplifying parameters and assumptions to determine the ESS condition, which can be explained as follows. If members of a population employ a particular strategy A (e.g. disperse from the population at a rate D), then A is an ESS if it is superior to any other strategy B that arises in the population (e.g. dispersal at a rate D'). The ESS can be defined much more precisely and rigorously in terms of mathematical expectation of fitness of the behavioral strategies (see 83 for a review), but such a definition is unnecessary for this discussion.

HAMILTON-MAY-COMINS MODELS The first formal ESS dispersal model was the inclusive fitness model developed by Hamilton & May (76). Previous verbal and mathematical models were usually concerned with dispersal in environments that fluctuated in time and space (see below). Hamilton & May (76) first demonstrated that it could be adaptive for parents to force the dispersal of a fraction of their offspring even if the environment was stable, and even when dispersal was very risky. Their basic model assumed a wholly parthenogenetic species in an environment with a fixed number of sites. Only one adult could occupy each site; each adult produced $m$ offspring and then died, leaving the site available for colonization. The mother's genotype determined the proportion of the offspring, $v$, that dispersed. A fraction, $p$, of those dispersers survived. All offspring competed equally for each site, i.e. philopatry did not convey to an offspring a potential advantage in assuming occupancy of its natal site. This assumption often differs from the assumptions of verbal models, where some advantage to a nondisperser in occupying its natal site is assumed.

Hamilton & May (76) found, given a large number of patches and a large number of offspring, that the ESS migration rate was $v^* = 1/(2-p)$. This remarkable and perhaps counterintuitive result means that parents should enforce the dispersal of one half or more of their offspring, even if the probability of survival of the dispersal episode is low. This result becomes more intuitively clear when the assumptions of the model are considered (76). Only one offspring was capable of occupying a site. If none of those offspring disperse, all would compete among themselves and with any dispersers for occupancy of the site. At best, an adult with nondispersing offspring can expect only one successful offspring. This would occur only if the offspring outcompetes dispersers for the site, which is not guaranteed under the assumptions of the model. A parent that forces some of her offspring to disperse could have several successful offspring, because her young could occupy her own site as well as other sites in the landscape. In effect, dispersal is positively selected because, all else being equal, it is better for a parent to have her offspring compete with nonsiblings at many sites than among themselves at a single site (88).

Hamilton & May (76) relaxed some of their assumptions to extend their deterministic model. Specifically, some adults were allowed to survive from one generation to the next, and some sites were allowed to remain vacant. They introduced stochasticity into the model by specifying that the number of offspring produced at each site was determined by a Poisson distribution with mean and variance $m$ (the same $m$ as used in the deterministic model). When $m$ was large, the results were no different from those of the deterministic model, indicating that large litter sizes tended to offset the effects of the possibility of vacant sites. When $m$ was small, $v^*$ was determined again by the probability of successful dispersal, and also by $m$ and the fraction of sites occupied. An interesting result from this analysis was that the ESS dispersal rate was always greater than the dispersal rate that maximized site occupancy for the population. Thus, an individual parent would force more of its offspring to disperse than would be in the best interests of the population. Finally, Hamilton & May (76) incorporated sexual reproduction in the analysis. The $v^*$ was expected to decrease, due to the introduction of a conflict of interest between offspring and parents. Parents, in an attempt to increase their own fitness, would be expected to force more of their offspring to disperse, while their offspring would resist dispersal, especially if the probability of survival was low. Analyzing the interesting case in which all male offspring dispersed, eliminating the possibility of inbreeding, Hamilton & May solved for the ESS $v^*$ of females and found that $v^*$ for females was lower in this case than with the parthenogenetic model,

$$v^* = 0 \qquad 0 < p < 1/2$$

$$v^* = (2p-1)/(4p-1-2p^2) \qquad 1/2 < p < 1.$$

Again, in this deterministic model, $v^*$ depends only on the probability of surviving the dispersal episode.

Comins et al (40) developed the model of Hamilton & May (76) a step further by introducing more complexity into the analysis. They produced a discrete generation island model in which stochasticity was introduced into the processes of migration, reproduction, and competition. Comins et al allowed for empty sites and for more than one adult per site, and they introduced environmental variability in the form of a probability of site destruction. Dispersal has two benefits in their model: colonization of empty sites, and competition for occupied sites in which the frequency of the dispersal gene is significantly lower than at the presently occupied site. The general results of their model were that the ESS dispersal rate increased with increased probability of site destruction as a result of the increased number of unoccupied sites. Also, dispersal increased with the increasing variance in gene frequency, a result of the large number of sites with a significantly lower gene frequency of the wild type. Additionally, the ESS $v^*$ decreased as the number of adults per site or the number of offspring per site increased, a result of the decreased number of unoccupied sites and a decrease in the variance in the gene frequency. Finally, $v^*$ increased with increasing probability of survival of the dispersal episode. Modifying the model to allow for sexual reproduction or stepping stone migration resulted in a lower $v^*$ in both cases. Sexual reproduction led to the conflict of interest discussed earlier, and stepping stone migration increased the spatial correlation in gene frequency and therefore lowered the variance in gene frequency. Additional analyses by Comins (39) indicated that increasing the spatial correlation in gene frequency, and consequently the lowering of the variance in gene frequency, could be compensated for by the benefits derived from colonization of empty sites at some distance from the presently occupied site.

These models assumed that the fecundity of all individuals was the same. Cohen & Motro (38) recognized that there could be a cost to dispersal in the form of reduced fecundity. They developed a model specifically for seed dispersal, where the cost to the plant was for the production of a physical structure for dispersal. However, their model could apply whenever a cost is involved in producing offspring that disperse. Their results collapsed to $v^* = 1/(2-p)$ when there was no cost to producing a dispersal mechanism. When a cost was inolved, the ESS $v^*$ fell, and therefore fewer dispersers were expected to be produced.

The Hamilton-May-Comins ESS approach was extended even farther by Frank (62) and Taylor (188). Frank (62) used Price's (155) method for the hierarchical analysis of natural selection to find a generalized solution for the problem of dispersal polymorphisms in stable habitats. Using essentially the same assumptions as Hamilton-May-Comins, Frank (62) arrived at the ESS dispersal rate $v^*$ as:

$$v^* = \frac{\rho - c}{\rho - c^2} \qquad 0 < c < \rho < 1,$$

where $\rho$ can be defined as the regression coefficient of relatedness of the controlling genotype (the genotype that determines who disperses) onto a member of the cohort of offspring chosen randomly before dispersal (i.e. the degree of relatedness among members of the population); $c$ was the cost of dispersal. Under the specific conditions of the Hamilton-May-Comins model, Frank's $v^*$ became exactly the same $v^* = 1/(2-p)$. Frank's model both simplified and generalized the results of Hamilton-May-Comins by incorporating the possibility of more than one individual per patch and inbreeding. Taylor (188) considered an inclusive fitness model under a variety of genetic systems and dispersal conditions. Again, his results provided a general solution that collapsed to the results of Hamilton-May-Comins and Frank (62) under the specific conditions of their models. The critical contribution of both Frank (62) and Taylor (188) was their explicit consideration of different levels of relatedness in haploid, diploid, or haplodiploid systems. Females dispersing from a patch in a stable habitat not only can colonize an empty site, they can avoid competing with related individuals that do not disperse. Even if the number of potential survivors in a dispersal episode is low, the gain in inclusive fitness can offset the loss. Inbreeding within subpopulations increases the dispersal rate regardless of the subpopulation size or cost of dispersal, because the gain in inclusive fitness is great. Outbreeding decreases dispersal, especially as the cost of dispersal increases.

Stenseth (185) generated another generalized ESS model based on a two-stage Leslie matrix with an explicit term for the gain in reproductive output in the new habitat after successful dispersal. His simple result was that:

$$v^* = 1/2[1/(1-p) - 1/(1-g)],$$

where $p$ = probability of successful dispersal, and $g$ = gain in reproductive output as a result of dispersal. For dispersal to be present at all ($v^* > 0$), it must be true that $g = p$ (set $v^* = 0$ in the equation above and solve), or the gain in reproductive output must offset the loss in fitness due to dispersal.

TEMPORAL AND SPATIAL VARIABILITY MODELS Additional ESS models have been developed to investigate dispersal in environments variable in space or in both space and time. Hastings (81) and Holt (88) found that variation over space, by itself, was not sufficient to select for dispersal, as the ESS $v^* = 0$. The reason for this was simple: Passive diffusion took individuals from more favorable to less favorable habitats more often than the reverse simply because more favorable habitats tended to have more individuals in the first place (88). The ESS models of Hastings (81) and Holt (88) represent population growth with differential equation models, which in effect assume that the population size in each habitat is large enough to be treated as a continuous variable. In this situation, the progeny of any single individual will compete with many non kin within each site. This is in contrast to the original Hamilton-May model, where the population size at each site is a discrete integer—one adult—and there is strong competition among kin at each site. Comins et al (40) demonstrated that as the number of adults at each site increases, the ESS dispersal rate $v^*$ approaches zero. So the Hastings-Holt result emerges as a limiting case of the Hamilton-May model (88). When population sites are large enough, dispersal is disadvantageous in a spatially variable but temporally constant environment. Yet dispersal does exist in nature, often under circumstances that do not meet the assumptions of the Hamilton-May models (e.g. competition among kin for limited sites), so there should be conditions that would favor the maintenance of an ESS $v^* > 0$. The obvious candidate to investigate is temporal variability. Levin, Cohen & Hastings (113) analyzed a model where environments could be variable in both space and time. Their model was constructed explicitly for seed dispersal and included seed dormancy as a possible complicating mechanism. However, under conditions of no dormancy, their model could apply to other taxa. Environments were locally unpredictable, and the variation was described by a probability distribution that remained constant. The number of patches was considered infinite, and each patch had a carrying capacity equal to $k$. Using a combination of analytical and simulation techniques, they found that when there was no cost to dispersal, the ESS dispersal rate was always one (as in the Hamilton-May-Comins models). As the probability of surviving a dispersal episode decreased, the ESS dispersal rate came to depend on the relativized variance in $k$, the carrying capacity of the patches. The ESS $v^*$ approached 0 as the survival rate of dispersers approached a critical level $a_{crit} = \hat{k}/\bar{k}$, the ratio of the harmonic mean of the carrying capacity to the arithmetic mean of the carrying capacity. Because $\hat{k} << \bar{k}$ when the variance in $k$ is $>>$ than 0, there was always some lower limit to the critical survival level that ensured a reasonable level of dispersal. As the variance in $k$ decreased, however, $\hat{k}$ approached $\bar{k}$, and $a_{crit}$ approached 1. In other words, when the environmental variability is slight, dispersal is selected against as soon as there is any substantial cost to dispersal. As the environmental variability increased, the ESS $v^*$ increased ($a_{crit}$ becomes lower) even when the probability of surviving a dispersal episode was low. Low environmental variability alone, or low variability coupled with high population size, resulted in lower dispersal, i.e. when the probability of finding a suitable environment elsewhere was low, the dispersal rate declined. Recent work by R. D. Holt & V. McPeck (personal communication) has shown that chaotic population dynamics can also select for dispersal. Even in a physically constant environment, such dynamics can lead to substantial temporal variation in fitness that is negatively correlated across space, thus setting up the condition needed to favor dispersal.

Bull et al (26) used a genetic model to address the question of whether spatial variation could lead to the evolution of migration (dispersal). They discovered that selection for migration depended on the geometric mean fitness of individuals involved in movements from each subpopulation to the other. It did not matter how large the gain was in fitness of individuals migrating from environment 1 to environment 2; if the gain in fitness of individuals migrating from environment 2 to environment 1 was low, there would be no selection for dispersal. Selection tended to reduce dispersal when individuals moved from an area where they realized higher fitness to an area where they realized lower fitness (as in the Hastings-Holt model).

Lomnicki (117–119) took a more mechanistic approach to modeling the evolution of dispersal by including social structure in his models. The ability of an individual to obtain resources was dependent on its rank in a dominance hierarchy. Limited resources were apportioned among the lowest $K$ ranks in the hierarchy, and individuals ranked above $K$ did not secure enough resources to reproduce unless they dispersed to a new location. Lomnicki (119) solved for the evolutionarily stable dispersal rank, the rank in the dominance hierarchy above which animals should disperse. Lomnicki (119) found in a stable environment (low probability of newly established empty habitats) that the ESS dispersal rank was very close to $K$, but as the environment became less stable, the ESS dispersal rank dropped below $K$. In other words, when an unstable environment translates to a high probability of empty habitats each generation, individuals with low ranks should disperse. Even if individuals could secure enough resources to reproduce, if they can reasonably expect to obtain more elsewhere, they should disperse.

The results of the ESS models are quite consistent, regardless of their details. In stable habitats, the dispersal rate (in the absence of complicating factors) appears to be $v^* = 1/(2-p)$. Consequently, the critical parameter is the probability of surviving a dispersal episode. If the environment is variable, the type and amount of variability is critical in determining the dispersal rate. Spatial variation alone tends to reduce dispersal rate, because on average an individual will disperse to a location that is not as good as the one presently occupied because there are more individuals in good than in bad locations. Temporal variability tends to increase dispersal rate, and when both temporal and spatial variation exist, the relative amounts of each determine whether the dispersal rate will increase or not.

## *Non-ESS Individual Selection Models*

GENERAL MODELS Early empirical studies of dispersal often documented a leptokurtic distribution of dispersal distances (e.g. 46, 90). Consequently, some of the earliest theoretical models attempted to explain this phenomenon. Murray (139), with a simple simulation model, was able to demonstrate that the skewed distribution of dispersal distances could be accounted for by subordinate individuals moving to the first available home range in response to aggression by dominant individuals. Individuals were moving only to maximize their own chance for reproduction when they were faced with the problem of not being able to occupy their natal site.

TEMPORAL AND SPATIAL VARIABILITY MODELS Selection for dispersal in response to fluctuations in the environment has been a common theme in the literature (see 205 for a general review of population responses to patchy environments). It is easy to envision an advantage to an individual that is able to escape deteriorating conditions and disperse to a more favorable location (e.g. 115, 116). Again, the problem arises as to whether the variability occurs over space or time, or both. As in ESS models, variability in space alone apparently tends to select for reduced dispersal, and variability in time tends to select for increased dispersal. The exact reasons for this vary with the model specifications but remain remarkably consistent.

In an analysis of seasonal migratory behavior, Cohen (37) developed a model that can be applied to dispersal between two habitats that are variable in both space and time. He found that the optimal migration fraction depended on the relative variances in the viability of individuals in the two habitats, and whether viabilities in the two habitats were positively or negatively correlated. When the viabilities in each habitat are positively correlated, there is a tendency to have a migration fraction of either 0 or 1. If the survival of the migratory and philopatric portions of the population are both low, whether an individual migrates or not depends on the variability in survival over time in the habitat into which migration occurs. High variability results in 100% migration, and low variability results in 0% migration. When the viabilities in the two habitats are negatively correlated, an intermediate fraction should disperse. The exact fraction again depends on the variances of the viabilities. Interestingly, in an attempt to explain dispersal in populations of *Microtus,* Grant (70) provided a mechanistic explanation for dispersal in a seasonally varying environment. Grant hypothesized that animals dispersed in response to cues given off by slowing vegetative growth and the corresponding increased difficulty in meeting nutritional needs, along with responding to increased aggression in environments crowded from natal recruitment. Later, Berger et al (16, 17; Negus & Berger 141) isolated secondary compounds from vegetation that signaled the beginning and the end of the growing season, indirectly lending support to Grant's hypothesis.

Gill (67) also implicated high density as a force behind the evolution of dispersal. He felt that the key was temporal variation in crowding in local habitats. If habitats were locally stable with respect to temporal variation in crowding, i.e. crowding would continue over time, selection would favor either increased tolerance of crowding or increased competitive ability. If, however, crowding did not persist and was locally unpredictable, dispersal was expected to evolve. Predictable temporal variation in crowding would lead to diapause, or some mechanism of avoiding the harsh period without incurring the risks of dispersal.

Kuno (110), and Metz et al (130), who amplified and clarified Kuno's original paper, analyzed a similar situation except that alternative behavioral strategies such as diapause were not possible. The environment consisted of several patches in which the growth rates were identically distributed and independent over years, but not over patches. Dispersers were at a selective advantage over nondispersers, and when the patches differed in some systematic manner over time, the advantage of dispersal over nondispersal was even greater, even if dispersing involved some mortality. Predictable variability in time rather than over space contributed to the selective advantage of dispersers over nondispersers. Gadgil (63) also analyzed the effect of environmental variability in space and time on dispersal. He focused on how that variability influenced the proximal mechanisms leading to the evolution of dispersal. However, his general findings were the same as in previous models, variability in space reduced dispersal while variability in time increased dispersal.

GENETIC MODELS Roff (164) formally coupled genetics to the problem of the evolution of dispersal in heterogeneous environments. He manipulated stability by varying the means and variances of the population growth rate and the carrying capacity. Through simulations, Roff found that it was impossible to predict the effect of changing environmental variability on the frequency of dispersers. What became critical to dispersal rate was the way the environmental variability was generated. Increasing environmental stability did not always decrease the proportion of dispersers in the population. For example, when the stability of the environment was increased by an increase in the mean growth rate of the population, the proportion of dispersers maintained in the population always increased. However, the general trend was for a decrease in the proportion of dispersers with increasing environmental stability.

Motro (136–138) examined optimal dispersal rates using a traditional genetic analysis of changes in gene frequency. Using a haploid population with discrete generations and only a single occupant per site, Motro found that the optimal dispersal rate was $v^* = 1/(2-p)$ where $p$ is the probability of surviving the dispersal episode. Consequently, if the environment is very harsh, and $p \rightarrow 0$, then $v^* = 1/2$, and if the environment is benign and $p \rightarrow 1$, then everybody should disperse, i.e. $v^* = 1$. Increasing the carrying capacity equally in each site depressed the dispersal rate because all individuals could acquire their natal site. Using a diploid population, Motro (137) examined the question of how an adult should divide its progeny into dispersed and nondispersed descendents, i.e. what conditions allowed for a stable dispersal polymorphism. He used a three genotype-two allele system, each genotype having its own dispersal rate (determined by its mother). Given that the optimal rate of dispersal was $v^* = 1/(2-p)$, Motro found that a polymorphic equilibrium could develop if the dispersal rate of the heterozygote was close to $v^*$, and was also closer to $v^*$ than the dispersal rate of either homozygote. Consequently, heterozygote advantage, in the form of dispersing its offspring at a rate closest to $v^*$, was the necessary condition for the maintenance of a polymorphism. In the final paper of the series, Motro (138) examined the optimal rate of dispersal when dispersal was determined by the offspring's genotype rather than the parent's. Motro (138) conjectured that this situation would lead to a possible conflict of interest between parent and offspring. Parents want more of their offspring to disperse because successful dispersers increase their inclusive fitness, while offspring should want to remain at their natal site, especially if the risk of mortality during dispersal is great. Motro was able to demonstrate that a genetic polymorphism for dispersal was possible, and that the optimal rate of dispersal was a function only of the probability of surviving the dispersal episode. The optimal dispersal rate was lower than when dispersal was controlled by the adult, as Motro surmised (see also 76).

Several other genetic models have been developed to investigate the evolution of dispersal under conditions of environmental variability. Balkau & Feldman (8) and later Karlin & McGregor (101) and Teague (189) demonstrated analytically that in either haploid or diploid populations, selection tended to reduce dispersal rates in environments that exhibited only spatial variation. This parallels the results of Hastings (81) and Holt (88). Gillespie first suggested (68) and later demonstrated (69) that in environments variable in time and space, selection could lead to either reduced or increased dispersal rates, depending on the relative amount of variation in each. Gillespie (69) found that selection maximized the geometric mean fitness of the population, which resulted in decreased dispersal rates when there was more spatial variation than temporal variation (see also 26). Selection led to an increase in dispersal rate when there was relatively more temporal than spatial variation. Moody (133) found similar results.

Asmussen (5) used the modifier approach to investigate analytically the evolution of dispersal in spatially variable populations where dispersal was density-dependent. She found that an allele that arises at a selectively neutral locus that influences dispersal will persist if and only if the allele reduces the migration rates at equilibrium population size. Even if migration rates were density independent, the conclusions were the same (e.g. 190)—selection again reduced the dispersal rate. Her justification was the same as Holt's (88). The genotype with the highest fitness in a given environment would be the most frequent in that environment and would comprise a majority of the dispersers to the new environment. However, these dispersers will be less fit in the new environment, and the modifier should act to reduce dispersal.

## DISPERSAL AND INBREEDING

The evolution of dispersal as a mechanism for the avoidance of inbreeding has been implicitly or explicitly suggested by a number of authors (e.g., 18, 25, 44, 47, 49, 50, 66, 71, 72, 78, 84–87, 89, 91, 93, 128a, 132, 149, 150, 153, 156–158, 170, 172, 209). Recently, empirical demonstrations of inbreeding depression have occurred in zoo populations (159–161, 171), and evidence for possible inbreeding depression has been provided by natural populations (74, 149, 150, 191a). Others, however, have questioned the assumption that inbreeding in natural populations is always deleterious (e.g. 182, 190). A major proponent of this view is Shields (175, 176; and see 134) who hypothesized that the leptokurtic dispersal distance distribution common in birds and mammals was a direct result of selection for inbreeding (contra 139, 197, 198).

Shields hypothesized a balance between inbreeding costs and benefits, predicated on the assumption that the environment is highly heterogeneous over small spatial scales. Selection and mild inbreeding result in highly linked gene complexes, such that movement over even short distances (10 home-range diameters) places the individual in an environment for which it is not adapted. Philopatry, in effect, is an intraspecific premating isolating mechanism for maintaining highly adaptive, tightly linked gene complexes. For empirical support, Shields cited three studies (6, 9, 167) as providing direct or indirect evidence that mild inbreeding (effective population size $<1000$) and local adaptation resulted in significant outbreeding depression (see also 10, 11).

Shields predicted that low fecundity species ($<10^5$ progeny per lifetime) could minimize the cost/benefit ratio of sexual reproduction by inbreeding intensely (but avoiding incest) in small demes ($<1000$). Species with high fecundity would be subjected to selection for outbreeding, because high fecundity organisms could afford the potential for outbreeding depression. More offspring are produced than can survive in the immediate vicinity, so some offspring could be sent to habitats some distance from the natal site, with the possibility that they could survive. As a test of his predictions, Shields evaluated the dispersal patterns of a wide range of taxa and found that the amount of inbreeding was associated with fecundity as he predicted. However, not all environments are heterogeneous, and Shields explained inbreeding in homogeneous environments as a way of maintaining individual mutations that are favorable when alone, but unfavorable in combination with other mutations. Different mutations arise in different populations, each conferring a different type of advantage on its carriers. Inbreeding maintains the mutations within the small groups and is a way for dealing with what Shields termed "genetic heterogeneity". Thus, Shields predicted reduced dispersal regardless of the amount of spatial variability in the environment.

Both Bengtsson (13) and Chesser & Ryman (33) analyzed the relative costs of inbreeding versus dispersal. Bengtsson found that in general, the strategy that provided the greatest fitness, either dispersal or inbreeding, was determined by the inequality:

$$(1 - m_1)\,[1 - i_1\,(1 + 2\,f_1)] < (1 - m_2)\,[1 - i_2\,(1 + 2\,f_2)],$$

where $m$ is the cost of migration, $i$ is the cost of inbreeding, and $f$ is the coefficient of consanguinity in the population. This expression simply states that the strategy 2 is favored over the strategy 1 provided the cost of migration of 2 and the inbreeding cost of 2 is less than the corresponding costs of strategy 1. If 2 is the strategy of no migration, leading to inbreeding, then $m_2 = 0$ and $f_2 = .25$ (the coefficient of consanguinity between sibs in an outbred population), and let $i_2 = i$. If strategy 1 is dispersal, leading to no inbreeding, then $f_1 = i_1 = 0$, and let $m_1 = m$. The general case expression above becomes simply $m > (3/2)i$, i.e. in order for migration to confer a selective advantage over inbreeding, the cost of migration must be less than 3/2 the cost of inbreeding. The immediate conclusion was that if $i > 2/3$, inbreeding could not evolve because the cost of migration cannot exceed 1. If there is no cost to dispersal, inbreeding should never occur (see also 151, 152). Unfortunately, this analysis does not directly address Shields' arguments, as Shields assumed that whatever dispersal occurred did so because it was adaptive.

Chesser & Ryman (33) expanded the analysis of the balance between inbreeding and dispersal by examining several types of breeding strategies in subdivided populations. They found that the conditions that favored inbreeding depended on the type of mating (e.g. full sib vs outbred matings), the cost of inbreeding, the relatedness of individuals in the population, and the cost of migration. After developing a general model, they analyzed specific cases that provided the same results as those of Bengtsson (13). Chesser & Ryman also demonstrated that in polygynous mating systems, it is genetically advantageous for males to disperse. Females should disperse only when the matings that occur after their male sibs disperse have a low probability of producing viable offspring (as would be the case if the population is already highly inbred). Male dispersal was always selected for unless the matings of female sibs with other males in the population resulted in no offspring. These were the first analytical results to provide a mechanism to account for sex-biased dispersal (see next section).

Despite all these models implicating dispersal as a means of avoiding inbreeding, Waser et al (199) in a general model were able to demonstrate that dispersal probably would not evolve primarily as a mechanism to avoid inbreeding. They found that if the fractional loss of offspring as a result of inbreeding was less than the cost of dispersal, dispersal would not evolve as the mechanism to avoid inbreeding. Waser et al (199) were also able to provide a mechanism for sex-biased dispersal. Their results did not rely on the mating system, but simply on the number of outbred matings forfeited when an individual mates with a relative.

One, somewhat controversial, paper suggests that inbreeding avoidance plays no role in the evolution of dispersal. Moore & Ali (134) proposed that male dispersal was solely the result of intrasexual competition for mates, and females dispersed to gain access to critical resources or to avoid infanticide by males. Moore & Ali were taken to task by Packer (150) for misinterpreting and misrepresenting evidence used to support their hypothesis; consequently this hypothesis has not received much serious consideration.

## AGE-BIASED OR SEX-BIASED DISPERSAL

The evolution of sex-biased dispersal has received as much attention in the last decade as has the evolution of dispersal itself, and yet it remains an arena of major controversy. Three major hypotheses have been proposed to explain sex-biased dispersal: (*a*) resource competition, (*b*) intrasexual competition for mates, and (*c*) inbreeding avoidance. These hypotheses are not mutually exclusive (50) and are often offered together as explanations for sex-biased dispersal. A fourth hypothesis is based on genetic asymmetry, where the disperser is the bearer of the heterogametic sex chromosome (203). However, this hypothesis has been largely discounted (71, 209). Other hypotheses (e.g. the male dominance hypothesis and the Oedipus hypothesis) can usually be categorized as combinations of the three.

The origins of the debate can be traced to a seminal paper by Greenwood (71, later amplified 72) who, after reviewing the literature, found that in birds, females tended to be the dispersing sex. In mammals, males tended to be the dispersing sex. Because the inbreeding avoidance hypothesis for dispersal predicts that either sex can disperse, Greenwood looked for an explanation that related dispersal to the mating system. Female-biased dispersal was hypothesized to be the result of monogamy, the predominant mating system of birds, and male-biased dispersal the result or polygyny, the predominant mating system of mammals. In monogamous systems, males must acquire and defend sufficient resources to attract females. Familiarity with their natal area could provide an advantage in acquiring those resources, and consequently, males should be philopatric. In polygynous mating systems, females invest more in the offspring than do males; therefore the acquisition of territory is critical to their reproductive success. Females would gain most from being philopatric and gaining access to their natal area, while males maximize their reproductive output by gaining access to many females. Male-biased dispersal is the result. However, Waser & Jones (200) questioned the validity of this hypothesis by pointing our several counterexamples.

Dobson (49) also assumed that inbreeding avoidance was responsible for the evolution of dispersal. He hypothesized that intrasexual competition for mates was responsible for sex-biased dispersal, and he generated predictions about dispersal in monogamous or polygynous mating systems from each of the three major hypotheses (intrasexual competition, inbreeding avoidance, and resource competition). Some predictions of the three models were the same. For example, for a monogamous mating system, both the intrasexual mate competition hypothesis and the resources competition hypothesis predicted that both sexes of juveniles should disperse. However, the prediction about dispersal in a polygynous mating system clearly separated the two competing hypotheses. The mate competition hypothesis predicted that only juvenile males should disperse, while the resource competition hypothesis predicted similar numbers of both sexes should disperse. He found that for 57 species of mammals, the evidence supported the mate competition hypothesis. His conclusions have since been questioned because avian taxa are not represented (e.g. 36). However, this hypothesis along with Greenwood's (71) hypothesis remains one of the more widely accepted.

Additional hypotheses have been proposed to explain sex-biased dispersal, but they seem to be variations of the three discussed above. The male dominance hypothesis (45, 66) was recognized by Woolfenden & Fitzpatrick (209) to be closely related to the resource competition hypothesis (but see 71). Males remain philopatric in monogamous systems to gain access to resources necessary to attract females, and through their aggressive behavior males force females to disperse. The Oedipus hypothesis (114) viewed dispersal from the point of view of the parents and was based on the premise that parents would evict offspring of the sex that would cost them the most to retain. In polygynous mammals or birds, adult males should evict sons because they represent potential competition for mates. In monogamous birds, female offspring are driven off because they could force their parents to raise the offspring's young through brood parasitism. In monogamous mammals, there is no opportunity for this type of cheating, so male and female offspring should disperse in equal proportions. Marks & Redmond (121) challenged this hypothesis, claiming that the assumptions underlying the model were incorrect. They also provided counterexamples that did not fit the predictions; these examples include leking birds, waterfowl, and numerous species of mammals. Finally, Johnson (94) proposed a hypothesis to explain sex-biased dispersal that combined aspects of Greenwood (71) and Dobson (49) as well as life history theory. He suggested that the philopatric sex would be the one in which dispersal would cause the greatest delay in breeding. For example, in polygynous systems, because of the extreme competition among males for females, juvenile males typically do not acquire mates; natal dispersal of males thus would not delay breeding. Juvenile females, or at least very young females, do breed, and consequently natal dispersal could delay reproduction for them. Females, therefore, should be philopatric. This hypothesis is intriguing, but more predictions need to be generated because the difficulty in measuring the "costs" of delayed breeding make this hypothesis very difficult to test.

Finally, while a great deal of effort has gone into formulating hypotheses about sex-biased dispersal, little work has been done to demonstrate that natal dispersal is adaptive. While the explanation of inbreeding avoidance almost necessitates dispersal before the physiological age at first reproduction, other hypotheses (e.g. mate competition) do not. Morris (135) used a life history approach analytically to demonstrate that when dispersal is adaptive for all age classes, prereproductives gain most by dispersing. There were even some circumstances when older age classes would not benefit from dispersal, and yet prereproductives would. Morris also was able to demonstrate that if inclusive fitness is added to the analysis, adults may gain benefits, in addition to any benefits the offspring themselves could gain, by forcing the offspring to disperse. However, when parents became too old, they benefited, through the gain in inclusive fitness, by dispersing and leaving their home range to their offspring. Consequently, dispersal was predicted to occur when animals were very young and very old.

## EMPIRICAL TESTS OF THE MODELS

Over the past two decades, a prodigious number of papers have contained information about dispersal of birds and mammals, an acknowledgment of the critical role of dispersal in the life history and population dynamics of most taxa. Unfortunately, few of these investigations addressed the evolution of dispersal. In fact, for birds and mammals, no empirical tests exist for any of the mathematical models for the evolution of dispersal. This is not unexpected for several reasons. First, only recently have techniques been developed to identify dispersers in the field. Consequently, initial empirical investigations of dispersal often do little more than document its presence in natural populations and identify the predominant age and sex class of dispersers. In fact, there is still debate over the proper techniques for identifying dispersers (43,

48, 51, 52, 109, 154, 169, 193). Second, the lack of testing is undoubtedly the result of the difficulty in meeting the assumptions or measuring the parameters in the models. For example, results as simple as those of Hamilton-May-Comins require a knowledge of the probability of surviving a dispersal episode, data that, with a few notable exceptions, are completely lacking. More involved models require knowledge of parameters such as the probability of site extinction, fraction of sites occupied, and carrying capacity. Measuring these parameters is a formidable task, and not surprisingly few data exist. The difficulty of following dispersers, especially small birds and mammals, after they leave their natal site can be insurmountable. They are impossible (or nearly so) to see, and trapping to relocate dispersers can be difficult. Radiotelemetry has been helpful, but the cost can be prohibitive if a large sample size is desired; small transmitters often require searches over large areas to relocate individuals (e.g. 19, 145, 206, 207). Large mammals move even greater distances, making dispersers just as difficult to find. Large animals are usually few, making large samples difficult to obtain. Finally, large animals can live for a long time, making data on survival and lifetime reproduction inaccessible. These problems make investigating the evolution of any life history trait difficult.

Additionally, assumptions of the models are very difficult to meet, especially when these deal with higher vertebrates. The assumptions that all parents die in each generation, and that parthenogenesis provides replacements obviously cannot be met for a majority of species. These problems leave two options. First is an attempt to test the theoretical models, even though all assumptions are not met, and parameters are not adequately measured. The second choice is not to test the models, but to use them for heuristic purposes.

Despite the problems, there have been some attempts to investigate the evolutionary basis of dispersal. Investigations usually fall into two categories: (*a*) direct field studies of the cost and benefits of dispersing, and (*b*) indirect evidence about dispersal, collected for other purposes and used as support for a hypothesis about the evolutionary basis of dispersal.

## *Tests of Group Selection Models*

Investigations of the first type are rare; in fact no tests exist of group selection using birds or mammals. Some laboratory investigations have demonstrated that group selection is possible. For example, Craig (41) was very effective in using group selection to establish populations of *Tribolium confusum* with either high or low emigratory activity. However, little evidence from natural populations supports the claim that dispersal evolved by group selection.

## *Tests of Individual Selection Models*

A number of recent studies attempted to measure the survival and reproduction of dispersing animals as a means of assessing costs and benefits of dispersal. While it has often been assumed that the survival of dispersers is very low (174), survival of a dispersal episode can vary greatly, depending on the species and conditions encountered by the dispersing individuals. Garrett & Franklin (65) found a 46% lower survival rate for dispersers than for resident black-tailed prairie dogs *(Cynomys ludovicianus)*. Leuze (111) found that one half of dispersing water voles *(Arvicola terrestris)* were lost to predation, an increase of 86 times over the predation rate on residents at the same time. Wiggett & Boag (206) found that 16% of dispersing male and 11% of dispersing female Columbian ground squirrels (*Spermophilus columbianus*), at a minimum, survived dispersal. Jones & Waser (100) were able to demonstrate that dispersal of banner-tailed kangaroo rats (*Dipodomys spectabilis*) was related to the degree of habitat saturation. Jones (99, Table 1) found that, at high density, survival of dispersers (animals moving >50 m) of both sexes was lower than the survival of nondispersers (30% for males and 76% for females). At low density, dispersers survived better than philopatric animals (27% for males and 49% for females). Wiggett et al (207) found 100% (6 of 6) survival of radiotagged dispersing Columbian ground squirrels (*Spermophilus columbianus*). Boyce & Boyce (22) also found good survival for a small sample of female *Microtus arvalis;* only 1 animal disappeared in 29 cases of breeding dispersal and 31 cases of natal dispersal. Berger (14, 15) found little mortality among dispersing feral horses. Krohne & Burgin (107) found no differences in survival between philopatric residents and successful dispersers of *Peromyscus leucopus*.

Unfortunately, measures of lifetime reprodutive success of dispersers and nondispersers are extremely difficult to obtain, and consequently, investigations usually determine the reproductive success of dispersers in the breeding season(s) immediately after dispersal. For example, Berger found that 63% of the females that dispersed produced at least one foal in two years postdispersal, and 90% produced at least one foal in three years postdispersal. Only 13% of male dispersers sired offspring by three years post-dispersal. Krohne & Burgin (107) found no differences in reproductive activity between male resident and dispersing *Peromyscus leucopus,* but a significantly greater proportion (28%) of female residents were reproductively active than were female dispersers. Production of independent young per bird was unrelated to dispersal in song sparrows (*Melospiza melodia*) (4), and dispersing kangaroo rats reproduced as successfully as philopatric rats (198).

Johnson & Gaines (95, 96) monitored the survival and reproduction of dispersing and nondispersing prairie voles (*Microtus ochrogaster*) in an attempt to assess experimentally the costs and benefits of dispersal. The first experiment (95) involved removing all animals from two live-trap grids and allowing dispersers to colonize. Dispersers survived slightly better than residents on the control grid, and their reproduction (as measured by external reproductive characters) was much higher than that of residents. In a second experiment (96), animals were captured as they attempted to disperse from an enclosed population, and one half were moved into an empty enclosure (dispersers) and one half were placed back into the original population (frustrated dispersers). Dispersers in the empty habitat experienced the highest fitness, as measured by the product of survival and reproductive activity. The fitnesses of the frustrated dispersers and nondispersers in the same enclosure were the same. However, frustrated dispersers had significantly lower survival than nondispersers during the first few weeks after they were placed back into the enclosure. This period was assumed to be one of transience, and the reduced survival reflected the problems involved in transience experienced by other species (e.g. 1, 29, 30, 56, 65, 106, 111, 131, 148, 178, 181, 189). However, once frustrated dispersers were able to establish a home range and "enter" the population, they experienced the same survival as the rest of the population, but had a much greater reproductive activity. In fact, their reproductive activity was as high as that of the dispersers in the empty enclosure. The reduced survival was compensated for by an increased reproductive activity, resulting in fitness comparable to that of nondispersers. Several problems exist with this experiment, including the placement of dispersers back into their original enclosure, but experiments such as this appear to be a step toward evaluating the costs and benefits of dispersal.

By far the most common types of investigation are those initiated for purposes other than testing hypotheses related to the evolutionary basis of dispersal. Enough data were collected that, when combined with the general features of the social system and life history of the species, some inferences about the ultimate causal factors of dispersal could be made (Table 2). As noted by several authors (e.g. 50) the hypotheses listed are not mutually exclusive, but more often than not, only a single hypothesis is offered. There appear to be few phylogenetic trends. Dispersal in rodents is explained using all five hypothesis; in fact, ground squirrels alone include four of the five. There is a strong trend for nonhuman primates: every study offers the avoidance of inbreeding as an explanation for the evolution of dispersal. Fewer studies document dispersal in birds, probably because of the greater vagility of this group. Most studies found dispersal of birds to be in accordance with Greenwood's resource competition hypothesis (71). Males are philopatric in order to increase their chances of acquiring a territory; females disperse to acquire males, and therefore the resources, necessary to reproduce successfully.

**Table 2** Species for which an ultimate cause for the evolution of dispersal has been hypothesized. IA = inbreeding avoidance, MC = mate competition, RC = resource competition, OI = optimal inbreeding, O = other

| Species | Mechanism | Reference |
|---|---|---|
| Mammals | | |
| *Equus caballus* (feral horse) | IA | 14, 15 |
| *Odocoileus virginianus* (white-tailed deer) | OI | 142, 143 |
| *Odocoileus hemionus* (black-tailed deer) | MC, IA | 27 |
| *Aepyceros melampus* (impala) | RC, IA | 140 |
| *Cervus elaphus* (red deer) | RC, IA(?) | 35 |
| *Helogale parvula* (dwarf mongoose) | MC, IAS | 166 |
| *Canus lupus* (wolf) | OI | 129 |
| *Vulpes vulpes* (red fox) | RC, IA | 193a |
| *Ursus americanus* (black bear) | MC | 165 |
| *Panthera leo* (lion) | MC | 77 |
| *Marmota flaviventris* (yellow-bellied marmot) | MC | 24 |
| *Spermophilus californicus* (California ground squirrel) | MC | 47, 49 |
| *Spermophilus columbianus* (Columbian ground squirrel) | IA | 60 |
| *Spermophilus beldingi* (Belding's ground squirrel) | MC, OI | 84, 86 |
| *Cynomys ludovicianus* (black-tailed prairie dog) | MC, RC | 65 |
| *Ondatra zibethicus* (muskrat) | IA | 28 |
| *Clethrionomys rufocanus* (grey-sided vole) | IA | 92 |
| *Clethrionomys rufocanus* (grey-sided vole) | MC | 102 |
| *Peromyscus leucopus* (white-footed mouse) | IA | 108, 208 |
| *Peromyscus maniculatus* (deer mouse) | MC, RC | 58, 104 |
| *Peromyscus maniculatus* (deer mouse) | O[1] | 104 |
| *Dipodomys spectabilis* (banner-tailed kangaroo rat) | RC | 98–100, 198 |
| *Ochotona princeps* (pika) | MC | 179, 180 |
| *Ochotona princeps* (pika) | O[2] | 204 |
| *Papio anubis* (olive baboon) | IA | 149 |
| *Pan troglodytes* (chimpanzee) | IA | 156 |
| *Gorilla gorilla* (gorilla) | IA | 78 |
| *Cercopithecus aethiops* (vervet monkey) | IA | 31 |
| **Antechinus* spp. (dasyurid marsupial) | UL, IA | 36 |
| **Trichosurus vulpecula* (brush-tailed possum) | MC | 34 |
| Birds | | |
| *Phalaropus lobatus* (Red-necked Phalarope) | RC | 163 |
| *Parus palustris* (Marsh Tit) | RC | 145 |
| *Numenius americanus* (Long-billed Curlew) | IA | 162 |
| *Melospiza melodia* (Song Sparrow) | RC | 4, 202 |
| *Gymnorhinus cyanocephalus* (Piñon Jay) | MC | 122 |
| *Ficedula hypoleuca* (Pied Flycatcher) | MC | 79 |
| *Melanerpes formicivorus* (Acorn Woodpecker) | IA | 105 |
| *Crotophage sulcirostris* (Groove-billed Ani) | RC | 21 |
| *Nucifraga columbiana* (Clark's Nutcracker) | RC | 191 |
| *Larus occidentalis* (Western Gull) | RC | 184 |
| *Centrocerus urophasianus* (Sage Grouse) | RC | 53 |

*Specifically for sex-biased dispersal
1. Sexual search hypothesis
2. Sibling competition/female dispersal hypothesis
? Indicates our interpretation of author's description

For dispersal to undergo the process of natural selection, it must have a genetic basis (55). Establishing this genetic basis has proved difficult. Familial resemblance in movement or a tendency for littermates to leave the nest has been documented (12, 20, 82, 104), indicating the possibility of a genetic basis. Attempts to measure the heritability of dispersal itself have resulted in estimates as high as 0.62 (2). More common are attempts to measure the heritability of dispersal distance, and Greenwood et al (75) found a heritability of 0.56–0.62 for great tits. However, such measurements are fraught with technical difficulties (e.g. 4, 191b), reducing the utility of the estimates. For example, after accounting for location of origin of nesting song sparrows, Arcese (4) found that sibs were not more similar in dispersal distance than nonsibs; this fact suggests a low heritability. Arcese (4) cautioned that this position effect could be the explanation for the high correlations of dispersal distances of sibs in spruce grouse (103) and mother-offspring dispersal distance in sparrowhawks (144). When Waser & Jones (201) accounted for sources of environmental variation, they found a heritability of 0 for both dispersal distance (see also 61) and the tendency to disperse for kangaroo rats. They assumed that some genetic variation for dispersal tendency existed in the population but concluded that its detection would be difficult because of sample size problems and the possibility that dispersal is a threshold trait (59). Obviously, the measurement of the heritability of dispersal in natural populations is still in its infancy.

## CONCLUSIONS

This review has attempted to provide a reasonably detailed description of theoretical models for the evolution of dispersal in order to illustrate that despite the broad range of modelling techniques and assumptions of the models, there are a number of common conclusions. It is also hoped that a somewhat detailed account of the models will help investigators who do empirical work to recognize models that could be applied to their research.

In general, dispersal should be a common phenomenon, even in stable habitats and even if the survival of dispersers is low. These results are common to a large number of models; in fact, the relationship that continually appears is an optimal dispersal rate (or the ESS dispersal rate) $v^* = 1/(2-p)$. Introducing more complexity into the models by relaxing some of the assumptions—e.g. about sexual reproduction, or increasing the carrying capacity of the patches—tended to reduce the optimal dispersal rate. These results also indicate that the critical parameter is the probability of surviving dispersal.

As was originally assumed in early speculations about the evolution of dispersal, temporal and spatial variability in the environment play a major role in determining the optimal dispersal rate. Temporal variation tends to result in increased dispersal, spatial variation in decreased dispersal.

There is considerable debate over the evolution of sex-biased dispersal. None of the hypotheses can be falsified by examining the literature of large numbers of taxa. Expanded reviews of the literature probably will not determine which hypothesis, if any, is correct. The hypotheses must generate additional predictions that can be tested in the field, not with the literature. Using data gathered for other purposes to test hypotheses may lead to conclusions that can be refuted simply by waiting for the next published data set. For example, published data on the survival of dispersing small mammals could be used to estimate the costs and benefits of dispersal and thereby indirectly to test hypotheses about the evolution of dispersal. However, this can lead to spurious conclusions because different techniques used to identify dispersing small mammals appear to provide different subsamples of the disperser population, each with its own probability of survival.

Empirical tests of models for the evolution of dispersal or sex-biased dispersal are totally lacking. Restrictive assumptions and unmeasurable parameters, along with the difficulty of designing experiments to investigate the evolution of a life history trait, have restricted investigations to measuring the survival and reproduction of dispersers within small areas, as a means of assessing costs and benefits (e.g. 95, 96). Empirical investigations aimed at measuring the costs and benefits of dispersal in environments that are variable over space and time appear necessary to evaluate many of the theoretical models. While such studies will involve a great deal of work, they offer the most promise for elucidating the mechanisms responsible for the evolution of dispersal.

ACKNOWLEDGMENTS

We thank Bob Holt for his insightful comments and for sharing some of the general results from his unpublished work. Our own work on dispersal has been supported by National Science Foundation grants DEB-8020343 and BSR-8314825 to M.S. Gaines and grants from the University of Kansas General Research Fund.

## *Literature Cited*

1. Ambrose, H. W. III. 1972. Effect of habitat familiarity and toe-clipping on rate of owl predation in *Microtus pennsylvanicus*. *J. Mammal* 53:909–12
2. Anderson, J. L. 1975. *Phenotypic correlations among relatives and variability in reproductive performance in populations of the vole* Microtus townsendii. PhD thesis. Univ. Br. Columbia, Vancouver, BC. 207 pp.
3. Aoki, K. 1982. A condition for group selection to prevail over counteracting individual selection. *Evolution* 36:832–42
4. Arcese, P. 1989. Intrasexual competition, mating system and natal dispersal in Song Sparrows. *Anim. Behav.* 38:958–79
5. Asmussen, M. A. 1983. Evolution of dispersal in density regulated populations: A haploid model. *Theor. Popul. Biol.* 23:281–99
6. Baker, M. C., Mewaldt, L. R. 1978. Song dialects as barriers to dispersal in White-crowned Sparrows, *Zonotrichia leucophrys* Nuttali. *Evolution* 32:712–22
7. Baker, R. R. 1978. *The Evolutionary*

*Ecology of Animal Migration*. London: Hodder & Stroughton

8. Balkau, B. J., Feldman, M. W. 1973. Selection for migration modification. *Genetics* 74:171–74
9. Bateson, P. P. G. 1978. Sexual imprinting and optimal outbreeding. *Nature* 273:659–60
10. Bateson, P. P. G. 1980. Optimal outbreeding and the development of sexual preferences in the Japanese quail. *Z. Tierpsychol*. 53:231–44
11. Bateson, P. P. G. 1983. Optimal outbreeding. In *Mate Choice,* ed. P. Bateson, pp. 257–77. Cambridge: Cambridge Univ. Press
12. Beacham, T. 1979. Dispersal tendency and duration of life of littermates during population fluctuations of the vole *Microtus townsendii*. *Oecologia* 42:11–21
13. Bengtsson, B. O. 1978. Avoiding inbreeding: at what cost? *J. Theor. Biol*. 73:439–44
14. Berger, J. 1986. *Wild Horses of the Great Basin: Social Competition and Population Size*. Chicago: Univ. Chicago Press
15. Berger, J. 1987. Reproductive fates of dispersers in a harem-dwelling ungulate: the wild horse. See Ref. 32, pp. 41–54
16. Berger, P. J., Negus, N. C., Sanders, E. H., Gardner, P. D. 1981. Chemical triggering of reproduction in *Microtus montanus*. *Science* 214:69–70
17. Berger, P. J., Sanders, E. H., Gardner, P. D., Negus, N. C. 1977. Phenolic compounds functioning as reproductive inhibitor in *Microtus montanus*. *Science* 195:575–77
18. Bischof, N. 1975. Comparative ethology of incest avoidance. In *Biosocial Anthropology,* ed. R. Fox, pp. 37–67. London: Malaby
19. Boag, D. A., Murie, J. O. 1981. Population ecology of Columbian ground squirrels in southwestern Alberta. *Can. J. Zool*. 59:2230–40
20. Boonstra, R., Craine, I. M. T. 1988. Similarity of residence times among *Microtus* littermates: importance of sex and maturation. *Ecology* 69:1290–93
21. Bowen, B. S., Koford, R. R., Vehrencamp, S. L. 1989. Dispersal in the communally breeding Groove-billed Ani (*Crotophaga sulcirostis*). *Condor* 91:52–64
22. Boyce, C. C. K., Boyce, J. L. III. 1988. Population biology of *Microtus arvalis*. II. Natal and breeding dispersal of females. *J. Anim. Ecol*. 57:723–36
23. Brereton, J. L. G. 1962. Evolved regulatory mechanisms of population control. In *The Evolution of Living Organisms,* ed. G. W. Leeper, pp. 81–93. Parkville, Victoria: Melbourne Univ. Press
24. Brody, A. K., Armitage, K. B. 1985. The effects of adult removal on dispersal of yearling yellow-bellied marmots. *Can. J. Zool*. 63:2560–64
25. Buechner, M. 1987. A geometric model of vertbrate dispersal: tests and implications. *Ecology* 68:310–18
26. Bull, J. J., Thompson, C., Ng, D., Moore, R. 1987. A model for natural selection of genetic migration. *Am. Nat*. 129:143–57
27. Bunnell, F. L., Harestad, A. S. 1983. Dispersal and dispersion of black-tailed deer: models and observations. *J. Mammal* 64:201–9
28. Caley, M. J. 1987. Dispersal and inbreeding avoidance in muskrats. *Anim. Behav*. 35:1225–33
29. Carl, E. A. 1971. Population control in arctic ground squirrels. *Ecology* 52:395–413
30. Cheney, D. L., Lee, P. C. 1981. Behavioral correlates of non-random mortality among free-ranging female vervet monkeys. *Behav. Ecol. Sociobiol*. 9:153–61
31. Cheney, D. L., Seyfarth, R. M. 1983. Non-random dispersal in free-ranging vervet monkeys: social and genetic consequences. *Am. Nat*. 122:392–41
32. Chepko-Sade, B. D., Halpin, Z. T., eds. 1987. *Mammalian Dispersal Patterns*. Chicago: Univ. Chicago Press
33. Chesser, R. K., Ryman, N. 1986. Inbreeding as a strategy in subdivided populations. *Evolution* 40:616–24
34. Clout, M. N., Efford, M. G. 1984. Sex differences in the dispersal and settlement of brushtailed possums (*Trichosurus vulpecula*). *J. Anim. Ecol*. 53:737–49
35. Clutton-Brock, T. H., Guiness, F. E., Albon, S. D. 1982. *Red Deer*. Chicago: Univ. Chicago Press
36. Cockburn, A., Scott, M. P., Scotts, D. J. 1985. Inbreeding avoidance and male-biased dispersal in *Antechinus* spp. (Marsupialia: Dasyuridae) *Anim. Behav*. 33:908–15
37. Cohen, D. 1967. Optimization of seasonal migratory behavior. *Am. Nat*. 101:5–17
38. Cohen, D., Motro, U. 1989. More on optimal rates of dispersal: taking into account the cost of the dispersal mechanism. *Am. Nat*. 134:659–63
39. Comins, H. N. 1982. Evolutionarily stable strategies for dispersal in two dimensions. *J. Theor. Biol*. 94:579–606
40. Comins, H. N., Hamilton, W. D., May, R. M. 1980. Evolutionarily stable dis-

persal strategies. *J. Theor. Biol.* 82: 205–30
41. Craig, D. M. 1982. Group selection versus individual selection: an experimental analysis. *Evolution* 36:271–82
42. Crow, J. F., Aoki, K. 1982. Group selection for a polygenic behavioral trait: A differential proliferation model. *Proc. Natl. Acad. Sci. USA* 79:2628–31
43. Danielson, B. J., Johnson, M. L., Gaines, M. S. 1986. An analysis of a method for comparing residents and colonists in a natural population of *Microtus ochrogaster*. *J. Mammal* 67:733–36
44. Demarest, W. J. 1977. Incest avoidance among human and nonhuman primates. In *Primate Bio-Social Development,* ed. S. Chevalier-Skolnikoff, F. E. Poirier, pp. 323–42. New York: Garland
45. Dhondt, A. A. 1979. Summer dispersal and survival of juvenile Great Tits in southern Sweeden. *Oecologia* 42:139–57
46. Dice, L. R., Howard, W. E. 1951. Distance of dispersal by prairie deermice from birthplace to breeding sites. *Contrib. Lab. Vertebr. Biol. Univ. Mich.* 50:1–15
47. Dobson, F. S. 1979. An experimental study of dispersal in the California ground squirrel. *Ecology* 60:1103–9
48. Dobson, F. S. 1981. An experimental examination of an artificial dispersal sink. *J. Mammal* 62:74–81
49. Dobson, F. S. 1982. Competition for mates and predominant juvenile male dispersal in mammals. *Anim. Behav.* 30:1183–92
50. Dobson, F. S., Jones, W. T. 1986. Multiple causes of dispersal. *Am. Nat.* 126:855–58
51. Dueser, R. D., Rose, R. K., Porter, J. H. 1984. A body-weight criterion to identify dispersing small mammals. *J. Mammal* 65:727–29
52. Dueser, R. D., Wilson, M. L., Rose, R. K. 1981. Attributes of dispersing meadow voles in open-grid populations. *Acta Theriol.* 26:139–62
53. Dunn, P. O., Braun, C. E. 1985. Natal dispersal and lek fidelity of Sage Grouse. *Auk* 102:621–27
54. Endler, J. 1977. *Geographic Variation, Speciation, and Clines*. Princeton, NJ: Princeton Univ. Press
55. Endler, J. A. 1986. *Natural Selection in the Wild*. Princeton, NJ: Princeton Univ Press
56. Errington, P. L. 1946. Predation and vertebrate populations. *Q. Rev. Biol.* 21:144–77, 221–45
57. Eschel, I. 1972. On the neighbor effect and the evolution of altruistic traits. *Theor. Popul. Biol.* 3:258–77
58. Fairbairn, D. J. 1978. Dispersal of deer mice *Peromyscus maniculatus:* proximal causes and effects on fitness. *Oecologia* 32:171–93
59. Falconer, D. S. 1981. *Introduction to Quantitative Genetics*. London: Longman. 2nd ed.
60. Festa-Bianchet, M., King, W. J. 1984. Behavior and dispersal of yearling Columbian ground squirrels. *Can. J. Zool.* 62:161–67
61. Fleischer, R. C., Lowther, P. E., Johnston, R. F. 1984. Natal dispersal in house sparrows: possible causes and consequences. *J. Field Ornithol.* 55: 446–56
62. Frank, S. A. 1986. Dispersal polymorphisms in subdivided populations. *J. Theor. Biol.* 122:303–9
63. Gadgil, M. 1971. Dispersal: population consequences and evolution. *Ecology* 52:253–61
64. Gaines, M. S., McClenaghan, L. R. Jr. 1980. Dispersal in small mammals. *Annu. Rev. Ecol. Syst.* 11:163–96
65. Garrett, M. G., Franklin, W. L. 1988. Behavioral ecology of dispersal in the black-tailed prairie dog. *J. Mammal* 69:236–50
66. Gauthreaux, S. A. Jr. 1978. The ecological significance of behavioural dominance. In *Perspectives in Ethology,* ed. P. P. G. Bateson, P. H. Klopfer, 3:17–54. London: Plenum
67. Gill, D. E. 1978. On selection at high population density. *Ecology* 59:1289–91
68. Gillespie, J. H. 1975. The role of migration in the genetic structure of populations in temporally and spatially varying environments. 1. Conditions for polymorphism. *Am. Nat.* 109:127–35
69. Gillespie, J. H. 1981. The role of migration in the genetic structure of populations in temporally and spatially varying environments. III. Migration modification. *Am. Nat.* 117:223–33
70. Grant, P. R. 1978. Dispersal in relation to carrying capacity. *Proc. Natl. Acad. Sci. USA* 75:2854–58
71. Greenwood, P. J. 1980. Mating systems, philopatry and dispersal in birds and mammals. *Anim. Behav.* 28:1140–62
72. Greenwood, P. J. 1983. Mating systems and the evolutionary consequences of dispersal. See Ref. 186, pp. 116–31
73. Greenwood, P. J., Harvey, P. H. 1982. Natal and breeding dispersal in birds. *Annu. Rev. Ecol. Syst.* 13:1–21
74. Greenwood, P. J., Harvey, P. H., Perrins, C. M. 1978. Inbreeding and dis-

persal in the Great Tit. *Nature* 271:52–54
75. Greenwood, P. J., Harvey, P. H., Perrins, C. M. 1979. The role of dispersal in the Great Tit (*Parus major*): the causes, consequences and heritability of natal dispersal. *J. Anim. Ecol.* 48:123–42
76. Hamilton, W. D., May, R. M. 1977. Dispersal in stable habitats. *Nature* 269:578–81
77. Hanby, J. P., Bygott, J. D. 1987. Emigration of subadult lions. *Anim. Behav.* 35:161–69
78. Harcourt, A. H. 1978. Strategies of emigration and transfer by primates, with particular reference to gorillas. *Z. Tierpsychol.* 48:401–20
79. Harvey, P. H., Greenwood, P. J., Campbell, B., Stenning, M. J. 1984. Breeding dispersal of the Pied Flycatcher (*Ficedula hypoleuca*). *J. Anim. Ecol.* 53:727–36
80. Harvey, P. H., Greenwood, P. J., Perrins, C. M. 1979. Breeding area fidelity of the Great Tit. *J. Anim. Ecol.* 48:305–13
81. Hastings, A. 1983. Can spatial variation alone lead to selection for dispersal? *Theor. Popul. Biol.* 24:244–51
82. Hilborn, R. 1975. Similarities in dispersal tendency among siblings in four species of voles (*Microtus*). *Ecology* 56:1221–25
83. Hines, W. G. S. 1987. Evolutionary stable strategies: a review of basic theory. *Theor. Popul. Biol.* 31:195–272
84. Holekamp, K. E. 1984. Natal dispersal in Belding's ground squirrels (*Spermophilus beldingi*). *Behav. Ecol. Sociobiol.* 16:21–30
85. Holekamp, K. E. 1984. Dispersal in ground-dwelling sciurids. In *The Biology of Ground Dwelling Sciurids,* ed. J. O. Murie, G. R. Michener, pp. 297–320. Lincoln: Univ. Nebraska Press
86. Holekamp, K. E. 1986. Proximal causes of natal dispersal in Belding's ground squirrels (*Spermophilus beldingi*). *Ecol. Monogr.* 56:365–91
87. Holekamp, K. E. 1989. Why male ground squirrels disperse. *Am. Sci.* 77: 232–39
88. Holt, R. D. 1985. Population dynamics in two-patch environments: some anomalous consequences of an optimal habitat distribution. *Theor. Popul. Biol.* 28:181–208
89. Hoogland, J. L. 1982. Prairie dogs avoid extreme inbreeding. *Science* 215:1639–41
90. Howard, W. E. 1949. Dispersal, amount of inbreeding, and longevity in a local population of prairie deermice on the George Reserve, southern Michigan. *Contrib. Lab. Vertebr. Biol. Univ. Mich.* 43:1–50
91. Howard, W. E. 1960. Innate and environmental dispersal of individual vertebrates. *Am. Midl. Nat.* 63:152–61
92. Ims, R.A. 1989. Kinship and origin effects on dispersal and space sharing on *Clethrionomys rufocanus*. *Ecology* 70: 607–16
93. Itani, J. 1972. A preliminary essay on the relationship between social organisation and incest avoidance in non-human primates. In *Primate Socialization,* ed. F. E. Poirer, pp. 165–71. New York: Random House
94. Johnson, C. N. 1986. Sex-biased philopatry and dispersal in mammals. *Oecologia* 69:626–27
95. Johnson, M. L., Gaines, M. S. 1985. Selective basis for emigration of the prairie vole, *Microtus ochrogaster:* open field experiment. *J. Anim. Ecol.* 54: 399–410
96. Johnson, M. L., Gaines, M. S. 1987. The selective basis for dispersal of the prairie vole, *Microtus ochrogaster*. *Ecology* 68:684–94
97. Johnston, R. F. 1961. Population movements of birds. *Condor* 63:386–89
98. Jones, W. T. 1984. Natal philopatry in bannertailed kangaroo rats. *Behav. Ecol. Sociobiol.* 15:151–55
99. Jones, W. T. 1988. Density-related changes in survival of philopatric and dispersing kangaroo rats. *Ecology* 88:1474–78
100. Jones, W. T., Waser, P. M., Elliot, L. F., Link, N. E., Bush, B. B. 1988. Philopatry, dispersal, and habitat saturation in the banner-tailed kangaroo rat, *Dipodomys spectabilis*. *Ecology* 88: 1466–73
101. Karlin, S., McGregor, J. 1974. Towards a theory of the evolution of modifier genes. *Theor. Popul. Biol.* 5:59–103
102. Kawata, M. 1989. Growth and dispersal timing in male red-backed voles *Clethrionomys rufocanus bedfordiae*. *Oikos* 54:220–26
103. Keppie, D. M. 1980. Similarity of dispersal among sibling male spruce grouse. *Can. J. Zool.* 58:2102–4
104. King, J. A. 1983. Seasonal dispersal in a seminatural population of *Peromyscus maniculatus*. *Can. J. Zool.* 61:2740–50
105. Koenig, W. D., Pitelka, F. A. 1979. Relatedness and inbreeding avoidance: counterplays in the communally nesting Acorn Woodpecker. *Science* 206:1103–05

106. Koford, C. B. 1965. Population dynamics of rhesus monkeys on Cayo Santiago. In *Primate Behavior,* ed. I. De Vore, pp. 160–74. New York: Holt, Rinehart & Winston
107. Krohne, D. T., Burgin, A. B. 1987. Relative success of residents and immigrants in *Peromyscus leucopus. Holarctic Ecol.* 10:196–200
108. Krohne, D. T., Dubbs, B. A., Baccus, R. 1984. An analysis of dispersal in an unmanipulated population of *Peromyscus leucopus. Am. Midl. Nat.* 112:146–56
109. Krohne, D. T., Miner, M. S. 1985. Removal trapping studies of dispersal in *Peromyscus leucopus. Can. J. Zool.* 63:71–75
110. Kuno, E. 1981. Dispersal and persistence of populations in unstable habitats: a theoretical note. *Oecologia* 49: 123–26
111. Leuze, C. C. K. 1980. The application of radio tracking and its effect on the behavioral ecology of the water vole, *Arvicola terrestris* (Lacepede). In *A Handbook on Biotelemetry and Radio Tracking,* ed. C. J. Amlamer, D.W. MacDonald, pp. 361–66. Oxford: Pergamon
112. Levin, B. R., Kilmer, W. L. 1974. Interdemic selection and the evolution of altruism: a computer simulation study. *Evolution* 28:527–45
113. Levin, S. A., Cohen, D., Hastings, A. 1984. Dispersal strategies in patchy environments. *Theor. Popul. Biol.* 26: 165–91
114. Liberg, O., von Schantz, T. 1985. Sex-biased philopatry and dispersal in birds and mammals: the Oedipus hypothesis. *Am. Nat.* 126:129–35
115. Lidicker, W. Z. Jr. 1962. Emigration as a possible mechanism permitting the regulation of population density below carrying capacity. *Am. Nat.* 96:29–33
116. Lidicker, W. Z. Jr. 1975. The role of dispersal in the demography of small mammals. In *Small Mammals: Their Production and Population Dynamics,* ed. F. B. Golley, K. Petrusewicz, L. Ryszkowski, pp. 103–28. London: Cambridge Univ. Press
117. Lomnicki, A. 1978. Individual differences between animals and the natural regulation of their numbers. *J. Anim. Ecol.* 47:461–75
118. Lomnicki, A. 1980. Regulation of population density due to individual differences and patchy environments. *Oikos* 35:185–93
119. Lomnicki, A. 1988. *Population Ecology of Individuals.* Princeton: Princeton Univ. Press
120. Deleted in proof
121. Marks, J. S., Redmond, R. L. 1987. Parent-offspring conflict and natal dispersal in birds and mammals: comments on the Oedipus hypothesis. *Am. Nat.* 129:158–64
122. Marzluff, J. M., Balda, R. P. 1989. Causes and consequences of female biased dispersal in a flock living bird, the Pinyon Jay. *Ecology* 70:316–28
123. Maynard Smith, J. 1974. The theory of games and the evolution of animal conflicts. *J. Theor. Biol.* 47:209–21
124. Maynard Smith, J. 1976. Group selection. *Q. Rev. Biol.* 51:277–83
125. Maynard Smith, J. 1976. Evolution and the theory of games. *Am. Sci.* 64:41–45
126. Maynard Smith, J. 1979. Game theory and the evolution of behavior. *Proc. R. Soc. London Ser. B* 205:475–88
127. Maynard Smith, J., Parker, G. A. 1976. The logic of asymmetric contests. *Anim. Behav.* 24:159–75
128. Maynard Smith, J., Price, G. R. 1973. The logic of animal conflicts. *Nature* 246:15–18
128a. McClean, I. G. 1982. The association of female kin in the arctic ground squirrel *Spermophilus parryii. Behav. Ecol. Sociobiol.* 10:91–99
129. Mech, L. D. 1987. Age, season, distance, direction, and social aspects of wolf dispersal from a Minnesota pack. See Ref. 32, pp. 55–74
130. Metz, J. A. J., de Jong, T. J., Klinkhamer, P. G. L. 1983. What are the advantages of dispersing; a paper by Kuno explained and extended. *Oecologia* 57:166–69
131. Metzgar, L. H. 1967. An experimental comparison of screech owl predation on resident and transient white-footed mice (*Peromyscus leucopus*). *J. Mammal.* 48:387–91
132. Michener, G. R., Michener, D. R. 1977. Population structure and dispersal in Richardson's ground squirrels. *Ecology* 58:359–68
133. Moody, M. 1981. Polymorphism with selection and genotype dependent migration. *J. Math. Biol.* 11:245–67
134. Moore, J., Ali, R. 1984. Are dispersal and inbreeding avoidance related? *Anim. Behav.* 32:94–112
135. Morris, D. W. 1982. Age-specific dispersal strategies in iteroparous species: who leaves when? *Evol. Theory* 6:53–65
136. Motro, U. 1982. Optimal rates of dispersal I. Haploid populations. *Theor. Popul. Biol.* 21:394–411
137. Motro, U. 1982. Optimal rates of dispersal. II. Diploid populations. *Theor. Popul. Biol.* 21:412–29

138. Motro, U. 1983. Optimal rates of dispersal. III. Parent-offspring conflict. *Theor. Popul. Biol.* 23:159–68
139. Murray, B. G. Jr. 1967. Dispersal in vertebrates. *Ecology* 48:975–78
140. Murray, M. G. 1982. Home range, dispersal and clan system of impala. *Afr. J. Ecol.* 20:253–69
141. Negus, N. C., Berger, P. J. 1977. Experimental triggering of reproduction in a natural population of *Microtus montanus*. *Science* 196:1230–31
142. Nelson, M. E., Mech, L. D. 1984. Home range formation and dispersal of deer in northeastern Minnesota. *J. Mammal.* 65:567–75
143. Nelson, M. E., Mech, L. D. 1987. Demes within a northeastern Minnesota deer population. See Ref. 32, pp. 27–40
144. Newton, I., Marquiss, M. 1983. Dispersal of Sparrowhawks between birthplace and breeding place. *J. Anim. Ecol.* 52:463–77
145. Nilsson, J-A. 1989. Causes and consequences of natal dispersal in the Marsh Tit, *Parus palustris*. *J. Anim. Ecol.* 58: 619–36
146. Deleted in proof
147. Deleted in proof
148. Otis, J. S., Froehlich, J. W., Thorington, R. W. 1981. Seasonal and age-related differential mortality by sex in the mantled howler monkey, *Alouatta palliata*. *Int. J. Primatol.* 2:197–205
149. Packer, C. 1979. Inter-troop transfer and inbreeding avoidance in *Papio anubis*. *Anim. Behav.* 27:1–36
150. Packer, C. 1985. Dispersal and inbreeding avoidance. *Anim. Behav.* 33:676–78
151. Parker, G. A. 1979. Sexual selection and sexual conflict. In *Sexual Selection and Reproductive Competition in Insects,* ed. M. B. Blum, N. A. Blum, pp. 123–66. New York: Academic
152. Parker, G. A. 1983. Mate quality and mating decisions. See Ref. 11, pp. 141–64
153. Pfeifer, S. L. R. 1982. Disappearance and dispersal of *Spermophilus elegans* juveniles in relation to behavior. *Behav. Ecol. Sociobiol.* 10:237–43
154. Porter, J. H., Dueser, R. D. 1989. A comparison of methods for measuring small-mammal dispersal by use of a Monte Carlo simulation model. *J. Mammal.* 70:783–93
155. Price, G.A. 1970. Selection and covariance. *Nature* 227:520–21
156. Pusey, A. E. 1980. Inbreeding avoidance in chimpanzees. *Anim. Behav.* 28:543–82
157. Pusey, A. E. 1987. Sex-biased dispersal and inbreeding avoidance in birds and mammals. *Trends Ecol. Evol.* 2:295–99
158. Pusey, A. E., Packer, C. 1987. Dispersal and philopatry. In *Primate Societies,* ed. B. B. Smuts, D. L. Cheney, R. M. Seyfarth, R. W. Wrangham, T. T. Struhsaker, pp. 250–66. Chicago: Univ. Chicago Press
159. Ralls, K., Ballou, J. 1982. Effects of inbreeding on infant mortality in captive primates. *Int. J. Primatol.* 3:491–505
160. Ralls, K., Brugger, K., Ballou, J. 1979. Inbreeding and juvenile mortality in small populations of ungulates. *Science* 206:1101–3
161. Ralls, K., Brugger, K., Glick, A. 1980. Deleterious effects of inbreeding in a herd of captive Dorcas gazelle. *Int. Zoo Yearb.* 20:137–46
162. Redmond, R. L., Jenni, D. A. 1982. Natal philopatry and breeding area fidelity of Long-billed Curlews (*Numenius americanus*): patterns and evolutionary consequences. *Behav. Ecol. Sociobiol.* 10:277–79
163. Reynolds, J. D., Cooke, F. 1988. The influence of mating systems on philopatry: a test with polyandrous red-necked phalaropes. *Anim. Behav.* 36:1788–95
164. Roff, D. A. 1975. Population stability and the evolution of dispersal in a heterogeneous environment. *Oecologia* 19: 217–37
165. Rogers, L. L. 1987. Factors influencing dispersal in the black bear. See Ref. 32, pp. 75–84
166. Rood, J. P. 1987. Dispersal and intergroup transfer in the dwarf mongoose. See Ref. 32, pp. 85–103
167. Ryman, N., Allendorf, F. W., Stahl, G. 1979. Reproductive isolation with little genetic divergence in sympatric populations of brown trout (*Salmo trutta*). *Genetics* 92:247–62
168. Deleted in proof
169. Schieck, J. O., Millar, J. S. 1987. Can removal areas be used to assess dispersal of red-backed voles? *Can. J. Zool.* 65:2575–78
170. Schwartz, O. A., Armitage, K. B. 1980. Genetic variation in social mammals: the marmot model. *Science* 207:665–67
171. Senner, J. W. 1980. Inbreeding depression and the survival of zoo populations. In *Conservation Biology,* ed. M. E. Soule, B. A. Wilcox, pp. 209–24. Sunderland: Sinauer
172. Sherman, P. W. 1977. Nepotism and the evolution of alarm calls. *Science* 197:1246–53
173. Sherman, P. W. 1980. The limits of ground squirrel nepotism. In *Sociobiology: Beyond Nature/Nuture?,* G. W. Bar-

low, J. Silverberg, pp. 505–44. Boulder: Westview

174. Sherman, P. W., Morton, M. L. 1984. Demography of Belding's ground squirrels. *Ecology* 65:1617–28
175. Shields, W. M. 1982. *Philopatry, Inbreeding, and the Evolution of Sex*. Albany: State Univ. NY Press
176. Shields, W. M. 1983. Optimal inbreeding and the evolution of philopatry. See Ref. 186, pp. 139–59
177. Slatkin, M., Wade, M. J. 1978. Group selection on a quantitative character. *Proc. Natl. Acad. Sci. USA* 75:3531–34
178. Smith, A. T. 1974. The distribution and dispersal of pikas: influences of behavior and climate. *Ecology* 55:1368–76
179. Smith, A. T. 1987. Population structure of pikas: dispersal versus philopatry. See Ref. 32, pp. 128–42
180. Smith, A. T., Ivins, B. L. 1983. Colonization in a pika population: dispersal vs philopatry. *Behav. Ecol. Sociobiol.* 13:37–47
181. Smith, C. C. 1968. The adaptive nature of social organization in the genus of tree squirrels *Tamiasciurus*. *Ecol. Monogr.* 30:31–63
182. Smith, R. H. 1979. On selection for inbreeding in polygynous animals. *Heredity* 43:205–11
183. Snyder, R. L. 1961. Evolution and integration of mechanisms that regulate population growth. *Proc. Natl. Acad. Sci. USA* 47:449–55
184. Spear, L. B. 1988. Dispersal patterns of Western Gulls from southeast Farallon Island. *Auk* 105:128–41
185. Stenseth, N. C. 1983. Causes and consequences of dispersal in small mammals. See Ref. 186, pp. 63–101
186. Swingland, I. R., Greenwood, P. J., eds. 1983. *The Ecology of Animal Movement*. Oxford: Clarendon
187. Tamarin, R. H. 1984. Body mass as a criterion of dispersal in voles: a critique. *J. Mammal.* 65:691–92
188. Taylor, P. D. 1988. An inclusive fitness model for dispersal of offspring. *J. Theor. Biol.* 130:363–78
189. Teague, R. 1977. A model of migration modification. *Theor. Popul. Biol.* 12:86–94
190. Templeton, A. R. 1987. Inferences on natural population structure from genetic studies on captive mammalian populations. See Ref. 32, pp. 257–72
191. Vander Wall, S. B., Hoffman, S. W., Potts, W. K. 1981. Emigration behavior of Clark's Nutcracker. *Condor* 83:162–70

191a. van Noordwijk, A. J., Scharloo, W. 1981. Inbreeding in an island population of the Great Tit. *Evolution* 35:674–88

191b. van Noordwijk, A. J. 1984. Problems in the analysis of dispersal and a critique on its 'heritability' in the Great Tit. *J. Anim. Ecol.* 53:533–44

192. Van Valen, L. 1971. Group selection and the evolution of dispersal. *Evolution* 25:591–98
193. Verts, B. J., Carraway, L. N. 1986. Replacement in a population of *Perognathus flavus* subject to removal trapping. *J. Mammal.* 67:201–5

193a. von Schantz, T. 1981. Female cooperation, mate competition, and dispersal in the red fox *Vulpes vulpes*. *Oikos* 37:63–68

194. Wade, M. J. 1982. Group selection: migration and the differentiation of small populations. *Evolution* 36:949–61
195. Wade, M. J., McCauley, D. E. 1984. Group selection: the interaction of local deme size and migration in the differentiation of small populations. *Evolution* 38:1047–58
196. Waser, P. M. 1985. Does competition drive dispersal? *Ecology* 66:1170–75
197. Waser, P. M. 1987. A model predicting dispersal distance distributions. See Ref. 32, pp. 251–56
198. Waser, P. M. 1988. Resources, philopatry, and social interactions among mammals. In *Ecology of Social Behavior*, ed. C. N. Slobodchikoff, pp. 109–30. New York: Academic
199. Waser, P. M., Austad, S. N., Keane, B. 1986. When should animals tolerate inbreeding? *Am. Nat.* 128:529–37
200. Waser, P. M., Jones, W. T. 1983. Natal philopatry among solitary mammals. *Q. Rev. Biol.* 58:355–90
201. Waser, P. M., Jones, W. T. 1989. Heritability of dispersal in banner-tailed kangaroo rats, *Dipodomys spectabilis*. *Anim. Behav.* 37:987–91
202. Weatherhead, P. J., Boak, K. A. 1986. Site infidelity in Song Sparrows. *Anim. Behav.* 34:1299–1310
203. Whitney, G. 1976. Genetic substrates for the initial evolution of human sociality 1. Sex chromosome mechanisms. *Am. Nat.* 110:867–75
204. Whitworth, M. R., Southwick, C. H. 1984. Sex differences in the ontogeny of social behavior in pikas: possible relationships to dispersal and territoriality. *Behav. Ecol. Sociobiol.* 15:175–82
205. Wiens, J. A. 1976. Population responses to patchy environments. *Annu. Rev. Ecol. Syst.* 7:81–120
206. Wiggett, D. R., Boag, D. A. 1989. Intercolony natal dispersal in the Col-

umbian ground squirrel. *Can. J. Zool.* 67:42–50

207. Wiggett, D. R., Boag, D. A., Wiggett, A. D. R. 1989. Movements of intercolony natal dispersers in the Columbian ground squirrel. *Can. J. Zool.* 67:1447–52

208. Wolff, J. O., Lundy, K. I., Baccus, R. 1988. Dispersal, inbreeding avoidance and reproductive success in white-footed mice. *Anim. Behav.* 36:456–65

209. Woolfenden, G. E., Fitzpatrick, J. W. 1986. Sexual asymmetries in the life history of the Florida Scrub jay. In *Ecological Aspects of Social Evolution,* ed. D. I. Rubenstein, R. W. Wrangham, pp. 87–107. Princeton: Princeton Univ. Press

210. Wynne-Edwards, V. C. 1962. *Animal Dispersion In Relation To Social Behaviour*. Edinburgh: Oliver & Boyd

*Annu. Rev. Ecol. Syst. 1990. 21:481–508*


# A MODEL SYSTEM FOR COEVOLUTION: AVIAN BROOD PARASITISM

*Stephen I. Rothstein*

Department of Biological Sciences and Marine Sciences, University of California, Santa Barbara, California 93106

KEY WORDS: cowbird, cuckoo, generalists, parasite-host interactions, social parasitism, specialists

---

## INTRODUCTION

Many putative examples of coevolution do not stand up to critical analysis. A rigorous definition of coevolution requires that a trait in one species has evolved in response to a trait of another species, which trait was itself evolved in response to the first species (50, 69). This type of intimate, reciprocal evolutionary relationship is hard to demonstrate because most species interact with many other species, all of which may affect their evolution. For example, a host species is likely to be affected by many types of parasitic helminths and protozoans. Accordingly, some of its defenses will be fairly general and not attributable to any particular species of parasite. Such situations are termed diffuse coevolution, as opposed to pairwise coevolution in which adaptations have a stepwise, reciprocal nature. Unfortunately, diffuse coevolution is difficult to document because additional species may need to be considered. Also, adapting to many species may compromise adaptations to any one species so much that coevolutionary traits are weakly expressed and hard to identify. Systems in which the interacting species are few (optimally only two) provide the clearest examples of coevolution. Such systems include many mutualistic relationships and some parasite-host associations (51). Among the latter, brood parasitism provides some of the most persuasive examples of coevolution because it often involves small numbers of species.



0066-4162/90/1120/0481$02.00

Brood or social parasitism is a form of breeding biology in which certain individuals, the parasites, receive parental care from unrelated individuals, the hosts. It is most prevalent in birds and hymenopterans (26, 106, 165). I concentrate here on avian parasites because the insect systems are less well known. Parasitism usually reduces a host's reproductive output which results in selection for host defenses. These in turn select for coevolved counter-defenses by parasites that may select for new host defenses and therefore result in an arms race (30). Payne (106) reviewed many evolutionary aspects of avian brood parasitism, but this was before the appearance of many recent studies, especially experimental ones.

Avian brood parasitism is an excellent system for studies of coevolution for several reasons (26, 63, 124). First, most hosts are parasitized by a single species since sympatric parasites usually show little overlap in host usage (44, 46, 48, 65, 99, 108). Second, parental care in birds is elicited by vocal and visual cues, the same sensory modalities humans stress, so parasitic adaptations for and host adaptations against parasitism involve features we can easily detect. This is not true with parasitic insects in which olfactory cues are critical (26).

Third, nearly all of the major adaptations related to parasite-host interactions are manifested in or near the nest. This provides such a strong central focus for both evolution and the efforts of researchers that there are few other contexts in which putative adaptations must be examined to assess overall effects on fitness. Thus, if a host removes a parasite's eggs from its nest, we can be reasonably sure that this behavior evolved in response to parasitism and currently has fitness consequences that apply only or mainly when parasitism occurs. This can be contrasted with adaptations one might envisage in other interactions between species. For example, while long horns may help a male antelope to defend itself against a predator, and might therefore be coevolved, the horns may also function in mate attraction or in fights with other males. It may be unclear which of these uses of horns is the primary reason for their evolution.

Lastly, a brood parasite's fitness typically is maximized when the host loses its entire brood, because this allows all of the host's parental effort to be directed toward the parasite. Unlike some types of parasitism (67), here there is no selection for benign effects (124). Thus, heavy host losses may impose very strong selection pressures on hosts, which enhance the likelihood that coevolution will occur and be detectable. Parasitism may have such a detrimental effect on hosts that it can be a significant factor in reducing host populations or even driving entire species to near extinction (10, 25, 57, 59, 88, 90, 112–114, 153)—which has prompted the development of parasite-control programs (6, 76, 91, 136).

Intraspecific as well as interspecific parasitism occurs in birds (2, 16, 80,

120, 173). Host defenses are less prominent in intraspecific parasitism (but see 42, 113a, 135a) because the parasitism is usually less costly. In addition, because the host and parasite are usually identical in appearance and behavior, it may be especially difficult for hosts to develop defenses and for researchers to identify any that do evolve. For these reasons I deal only with obligate interspecific parasitism which has evolved independently at least seven times, as it occurs in seven taxa (75). These taxa are two subfamilies of cuckoos (Cuculinae and Neomorphinae), two types of finches (the cuckoo-finch *Anomalospiza imberbis* and Whydahs in the Viduinae), the honeyguides (Indicatoridae), the cowbirds (Icterinae), and the black-headed duck *(Heteronetta atricapilla)*. These seven taxa add up to about 80 species or only 1% of all bird species, but because some parasites use many host species, brood parasitism affects a large proportion of all birds, especially passerines. In addition to these obligate parasites, some ducks commonly practice both intra- and interspecific parasitism and also tend their own nests (120, 166).

## GENERALISTS VERSUS SPECIALISTS AMONG THE PARASITES

A few brood parasites such as the brown-headed *(Molothrus ater)* and shiny cowbirds *(M. bonariensis)* are extreme generalists and parasitize most passerines with which they are sympatric. The host list for the former cowbird includes 220 species (47), few of which are utilized by any other parasitic bird. The only conclusive study of individual female brown headed cowbirds showed that they use more than one host species over a span of a few days (35). At the other extreme, the screaming cowbird *(M. rufoaxillaris)* usually parasitizes only one host species (43), as do nearly all viduine species. The remaining parasites specialize to varying degrees. Some primarily use only one host species, and most normally use no more than five to ten. The host list for some of these with extensive ranges, such as the common cuckoo *(Cuculus canorus)*, is very long. Only a few species are used regularly in any region (153, 172), however, and the use of rare hosts may represent a best-of-a-bad-situation strategy (19, 74, 117).

## HOST LOSSES DUE TO PARASITISM

Parasites lower host reproductive output in three major ways. In most species, females usually remove one or more host eggs from nests they parasitize (106). Parasitism often reduces the hatching success of those host eggs that remain (45, 122a). Lastly, parasitic nestlings often inflict severe host losses. In most cuckoos, nestlings push all of the host eggs and young out of the nest, and adult hosts never attempt to stop the destruction of their own young (75, 172). At least one cuckoo (97) and probably all honeyguides (44, 160) stab host nestlings to death with special mandibular hooks. Cowbird nestlings do not attack host young directly but usually cause the loss of all or some of the host young by outcompeting them for food (45). Viduine nestlings may have little effect on hosts (106). The least costly parasitism occurs in the precocial black-headed duck whose hatchlings leave host nests after 1–2 days and develop with no further parental care (167). Parasitism has costs even when it does not depress a host's current reproductive output because parental effort may depress future reproductive output (77). In one system, hosts may sometimes receive net benefits from parasitism. Giant cowbirds *(Scaphydura oryzivora)* depress host reproductive output but seem to do so to a lesser extent than do the parasitic flies which they remove from the host young (149–151). However, critical details about this system still need to be published.

## FREQUENCY OF PARASITISM

The frequency of parasitism is a major determinant of the potential selection for a host defense and varies greatly according to parasite species. Cuckoos are relatively uncommon birds and lay 10–25 eggs per season (106, 172). Although local parasitism rates sometimes approach 50% (11, 49, 75, 153), "global" rates are usually much lower. Britain's nest record system shows only six hosts of the common cuckoo with rates of parasitism above 1% and only one above 5% (11, 54; see also Ref. 94). However, some other cuckoo species are limited to one or a few hosts and often inflict parasitism rates of 20% or more (14, 52, 53, 71, 108, 159). Most parasitic cowbirds are common and the best-studied species, the brown-headed cowbird, is more abundant than many if not most of its hosts. Females lay 40 or more eggs a year (3, 137, 139). Local parasitism rates by cowbirds sometimes approach 100% (32, 45, 56, 169). A nest record system from Ontario shows 47 species with parasitism rates of at least 5% and 20 with rates of at least 20% (counting only species with more than 24 nests in Ref. 110). Quantitative data are fewer for other parasite-host systems, but some systems also appear to have high rates of parasitism, as in cowbirds (106).

Since predation occurs at roughly half of all noncavity nests of passerines (115b), it might seem to be a much stronger selection pressure on avian breeding biology than parasitism, which often occurs at 10% or less of nests. However, rearing a parasite like a cuckoo can occupy a host for its entire breeding season whereas birds readily renest after predation (101, 110, 140). Birds that can renest three times have only a 6% chance of producing no young if each of their four nests has a 50% chance of predation ($0.5^4 = 0.06$).

A parasitism rate of 10% can also cause 5–6% of birds to produce no young if half of the parasitized nests are destroyed by predators and 10% of renests are parasitized. So even when it is five times less frequent, parasitism may exert selection pressures similar to those due to predation.

## HOST DEFENSES AND PARASITIC COUNTERDEFENSES

In addition to the losses incurred by parasitism and the probability that parasitism will occur, the adaptive value of a host defense is also determined by its cost. Some defenses seem essentially cost free while others clearly inflict costs. Since we are dealing with coevolution and arms races, it is proper to call a host feature a defense only if it both reduces the impact of parasitism and has evolved in response to, or is currently maintained by, selection pressures arising from parasitism. Similarly, adaptations by the parasite should be called counterdefenses only if they evolved in response to host defenses. It is useful to consider potential adaptations in the sequence over which they could be manifested.

### *Avoiding Versus Facilitating Parasitism*

Certain nest locations can make it difficult for parasites to find nests (23, 56), but predation is also important in determining nest sites. Hosts try to defend nests and are especially aggressive toward parasites, as is shown by observations of wild birds (23, 171, 172) and by experiments with mounts (9, 27, 36, 66, 96, 109, 118, 119) and live caged birds (17). When parasites are larger than hosts, which is usually the case, aggression is often ineffective (118, 172; Rothstein & O'Loghlen, in preparation). Aggression in some or all hosts may reflect general responses to nest intruders, rather than evolved host defenses, e.g. birds may attack parasites because they remember having seen them near nests or because they mistake parasites for predators. This could explain why host aggression seems to be maladaptive in some cases, with parasites using it as a nest finding cue (92, 143), and why the most experienced and aggressive birds are the ones most likely to be parasitized (147, 148).

Some features of parasites seem to incite host aggression, e.g. hosts may attack some cuckoo species because they are hawk-like in plumage and flight behavior, thus eliciting a mobbing response (172). In one cuckoo, the koel *(Eudynamis scolopacea)*, males are black, as are the crows that are their hosts. The koel male decoys the host into chasing him from the nest while the less conspicuous female sneaks in to lay her egg (75, 172). Similar decoying behavior has been reported for other cuckoo species (49) and at least one honeyguide (145, 146). Other features seem designed to thwart host aggression, e.g. parasites lay eggs much more quickly than do other birds (15, 19, 20, 61, 172). One cowbird often waits for hosts to depart before visiting a nest (170a). Parasitic females tend to be especially dull in color, which may make them inconspicuous (103). Many cuckoo species also have plumage polymorphisms or high levels of continuous variation, especially among females, which may make it harder for hosts to recognize cuckoos (103). By making themselves inconspicuous to large hosts and conspicuous to small ones, parasites may benefit from both avoiding and provoking aggression. At least one parasite may use joint action to overwhelm host aggression. Screaming cowbirds travel in mated pairs, and up to five pairs may simultaneously visit nests of their usual host, the bay-winged cowbird (*Molothrus badius;* 41, 85). An unusual behavior in cowbirds was originally suggested to be an appeasement display that lessens host aggression (142). However, the display functions in other contexts and has never been seen in response to host attacks (129, 132) although, paradoxically, it probably would serve to lessen host aggression (118). Hosts may get some protection from parasitism by nesting at high densities, by practicing joint defense, and by nesting near larger, more aggressive host species (21, 23, 42a, 168, 171).

## *Adaptations Related to Eggs*

HOST EGG REJECTION Experimentation has shown that many species eject parasitic eggs from their nests (1, 18, 23, 28, 34, 83, 95, 124, 156). Such ejection is essentially cost-free for hosts that have eggs easily distinguishable from those of the parasite and that can grasp parasitic eggs in their bills and lift them without damaging their own eggs. But ejection can incur costs if a host mistakenly ejects its own egg (a "recognition cost"; 27, 28, 134) or has a bill too small to grasp eggs so that it must puncture parasitic eggs to remove them (27, 28, 127). Puncture ejections cause occasional breakage of the host's own eggs, either when the bird's bill or the parasitic egg bangs into them. But the loss is only 0.18 to 0.42 host eggs per ejection for two cuckoo and two cowbird hosts (27, 28, 122, 127). There is no evidence that these costs make it more adaptive to accept parasitic eggs than to eject them, except when parasitic eggs are laid more than several days after the host's laying period and receive insufficient incubation to hatch. In such cases, cowbird hosts that pay ejection costs show a decline in rejection tendency, but other hosts do not (127, 129a). By contrast, cuckoo hosts that pay rejection costs do not show increased acceptance when experimentally parasitized during incubation (28). This contrast may occur because cuckoos parasitize nests in the incubation stage less often than do cowbirds (27, 45, 172).

Cowbirds and cuckoos have unusually thick-shelled eggs (5, 7, 68, 154), and cowbird eggs are especially spherical (111). These features make the eggs unusually strong (111, 115, 154) and could be counterdefenses to puncture ejections (121), but a very small passerine was able to pierce cowbird eggs (154). It is likely, therefore, that strong eggshells evolved because they keep parasitic eggs from being broken when laid hurriedly in host nests and because they break some host eggs thereby reducing the number of host nestlings that compete with the parasite (7, 75).

Ejection may not be a viable defense for species with small bills because it is either physically impossible or too costly. Such species are more likely to reject by desertion and renesting and less likely to use ejection than are larger hosts (28, 58, 124, 127). One cowbird host often reduces the costs of renesting by burying a parasitized clutch with nesting material and laying a replacement clutch on the new nest floor (22). While desertion in response to experimentally introduced parasitic eggs is frequent in only one cowbird host (127, but in several cuckoo hosts—28, 64, 95, 96), it is common in many species in apparent response to natural cowbird parasitism (22, 58, 101, 124, 141, 169). This contrast may be due to the timing of natural versus experimental parasitism (58, but see Ref. 135) or to naturally parasitized hosts deserting after seeing adult cowbirds near nests (58, 124). One experimental study found that only the timing of egg insertion, not the sight of an adult cowbird, contributed to desertion, but this result was based on small sample sizes (17; see also 8). Experiments on two cuckoo hosts show that both the presence of a cuckoo egg and the sight of a cuckoo mount contribute to rejection behavior (27, 96).

If birds show any ejection behavior, they always remove foreign eggs regardless of whether these are the majority or the only egg type present (126, 129a, 133). At least some hosts seem to learn the appearance of their own eggs in an imprinting-like process (123, 131), and rejection is directed toward any dissimilar egg, not just the eggs of parasites (1, 133, 134). Learned egg recognition facilitates the evolution of certain host defenses because it means that a host can evolve a new egg type (see below) if a mutation for a new egg coloration appears in a single individual. By contrast, if the recognition were innate, a new egg type could evolve only if mutations for a new egg color and for the neural mechanism needed to recognize that new color appeared in a single individual (123).

Species that reject cowbird eggs exhibit degrees of tolerance toward foreign eggs proportional to the divergence between their eggs and those of the parasite (134). Species with cowbird-like eggs must exercise fine discrimination to ensure that cowbird eggs are rejected. But the American robin's *(Turdus migratorius)* eggs and those of the cowbird are highly divergent. The former are larger, are blue instead of white, and are immaculate instead of spotted. Eggs differing from robin eggs by one of these parameters are easily distinguishable but are normally accepted, while eggs that differ in two ways are usually rejected. This built-in tolerance may be an adaptation that ensures that cowbird eggs are rejected but that occasional deviant robin eggs are not (134).

THE ORIGIN OF EGG REJECTION There seems little doubt that avian egg recognition evolved in response to brood parasitism, except in a few species that nest on the ground in dense colonies and risk having their eggs mixed up (157). Since this context and brood parasitism are the only ones known in which it is adaptive to reject deviant eggs, parsimony suggests that egg rejection is a true host defense. Also, passerine species that are unsuitable as hosts because of nest placement or diet show little or no clear egg rejection (28). Even more convincing are intrageneric and intraspecific tests showing that populations that are not parasitized by cuckoos are significantly more likely to accept nonmimetic eggs than are parasitized populations (16a, 28a, 152). An especially revealing intraspecific example involves the village weaver *(Ploceus cucullatus)* which rejects dissimilar eggs in Africa where it is parasitized by cuckoos (161). Weavers introduced sometime before 1797 to Hispaniola, where no parasites occurred, have lost most of their rejection behavior (24). Experiments in the future may show a return to higher rejection rates because the shiny cowbird colonized Hispaniola in 1972 and now parasitizes the weaver.

Another trend suggesting that parasitism is the factor responsible for the evolution of egg rejection comes from comparisons of entire avifaunas (135b). North America has only one widespread parasitic bird whereas Africa has about 30 parasites. Only a small proportion of North American species reject divergent eggs (124, 133), but Swynnerton's (156) experiments in southern Africa indicate that nearly all passerines there reject. Lastly, certain features of rejection seem designed to maximize fitness in the context of brood parasitism, e.g. declines in rejection of cowbird eggs during incubation by hosts that incur ejection costs, and the American robin's built-in tolerance.

PARASITIC COUNTERDEFENSES TO HOST EGG REJECTION An obvious counterdefense to host rejection is for a parasite to mimic the host's eggs. The brown-headed cowbird of North America has eggs like those of some of its major hosts, but the egg type, whitish with brown and gray spots, is a common general type among passerines. These spotted eggs vary greatly, and some may closely match the spotted eggs of certain hosts in the Great Plains (31), the arena where this cowbird has been abundant for the longest period (43, 89). A better case for mimicry occurs in the shiny cowbird, which has both spotted and immaculate white eggs, in both Uruguay and northern Argentina. Most hosts accept or reject both morphs, but two accept the spotted morph and reject the white one (49, 87). No known host shows the reverse, which may explain a possible recent decline in the white morph.

The clearest egg mimicry occurs among cuckoos, and one species, the common cuckoo, has eggs that span nearly the entire range shown by passerines. Each egg type is generally adapted to a particular host species (5, 12), and in some cases the mimicry as regards color, but not size (172), is nearly perfect by human standards. The eggs are slightly larger than those of most hosts but are still remarkably small given the size of a common cuckoo. Significantly, hosts disdriminate against oversized eggs (27). Cuckoo species that parasitize hosts larger than themselves have eggs that are appropriate to their body size (172). Females laying different egg morphs are members of different "gentes" (singular = gens). Over its huge range in Eurasia and Africa, the common cuckoo probably has several dozen gentes, but any one region, such as England, has from one to a few (5, 12, 153). Host choice is thought to be learned by females imprinting on the hosts that reared them, but there is no relevant evidence. The determination of egg coloration probably has a large genetic component, as in other birds. Thus, the maintenance of gentes may require reproductive isolation because males in other birds carry, but of course do not express, genes for egg coloration. Gentes are usually separated by habitat, and cuckoos may have limited dispersal (106)—features that may facilitate reproductive isolation. Alternatively, the gene(s) for egg color may be on the female chromosome, thereby allowing for interbreeding among gentes (19, 70).

There is one case in which a host discriminates amongst eggs solely by size. The rufous hornero *(Furnarius rufus)* nests in mud domes whose dark interiors may allow only tactile discrimination. Horneros usually eject shiny cowbird eggs whose width is less than 88% that of their own eggs, but they accept larger eggs (86). Cowbird eggs have unusually large widths in Uruguay, where the cowbird has parasitized the hornero for a long period, but not in Argentina, where the interaction has been frequent for less than 300 years. The more spherical eggs in Uruguay are not an allometric consequence of increased egg size because neither the Uruguayan or Argentine cowbirds have eggs that become more spherical as they get larger. Since cowbirds in Uruguay parasitize many species, why have they evolved a response to the hornero? The hornero is the highest quality host when it does not reject because its domed nests have remarkably low predation rates (83). Although many more eggs are laid in the nests of other hosts, horneros contribute disproportionately to recruitment to the parasite's population and probably also have a similarly disproportionate effect on the parasite's evolution. Some features of generalist parasites may thus make them particularly well suited to certain high quality hosts. Such parasites might then be generalists and specialists at the same time.

Davies & Brook (27) pointed out that instead of host discrimination and coevolution, parasites themselves could select for mimicry if they preferentially remove parasitic eggs when laying in already parasitized nests.

Brooker & Brooker (14) endorsed this alternative because they found that a host of an Australian cuckoo was parasitized with mimetic eggs yet never rejected nonmimetic eggs. However, this cuckoo parasitizes 20 other species that have similar eggs (14), one or more of which may have selected for egg mimicry via host rejection. So this may be another example of a generalist evolving a major adaptation that applies to one or only a few host species. Furthermore, it is not clear that cuckoos selectively remove cuckoo eggs from already parasitized nests (27), and even if they did, the incidence of multiple parasitism may usually be too low to be an important selection pressure. Multiple parasitism is very common in cowbird parasitism (45), but there is no evidence that cowbirds selectively remove cowbird eggs. Another alternative to host egg rejection is predation, with selection favoring the same egg colors in parasites as in their hosts because of crypsis (62). But this hypothesis has no support (27, 87).

HOST COUNTERDEFENSES TO EGG MIMICRY A host whose egg is mimicked may escalate the coevolution or arms race. First, as some cuckoo hosts do, it can desert its nest if there is any indication of parasitism such as eggs that are slightly different from one another or an adult parasite detected near the nest (27, 96). Secondly, the host could evolve a new egg type and thereby escape the parasite's egg mimicry. No definite examples of such directional selection are known. However, the bright immaculate blue eggs of the American robin may be an example; these are much more distinct from cowbird eggs than are the marked eggs of nearly all of its close relatives (73a). Another directional change might be the evolution of egg markings complex enough to serve as a signature for each host individual. A survey of accepting and rejecting cuckoo hosts, however, showed no overall differences in the complexity of egg markings (29).

A third host response to egg mimicry involves frequency dependent selection and results in polymorphic or continuously variable eggs (155, 156, 161). While the parasitic species might evolve a parallel range of egg variation, each individual could lay only one egg type. Errors by parasites (i.e. mismatched parasite and host eggs) would be common because parasites apparently commit themselves to host nests before eggs appear (19, 61, 170a), presumably to synchronize their egg laying with that of the host. Parasites thus have no way to predict whether their egg will match those in a particular host nest. Some African glossy cuckoos (*Chrysococcyx* spp.) and their weaver (*Ploceus* spp.) hosts show extreme but parallel intraspecific variation in egg color of the sort coevolution could produce (103). The situation may be even more complex because some weavers practice intraspecific parasitism. This may have led to the evolution of egg polymorphisms (42), meaning that the cuckoo polymorphisms could be an adaptation to a condition that predated interspecific parasitism. Other African cuckoos and some Asian ones also have extreme egg variation that parallels that of individual host species (5, 49).

## *Adaptations During the Nestling Stage*

The disparity between host and parasitic young is sometimes ludicrous, as anyone who has seen a small songbird feeding a cuckoo more than five times its size can attest. This anomaly occurs even with hosts that exhibit finely tuned discrimination among egg types! Yet no known host routinely shows outright rejection of parasitic nestlings. The evolution of nestling rejection may be unlikely if parasitic nestlings are supernormal stimuli for parental care due to their large size or to some other feature. But data on feeding rates of parasitized and nonparasitized nests (13) and on choice tests involving cuckoo and various nonparasitic nestlings provided no evidence for a supernormal stimulus effect (27). Similarly, nestlings of nonparasitic species fare as well as cowbirds when placed in nests of other species (30a).

Davies & Brooke (27) suggested that cuckoo nestlings are less likely to be rejected than are cuckoo eggs because: (*a*) It is too late for a host to save its young if a cuckoo nestling is present, and (*b*) it is easier to discriminate among eggs than among nestlings. Nevertheless, nestling acceptance is puzzling since simple recognition rules (e.g. never feed a nestling with an orange colored mouth) could result in nestling rejection (27, 133). Furthermore, even if a host cannot save its own young, it can gain no benefits from caring for an unrelated individual for a month or more. Perhaps nestlings are accepted because brood parasites have capitalized on a very fundamental aspect of parental care. Namely, in the absence of parasitism, selection may have usually favored learned recognition of the nestling features that prevail just before fledging because nestling recognition has no value while young are still in the nest.

If no hosts showed nestling rejection, one could suggest that there is some intrinsic feature of birds that absolutely bars its evolution. But two unrelated systems show nestling rejection of a sort. Nestlings of parasitic viduines mimic the intricate mouth colorations of their particular species of estrildid host(s), which usually do not feed nestlings with mismatched mouth patterns (99, 100, 107a). Although such nestlings starve, there is no outright rejection (i.e. removal). Reproductive isolation among viduines is achieved by imprinting, with males incorporating the song of the estrildid host that reared them into their own song and females assortatively mating with males that give the song of the host that reared them (104, 105, 107). Intricate mouth patterns occur in all estrildids, even on continents where they are not parasitized (100), so it is unclear whether the feature evolved in response to parasitism by African viduines. Furthermore, the only Australian estrildid that has been tested shows preferential feeding of mimetic young (68a).

Nearly perfect nestling mimicry also occurs between the parasitic screaming cowbird and its host, the bay-winged cowbird (38a, 43). Bay-wings are also parasitized by the generalist shiny cowbird which is not mimetic. Shiny cowbird nestlings receive normal care, but bay-wings give them little or no care after they fledge (41). Bay-wing nests often have dark interiors (41a) which may make it difficult for them to distinguish among nestlings until the young fledge. As would be expected under this interpretation, bay-wings do not reject nonmimetic eggs (39, 83). Some hosts of at least one honeyguide species also nest in cavities and may routinely drive away the nonmimetic young honeyguides as soon as the parasites fledge from their nests (145, 146).

It may be no coincidence that the two cases of nestling mimicry occur in systems in which parasites are normally limited to one host species. This extreme specialization may have made the selection pressures on the hosts especially high (screaming cowbirds parasitize at least 80%–90% of all bay-wing nests—41, 41a), thereby facilitating the appearance of an adaptation that is especially hard to evolve. In addition, it is possible that some or even many hosts give slightly better care to nestlings similar to their own, but that selection for nestling mimicry becomes diffused when a parasite uses more than one host species. Possible host selection on parasitic nestlings is suggested by the variable nestling mouth colors of two parasitic cowbird species (38, 130) because such variation is rare in nonparasitic passerines (33). Also, the nestling mouth-colors and feathers of parasitic cuckoos resemble those of passerines more than those of closely related nonparasitic cuckoos (106).

Most of the few cuckoo species that coexist with host nestlings have nestling plumages somewhat similar to those of their chief hosts (27, 75; but see Ref. 52). As the similarities normally involve the dorsal area rather than the face and bill, they may relate more to predation than host care. A predator is most likely to see a nestling's back, and nestlings with contrasting appearances may make a nest conspicuous. The begging calls of some cuckoo and host nestlings are similar; this raises the possibility of vocal mimicry (93, 97, 98, 115a).

## SOME ADAPTATIONS THAT DO NOT SEEM TO REPRESENT COEVOLUTION

Some parasitic traits are not adaptations to host defenses evolved in response to parasitism, e.g. the tendency for parasites to steal a host egg for each egg they lay could be an adaptation to hosts that count eggs to detect parasitism (60, 94). But cowbird and cuckoo hosts show no differences in responses whether or not a host egg is removed when a parasitic egg is added (27, 96, 124). Because host nestlings are usually present with cowbird nestlings, egg removal in that system may function to return the host's brood to a size it can feed. However, this function cannot apply to most cuckoos, but they and cowbirds usually eat stolen host eggs (20, 79, 172) which may provide energy and materials for forming eggs. Neither the brood reduction nor egg eating functions are responses to evolved host defenses, so neither represents coevolution.

Parasites must lay before the host is well into its incubation period or their eggs will not receive enough incubation to hatch. They appear to watch hosts and to estimate when the laying stage will occur (19, 61, 170d). This is a crucial adaptation but not a response to a host defense. Similarly, parasitic eggs hatch after relatively short incubation periods (106), and some even undergo embryonic development while still in the female's oviduct (158). This headstart at hatching gives parasitic nestlings an advantage over host young but would be selected for even in the absence of host defenses; further, some nonparasitic relatives of cuckoos and cowbirds also have very short incubation periods (60). It is probably also correct to exclude from coevolution those parasitic adaptations that allow nestlings to kill host young or to beg with unusual intensity (55). Reducing competition with host nestlings is adaptive for parasites regardless of whether hosts have evolved defenses. Of course some of these features may become more extreme due to coevolution, e.g. host and parasite incubation periods may become shorter in response to one another. Among the features of hosts, it is likely that recognition and aggression toward adult parasites is often not a coevolved adaptation—a point discussed previously.

## THE ABSENCE OF SOME ADAPTATIONS

Although some of the adaptations for and against parasitism are remarkable, some of the adaptations one might expect to see are surprisingly absent. The nearly total lack of nestling rejection has already been discussed. Other examples exist.

### *Host Choice by Cowbirds*

Actual and potential passerine hosts of cowbirds in the Neotropics and the Nearctic fall into two discrete groups, rejecter and accepter species. The former show nearly 100% rejection of real or model cowbird eggs experimentally placed in their nests, whereas the latter show nearly 100% acceptance (22, 34, 83, 125). Some of the accepters show low to high levels of nest desertion in response to natural parasitism for proximate reasons that are not wholly understood (as described above). Since rejecters usually mean failure, cowbirds should avoid them. But at least some rejecters are parasitized (84, 128, 138) although most birds (especially nonpasserines) with incompatible diets are not parasitized (45, 170a). Parasitism of rejecters is difficult to determine because they may remove cowbird eggs so quickly that researchers rarely see them. By visiting gray catbird *(Dumetella carolinensis)* nests several times during the early morning when cowbirds lay, Scott (138) found at least 44% to be parasitized, whereas more casual nest checks show only 1.5% parasitism for the same region (110). Parasitism of rejecter species occurs even though many nests of accepters go unparasitized (128). This "shotgun" approach to host use may occur because accepters outnumber rejecters and both groups are so diverse that no general feature separates all rejecters from all accepters (128). Unlike generalist cowbirds, gentes of the common cuckoo are thought rarely to parasitize hosts likely to reject their eggs (19, 20). But estimates of the extent to which cuckoos choose the correct host are probably at least slightly inflated.

## *The Lack of Egg Rejection in Some Hosts*

Two general views, equilibrium versus evolutionary lag, can explain the acceptance by many hosts parasitized with nonmimetic eggs. Under the equilibrium view, rejection incurs costs greater than acceptance and is therefore less adaptive (121, 174). Under the lag view, rejection would be adaptive but has not yet become common or even detectable because it takes time for new genetic variants to appear and to increase due to selection (12, 124, 133).

One equilibrium hypothesis argues that adult parasites destroy all of their host's young if their egg is removed, thus negating the benefits of host egg recognition (173). But removal of cowbird and cuckoo eggs by researchers produced no evidence for such behavior (133; A. Lotem, personal communication). Nor are recognition costs likely to explain the absence of rejection in many accepter cowbird hosts because they have eggs easily distinguishable from those of their parasite (124). The same is true for the dunnock *(Prunella modularis)*, the only major British host of the common cuckoo that lacks rejection behavior and is parasitized with nomimetic eggs (12). Recognition errors are more likley if the parasite has mimetic eggs. Reed warblers *(Acrocephalus scirpaceus)* ejected one or more of their own eggs at 6 of 19 nests that received models of mimetic cuckoo eggs but never did so at nests that received nonmimetic egg models or no experimental eggs (27). If recognition errors occur only at nests parasitized with mimetic eggs, they are not really costs for cuckoo hosts because the birds lose all of their brood whether they exert no rejection behavior or mistakenly reject their own egg. For such hosts, only recognition errors at unparasitized nests are costs, and as these have not been demonstrated, there is no evidence that recognition costs explain the absence of rejection. While recognition errors at parasitized nests cannot make acceptance more adaptive than rejection, they do reduce the relative adaptive value of rejection to the extent that hosts remove their own egg and leave the cuckoo egg.

Another type of cost occurs when a host with a small bill must reject by puncturing the parasitic egg or by renesting. As discussed above, puncture ejections occasionally cause breakage of host eggs, and renesting incurs a time lag and the costs of a new nest and eggs. Neither of these costs is relevant if parasitized hosts always lose all of their own young, as with most honeyguide and cuckoo hosts. In such cases, any response, even a cessation of reproductive effort during the current season, is better than rearing a parasite. Ejection costs might explain the status of some acceptors of cowbird parasitism (121), but this seems most unlikely for at least 10 species that usually lose all of their young when parasitized (8, 45, 57, 59, 82, 125, 135, 141, 163). Some cowbird hosts experience little or no decline in fledging rate if a cowbird nestling is present, but it is unlikely that such hosts, which tend to be large, incur any ejection costs (102). Even for a small species that must use puncture ejections, the known costs are slight (above) and are below the usual losses incurred by raising a cowbird.

Consideration of nest desertion also suggests that it should often, perhaps even usually, be favored over acceptance. First, the energy expenditure needed to lay a new clutch is less than a third that required to rear a cowbird (S. I. Rothstein, in preparation) because egg laying is much less costly than feeding nestlings and fledglings (116). Second, renesting incurs only a 5–7 day time lag, a period over which measures of productivity such as clutch size show little or no change (73). Lastly, in a model that contrasts the losses due to accepting a cowbird egg to those of renesting and the risk that a renest will be parasitized, it appears that parasitism of renests must be unrealistically high for selection to favor acceptance over renesting (S. I. Rothstein, in preparation). Thus overall, evolutionary lag remains the best explanation for the accepters among cowbird hosts. Lag is especially likely because cowbird parasitism in North America seems to be of recent vintage. Cowbirds have a Neotropical origin (44) and have enormously increased their range and abundance in North America in the last 200 years (44, 89).

The stochastic generation of genetic variants coding for rejection is probably a major factor determining a species' status as an accepter or a rejecter. It is not the only factor, however, because accepters and rejecters are not completely random assemblages of species. Although there is overlap, rejecters tend to be larger (121, 124). How can lag explain this trend? Large species can undoubtedly grasp and manipulate cowbird eggs more easily than small ones, and this probably facilitated the evolution of grasp ejections. Small species may require a relatively rare mutation that codes for an especially high amount of effort directed toward grasp ejection or for ejection by puncturing. The latter may be less likely to evolve because nest sanitation, a possible preadaptation for ejection (124), involves birds grasping but not puncturing fecal sacs and empty eggshells. Large species might also have been more likely to evolve rejection because they were preferred hosts in the past, due to their relatively low rates of nest predation and greater ability to feed cowbird nestlings (83, 170, but see Ref. 81).

## *The Long-Range Evolution of Hosts and Parasites, or, Which Side Wins?*

It has long been assumed that parasite-host coevolution will become more refined over time and that a parasite will tend to become specialized as more host species develop defenses (5, 60, 153, 156). Specialization on only one to a few hosts should occur because different counterdefenses will be needed for each host species, and limits probably exist as to the number of different specialists that can cooccur in one gene pool without a parasite undergoing speciation. For example, the three common egg types or gentes that occur in British cuckoos may be close to the maximum that can be maintained in this one population (or conceivably the gentes could be sibling species). On the other hand, some parasitic species may specialize on a single host from their inception, since one of the likely routes for the evolution of brood parasitism involves an incipient parasite laying in the nests of a close relative (60).

Davies & Brooke (29; see also Ref. 4) have formalized the likely sequence of events leading to heightened parasite-host arms races as follows: (*a*) Before being parasitized, a host shows no egg rejection. (*b*) Once parasitism begins, selection favors egg rejection; the time needed for the defense to increase depends on its adaptive value. (*c*) When hosts begin to reject eggs, selection begins to favor mimetic parasitic eggs. (*d*) If parasitism with mimetic eggs becomes too frequent, the host may go extinct. Otherwise, the host evolves egg discrimination that becomes more widespread and/or more refined. This heightened host discrimination could cause the parasite to switch to previously unused and therefore accepter hosts which would start the sequence over again. Heightened egg mimicry by the parasite is also possible, especially if the parasite-host community is relatively old and all or most potential hosts have evolved some degree of egg discrimination. I would suggest some additional possibilities at this stage. Even if egg mimicry becomes perfect, the host may evolve new or more variable egg colorations or preferential care of its own young. Either possibility should elicit new parasitic adaptations, e.g. the evolution of new egg types or mimetic nestlings. (*e*) If a rejecter host is freed from parasitism, Davies & Brooke suggest that selection produces a slow reversion to acceptance because of occasional recognition errors. However, as recognition errors have only been demonstrated in parasitized nests (above), rejection might become a neutral trait. Thus, step *a,* a lack of host rejection before parasitism begins, may not always apply because many species that have only recently become hosts may be descended from lineages that were parasitized in the past and that retain some atavistic rejection over successive speciation events, even in the absence of parasitism. Also, intraspecific parasitism may select for egg rejection (42). Whatever its basis, egg rejection occurs at high levels in some populations or even in species that have had no known contact with interspecific parasites (23). Rejection sometimes wanes slowly; thus three passerines show moderate to high levels of rejection of strongly dissimilar eggs in Iceland, even though they are allopatric with parasites (29, but see Ref. 24).

Davies & Brooke's proposed sequence assumes that genetic variation is always present for host discrimination and that this defense begins to increase when parasitism begins. Although some species may have atavistic retention of rejection, evolutionary lag may occur. Evolutionary lag has three components. First, it may take time after the selection pressure is first exerted for the appearance in the host of random genetic variation (mutations or recombinations) that codes for incipient rejection behavior. While artificial selection experiments always show that natural populations have considerable genetic variation for continuously variable traits such as size, the genetics may be very different for discrete traits that are either present or absent, i.e. rejection versus no rejection (124). Second, many beneficial mutations are lost soon after they appear due to independent assortment and drift (37). Third, even after an adaptation becomes common enough that its loss is unlikely, time is needed before the adaptation is frequent enough for us to detect and to identify it as an evolutionary response. This last component was the only one considered in Davies & Brooke's (29) sequence, but in reality it may be nearly impossible to distinguish among these three components.

Species that are probably in a phase of waiting for the right genetic variation are the dunnock in Britain, bulbuls (*Pycnonotus* spp.) in Africa, and many cowbird hosts in North America (12, 78, 124, 156). Although some of these species may show occasional rare rejections of experimentally placed parasitic eggs, these apparent rejections are not reliable indicators of responses to parasitic eggs (125). One could say that the parasite is winning the arms race while a host is "waiting" for egg rejection to appear. Although a magpie *(Pica pica)* population showed a high (74%) rejection rate of nonmimetic eggs only 25–30 years after becoming sympatric with a cuckoo (152), one cannot assume that the time lag was short. The population could have been receiving "rejection alleles" via gene flow for a long time from a nearby population in sympatry with the cuckoo. Comparative studies of cowbird hosts on Caribbean islands found no indication that hosts had developed defenses after 40 years of sympatry with cowbirds (111a).

A time lag may also apply to parasitic egg mimicry in response to host egg recognition. In such cases, the host is clearly winning the arms race. Examples are provided by rejecter hosts of the brown-headed cowbird (34, 124) and ecologically suitable hosts that have moderate levels of rejection but are almost never parasitized by cuckoos (28, 29). Rejection by the latter may have caused the cuckoo to switch to other host species. Whether a parasite's defeat is temporary depends on whether it switches back to these hosts and evolves egg mimicry.

Once a parasite evolves strong egg mimicry, it becomes the current winner until the host evolves new egg types or a response to nonmimetic nestlings. If the parasite then evolves nestling mimicry it might become a permanent winner unless the host evolves a new egg or nestling type or variable eggs or nestlings. When host variation forces matching errors on parasites, it seems best to declare a draw because both interactants sometimes neutralize the other's adaptations. Lastly, hosts could undergo directional selection and evolve a new egg type with parasitic egg mimicry usually lagging somewhat behind. This could result in a neverending arms race, or it could stabilize in parallel egg polymorphisms in parasites and hosts if new egg types in hosts are favored only when uncommon (because they are mimicked by the time they become common).

Parasites may possess some features that tend to shift arms races in their favor because their evolutionary histories have been shaped by parasitism more than have those of hosts. Parasites with mimetic eggs may for example have genetic mechanisms that allow minor genetic changes to result in major shifts in egg coloration. Such genetic processes occur in mimicry complexes involving butterfly coloration (144). So evolutionary lag may be less of a factor when a parasite develops a mimetic egg than when a host develops rejection.

Davies & Brooke (29) suggested that parasitic egg mimicry spreads faster than host egg rejection when both are increasing because every parasite encounters hosts, whereas only some hosts encounter parasites, i.e. selection on parasites should be stronger. However, not all hosts are rejecters unless lag has kept the parasite from evolving egg mimicry until host rejection has reached fixation. The adaptive value of host rejection depends largely on the frequency of parasitism, whereas that of parasitic egg mimicry depends on the frequency of host rejection. Thus, selection on parasites for egg mimicry will be weaker than selection on hosts until the frequency of host rejection exceeds the frequency of parasitism. But if parasitic egg mimicry is subject to evolutionary lag and appears after the frequency of host rejection has exceeded that of parasitism, mimicry will be selected for more strongly than host rejection as soon as it appears. Another complication is that mimicry will always lower the selective advantage of host rejection because some or even all parasitized rejecter individuals will not detect the mimetic parasitic eggs and will therefore have the same fitness as accepter individuals (72). Thus, as mimicry becomes more common, the increase in host rejection will slow down and may even stall completely if mimicry reaches fixation. A possible example of a near or complete stall is the redstart *(Phoenicurus phoenicurus)*. Although its immaculate blue eggs are mimicked perfectly by a cuckoo gens in Scandinavia, fewer than 30% of redstarts reject strongly dissimilar eggs (28, 162). A stall in the increase of egg rejection by a host might place a special premium on desertion in response to adult cuckoos.

## COMPARISONS BETWEEN COEVOLUTION IN THE CUCKOO AND COWBIRD SYSTEMS

Experimental introductions of parasitic eggs have shown three major differences between the defenses of cowbird and cuckoo hosts. First, while experimental data divide actual and potential cowbird hosts into two discrete categories, accepters and rejecters, they show intermediate response levels in many actual and potential hosts of the common cuckoo (Figure 1). Possible explanations are: (*a*) Few cowbird hosts may show intermediate rejection levels because selection for host rejection is so strong that species are rapidly transformed from accepters to rejecters once rejection appears (125). But cuckoo parasitism is a much weaker selection pressure because it is a relatively rare event, as discussed above. Thus, rejection would approach fixation more slowly in a cuckoo host (72). (*b*) Only the cuckoo system involves egg mimicry, which as stated above, can slow or even stall the increase in host rejection. (*c*) Cuckoos rarely parasitize many species that develop moderate to high rates of rejection, which would cause host rejection to stay at a frequency below 100% or to decline slowly. By contrast, cowbirds frequently parasitize at least some rejecter species, causing rejection in these species to approach and remain near fixation. (*d*) European birds are regularly parasitized only in part of their areas of sympatry with the cuckoo (153, 172), whereas cowbirds tend to be more geographically consistent in their use of host species (some geographic trends, however, do exist—45). Thus, gene flow between parasitized and unparasitized populations may slow movement toward fixation for acceptance or rejection in cuckoo hosts. (*e*) A. Lotem and H. Nakamura, (pers. commun.) suggest that intermediate rates of rejection by some cuckoo hosts reflect permanent equilibrium states. Because of the cuckoo's egg mimicry, hosts may need a prolonged period of learning to distinguish reliably between host and cuckoo eggs. This need may select for hosts that show little rejection during the first nesting of their lives and which instead use this period to learn the range of variation of their own eggs. This "slow learning" strategy results in acceptance of parasitic eggs by many first time breeders and could not be adaptive if rates of parasitism were high. Cowbird hosts appear to learn quickly from their first egg, because rates of parasitism

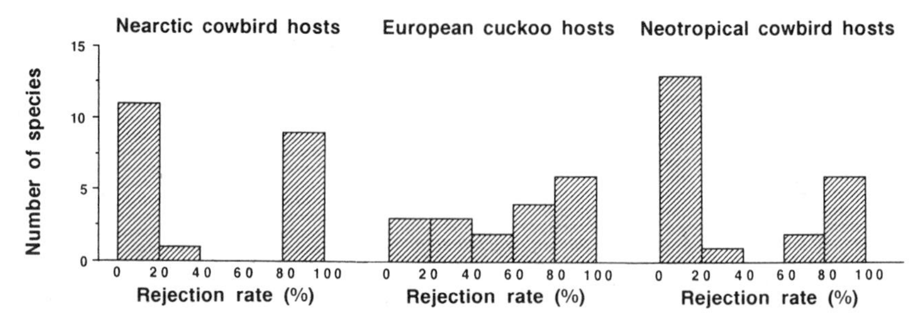


*Figure 1* Numbers of species showing various rejection rates of experimentally introduced nonmimetic eggs (models or real ones) in three parasite-host systems. Rejection rate is the proportion of nests tested at which rejection occurred. The graph includes only species for which experiments have been done on five or more nests. N's for most species are more than 10 nests and for several, 50 or more. Data on European Cuckoo hosts (28, 95, 115a, 162) include only passerines deemed to be suitable hosts (28) because cuckoos very rarely parasitize species other than their usual hosts. Data on cowbird hosts (mostly from Refs. 8, 23, 83, 124. See Ref 135a for a complete tabulation.) include all passerines that have been tested because cowbirds parasitize virtually all passerines including some that are often or usually unsuitable because of diet (110, 128).

are high and cowbird eggs are usually so divergent that even one host egg enables an individual to learn to distinguish reliably between egg types. Most acceptances by great reed warblers *(Acrocephalus arundinaceus)* in Japan are by yearlings (A. Lotem and H. Nakamura, personal communication) as the slow learning model predicts, whereas the reed warbler in England shows no clear relation between age and response (27).

The second difference between cowbird and cuckoo hosts is the proportion of species that show some rejection. Most potentially suitable cuckoo hosts in England show at least low levels (i.e. $\geq$ 20%) of unambiguous rejection toward experimentally introduced nonmimetic eggs (28, Fig. 1), but most potentially suitable cowbird hosts show no clear rejection (124). There are about 50 species of parasitic cuckoos but only 5 parasitic cowbirds, which suggests that the cuckoos have been parasites for much longer. Thus cuckoo-host arms races are older than cowbird-host arms races. Alternatively, while rejection is always adaptive when a cuckoo host is parasitized, acceptance may be more adaptive for some cowbird hosts that do not lose all of their young when parasitized and that must pay costs to reject (121). However, as stated previously there is no evidence that rejection is more costly than acceptance in any cowbird host.

Lastly, while experiments show that only one or two cowbird hosts regularly desert in response to parasitic eggs (17, 18, 23, 34, 83, 124, 127), many potential cuckoo hosts show frequent desertion in comparable experiments (28, 64, 95, 96). This contrast may exist because any rejection mode, regardless of costs, is adaptive for parasitized cuckoo hosts which will otherwise always lose all of their young. This is not true, however, for cowbird hosts, many of which rear some of their own young even if they accept parasitism. However, some cowbird hosts lose all of their young almost as consistently as do cuckoo hosts. The contrast may thus also exist because desertion in response to a parasitic egg is less likely to evolve, due to an absence of preadaptations, than is ejection (124) and because cuckoo hosts have had more time to evolve host defenses.

## CONCLUSIONS

Many researchers have recently addressed the significant and clear evolutionary questions posed by brood parasitism by carrying out original field studies, often with experimental approaches. This has resulted in considerable progress in the 13 years since Payne's review (106), which found little quantitative evidence that hosts of even the most famous parasite, the common cuckoo, reject nonmimetic eggs, although such rejection had been hypothesized for over a century. We now know that birds in numerous taxa and from many parts of the world exhibit egg recognition.

However, the recent findings have left some longstanding questions unanswered and have created new ones as well. In the former category, the scarcity of nestling rejection remains especially intriguing (and bothersome!). The lack of egg rejection in many hosts is at least as intriguing; we know that it is feasible since many species possess this defense. Despite a century of speculation, we still do not know whether specialized parasites other than the viduines imprint on their hosts. Among the new questions posed by the recent research is the unexpected contrast cuckoo and cowbird hosts show in intraspecific variation in rejection rates of nonmimetic eggs. Also unexpected is the discovery that some hosts whose eggs seem to be mimicked by the parasites express little or no rejection of nonmimetic eggs (14, 162). Despite the expectation that nestling recognition is less likely to evolve than egg recognition, there is no egg recognition in one of the two systems that show clear nestling recognition (39). There are answers to some of these problems, as described above, but perhaps the clearest generalization to be made about coevolution and brood parasitism is that the latter is a curious mosaic of adaptation and counteradaptation. Although parasites and hosts often express effective reciprocal adaptations much more clearly than do other putatively coevolving species, they sometimes lack seemingly beneficial and feasible adaptations. This has prompted some workers to endorse evolutionary lag as an explanation for certain aspects of the system, after all reasonable adaptive explanations have been tested and rejected (12, 29, 133).

If lag is important and if different parasite-host systems have been evolving for different periods of time, as seems all but certain (135b), many variations should appear on the themes of adaptations for and against parasitism. This is precisely what we see, and attempts to develop a general model of parasite-host coevolution are further exacerbated by the extreme variation that occurs in key aspects of the system. Parasitism can be rare or common. It can result in a minimal or a complete loss to the host. Parasites use from one to hundreds of host species. Some continents have only one or two parasitic species while others have over 20. Nevertheless, the general sequence of coevolutionary events proposed by Davies & Brooke and modified here seems to be a reasonable description of at least the two best-known systems—those of the common cuckoo and the brown-headed cowbird. These two parasites may well be at different points on somewhat similar coevolutionary progressions with the cuckoo at a later stage. But we need data from other parasites, especially ones that are sympatric with many other parasites and that use only one or a few host species. Some of these more specialized parasites may be even further along the progression than the common cuckoo, while others may have specialized on single hosts early in their evolutionary histories (104).

ACKNOWLEDGMENTS

I am grateful to N. B. Davies, R. B. Payne, and R. R. Warner for their valuable comments on this paper. Preparation of the manuscript was supported by NSF grant BNS-8616922.

## *Literature Cited*

1. Alvarez, F., Arias de Reyna, L., Segura, M. 1976. Experimental brood parasitism of the magpie *(Pica pica)*. *Anim. Behav.* 24:907–16
2. Andersson, M. 1984. Brood parasitism within species. In *Producers and Scroungers*, ed. C. J. Barnard, pp. 195–228. London: Chapman & Hall
3. Ankney, C. D. 1985. Variation in weight and composition of brown-headed cowbird eggs. *Condor* 87:296–99
4. Arias de Reyna, L., Hidalgo, S. J. 1982. An investigation into egg-acceptance by azure-winged magpies and host-recognition by great spotted cuckoo chicks. *Anim. Behav.* 30:819–23
5. Baker, E. C. S. 1942. *Cuckoo Problems*. London: Witherby. 207 pp.
6. Beezley, J. A., Rieger, J. P. 1987. Least Bell's vireo management by cowbird trapping. *West. Birds* 18:55–61
7. Blankespoor, G. W., Oolman, J., Uthe, C. 1982. Eggshell strength and cowbird parasitism of Red-winged Blackbirds. *Auk* 99:363–65
8. Briskie, J. V., Sealy, S. G. 1987. Responses of least flycatchers to experimental inter- and intraspecific brood parasitism. *Condor* 89:899–901
9. Briskie, J. V., Sealy, S. G. 1989. Changes in nest defense against a brood parasite over the breeding cycle. *Ethology* 82:61–67
10. Brittingham, M. C., Temple, S. A. 1983. Have cowbirds caused forest songbirds to decline? *BioScience* 33:31–35
11. Brooke, M. de L., Davies, N. B. 1987. Recent changes in host usage by cuckoos *Cuculus canorus* in Britain. *J. Anim. Ecol.* 56:873–83
12. Brooke, M. de L., Davies, N. B. 1988. Egg mimicry by cuckoos *Cuculus canorus* in relation to discrimination by hosts. *Nature* 335:630–32
13. Brooke, M., de L., Davies, N. B. 1989. Provisioning of nestling Cuckoos *Cuculus canorus* by Reed Warbler *Acrocephalus scirpaceus* hosts. *Ibis* 131:250–56
14. Brooker, M. G., Brooker, L. C. 1989.

The comparative breeding behavior of two sympatric cuckoos, Horsfield's bronze-cuckoo *Chrysococcyx basilis* and the shining bronze-cuckoo *C. lucidus*, in Western Australia: a new model for the evolution of egg morphology and host specificity in avian brood parasites. *Ibis* 131:528–47

15. Brooker, M. G., Brooker, L. C., Rowley, I. 1988. Egg deposition by the bronze-cuckoos *Chrysococcyx basalis* and *Ch. lucidus*. *Emu* 88:107–9
16. Brown, C. R., Brown, M. B. 1988. A new form of reproductive parasitism in cliff swallows. *Nature* 331:66–68

16a. Brown, R. J., Brown, M. N., Brooke, M. de L., Davies, N. B. 1990. Reactions of parasitized and unparasitized populations of *Acrocephalus* warblers to model cuckoo eggs. *Ibis* 132:109–11

17. Burgham, M. C. J., Picman, J. 1989. Effect of brown-headed cowbirds on the evolution of yellow warbler antiparasite strategies. *Anim. Behav.* 38:298–308
18. Carter, M. D. 1986. The parasitic behavior of the bronzed cowbird in south Texas. *Condor* 88:11–25
19. Chance, E. P. 1940. *The Truth About the Cuckoo*. London: Country Life. 207 pp.
20. Chance, E. P., Hann, H. W. 1942. The European cuckoo and the cowbird. *Bird-Banding* 13:99–103
21. Clark, K. L., Robertson, R. J. 1979. Spatial and temporal multi-species nesting aggregations in birds as anti-parasite and anti-predator defenses. *Behav. Ecol. Sociobiol.* 5:359–71
22. Clark, K. L., Robertson, R. J. 1981. Cowbird parasitism and evolution of anti-parasite strategies in the yellow warbler. *Wilson Bull.* 92:244–58
23. Cruz, A., Manolis, T., Wiley, J. W. 1985. The shiny cowbird: a brood parasite expanding its range in the Caribbean region. In *Neotropical Ornithology, Ornithol. Monogr.* 36, ed. P. A. Buckley, M. S. Foster, E. S. Morton, R. S. Ridgely, F. G. Buckley, pp. 607–20. Washington, DC: Am. Ornithol. Union. 1041 pp.
24. Cruz, A., Wiley, J. W. 1989. The decline of an adaptation in the absence of a presumed selection pressure. *Evolution* 43:55–62
25. Cruz, A., Wiley, J. W., Nakamura, T. K., Post, W. 1989. The shiny cowbird *Molothrus bonariensis* in the West Indian region—biogeographical and ecological implications. In *Biogeography of the West Indies—Past, Present, and Future*, ed. C. A. Woods, pp. 519–41. Gainesville, Fla: Sandhill Crane
26. Davies, N. B., Bourke, A. F. G., Brooke, M. de L. 1989. Cuckoos and parasitic ants: interspecific brood parasitism as an evolutionary arms race. *Trends Ecol. Evol.* 4:274–78
27. Davies, N. B., Brooke, M. De L. 1988. Cuckoos versus reed warblers: adaptations and counteradaptations. *Anim. Behav.* 36:262–84
28. Davies, N. B., Brooke, M. de L. 1989. An experimental study of co-evolution between the cuckoo, *Cuculus canorus*, and its hosts. I. Host egg discrimination. *J. Anim. Ecol.* 58:207–24
29. Davies, N. B., Brooke, M. de L. 1989. An experimental study of co-evolution between the cuckoo, *Cuculus canorus*, and its hosts. II. Host egg markings, chick discrimination and general discussion. *J. Anim. Ecol.* 58:225–36
30. Dawkins, R., Krebs, J. R. 1979. Arms races between and within species. *Proc. R. Soc. London B Ser.* 205:489–511

30a. Eastzer, D., Chu, P. R., King, A. P. 1980. The young cowbird: average or optimal nesting? *Condor* 82:417–25

31. Elliott, P. F. 1977. Adaptive significance of cowbird egg distribution. *Auk* 94:590–93
32. Elliott, P. F. 1978. Cowbird parasitism on the Kansas tallgrass prairie. *Auk* 95:161–67
33. Ficken, M. S. 1965. Mouth color of nestling passerines and its use in taxonomy. *Wilson Bull.* 77:71–75
34. Finch, D. M. 1982. Rejection of cowbird eggs by crissal thrashers. *Auk* 99:719–21
35. Fleischer, R. C. 1985. A new technique to identify and assess the dispersion of eggs of individual brood parasites. *Behav. Ecol. Sociobiol.* 17:91–99
36. Folkers, K. L., Lowther, P. E. 1985. Responses of nesting red-winged blackbirds and yellow warblers to brown-headed cowbirds. *J. Field Ornithol.* 56:175–77
37. Ford, E. B. 1964. *Ecological Genetics*. London: Methuen. 335 pp.
38. Fraga, R. M. 1978. The rufous-collared sparrow as a host of the shiny cowbird. *Wilson Bull.* 90:271–84

38a. Fraga, R. M. 1979. Differences between nestlings and fledglings of screaming and baywinged cowbirds. *Wilson Bull.* 90:151–54

39. Fraga, R. M. 1983. The eggs of the parasitic screaming cowbird *(Molothrus rufoaxillaris)* and its host, the baywinged cowbird *(M. badius)*: is there evidence for mimicry? *J. Ornithol.* 124:187–93
40. Fraga, R. M. 1985. Host-parasite in-

teractions between chalk-browed mockingbirds and shiny cowbirds. See Ref. 23, pp. 829–44
41. Fraga, R. M. 1986. *The bay-winged cowbird* (Molothrus badius) *and its brood parasites: interactions, coevolution and comparative efficiency*. PhD thesis. Univ. Calif., Santa Barbara
41a. Fraga, R. M. 1988. Nest sites and breeding success of baywinged cowbirds *(Molothrus badius). J. Ornithol.* 129:175–83
42. Freeman, S. 1988. Egg variability and conspecific nest parasitism in *Ploceus* weaverbirds. *Ostrich* 59:49–53
42a. Freeman, S., Gori, D. F., Rohwer, S. 1990. Red-winged blackbirds and brown-headed cowbirds: some aspects of a host-parasite relationship. *Condor* 92:336–40
43. Friedmann, H. 1929. *The Cowbirds, A Study in the Biology of Social Parasitism*. Springfield, Ill: C. C. Thomas. 421 pp.
44. Friedmann, H. 1955. The honey-guides. *US Natl. Mus. Bull.* 208. 292 pp.
45. Friedmann, H. 1963. Host relations of the parasitic cowbirds. *US Natl. Mus. Bull.* 233. 273 pp.
46. Friedmann, H. 1967. Alloxenia in three African species of *Cuculus*. *Proc. US Natl. Mus.* 124:1–13
47. Friedmann, H., Kiff, L. F. 1985. The parasitic cowbirds and their hosts. *Proc. West. Found. Zool.* 2:226–304
48. Fry, C. H., Keith, S., Urban, E. K. 1985. Evolutionary expositions from "The Birds of Africa": *Halcyon* song phylogeny, cuckoo host partitioning; systematics of *Aplopelia* and *Bostrychia*. In *African Vertebrates: Systematics, Phylogeny and Evolutionary Ecology*, ed. K. L. Schuckmann, pp. 163–80. Bonn: Selbstverlag
49. Fry, C. H., Keith, S., Urban, E. K. 1988. *The Birds of Africa,* Vol. III. London: Academic
50. Futuyma, D. J., Slatkin, M. 1983. Introduction. In *Coevolution,* ed. D. J. Futuyma, M. Slatkin, pp. 1–13. Sunderland, Mass: Sinauer. 555 pp.
51. Futuyma, D. J., Slatkin, M. 1983. Epilogue: The study of coevolution. See Ref. 50, pp. 459–64
52. Gaston, A. J. 1976. Brood parasitism by the pied crested cuckoo *Clamator jacobinus*. *J. Anim. Ecol.* 45:331–48
53. Gill, B. J. 1983. Brood parasitism by the shiny cuckoo *Chrysococcyx lucidus* at Kaikoura, New Zealand. *Ibis* 125:40–55
54. Glue, D., Morgan, R. 1972. Cuckoo hosts in British habitats. *Bird Study* 19:187–92
55. Gochfeld, M. 1979. Begging by nestling shiny cowbirds: adaptive or maladaptive. *Living Bird* 17:41–50
56. Gochfeld, M. 1979. Brood parasite and host coevolution: interactions between shiny cowbirds and two species of meadowlarks. *Am. Nat.* 113:855–70
57. Goldwasser, S., Gaines, D., Wilbur, S. R. 1980. The least Bell's vireo in California: a de facto endangered race. *Am. Birds* 34:742–45
58. Graham, D. S. 1988. Responses of five host species to cowbird parasitism. *Condor* 90:588–91
59. Grzybowski, J. A., Clapp, R. B., Marshall, J. T. Jr. 1986. History and current population status of the black-capped vireo in Oklahoma. *Am. Birds* 40:1151–61
60. Hamilton, W. J. III, Orians, G. H. 1965. Evolution of brood parasitism in altricial birds. *Condor* 67:361–82
61. Hann, H. W. 1941. The cowbird at the nest. *Wilson Bull.* 53:211–21
62. Harrison, C. J. O. 1968. Egg mimicry in British cuckoos. *Bird Study* 15:22–28
63. Harvey, P. H., Partridge, L. 1988. Of cuckoo clocks and cowbirds. *Nature* 335:586–87
64. Higuchi, H. 1989. Responses of the bush warbler *Cettia diphone* to artificial eggs of *Cuculus cuckoos* in Japan. *Ibis* 131:94–98
65. Higuchi, H., Sato, S. 1984. An example of character release in host selection and egg colour of cuckoos *Cuculus* spp. in Japan. *Ibis* 126:398–404
66. Hobson, K. A., Sealy, S. G. 1989. Responses of yellow warblers to the threat of cowbird parasitism. *Anim. Behav.* 38:510–19
67. Holmes, J. C. 1983. Evolutionary relationships between parasitic helminths and their hosts. See Ref. 50, pp. 161–85
68. Hoy G., Ottow, J. 1964. Biological and oological studies of the molothrine cowbirds (Icteridae) of Argentina. *Auk* 81:186–203
68a. Immelmann, K., Piltz, A., Sossinka, R. 1977. Experimentelle untersuchungen zur bedeutung der rachenzeichnung junger zebra finken. *Zeit. Teirpsychol.* 45:210–18
69. Janzen, D. H. 1980. When is it coevolution? *Evolution* 34:611–12
70. Jensen, R. A. 1966. Genetics of cuckoo egg polymorphism. *Nature* 209:827
71. Jensen, R. A., Clinning, C. F. 1974. Breeding biology of two cuckoos and their hosts in South West Africa. *Living Bird* 13:5–50
72. Kelly, C. 1987. A model to explore the rate of spread of mimicry and rejection

in hypothetical populations of cuckoos and their hosts. *J. Theor. Biol.* 125:283–99
73. Klomp, H. 1970. The determination of clutch size in birds. A review. *Ardea* 58:1–124
73a. Lack, D. 1958. The significance of the colour of turdine eggs. *Ibis* 100:145–66
74. Lack, D. 1963. Cuckoo hosts in England. *Bird Study* 10:185–201
75. Lack, D. 1968. *Ecological Adaptations for Breeding in Birds*. London: Methuen. 409 pp.
76. Laymon, S. A. 1987. Brown-headed cowbirds in California: historical perspectives and management opportunities in riparian habitats. *West. Birds* 18:63–70
77. Linden, M., Møller, A. P. 1989. Cost of reproduction and covariation of life history traits in birds. *Trends Ecol. Evol.* 4:367–71
78. Liversidge, R. 1971. The biology of the Jacobin cuckoo *Clamator jacobinus*. *Proc. 3rd Pan-Afr. Ornithol. Congr., Ostrich Suppl.* 8:117–37
79. Lohrl, H. 1979. Untersuchungen am Kuckuck, *Cuculus canorus* (Biologie, Ethologie, und Morphologie). *J. Ornithol.* 120:139–73
80. MacWhirter, R. B. 1989. On the rarity of intraspecific brood parasitism. *Condor* 91:485–92
81. Manolis, T. D. 1982. *Host relationships and reproductive strategies of the shiny cowbird in Trinidad and Tobago*. PhD thesis. Boulder, Colo: Univ. Colo.
82. Marvil, R. E., Cruz, A. 1989. Impact of brown-headed cowbird parasitism on the reproductive success of the solitary vireo. *Auk* 106:476–80
83. Mason, P. 1986. Brood parasitism in a host generalist, the shiny cowbird: I. The quality of different species as hosts. *Auk* 103:52–60
84. Mason, P. 1986. Brood parasitism in a host generalist, the shiny cowbird: II. Host selection. *Auk* 103:61–69
85. Mason, P. 1987. Pair formation in cowbirds: evidence found for screaming but not shiny cowbirds. *Condor* 89:349–56
86. Mason, P., Rothstein, S. I. 1986. Coevolution and avian brood parasitism: Cowbird eggs show evolutionary response to host discrimination. *Evolution* 40:1207–14
87. Mason, P., Rothstein, S. I. 1987. Crypsis versus mimicry and the color of shiny cowbird eggs. *Am. Nat.* 130:161–67
88. May, R. M., Robinson, S. K. 1985. Population dynamics of avian brood parasitism. *Am. Nat.* 126:475–94
89. Mayfield, H. 1965. The brown-headed cowbird with old and new hosts. *Living Bird* 4:13–28
90. Mayfield, H. 1977. Brown-headed Cowbird: agent of extermination. *Am. Birds* 31:107–13
91. Mayfield, H. 1978. Brood parasitism: Reducing interactions between Kirtland's warblers and brown-headed cowbirds. In *Endangered Birds: Management Techniques for Preserving Threatened Species*, ed. S. A. Temple, pp. 85–91. Madison: Univ. Wisc. Press
92. McLean, I. 1987. Response to a dangerous enemy: Should a brood parasite be mobbed? *Ethology* 75:235–45
93. McLean, I., Waas, J. R. 1987. Do cuckoo chicks mimic the begging calls of their hosts? *Anim. Behav.* 35:1896–98
94. Moksnes, A., Røskaft, E. 1987. Cuckoo host interactions in Norwegian mountain areas. *Ornis Scand.* 18:168–72
95. Moksnes, A., Røskaft, E. 1988. Responses of fieldfares, *Turdus pilaris* and bramblings *Fringilla montifringilla* to experimental parasitism by the cuckoo *Cuculus canorus*. *Ibis* 130:535–39
96. Moksnes, A., Røskaft, E. 1989. Adaptations of meadow pipits to parasitism by the common cuckoo. *Behav. Ecol. Sociobiol.* 24:25–30
97. Morton, E. S., Farabaugh, S. M. 1979. Infanticide and other adaptations of the nestling striped cuckoo *Tapera naevia*. *Ibis* 121:212–13
98. Mundy, P. J. 1973. Vocal mimicry of their hosts by nestlings of the great spotted cuckoo and striped crested cuckoo. *Ibis* 115:602–4
99. Nicolai, J. 1964. Der brutparasitismus der viduinae als ethologisches problem. *Z. Tierpsychol.* 21:129–204
100. Nicolai, J. 1974. Mimicry in parasitic birds. *Sci. Am.* 231:92–98
101. Nolan, V. Jr. 1978. *The Ecology and Behavior of the Prairie Warbler* Dendroica discolor. *Ornithol. Monogr.* 26. Washington, DC: Am. Ornithol. Union. 595 pp.
102. Ortega, C. P., Cruz, A. 1988. Mechanisms of egg acceptance by marsh-dwelling blackbirds. *Condor* 90:349–58
103. Payne, R. B. 1968. Interspecific communication signals in parasitic birds. *Am. Nat.* 101:363–76
104. Payne, R. B. 1973. *Behavior, Mimetic Songs and Song Dialects, and Relationships of the Parasitic Indigobirds (Vidua) of Africa*. *Ornithol. Monogr.* 11. Washington, DC: Am. Ornithol. Union. 333 pp.
105. Payne, R. B. 1973. Vocal mimicry of the paradise whydahs *(Vidua)* and response of female whydahs to the songs

of their hosts *(Pytilia)* and their mimics. *Anim. Behav.* 21:762–71
106. Payne, R. B. 1977. The ecology of brood parasitism in birds. *Annu. Rev. Ecol. Syst.* 8:1–28
107. Payne, R. B. 1980. Behavior and songs in hybrid parasitic finches. *Auk* 97:118–34
107a. Payne, R. B. 1982. Species limits in the indigobirds (Ploceidae, *Vidua*) of West Africa: mouth mimicry, song mimicry, and description of new species. *Misc. Publ. Univ. Mich. Museum Zool.* 102
107b. Payne, R. B. 1985. Behavioral continuity and change in local song populations of village indigobirds *Vidua chalybeata*. *Z. Tierpsychol.* 70:1–44
108. Payne, R. B., Payne, K. 1967. Cuckoo hosts in southern Africa. *Ostrich* 38:135–43
109. Payne, R. B., Payne, L. L., Rowley, I. 1985. Splendid wren *Malurus splendens* response to cuckoos: an experimental test of social organization in a communal bird. *Behaviour* 94:108–27
110. Peck, G. K., James, R. D. 1987. *Breeding Birds of Ontario, Nidiology and Distribution.* Vol. 2. *Passerines.* Toronto: Royal Ontario Mus.
111. Picman, J. 1989. Mechanisms of increased puncture resistance of eggs of brown-headed cowbirds. *Auk* 106:577–83
111a. Post, W., Nakamura, T. K., Cruz, A. 1990. Patterns of shiny cowbird parasitism in St. Lucia and southwestern Puerto Rico. *Condor* 92:461–69
112. Post, W., Wiley, J. W. 1976. The yellow-shouldered blackbird—present and future. *Am. Birds* 30:13–20
113. Post, W., Wiley, J. W. 1977. Reproductive interactions of the shiny cowbird and the yellow-shouldered blackbird. *Condor* 79:176–84
113a. Power, H. W., Kennedy, E. D., Romagnano, L. C., Lombardo, M. P., Hoffenberg, A. S. et al. 1989. The parasitism insurance hypothesis: why starlings leave space for parasitic eggs. *Condor* 91:753–65
114. Pulich, W. M. 1976. *The Golden-cheeked Warbler, A Bioecological Study.* Austin, Texas: Texas Parks & Wildlife Dept. 172 pp
115. Rahn, H., Curran-Everett, L., Booth, D. T. 1988. Eggshell differences between parasitic and nonparasitic Icteridae. *Condor* 90:962–64
115a. Redondo, T., Arias de Reyna, L. 1988. Vocal mimicry of hosts by great spotted cuckoo *Clamator glandarius:* Further evidence. *Ibis* 130:540–44
115b. Ricklefs, R. E. 1969. An analysis of nesting mortality in birds. *Smithson. Contrib. Zool.* 9:1–48
116. Ricklefs, R. E. 1974. Energetics of reproduction in birds. In *Avian Energetics,* ed. R. A. Paynter Jr., pp. 152–292. Cambridge, Mass: Nuttall Ornithol. Club
117. Riddiford, N. 1986. Why do cuckoos *Cuculus canorus* use so many species of hosts? *Bird Study* 33:1–5
118. Robertson, R. J., Norman, R. F. 1976. Behavioral defenses to brood parasitism by potential hosts of the brown-headed cowbird. *Condor* 78:167–73
119. Robertson, R. J., Norman, R. F. 1977. The function and evolution of aggressive host behavior towards the brown-headed cowbird *(Molothrus ater). Can. J. Zool.* 55:508–18
120. Rohwer, F. C., Freeman, S. 1989. The distribution of conspecific nest parasitism in birds. *Can. J. Zool.* 67:239–53
121. Rohwer, S., Spaw, C. D. 1988. Evolutionary lag versus bill-size constraints: a comparative study of the acceptance of cowbird eggs by old hosts. *Evol. Ecol.* 2:27–36
122. Rohwer, S., Spaw C. D., Røskaft, E. 1989. Costs to northern orioles of puncture-ejecting parasitic cowbird eggs from their nests. *Auk* 106:734–38
122a. Røskaft, E., Orians, G. H., Beletsky, L. D. 1990. Why do red-winged blackbirds accept eggs of brown-headed cowbirds. *Evol. Ecol.* 4:35–42
123. Rothstein, S. I. 1974. Mechanisms of avian egg recognition: possible learned and innate factors. *Auk* 91:796–807
124. Rothstein, S. I. 1975. An experimental and teleonomic investigation of avian brood parasitism. *Condor* 77:250–71
125. Rothstein, S. I. 1975. Evolutionary rates and host defenses against avian brood parasitism. *Am. Nat.* 109:161–76
126. Rothstein, S. I. 1975. Mechanisms of avian egg recognition: Do birds know their own eggs? *Anim. Behav.* 23:268–78
127. Rothstein, S. I. 1976. Experiments on defenses cedar waxwings use against cowbird parasites. *Auk* 93:675–91
128. Rothstein, S. I. 1976. Cowbird parasitism of the cedar waxwing and its evolutionary implications. *Auk* 93:498–509
129. Rothstein, S. I. 1977. The preening invitation or head-down display of parasitic cowbirds: I. Evidence for intraspecific occurrence. *Condor* 79:13–23
129a. Rothstein, S. I. 1977. Cowbird parasitism and egg recognition of the northern oriole. *Wilson Bull.* 89:21–32
130. Rothstein, S. I. 1978. Geographical

variation in the nestling colorations of parasitic cowbirds. *Auk* 95:152–60
131. Rothstein, S. I. 1978. Mechanisms of avian egg recognition: additional evidence for learned components. *Anim. Behav.* 26:671–77
132. Rothstein, S. I. 1980. The preening invitation or head-down display of parasitic cowbirds: II. Experimental analyses and evidence for behavioural mimicry. *Behaviour* 75:148–84
133. Rothstein, S. I. 1982. Successes and failures in avian egg recognition with comments on the utility of optimality reasoning. *Am. Zool.* 22:547–60
134. Rothstein, S. I. 1982. Mechanisms of avian egg recognition: Which egg parameters elicit responses by rejector species? *Behav. Ecol. Sociobiol.* 11:229–39
135. Rothstein, S. I. 1986. A test of optimality: egg recognition in the eastern phoebe. *Anim. Behav.* 34:1109–19
135a. Rothstein, S. I. 1990. Brood parasitism and clutch-size determination in birds. *Trends Ecol. Evol.* 5:101–2
135b. Rothstein, S. I. In press. Brood parasitism, the importance of experiments and host defenses of avifaunas on different continents. *Proc. 7th Pan-Afr. Ornithol. Congr.*
136. Rothstein, S. I., Verner, J., Stevens, E., Ritter, L. V. 1987. Behavioral differences among sex and age classes of the brown-headed cowbird and their relation to the efficacy of a control program. *Wilson Bull.* 99:322–37
137. Rothstein, S. I., Yokel, D. A., Fleischer, R. C. 1986. Social dominance, mating and spacing systems, female fecundity, and vocal dialects in captive and free ranging brown-headed cowbirds. In *Current Ornithology,* ed. R. F. Johnston, 3:127–85. New York: Plenum. 522 pp.
138. Scott, D. M. 1977. Cowbird parasitism on the gray catbird at London, Ontario. *Auk* 94:18–27
139. Scott, D. M., Ankney, C. D. 1983. The laying cycle of brown-headed cowbirds: passerine chickens? *Auk* 100:583–92
140. Scott, D. M., Lemon, R. E., Darley, J. A. 1987. Relaying interval after nest failure in gray catbirds and northern cardinals. *Wilson Bull.* 99:708–12
141. Sedgewick, J. A., Knopf, R. L. 1988. A high incidence of brown-headed cowbird parasitism of willow flycatchers. *Condor* 90:253–56
142. Selander, R. K., LaRue, C. J. Jr. 1961. Interspecific preening invitation display of parasitic cowbirds. *Auk* 78:473–504
143. Seppa, V. 1969. The cuckoo's ability to find a nest where it can lay an egg. *Ornis Fenn.* 46:78–80
144. Sheppard, P. M., Turner, J. R. G., Brown, K. S., Benson, W. W., Singer, M. C. 1985. Genetics and the evolution of muellerian mimicry in *Heliconius* butterflies. *Philos. Trans. R. Soc. London Ser. B* 308:433–613
145. Short, L. L., Horne, J. F. M. 1985. Behavioral notes on the nest-parasitic Afrotropical honeyguides (Aves: Indicatoridae). *Am. Mus. Novit.* 2825:1–46
146. Short, L. L., Horne, J. F. M. 1988. Lesser honeyguide interactions with its barbet hosts. *Proc. 6th Pan-Afr. Ornithol. Congr.* pp. 65–75
147. Smith, J. N. M. 1981. Cowbird parasitism, host fitness, and age of the host female in an island song sparrow population. *Condor* 83:152–61
148. Smith, J. N. M., Arcese, P., McLean, I. G. 1984. Age, experience and enemy recognition by wild song sparrows. *Behav. Ecol. Sociobiol.* 14:101–6
149. Smith, N. G. 1968. The advantage of being parasitized. *Nature* 219:690–94
150. Smith, N. G. 1979. Alternate responses by hosts to parasites which may be helpful or harmful. In *Host-Parasite Interfaces,* ed. B. B. Nickel, pp. 7–15. New York: Academic
151. Smith, N. G. 1980. Some evolutionary, ecological, and behavioural correlates of communal nesting by birds with wasps or bees. In *Acta XVII Congressus Internationalis Ornithologici,* ed. R. Nohring, 2:1199–1205. Berlin: Deutschen Ornithol.-Gessellschaft
152. Soler, M., Moller, A. P. 1990. Duration of sympatry and coevolution between the great spotted cuckoo and its magpie host. *Nature* 343:748–50
153. Southern, H. N. 1954. Mimicry in cuckoo's eggs. In *Evolution as a Process,* ed. J. Huxley, A. C. Hardy, E. B. Ford, pp. 219–32. London: Allen & Unwin
154. Spaw, C. D., Rohwer, S. 1987. A comparative study of eggshell thickness in cowbirds and other passerines. *Condor* 89:307–18
155. Swynnerton, C. F. M. 1916. On the coloration of the mouths and eggs of birds. II. On the coloration of eggs. *Ibis* 4:529–606
156. Swynnerton, C. F. M. 1918. Rejection by birds of eggs unlike their own: with remarks on some of the cuckoo problems. *Ibis* 6:127–54
157. Tschantz, B. 1959. Zur brutbiologie der trottellumme *(Uria aalge aalge). Behaviour* 14:1–100
158. Vernon, C. J. 1970. Pre-incubation

embryonic development and egg "dumping" by the Jacobin cuckoo. *Ostrich* 41:259–60
159. Vernon, C. J. 1984. The breeding biology of the thickbilled cuckoo. *Proc. 5th Pan-Afr. Ornith. Congr.*, pp. 825–40
160. Vernon, C. J. 1987. Bill hooks of *Protodiscus* nestlings. *Ostrich* 58:187
161. Victoria, J. K. 1972. Clutch characteristics and egg discriminative ability of the African village weaverbird *(Ploceus cucullatus). Ibis* 114:367–76
162. von Haartman, L. 1981. Co-evolution of the cuckoo *Cuculus canorus* and a regular cuckoo host. *Ornis Fenn.* 58:1–10
163. Walkinshaw, L. H. 1961. The effect of parasitism by the brown-headed cowbird on *Empidonax* flycatchers in Michigan. *Auk* 78:266–68
164. Walkinshaw, L. H. 1983. *Kirtland's Warbler; The Natural History of an Endangered Species*. Bloomfield Hills, Mich: Cranbrook Inst. Science. 207 pp.
165. Wcislo, W. T. 1989. Behavioral environments and evolutionary change. *Annu. Rev. Ecol. Syst.* 20:137–69
166. Weller, M. W. 1959. Parasitic egg laying in the redhead *(Aythya americana)* and other North American Anatidae. *Ecol. Monogr.* 29:333–65
167. Weller, M. W. 1968. The breeding biology of the parasitic black-headed duck. *Living Bird* 7:169–208
168. Wiley, J. W. 1982. *Ecology of avian brood parasitism at an early interfacing of host and parasite populations*. PhD thesis. Coral Gables, Fla: Univ. Miami
169. Wiley, J. W. 1985. Shiny cowbird parasitism in two avian communities in Puerto Rico. *Condor* 87:167–76
170. Wiley, J. W. 1986. Growth of shiny cowbird and host chicks. *Wilson Bull.* 98:126–31
170a. Wiley, J. W. 1988. Host selection by the shiny cowbird *Condor* 90:289–303
171. Wiley, R. H., Wiley, M. S. 1980. Spacing and timing in the nesting ecology of a tropical blackbird: comparison of populations in different environments. *Ecol. Monogr.* 50:153–78
172. Wyllie, I. 1981. *The Cuckoo*. New York: Universe. 176 pp.
173. Yom Tov, Y. 1980. Intraspecific nest parasitism in birds. *Biol. Rev.* 55:93–108
174. Zahavi, A. 1979. Parasitism and nest predation in parasitic cuckoos. *Am. Nat.* 113:157–59

*Annu. Rev. Ecol. Syst. 1990. 21:509–39*


# THE GEOLOGIC HISTORY OF DIVERSITY

*Philip W. Signor*

Department of Geology, University of California, Davis, California 95616

KEY WORDS: species richness, taxonomic diversity, fossil record, mass extinctions, paleodiversity

---

## INTRODUCTION

A spirited debate occurred amongst the founders of geology over the geologic history of diversity (44, 104). Lyell (79) and Agassiz (1), among other adherents to strict uniformitarianism, maintained that the Earth's biota had been at a steady state for uncounted millennia. Against this view were the progressionists, who found evidence for directional change in the fossil record (e.g. 93). The debate has persisted, with occasional lulls, to the present. As recently as a decade ago, several eminent paleobiologists argued pursuasively that the taxonomic richness of marine Metazoa has been at equilibrium for much of the past 600 million years (46, 94, 96, 112). Strong arguments in support of a global increase in diversity were raised in response (139, 142). This dichotomy, equilibrium versus directional change through time, is a central theme in evolutionary paleontology (44).

The precise pattern of taxonomic richness in geologic time remains the subject of considerable debate, but the broad outline of that history is now generally accepted (14, 15, 72, 73, 85–88, 91, 118, 122, 123). There were relatively few species during the Paleozoic and early Mesozoic, and diversity increased substantially in the past hundred million years. The biosphere reached the zenith of the longest sustained period of taxonomic diversification in the Earth's history in the Pliocene and Pleistocene, when climatic change and the advent of organized human activity then checked that diversification. Continuing unabated for nearly one hundred million years, this diversification



0066-4162/90/1120-0509$02.00

proceeded through one major mass extinction (at the end of the Cretaceous Period) and, probably, two minor extinction events in the Cenozoic (100–102). There were more species and higher taxa of plants and animals, marine and terrestrial, in the Pliocene and Pleistocene world than at any time in the geologic past.

While the patterns of taxonomic diversity through time have become increasingly evident, the processes underlying those patterns, and their ultimate causes, remain obscure. A number of physical and biological processes likely influence diversity, including continental drift, changing sea level, mass extinctions, evolutionary innovation, and others. Attempts to single out particular processes as the primary control on diversity have not been successful. Temporal trends in diversity most likely result from a complex interaction of physical and biological processes that operate at different hierarchical levels of the biosphere.

This review first considers the strength and weaknesses of the fossil record as a chronicle of biological diversity. It reviews the record of taxonomic richness at several scales of analysis in the marine and terrestrial realms and examines possible links between these different levels and some of the hypotheses advanced to account for the observed trends. Lastly, the review considers the tempo and structure of the diversification of life through time.

## THE FOSSIL RECORD AND THE HISTORY OF LIFE

The shortcomings of the fossil record are legion and legendary. These inadequacies have been accepted as sufficient justification for not seeking direct tests of hypotheses in the fossil record and for the failure of certain predicted patterns to emerge from the record's data (34, 79). The published literature on biases in the fossil record now runs to hundreds of papers spanning over a century of research. A review of the shortcomings of the fossil record is beyond the scope of this chapter. Nevertheless, some understanding of the biases is a necessary preliminary to studies of diversity through time.

The fossil record of marine invertebrates is generally superior to that of terrestrial plants or animals. But even the marine record includes only a small fraction of the species that have existed in the geological past. Something between 1 and 10% of extinct skeletogenous animals are represented in the fossil record (123, 139). These are mostly shallow water benthic species that construct durable, heavily mineralized skeletons. In the modern oceans, species that possess heavily mineralized skeletons constitute only a minority of benthic communities. Estimates of the proportion of skeletogenous species in modern marine benthic communities run from approximately 10 to 70%, with a mean near 30% (67, 75, 108). That fraction might have been still lower in the geological past. In the famous Middle Cambrian Burgess Shale, where unique conditions led to the preservation of soft-bodied organisms, only 14% of the known species possess durable skeletons (27). Discoveries of similar faunas in Cambrian sediments of China (58–59, 153), Greenland (28), and the western United States (29, 103) indicate that the Burgess Shale fauna is not atypical of its time.

## *Bias in the Fossil Record*

The fossil record is dominated by a series of nonrandom biases. Studies that employ data based on the record must allow for these shortcomings. Biological processes exert a significant influence on the composition of the fossil record, but their action is often not appreciated. Soft tissues are quickly destroyed by scavengers and decomposers, and are rarely preserved. Mineralized tissues are also subject to attack by a variety of organisms. These processes act from the time of death of a potential fossil (perhaps including the death event, which could eliminate any chance of fossilization) and continue until the potential fossil is destroyed or buried beyond the depth of bacterial or fungal activity. The survival of mineralized tissues depends upon the structure and composition of the skeleton, the time of death, the habitat of death, and the other members of the local community that might damage or destroy the potential fossil (75, 82). In fact, most losses of potential fossils prior to burial and fossilization are the result of biological activity. Even the actual fossilization process is partially dependent upon biologically mediated, syndepositional biogeochemical processes (4, 16). Very rapid fossilization is often a key to preservation of soft tissues (5, 110).

Among the primary physical influences on the fossil record are the patterns and processes of sediment accumulation in the sedimentary record, the total aggregate of sedimentary deposits on earth. These deposits entomb and preserve the fossil record. The shortcomings of the sedimentary record are necessarily overprinted on the history of life. Many environments, especially terrestrial habitats, lack any significant sedimentation or are dominated by erosion, and consequently lack a fossil record. Where sediments do accumulate, their accumulation is unsteady and inconsistent, even capricious, in geological time (2, 3, 105, 106). Some environments occasionally disappear from the sedimentary record while other environments sometimes dominate the global record (e.g. the Cretaceous chalks that occur on most continents in the Northern Hemisphere). Sea level is one of the most important controls on marine sedimentation; changes in sea level control the accumulation of sediment on the continental shelves throughout the world. Sediments deposited in ocean basins are destroyed by subduction; those that survive and are commonly accessible for study are deposited on the continents and continental shelves during periods of high sea stand. Low sea stands produce few or no fossiliferous rocks that survive for future study. Hence, the accumulation of sediments in different parts of the world are not independent events, and one often cannot look elsewhere for the record that is missing in one region. No temporal trends are apparent in the accumulation of sediments; while the accumulation of sediment is certainly not steady in geological time, there is no reason to suppose that the nature of processes that lead to the accumulation of sediment have changed through geologic time.

The survival of fossils, once entombed in sediments, is less than certain. Sediments and their entombed fossils are subject to destruction by erosion and diagenesis (post-depositional mechanical or chemical alteration), and the chances of such loss increase with time and the age of the sediments (17, 48, 49, 94). This loss of sediments through time imposes a strong temporal bias on the quality of the fossil record, with the younger portion of the record providing a more complete history of ancient life (94).

Valentine (141) employed the Pleistocene record of Californian mollusks to calibrate the quality of the fossil record. He discovered that a large majority (77%) of the modern marine Californian species have a Pleistocene fossil record. Species lacking a Pleistocene fossil record are mostly small, fragile, or rare. Also, the Pleistocene record is biased toward shallow water species. These biases notwithstanding, the Pleistocene record of marine mollusks is much better than one would expect from overall estimates of the percentage of skeletogenous species preserved in the record (from 1 to 10%; 123). Valentine (141) concluded that the record, as initially formed, retains a large proportion of the skeletogenous species, but subsequent loss of fossils and sedimentary rock through erosion or diagenesis quickly degrades the quality of the fossil record. His results imply that the potential resolution of the fossil record is very good, and erosion and diagenesis are the primary agents reducing the quality of that record.

## EMPIRICAL PATTERNS: THE MARINE REALM

### *Trends in Global Taxonomic Diversity*

THE FIRST THREE BILLION YEARS The oldest rocks on earth (3.8 billion years) are not much younger than the earth itself (about 4.6 billion years). Stromatolites, laminated sedimentary structures formed by bacterial mats, are found in rocks dated at about 3.6 billion years (78, 149). Fossil cells preserved in cherts also testify to the presence of bacteria and, beginning about 1.6 billion years ago, eukaryotic algae in the Precambrian. Enigmatic fossils such as *Tawuia* (43), interpreted as macroalgae, are known from the late Proterozoic. More importantly, evidence of animal life is absent from the fossil record until about 650 million years ago. There are no undisputed burrows, tracks, or body fossils known until the late Precambrian (23, 33), when the famed Ediacaran (or Vendian) soft-bodied fauna appears in the fossil record (although occasional claims of earlier trace fossils appear in the literature—e.g, 68). The reasons for the apparent delay in the evolution of animals or the processes responsible for triggering the radiation are not understood, although a number of imaginative hypotheses have been proposed (21, 43, 130, 143).

THE CAMBRIAN DIVERSIFICATION Following the appearance of animals, the Metazoa embarked upon a diversification that has not been matched since in tempo or scope. Within a period of no more than a few tens of millions of years, every phylum of skeletogenous invertebrate (with the possible exception of the Bryozoa) and many of the classes and orders known today appeared in the fossil record (21, 43, 112, 130, 143). The numbers of species known from the early Phanerozoic are still relatively small, but the numbers of phyla, classes, and orders are impressive. The number of orders present in the fossil record climbed steadily through the Cambrian and Ordovician, reaching a plateau of between 125 and 140 that was maintained through the remainder of the Phanerozoic (112) (Figure 1).

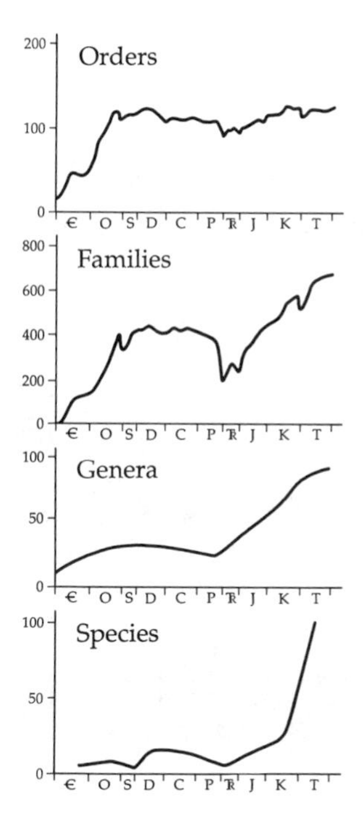


*Figure 1* Marine orders, families, genera, and species through geologic time. Species data are normalized to length of the geologic interval. Data from (11–13, 116, 118, 123)

METAZOAN DIVERSITY IN THE PHANEROZOIC Direct counts of the numbers of animal species from the fossil record show relatively low levels of diversity until the Cenozoic (95) (Figure 2). Beginning in the mid-Cretaceous, the number of marine skeletogenous species increased by an order of magnitude (Figures 1, 2). Indeed, more species are known from the modern world than from the entire fossil record. Accepted at face value, these data suggest a strong Cenozoic increase in biotic diversity. But that approach neglects the biases inherent in the fossil record (50, 94). The loss of older rocks through erosion or subduction, and the destruction of fossils through metamorphism, surely reduces the numbers of species known from more ancient times. There is a strong statistical correlation between the number of species known from the geologic periods and the area (r = .84) and volume (r = .88, N = 10) of sedimentary rocks deposited during those intervals (Figure 2)(96). The distribution of paleontological research effort also correlates with the numbers of species known (r = .94), and with sedimentary area and volume (120). Similar correlations have been obtained for counts of plant taxa and terrestrial sediments (73). These correlations suggest that rock area or volume controls the number of species available for discovery and so are strong support for

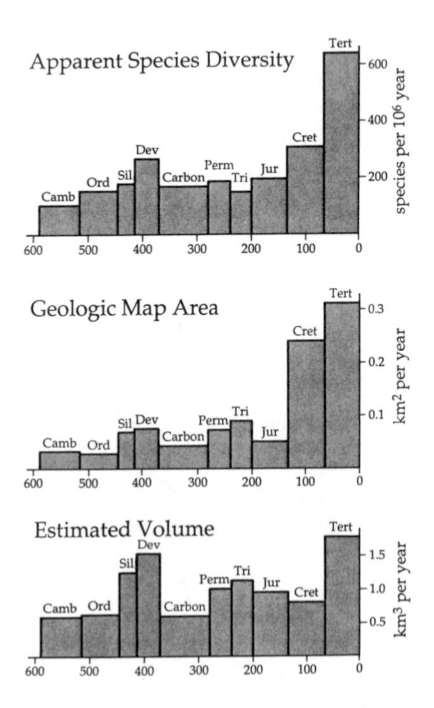


*Figure 2* Marine species, rock area, and rock volume deposited during the geologic periods. Modified from (95, 96).

claims that simple counts of species cannot be interpreted directly as the history of taxonomic diversity (96, 121).

Debates over trends of species richness through time have centered on the quality of the fossil record. Those claiming constant levels of diversity through time (46, 94, 112) have argued that the imperfections of the fossil record prevent direct evaluation of changes in the numbers of species through time; they have sought alternative methods to document diversity in geological time. Among these methods are counts of higher taxa (94, 96, 112). Presumably, higher taxa are less vulnerable to sampling bias because it is (generally) a simpler task to document the presence of an order or a family, rather than each of their constituent species. But this approach presumes that higher taxa have represented roughly constant numbers of subtaxa through time, an assumption that lacks philosophical or substantive justification. In fact, the model number of subtaxa within higher taxa is one (e.g. 6, 112, 145), and classes or orders with large numbers of species are rare. A further assumption is that the higher taxa are, in fact, valid clades. Preliminary efforts to corroborate patterns of diversity using holophyletic taxa have produced distinctly different results (92, 127), although the general consensus holds that incorporation of paraphyletic taxa into tabulations of taxonomic richness involves no systematic biases.

Do higher taxa accurately reflect variations in the number of subtaxa? There are strong arguments to the contrary. The first argument is fairly direct: Patterns of variation at different levels of the Linnaean hierarchy should be consistent if numbers of higher taxa reflect variation at the species level. This is clearly not the case, as Sepkoski (112–113, 116) has demonstrated for orders and families, and Raup & Sepkoski (102) documented for genera (Figure 1). A more complex argument can be found in Raup's (97) estimation of the severity of the mass extinction at the end of the Permian. The reduction in the number of orders was only 16.8%, while families were reduced by 52%. These reductions were generated by a loss of 88% or more of the marine species on earth. Clearly, higher taxa are suboptimal metrics for species diversity, and classes and orders are inferior to families and genera. Niklas has reached similar conclusions in his studies of plant diversity (85). Comparison of the numbers of higher taxa to the standing diversity of species, calculated by Signor (123), demonstrates the decreasing utility of increasingly higher level taxa as metrics for species richness. The numbers of phyla, classes, and orders are fairly constant through the Phanerozoic (112). But lower levels of the taxonomic hierarchy show quite different patterns (123)(Figure 1). The diversification attained a mid-Paleozoic plateau of nearly 500 families that was maintained until the Mesozoic (116). At the species level, the standing diversity of marine skeletogenous species remained low throughout the first 500 million years of metazoan life. But over the past 100 million years, the number of marine skeletogenous species probably increased tenfold.

The second argument in support of equilibrium diversity through geological time comes from a theoretical analysis of clade "shape." Gould et al (45–6) employed simulations of evolving lineages to demonstrate that evolving clades within a system at equilibrium tend to have a symmetrical shape—Center of Gravity (CG) = .5, while clades in an expanding (pre-equilibrium) universe are bottom-heavy (reach their peak diversity prior to the midpoint of their temporal duration (CG<.5)). Applying this clade "statistic" to the fossil record, they demonstrated that Cambro-Ordovician clades of marine invertebrates have a mean CG of .482 while post-Ordovician clades have an average CG of .499. This result was confirmed through analysis of mammalian clades, which are bottom heavy early in the Cenozoic (CG = .474 during the radiation of mammals) but have CGs of .508 later in the Cenozoic.

Predictions about the behavior of CG are predicted on the assumption that rates of origination and extinction are stochastically constant and equal. However, Gilinsky & Bambach (41) have demonstrated that there are systematic variations in rates of origination and extinction within clades in the fossil record. Rates of origination tend to decline while rates of extinction generally increase with time. Whether this alters the shape of simulated clades remains to be demonstrated. In addition, Kitchell & MacLeod have found that the values of CG documented in the fossil record fall within the range of variation observed among randomly generated clades. Their results suggest that the small differences in CG observed by Gould et al may be meaningless.

The recent debate on diversity patterns of marine organisms closed in 1981 with the publication of a consensus paper coauthored by the principles in the controversy. Sepkoski et al (118) examined parallel trends in five different data sets: animal families and genera, species per million years, trace fossils, and within-habitat diversity (Figure 3). After accounting for the temporal autocorrelations present in each data set, each of the five data sets are still strongly intercorrelated. Sepkoski et al (118) concluded that the underlying similarity between these different data sets represents a powerful primary signal of diversity change through time. It isn't clear why these authors were convinced at the time that the pattern was real, and not an equally powerful artifact of the fossil record's biases, as several had previously argued (46, 94, 96, 112). Later work has tended to confirm their conclusion (122, 123).

CONTROLS ON GLOBAL DIVERSITY The controls on global diversity include both processes acting at global scales and those limited in their action to individual communities or regions. In the latter case, any processes that modify diversity at local or regional scales must necessarily affect global diversity, unless the effects are somehow offset by corresponding coun-

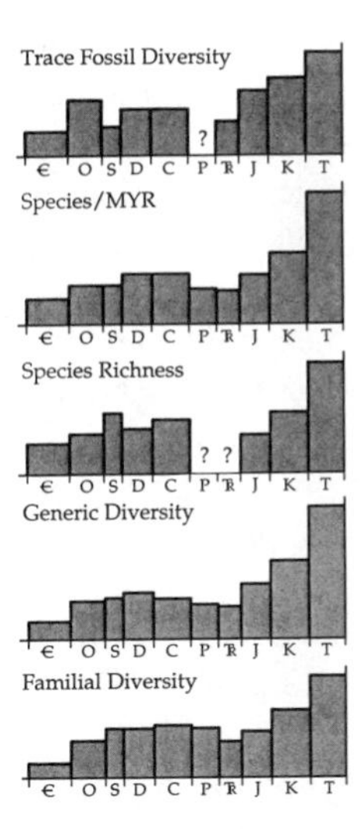


*Figure 3* Numbers of trace fossils, species, within-habitat species richness, generic diversity and familial diversity in the Phanerozoic. Data from Sepkoski et al (118).

tereffects operating at larger scales. Processes acting at more limited scales are addressed in a later section. Processes that act at global scales include changing levels of provinciality (mediated by continental drift), changing climates, variation in sea level, and mass extinctions.

The dominant factor determining long-term variation in the taxonomic diversity of marine organisms is the arrangement of continental land masses on the earth's surface (109, 123, 142, 144). When the continents are gathered together into a single land mass, as at the close of the Paleozoic Era, there are relatively few faunal provinces. In contrast, large numbers of provinces form when the land masses are widely dispersed. Land masses restrict the ability of marine organisms to disperse or migrate, depending upon their position, especially when the continents interrupt the flow of east/west currents. Similarly, dispersed land masses allow the independent evolution of terrestrial faunas and floras, such as the unique biotas of Australia or Madagasgar (or South America prior to the formation of the Central American land bridge) (35, 126).

Estimates of marine faunal diversity track the number of marine faunal provinces through time (142). The number of provinces was uniformly low during the Paleozoic and early Mesozoic, and spurted to a peak in the Recent (Figure 4). The additional faunal provinces multiply the numbers of shelf-dwelling benthic species, yielding much larger numbers of species without necessarily increasing the numbers of taxa within individual communities or regional (beta) diversity. Obviously, estimates based on the numbers of provinces are crude at best, given the striking variation in taxonomic diversity among the modern marine provinces and the errors inherent in counting the numbers of ancient provinces. Also, the increase in provinciality is less than the resulting increase in Cenozoic species richness, so it is difficult to attribute the entire change to plate tectonics. But a large share of the striking increase in provinciality among marine faunas in the past 100 m.y. can reasonably be ascribed to the action of continental drift and the developing configuration of continents on the surface of the earth (109, 123, 142, 144).

Climatic variation is generally not recognized as a process responsible for long-term variation in biotic diversity, but global refrigeration is frequently cited as a factor, or possibly the primary cause, in mass extinctions (30, 89, 132, 133). Tropical shelf faunas, including reef and reef-associated communities, appear more vulnerable to mass extinctions than are deep-dwelling or temperate faunas (83, 84, 132, 133, but see 99). These tropical faunas typically include very diverse communities, and their loss has a proportionally large impact on the diversity of the biosphere. However, Jablonski (62) rejects this hypothesis, arguing that not all extinctions correspond to periods of cooling and that the relatively recent Quaternary glaciations are matched by only minor extinctions in the fossil record (but see 132, 133).

It seems likely that long-term cooling or, in the extreme, refrigeration of the earth would have a deleterious effect on diversity (53, 54, 83, 84). Cooling would reduce rates of productivity, reduce population sizes, and increase rates of extinction. Cooler temperatures and reduced rates of photosynthesis would likely preclude the formation of tropical organic reefs. Stanley (132, 133) has argued that cooling will not harm temperate faunas, as they can migrate to follow their shifting habitat, but tropical faunas have no geographic escape and will become extinct.

Hansen (53, 54) demonstrated a general correspondence between thermal

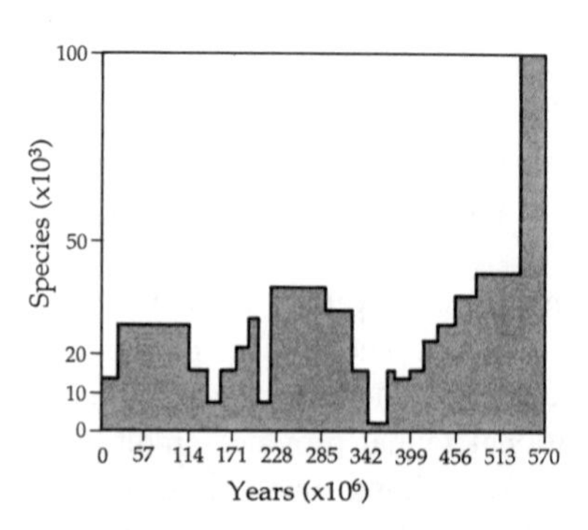


*Figure 4* Changing numbers of species estimated from the numbers of marine faunal provinces during the Phanerozoic (142).

trends and the diversity of early Cenozoic benthic mollusks on the US Gulf coast. The benthic fauna was generally species-poor through the early Cenozoic, slowly rebounding from the Cretaceous-Tertiary extinction (Figure 5). Local diversity maxima were reached concurrently with thermal maxima. However, each subsequent diversity peak was higher than the previous one, while thermal peaks showed no such trend. No long-term increase in global temperature occurred concurrent with the Cretaceous-Cenozoic diversification of marine species. An alternative scenario suggests that diversity maxima would be obtained when strong global thermal gradients lead to the formation of thermally restricted provinces (138). Global, warm seas would produce few provinces and a lower global diversity. Debates about the potential effects of climatic variation aside, long-term thermal trends that might drive diversity change, especially the Cretaceous-Cenozoic diversification, remain to be demonstrated.

Sea level is another factor that might plausibly affect global marine diversity (e.g. 2, 38, 51, 83, 84, 111, 152). The large majority of marine species known from the fossil record inhabited the continental shelves. A relatively small drop in sea level, on the order of a hundred meters, significantly reduces the shelf area flooded by the oceans. In turn, the reduction in shelf area would reduce the area available for habitation by benthic communities, thus reducing the standing crop and increasing rates of extinction. In many instances, the number of taxa known from the fossil record tracks the area flooded by the sea. The diversities of Cambrian trilobites (22), Mesozoic ammonites (12, 52, 69), Mesozoic brachiopods (52), and other taxa, correlate with the submerged areas of the continents inhabited by these groups. Global species richness through time also correlates with the flooded area of the continents, although that correlation is not strong (40, 111). This effect will be compounded by

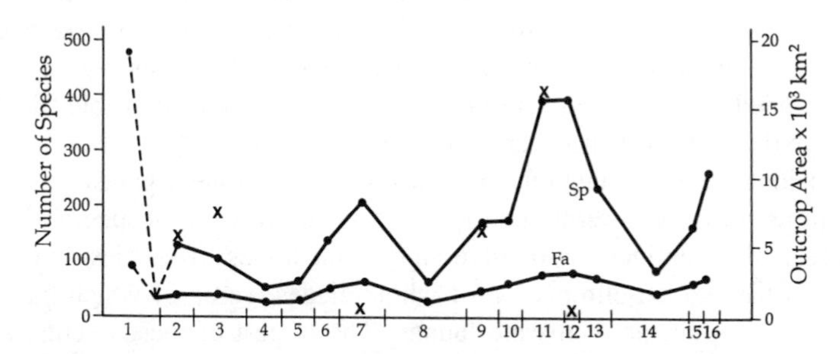


*Figure 5* Rebounding molluscan diversity along the gulf coastal plain of North America following the Cretaceous-Tertiary mass extinction (53, 54). Large "X" indicate outcrop areas of selected fossiliferous formations.

sampling biases, whereby stratigraphic intervals represented by more broadly exposed sediments will yield more fossil taxa and appear more diverse (60).

The long-term consequences of variation in sea level are uncertain. While lower sea levels reduce the flooded area on the continental shelves, they actually increase the shoreline and submerged shallow areas of oceanic islands (40, 61, 62, 131). Oceanic islands in the Indo-Pacific region support some of the most diverse marine faunas known today, and more than 85% of the extant families of shallow water marine mollusks have representatives in these faunas (61). Other clades are represented in oceanic faunas by a somewhat lower percentage, but in no case was the percentage less than 70. These faunas should be immune to the effects of lowered sea level, and the families and higher taxa represented among the island-dwelling faunas should survive. Over the longer term, dynamic fluctuations in sea level have proven to be common (e.g. 55), but no long-term increases in sea level have been recognized. Thus, variation in shelf area might influence short-term fluctuations in diversity but could not be responsible for the major changes in diversity through time.

A sudden and dramatic loss of shelf area is a possible cause of the great Permo-Triassic mass extinction (107, 125), and some of the other extinctions occurred during regressions (2). However, not all regressions correspond with extinctions (53, 54). A late Oligocene low stand of sea level does not correspond to a significant extinction event, and a late Eocene extinction event was not paired to a regression. Furthermore, the broad distribution of marine shallow water families demonstrates that a reduction in sea level would not result in mass extinction (61, 62, 64, 152).

It has been suggested that the effects of temperature and sea level could combine to produce pronounced extinctions (e.g. 90). Fluctuations in sea-level and temperature are very likely linked as cause and effect. A rise in sea-level, flooding the continental lowlands, would reduce the Earth's albedo and cause a general warming. A regression would induce a reduction in the mean surface temperature. Therefore, the effects of these two processes are probably inseparable (54). Vermeij (147), among others, has suggested that the combined effects of temperature and regression are responsible for Pleistocene extinctions of Caribbean mollusks.

Occasional severe reductions in the diversity of marine organisms, or mass extinctions, have long-term consequences in the marine biosphere (36, 54, 98, 116). Mass extinctions involve significant losses of species and higher taxa, and the sudden elimination of whole clades, some of which had been important components of marine faunas. The largest of these events, at the end of the Permian, included the extinction of 13.5% of the classes, 16.8% of the orders, 52% of the families, 64.8% of the genera, and as many as 96% of the marine species on Earth (97). Recovery from mass extinctions is prolonged, requiring ten million years or more depending upon the severity of the extinction (36, 54, 98, 116).

Hansen's (54) study of the recovery of marine mollusk diversity following the Cretaceous-Tertiary mass extinction indicates that molluscan diversity rebounded rapidly following the extinction, but the rate of origination quickly slowed. Diversity had not regained Late Cretaceous levels by the time of the next extinction event, in the Late Eocene (Figure 5). These results suggest a dynamic balance between diversification and mass extinction (54), with diversity varying in response to several physical processes (continental drift, temperature, sea-level) and subject to occasional resetting by mass extinctions.

There is no reason to suppose that diversity is controlled by any single process, although we are bound by scientific convention first to seek simple explanations. Hansen's results suggest it is the interaction of several physical processes that influences global diversity, a view shared by Ager (3), Cracraft (31), Officer et al (90), and others. Biological processes undoubtedly play a role also, perhaps through the evolution and increasing dominance of species-rich clades (e.g. bivalves, gastropods—148).

## *Trends in Within-Habitat Diversity*

HISTORICAL PATTERNS The numbers of species occurring within ancient faunal assemblages have increased episodically through time (9). Nearshore, physically stressed benthic communities include approximately constant numbers of constituent species through time, but the number of species doubled in Mesozoic communities inhabiting open marine environments. Benthic communities occupying intermediate, physically variable environments showed intermediate increases in diversity (Figure 6).

The significance of this pattern is the subject of some dispute (13, 56, 57). There are several systematic biases in the fossil record that could yield the observed result, and there is insufficient evidence at the present time to discard alternative hypotheses invoking these biases. Comparisons among tabulations of the numbers of skeletogenous species within fossil marine benthic communities presume that the proportion of skeletogenous species within such communities has remained constant through time (56, 57). However, some evidence indicates that this is not true. The increase in diversity noted by Bambach occurs coincident with the great radiation of higher gastropods (128, 134). These advanced snails are primarily predators, a trophic group largely missing from earlier fossil communities. In modern (108) and Cambrian (27) fossil communities, the soft-bodied component of the fauna includes disproportionate numbers of predators. Furthermore, Hoffman (57) contends that the Bambach study lacks controls on the numbers and types of benthic communities included in the study. Are all

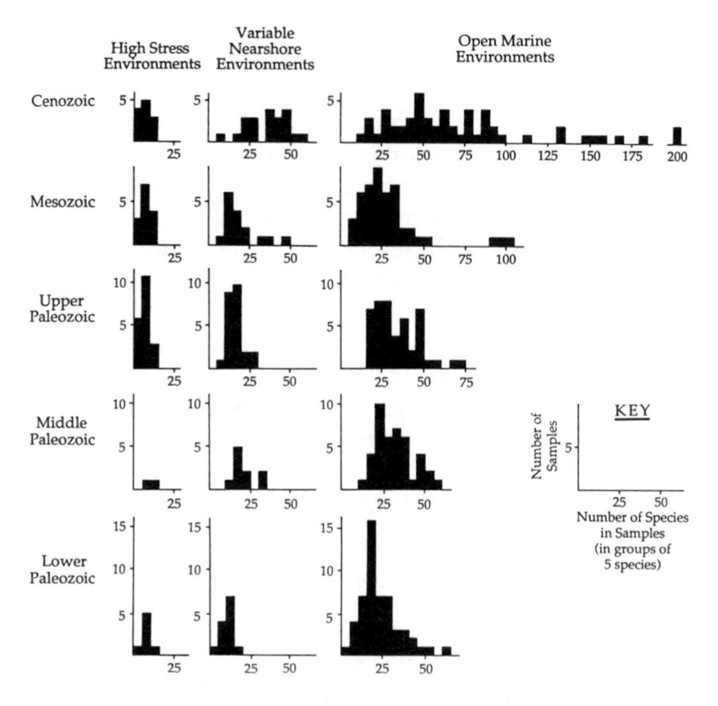


*Figure 6* Within-habitat species richness for marine benthic invertebrates (9).

open water marine communities, for example, sufficiently similar to justify such straightforward comparison?

One bias that could account for the increase in within-habitat diversity results from variable sample sizes. Bambach's compilation of within-habitat species richness through time is derived from a large number of studies of community paleoecology, the work of many researchers over more than 15 years. There is no standardization of sampling design or effort among these studies. Whether this might lead to any systematic bias in Bambach's data is unknown. A closely related issue is time-dependent variation in the preservation of fossils. Prior to the Late Cretaceous, fossils preserved in unconsolidated sediments are uncommon. The increase in within-habitat species richness documented by Bambach corresponds to the first occurrence of well-preserved fossils in unconsolidated sediments, especially the diverse faunas from the Gulf Coast region of North America. The absence of lithification greatly simplifies collection and preparation of fossils, thus increasing the efficiency of sampling. Another possible bias is taphonomic in nature. Behrensmeyer & Kidwell (13) suggest that the increase in species richness might be related to post-mortem mixing of separate communities in complex shell beds. These and other (57) potential biases notwithstanding, the Cretaceous increase in within-habitat diversity is generally accepted as a genuine biological phenomenon (118, 123, 124, 147, 148).

## *Mechanisms of Change*

SPECIES PACKING WITHIN COMMUNITIES It is difficult to envision biological processes that could lead to an episodic doubling of within-habitat species richness in some habitats while maintaining constant numbers of species in others. One hypothesis proposed to account for the pattern is that the number of species increased through occupation of previously unfilled roles within ancient communities [Bambach (10) employs the term "guilds"] (10, 11). These guilds are defined by trophic role within the community, by the species' utilization of space, and by the general biological characteristics of the species. Surprisingly, Bambach (10) discovered no increase in resource partitioning through time (the number of species within guilds in each community); the increase in diversity resulted solely from the invention of new guilds. But the processes controlling the number of guilds within communities, and the episodic nature of that increase, are not well understood.

TIERING The spatial structure of marine benthic communities has changed over geologic time (7, 8, 18). Suspension-feeding invertebrates have increased the maximum depth of dwelling burrows below the sediment surface through time, with the major increase in depth coming at the end of the Paleozoic. The maximum height of epifaunal suspension-feeders has also varied through time, reaching peaks in the mid-Paleozoic and mid-Mesozoic. Thus, the ranges of tiers above and below the sediment-water interface has varied through time in a nonsystematic fashion and shows a general increase in spatial complexity (Figure 7). Ausich & Bottjer (7, 8, 18) analogized this change to the vertical development of forest communities, although the mechanism responsible for variation in tiering through time remains to be demonstrated.

Ausich & Bottjer (7, 8, 18) argue that variations in tiering, or increasing spatial complexity of benthic communities, was responsible, in part, for changing global species richness through time. Variation in tiering is very much a community-level phenomenon and should be reflected in Bambach's compilation of within-habitat diversity. Comparison of Figures 6 and 7 indicates no such correspondence, in part because the increase in burrow depth primarily reflects activity by soft-bodied organisms that are generally not preserved in the fossil record. Nevertheless, variations in tiering through time are unmistakable to those familiar with the fossil record, and provide opportunities for increasing within-habit at diversity.

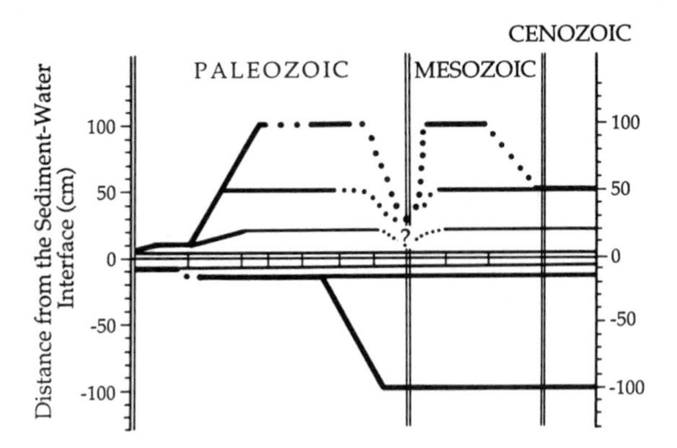


*Figure 7* Tiering, or vertical expansion of marine benthic communities above and below the sediment-water interface through geologic time (7, 8, 18).

ROLE OF MASS EXTINCTIONS The most interesting aspect of mass extinctions, besides the remarkable loss of taxa, is the environmental breadth of the extinctions. Diversity in terrestrial and marine animal communities alike was severely reduced by the mass extinction at the end of the Cretaceous Period, 65 million years ago. Perhaps as many as 95% of all species on earth became extinct during the mass extinction at the end of the Permian Period (97). These extinctions severely reduced diversity at the community level (53, 54, 80), although there is an ongoing debate regarding the relative influence of extinctions on different types of marine communities (e.g., 99, 132, 133). Recovery to preextinction levels of diversity following mass extinction is a prolonged process requiring millions of years (e.g. 54). Hansen found the recovery time to exceed the waiting time to the next extinction, indicating that mass extinctions might control maximum diversity within marine benthic communities.

## *Trends in Regional Species Diversity*

A third pathway to increased diversity is intermediate between increasing the numbers of species within communities and increased global diversity. Diversity can increase by decreasing the similarity between communities within a province (e.g. 150). The possibility of temporal variation in the differentiation of faunal assemblages, or beta diversity, has only recently been evaluated by Sepkoski (117). Sepkoski compiled data on the environmental distribution of genera through the Paleozoic. He employed a six-fold classification of marine benthic habitats, classifying them as (*a*) peritidal, (*b*) nearshore protected shallow low-energy, (*c*) offshore high-energy, (*d*) shallow open-shelf, (*e*) deep open shelf, and (*f*) deepwater slope/basin deposits (note that reefs and mounds were excluded from the data base). Environments were classified by sedimentological criteria, to avoid circularity in interpretation. Faunal similarity was calculated with Jaccard's Coefficient of Similarity (66).

Sepkoski's data indicate that within-habitat diversity doubled from the Cambrian to the Ordovician, and remained relatively steady thereafter; this partially corroborates Bambach's earlier results. Beta diversity also increased subsequent to the Cambrian and remained approximately steady thereafter. Among the six environments, Sepkoski found the greatest similarity between offshore high-energy and shallow open-shelf environments, with faunal similarity declining both seaward and landward.

The increase in alpha diversity only accounts for about a sixth of the early Paleozoic increase in global diversity, and changing beta diversity accounts for another sixth (117). Sepkoski envisions no concurrent changes in provinciality, nor do Valentine et al (142) find any evidence of early Paleozoic increases in provinciality. Sepkoski (117) suggests that the remaining increase in global diversity might come from evolutionary change not incorporated in his data base (e.g. the evolution of reef communities).

CONTROLS ON BETA DIVERSITY The processes that control beta diversity in geologic time are not known. Sepkoski (117), following Valentine (140), suggests that the variation in beta diversity results from increased ecological specialization of marine benthic species. A subjective, but commonly held, interpretation of Cambrian faunas is that they are simple in structure and broadly distributed across many environments (117, 140). These early faunas are replaced by more specialized, and ecologically restricted, faunas beginning in the Ordovician. Specialization of marine arthropods, as inferred from limb morphology, has increased through time and showed the strongest increase through the Paleozoic (26, 39). However, these patterns have not been rigorously substantiated for other clades. Post-Paleozoic patterns of beta diversity have not been documented, and the contribution of beta diversity to the Cretaceous-Cenozoic diversification is not known.

## EMPIRICAL PATTERNS: THE TERRESTRIAL REALM

### *Vascular Plants*

The history of plant diversity has been examined recently in a series of excellent publications on the taxonomic history of floras of the northern hemisphere (71–73, 76, 85–88, 136, 137, 151), which are summarized here. Debates about the age of the first vascular plants linger, with most paleobotanists citing a Silurian age for the first plants, but a minority finding convincing evidence for terrestrial vascular plants in the Late Ordovician (47). Regardless of the precise time of origin, the diversification of terrestrial

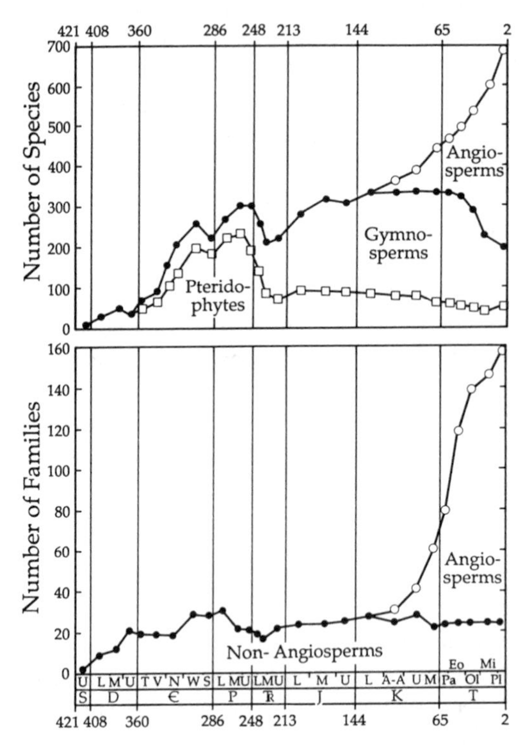


*Figure 8* Taxonomic richness of vascular plants, in the Northern Hemisphere, through geologic time (85–8). The diagram is subdivided to indicate the relative contribution of different clades to the overall diversity.

plants was well underway by the Late Silurian (73, 86–88) (Figure 8). The Devonian appearance of seed plants accelerated the diversification to a Late Devonian peak of over 40 genera. After a slight decline, the diversification resumed in the Carboniferous, and the number of species quickly increased fivefold, to over 200. Arborescent lycopods constitute the largest portion of this increase, although ferns and other groups constitute a significant part of the new flora (86, 88).

The number of plant species increased only gradually between the mid-Carboniferous and the end of the Permian. There was a minor reduction (20%) in diversity at the end of the Permian, followed by a rapid rebound to pre-Mesozoic levels. Diversity then resumed its slow rate of increase, reaching approximately 250 species in the Early Cretaceous. However, profound changes in the composition of the flora underlie this minor trend in diversity. The previously dominant groups became extinct or were severely reduced in diversity and were replaced by coniferophytes, cycads, and cycadeoids, which dominated the plant biotas until the middle Cretaceous (88).

The final phase of plant diversification opened in the mid-Cretaceous, with the appearance of angiosperms. Angiosperms began to diversify, slowly at first but more rapidly in the Cenozoic, and they now constitute more than 80% of the plant species on earth (76, 88). The angiosperms have the highest rate of diversification of the major plant groups and appear immune to extinction events. Furthermore, there is no evidence of an equilibrium plant diversity in geologic time.

How are the increased numbers of plant species accommodated? Niklas et al (86) and Knoll (71) found that the average number of species in plant floras, or alpha diversity, increased episodically through time. Within-flora species richness doubled in the Early Carboniferous and again in the Cretaceous (85, 88). But the magnitude of the change in alpha diversity is insufficient to account for the overall increase in global diversity. The same authors document striking increases in angiosperm gamma diversity through the Cretaceous and Cenozoic, suggesting that there are significant components of diversification at other levels contributing to the overall pattern.

The nature of processes controlling plant diversity through geologic time has not received the same attention devoted to the marine fauna, but recent work by Tiffney & Niklas (137) has produced interesting results. In particular, poor correlations between the dispersal of land masses and plant diversity indicate that changing biogeographic provinciality is not a significant component of plant diversity, in striking contrast to the marine fauna. Plant diversity correlates strongly with land area and uplands area (137), but it is possible that these correlations reflect sampling biases (73, 86, 94). However, Tiffney & Niklas (137) observe that the correlation between upland area, which is poorly represented in the fossil record, and plant diversity cannot result from sampling bias alone. In addition, the increases in within-flora diversity through time cannot be explained by variation in land area (72, 137).

The role of evolutionary innovation cannot be overlooked in the history of plant diversity (72, 85a, 86–88, 136, 151). The evolution of new modes of reproduction (seeds, flowers) and other innovations (wood) surely have enhanced the capacity of plants to invade new habitats and increase the numbers of species within plant communities (85a, 86–88, 136). Niklas (85a) concludes that the increases in plant diversity through time can be explained largely through innovations within plants (mainly in reproductive organs—flowers and seeds) and geological processes (tectonics).

## *Terrestrial Animals*

In terms of biological diversity, the history of terrestrial animals is the history of insects and arachnids. However, biases of the fossil record and human interest conspire against us, and the published literature is not large. The earliest insects are Carboniferous in age (24), and advanced insects (Neoptera) are known from the Late Carboniferous. Carpenter & Burnham (24) estimate that by the mid-Tertiary about 60% of the insect genera were extant. Niklas (85) presents an estimate of the numbers of insect genera through geologic time that shows relatively few genera in the Paleozoic and Mesozoic, followed by a post-Mesozoic doubling of diversity (Figure 9). The Cenozoic increase in insect diversity coincides with the radiation of angiosperms, and Niklas (85) envisions a coevolutionary relationship between the two clades.

Perhaps unsurprisingly, the fossil record of terrestrial vertebrates has received considerably more attention (e.g. 14, 15, 25, 42, 77, 85, 91, 135), and the diversity patterns of nonmarine tetrapods are relatively well documented. The fossil record of birds is much less well known, probably as a result of their light skeletons (91).

The fossil record of terrestrial vertebrates began in the Late Devonian; Paleozoic diversity was never large (>50 families) and declined during the early Mesozoic. Beginning in the mid-Cretaceous, the number of nonmarine families began to increase rapidly to a Recent peak of approximately 340 (Figure 9). The generic pattern mimics the familial trend but is much more exaggerated (91). There is a late Paleozoic peak in the numbers of genera (approximately 310/myr), followed by an early Mesozoic downturn and a Cretaceous-Cenozoic increase to modern levels. Excepting the Permian peak, the trend for terrestrial vertebrates is very similar to the diversity trends of marine animals and plants.

Changing patterns of geographic isolation, driven by continental drift, have unquestionably played a major role in forming global patterns of vertebrate

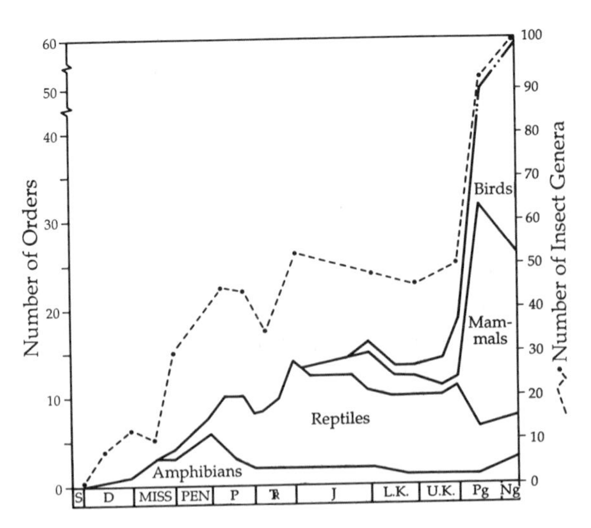


*Figure 9* Diversity of terrestrial animals through geologic time. Insect data from (85), vertebrate data from (14, 15, 85, 91).

diversity (15, 35, 37, 74, 85, 91, 126). Each land mass developed its own indigenous fauna, some coming to be dominated by placentals and others by marsupials. The opening of migration routes that foster faunal interchanges, such as across the Bering Strait or the Isthmus of Panama, has resulted in occasional reductions in the numbers of species and higher taxa (37). Also, the morphological diversification of angiosperms probably fostered a parallel diversification among terrestrial animals (151).

## CENOZOIC INCREASE IN DIVERSITY: A COMMON PATTERN?

The tabulations of plant and animal diversity through time, presented above, document a spectacular diversification over the past 100 million years. There were more species and higher taxa extant in the Pliocene and Pleistocene than at any time in the past, and the trend, at least until the intervention of climatic change and organized human activity, was toward still more species (118, 122, 123, 139, 142). The striking similarity of diversity patterns among terrestrial and marine organisms, particularly the Cretaceous-Cenozoic diversification, is a salient feature of the fossil record. Can the patterns be trusted, or does the problem of bias in the fossil record still linger here? The asynchronous increases in marine metazoan and terrestrial plant alpha diversities suggest that the patterns are not entirely artifactual, although the potential biases introduced by the fossil record cannot be completely discounted. If the patterns are a biological signal, and not artifactual, why would diversity trends from the two realms be so similar? Is there a common causation or a causal link between the realms? The answers are elusive, but the parallel trends in the diversities of marine animals and terrestrial arthropods, vertebrates, and plants suggest a common underlying process that is global in its action.

## CONTROLS ON GLOBAL DIVERSIFICATION

### *Plate Tectonics*

The single process that is global in scope and historically corresponds in activity to patterns of diversity is continental drift. Its impact on the development of provincial floral and faunal distributions is undisputed. There is an extensive body of evidence derived from the phylogenetic relationships of faunas inhabiting the different continents, particularly for terrestrial vertebrates, indicating that plate tectonics has a profound impact on biological diversity. Also, few other processes act on such a broad geographic scale. Cracraft (31) has made a general case for the importance of lithospheric complexity as a control on diversification. With the exception of terrestrial plants (137), most authors see a primary role for plate tectonics in controlling diversity through time. And, tectonics controls the numbers of uplands and environmental complexity, cited by Tiffney & Niklas (137) as the factors best correlated with plant diversity.

### *Mass Extinctions*

Mass extinctions impact both the terrestrial and marine realms, reducing diversity and eliminating many clades (98). As noted previously, the recovery of faunas following extinction events can extend over geologically significant intervals (54, 116). While mass extinctions reduce the number of extant taxa, they do not engender diversification (beyond the expected post-extinction rebound). The role of extinctions in diversification must be as a second order effect, perhaps by eliminating some clades to allow diversification of others. Mass extinctions are also opportunities for the study of diversity. They are relatively brief intervals of time in which potential controls on diversity such as sea level or climate can be studied and evaluated (e.g. 54, 132, 133).

### *Escalation*

Vermeij (148) hypothesized that escalation, or predator-prey interactions, has led to biological diversification over geological time. Greatly simplified, his argument suggests that the evolution of a predator provides evolutionary opportunities for predator avoidance. As prey adopt different strategies for avoiding predation, they create new opportunities for evolving predators. This creates an evolutionary feedback leading to sustained diversification. The increasing prevalence in geological time of morphological features associated with resisting or avoiding predation among marine invertebrates (Vermeij, 146–148) and the apparent coevolution of plants and insects (85) lend considerable support to this view.

## THREE GREAT MARINE FAUNAS

The fossil record is a loose composite of enormously complex and voluminous data of varying accuracy that has accumulated over several centuries of scientific inquiry. The possible existence of patterns or simplifying themes within the record has been an important theme in paleontological research. Sepkoski (114) applied a Q-mode factor analysis to his data base on marine families (115) to search for structure within the composite record of marine families. His approach followed that employed by Flessa & Imbrie (38), and their results are somewhat similar, although they employed different data bases. The factor analysis reveals that just three factors account for more than 90% of the data.

The first of these factors is heavily loaded on early families, including trilobites, primitive mollusks, inarticulate brachiopods, and a few other taxa. Trilobites are the largest contributors to the first factor, which is not surprising given that trilobites include 75% of all described Cambrian species (95). Sepkoski interpreted this first factor to represent a Cambrian Fauna. The second factor is heavily loaded on Paleozoic families, including articulate brachiopods, ostracodes, cephalopods, rugose and tabulate corals, and other typically Paleozoic taxa. This group of taxa was designated the Paleozoic fauna (Figure 10). The third factor loads heavily on gastropods, bivalves, osteichthian fish, crustaceans, and echinoids and other taxa characteristic of the modern world. Sepkoski labeled this last association the Modern Fauna. On the basis of these results, Sepkoski (114:44) suggests "there is a fundamental simplicity to all the faunal change we see in the fossil record; this seems true in spite of the almost chaotic variation initially apparent. . . ."

Sepkoski provided no critical test of his results; could a factor analysis force pattern where none exists? In a subsequent paper, Kitchell & MacLeod (70) developed an appropriate null model. They generated large numbers of random clades and assembled them into an artificial record. Application of a factor analysis to the artificial record failed to generate factors with the relatively even loadings produced in Sepkoski's study. Kitchell & MacLeod (70) conclude that the null model fails to replicate the patterns produced in Sepkoski's work.

## *Ecological Integrity of Faunas*

Quantitative analyses revealed structure in the composite stratigraphic record of marine families but not the ecological or evolutionary significance of that

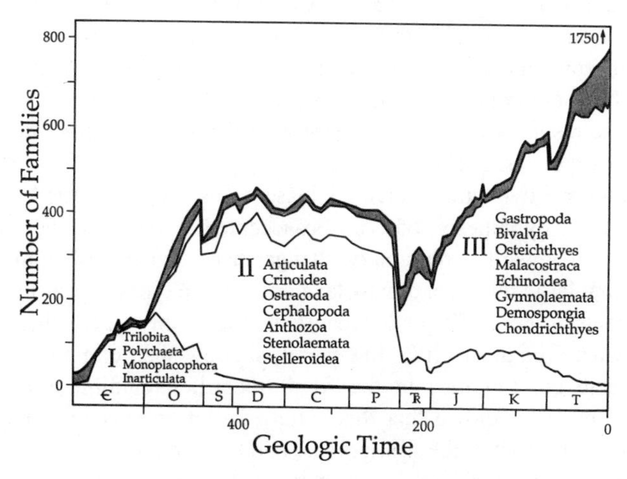


*Figure 10* Three great faunas in geologic time and the major components of each fauna (116).

structure. To examine this aspect of Phanerozoic evolutionary history, Sepkoski & Sheehan (119) compiled reports of the paleoecological distribution of orders within Paleozoic communities. They assigned each community to a five-fold classification of marine environments on the basis of sedimentological criteria, to avoid circularity. Then, applying cluster analysis to the communities, they were able to show that the Cambrian Fauna initially dominated all benthic environments, but was quickly displaced from (or replaced in) nearshore environments by communities dominated by the Paleozoic Fauna. In turn, the Paleozoic Fauna was later displaced (or replaced) from nearshore environments by the Modern Fauna. These results indicate that the communities of the successive faunas were formed in nearshore environments and subsequently expanded into offshore habitats. Furthermore, they indicate an ecological and evolutionary coherence to the three Great Faunas that has persisted over geological time (119).

### *Onshore-Offshore Patterns of Origination*

Jablonski & Bottjer (63) observed that adaptive types new to the fossil record tended to appear first in nearshore environments. Antecedents of this observation can be found in the literature (e.g. trace fossils, 19, 32; bivalve mollusks, 129), but the generality of the pattern had not been appreciated. Joining with Sepkoski & Sheehan, they (65) proposed that significant evolutionary innovations tended to appear first in nearshore environments. They were unable to demonstrate the mechanism that would generate such a pattern, but subsequent work has tended to support their conclusions (20).

Mount & Signor (81) examined the habitats of first occurrence of marine families during the Cambrian radiation of Metazoa. Working in the late Precambrian and Early Cambrian of western North America, they demonstrated a strong statistical bias toward first occurrences in shallow, subtidal marine environments that lay below fairweather wave base. While this is not the nearshore environment of Jablonski et al (65) or Sepkoski & Sheehan (119), these results are not inconsistent with the general pattern documented by those workers (see 20).

Nevertheless, two other potential biases must be discounted before the Jablonski et al hypothesis (65) can be accepted. First, there is a statistical predisposition toward evolutionary change in particular environments (20). To show that large numbers of new taxa or adaptive types appear in nearshore environments is not sufficient, because it might only prove that evolution happens where organisms live. In general, shallow shelf communities are the most diverse communities in the fossil record, excluding tropical reefs. Second, it must be shown that this pattern is not another bias of the fossil record. As noted above, the stratigraphic and fossil record consist primarily of shallow water deposits laid down on the continental shelves or in epicontinental seas. Valentine (141) found the Pleistocene record of California to be biased toward shallow water species. In this situation, first appearances in the fossil record would be expected to occur in shallow water sediments, regardless of the actual habitat of origination. Further work is needed to resolve these doubts.

## MODELING DIVERSITY THROUGH TIME

### *Logistic Model*

Sepkoski (112) proposed that a kinetic model of Phanerozoic diversity is consistent with the observed pattern of variation of marine orders through time. His kinetic model, the well-known logistic model for population growth, also assumes that diversification is initially exponential but that rates of origination subsequently decline to balance rates of extinction, thus maintaining an equilibrium diversity. That the model produces patterns so similar to the history of ordinal diversity suggests some sort of diversity-dependent equilibrium following a period of exponential diversification. Several criticisms of this model have been raised. The most important criticism is cladistic; the orders employed by Sepkoski are largely paraphyletic; holophyletic units might produce other patterns. Also, the numbers of orders through time do not parallel the numbers of families, genera, or species through time (123), leaving open the question of what changing numbers of orders might represent. Lastly, there is no convincing evidence that the rates of ordinal origination are diversity dependent, as implied in the model (57). Cracraft (31) concludes that speciation is probably diversity-independent over geological time.

MARINE ORDERS Sepkoski's compilation of the numbers of marine orders through time matches the predictions of the logistic model rather well, showing a rapid increase in the numbers of orders in the late Precambrian and Cambrian followed by approximately constant numbers of orders through the remainder of the Phanerozoic (Figure 11). Detailed examination of the radiation of orders early in the Phanerozoic confirmed the exponential character of the diversification. However, more recent work indicates that the pattern cannot be explained as a simple exponential diversification (143).

MARINE FAMILIES The history of familial diversity is indeed complex and cannot be described by a simple logistic model. Sepkoski (113) introduced a more complex version of the kinetic model, similar in structure to logistic models of two- and three-population competition. He differentiated two populations of families, a Cambrian fauna and a Paleozoic fauna (these

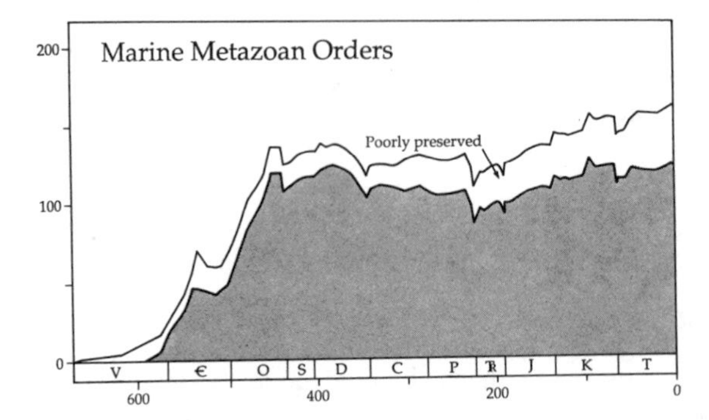


*Figure 11* Sepkoski's (112–3, 116) kinetic model of taxonomic diversification. The figure shows the model's prediction and the corresponding change in numbers of orders through time (112).

faunas are similar but are not identical to the faunas defined in Sepkoski's later factor analytic study of the fossil record). The Cambrian families radiate quickly but fail to sustain their radiations and are quickly overhauled by the Paleozoic fauna. Sepkoski (116) later extended this study with a third population of families, the Modern Fauna. The three-phase model provides an excellent description of the observed pattern of Phanerozoic familial diversity.

Hoffman (57) has criticized Sepkoski's models, arguing that the underlying assumptions of equilibrium and competition among populations of families are unsupported and inappropriate. Furthermore, Hoffman argues that families are insensitive to changing levels of species richness. This latter criticism is undoubtedly valid, as the documented trends in numbers of marine families and species through time are not closely similar (123). It is difficult to envision a macroevolutionary version of competition between populations of families, regardless of the apparent ecological integration of the Three Great Faunas (114).

## SUMMARY

The dynamic behavior of taxonomic diversity reflects the complex interaction of physical and biological factors over geological time. Evolutionary change leading to increased alpha diversity in marine and benthic communities contributed directly to increased diversity on regional and global scales. Coevolutionary interactions between terrestrial plants and animals led to increased species richness in terrestrial communities. At the global scale, continental drift has been a major force controlling global diversity, both in the marine and terrestrial realms, over geologic time. Climate and sea level undoubtedly play a role in the short term but are unlikely to control diversity in the long term.

While the general patterns are now apparent, many details remain to be resolved. In particular, the nature of controls on the alpha (within-habitat) and beta (regional) diversity of marine organisms in geologic time need to be identified.

ACKNOWLEDGMENTS

The ideas presented here were improved by discussions with Jack Sepkoski and Geerat Vermeij. This work was supported by NSF EAR 88-04798.

## *Literature Cited*

1. Agassiz, L. 1854. The primitive diversity and number of animals in geological times. *Am. J. Sci. Arts. Sec. Ser.* 17:309–24
2. Ager, D. V. 1976. The nature of the fossil record. *Proc. Geol. Ass.* 87:131–59
3. Ager, D. V. 1981. *The Nature of the Stratigraphical Record*. London: Halstead. 122 pp. 2nd ed.
4. Aller, R. C. 1982. The effects of macrobenthos on chemical properties of marine sediment and overlying water. In *Animal-Sediment Relations,* ed. P. L. McCall, M. J. S. Tevesz, pp. 53–102. New York: Plenum
5. Allison, P. A. 1988. The role of anoxia in the decay and mineralization of proteinaceous macro-fossils. *Paleobiology* 14:139–54
6. Anderson, S. 1974. Patterns of faunal evolution. *Q. Rev. Biol.* 49:311–32
7. Ausich, W. I., Bottjer, D. J. 1982. Tiering in suspension-feeding communities on soft substrata throughout the Phanerozoic. *Science* 216:173–74
8. Ausich, W. I., Bottjer, D. J. 1985. Phanerozoic tiering in suspension-feeding communities on soft substrata: implications for diversity. In *Phanerozoic Diversity Patterns: Profiles in Macroevolution,* ed. J. W. Valentine, pp. 255–274. Princeton NJ: Princeton University Press. 435 pp.
9. Bambach, R. K. 1977. Species richness in marine benthic habitats through the Phanerozoic. *Paleobiology* 3:152–67
10. Bambach, R. K. 1983. Ecospace utilization and guilds in marine communities through the Phanerozoic. In *Biotic Interactions in Recent and Fossil Benthic Communities*. ed. M. J. S. Tevesz, P. L. McCall, pp. 719–746. New York: Plenum. 837 pp.
11. Bambach, R. K. 1986. Phanerozoic marine communities. In *Patterns and Processes in the History of Life,* ed. D. M. Raup, D. Jablonski, ed., pp. 407–428. Berlin: Springer Verlag. 447 pp.
12. Bayer, U., McGhee, G. R., Jr. 1985. Ammonite replacements in the German Lower and Middle Jurassic. *Lect. Notes Earth Sci.* 1:164–220
13. Behrensmeyer, A. K., Kidwell, S. M., 1985. Taphonomy's contribution to paleobiology. *Paleobiology* 11:105–19
14. Benton, M. J. 1985. Mass extinction among non-marine tetrapods. *Nature* 316:811–14
15. Benton, M. J. 1985. Patterns in the diversification of Mesozoic non-marine tetrapods and problems in historical diversity analysis. *Spec. Pap. Palaeont.* 33:185–202
16. Berner, R. A. 1980. *Early Diagenesis—A Theoretical Approach*. Princeton, NJ: Princeton Univ. Press. 341 pp.
17. Blatt, H., Jones, R. L. 1975. Proportions of exposed igneous, metamorphic, and sedimentary rocks. *Geol. Soc. Am. Bull.* 86:1085–88
18. Bottjer, D. J., Ausich, W. I. 1986. Phanerozoic development of tiering in soft substrata suspension-feeding communities. *Paleobiology* 12:400–420
19. Bottjer, D. J., Droser, M. L., Jablonski, D. 1988. Paleoenvironmental trends in the history of trace fossils. *Nature* 333: 252–55
20. Bottjer, D. J., Jablonski, D. 1988. Paleoenvironmental patterns in the evolution of post-Paleozoic benthic marine invertebrates. *Palaios* 3:540–60
21. Brasier, M. D. 1979. The Cambrian radiation event. *Syst. Ass. Sp. Vol.* 12:103–59
22. Burrett, C. F., Richardson, R. G. 1978.

Cambrian trilobite diversity related to cratonic flooding. *Nature* 272:717–19
23. Byers, C. W. 1982. Geological significance of marine biogeneic sedimentary structures. In *Animal-Sediment Relations,* ed. P. L. McCall, M. J. S. Tevesz, pp. 221–56. New York: Plenum. 336 pp.
24. Carpenter, F. M., Burnham, L., 1985. The geological record of insects. *Annu. Rev. Earth Planet. Sci.* 13:297–314
25. Carroll, R. L. 1977. Patterns of amphibian evolution: An extended example of the incompleteness of the fossil record. In *Patterns of Evolution,* ed. A. Hallam, pp. 405–437. Amsterdam: Elsevier. 591 pp.
26. Cisne, J. L. 1975. Evolution of the world fauna of aquatic free-living arthropods. *Evolution* 22:337–66
27. Conway Morris, S. 1986. The community structure of the Middle Cambrian phyllopod bed (Burgess Shale). *Palaeontology* 29:423–67
28. Conway Morris, S., Peel, J. S., Higgins, A. K., Soper, N. J., Davis, N. C. 1987. A Burgess shale-like fauna from the Lower Cambrian of North Greenland. *Nature* 326:181–83
29. Conway Morris, S., Robison, R. A. 1988. More soft-bodied animals and algae from the Middle Cambrian of Utah and British Columbia. *Univ. Kans. Paleont. Cont.* 121. 48 pp.
30. Copper, P. 1977. Eustacy during the Cretaceous: its implications and importance. *Palaeogeog. Palaeoclimat. Palaeoeco.* 22:1–60
31. Cracraft, J. 1985. Biological diversification and its causes. *Ann. Mo. Bot. Gard.* 72:794–822
32. Crimes, T. P. 1974. Colonisation of the early ocean floor. *Nature* 248:328–30
33. Crimes, T. P. 1989. Trace fossils. *In The Precambrian-Cambrian Boundary,* ed. J. W. Cowie, M. D. Brasier, pp. 166–185. Oxford: Clarendon. 213 pp.
34. Darwin, C. 1859. *On the Origin of Species*. London: John Murray. 513 pp.
35. Eisenberg, J. F. 1981. *The Mammalian Radiations*. Chicago: Univ. Chicago Press. 610 pp.
36. Erwin, D. H., Valentine, J. W., Sepkoski, J. J. Jr. 1987. A comparative study of diversification events: The early Paleozoic versus the Mesozoic. *Evolution* 41:1177–86
37. Flessa, K. W. 1975. Area, continental drift, and mammalian diversity. *Paleobiology* 1:189–94
38. Flessa, K. W., Imbrie, J. 1973. Evolutionary pulsations: Evidence from Phanerozoic diversity patterns. In *Implications of Continental Drift to the Earth Sciences,* ed. D. H . Tarling, S. K. Runcorn, pp. 247–94. London: Academic. 622 pp.
39. Flessa, K. W., Powers, K. V., Cisne, J. L. 1975. Specialization and evolutionary longevity in the Arthropoda. *Paleobiology* 1:71–81
40. Flessa, K. W., Sepkoski, J. J. Jr. 1978. On the relationship between Phanerozoic diversity and changes in habitable area. *Paleobiology* 4:359–66
41. Gilinsky, N. L., Bambach, R. K. 1987. Asymmetrical rates of origination and extinction in higher taxa. *Paleobiology* 13:446–64
42. Gingerich, P. D. 1977. Patterns of evolution in the mammalian fossil record. See Ref. 25, pp. 469–500
43. Glaessner, M. F. 1984. *The Dawn of Animal Life*. Cambridge: Cambridge Univ. Press. 244 pp.
44. Gould, S. J. 1977. Eternal metaphors of palaeontology. See Ref. 25, pp. 1–26
45. Gould, S. J., Gilinsky, N. L., German, R. Z. 1987. Asymmetry of lineages and the direction of evolutionary time. *Science* 236:1437–41
46. Gould, S. J., Raup, D. M., Sepkoski, J. J. Jr., Schopf, T. J. M., Simberloff, D. S. 1977. The shape of evolution: A comparison of real and random clades. *Paleobiology* 3:23–40
47. Gray, J., Massa, D., Boucot, A. J. 1982. Caradocian land plant microfossils from Libya. *Geology* 10:197–201
48. Gregor, B. 1970. Denudation of the continents. *Nature* 228:273–75
49. Gregor, C. B. 1985. The mass-age distribution of Phanerozoic sediments. *Geol. Soc. Lond. Mem.* 10:294–89
50. Gregory, J. T. 1955. Vertebrates in the geologic time scale. *Geol. Soc. Am. Spec. Pap.* 62:593–608
51. Hallam, A. 1981. *Facies Interpretation and the Stratigraphic Record.* New York: Freeman. 291 pp.
52. Hallam, A. 1987. Radiations and extinctions in relation to environmental change in the marine Lower Jurassic of northwest Europe. *Paleobiology* 13: 152–68
53. Hansen, T. A. 1987. Extinction of Late Eocene to Oligocene molluscs: relationship to shelf area, temperature changes, and impact events. *Palaios* 2:69–75
54. Hansen, T. A. 1988. Early Tertiary radiation of marine molluscs and the long-term effects of the Cretaceous-Tertiary extinction. *Paleobiology* 14:37–51
55. Haq, B. U., Hardenbol, J., Vail, P. R. 1987. Chronology of fluctuating sea

levels since the Triassic. *Science* 2325: 1156–67
56. Hoffman, A. 1983. Paleobiology at the crossroads: A critique of some modern paleobiological research programs. In *Dimensions of Darwinism,* ed. M. Grene, pp. 241–271. Cambridge: Cambridge Univ. Press. 471 pp.
57. Hoffman, A. 1989. *Arguments on Evolution*. Oxford: Oxford Univ. Press. 274 pp.
58. Hou, X.-g. 1987a. Two new arthropods from Lower Cambrian, Chengjian, Eastern Yunnan. *Acta Palaeont. Sin.* 26: 236–56
59. Hou, X.-g. 1987b. Early Cambrian large bivalved arthropods from Chengjiang, Eastern Yunnan. *Acta Palaeont Sin.* 26:286–98, 2 pl.
60. Jablonski, D. 1980. Apparent versus real biotic effects of transgressions and regressions. *Paleobiology* 6:397–407
61. Jablonski, D. 1985. Marine regressions and mass extinctions: A test using the modern biota. See Ref. 8, pp. 335–53.
62. Jablonski, D. 1986. Causes and consequences of mass extinctions: a comparative approach. In *Dynamics of Extinction,* ed. D. K. Elliott, pp. 183–229. New York: Wiley. 312 pp.
63. Jablonski, D., Bottjer, D. J. 1983. Soft-bottom epifaunal suspension-feeding assemblages in the Late Cretaceous. See Ref. 10, pp. 747–812
64. Jablonski, D., Flessa, K. 1984. The taxonomic structure of shallow-water marine faunas: Implications for Phanerozoic extinctions. *Malacologia* 27:43–66
65. Jablonski, D., Sepkoski, J. J. Jr., Bottjer, D. J., Sheehan, P. M., 1983. Onshore-offshore patterns in the evolution of Phanerozoic shelf communities. *Science* 222:1123–25
66. Jaccard, P. 1980. Nouvelles recherces sur la distribution florale. *Bull. Soc. Vaud. Sci. Nat* 44:223–70
67. Johnson, R. G. 1964. The community approach to paleoecology. In *Approaches to Paleoecology,* ed. J. Imbrie, N. D. Newell, pp. 107–134. New York: Wiley. 432 pp.
68. Kauffman, E. G., Steidtmann, J. R. 1981. Are these the oldest metazoan trace fossils? *J. Paleontol.* 55:923–48
69. Kennedy, W. J. 1977. Ammonite evolution. See Ref. 25, pp. 251–304
70. Kitchell, J. A., MacLeod, N. 1988. Macroevolutionary interpretations of symmetry and synchroneity in the fossil record. *Science* 240:1190–93
71. Knoll, A. H. 1984. Patterns of extinction in the fossil record of vascular plants. In *Extinctions,* ed. M. Nitecki, pp. 21–67. Chicago: Univ. Chicago Press. 487 pp.
72. Knoll, A. H. 1986. Patterns of change in plant communities through geologic times. In *Community Ecology,* ed. J. Diamond, T. J. Case, pp. 125–141. New York: Harper & Row. 523 pp.
73. Knoll, A. H., Niklas, K. J., Tiffny, B. H. 1979. Phanerozoic land-plant diversity in North America. *Science* 206:1400–1402
74. Kurtén, B. 1969. Continental drift and evolution. *Sci. Am.* 220(3):54–64
75. Lawrence, D. R. 1968. Taphonomy and information losses in fossil communities. *Bull. Geol. Soc. Am.* 79:1315–30
76. Lidgard, S., Crane, P. R. 1988. Quantitative analyses of the early angiosperm radiation. *Nature* 331:344–46
77. Lillegraven, J. A. 1972. Ordinal and familial diversity of Cenozoic mammals. *Taxon* 21:261–74
78. Lowe, D. R. 1980. Stromatolites 3400-Myr old from the Archean of Western Australia. *Nature* 284:441–43
79. Lyell, C. 1830–3. *Principles of Geology*. London: John Murray. (1970 reprint).
80. McGhee, G. R. Jr. 1988. The Late Devonian extinction event: evidence for abrupt ecosystem collapse. *Paleobiology* 14:221–34
81. Mount, J. F., Signor, P. W. 1985. Early Cambrian innovation in shallow subtidal environments: Paleoenvironments of Early Cambrian shelly fossils. *Geology* 13:730–33
82. Müller, A. H. 1979. Fossilization (Taphonomy). In *Treatise on Invertebrate Paleontology,* ed. R. A. Rubison, C. Teichert, pp A2–A78. Lawrence: Geol. Soc. Am. & Univ Kans Press. 569 pp.
83. Newell, N. D. 1967. Revolutions in the history of life. *Geol. Soc. Am. Spec. Pap.* 89:63–91
84. Newell, N. D. 1971. An outline history of tropical organic reefs. *Am. Mus. Novit.* 2465. 37 pp.
85. Niklas, K. J. 1986. Large-scale changes in animal and plant terrestrial communities. In *Patterns and Processes in the History of Life,* ed. D. M. Raup, D. Jablonski, pp. 383–405. Berlin: Springer Verlag. 447 pp.
86. Niklas, K. J., Tiffney, B. H., Knoll, A. H. 1980. Apparent changes in the diversity of fossil plants. *Evol. Bio.* 12:1–89
87. Niklas, K. J., Tiffney, B. H., Knoll, A. H. 1983. Patterns in vascular plant diversification. *Nature* 303:614–16

88. Niklas, K. J., Tiffney, B. H., Knoll, A. H. 1985. Patterns in vascular land plant diversification: an analysis at the species level. See Ref. 8, pp. 97–128.
89. Officer, C. B., Drake, C. L. 1985. Terminal Cretaceous environmental effects. *Science* 227:1161–67
90. Officer, C. B., Hallam, A., Drake, C. L., Devine, J. D. 1987. Late Cretaceous and paroxysmal Cretaceous/Tertiary extinctions. *Nature* 326:143–49
91. Padian, K., Clemens, W. A. 1985. Terrestrial vertebrate diversity: Episodes and insights. See Ref. 8, pp. 41–96
92. Patterson, C., Smith, A. B. 1987. Is the periodicity of extinctions a taxonomic artefact? *Nature* 330:248–51
93. Phillips, J. 1860. *Life on Earth, Its Origin and Succession*. Cambridge: MacMillan. 224 pp. (1980 reprint).
94. Raup, D. M. 1972. Taxonomic diversity during the Phanerozoic. *Science* 177: 1065–71
95. Raup, D. M. 1976. Species diversity in the Phanerozoic: A tabulation. *Paleobiology* 2:279–88
96. Raup, D. M. 1976. Species diversity in the Phanerozoic: An interpretation. *Paleobiology* 2:289–97
97. Raup, D. M. 1979. Size of the Permo-Triassic bottleneck and its evolutionary implications. *Science* 206:217–18
98. Raup, D. M. 1986. Biological extinction in Earth History. *Science* 231:1528–53
99. Raup, D. M., Boyajian, G. E. 1988. Patterns of generic extinction in the fossil record. *Paleobiology* 14:109–25
100. Raup, D. M., Sepkoski, J. J. Jr. 1982. Mass extinctions in the fossil record. *Science* 215:1501–03
101. Raup, D. M., Sepkoski, J. J. Jr. 1984. Periodicity of extinctions in the geologic past. *Proc. Nat. Acad. Sci. USA* 81: 801–805
102. Raup, D. M., Sepkoski, J. J. Jr. 1986. Periodic extinction of families and genera. *Science* 231:833–36
103. Robison, R. A. 1985. Affinities of *Aysheaia* (Onychophora) with description of a new Cambrian species. *J. Paleontol.* 59:226–35
104. Rudwick, M. J. S. 1976. *The Meaning of Fossils*. Chicago: Univ. Chicago Press. 287 pp. 2nd ed.
105. Sadler, P. M. 1981. Sediment accumulation rates and the completeness of stratigraphic sections. *J. Geol.* 89: 569–84
106. Schindel, D. E. 1980. Microstratigraphic sampling and the limits of paleontologic resolution. *Paleobiology* 6:408–26
107. Schopf, T. J. M. 1974. Permo-Triassic extinctions: Relation to sea-floor spreading. *J. Geol.* 82:129–39
108. Schopf, T. J. M. 1978. Fossilization potential of an intertidal fauna, Friday Harbor, Washington. *Paleobiology* 4: 261–70
109. Schopf, T. J. M. 1979. The role of biogeographic provinces in regulating marine faunal diversity through geologic time. In *Historical Biogeography, Plate Tectonics, and the Changing Environment* ed. J. Gray, A. J. Boucot, pp. 449–57. Corvallis: Oregon State Univ. Press
110. Seilacher, A., Reif, W.-E., Westphal, F. 1985. Sedimentological, ecological and temporal patterns of fossil *Lagerstaetten*. *Philos. Trans. R. Soc. Lond.* B311:5–23
111. Sepkoski, J. J. Jr. 1976. Species diversity in the Phanerozoic: species-area effects. *Paleobiology* 2:298–303
112. Sepkoski, J. J. Jr. 1978. A kinetic model of Phanerozoic taxonomic diversity I. Analysis of marine orders. *Paleobiology* 4:223–51
113. Sepkoski, J. J. Jr. 1979. A kinetic model of Phanerozoic taxonomic diversity II. Early Phanerozoic families and multiple equilibria. *Paleobiology* 5:222–51
114. Sepkoski, J. J. Jr. 1981. A factor analytic description of the Phanerozoic marine fossil record. *Paleobiology* 7:36–53
115. Sepkoski, J. J. Jr. 1982. *A Compendium of Fossil Marine Families*. *Milwauk. Pub. Mus. Cont. Biol. Geol.* 51. 125 pp.
116. Sepkoski, J. J. Jr., 1984. A kinetic model of Phanerozoic taxonomic diversity III. Post-Paleozoic families and mass extinctions. *Paleobiology* 10:246–67
117. Sepkoski, J. J. Jr. 1988. Alpha, beta, or gamma: where does all the diversity go? *Paleobiology* 14:221–34
118. Sepkoski, J. J. Jr., Bambach, R. K., Raup, D. M., Valentine, J. W. 1981. Phanerozoic marine diversity and the fossil record. *Nature* 293:435–537
119. Sepkoski, J. J. Jr., Sheehan, P. M. 1983. Diversification, faunal change, and community replacement during the Ordovician radiations. See Ref. 10, pp. 673–717
120. Sheehan, P. M. 1977. Species diversity in the Phanerozoic: A reflection of labor by systematists? *Paleobiology* 3:325–28
121. Signor, P. W. 1978. Species richness in the Phanerozoic: An investigation of sampling effects. *Paleobiology* 4:394–406
122. Signor, P. W. 1982. Species richness in the Phanerozoic: Compensating for sampling bias. *Geology* 10:625–28
123. Signor, P. W. 1985. Real and apparent

trends in species richness through time. See Ref. 8, pp. 129–50
124. Signor, P. W., Brett, C. E. 1984. The mid-Paleozoic precursor to the Mesozoic marine revolution. *Paleobiology* 10: 229–45
125. Simberloff, D. S. Permo-Triassic extinctions: Effects of area on biotic equilibrium. *J. Geol.* 82:267–74
126. Simpson, G. G. 1980. *Splendid Isolation*. New Haven: Yale Univ. Press
127. Smith, A. B., Patterson, C., 1988. The influence of taxonomic method on the perception of patterns of evolution. *Evol. Biol.* 23:127–216
128. Sohl, N. D. 1987. Cretaceous gastropods: contrasts between tethys and the temperate provinces. *J. Paleontol.* 61: 1085–1111
129. Stanley, S. M. 1972. Functional morphology and evolution of bysally attached bivalve mollusks. *J. Paleontol.* 46:165–213
130. Stanley, S. M. 1976. Fossil data and the Precambrian-Cambrian evolutionary transition. *Am. J. Sci.* 276:56–76
131. Stanley, S. M. 1979. *Macroevolution: Pattern and Process*. San Francisco: Freeman. 332 pp.
132. Stanley, S. M. 1984. Temperature and biotic crises in the marine realm. *Geology* 12:205–208
133. Stanley, S. M. 1986. Anatomy of a regional mass extinction: Plio-Pleistocene decimation of the western Atlantic bivalve fauna. *Palaios* 1:17–36
134. Taylor, J. D., Morris, N. J., Taylor, C. N. 1980. Food specialization and the evolution of predatory prosobranch gastropods. *Palaeontology* 23:375–409
135. Thomson, K. S. 1977. The pattern of diversification among fishes. See Ref. 25, pp. 377–404
136. Tiffney, B. H. 1981. Diversity and major events in the evolution of land plants. In *Paleobotany, Paleoecology, and Evolution*. ed. K. J. Niklas, pp. 193–230. New York: Praeger
137. Tiffney, B. H., Niklas, K. J. 1990. Continental area, dispersion, latitudinal distribution and topographic variety: A test of correlation with terrestrial plant diversity. In *Biotic and Abiotic Factors in Evolution,* ed. W. Allmon, R. D. Norris, Chicago: Univ. Chicago Press. In press
138. Valentine, J. W. 1968. Climatic regulation of species diversification and extinction. *Geol. Soc. Am. Bull.* 79:273–76
139. Valentine, J. W. 1970. How many marine invertebrate fossils? *J. Paleontol.* 44:410–15
140. Valentine, J. W. 1973. *Evolutionary Paleoecology of the Marine Biosphere*. Englewood Cliffs: Prentice-Hall. 511 pp.
141. Valentine, J. W. 1989. How good was the fossil record? Clues from the California Pleistocene. *Paleobiology* 15:83–94
142. Valentine, J. W., Foin, T. C., Peart, D. 1978. A provincial model of Phanerozoic marine diversity. *Paleobiology* 4:55–66
143. Valentine, J. W., Awramik, S. M., Signor, P. W., Sadler, P. M. 1990. The biological explosion at the Precambrian-Cambrian boundary. *Evol. Biol.* In press
144. Valentine, J. W., Moores, E. M. 1972. Plate-tectonic regulation of faunal diversity and sea level: a model. *Nature* 228:657–59
145. Van Valen, L. 1972. Are categories in different phyla comparable? *Taxon* 22: 333–73
146. Vermeij, G. J. 1977. The Mesozoic marine revolution: evidence from snails, predators, and grazers. *Paleobiology* 3:245–58
147. Vermeij, G. J. 1978. *Biogeography and Adaptation*. Cambridge Mass: Harvard Univ. Press. 332 pp.
148. Vermeij, G. J. 1987. *Evolution and Escalation*. Princeton NJ: Princeton Univ. Press. 527 pp.
149. Walter, M. R., Buick, R., Dunlop, J. S. R. 1980. Stromatolites 3400–3500 m.y. old from the North Pole area, Western Australia. *Nature* 284:443–45
150. Whittaker, R. H. 1977. Evolution of species diversity in land communities. *Evol. Biol.* 10:1–67
151. Wing, S. L., Tiffney, B. H. 1987. The reciprocal interaction of angiosperm evolution and tetrapod herbivory. *Rev. Palaeobot. Palynol.* 50:179–210
152. Wise, K. P., Schopf, T. J. M. 1981. Was marine faunal diversity in the Pleistocene affected by changes in sea level? *Paleobiology* 7:394–99
153. Zhang, W.-t., Hou, X.-g. 1985. Preliminary notes on the occurrence of the unusual trilobite *Naraoia* in Asia. *Acta Palaeontol. Sin.* 24:591–95, 4 pl.

*Annu. Rev. Ecol. Syst. 1990. 21:541–66*


# RIBOSOMAL RNA IN VERTEBRATES: EVOLUTION AND PHYLOGENETIC APPLICATIONS

*David P. Mindell*

Department of Biological Sciences, University of Cincinnati, Cincinnati, Ohio 45221

*Rodney L. Honeycutt*

Department of Wildlife and Fisheries Science, Texas A&M University, College Station, Texas 77843

KEY WORDS: ribosomal RNA, molecular systematics, vertebrates, rates of evolution

## INTRODUCTION

Many evolutionary biologists seek to learn about genealogy among life forms. Toward this end, researchers have sought out informative systems or characters from fields as disparate as behavior, morphology, physiology, biogeography, and molecular genetics. DNA sequences in particular appear promising for resolving evolutionary questions, due in part to the fact that they constitute the physical material of inheritance in its most particulate form. However, the promise of DNA rests on the assumption that historical change in nucleotide sequences can be inferred unambiguously from sequences of extant (or recently extinct) forms, and this task remains as a challenge.

It is important to recognize that the pursuits of molecular systematics (history of descent among organisms) and molecular evolution (history of change among and within molecules) are mutually informing. Molecular systematic studies have been instrumental in studies of molecular evolution because patterns of sequence change over thousands and millions of years can be inferred only by comparing sequences among taxa whose divergences span



0066-4162/90/1120-0541$02.00

such periods of evolutionary time. Conversely, our increasing understanding of the nature of DNA sequence change informs our attempts to determine phylogenetic relationships among organisms, in that we can emphasize or de-emphasize various types of sequence change in phylogenetic analyses, depending on the age of the evolutionary events of interest and the rate at which particular sequences change over time. This sounds like the systematists' age-old problem of weighting (determining the phylogenetic information content of characters), and it is, albeit with a new twist; a better understanding of the relationship between change and time. Given that DNA sequence comparisons among taxa comprise the same comparative approach that has served systematists for the past 180 years (if we go back to Lamarck's time), it is not surprising that molecular phylogeneticists face similar decisions regarding treatment of variable characters. The mutually informing nature of molecular phylogenetics and molecular evolution requires constant reevaluation of methods used in both pursuits.

Ribosomal RNAs (rRNAs) have become one of the best known groups of genes, as their important functional role and near-ubiquitous distribution in organisms have drawn the attention of a wide range of biologists (see review by Gerbi—35). In this article we review and investigate variation in large subunit and small subunit rRNAs and the implications of this variability in studies of phylogeny, with particular emphasis on vertebrate animals.

## STRUCTURE AND ORGANIZATION OF RIBOSOMAL RNA GENES

### *Origin and Primary Sequence Organization*

Ribosomal RNA genes, variously called rRNAs or rDNAs, code for RNAs and can be found in nuclear, mitochondrial (mt), and chloroplast genomes. Each RNA molecule is bonded to 30 to 100 smaller protein molecules to form a compact subunit of the ribosome (8, 79). Two such subunits combine to form a ribosome, providing the apparatus for synthesizing proteins from messenger RNAs. Functional rRNAs are crucial to the survival of individuals, as ribosomally synthesized proteins are the basic constituents of organismal phenotypes. Thus, rRNA sequences are relatively resistant to change over time. The number of ribosomes in any cell is directly related to the level of protein-synthesizing activity. Higher sequence similarity between mt rRNAs and bacterial rRNAs than between mt and nuclear rRNAs within any given species (39) supports the notions of a very early origin of ribosomes in the history of life and of mitochondria arising from bacteria (70).

Nuclear rRNA complexes in eukaryotes include coding and noncoding sequences arranged tandemly in repeated units in the nucleolus organizer region on one to several chromosomes (Figure 1A). In vertebrates the three coding regions are the 18S, 5.8S, and 28S genes (S = sedimentation values based on an index of weight and shape). Noncoding regions include an external transcribed spacer, two internal transcribed spacers, and a nontranscribed spacer found between each pair of repeat units. The number of repeat copies per cell varies widely among taxa, ranging from one copy in some protozoa to 6400 copies in wheat (3). Most animals, however, have less than 1000 copies [e.g. 200 in the chicken (*Gallus gallus*; 71), and 100 in the mouse (*Mus musculus;* 44)]. Like bacterial rRNAs, animal mt rRNAs (12S and 16S genes in vertebrates) occur in one copy per genome and have no associated spacers. In most animals these genes are adjacent to each other, with each being flanked by transfer RNAs.

## *Secondary Structure*

Ribosomal RNAs contain considerable amounts of within-molecule base pairing to form specific secondary structures. A variety of approaches has

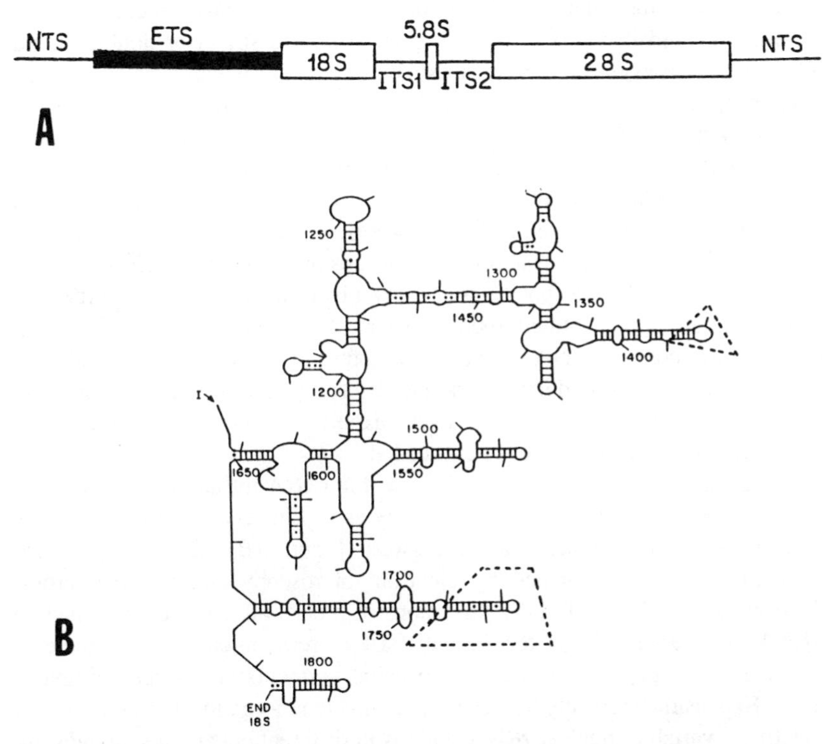


*Figure 1* (A) Diagram illustrating components of the vertebrate nuclear rRNA repeat unit, including three genes (18S, 5.8S, 28S), a nontranscribed spacer (NTS), an external transcribed spacer (ETS), and two internal transcribed spacers (ITS). (B) Secondary structure for the 3' half of *Xenopus laevis* 18S rRNA, after Atmadja et al (4a). Reproduced with permission from Gerbi (35). Dashed boxes denote expansion segments.

been used to estimate rRNA secondary structure from the nucleotide sequence; however, in molecules as large as rRNAs, many configurations are possible. Graphical, dot matrix methods are widely used, in which all pairwise nucleotide comparisons are made within a matrix, and complementary regions show as pairs of short diagonal lines (67, 83). "Minimum energy structure" estimates seek configurations that are energetically most favorable and/or maximize paired bases (66, 74, 109, 125, 126), given certain thermodynamic rules (90; see 124). In most cases, modeling results should be considered heuristic. Minor sequence changes in closely related species can yield different secondary structures (106), and many details are unaccounted for or poorly understood (see review in 109). For example, in their functional state, rRNAs are bound to proteins, and location of the binding sites will constrain secondary structure in ways that are not known. Ribosomal RNA molecules can start to fold as they are being synthesized, and the relative proximity of complementary regions likely plays a role in determining structure.

Direct experimental approaches to determining secondary structure include determining sensitivity of various regions to single strand or double strand specific nucleases, use of chemicals such as kethoxal, which attacks unpaired guanines, and bisulfite, which modifies unpaired cytosines, and intra-RNA cross-linking experiments (78). Most direct experimental evidence for deducing secondary structure has been done for *Escherichia coli* (68, 114, 127). Structural hypotheses for other species are often based on or directly overlaid on the *E. coli* model. Gutell & Fox (40) present such diagrams for numerous species. These are useful as approximations, as one can quickly place variable or conserved nucleotide positions for any given taxon in the context of an experimentally corroborated model. There does appear to be a common architectural core in rRNA secondary structure across tremendously diverse taxa, based on both modeling work and determination of a common melting curve component (20, 24). However, experimentally deduced secondary structures for other taxa are much needed.

Long-range base pairing interactions within rRNA influence the folding of the molecule into domains, which are series or clusters of stems and loops interconnected by single stranded regions (Figure 1B). Existence of such domains has been indicated by electron microscopy and psoralen cross-linking studies (54, 104, 117). Insertions of blocks of sequence into nuclear rRNAs are responsible for the dramatic size increase in eukaryotic structures relative to *E. coli,* and are called expansion segments or divergent domains (17). Expansion segments of different lengths tend to be located in the same, relatively variable, nuclear rRNA regions in different eukaryotes, suggesting that functionally important regions are not being disrupted by the insertion. Vertebrate mt rRNAs are smaller in size relative to *E. coli* rRNAs, yet, in some cases, the sequence block losses are from the same positions at which eukaryotic nuclear rRNA expansion segments are found (e.g. 69). Expansion segments have been hypothesized to be remnants of mobile elements (17) or remnants of "linkers" involved in assembly of the ribosome (16). Larson & Wilson (62) found nonrandom distribution of substitutions within salamander expansion segments, suggesting the presence of at least some structural or functional constraints.

# METHODS OF ANALYSIS

## *Molecular Techniques*

RESTRICTION ENDONUCLEASE SITE MAPPING Restriction site mapping of nuclear rRNA involves the isolation of high molecular weight DNA followed by the digestion of a DNA aliquot with specific restriction endonucleases. Digested fragments are then separated based on size via electrophoresis and are transferred to a membrane filter using the Southern blot method (99). Filters are then hybridized to radioactive, often heterologous, probes [e.g. *Mus musculus* genes (4) for comparisons among tetrapods], and exposed to X-ray film. Restriction site maps are constructed by first determining the rRNA repeat length, and then using a combination of single and double restriction endonuclease digestions to locate specific sites relative to each other. The repetitive nature of rRNA facilitates cloning or isolation experiments, as well as direct visualization of cleaved fragments with autoradiography. Restriction maps may cover all regions of the nuclear rRNA repeat unit, including several different levels of divergence. Restriction site mapping in mt rRNAs, which lack spacers, may also cover a wide range of divergence rates.

Drawbacks of restriction site analysis are that it is time consuming and subject to errors of gel interpretation, and studies from different labs are often difficult to compare, due to use of different electrophoretic techniques and restriction enzymes. Regions not directly covered by heterologous probes, especially variable nontranscribed spacers, will be difficult to map (113). Partial digestions can help determine the existence of sites that are distant from the probe (49); however, for taxa with large nontranscribed spacers, such as mammals (3) this strategy is limited. An approach that can provide more phylogenetically informative characters is the use of four-base, rather than six-base, recognition enzymes, combined with partial digestions and smaller probes to provide more detailed maps of specific regions. Amplification of doublestranded rDNA fragments, using oligonucleotide primers and the polymerase chain reaction (PRC) combined with partial digestions and four-base recognition enzymes, would allow resolution of greater variability without development and use of heterologous probes.

NUCLEOTIDE SEQUENCING There are three basic procedures for sequencing both nuclear and mitochondrial rRNA genes: direct sequencing of isolated

RNA, cloning and sequencing DNA, or direct sequencing of amplified DNA. The traditional approach has been cloning of rDNA segments into plasmids or bacteriophages followed by sequencing. Direct sequencing of RNA through reverse transcriptase extension of 5' end-labeled DNA primers in the presence of dideoxynucleotides is becoming a common method in systematic studies as well (31, 82). No cloning is required, and synthetic oligonucleotides complementary to specific segments of rDNA can be constructed for a particular comparative study. The potentially most efficient approach involves PCR. In vitro DNA amplification is done, in which specific segments of DNA can be amplified as much as $10^5$-fold in a few hours (e.g. 88). Direct sequencing of single-stranded DNA amplified via PCR with an asymmetrical primer ratio eliminates the need for RNA purification or DNA subcloning prior to sequencing (41, 118).

## *Phylogenetic Techniques*

A primary challenge facing systematists working with nucleotide sequence characters is to distinguish phylogenetically informative change from a larger set of character change that has been determined. Sequence informativeness is variable because the rate of evolutionary change within and among sequences is variable. Nucleotide positions that have changed many times since divergence of the species in question will be saturated with change and phylogenetically uninformative. Thus, the challenge stated above requires systematists to learn about the variable nature and rate of change for DNA sequences. Recognized tendencies for nucleotide sequence change include: faster turnover in noncoding sequences (e.g. pseudogenes) versus protein coding genes and rRNAs (e.g. 63); faster turnover at third codon positions relative to first and second positions in protein-coding genes (due to frequency of synonymous substitutions at that position) (e.g. 73); and faster rate for transitions (T'C, A'G) versus transversions (T'G, T'A, A'C, C'G) both in nuclear and mitochondrial genes (10, 11, 32, 108). The difference in relative rate in base-paired (stem) versus unpaired (loop) regions of rRNAs is still unclear. Change in single-stranded loop regions does not "require" a compensatory change as in a stem region; however, selection will likely favor and perhaps accelerate such compensatory changes within stems. Counting such changes as independent events can bias phylogenies by placing too much weight on covarying characters as seen in 5S rRNA sequences (111), but see (97). Finally, some taxa, generally characterized by shorter generation time or high number of germ-line divisions per unit time, tend to have faster rates of sequence change than others (9, 36, 58, 60, 64, 119).

Alignments are crucial to phylogenetic reconstruction because they determine which nucleotide positions will be considered homologous in phylogenetic analyses. Computer algorithms, with varying assumptions and optimization schemes, are available for both pairwise (51, 75, 98, 112) and multiple (5, 6, 46, 91) alignments; however, they are best considered as heuristic, because they are often sensitive to small alterations in search parameters and inclusion or exclusion of taxa. Multiple alignment of rRNAs is particularly difficult, as they do not have reading frames as protein-coding genes do, with relatively conserved first and second codon positions to provide guidelines, and because the abundant 1-2 base pair insertions and deletions could be parsimoniously placed at multiple locations. In a multiple alignment bout, DNA sequences from various taxa should be included (input) in decreasing order of phylogenetic relationship (30a, 71a, 92). Homology denotes similarity of features due to shared inheritance from a common ancestor, and may be defined on the basis of continuity of information. As such, that information is primarily genetic, and the "continuity" is provided by genealogy. Alignment of sequences from distant relatives, prior to alignment of more closely related taxa would tend to disrupt the continuity of information in those DNA sequences. Similarly, equal treatment, via simultaneous alignment, of sequences from all taxa would also remove the genealogical basis of the homology assessment (unless all the taxa involved experienced simultaneous origins from a common ancestor). Thus, multiple sequence alignments should be informed by phylogeny, to whatever extent possible. Alignments may also be improved by constraining the alignment in regions of conserved secondary structure (39).

Weighting schemes enable emphasis or deemphasis of different types of nucleotide character change in phylogenetic analysis. For example, transversions may be given additional weight in cases where transitions are thought to be relatively uninformative. The computer program PAUP (101) has a "matrix weighting" capability suitable for such applications. Alternatively, Penny & Hendy (81) have developed a compatibility approach, whereby internally supported subsets of characters are identified for subsequent weighting in parsimony analyses. Another approach is that of successive weighting (12, 26) in which weights are based on consistency, or agreement, within the data set, and no prior decisions regarding weighting are required. The program Hennig86 (29) has such a facility (*xsteps w* option) for use in tree building. No treatment is still a form of weighting, albeit equal across all characters, and use of the increasing knowledge of molecular evolution via unequal weighting is preferred to equal weighting, given known rate heterogeneity among characters (see 28 and 55 for discussions of weighting). In regard to morphological characters, Donoghue (22) has pointed out that the likelihood of various character changes probably varies over time, and a similar argument could be made for some molecular characters. Weighting should thus be carefully considered.

The earliest phylogenetic analyses of molecular data generally involved equal weighting of all characters in calculating similarity measures (distances) between pairs of taxa, often with divergence times spanning a wide range (59, 76, 93). Criticism of this approach has centered on the assumption of rate constancy of character change inherent in some, but not all, of its applications, and the loss of information and the rationale involved in submerging discrete differences into genetic distances (27, 47, 105). Discrete character analyses (45) for nucleotide bases allow one to retain information on specific mutation events that may be used in subsequent studies of character evolution, and to avoid assumptions of rate constancy. For a recent discussion of parsimony, likelihood, and distance methods for inferring phylogeny from sequences, see Felsenstein (30).

## VARIATION IN EVOLUTIONARY RATE

A correlation between time and increasing molecular divergence was first deduced from comparative amino acid replacement data among species (122, 123). We now recognize, within this general correlation, continuous variability in degree of correlation, both among sections of homologous nucleotide sequence and among species. Our objective in the following section is to assess aspects of evolutionary rate variability in rRNAs, as they pertain to phylogenetic studies.

### *Rate Heterogeneity Among Taxa*

Rates of nucleotide change in protein coding genes are faster in some taxa characterized by shorter generation-times, or more germ-line cell divisions per unit time (7, 9, 15, 33, 60, 96, 64, 65). We would like to know if such heterogeneity in rate of change among taxa occurs in rRNAs as well. One approach used in assessing such rate variability involves DNA hybridization experiments. This approach provides an informative average across large portions of genomes; however, information on specific substitutions and rearrangement events from identifiable homologous sites is lost in the averaging process, and in some applications, divergence times for species pairs must be estimated from fossil or biogeographic evidence. Another approach involves comparing discrete nucleotide sequence characters among taxa, determining genetic distances from those comparisons, and assessing those distances, or their variances, across numerous genes for significant differences.

Here, we assess rate differences among species by comparing amounts of discrete nucleotide character change between species and an outgroup, or reference, species. This is essentially a "relative rate test" (94), in which number of discrete differences between each of two ingroup species and the outgroup are compared. Because the two ingroup species are more closely related to each other than to the outgroup, they are of equal age relative to the outgroup, and their respective rates of change may be directly compared. Advantages of this approach are that estimated divergence dates based on fossils are not required, and amounts of discrete change are compared, rather than averaged genetic distances. We proceeded as follows: (*a*) entire homologous rRNAs for the two ingroup species were aligned using the algorithm of Myers & Miller (75); (*b*) a consensus sequence was made from that alignment in which all variable nucleotide sites were overwritten as "N" (ambiguous); (*c*) the consensus sequence was aligned with the outgroup sequence; (*d*) for each ingroup species, unique substitutions relative to the outgroup were tallied; and (*e*) relative numbers of substitutions were compared and analyzed for degree of difference with a binomial test. The alignments determine character homology, and by aligning the ingroup species first, and aligning the outgroup to their conserved positions only, we use (in this case obvious) phylogenetic information to constrain the homology assignments. Insertions and deletions were ignored, because alignments are the most unreliable in those areas. In using a binomial test, we assume that equal rates of change would mean that 50% of the total number of substitutions (relative to the outgroup) would be found in each of the ingroup species. We are testing for departure from the expected 50% in the observed distribution of substitutions. The binomial test was done according to the following formula:

$$\left[\sum \frac{m!}{(m-n)!n!}\right].5^{m}$$

where $m$ is the total number of substitutions between the two ingroup species and the outgroup, and $n$ is the number of substitutions actually observed in any one of the ingroup species (a Cray computer was used, due to the large numbers associated with factorials). The binomial test assumes independence of substitution events, which will not hold if selection favors compensatory changes within rRNA stem regions. However, if secondary structure for ingroups is as similar as it appears to be for those species analyzed here (primarily *Homo* and *Mus*) this bias will be similar in the two lineages.

Various substitution rate comparisons were made between *Homo sapiens, Mus musculus,* or *Gallus gallus* with *Gallus gallus* and/or *Xenopus laevis* as an outgroup, for the mt rRNAs, 12S and 16S, and nuclear 28S rRNAs (see Table 1 for sequence references). In each comparison *Mus* showed more substitutions relative to the outgroup than did *Homo* (12S, 16S, and 28S genes) or *Gallus* (12S), and in the only comparison between *Gallus* and *Mus* (12S), *Mus* also showed more substitutions relative to the outgroup. Increased substitutions in *Mus* ranged from 10 to 16 bases with probabilities between 0.087 and 0.319 (Table 2). Although these differences are not statistically

**Table 1** Ribosomal RNA sequences compared in this review. Abbreviations: SSU = small subunit, LSU = large subunit, mt = mitochondrial. In vertebrates: mt SSU = 12S, mt LSU = 16S, nuclear SSU = 18S, and nuclear LSU = 28S.

| | Gene length in bases (reference) | | | |
|---|---|---|---|---|
| | mt | | nuclear | |
| | SSU | LSU | SSU | LSU |
| *Saccharomyces cerevisiae* (yeast) | | | 1798(86) | 3391(34) |
| *Artemia salina* (brine shrimp) | | | 1810(77) | |
| *Drosophila yakuba* (fruit fly) | 788(18) | 1326(18) | | |
| *Xenopus laevis* (frog) | 819(23, 85) | 1640(85) | 1824(89) | 4081(110) |
| *Bos taurus* (cow) | 958(2) | 1571(2) | | |
| *Giraffa camelopardalis* (giraffe) | 955(102) | | | |
| *Sus scrofa* (pig) | 956(102) | | | |
| *Mus musculus* (house mouse) | 954(107) | 1582(107) | 1869(84) | 4712(43) |
| *Rattus norvegicus* (rat) | 953(56) | 1559(87) | | |
| *Homo sapiens* (human) | 954(1) | 1559(1) | 1869(38) | 5025(37) |
| *Pan troglodytes* (chimp) | 948(52) | | | |
| *Pongo pygmaeus* (orangutan) | 950(52) | | | |
| *Gallus gallus* (chicken) | 975(20d) | 1619(20d) | | |

significant, they are consistent with a rate acceleration in various protein coding genes in rodents as compared to *Homo* noted by others (15, 119), and indicate that, despite their relative conservation, rRNAs also tend to show evolutionary rate heterogeneity among taxa. That 12S and 16S rRNAs show limited evidence of rate heterogeneity extends the phenomenon to the mitochondrial genome, and suggests that the proposed cause of the rate increase (mutation associated with cell division of gametes) is not limited to the paternal lineage. Findings were the same despite use of two different outgroups in 12S comparisons, strengthening the result. The number of substitutions involved in similar 18S comparisons was too small to analyze.

**Table 2** Nucleotide substitution rate comparisons in species pairs relative to an outgroup.

| Outgroup | Ingroup species | (Number of substitutions) | Probability |
|---|---|---|---|
| *12S* | | | |
| *Gallus* | *Homo/Mus* | (129/142) | 0.2330 |
| *Xenopus* | *Homo/Mus* | (109/119) | 0.2756 |
| *Xenopus* | *Homo/Gallus* | (122/129) | 0.3525 |
| *Xenopus* | *Gallus/Mus* | (127/137) | 0.2899 |
| *16S* | | | |
| *Gallus* | *Homo/Mus* | (219/230) | 0.3185 |
| *28S* | | | |
| *Xenopus* | *Homo/Mus* | (53/69) | 0.0871 |

## *Rate Heterogeneity Among Genes*

To assess rate heterogeneity among genes, we aligned homologous rRNAs among species pairs, as above, and determined percentage sequence divergence (% divergence). We also determined the percentage of all substitutions that were transitions (% transitions) for consideration in the Rate Heterogeneity Within rRNAs section, below. To facilitate comparisons among genes we have plotted averages of these percentage values against estimated divergence time for multiple species pairs in four rRNAs (Figure 2, Table 1). Actual divergence times are unknown, and estimates used here (see 9, 14, 21, 42, 95 and references therein) should be considered approximate, relative values. Percent divergence scores were calculated as: 100(1-[number identical nucleotides/{number shared base positions + total number gaps}]). Coding of gaps in rRNAs in phylogenetic and divergence analyses is controversial, and ranges from the counting of each nucleotide positional gap as a separate character to discounting gaps entirely. We take a relatively conservative approach here in treating each gap equally, as a single character, regardless of its length in base positions.

As with other mitochondrial genes, mt rRNAs are changing at a faster rate than are their nuclear counterparts (Figure 2). For example, sequence comparisons between different mammalian orders, as represented by *Homo sapiens* and *Mus musculus*, had % divergence values of 25, 24, 8, and 1 in 16S, 12S, 28S, and 18S, respectively. Comparisons between *Drosophila yakuba* and various vertebrates had % divergence values ranging from 35 to 41 in 12S and 16S mt rRNAs, whereas comparisons involving similar divergence times between *Artemia salina* and vertebrates show 16% divergence in nuclear 18S rRNA. Comparisons involving *Xenopus laevis* follow this same trend. Accelerated rates of change in mt DNAs may be due to several factors, including decreased polymerase efficiency, or less efficient mechanisms for repair of mutations (10, 19, see 53).

Rates of change are similar in the two mt rRNAs (Figures 2a, b). For example, *Homo* and *Bos* are 21% divergent in comparisons of both 12S and 16S sequences. Similarly, *Mus* and *Drosophila* are 35% divergent in comparison of both 12S and 16S genes. Figures 2c and 2d illustrate the generally faster rate of change in 28S as compared to 18S sequences that is largely due to faster change in the 28S expansion segments, a point discussed below. Average % divergence scores for species pairs with estimated divergence times of 85 and 350 million years ago are 0.7 and 4.9 versus 8.1 and 10.5 in 18S and 28S, respectively.

Given continuing nucleotide change over time, % divergence values will eventually level off as species reach a maximal sequence divergence, with some portion of sites being identical due either to strong selective constraints or to chance. This saturation with change will occur faster over time as a

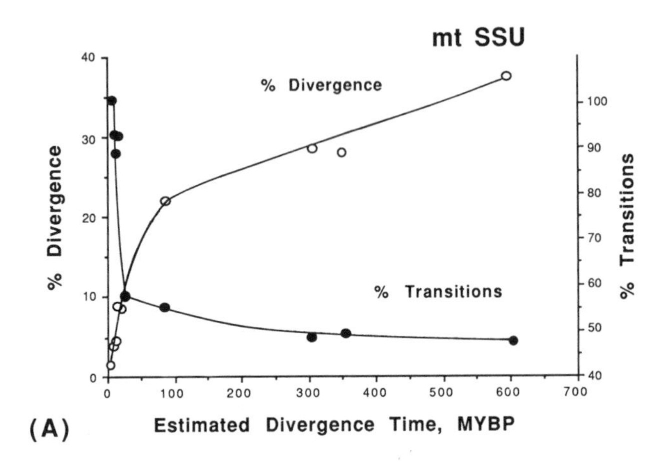
mt SSU
% Divergence
% Transitions
% Divergence
% Transitions
0
10
20
30
40
40
50
60
70
80
90
100
0
100
200
300
400
500
600
700
(A)
Estimated Divergence Time, MYBP

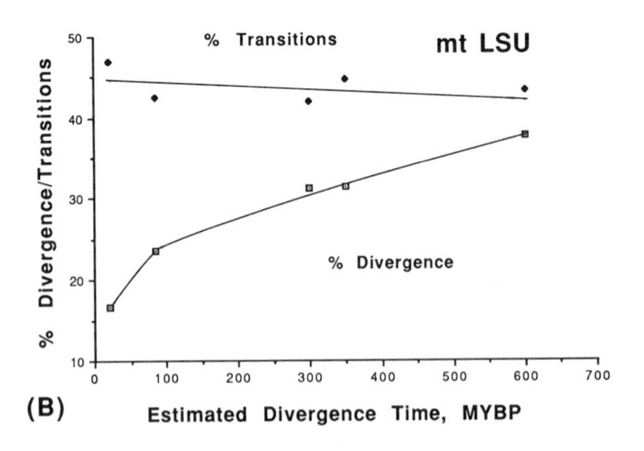
% Transitions
mt LSU
% Divergence
% Divergence/Transitions
10
20
30
40
50
0
100
200
300
400
500
600
700
(B)
Estimated Divergence Time, MYBP

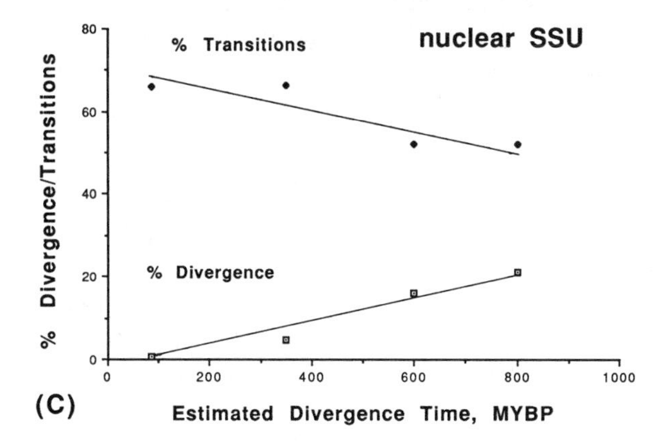


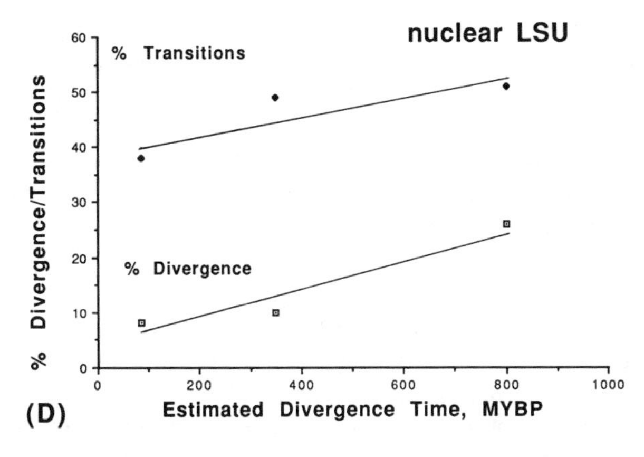


*Figure 2* Percent nucleotide sequence divergence and percent of all substitutions that are transitions versus estimated divergence time for species pairwise comparisons in: (*A*) mitochondrial (mt) small subunit (SSU) ribosomal RNA (rRNA), 12S gene in vertebrates; (*B*) mt large subunit (LSU) rRNA, 16S gene in vertebrates; (*C*) nuclear SSU rRNA, 18S in vertebrates; and (*D*) nuclear LSU rRNA, 28S in vertebrates. Sequence data sources are as listed in Table 1. Points represent averages for species comparisons having the same estimated divergence times. Percent sequence divergence was calculated as: 100(1-[no. identical nucleotides/{no. shared base positions+total number of gaps}]). Estimated times of divergence in millions of years before present (MYBP) for various species pairs are: > 800 for *Saccharomyces* and *Artemia* with others; 600 for *Drosophila* with vertebrates; 350 for *Xenopus* with other vertebrates; 300 for birds with mammals; 85 for comparisons among mammalian orders; 20 for within mammalian orders; and variously < 20 for primate divergences as in Hixson & Brown (52).

function of the overall rate of change. The initial portions of the % divergence curves (Figure 2) rise quickly, being relatively free of convergent sequence similarity due to multiple substitutions at a given site. 12S has the largest number of sequences available for comparison and, hence, points for fitting a curve. Saturating or dampening of 12S sequence divergence values begins under 100 million years ago (MYA), or around the time of origin for many mammalian orders (14). Divergence continues, however, at least as far back as 600 MYA, although much less quickly. Complete saturation and uninformativeness of nucleotide change would result in no increase in divergence over time and horizontal % divergence "lines" in Figure 2; such is not observed. In 16S comparisons (Figure 2b), the dampening of the % divergence curve is less visible because fewer sequence comparisons are possible between species whose divergences occurred within the past 100 million years. Reduction in slope of divergence curves will occur further back in time for 28S and 18S, in light of their slower rate of change. Percent divergence plots for 18S and 28S (Figures 2c,d) do not appear to dampen even with inclusion of comparisons between *Saccharomyces* and vertebrates with an estimated divergence time of over 800 MYA. 18S sequence comparisons of *Saccharomyces* with *Mus*, *Homo*, and *Artemia* are all less than 22% divergent. These rate differences among rRNAs are based on averages including all types of nucleotide change within genes and, given within-gene variability in rates, they provide only a first-level approximation of the variability present. This approximation, however, can indicate relative suitability of the genes for particular phylogenetic issues.

## *Rate Heterogeneity within rRNAs*

Although only average % transition values are plotted in Figure 2, we found the % transition values to be more variable, for divergences of a given age than are the % divergence values discussed above. For 12S sequences, percent transitions for comparisons between representatives of different mammalian orders (estimated age of divergence is 85 MYA) ranges from 64 (*Bos/Homo*) to 43 (*Bos/Mus*), whereas the % divergence values for the same comparisons are 21 each. Similarly, % transitions for conordinal comparisons (estimated age of divergence is 20 MYA) range from 79 (*Bos/Giraffa*) to 57 (*Mus/Rattus*), whereas the corresponding % divergence values of 12 and 9 have a smaller range. Thus, the transition substitution rate seems less predictable than the overall rate of change as indicated by the % divergence figures. Saturation of transitional change precedes whole sequence saturation due to the faster rate of transition substitutions. As seen in 12S (Figure 2a), there is a levelling-off at minimal values of about 50%, beginning about 20 MYA, though the decrease continues gradually as far back in time as 600 MYA. Percent transitions for 16S is less than 50 for divergences of 20 MYA and remains over 40 for divergences of 600 MYA (Figure 2b). Percent transitions in 18S comparisons (Figure 2c) do not reach 50 until much later in time, about 600 MYA, as might be expected with its slower overall rate of change. Percent transitions in a *Homo/Xenopus* comparison are 67 and 47 in 18S and 12S, respectively. Percent transition values for comparisons of 28S sequences (Figure 2d) are closer to the mt rRNA values, due to more rapid change in expansion segments and small numbers of substitutions in other gene regions. Lower % transitions in 28S for divergences of 85 MYA compared to later divergences are most likely an artifact of the relatively small number of 28S comparisons made and substitutions inferred.

Considering the percentage of identical nucleotide base positions for 28S comparisons between species pairs, we find that regions outside of the expansion segments (ES), exact positions defined for *Mus musculus* in Hassouna et al (94)—are as conserved as is nucleotide change in the 18S gene. The 28S expansion segment sequences, however, are changing more quickly than both 18S sequences and 28S nonexpansion segments, and more slowly than the mt rRNAs. Illustrating this, we found *Xenopus/Mus* comparisons of % identical nucleotide positions to be: 12S/61, 16S/68, 18S/94, 28S ES/91, and 28S non-ES/97. Similarly *Homo/Mus* comparisons are: 12S/76, 16S/75, 18S/99, 28S ES/89, and 28S non-ES/99.7. Detailed 28S expansion segment analyses by Larson & Wilson in salamanders (62) showed a mean substitution rate of 0.32 within expansion segments and only 0.05 outside them. Gonzalez et al (37) found as much variation in the D6 28S expansion segment among six humans as between humans and a chimpanzee.

To illustrate regional, within-gene rate heterogeneity, we compare divergence rates for a 140 base pair segment of 12S rRNA with rates for the whole gene. This segment corresponds to positions 1135 through 1275 of the complete human mtDNA sequence (1), and comparisons made include five species not appearing in Table 1: *Sphenodon punctatus* (tuatara), *Rhea americana* (rhea), *Struthio camelus* (ostrich), *Spheniscus humbolti* (penguin) (D. Mindell, T. Quinn, A. C. Wilson, unpublished data), and *Cichlasoma citrinellum* (cichlid) (A. Meyer & A. C. Wilson, submitted). These sequences were obtained using polymerase chain reaction primers for 12S as described in Kocher et al (57). Overall, rate of change is slower across species in this gene region (Figure 3) than in the 12S sequence as a whole (Figure 2a). Comparison of *Mus* and *Homo* % divergence values across the entire 12S gene versus the 140 base pair segment is 24 and 10, respectively, and in *Xenopus/Mus* comparisons, the corresponding % divergence values are 28 and 18, respectively. Most of this 140 base pair segment is highly conserved in primary sequence and in apparent secondary structure in all species that have been sequenced (see Ref. 102, pp. 68–69). Figure 3 illustrates that the same relative divergence patterns seen in Figure 2a hold in the representatives

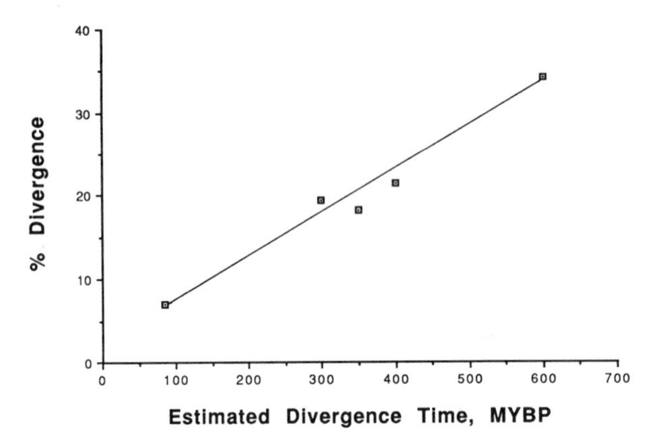


*Figure 3* Percent nucleotide sequence divergence versus estimated divergence time in millions of years before present (MYBP) for pairwise species comparisons based on published (Table 1) and unpublished sequences (see text) for a relatively conserved 140 base pair region of the 12S mitochondrial ribosomal RNA gene. Percent divergence calculations and divergence time estimates are as in Figure 2.

of two additional vertebrate classes, Aves and Reptilia (divergence time among birds, mammals, and reptiles is estimated to be 300 MYA). Representatives of avian orders had lower % divergence values than did members of different mammalian orders in these 12S segment comparisons, and this could be due to differences in age or rate of change across taxa. These partial 12S gene comparisons demonstrate that sequence position effects can yield substantial rate heterogeneity.

Ribosomal RNA sequences cover a broad spectrum of evolutionary change rates, making them potentially applicable to most phylogenetic issues among organisms. We summarize recent phylogenetic studies of vertebrates using rRNAs in Table 3. Relative success of these studies in accurately determining phylogeny is difficult to assess. Phylogenetic congruence among independent data sets, however, provides the best indication of accuracy, and on this basis it appears that rRNAs can be informative (see references in Table 3), particularly where knowledge of rate variability is applied to phylogenetic analyses. Analysis of bacterial rRNAs, particularly conserved core regions of the small subunit, have provided significant advances in understanding early branching events among prokaryotes and eukaryotes (31, 61, 115). Within eukaryotes, nuclear rRNA sequences have been used in phylogenetic studies of protoctista (25), of plant origins (116, 120), and within and among vertebrate classes (e.g. 50, 62, 72). Rapidly evolving transcribed and nontranscribed spacer regions have been used in analysis of more closely related taxa including populations within species (48, 100). Mitochondrial 12S rRNA sequences have been used less often in phylogenetic analyses; however, recent analyses

**Table 3** Comparative studies using ribosomal RNA (rRNA) genes as phylogenetic characters in vertebrates and in representatives of other eukaryotes. Abbreviations: nucl, nuclear; mt, mitochondrial; cp, chloroplast; rsa, restriction site analysis; seq, nucleotide sequence; tr initiat, transcription initiation; NTS, nontranscribed spacer; ITS, internal transcribed spacer; SSU, small subunit rRNA; LSU, large subunit rRNA.

| Taxonomic group | Source | Gene/region | Type of comparison | Reference |
|---|---|---|---|---|
| VERTEBRATES | | | | |
| Multiple classes | nucl | 28S | seq | 50, 50a |
| Amphibia | | | | |
| Caudata | nucl | 28S | seq | 62 |
| Anura | nucl | 28S | rsa | 48a |
| *Rana* | nucl | 18S, 28S | rsa | 48 |
| *Xenopus* | nucl | ITS, 28S | seq | 33a, 20a |
| Reptilia | | | | |
| *Sceloporus* | nucl | 18S, 28S | rsa | 96a |
| Mammalia | | | | |
| Marsupials | mt | 12S | seq | 103 |
| Primates | mt | 12S | seq | 52 |
| Primates | nucl | 28S | seq | 38a, 94a |
| Primates | ncul | tr initiat | rsa | 113a |
| Artiodactyla | mt | 12S, 16S | seq | 72a, 72b |
| *Geomys* | nucl | 18S, 28S, ITS | rsa | 20b, 5a |
| *Spalax* | nucl | NTS | rsa | 100, 77a |
| *Mus* | nucl | NTS | rsa | 99a |
| Aves | | | | |
| Multiple orders | nucl | 18S, 28S, ITS | rsa | 72 |
| OTHER EUKARYOTES | | | | |
| Multiple kingdoms | nucl, mt, cp | SSU, LSU | seq | 15a |
| Multiple animal phyla | nucl | SSU | seq | 61a |
| Seed plants | nucl | SSU, LSU | seq | 120 |
| Angiosperms | nucl, cp | SSU, LSU | seq | 116 |

of 12S genes have been made among vertebrates such as frogs (13), primates (52), and marsupials (103). Miyamoto & Boyle (72a) and Miyamoto et al (72b) have found 12S and 16S mt rRNAs to provide a single most parsimonious tree for intraordinal and interordinal relationships of select eutherian mammals that is generally, though not uniformly, congruent with phylogenetic hypotheses based on morphology and other molecular data sets. Analysis of 28S and internal transcribed spacer sequences has indicated a sister relationship for humans and chimpanzees among hominoids, in agreement with analyses based on DNA hybridization and globin gene nucleotide sequences (38a).

We will describe here an approach to applying the rRNA rate heterogeneity

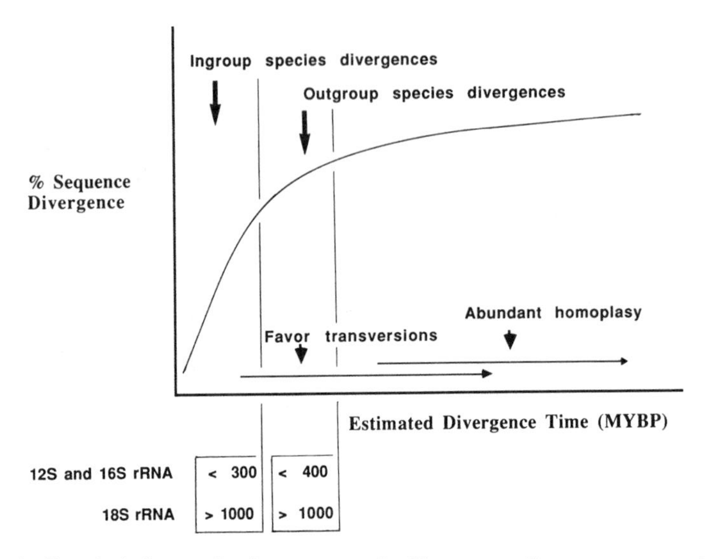


*Figure 4* Hypothetical curve showing percent nucleotide sequence divergence versus estimated divergence time for pairwise species comparisons with phylogenetic application regions noted (see text). Numbers at bottom are estimated divergence time estimates in millions of years before present (MYBP) for three ribosomal RNAs, based on Figure 2.

characterizations in phylogenetic analyses. We do not intend to make specific recommendations for particular phylogenetic questions, however, as that requires more detailed knowledge of rRNA evolutionary rate variation within and among taxa. Figure 4 includes a hypothetical curve showing percent nucleotide sequence divergence between species pairs plotted against estimated time since divergence (as in Figures 2 and 3), with salient features regarding phylogenetics noted. To match nucleotide sequences to appropriate phylogenetic questions, ingroup species divergence times should fall under the initial portion of the curve (approximately during the past 150 million years [300 million years at most] for 12S and 16S rRNA analyses; see Figure 2), where multiple substitutions at a given site and subsequent homoplasy are reduced. This will reduce the amount of phylogenetically uninformative, or misleading, nucleotide character change included in analyses. In the Phylogenetic Techniques section we discussed the need to maintain continuity of information in nucleotide sequences by attempting to align them in decreasing order of phylogentic relationship, lest the alignments and subsequent phylogenetic analyses be adversely affected by homoplasious similarity between distant relatives. Conservatively, taxa whose divergences fall within the region of abundant homoplasy in Figure 4 could be excluded from both phylogenetic and alignment analyses for this same reason. If species with divergences not falling on the initial steepest portion of the curve are included, alignment and phylogenetic analyses will need to take increasing levels of homoplasy into account.

Outgroups are used in phylogenetic analyses to estimate ancestral character states. The divergence times for the outgroups must precede ingroup divergences, yet they should also avoid falling under that portion of the curve where comparisons include abundant homoplasy for reasons stated above. Multiple phylogenetic analyses using different individual and combinations of outgroup species can lend credence to the result if the same relationships are found in each trial. Outgroup selection is crucial in assessing phylogenetic relationships, yet criteria for making that selection have been difficult to identify and apply. Choosing outgroup taxa whose divergences from the ingroup taxa predate ingroup divergences yet still occur along the initial portion of the % divergence curve is one criterion that may be applied, if sufficient information from other species is available.

Transitions occur at a faster rate than transversions, and for species whose divergences fall under the shoulder and relatively horizontal portions of the % divergence and % transition curves (Figures 2 and 4), transversions may be favored in phylogenetic analyses by giving them more weight. Transitions begin to saturate with change when they represent about 50% of the substitutions, which is often the case for species divergences of over 50 MYA in 12S and 16S rRNAs (Figure 2). In considering relationships among species with divergences not occurring on the steepest portion of % transition curves (Figure 2), transitions may be omitted or transversions may be given additional weight. Because rates of change vary between sequence regions, greater phylogenetic resolution may be attained by considering transition to transversion ratios in various sequence regions before assigning weights for phylogenetic analyses. Phylogenetic analysis of rRNA sequences remains problematic. We stress that the % divergence and % transition curves are only averages, providing estimated guidelines based on a small number of species for whom comparable sequences are available. Species differences in abundance of the different nucleotide bases may significantly affect % divergence and % transition figures (20c), and should be considered in future analyses. The incidence and confounding effects of sequence polymorphism, especially for nuclear rRNAs present in repeat copies, are little known and need to be investigated.

## CONCLUSIONS

Ribosomal RNAs are potentially applicable to most issues of organismal phylogeny. This is due to their functionally important role in protein synthesis, which makes them relatively resistant to evolutionary change, and the presence of a broad range of rate variability both among and within regions.

Based on species comparisons and % divergence figures, 12S and 16S mt rRNAs are potentially informative of genealogy among taxa that diverged as far back as 300 MYA, although they seem best suited for divergences of about 150 MYA or less. Nuclear 18S is the most slowly changing of the four rRNAs assessed; sequences for a frog and a human are less than 5% divergent. Nuclear 28S rRNA changes faster than 18S, due to variable expansion segments, but more slowly than the mt rRNAs. Transition substitutions appear less predictable in their initial rate of accumulation compared to all substitutions, and in the 12S gene, transitions begin to saturate with change for divergences of about 50 MYA, and upon accounting for about 50% of all substitutions. Slower rate of character change in a 140 base pair segment of the 12S gene demonstrates the capability for rRNA sequence position or location to affect evolutionary rate. In comparing 12S, 16S, and 28S nucleotide character change between *Mus musculus* and *Homo sapiens* or *Gallus gallus* relative to two different outgroups, *Mus* tended to show a faster rate of change. Information on variable rates of nucleotide character change should be used, to avoid incorporating misinformation in both sequence alignments and subsequent phylogenetic analyses, and thereby increase our precision in phylogenetic analyses.

ACKNOWLEDGMENTS

We would like to thank Thomas Quinn, Kelley Thomas, Wen-Hsiung Li, Thomas Beck, Joel Cracraft, Brian Hanks and Marc Allard for thoughtful and valuable discussions pertaining to this manuscript. Réjean Morais and Paul Desjardins kindly provided a preprint of chicken 12S and 16S rRNA sequences. We also thank Thomas Quinn, Axel Meyer, and Allan Wilson for allowing us to include various unpublished 12S mitochondrial sequences. Christopher Dick and Steven Sawchuk helped with aspects of nucleotide sequencing. This work was facilitated by a University Research Council grant from the University of Cincinnati to D. P. Mindell.

*Literature Cited*

1. Anderson, S., Bankier, A. T., Barrell, B. G., de Bruijn, M. H. L., Coulson, A. R., et al. 1981. Sequence and organization of the human mitochondrial genome. *Nature* 290:457–65
2. Anderson, S., de Bruijn, M. H. L., Coulson, A. R., Eperon, E. C., Sanger, R., Young, I. G. 1982. Complete sequence of bovine mitochondrial DNA: conserved features of the mammalian mitochondrial genome. *J. Mol. Biol.* 156:683–717
3. Appels, R., Honeycutt, R. L. 1986. rDNA: evolution over a billion years. In *DNA Systematics*, ed. S. K. Dutta, pp. 81–135. Boca Raton: CRC
4. Arnheim, N. 1979. Characterization of mouse ribosomal gene fragments by molecular cloning. *Gene* 7:83–96

4a. Atmadja, J., Brimacombe, R., Maden, B. E. H. 1984. *Xenopus laevis* 18S ribosomal RNA: Experimental determination of secondary structure elements and locations of methyl groups in the sec-

ondary structure model. *Nucl. Acids Res.* 12:2649–67
5. Bacon, D. J., Anderson, W. F. 1986. multiple sequence alignment. *J. Mol. Biol.* 191:153–61
5a. Baker, R. J., Davis, S. K., Bradley, R. D., Hamilton, M. J., Van Den Bussche, R. A. 1989. Ribosomal-DNA, mitochondrial-DNA, chromosomal, and allozymic studies on a contact zone in the pocket gopher, *Geomys*. *Evolution* 43:63–75
6. Bains, W. 1986. MULTAN: A program to align multiple DNA sequences. *Nucleic Acids Res.* 14:159–77
7. Benveniste, R. E., Todaro, G. J. 1975. Evolution of type C viral genes: preservation of ancestral murine type C viral sequences in pig cellular DNA. *Proc. Natl. Acad. Sci. USA* 72:4090–94
8. Bielka, H., ed. 1982. *The Eukaryotic Ribosome*. New York: Springer-Verlag
9. Britten, R. J. 1986. Rates of DNA sequence evolution differ between taxonomic groups. *Science* 231:1393–98
10. Brown, W. M., George, M. J., Wilson, A. C. 1979. Rapid evolution of animal mitochondrial DNA. *Proc. Natl. Acad. Sci. USA* 76:1967–71
11. Brown, W. M., Prager, E. M., Wang, A., Wilson, A. C. 1982. Mitochondrial DNA sequences of primates: tempo and mode of evolution. *J. Mol. Evol.* 18: 225–39
12. Carpenter, J. M. 1988. Choosing among multiple equally parsimonious cladograms. *Cladistics* 4:291–96
13. Carr, S. M., Brothers, A. J., Wilson, A. C. 1987. Evolutionary inferences from restriction maps of mitochondrial DNA from nine taxa of Xenopus frogs. *Evolution* 41:176–8
14. Carroll, R. L. 1988 *Vertebrate Paleontology and Evolution*. 1988. New York: W. H. Freeman
15. Catzeflis, F. M., Sheldon, F. H., Ahlquist, J. E., Sibley, C. G. 1987. DNA-DNA hybridization evidence of the rapid rate of muroid rodent DNA evolution *Mol. Biol. Evol.* 4:242–53
15a. Cedergren, R., Gray, M. W., Abel, Y., Sankoff, D. 1988. The evolutionary relationships among known life forms. *J. Mol. Evol.* 28:98–112
16. Clark, C. G. 1987. On the evolution of ribosomal RNA. *J. Mol. Evol.* 25:343–50
17. Clark, C. G., Tague, B. W., Ware, V. C., Gerbi, S. A. 1984. *Xenopus laevis* 28S ribosomal RNA: a secondary structure model and its evolutionary and functional implications. *Nucleic Acids Res.* 12:6197–20
18. Clary, D. O., Wolstenholme, D. R. 1985. The ribosomal RNA genes of *Drosophila* mitochondrial DNA. *Nucleic Acids Res.* 13:4029–45
19. Clayton, D. A., Doda, J. N., Friedberg, E. C. 1974. The absence of a pyrimidine dimer repair mechanism in mammalian mitochondria. *Proc. Natl. Acad. Sci. USA* 71:2777–81
20. Cox, R. A., Hüvös, P., Godwin, E. A. 1973. Unusual features of the structure of the major RNA component of the larger subribosomal particle of the rabbit reticulocyte. *Israel J. Chem.* 11:407–22
20a. Cutruzzola, F., Loreni, F., Bozzoni, I. 1986. Complementarity of conserved sequence elements present in 28S ribosomal RNA and in ribosomal protein genes of *Xenopus* laevis and X. tropicalis. *Gene* 49:371–76
20b. Davis, S. K. 1986. *Population structure and patterns of speciation in Geomys (rodentia: Geomyidae): An analysis using mitochondrial and ribosomal DNA*. PhD thesis. Washing. Univ., St. Louis
20c. DeSalle, R. T., Freedman, E. M., Wilson, A. C. 1987. Tempo and mode of sequence evolution in mitochondrial DNA of Hawaiian *Drosophila*. *J. Mol. Evol.* 26:157–64
20d. Desjardins, P., Morais, R. 1990. Sequence and gene organization of the chicken mitochondrial genome. *J. Mol. Biol.* 212:599–634
21. Dickerson, R. E., Geis, I. 1983. *Hemoglobin*. 1983. Menlo Park, Calif: Benjamin/Cummings
22. Donoghue, M. J. 1989. Phylogenies and the analysis of evolutionary sequences, with examples from seed plants. *Evolution* 43:1137–56
23. Dunon-Bluteau, D., Brun, G. 1986. The secondary structures of the *Xenopus laevis* and human mitochondrial small ribosomal subunit RNA are similar. *FEBS* 198:333–38
24. Ebel, J. P., Branlant, C., Carbon, P., Ehresmann, B., Ehresmann, C., et al. 1983. Sequence and secondary structure conservation in ribosomal RNAs in the course of evolution. In *Nucleic Acids: The Vectors of Life,* ed. B. Pullman, J. Jortner, pp. 387–401. Durdrecht: D. Reidel
25. Elwood, J. J., Olsen, G. H., Sogin, M. L. 1985. The small-subunit ribosomal RNA gene sequences from the hypotrichous ciliates *Oxytricha nova* and *Stylogychia pustulata*. *Mol. Biol. Evol.* 2:399–410
26. Farris, J. S. 1969. A successive approx-

imations approach to character weighting. *Syst. Zool.* 18:374–85

27. Farris, J. S. 1981. Distance data in phylogenetic analysis. In *Advances in Cladistics: Proceedings of the First Meeting of the Willi Hennig Society*, ed. W. A. Funk, D. R. Brooks, pp. 3–23. Bronx: NY Bot. Gard.
28. Farris, J. S. 1983. The logical basis of phylogenetic analysis. In *Advances in Cladistics*, ed. N. W. Platnick, V. A. Funk, pp. 7–36. New York: Columbia Univ. Press
29. Farris, J.S. 1988. Hennig86, version 1.5 Stony Brook, NY
30. Felsenstein, J. 1988. Phylogenies from molecular sequences: inference and reliability. *Annu. Rev. Genet.* 22:521–65

30a. Feng, D., Doolittle, R. F. 1987. Progressive alignment as a prerequisite to correct phylogenetic trees. *J. Mol. Evol.* 25:351–60

31. Field, K. G., Olsen, G. J., Lane, D. J., Giovannoni, S. J., Ghiselin, M. T., et al. 1988. Molecular phylogeny of the animal kingdom. *Science* 239:748–53
32. Fitch, W. M. 1967. Evidence suggesting a non-random character to nucleotide replacements in naturally occurring mutations. *J. Mol. Biol.* 26:499–507
33. Fitch, W. M. 1986. Commentary. *Mol. Biol. Evol.* 3:296–98

33a. Furlong, J. C., Maden, B. E. H. 1983. Patterns of major divergence between the internal transcribed spacers of ribosomal DNA in *Xenopus borealis* and *X. laevis*, and of minimal divergence within ribosomal coding regions. *EMBO* 2:443–48

34. Georgiev, O. I., Nikolaev, N., Hadjiolov, A. A., Skryabin, K. G., Zakharyev, V. M., Bayev, A. A. 1981. The structure of the yeast ribosomal RNA genes. 4. Complete sequence of the 25S rRNA gene from Saccharomyces cerevisiae. *Nucleic Acids Res.* 9:6953–58
35. Gerbi, S. A. 1985. Evolution of Ribosomal DNA. In *Molecular Evolutionary Genetics*, ed. R. J. MacIntrye, pp. 419–517. New York: Plenum
36. Gillespie, J. H. 1986. Rates of molecular evolution. *Annu. Rev. Ecol. Syst.* 17:637–65
37. Gonzalez, I. L., Gorski, J. L., Campen, T. J., Dorney, D. J., Erickson, J. M., et al. 1985. Variation among human 28S ribosomal RNA genes. *Proc. Natl. Acad. Sci. USA* 82:7666–70
38. Gonzalez, I. L., Schmickel, R. D. 1986. The human 18S ribosomal RNA gene: evolution and stability. *Am. J. Hum. Genet.* 38:419–27

38a. Gonzalez, I. L., Sylvester, J. E., Smith, T. F., Stambolian, D., Schmickel, R. D. 1990. Ribosomal RNA gene sequences and Hominoid Phylogeny. *Mol. Biol. Evol.* 7:203–19

39. Gray, M. W., Sankoff, D., Cedergren, R. J. 1984. On the evolutionary descent of organisms and organelles: a global phylogeny based on a highly conserved structural core in small subunit ribosomal RNA. *Nucleic Acids Res.* 12:5837–52
40. Gutell, R. R., Fox, G. E. 1988. A compilation of large subunit RNA sequences presented in a structural format. *Nucleic Acids Res.* 16 Supplement: r175–r270
41. Gyllensten, U. B., Erlich, H. A. 1988. Generation of single-stranded DNA by the polymerase chain reaction and its application to direct sequencing of the HLA-DQA locus. *Proc. Natl. Acad. Sci. USA* 85:7652–56
42. Hasegawa, M., Yano, T., Kishino, H. 1984. A new molecular clock of mitochondrial DNA and the evolution of hominoids. *Proc. Jpn. Acad. (Nihon Gakushiin)* [B]. 60:95–98
43. Hassouna, N., Michot, B., Bachellerie, J. 1984. The complete nucleotide sequence of mouse 28S rRNA gene. Implications for the process of size increase of the large subunit rRNA in higher eukaryotes. *Nucleic Acids Res.* 12:3563–83
44. Henderson, A. S., Eicher, E. M., Yu, M. T., Atwood, K. C. 1976. Variation in ribosomal RNA gene number in mouse chromosomes. *Cytogen. Cell. Genet.* 17:307–16
45. Henning, W. 1966. *Phylogenetic Systematics*. Urbana: Univ. Ill. Press
46. Higgins, D. G., Sharp, P. M. 1989. Fast and sensitive multiple sequence alignments on a microcomputer. *CABIOS* 5: 151–53
47. Hillis, D. M. 1984. Misuse and modification of Nei's genetic distance. *Syst. Zool.* 33:238–40
48. Hillis, D. M., Davis, S. K. 1986. Evolution of ribosomal DNA: fifty million years of recorded history in the frog genus *Rana*. *Evolution* 40:1275–88

48a. Hillis, D. M., Davis, S. K. 1987. Evolution of the 28S ribosomal RNA gene in Anurans: regions of variability and their phylogenetic implications. *Mol. Biol. Evol.* 4:117–25

49. Hillis, D. M., Davis, S. K. 1988. Ribosomal DNA: intraspecific polymorphism, concerted evolution, and phylogeny reconstruction. *Syst. Zool.* 37:63–66
50. Hillis, D. M., Dixon, M. T. 1989. Vertebrate phylogeny: evidence from 28S ribosomal DNA sequences. In *The*

*Hierarchy of Life*, B. Fernholm, K. Bremer, H. Jörnvall, pp. 355–67. Amsterdam: Elsevier.

50a. Hillis, D. M., Dixon, M. T., Ammerman, L. K. 1990. The relationships of coelacanths: evidence from sequences of vertebrate 28S ribosomal RNA genes. In *The Biology and Evolution of Coelacanths*, J. A. Musiak, M. Bruton. Environmental Biol. of Fishes, Special Vol., in press

51. Hirschberg, D. S. 1975. A linear space algorithm for computing longest common subsequences. *Commun. Assoc. Comput. Mach.* 18:341–43

52. Hixson, J. E., Brown, W. M. 1986. A comparison of the small ribosomal RNA genes from the mitochondrial DNA of the great apes and humans: sequence, structure, evolution, and phylogenetic implications. *Mol. Biol. Evol.* 3:1–18

53. Holland, J., Spindler, K., Horodyski, F., Grabau, E., Nichol, S., VandePol, S. 1982. Rapid evolution of RNA genomes. *Science* 215:1577–85

54. Klein, B. K., Forman, P., Shiomi, Y., Schlessinger, D. 1984. Electron microscopy of secondary structure in partially denatured *Escherichia coli* 16S-rRNA and 30S subunits. *Biochemistry* 23: 3927–33

55. Kluge, A. G., Farris, J. S. 1969. Quantitative phyletics and the evolution of anurans. *Syst. Zool.* 18:1–32

56. Kobayashi, M., Seki, T., Yaginuma, K., Koike, K. 1981. Nucleotide sequences of small ribosomal RNA and adjacent transfer RNA genes in rat mitochondrial DNA. *Gene* 16:297–307

57. Kocher, T. D., Thomas, W. K., Meyer, A., Edwards, S. V., Pääbo, S., Villablanca, F. X. 1989. Dynamics of mitochondrial DNA evolution in animals: amplification and sequencing with conserved primers. *Proc. Natl. Acad. Sci. USA* 86: 6196–6200

58. Kohne, D. E. 1970. Evolution of higher-organism DNA. *Q. Rev. Biophys.* 3: 327–75

59. Kohne, D. E., Chiscon, J. A., Hoyer, B. H. 1972. Evolution of primate DNA sequences. *J. Human Evol.* 1:627–44

60. Laird, C. D., McConaughy, B. L., McCarthy, B. J. 1969. Rate of fixation of nucleotide substitutions in evolution. *Nature* 224:149–54

61. Lake, J. A. 1988. Origin of the eukaryotic nucleus determined by rate-invariant analysis of rRNA sequences. *Nature* 331:184–86

61a. Lake, J. A. 1990. Origin of the metazoa. *Proc. Natl. Acad. Sci. USA* 87:763–66

62. Larson, A., Wilson, A. C. 1989. Patterns of ribosomal RNA evolution in salamanders. *Mol. Biol. Evol.* 6:131–54

63. Li, W., Gojobori, T., Nei, M. 1981. Pseudogenes as a paradigm of neutral evolution. *Nature* 292:237–39

64. Li, W., Tanimura, M. 1987. The molecular clock runs more slowly in man than in apes. *Nature* 326:93–96

65. Li, W., Wu, C., Luo, C. 1985. A new method for estimating synonymous and non-synonymous rates of nucleotide substitution considering the relative likelihood of nucleotide and codon changes. *Mol. Biol. Evol.* 2:150–74

66. Lockard, R. E., Currey, K., Browner, M., Lawrence, C., Maizel, J. 1986. Secondary structure model for mouse B major globin mRNA derived from enzymatic digestion data, comparative sequence and computer analysis. *Nucleic Acids Res.* 14:5827–5841

67. Maizel, J. V., Lenk, R. P. 1981. Enhanced graphic matrix analysis of nucleic acid and protein sequences. *Proc. Natl. Acad. Sci. USA* 78:7665–69

68. Maly, R., Brimacombe, R. 1982. Refined secondary structure models for the 16S and 23S ribosomal RNA of *Escherichia coli. Nucleic Acids Res.* 11:7263–86

69. Mankin, A. S., Kopylov, A. M. 1981. A secondary structure model for mitochondrial 12S rRNA: an example of economy in rRNA structure. *Biochem. Int.* 3: 587–93

70. Margulis, L. 1981. *Symbosis in Cell Evolution: Life and Its Environment on the Early Earth*. San Francisco: Freeman

71. McClements, W., Skalka, A. M. 1977. Analysis of chicken ribosomal RNA genes and construction of lambda hybrids containing gene fragments. *Science* 196:195–97

71a. Mindell, D. P. 1991. Aligning DNA sequences: homology and phylogenetic weighting. In *Phylogenetic Analysis of DNA Sequences*, ed. M. M. Miyamoto, J. Cracraft. London: Oxford Univ. Press. In press

72. Mindell, D. P., Honeycutt, R. L. 1989. Variability in transcribed regions of ribosomal DNA and early divergences in birds. *Auk* 106:539–48

72a. Miyamoto, M. M., Boyle, S. M. 1989. The potential importance of mitochondrial DNA sequence data to eutherian mammal phylogeny. In *The Hierarchy of Life*, B. Fernholm, K. Bremmer, H. Jornvall, pp. 437–50. Amsterdam: Elsevier.

72b. Miyamoto, M. M., Tanhauser, S. M., Laipis, P. J. 1989. Systematic relationships in the artiodactyl tribe Bovini (family Bovidae), as determined from mitochondrial DNA sequences. *Syst. Zool.* 38:342–49

73. Miyata, T., Yasunaga, T., Nishida, T. 1980. Nucleotide sequence divergence and functional constraint in mRNA evolution. *Proc. Natl. Acad. Sci. USA* 77:7328–32
74. Mount, D. W. 1984. Modeling RNA structure. *Bio/Technology*. September: 791–95
75. Myers, W. W., Miller, W. 1988. Optimal alignments in linear space. *CABIOS* 4:11–17
76. Nei, M. 1971. Interspecific gene differences and evolutionary time estimated from electrophoretic data on protein identity. *Am. Nat.* 105:385–98
77. Nelles, L. G., Fang, B., Volckaert, G., Vandenberghe, A., De Wachter, R. 1984. Nucleotide sequence of a crustacean 18S ribosomal RNA gene and secondary structure of eukaryotic small subunit ribosomal RNAs. *Nucleic Acids Res.* 12:8749–68

77a. Nevo, E., Beiles, A. 1988. Ribosomal DNA non-transcribed spacer polymorphism in subterranean mole rats: genetic differentiation, environmental correlates and phylogenetic relationships. *Evol. Ecol.* 2:139–56

78. Noller, H. F. 1984. Structure of ribosomal RNA. *Annu. Rev. Biochem.* 53:119–62
79. O'Brien, T. W., Denslow, N. D., Harville, T. O., Hessler, R. A., Matthews, D. E. 1980. Functional and structural roles of proteins in mammalian mitochondrial ribosomes. In *Organization and Expression of the Mitochondrial Genome,* ed. C. Saccone, A. M. Kroon, pp. 301–5. New York: Elsevier/North-Holland Biomedical Press.
80. Patterson, C. 1987. Introduction. In *Molecules and Morphology in Evolution: Conflict or Compromise?,* ed. C. Patterson, pp. 1–22. London: Cambridge Univ. Press
81. Penny, D., Hendy, M. D. 1985. Testing methods of evolutionary tree construction. *Cladistics* 1:266–78
82. Qu, L., Nicoloso, M., Bachellerie, J. 1988. Phylogenetic calibration of 5' terminal domain of large rRNA achieved by determining twenty eukaryotic sequences. *J. Mol. Evol.* 28:113–24
83. Quigley, G. J., Gehrke, L., Roth, D. A., Auron, P. E. 1984. Computer-aided nucleic acid secondary structure modeling incorporating enzymatic digestion data. *Nucleic Acids Res.* 12:347–66
84. Raynal, R., Michot, B., Bachellerie, J. 1984. Complete nucleotide sequence of mouse 18S rRNA gene: Comparison with other available homologs. *FEBS Lett.* 167:263–68
85. Roe, B. A., Ma, D., Wilson, R. K., Wong, J. F. 1985. The complete nucleotide sequence of the *Xenopus laevis* mitochondrial genome. *J. Biol. Chem.* 260:9759–74
86. Rubstov, P. M., Musakhanov, M. M., Zakharyev, V. M., Krayev, A. S., Skryabin, K. G., Bayev, A. A. 1980. The structure of the yeast ribosomal RNA genes. I. The complete nucleotide sequence of the 18S ribosomal RNA gene from *Saccharomyces cerevisiae*. *Nucleic Acids Res.* 8:5779–94
87. Saccone, C., Cantatore, P., Gadaleta, G., Gallerani, R., Lanave, C., et al. 1981. The nucleotide sequence of the large ribosomal RNA gene and the adjacent tRNA genes from rat mitochondria. *Nucleic Acids Res.* 9:4139–48
88. Saiki, R. K. 1985. Enzymatic amplification of beta-globin genomic sequences and restriction site analysis for diagnosis of sickle cell anemia. *Science* 230:1350–54
89. Salim, M., Maden, B. E. H. 1981. Nucleotide sequence of *Xenopus laevis* 18S ribosomal RNA inferred from gene sequence. *Nature* 291:205–8
90. Salser, W. 1977. Globin messenger-RNA sequences: analysis of base-pairing and evolutionary implications. *Cold Spring Harbor Symp. Quant. Biol.* 42:985–1002
91. Sankoff, D. D., Cedergren, R. J. 1983. Simultaneous comparison of three or more sequences related by a tree. In *Time Warps, String Edits, and Macro-Molecules: the Theory and Practice of Sequence Comparison,* ed. D. Sankoff, J. B. Kruskal, pp. 253–63. Reading, Mass: Addison-Wesley
92. Sankoff, D. D., Morel, D., Cedergren, R. J. 1973. Evolution of 5S RNA and the nonrandomness of base replacement. *Nature New Biol.* 245:232–34
93. Sarich, V., Wilson, A. C. 1967a. Rates of albumin evolution in primates. *Proc. Natl. Acad. Sci. USA* 58:142–47
94. Sarich, V. M., Wilson, A. C. 1967b. Immunological time scale for hominid evolution. *Science* 158:1200–2

94a. Schmickel, R. D., Sylvester, J., Stambolian, D., Gonzalez, I. L. 1990. The ribosomal DNA sequence for the study of the evolution of human, chimpanzee, and gorilla. In *DNA Systematics,* Vol. III, S. K. Dutta, W. P. Winter; pp. 11–32. Boca Raton: CRC Press

95. Schopf, J. W., Hayes, J. M., Walter, M. R. 1983. Evolution of Earth's earliest ecosystems: Recent progress and unsolved problems. In *Earth's Earliest Biosphere: Its Origin and Evolution,* J. W. Schopf, 361–84. Princeton: Princeton Univ. Press
96. Sheldon, G. H. 1987. Rates of single-copy DNA evolution in herons. *Mol. Biol. Evol.* 4:56–69
96a. Sites, J. W., Davis, S. K. 1989. Phylogenetic relationships and molecular variability within and among six chromosome races of *Sceloporus grammicus* (Sauria, Iguanidae), based on nuclear and mitochondrial markers. *Evolution* 43:296–317
97. Smith, A. B. 1989. RNA sequence data in phylogenetic reconstruction: testing the limits of its resolution. *Cladistics* 5:321–44
98. Smith, T. F., Waterman, M. S. 1981. Identification of common molecular subsequences. *J. Mol. Biol.* 147:195–97
99. Southern, E. M. 1975. Detection of specific DNA fragments separated by gel electrophoresis. *J. Mol. Biol.* 98: 503–17
99a. Suzuki, H., Miyashita, N., Moriwaki, K., Kominami, R., Muramatsu, M., Kanehisa, T., Bonhomme, F., Petras, M., Yu, Z., Lu, D. 1986. Evolutionary implication of heterogeneity of the non-transcribed spacer region of ribosomal DNA repeating units in various subspecies of *Mus musculus*. *Mol. Biol. Evol.* 3:126–37
100. Suzuki, H., Moriwaki, K., Nevo, E. 1987. Ribosomal DNA (rDNA) spacer polymorphism in mole rats. *Mol. Biol. Evol.* 4:602–10
101. Swofford, D. L. 1989. Phylogenetic analysis using parsimony, Version 3. Oh. Champaign: Ill. Nat. Hist. Survey
102. Tanhauser, S. M. 1985. *Evolution of mitochondrial DNA: patterns and rate of change*. Ph.D. thesis. Univ. Florida, Gainesville
103. Thomas, R. H., Schaffner, W., Wilson, A. C., Pääbo, S. 1989. DNA phylogeny of the extinct marsupial wolf. *Nature* 340:465–67
104. Thompson, J. F., Hearst, J. E. 1983. Structure of *E. Coli* 16S RNA elucidated by psoralen crosslinking. *Cell* 32:1355–65
105. Thorpe, J. P. 1982. The molecular clock hypothesis: Biochemical evaluation, genetic differentiation and systematics. *Annu. Rev. Ecol. Syst.* 13:139–68
106. Troutt, A., Savin, T. J., Curtiss, W. C., Celentano, J., Vournakis, J. N. 1982. Computer-aided prediction of RNA secondary structures. *Nucleic Acid Res.* 10:403–19
107. Van Etten, R. A., Walberg, M. W., Clayton, D. A. 1980. Precise localization and nucleotide sequence of the mouse mitochondrial rRNA genes and three immediately adjacent novel tRNA genes. *Cell* 22:157–70
108. Vogel, F. 1972. Non-randomness of base replacement in point mutation. *J. Mol. Evol.* 1:334–67
109. von Heijne, G. 1987. *Sequence Analysis in Molecular Biology*. San Diego: Academic
110. Ware, V. C., Taque, B. W., Clark, C. G., Gourse, R. L., Brand, R. C., Gerbi, S. A. 1983. Sequence analysis of 28S ribosomal DNA from the amphibian *Xenopus laevis*. *Nucleic Acids Res.* 11:7795–18
111. Wheeler, W. C., Honeycutt, R. L. 1988. Paired sequence difference in ribosomal RNAs: evolutionary and phylogenetic implications. *Mol. Biol. Evol.* 5:90–96
112. Wilbur, W. J., Lipman, D. J. 1983. Rapid similarity searches of nucleic acid and protein data bands. *Proc. Natl. Acad. Sci. USA* 80:726–30
113. Williams, S. M., DeBry, R. W., Feder, J. L. 1988. A commentary on the use of ribosomal DNA in systematic studies. *Syst. Zool.* 60–62:
113a. Wilson, G. N., Knoller, M., Szura, L., Schmickel, R. D. 1984. Individual and evolutionary variation of primate ribosomal DNA transcription initiation regions. *Mol. Biol. Evol.* 1:221–37
114. Woese, C. J., Magrum. L. J., Gupta, R., Siegel, R. B., Stahl, D. A., et al. 1980. Secondary structure model for bacterial 16S ribosomal RNA: phylogenetic, enzymatic and chemical evidence. *Nucleic Acids Res.* 8:2275–93
115. Woese, C. R. 1987. Macroevolution in the microscopic world. In *Molecules and Morphology in Evolution: Conflict or Compromise?* ed. C. Patterson, pp. 177–202. Cambridge: Cambridge Univ. Press
116. Wolfe, K. H., Gouy, M., Yang, Y., Sharp, P. M., Li, W. 1989. Date of the monocot-dicot divergence estimated from chloroplast DNA sequence data. *Proc. Natl. Acad. Sci. USA* 86:6201–05
117. Wollenzein, P. L., Hearst, J. E., Thammana, P., Cantor, C. R. 1979. Base-pairing between distant regions of the *Escherichia coli* 16S ribosomal RNA in solution. *J. Mol. Biol.* 135:255–69
118. Wrischnik, L. A., Higuchi, R. G., Stoneking, M., Erlich, H. A., Arnheim, N. 1987. Length mutations in human

mitochondrial DNA: direct sequencing of enzymatically amplified DNA. *Nucleic Acids Res.* 15:529–42

119. Wu, C., Li, W. 1985. Evidence for higher rates of nucleotide substitution in rodents than in man. *Proc. Natl. Acad. Sci. USA* 82:1741–45
120. Zimmer, E. A., Hamby, R. K., Arnold, M. L., Leblanc, D. A., Theriot, E. C. 1989. Ribosomal RNA phylogenies and flowering plant evolution. In *The Hierarchy of Life,* ed. B. Fernholm, K. Bremer, H. Jörnvall, pp. 205–14. Amsterdam: Elsevier
121. Deleted in proof
122. Zuckerkandl, E., Jones, R. T., Pauling, L. 1960. A comparison of animal hemoglobins by tryptic peptide pattern analysis. *Proc. Natl. Acad. Sci. USA* 46:1349–60
123. Zuckerkandl, E., Pauling, L. 1962. Molecular disease, evolution, and genic heterogeneity. In *Horizons in Biochemistry,* ed. M. Kasha, B. Pullman. New York: Academic
124. Zuker, M. 1986. RNA folding prediction: the continued need for interaction between biologists and mathematicians. *Lect. Math. Life Sci.* 17:87–123
125. Zuker, M., Sankoff, D. 1984. RNA secondary structures and their prediction. *Bull. Math. Biol.* 46:591–621
126. Zuker, M., Stiegler, P. 1981. Optimal computer folding of large RNA sequences using thermodynamics and auxiliary information. *Nucleic Acids Res.* 9:133–48
127. Zweib, C., Glotz, C., Brimacombe, R. 1981. Secondary structure comparisons between small subunit ribosomal RNA molecules from six different species. *Nucleic Acids Res.* 9:3621–40

# SUBJECT INDEX

F

## Q



## R



## S

# CUMULATIVE INDEXES

## CONTRIBUTING AUTHORS, VOLUMES 17–21

# CHAPTER TITLES, VOLUMES 17–21